Separations Division 2016

Core Programming Area at the 2016 AIChE Annual Meeting

San Francisco, California, USA
13 - 18 November 2016

ISBN: 978-1-5108-3447-7

Printed from e-media with permission by:

Curran Associates, Inc.
57 Morehouse Lane
Red Hook, NY 12571

CURRAN ASSOCIATES INC. proceedings.com

Printed by Curran Associates, Inc. (2017)

AIChE
120 Wall Street, FL 23
New York, NY 10005-4020

Phone: (800) 242-4363
Fax: (203) 775-5177

www.aiche.org

Additional copies of this publication are available from:

Curran Associates, Inc.
57 Morehouse Lane
Red Hook, NY 12571 USA
Phone: 845-758-0400
Fax: 845-758-2633
Email: curran@proceedings.com
Web: www.proceedings.com

TABLE OF CONTENTS

461236 Bioinspired Separation Materials for Environmental and Energy Applications

Abstract:

Research Interests:

My research motivation in membrane science and technology is based on an acute awareness of current environmental issues and the potential of separations to address some of these issues. I have a strong interest in membrane technologies because they have been widely used to supply clean water and there is a growing interest in applications to clean energy and pollution prevention. My Ph.D. program has opened the door of a completely new membrane research field to meâ€"biomimetic membranes and I have helped in bringing the new area of artificial water channels for separations into focus through my PhD work. ***Biomimetic membranes, utilizing biological elements or borrowing concepts from biological systems, can take advantage of processes evolved by nature over billions of years for improving transport efficiency and specificity.*** My Ph.D. study started with aquaporins (AQPs) based biomimetic desalination membranes. Aquaporins are biological water channel proteins that facilitate extremely fast water transport while rejecting all the other ions and solutes. The later part of my Ph.D. work conducted some of the initial work in the exciting area of bioinspired artificial channels. My work showed channels that mimic natural channel proteins self-assemble in lipid and block copolymer bilayers and can approach the performance of biological water and proton channels.

In my future academic career, I am devoted to developing bioinspired energy efficient functional separation materials that can be used in the field of separations important for environmental and energy applications. I want to use artificial channels to fabricate energy-efficient separation materials for liquid and gas separations, and develop environmental sensors for environmental applications. I am also interested in utilizing the bioinspired concepts to mitigate membrane fouling issues in water and wastewater treatment, with a combination of biomimetic topography, controlled surface chemistry and antifouling signals.

Teaching Interests:

During my Ph.D. program, I have volunteered to instruct different classes for 2 years to learn how to teach and develop my own teaching style. Because I have both chemical and environmental engineering backgrounds, I am qualified to teach undergraduate or graduate core courses (such as transport phenomenon, fluid mechanics and reaction engineering), specialized courses (membrane separations, water and wastewater treatment) and experimental courses (environmental or chemical engineering unit operations). ***My teaching will be focused on how to utilize the basic skills acquired in classroom to solve practical problems as this leads to the greatest retention of core concepts. All of the courses I will instruct will be designed to train the students in extending their learning abilities (how to find problems and where to get necessary resources to solve them), communication and collaboration, writing and presenting skills.***

Representative publications:

1. **Shen, Y.**, et al. Highly permeable artificial water channels in block copolymer membranes. (in preparation)
2. Licsandru, E., Kocsis, I., **Shen, Y.**, at al. Salt excluding artificial water channels reveal enhanced dipolar water and proton translocation. *J. Am. Chem. Soc.* 138, 5403-5409, (2016). (**co-first author**)
3. **Shen, Y.**, et al. Highly permeable artificial water channels that can self-assemble into two-dimensional arrays. *Proc. Natl. Acad. Sci. U.S.A.* 112, 9810-9815, (2015).
4. **Shen, Y.**, et al. Biomimetic membranes: A review. *J. Membr. Sci.* 454, 359-381 (2014).
5. Sun, Y., **Shen, Y.**, et al. Linkages between microbial functional potential and wastewater constituents in large-scale membrane bioreactors for municipal wastewater treatment. *Water Res.* 56, 162-171 (2014). (**co-first author**)
6. **Shen, Y.**, et al. Improvement on Modified Lowry Method against Interference by divalent cations for soluble protein measurement in wastewater systems. *Appl. Microbiol. Biotechnol.* 97, 4167-4178 (2013).
7. **Shen, Y.**, et al. Characterization of soluble microbial products in 10 large-scale membrane bioreactors for municipal wastewater treatment in China. *J. Membr. Sci.* 415â€“416, 336-345 (2012).
8. **Shen, Y.**, et al. A systematic insight into fouling propensity of soluble microbial products in membrane bioreactors based on hydrophobic interaction and size exclusion. *J. Membr. Sci.* 346, 187-193 (2010).

Contact Information

Yuexiao Shen, 162 Fenske Laboratory

The Pennsylvania State University

University Park, PA, 16802

Phone: (814)777-8181 Email: yxs174@psu.edu

475849 Demulsification of the Phosphoric Acid-Tributyl Phosphate (W/O) Emulsion By Hydrocyclone

Abstract:

Research Interests: separation by hydrocyclone,demulsification of water-in-oil emulsion, drop coalescence behavior, liquid-liquid dispersion

Teaching Interests: mechanical separation

Emulsions are often generated as unwanted byproducts during the mixing of organic and aqueous phases, but they are required in the emulsification extraction to greatly improve the extraction rate by increasing the contact interfacial area between the two phases. A rapid, convenient, efficient and economic demulsification method is subsequently needed to separate the two phases whether the emulsions are required or not. Liquidâ€“liquid hydrocyclone is efficient for the two-phase separation of emulsion systems with the aid of strong centrifugal force created by the swirling flow. In this study, a demulsification process of phosphoric acidâ€“tributyl phosphate (W/O) emulsion was investigated by hydrocyclone. The drop size distributions and average drop sizes of the emulsions pre and post demulsification were measured. The average drop size differences were used to evaluate the demulsification performance. The force condition and coalescence process of drops in the hydrocyclone were analyzed. The effects of density difference between the two phases, inlet flow rate and structural parameters (large cone part angle, small cone part angle, cylindrical part height, inlet and overflow pipe inner diameter) of the hydrocyclone on the demulsification performance were investigated. Larger density difference between the two phases led to better demulsification performance. The demulsification performance firstly became better and then got worse with the increase of inlet flow rate. The decrease in large and small cone part angle, as well as the increase in cylindrical part height, inlet and overflow pipe inner diameter of the hydrocyclone had positive effects on the demulsification performance. The influence degree of structural parameters for the demulsification performance decreased in the order: the cylindrical part height > the large cone part angle > the inlet inner diameter > the overflow pipe inner diameter > the small cone part angle. The optimum density difference, inlet flow rate and structural parameters were obtained. Under the optimum conditions, the average drop size of emulsion increased from 4.78 to 55.07 Î¼m, and the difference was 50.29 Î¼m. The average drop size differences were associated with these above-mentioned parameters through a dimensionless model. The comparison between the experimental and calculated data showed that this model was reliable to predict the average drop size difference in most demulsification conditions by hydrocyclone.

475710 Membranes for Liquid and Gas Separations

Abstract:

Membranes for liquid and gas separations

Oishi Sanyal

1st year Postdoctoral Researcher

Georgia Tech

Research Interests:

My broad research interest lies in the area of membrane design with an equal focus on both liquid and gas separations. For both liquid and gas feeds, membrane material selection plays a key role in engineering more energy efficient separations based on environmentally benign and energy efficient processes. Conventional thermally driven techniques like distillation have dominated large scale industrial separations; however, membrane-based processes are emerging to replace or debottleneck conventional techniques.

In the water industry, reverse osmosis and nanofiltration are already dominant for desalination and wastewater purification. The major challenge faced by water treatment membranes is their fouling propensity, and I am specifically interested in working on membrane modification techniques to lower fouling tendencies of commercial membranes.In the area of gas separation, I am primarily focused on creating membranes using advanced materials to surpass the current performance "upper bound" for polymeric membranes. Such membranes pose many opportunities for applications such as olefin-paraffin separation, natural gas and flue gas separations.

Teaching Interests:

Having done both my undergraduate and graduate studies in chemical engineering, I am familiar with the usual chemical engineering courses. Specifically, I would be interested in teaching courses like Mass Transfer, Transport Phenomena and Reaction Engineering. I would also like to offer a graduate level course on membrane separations as I personally benefited a lot from a similar course taught during my graduate school. I believe I will also be comfortable in teaching some environmental engineering courses or some electives related to environmental engineering.

Postdoctoral project: Understanding the engineering fundamentals of carbon molecular sieve (CMS) membranes Advisor: Prof. William J. Koros; School of Chemical and Biomolecular Engineering, Georgia Institute of Technology, Atlanta, Georgia, USA.

PhD Dissertation: Design of polyelectrolyte multilayer (PEM) membranes for ion rejection and wastewater effluent treatment

Advisor: Prof. Ilsoon Lee; Department of Chemical Engineering and Materials Science, Michigan State University, East Lansing, MI, USA.

Research experience:

Through my PhD and postdoctoral research, I have gained experience in water treatment and gas separation membranes. For my doctoral research, I worked on the surface modification of membranes using layer-by-layer technology to form polyelectrolyte multilayer (PEM) membranes. In collaboration with another research group in MSU, we extended the usual application of PEM membranes from specific ion rejection to treating a real wastewater effluent. My postdoctoral research focuses on designing asymmetric carbon-molecular sieve (CMS) fibers for advanced gas separation applications. These CMS membranes are produced via controlled high temperature pyrolysis of polyimide hollow fiber polymer precursors. My specific project is about the fundamentals of CMS formation and the effects of different pyrolysis parameters on final separation performance for different gas-pairs.

Teaching experience

During my PhD, I worked as a Teaching Assistant for the course on Transport Phenomena. I was responsible for grading assignments and tests as well as preparing solutions for the same. I hope to further my teaching experience during my stay at Georgia Tech. During my years as a PhD student, I had the opportunity to mentor 5 undergraduate students to perform research in the lab.

Future direction

As a faculty member, I will continue working in the area of membrane-based separation. In the water-treatment area, I will focus on membrane modification but more importantly, I will evaluate the performance of modified membranes under "real" conditions, to extend my doctoral research. Besides traditional water purification, I am interested in other liquid-based separations where the potential of membranes has not been explored in great detail. In the area of gas separation, I will continue my efforts in developing membranes for important separations in the chemical/petrochemical industries like olefin-paraffin mixtures and xylene isomers etc. which are conventionally carried out by highly energy-intensive separation methods.

Selected Publications

1. **Sanyal, O.;** Liu, Z.; Yu, J; Meharg, B. M.; Hong J. S; Liao, W.; Lee, **I.** " Design of fouling-resistant clay-embedded polyelectrolyte multilayer membranes for wastewater effluent treatment," ***Journal of Membrane Science***, **512**, 21-28, (2016).
2. **Sanyal, O.;** Liu, Z.; Meharg, B. M.; Liao, W.; Lee, **I.** "Development of polyelectrolyte multilayer membranes to reduce the COD level of electrocoagulation treated high-strength wastewater," ***Journal of Membrane Science,*** **496**, 259-266, (2015).
3. **Sanyal, O.;** Sommerfeld, A.N.; Lee, **I.** "Design of ultrathin nanostructured polyelectrolyte-based membranes with high perchlorate rejection and high permeability," ***Separation and Purification Technology*** **145**, 113-119, (2015).
4. **Sanyal, O**.; Lee, **I.** "Recent progress in the application of layer-by-layer assembly to the preparation of nanostructured ion-rejecting water purification membranes," ***Journal of Nanoscience and Nanotechnology*** **14**, 2178-2189, (2014). (review)

476028 Elucidating Diffusion, Adsorption, and Catalytic Processes in Oxide Materials

Abstract:

Research Interests:

A conceptual and molecular level understanding of diffusion, adsorption, and catalytic processes forms the basis of most separations and catalytic processes. Herein, I present two case studies in which an improved understanding of physical and chemical rates processes in porous oxide materials enables the design of more effective materials: firstly, amine-modified mesoporous oxide materials for use in large-scale CO_2capture applications, and secondly, aliovalently doped non-reducible bulk oxides for alkane dehydrogenation.

A combination of advanced synthetic techniques, thorough physicochemical characterization, batch and breakthrough adsorption uptake measurements, and mass and heat transfer modeling was used to determine critical factors governing rates of CO_2 adsorption onto supported amine materials. At amine coverages significantly exceeding monolayer coverage, diffusion through the amine-polymer phase was identified as the key step limiting CO_2 adsorption rates. Heteroatom incorporation into the silica framework was used to ameliorate the adverse effect of amine loading on diffusion rates, presumably by resulting in a more open amine-polymer phase. The modified oxide materials thus developed represent *effective*adsorbents exhibiting a combination of both high equilibrium adsorption capacities and rapid diffusion through the amine-polymer network.

In my postdoctoral research work, kinetic and mechanistic details of the non-oxidative conversion of propane to propylene over non-reducible oxides were investigated. High surface area (127m^2/g) bulk yttria doped zirconia materials were demonstrated to be highly active, selective, and regenerable propane dehydrogenation catalysts with

propylene rates comparable to Pt and Cr-based catalysts used commercially. The kinetics and mechanism of alkane dehydrogenation over this new class of catalysts was studied, and methods for assessing the number of catalytic active sites under reaction conditions, developed. Ongoing and future research efforts will focus on extending this concept of highly selective acid-base pair mediated C-H activation on non-reducible oxides to other hydrocarbon conversion reactions.

Building off of my training in separations, catalysis, and industrial R&D, I envisage that my research program will emphasize a molecular level understanding of diffusion, adsorption, and reaction processes, with wide-ranging applications in chemical and energy conversion processes.

Teaching Interests:

Kinetics & Reaction Engineering, Transport Phenomena, Mass Transfer

476129 Advanced Crystallization Technologies for (Bio)Manufacturing

Abstract:

Research Interests: To create novel crystallization strategies and processes for the purification and separation of proteins and other biologically derived molecules.

Teaching Interests: All core courses of chemical engineering. My goal is to help students build a strong background in chemical engineering, sharpen their analytical skills, and apply their knowledge and skills creatively.

Crystallization from solution is a key unit operation for purification, separation, and manufacturing in the environmental, food, pharmaceuticals, chemicals, and materials industries. Challenges that can arise during crystallization include (1) high sensitivities of crystal nucleation and growth kinetics to small variations in the concentrations of contaminating chemicals, (2) high sensitivities of aggregation and agglomeration kinetics to flow fields and particle size, (3) spatial inhomogeneity, (4) sensor limitations, and (5) many orders of magnitude variation in length scales (0.1 nm to 100s of microns) and time scales (20 ps to many hours).

New process designs are described that employ process intensification strategies to greatly enhance the control of the crystal size distribution. These strategies include using noncontact focused ultrasonication, exploiting differences in molecular and thermal diffusion rates at fluid-fluid interfaces to induce nucleation, and selecting crystallizer designs and operations to spontaneously induce multiphase flows that decouple nucleation and growth to enable their individual control. My experiments are supported by mathematical modeling and theoretical analysis, which collectively are also relevant to other chemical process systems.

The growth in demand for new therapeutics and fine chemicals manufactured in bioreactors motivates the invention of technologies to reduce purification and storage costs by at least an order of magnitude compared to existing technologies. A research program is proposed that employs fundamental understanding, mathematical models, and cleverly designed experiments to create novel crystallization strategies and processes for the purification and separation of proteins and other biologically derived molecules.

Keywords: Process intensification, kinetics, separation, biomanufacturing, process design and control, crystallization, purification, protein

479778 Membrane Contactors for Reactive Absorption Process

Abstract:

Research Interests:

I have wide skill set, ranging from membrane science, to reaction engineering, to analytical chemistry. My bachelor's degree is in chemistry, and I work as an antibody chemist and analytical chemist for a number of years prior to graduate school. This experience helped propel my research forward when I decided to go back to school for a doctorate in chemical engineering. My main PhD research focused on developing membrane contract reactors for 3-phase reactions; however, I was also part of an interdisciplinary NSF-IGERT Traineeship, exposing me to a number of collaborations.

Membrane contact reactors: Design and fabricate membranes and membrane reactors. Characterize liquid and gas transport properties of membranes. Characterize structural properties of polymeric and ceramic membranes. Surface modifications of membranes. Reaction engineering: rate laws, reaction order, kinetics, selectivity, and catalyst characterizations.

Liquid-liquid equilibria (LLE): Interdisciplinary project in which phase diagrams of four ternary systems of water-alcohol(s)-methoxycyclopentane (CPME) were determined. Binodal curves were experimentally determined using tie-line measurements and the data was correlated using the universal quasichemical (UNIQUAC) activity coefficient model.

Single cell oil from oleaginous yeast: Interdisciplinary project in which oleaginous yeast was investigated as a renewable alternative to crop-based oil. This project involved investigating scale-up of the fermentation, including cell lysis, separation, purification and economic analysis of the endeavor. Mentoring an undergraduate group involved in the process. And using the above mentioned LLE data to developed a green solvent alternative to the Bligh and Dyer lipid extraction.

Teaching Interests:

I have experience teaching undergraduate classes ranging from freshman to senior levels. In addition to teaching chemical engineering classes, I have also taught Material Science, which includes students from a variety of majors. I am comfortable teaching in both large lecture halls (Material Science), and smaller senior level chemical engineering classes (Chemical Reaction Engineering). My experiences mentoring undergraduates in the labs and coaching the undergraduate rugby team has allowed me to relate very easily to undergraduates and foster fun and welcoming atmospheres in the classroom, laboratories, and office.

Courses taught:

- Current Topics in Chemical Engineering (CHE110)
- Basic Concepts in Materials Science and Engineering (CHE354)
- Fundamentals of Mechanical Properties (CHE355)
    - Fundamentals of Electrical Properties (CHE356)
    - Chemical Engineering Thermodynamics 1 (CHE520)
    - Chemical Reaction Engineering (CHE550)

Professional Preparation:

Ph.D. 2016 Chemical Engineering Kansas State University

B.A. 2011 Chemistry University of California, Santa Cruz

Graduate Certificate- Biobased Products and- Kansas State University

2016 Bioenergy Experience:

2016-Curent Kansas State University, Department of Chemical Engineering

- Postdoctoral Researcher

- Instructor (two classes per semester)

2011-2015: Kansas State University, Department of Chemical Engineering

- NSF IGERT Trainee in Biorefining

- Graduate Research Assistant (PhD Student)

2008-2010 Tainted Love (80's cover band)

- Sound Technician (Roadie)

I was a roadie for a traveling band while I was taking community college classes. I was taking community college classes b/c I was switching from chemistry to engineering.

2004 – 2008: Curtis and Tompkins, LTD. Analytical Laboratories

- GC-MS Chemist

- Total Extractable Hydrocarbon Chemist

- Mercury/Inorganic Chemist

2004 – 2004: Protein Design Labs, Inc

- Antibody Chemist (Internship) PhD Dissertation:

Title: "Membrane Contact Reactors for Three-Phase Catalytic Reactions"

Advisors: Mary E. Rezac, Peter H. Pfromm

Department of Chemical Engineering, Kansas State University

Future Direction:

As faculty I would like to continue applying **Membrane Contactors** for gas-liquid reactions in reactive absorption processes, as well as conducting membrane research in interdisciplinary endeavors. Inorganic nanoparticle (NPs) synthesis is one such process that offers considerable opportunities for future research. These reactions involve a gas being brought into contact with a liquid that contains a dissolved metal salt. The gas must be absorbed into the liquid in order for the gas-metal reaction to take place. In traditional systems, such as stirred tanks or bubble columns, when the gas phase is dispersed into the liquid phase the gas-liquid interfacial area is small and varies with operating conditions, leading to difficulty in control of NP formation and scale-up. However, it is possible to substantially increase the gas-liquid interfacial area and control the gas-liquid mass transfer rate using membrane contactors. By employing a micro porous membrane between the gas and liquid phase, the gas is introduced to the liquid phase through the pores of the membrane. The membrane pores provide a large and stable interfacial surface area, favoring increased absorption rates and faster reaction times. One of the few studies using membrane contactors for NP production was by Jia et al. (J. Membr. Sci., 342 (2009) 1-5). Jia et al. produced $CaCO_3$ NPs by reacting aqueous $Ca(OH)_2$ with CO_2 gas using a membrane contactor described above; additionally, they used a surface renewal model to calculate CO_2absorption per membrane surface area, with the model and empirical data matching well. Proposed future work in this field includes: (1) more examples of noble metal chemistry, (2) systematically varying operating conditions to control size distributions, and (3) experimenting with different NP shapes.

In addition to nanoparticle production and membrane reactors, my interest includes separations and chromatography. With my extensive analytical chemistry background, I am naturally curious about solvent/polymer interactions, as well as physical/chemical interactions in relation to chromatography and separations. Membrane research is sufficiently flexible to adapt to the changes in the funding winds, and I foresee myself using membranes on collaboration research, both intra and interdisciplinary. Selected Awards

- NSF IGERT Trainee 2012-2015

- Carl Storm Underrepresented Minority Fellowships, July 2014

- Accepted to present, and awarded a travel grant to: New York Academy of Science and PepsiCo, Journey through Science Day, December 2014

- Elias Klein Founders' Travel Awards, North American Membrane Society, June 2013, Selected Publications

Wales, M.D.; Huang, C.; Joos, L.B.; Probst, K.V.; Vadlani, P.V.; Anthony, J.; Rezac, M.E.; "Liquid-Liquid Equilibria for Ternary Systems of Water + Methoxycyclopentane (CPME) + Alcohol; Methanol, Ethanol, 1-Propanol, or 2-Propanol"; Journal of Chemical & Engineering Data, 2016, 61 (4), pp 1479–1484, **DOI:** 10.1021/acs.jced.5b00803

Wales, M.D.; Joos, L.B.; Traylor, W.A.; Pfromm P.H.; Rezac, M.E.; "Composite Catalytic Tubular Membranes for Selective Hydrogenation in Three-Phase Systems," Catalysis Today, 2016, 268, pp 12-18, DOI :10.1016/j.cattod.2015.12.011

Zheng, Q.; **Wales, M.D**.; Heidlage, M.G.; Rezac, M.E.; Wang H., Bossmann, S.H.; and Hohn K.L.;

"Conversion of 2,3-Butanediol to Butenes over Bifunctional Catalysts in a Single Reactor," Journal of Catalysis, 2015, 330, pp 222-237, DOI: 10.1016/j.jcat.2015.07.004

Wales, M.D.; Joos, L.B.; Traylor, W.A.; Pfromm P.H.; Rezac, M.E; "Platinum and Palladium Composite Catalytic Membrane for Three-Phase Hydrogenation", Chemical Engineering Journal, (**In Prep**)

Probst, K.V.; **Wales, M.D.**; Rezac, M.E.; Vadlani, P.V.; "Evaluation of Green Solvents: "Oil Extraction from Oleaginous Yeast Lipomyces Starkeyi using Cyclopently Methyl Ether (CPME)", Biotechnology Progress, (**In Prep**)

481045 Development of Green Technologies using Fluid Simulation Knowledge

Abstract:

Research statement

My research interests can be categorized in the broad area of complex fluid simulation and its applications in industrial and environmental aspects. I am predominantly concerned with phase equilibrium and understanding the application of complex systems such as micro-emulsions, aqueous two phase systems, Ionic liquids as green solvents. Theoretical and simulation techniques are utilized to generate important insights into how the inter particle forces influence the behavior of such systems over a broad range of temperature and pressure. Phase equilibrium can play great role in various fields such as polymer synthesis, reservoir engineering, nano materials...

Research Interests:

- **Application of micro-emulsions for synthesis of nanomaterials and nano catalysts**

Microemulsions are colloidal 'nano-dispersions' which stabilized by a surfactant film. These thermodynamically stable dispersions can be considered like a Nano-reactors which can be utilized to carry out chemical reactions to synthesize nanomaterials. In fact using micro emulsions synthesis parameters are easily controlled to produce tailor-made products down to a nanoscale level with new and special properties. Due to their unique properties, namely, ultralow interfacial tension, large interfacial area, thermodynamic stability and the ability to solubilize otherwise immiscible liquids. The uses and applications of microemulsions are numerous in chemical and biological fields.

- **Application of aqueous two phase systems(ATPS) in separation of bio molecules**

Nowadays demands for purified bio-molecules are growing up. For many years chromatography was known as the most common way for purification of both single and double-stranded nucleic acids in industrial scale . But because of its high cost and operational difficulties in scale up, attempts for replacing such systems have been in progress in last decades. Aqueous two-phase systems (ATPSs) as a proper alternative have several preferences in comparison with chromatography technics especially easiness in scale up and lower costs. Furthermore ATPSs have the ability to combine two or more steps of purification process such as mixing, concentration and separation into one which makes the method more applicable.

- **Application of Ionic liquids as a green solvent**

During recent years considerable attention has been paid to the application of IL's as green solvent in various processes such as gas sweetening, CO2 capturing, absorption VOCs and solvent extraction processes. My research in this area can be divided into three main categories: a considerable effort have been focused on the synthesis of IL's for specific applications. In the second category, I am trying to report experimental data on solubility of gases in IL's. finally, modeling of gas solubility in ionic liquids.

- **Phase equilibrium of reservoir fluids in the presence of Surfactant and brine**

Reservoir fluid under extreme conditions of temperature and pressure are of fundamental importance in oil industry, as well as being ubiquitous in nature. Usually surfactants are injected to the reservoir for enhanced oil recovery. Modeling of phase equilibrium in the presence of electrolytes as well as surfactants is really complicated and needs a lot of research in this field. Conventional equations of state need improvement for application in this field. I have focused on theoretical as well as experimental study on phase behavior of reservoir fluids in the presence of surfactants, electrolytes.

- **Thermodynamics of electrolyte solution**

Industrial technologies make use of semi-empirical thermodynamic models extrapolated to extreme conditions (high temperatures, high (azeotropic) concentrations and high pressures). Molten salts are proposed as heat carriers. Pharmaceutical companies are designing processes with organic electrolytes and proteins. Nowadays, the development of accurate correlations and predictive models for thermo-physical properties of electrolyte solutions is highly desirable. Electrolyte thermodynamics concepts can be extended to describe the properties of ionic liquids, radioactive and biological materials. Theories of electrostatics and thermodynamics have to be combined and extended for modeling these novel materials at extreme conditions. Parameter estimation for thermodynamic models can be formulated through the use of advanced optimization algorithms constrained so that they predict accurately the observed number of phase splits, phases and azeotropes.

Teaching Interests:

I have taught several chemical engineering courses during last 10 years at the *Iran University of Science and technology* which gave me teaching ability and a variety of educational skills. At the undergraduate level, I have enough experience in teaching of following courses: *Material and Energy Balances, Thermodynamics I&II, Mass Transfer, Computer aided process engineering (using ASPEN HYSYS simulation environment, Petrochemical processes (3 semesters, Process safety.*

At the graduate level I have taught following courses: *Advanced Thermodynamics, Advanced kinetic & Reaction, PVT simulation, Thermodynamics of reservoir fluids.*

Future Plan

As a long-term plan, I am planning to apply my thermodynamic knowledge in technology development. Knowledge about phase behavior of mixture is an essential tools in process design, material synthesis and so on. In short-term I have planned to commercialize a new oxidative-extractive desulfurization process based on Ionic liquids.

Selected Publications:

1. Partitioning of pyrimidine single stranded oligonucleotide using polyethylene glycol sodium sulfate aqueous two-phase systems; experimental and modeling, **Fluid Phase Equilibria 432 (2017)45-53**
2. Removal of thiophene from model diesel oil with nitrate based ionic liquids at several temperatures, **Journal of Molecular Liquids 221 (2016) 1104-1110**
3. An Efficient Workflow for Production Allocation During Water Flooding, **Journal of Energy Resources Technology 139 (3), 032902 (2016)**
4. A simple accurate model for prediction of Petroleum Fractions physical Properties**, Korean Journal of Chemical Engineering 33: (2016) 2930.**
5. Oxidative desulfurization of Thiophenes using supported Dawson catalyst. **Accepted for publication in Petroleum Research Journal. (2016)**
6. Experimental measurement and thermodynamic modeling of CO2 solubility in aqueous solutions of morpholine, **Journal of Molecular Liquids, 214 (2016) 411-417**
7. Novel liquid–liquid equilibrium data for six ternary systems containing IL, hydrocarbon and thiophene at 25 °C, **Fluid Phase Equilibria, 412 (2016) 21-28**
8. Phase behavior of glycol ether surfactant systems in the presence of brine and hydrocarbon: experiment and modeling, **Fluid Phase Equilibria, 414 (2016) 101-110**
9. Nanostructures in clear and homogeneous mixtures of rapeseed oil and ethanol in the presence of green additives, **Colloid and Polymer Science 293, Issue 11(2015) 3225-3235**
10. Effects of salts and sucrose on the phase behavior of ternary mixtures of water, decane, and mono-ethylene glycol butyl ether, **Colloids and Surfaces A: Physicochemical and Engineering Aspects, 477(2015) 19–25**
11. An investigation of the fish diagrams of water or brine/decane or dodecane/propylene glycol ether (C3P1 or C3P2) systems**, Journal of Molecular Liquids, 206 (2015) 170-175**

12. Modeling adsorption in binary associating solvents using the extended MPTA model, **Adsorption 2014, 1-9**

13. Thermodynamic Modeling of Aqueous Ionic Liquid Solutions Using PC-SAFT Equation of State, **Ind. Eng. Chem. Res., 2012, 51 (30), pp 10274–10282**

14. Application of PHSC Equation of State in Prediction of Gas Hydrate Formation Condition, **Fluid Phase Equilibria, 2012, 333, p.27-37**

15. Modeling CO2 Solubility in Aqueous N-methyldiethanolamine Solution by Electrolyte Modified Peng–Robinson Plus Association Equation of State, **Ind. Eng. Chem. Res., 2012, 51 (29), pp 9875–9885**

1. 16. Prediction of gas hydrate formation condition in the presence of thermodynamic inhibitors with the Elliott-Suresh-Donohue Equation of State , **Journal of Petroleum Science and Engineering , 77 (2011) 93-103**
2. 17. Simplified Local Density Model for Adsorption of Pure Gases on Activated Carbon using Sutherland and Kihara Potentials, **Microporous and Mesoporous Materials 136 (2010) 1–9**

477820 Separation and Reaction Processes in Functional Nanostructured Porous Materials

Abstract:

Research Interests:

The development of new-generation materials that extend the industrial applications of reactionseparation processes is being addressed in some applications like water treatment or fuel cells. The development of functionalized surfaces in highly porous materials can led to stimuli responsiveness such as ionization, electron and ion transfer, swelling, hydrophobicity/hydrophilicity or sorption. These new combination of techniques and materials prove to be effective in selective separations in various environmental conditions, enhancing reactivity, durability and permeation.

The main idea is how to increase the number of functional groups by unit of area on the material used in order to give and augment sensibility, reactivity, selectivity or sorptive characteristics and, at the same time not affecting or even improve the application performance. Polymers and biopolymers may display these different surface functionalities and due to their versatility, they can be applied in diverse research fields and
simultaneously be sustainable technologies.

Current Research

Currently, we are using "green" techniques to create nanocomposites based on polyvinylidene fluoride membranes. This post-manufacture modification relates to the incorporation of various nanomaterials into the structure of the membrane, usually a commercial one. This feature allows better conformation and distribution of the nanomaterial, avoiding aggregation, mixing problems and losses. Membrane pores can be functionalized with hydrophilic materials such as polyvinylpyrrolidone and poly(acrylic acid). Then, by double ion exchange and subsequent reduction, metallic and bimetallic nanoparticles can be synthesized for metal capture or reduction/oxidation of organic contaminants. Through layer-by-layer functionalization, membranes can also be loaded with protein channels (porins) and enzymes. These porins permit the separation of different molecular weights. In the case of enzymes, the objective is the oxidation of aryl halides, avoiding enzyme inactivity.

Dibakar Bhattacharyya, PhD leads this research and the grants associated are the Superfund Research Program of the NIEHS and the KY NSF EPSCoR.

Research Experience

To prepare for the career in separation-reaction process design and engineered materials for different treatment applications, I have worked for more than 10 years in elimination of hard microbial contamination from water, remediation of recalcitrant and harmful pollutants, combination of physicochemical and biological treatments, and plant design for different industries ranging from drinking water to foods to pharmaceutical. In addition, I have good experience in polymeric materials because it is a personal interest, the backbone support of my current research and I had taught this subject during my work as an adjunct professor in Chemical Engineering. My research work (MS, Faculty and PhD) in advanced oxidation processes and membrane functionalization with nanoparticles, respectively, used different platforms in both lab-scale and full-scale for environmentally benign syntheses and manufacture, and the establishment of individual and combined technology strategies to design suitable processes. As an adjunct professor, we succeeded in immobilizing Titania onto activated carbon and polyester for the photocatalytic degradation of wastewater of the printer cartridges reuse process. We advised, assessed equipment, engineered and designed processes for a Drinking Water Treatment and Supply in an Emergency and/or Disaster Scenario in Bogotá D.C., Colombia. In addition, we developed a pilot software Excel app for configuration of water treatment systems in pharmaceutical, food and soft drink applications.

Teaching Interests:

I have an extensive teaching experience. I spent 7 years teaching science and technology to a gifted youngsters in a special education institution, from middle to high school. I was also an adjunct professor in the Chemical Engineering Department at the Universidad de los Andes (Colombia) for 3 years after my master's degree. At los Andes, I taught Separation Processes, Material an Energy Balances, Physicochemical Aspects of
Polymers and Transport Phenomena. At University of Kentucky I was TA and have been guest-lecturer of undergraduate courses in Chemical Engineering such as Separation Processes, Chemical Reactor Design and Chemical Engineering Process Design. To finish, I have mentored graduate and undergraduate students in my group and overseas, including a master's student, REU-NSF and NSF Graduate Fellowship awardees.

Future plans

My research focus on the development of integrated, cost-effective technologies that allow targeted reaction, purification and selective separation of mixtures in different qualities. Likewise, the development of solutions for specific separations and reactions needs intensive research in functional polymers and bioinspired materials at micro and nano scale levels. Specifically I am interested in enabling and designing water treatment systems for specific needs without the need to build conventional, often unsustainable and expensive treatment systems. During the graduate studies and the research as an adjunct professor, I realized that much of the research in water applications focuses on development and study of nano and micro-materials, and how to apply these in current treatment processes. Although this approach has produced excellent developments such as aquaporins, metal nanoparticles or carbon nanotubes inserted into membranes, it has posed also new types of problems like contamination related to nanoparticle activity. In order to address this, it is necessary to develop new technologies based on responsive and immobilized nanostructured materials and to combine reaction and separation processes for each of the applications to increase their selectivity and effectivity. Specifically, start by understanding that the importance of the supporting material and how is functionalized determine the performance of the combined process and the materials used.

Selected Publications

Sebastián Hernández, Anthony Saad, Lindell Ormsbee and Dibakar Bhatacharyya: Nanocomposite and Responsive Membranes for Water Treatment, In Emerging Membrane Technology for Sustainable Water Treatment,
edited by Nicholas Hankins and Rajindar Singh, Elsevier, Boston, 2016, pages 389-431,
ISBN 9780444633125. DOI:10.1016/B978-0-444-63312-5.00016-4

Sebastián Hernández, Shi Lei, Wang Rong, Lindell Ormsbee and Dibakar Bhattacharyya: Functionalization of Flat Sheet and Hollow Fiber Microfiltration Membranes for Water Applications. ACS Sustainable Chemistry & Engineering. 2016; 4 (3): 907-918. DOI: 10.1021/acssuschemeng.5b01005

Sebastián Hernández, Joseph K. Papp and Dibakar Bhattacharyya: Iron-Based Redox Polymerization of Acrylic Acid for Direct Synthesis of Hydrogel/Membranes and Metal Nanoparticles for Water Treatment. Industrial &
Engineering Chemistry Research 2014; 53(3):1130-1142. DOI: 10.1021/ie403353g

Jenny Dussan Garzón, Martha Josefina Vives-Florez, Víctor Manuel Sarria, Oscar Fernando Sánchez Medina, Luis Fernando Delgado, Andrés Fernando González Barrios and Sebastián Hernández Sierra: Aproximaciones Biológicas y Fisicoquímicas en el tratamiento de contaminantes: un resumen del aporte de la Universidad de Los Andes. 2009; DOI: 10.16924/riua.v0i30.233

Sebastián Hernández Sierra, Maryhan Paola Villamil Rodríguez and Lady Liliana Miranda Gómez: Agua Potable. Evaluación de Tecnologías para Dotación de Servicios Vitales en Alojamientos Institucionales, edited by Universidad de Los Andes, Alcaldía Mayor de Bogotá D.C., 07/2010

472251 Viability of Low Pressure Ammonia Synthesis Via Reactive Separation

Abstract:

Viability of Low Pressure Ammonia Synthesis via Reactive Separation

Mahdi Malmali, Alon McCormick, and Edward L. CusslerDepartment of Chemical Engineering and Materials Science, University of Minnesota, Minneapolis, MN 55455

Ammonia is a crucial chemical primarily used as an agricultural fertilizer; it is the second most manufactured chemical in the world with more than 160 million tons produced per year[1]. Ammonia is produced via the Haber-Bosch process, where hydrogen (from thermochemical steam-reforming of methane) and nitrogen are combined and react over a promoted iron-based catalyst at temperatures ranging from 380° to 520° C and pressures from 150 to 300 bar[2]. Reaction at such an elevated pressures only achieves partial conversions (typically less than 30%), which necessitates the recycle of unreacted gas using costly phase separation processes. After more than a century of research and optimization, this process is still highly capital- and energy-intensive[3], hence only economically viable in large-scale centralized plants.

We seek to develop technologies for small-scale synthesis of ammonia with zero-carbon footprint[4,5]. This process is powered by electricity produced from stranded intermittent wind or solar energy and can make hydrogen from electrolysis of water and nitrogen from pressure swing adsorption of air. A major drawback for widespread application of such a process is the high cost of the synthesized ammonia. The pressure in this process can be reduced to improve the economics of the process. This is achieved by simultaneous removal of the ammonia from the reaction system and circumventing the reaction equilibrium limitation.

We will initially present the ammonia reaction rate kinetics at low pressures. Then we discuss our findings of the reaction-separation process, which consists of a catalytic reactor and absorber unit; we report the absorption of ammonia on $CaCl_2$, which is utilized to selectively remove the ammonia from the system at high temperatures. Our experimental results demonstrate the high capacity of the absorption material at close-to-reaction temperature, which suppresses the equilibrium and increases production rates in our reaction-separation system; conversions as high as 80% at significantly reduced pressure (< 20 bar) are achieved. The effect of different operating conditions, including reaction temperature, reaction pressure, absorption temperature, gas transport, etc. on the kinetics of ammonia synthesis will be discussed in detail. For proof-of-concept, we compare the results obtained from high pressure reaction and low pressure reaction-separation process.

(1) http://minerals.usgs.gov/minerals/pubs/commodity/nitrogen/mcs-2011-nitro.pdf. (2) *Catalytic Ammonia Synthesis: Fundamentals and Practice*; Springer Science & Business Media, 2013. (3) Vojvodic, A.; Medford, A. J.; Studt, F.; Abild-Pedersen, F.; Khan, T. S.; Bligaard, T.; Nørskov, J. K. Exploring the Limits: A Low-Pressure, Low-Temperature Haberâ€"Bosch Process. *Chem. Phys. Lett.* **2014**, *598*, 108.

Â (4) Reese, M.; Marquart, C.; Malmali, M.; Wagner, K.; Buchanan, E.; McCormick, A.; Cussler, E. L. Performance of a Small-Scale Haber Process. *Ind. Eng. Chem. Res.* **2016**, *55*(13), 3742.

Â (5) Himstedt, H. H.; Huberty, M. S.; McCormick, A. V.; Schmidt, L. D.; Cussler, E. L. Ammonia Synthesis Enhanced by Magnesium Chloride Absorption. *AIChE J.* **2015**, *61* (4), 1364.

472974 Production of Acetic Acid from Natural Gas with Zero Carbon Emissions

Abstract:

In this work, the design of a process plant flow sheet that produces acetic acid from natural gas with zero carbon emissions is developed. Two such cases are developed - the production of acetic acid alongside power generation and hydrogen respectively.

The UNISIM software (Honeywell trademark) is used to create the flow sheet for the various modules of each plant based on established technologies of industrial acetic acid and hydrogen production processes. The various modules, when coupled for each process plant, create the net effect of zero-carbon emission.

Energy self-sufficiency, feasibility and operability limits of both cases are established via a thermodynamic analysis, while profitability is ascertained via economic and revenue studies based on current prices of the various commodities. The results are compared to stand-alone power, hydrogen and acetic production plants.

469218 Ionic Liquid Incorporation for Self Healing Purposes: New Measuring Methods for Bitumen Binder, Applications to Asphalt Concrete and Understanding of Chemical Interactions

Abstract:

Asphalt concrete is the most commonly used material in pavement industry. It is prepared by mixing bitumen, which is the bottom product of crude oil distillation, and aggregates at temperatures around 180 â°C. Asphalt concrete, once paved in the roads, experiences traffic loading along with changes in the ambient temperature throughout its service life. Thereby, failures in asphalt pavement such as fatigue cracking, rutting and thermal cracking, occur due to these continuous loading and climatic conditions. The mentioned failures in the roads give rise to building of new roads or maintaining the damaged roads. Building new roads or maintaining the already built ones requires preparation of asphalt mixtures, which consume large amounts of hydrocarbons, bitumen, together with excess energy due to high mixing temperatures. In addition to the large resource and energy consumptions, heating the mixture to high temperatures results in emission of greenhouse gases and bitumen fumes into the air. Therefore, extending the service life of asphalt pavements is very important both from an economic and environmental point of view. Such an improvement would also mean more efficient usage of the fossil resources.

Bitumen, which is the binder of asphalt concretes, is known to have intrinsic self- healing properties. It has the ability to restore its mechanical strength by closing the cracks at high temperatures or during long rest periods. However, during its service life, the asphalt pavement may not always have enough resting periods or the high temperatures which are needed to heal itself. Therefore, improving the self- healing capabilities of bitumen/asphalt is a necessity to minimize the mentioned environmental and economic problems.

In this study, the effects of ionic liquids, which are also known as green solvents, on the self- healing properties of bitumen is studied. 50/70 penetration grade bitumen from Tupras Ä°zmit Refinery is used in the studies. As part of the ionic liquid characterization, Thermogravimetric Analysis (TGA), Differential Scanning Calorimetry and Fourier Transform Infrared Spectroscopy are carried out. The bitumen is characterized by means of TGA, FTIR, DSC, Elemental Analysis and Saturate- Aromatic- Resin- Asphalthene (SARA) fractionation. Two new test methods are developed in order to measure the self- healing capability of bitumen. One of these methods is developed to see the effects of rest periods when intermittent loading is applied on the sample at high temperatures. The other method is developed to investigate the effects of long rest periods on the self- healing capability of bitumen at low temperatures. Measurements are carried out both on unmodified and ionic liquid modified bitumen to investigate the effects of ionic liquids with different side chain lengths on the self- healing properties of the bitumen. In addition to the measurements by the newly-developed techniques, stripping tests are carried out to find out about the effects of different ionic liquids on bitumen- aggregate interactions. Zeta potential measurements are done to quantify the electrostatic interactions between bitumen and aggregates. The asphalt performance of the modified bitumen is measured by Universal Testing Machine with Fatigue test, in which a loading- rest period cycle is applied to the sample until the sample is cracked, and Stiffness tests, which measures the hardness of the asphalt sample.

The results indicated that different ionic liquids have different effects on the bitumen/asphalt properties. With the addition of ionic liquids into bitumen, a 25% improvement in stripping properties, 30% improvement in high temperature self-healing performance, 200% improvement in low temperature performance and 120% improvement in asphalt fatigue life is obtained, meaning that different modification recipes can be used in road pavements depending on the climatic conditions and traffic density.

470079 Environmentally Responsible Fabrication of Efficient Perovskite Solar Cells from Recycled Car Batteries

Abstract:

Environmentally Responsible Fabrication of Efficient Perovskite Solar Cells

from Recycled Car Batteries

Po-Yen Chen[1,2]*, Paula T. Hammond[1], Angela M. Belcher[3]

[1] *Department of Chemical Engineering, Massachusetts Institute of Technology, Cambridge, MA, 02139*

[2] *School of Engineering, Brown University, Providence, RI 02912*

[3] *Department of Materials Science and Engineering, Massachusetts Institute of Technology, Cambridge, MA, 02139*

Email: pychen@mit.edu

Organolead halide perovskite solar cells (PSCs) show great promise as a new large-scale and cost-competitive photovoltaic technology. Power conversion efficiencies over 15% to over 20% have been achieved within 18 to 36 months of development, and thus perovskite materials have attracted great attention in photovoltaic research. However, the manufacture of PSCs raises environmental concerns regarding the over-production of raw lead ore, which has harmful health and ecological effects. In this study, we have demonstrated an environmentally responsible synthetic pathway to reuse car batteries for the fabrication of efficient PSCs (**Figure 1**). Both the anodes and cathodes of car batteries serve as material sources for the synthesis of lead iodide perovskite materials. In contrast to the traditional lead extraction process, our synthesis pathway from recycled battery materials occurs at a lower temperature (600 degree C) and does not include the hazardous emission of lead vapor/dust and CO_2 to the environment. The lead iodide perovskite materials synthesized from car batteries and high-purity reagents demonstrate identical material characteristics. The photovoltaic performance of the PSCs synthesized by each route is the same, which demonstrates that the device quality does not suffer from the materials sourced from spent car batteries. Also, electrochemical impedance spectra (EIS) measurements reveal that each device type displays the same resistances of electron recombination, indicating that the electron-trans- port properties of the lead iodide perovskite are identical. Finally, a simple economic analysis reveals that a single lead- acid car battery can supply enough lead material for the fabrication of 709-m^2 PSCs, which can provide enough electricity to power 30.2 US residential units in Las Vegas, Nevada. With these techniques, the time required to find a lead replacement for PSCs can be further increased. The environmentally responsible fabrication is expected to be broadly applicable not only to the PSC technology but also other applications.

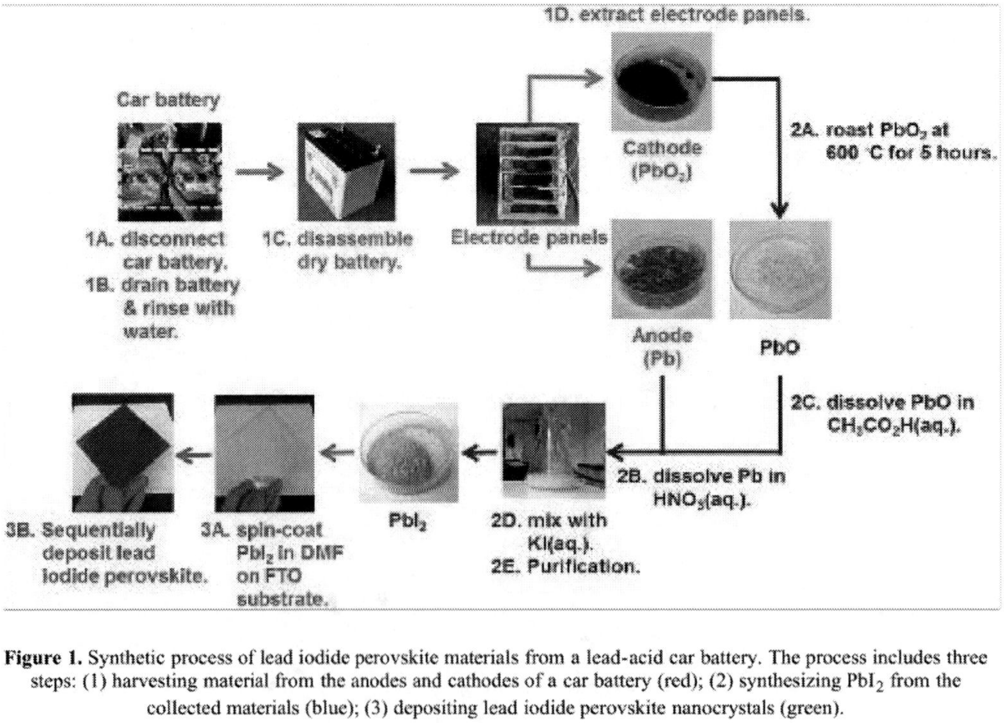

Figure 1. Synthetic process of lead iodide perovskite materials from a lead-acid car battery. The process includes three steps: (1) harvesting material from the anodes and cathodes of a car battery (red); (2) synthesizing PbI_2 from the collected materials (blue); (3) depositing lead iodide perovskite nanocrystals (green).

471578 Recycling & Garbage - a Separations & Materials Challenge

Abstract:

The field of recycling and garbage is largely overlooked by the Chemical Engineering community and has been the realm of mechanical, civil and even electrical engineers. But there is much to learn and apply in terms of Chemical Engineering from this field. Some surprising facts is that 99.9% of the separations at recycling facilities are done by machines - using the densities and shapes of materials to perform the separation. There are also major opportunities to increase the profit in garbage and recycling through better separation techniques and innovative material applications.

San Francisco has the goal of not sending anything to the landfill by 2020 - only 4 years away. Right now, they are currently around a 70% diversion rate. As the deadline approaches, innovations in this field cannot be just limited to the realm of mechanical engineering. For instance, in the past 10 years, composting technology has increased the amount of materials which are able to be composted. First only yard trimmings were accepted, then food scraps, now San Francisco is able to accept food stained paper products. The ultimate goal of composting is â€œanything which was a plant.â€

The garbage stream is also seeing innovations. Earlier this year, San Francisco adopted an organic press, which squeezes all the liquids from the garbage with a hydraulic press. Liquid from the process is sent to a biogas plant to convert into methane & then electricity.

On the recycling side, more materials and separation advances are required to keep the recycling industry afloat.

Currently, the driving force for places where recycling makes sense is areas with not a large amount of space - therefore landfill tipping rates (the cost to dump at a landfill) increase. The Pacific Northwest is home to the highest tipping fee in the nation ($142/ton) and land rich Texas has the lowest at $5/ton. Recycling programs therefore follow the incentives of the state they reside. Diversion from the landfill makes economic sense when instead of paying

$40/ton (Bay Area price), separating and giving away material for free. The $40 gap is saved but no profit is made.

The real profit from recycling comes from one material - aluminum. It accounts for 2% of the weight of the recyclate but 40% of the revenue. And paper and plastic are shipped overseas to be manufactured. Advances in material separations and new applications of recycled material is required to make recycling a truly sustainable business model.

Glass is the major loser of recycling. It is such a heavy material so the cost of transportation and disposal is very high. Additionally, remanufacturing options are limited - making new glass bottles is much cheaper than using recycled glass. New, business-sustainable applications of glass are required to make recycling of glass not a losing proposition. Companies are putting recycled glass into concrete and pavement - which is good - but these are not high margin applications. So, glass needs to be and is heavily subsidized by the State of California - which charges a fee to distributors (and then end customers) for glass products.

These opportunities in the garbage and recycling industries are ripe for the chemical engineering community. The focus must turn towards novel applications of materials and advanced chemical separation techniques â€“ both well within the chemical engineering realm.

463137 John Quinn: Master Weaver of a Remarkable Professional Tapestry

Abstract:

Professor John A. Quinn first came to prominence in chemical engineering circles for his fundamental research on mass transfer and interfacial phenomena. His work featured experiments thoughtfully conceived to permit accurate determination and rigorous analysis of interphase mass transfer rates at freshly created interfaces – an achievement made possible by his design of apparatus that brought immiscible fluids together in well-defined geometries at unprecedentedly short phase contact times. These experiments enabled the intrinsic mass transfer resistance of the interface itself to be explored, thus providing John with the opportunity to test the prevailing assumption that solutes diffusing from one immiscible phase into another would always experience equilibrium across the infinitely thin gas/liquid and liquid/liquid interfaces separating them.

While in the process of elucidating the role of the interface in mass transport, John Quinn and his students would initially be plagued by -- and then later intrigued with -- the thin films of surface-active materials that invariably concentrated at phase boundaries. Exploration of these insoluble monolayers revealed that their influence frequently extended well beyond the immediate vicinity of the interface; under some circumstances these interfacial films promoted convective instabilities while at other times they exerted a stabilizing effect on the bulk phases on either side of them.

Given his preoccupation with these thin films or interfacial "membranes" that spontaneously accumulated at phase boundaries, it was inevitable that John Quinn's attention would turn to synthetic membranes, inasmuch as these afforded controllable properties (e.g., porosities and thicknesses) that made them compelling platforms for fundamental transport studies. Track-etched mica membranes provided particularly fertile ground for the investigation of various steric, hydrodynamic, and electrodynamic effects key to the transport of solutes and colloids through submicron pores. Out of this work would come a much-improved

theoretical description of sterically hindered diffusion, as well as critical insights on filtration and adsorption in finely porous media. Synthetic membrane composites also informed studies in his group of gas permeation across the skin.

By virtue of his fundamental mass transfer work and his keen observations of transport across thin films, John Quinn was well-positioned to witness and participate in the birth and subsequent growth of a membrane separation industry initially focused on simple polymer films. But John's vision extended far beyond the capabilities of those conventional synthetic membranes to what might be possible if engineers were to mimic elements of the structure and function of vastly more powerful biological membranes. For instance, John noted that cell membranes achieved unparalleled selectivity by exploiting reversible reactions between permeants and membrane-localized carriers, and he was fascinated by the ability of membrane-bound biomolecules to deftly orchestrate key biochemical reactions. He responded by materially advancing the theory of carrier-mediated or facilitated transport and by inspiring the development of enzyme membrane bioreactors capable of chiral separations important to the pharmaceutical industry.

More significantly, John Quinn's enthusiasm for the intellectual interface between biology and engineering – and his influence on the direction that his department at Penn and his chosen field of chemical engineering would take -- amply justify his recognition as a pioneer in the then-emerging field of biomolecular engineering. John and his students would go on to explore several phenomena mediated by biomolecule-rich cell membranes, including, for instance, the attachment of cells to surfaces, the motility of cells on those surfaces, and the directed transport of motile bacteria.

This presentation traces the logic and serendipity that underlie the evolution of John Quinn's varied research interests, identifying the threads that tie together the body of his work into a coherent whole – a truly remarkable "professional tapestry". It documents the considerable impact that his scientific and technical contributions have had on our profession – and it provides an occasion to honor the thoughtful guidance and warm friendship that John so generously offered to so many, his students and colleagues alike.

457447 John Quinn: Advisor, Mentor, and Friend

Abstract:

Much of John Quinn's interest and curiosity centered on surfaces and interfacial phenomena. My thesis on interfacial resistance to mass transfer at liquid/liquid interfaces was based on a most unlikely experiment that somehow John knew had a chance of working. This work inspired much of my career. For example, early in my career I made membranes as thin as 10nm, not much more than two surfaces. Later I helped find a catalytic needle in a haystack that saved a business based on a reaction at a solid surface.

456402 Facilitated Transport Membranes for CO_2 Separation and Capture

Abstract:

This presentation is in honor of Dr. John Quinn for his pioneering work on facilitated transport membranes including those for CO_2 separation, and it covers CO_2 facilitated transport membranes for: (1) hydrogen purification for fuel cells and (2) CO_2 capture from flue gas in power plants and from <1% CO_2 concentration sources. We have synthesized CO_2-selective membranes by incorporating amino groups into polymer networks. The membranes have shown high CO_2 permeability and selectivity vs. hydrogen, carbon monoxide and nitrogen. We have elucidated the effect of amine steric hindrance in the solid membrane, showing significant enhancement for CO_2 transport. The membranes were scaled up to 14 inches in width by using a continuous roll-to-roll machine equipped with a thin-film-coating assembly. For the CO_2 separation from a synthesis gas stream containing CO_2 and H_2S, i.e., hydrogen purification for fuel cells or pre-combustion CO_2 capture in IGCC (integrated gasification combined cycle) technology, thicker amine-containing membranes with around 15 ± 1 microns of selective layer thicknesses were fabricated. For the post-combustion CO_2 capture from flue gas, thinner amine-containing membranes with < 200 ± 20 nm of selective layer thicknesses were scaled up. From gas transport measurements, the flat-sheet samples of the scale-up membranes for both applications exhibited similar performances compared to the membranes synthesized in lab scale. CO_2 permeance of approximately 200 GPU and CO_2/H_2 selectivity > 200 were for the thicker membranes for the purification of hydrogen and synthesis gas along with CO_2 capture (e.g., pre-combustion capture), and CO_2 permeance of about 800 GPU and CO_2/N_2 selectivity > 140 were for the thinner membranes for CO_2 capture from flue gas. The scale-up membrane has been fabricated into spiral-wound membrane modules. Techno-economic analysis has shown that the post-combustion capture process using the thin membrane has approached DOEâ€™s

capture cost target for 2025.

455619 The Force on a Body in Active Matter

Abstract:

The statistical mechanics and microhydrodynamics of active matter systems have seen a dramatic increase in interest in the past several years. Much attention has focused on the fascinating nonequilibrium behaviors of active matter not observed in equilibrium thermodynamic systems, such as spontaneous collective motion and swarming. Even minimal kinetic models of active Brownian particles exhibit self-assembly that resembles a gas-liquid phase separation. Self propulsion allows active systems to generate internal stresses that enable them to control and direct their own behavior and that of their surroundings. Recent work involving the unique 'swim pressure' exerted by active systems offers a perspective on the basic underlying physical mechanism responsible for self-assembly and pattern formation in active matter [1-3].

In this work we take a more microscopic view and present a general theory for determining the force (and torque) exerted on a body (or boundary) in active matter [4]. The theory extends the description of passive Brownian colloids to self-propelled active particles and applies for all ratios of the thermal energy k_BT to the swimmer's activity $k_sT_s = \zeta U^2{}_0 \tau R/6$, where ζ is the Stokes drag coefficient, U_0 is the swim speed and τR is the reorientation time of the active particles. The theory, which is valid on all length and time scales, has a natural microscopic length scale over which concentration and orientation distributions are confined near boundaries, but the microscopic length does not appear in the force. The swim pressure emerges naturally and dominates the behavior when the body size is large compared to the swimmer's run length $l = U_0 \tau R$. The theory is used to predict the motion of bodies of all sizes immersed in active matter.

References
[1] S.C. Takatori, W. Yan and J.F. Brady, Phys. Rev. Lett. 113, 028103 (2014)
[2] S.C. Takatori and J.F. Brady, Phys. Rev. E 91, 032117 (2015)
[3] S.C. Takatori and J.F. Brady, Current Opinion in Colloid & Interface Science, 21, 24 (2016)
[4] W. Yan and J.F. Brady, J. Fluid Mech. 785, R1 (2015)

459226 A Talk on Bijels, in Honor of John A. Quinn

Abstract:

John A. Quinn was a pioneer who advanced our fundamental understanding of processes that occur at fluid interfaces in the field of chemical engineering with practical implications in separations and reaction processes. John was also an outstanding colleague who formed and nurtured at the University of Pennsylvania our extraordinary department culture of intellectual exchange and interaction. In this talk, I briefly highlight a few examples of his work in the field. Thereafter, I discuss an emerging vehicle for study by chemical engineers in these contexts: Bijel membranes. Bijels, first predicted in simulation (1) and realized by a delicate experimental process, (2) comprise bicontinuous streams of immiscible fluids stabilized by jammed layers of nanoparticles at their interface. We have recently developed a robust and versatile means of fabricating these membranes that exploits a suspension of nanoparticles in a ternary aqueous/oil/co-solvent system. (3) This platform allows a broad range of particles and fluid pairs to be formed into fibers or membranes with tremendous promise in the fields of separations and interphase catalysis. Further, the potential importance of these vehicles in practical application will spur fundamental questions of trapped states and interfacial transport worthy of detailed analysis in the style of Quinn. Finally, this work, performed with colleague Daeyeon Lee at the University of Pennsylvania, embodies Johnâ€™s spirit of collaborative endeavor.

(1) **Colloidal Jamming at Interfaces: A Route to Fluid-Bicontinuous Gels, K. Stratford et al** *ScienceÂ* 2005,**309**, 2198-2201.

Â (2) **Bijels: a new class of soft materials, Michael E. Cates and Paul S. Clegg** *Soft Matter*, 2008, **4**, 2132-2138.

Â (3) **Continuous Fabrication of Hierarchical and Asymmetric Bijel Microparticles, Fibers, and Membranes by Solvent Transfer-Induced Phase Separation (STRIPS)**Â M.F. **Continuous Fabrication of Hierarchical and Asymmetric Bijel Microparticles, Fibers, and Membranes by Solvent Transfer-Induced Phase Separation**

(STRIPS) Haase, D. Lee, K.J. Stebe, Advanced Materials 2015, **27**, 7065-7071.

466714 Reverse Osmosis As a Tool in Water Treatment Strategies

Abstract:

Reverse osmosis is a workhorse tool commonly leveraged in water treatment flowsheets to remove highly soluble salts. This paper will briefly review state of the art reverse osmosis membrane chemistry and the manufacturing process, and then discuss how RO spiral wound modules are employed in conjunction with suitable pre-treatment technologies to create very high quality water for a wide variety of municipal and industrial applications.

463011 Recombinant Oleosin As a Functional Surfactant

Abstract:

Our laboratory is interested in novel surfactants that can be self-assembled into technologically useful suprastructures and can be equipped with novel biofunctionality. Oleosin is a naturally occurring surfactant protein found in plants. The molecule has three domains – two terminal hydrophilic domains, and a central hydrophobic domain with an embedded proline knot, which forces the molecule into a turn. We have obtained the gene for oleosin and expressed it in E.coli, and, using the tools of molecular biology, made hundreds of variants of the molecule, including numerous truncations and additions of functional or targeting domains. Depending on the truncation, oleosin can assemble into vesicles, worm-like micelles, sheets, or spherical micelles – assemblies which have utility in drug delivery and tissue engineering. In one application, we show how spherical micelles equipped with receptor targeting domains can be activated by a protease by protecting receptor-targeting domains with protease cleavable domains. The corresponding spherical micelles bind to target cells, but only when both the protease and the receptor are present, effectively creating an "and" switch for cellular recognition. We also show how oleosin's properties as a surfactant can be exploited to stabilize mondisperse air bubbles for ultrasound imaging. We will also provide a further update on ongoing efforts to incorporate other functional domains into oleosin for a variety of applications.

473326 Driving Change in Pharmaceutical Development & Manufacturing through Innovation and Standards

Abstract:

We are all familiar with the old adage â€œbetter-faster-cheaper, Pick Twoâ€. However in todayâ€™s world of product development and manufacturing, all three are needed and needed now. Organizations are tasked with moving diverse and complex pipelines forward in record time with undivided attention required for product quality, safety and reliability of manufacturing. These challenges do not require incremental change in the operating model but revolutionary change in our model. In this talk I will discuss activities, specifically process and product innovation geared to customer value and standards-based thinking for product lifecycle management and partnership models. I will discuss the implications for the processes, plants and people to ensure successful commercial products.

461582 Distillation Tray Design through a Better Understanding of Operating Windows

Abstract:

Operational capacity of trays is quite detailed and there are numerous hydraulics parameters that have been offered for decades to help the designer understand both turnup and turndown of such equipment. Understanding all of the terms available such as; Flood, Spray, System Limit, Backup, Max velocity, Min velocity, Weir loading, entrainment, weeping, dumping, stability, balancing, plug flow, downcomer seal, etc. can be a daunting task for even an experience distillation person. All of these hydraulic parameters can be plotted on an Operating Window (or sometimes referred to as a performance diagram) and once plotted, the designer can/should have a better understanding of each hydraulic parameterâ€™s behavior. The author has written a previous paper on this topic, this paper will take the performance diagram one step further and show some of the practical aspects of its usefulness in industry.

461594 Perspectives on the Impact of Environmental Factors on Performance of Metal-Organic Frameworks in Adsorption Systems

Abstract:

Metal-organic frameworks (MOFs) are permanently porous coordination polymers that have emerged as an important new class of porous materials with the potential to make an immediate impact in adsorption separation technologies. MOFs are synthesized by a self-assembly process in which metal or metal-oxide vertices are connected by rigid or semi-rigid organic molecules. The rational synthesis approach opens up the possibility for new advances in the development of new and improved adsorbents for air purification and the removal of toxic chemicals. MOFs have shown the potential to be designed for adsorbate-specific interactions, creating effective and selective â€œmolecular spongesâ€. However, the density of active sites is often low, and the commercial use of MOFs is still limited by issues with chemical stability and synthesis scale-up. This perspective will discuss our comprehensive work and recent progress in this area and will demonstrate the applicability of MOFs for adsorption separations and catalysis.

461596 Using All Available Tools for Advanced Membranes for Gas Separations

Abstract:

Gas separations will be factors in determining long term winners from losers in the competitive global chemical industry, where large costs and savings are at stake in a competitive market. As highly energy-intensive processes, gas separations benefit from strategies that reduce energy use while minimizing capital costs needed to deliver each unit of product. Membranes are tools to implement such strategies in capacity expansions and new processes, which are enabled by approaches that were less attractive even a few decades ago. The ability to naturally incorporate molecularly-selective partitioning and diffusive discrimination between components in a single device causes membranes to be tools for process-intensification. The selective partitioning feature is understandable in terms of well-known thermodynamic factors. On the other hand, engineering the diffusion selectivity feature is a more recent tool. Examples of advanced materials, novel membrane structures and processes displaying high productivity and economical energy-efficient gas separations will be discussed. Not only engineering of diffusion activation energies but also of diffusion activation entropies associated with size-similar penetrants will be shown to be responsible for the development of these advanced tools.

461600 Improving Energy Efficiency of Distillation

Abstract:

Distillation is the most widely used separation process. It is always important to design and build an energy efficient process within the capital constraint. We will discuss some of our recent work on how to improve efficiency of multicomponent distillation through proper choice of column arrangement. We will also present new heat and mass integrated distillation columns that are not only most energy efficient, but also use reduced number of distillation sections leading to lower capital cost. Finally, we will briefly discuss novel dividing wall column arrangements that do not have the conventional constraint of splitting vapor between the two sides of a dividing wall. This enables multicomponent distillation with process intensification while providing high energy efficiency.

These examples illustrate that multicomponent distillation research is vibrant and provides rewarding opportunities for performance improvements.

461603 Towards MOF Membrane Technology: High-Performance Separations via Mixed-Linker ZIFs and Membrane Processing on Polymeric Hollow Fibers

Abstract:

This talk will describe our progress in the synthesis and properties of ZIF-type MOF materials and their subsequent use in scalable fabrication processes to obtain molecular sieving membranes on polymeric hollow fibers. Membrane separation is a key component of strategies targeting step changes (rather than incremental ones) in cost and energy-efficiency during production of fuels and chemicals from conventional and renewable feedstocks. Two important barriers to the development of MOF membrane separation technology are (i) the difficulty in obtaining tunable, high-performance, and robust molecular sieving materials; and (ii) the lack of low-cost, scalable membrane fabrication processes. A multiscale approach to membrane fabrication process engineering is required to overcome these barriers. We will demonstrate that the mixed-linker approach to synthesis of ZIF-type MOFs offers a path towards fine (sub-Ångstrom) control of pore structure and separation properties. Furthermore, we will discuss how such materials can be processed into polycrystalline membranes on a scalable hollow fiber platform, with control over nanometer-scale defects as well as micron-scale transport and film crystallization phenomena within the hollow fibers. These membranes allow high-performance separations of molecules such as hydrogen, light olefins, and isomers. The talk will conclude with a proposal to generalize the above strategy towards a versatile MOF membrane technology that addresses a wide range of molecular separation targets.

481236 Non-Classical Mechanisms of Nucleation of Crystals in Solution

Abstract:

Crystal nucleation is a central problem in numerous areas of science, technology, and medicine and one of the last scientific fields, in which theoretical predictions diverge from careful measurements by many orders of magnitude. Recent studies have demonstrated that the discrepancies between theory and experiment are due to behaviors that go beyond the assumptions of classical model structures. Three non-classical nucleation behaviors have received particular attention. Crystals of several substances were shown to nucleate form within pre-existing precursors, and not directly in the solution. The nucleation precursors are clusters consisting of dense liquid, which are metastable with respect to the host solution. Additional work has revealed that the shape of the nucleus is determined not by minimization of its free energy, but by the kinetics of aggregation. The non-equilibrium nucleus shape results in higher nucleation barriers and significantly slower nucleation. At high supersaturations, the size of the nucleus shrinks to just one molecule. In contrast to common beliefs , the nucleation rate in this spinodal regime is not infinite but is limited by the rate of protein association. The investigations of the non-classical nucleation behaviors are still in their infancy. Results on direct imaging of clusters and characterization of cluster mechanisms with a variety of compounds will soon lead to major breakthroughs in bottom-up design of new materials.

481227 Recent Advances in Crystal Structure Prediction: Perspectives and Opportunities

Abstract:

The prediction of the possible crystal structure(s) of organic molecules is an important activity for the pharmaceutical and agrochemical industries, among others, due to the prevalence of crystalline products [1]. The specific crystal structure adopted has implications for product performance as well as manufacturing. In this talk, recent progress in the area of crystal structure prediction is reviewed, focusing on the general requirements that crystal structure prediction (CSP) methodologies need to fulfil in order to achieve reliable predictions over a wide range of organic systems [2]. The current status of a multistage CSP methodology that has proven successful for a number of systems of practical interest, including pharmaceutical compounds and co-crystals, is discussed. The approach is based on a global search stage, with the *CrystalPredictor* algorithm [3], and a refinement stage based on local minimization with more accurate models, using the *CrystalOptimizer* algorithm [4]. The application of the approach to specific compounds is discussed, with particular emphasis on reliability and computational tractability. Further development needs are highlighted and ways in which one might progress from a description of the most likely solid phases to the prediction of properties of relevance to process and product development, such as solubility, are discussed. This requires the prediction, from a computed crystalline energy landscape, of the free energies of various solid phases, and the use of recent advances in the predictive modelling of the free energy of liquid phases [5].

[1] Storey R.A., Ymén I. (2011). Solid State Characterization of Pharmaceuticals. Wiley, Chichester, U.K.
[2] Pantelides, C.C., Adjiman, C.S., Kazantsev, A.V., 2014. General Computational Algorithms for Ab Initio Crystal Structure Prediction for Organic Molecules, in: Atahan-Evrenk, S., Aspuru-Guzik, A. (Eds.), Topics in Current Chemistry. Springer, pp. 25-58.
[3] Karamertzanis, P.G., Pantelides, C.C., 2005. Ab initio crystal structure prediction â€" I. Rigid molecules. Journal of Computational Chemistry 26, 304-324 ; Karamertzanis, P.G., Pantelides, C.C., 2007. Ab initio crystal structure

prediction. II. Flexible molecules. Molecular Physics 105, 273-291; Habgood, M., Sugden, I.J., Kazantsev, A., Adjiman, C.S., Pantelides, C.C., 2015. Efficient Handling of Molecular Flexibility in Ab Initio Generation of Crystal Structures, JCTC, 11, 1957-1969; Sugden, I.J., Adjiman, C.S., Pantelides, C.C., Accurate and efficient representation of intramolecular energy in ab initio generation of crystal structures. Part I: Adaptive local approximate models, Acta Cryst. B, accepted (2016).
[4] Kazantsev, A.V., Karamertzanis, P.G., Adjiman, C.S., Pantelides, C.C., 2011a. Efficient Handling of Molecular Flexibility in Lattice Energy Minimization of Organic Crystals. Journal of Chemical Theory and Computation 7, 1998-2016.
[5] 64. Papaioannou, V., Lafitte, T., AvendaÃ±o, C., Adjiman, C.S., Jackson, G., MÃ¼ller, E.A., Galindo, A., 2014. Group contribution methodology based on the statistical associating fluid theory for heteronuclear molecules formed from Mie segments, J. Chem. Phys., 14 , 054107.

466412 Electrophoretic Separation of Biomolecules in Nano-Lipid Membrane

Abstract:

The high performance separation of membrane proteins has not yet been attained in despite of the significance, and been demanded. For that, we developed a novel separation method in a nano-lipid membrane, which utilizes the generated function of nano-lipid membranes.The nano-lipid membranes were constructed on the glass slide by self-assembly of lipid molecules. At first we investigated the factors affecting the separation in the nano-lipid membrane. In the nano-lipid membrane, membrane proteins kept its original structures as like in the cell membranes. In addition, the diffusion of membrane proteins was suppressed compared that in free solution. These specific properties of the nano-lipid membrane enable non-denaturing separation and high separation efficiency. Furthermore, we clarified that the separation efficiency of membrane proteins could be easily controlled by varying the composition and phase state of lipid membrane. Finally, we attempted the electrophoretic separation of membrane proteins and easily attained it in the nano-lipid membrane. Thus, we could successfully demonstrate the possibility of high performance separation in nano-lipid membranes and developed a new field of nano-separation sciences.

457175 Immobilization and Orientation Controll of Anti-CRP Scfv on a Seuface of Polystyrene Latex Beads and Its Application to Latex-Turbidimetoric Assay

Abstract:

In this study, we investigated immobilization of anti-C-Peactive Protein (CRP) single-chain Fv antibodies onto a surface of polystyrene latex beads for use in latex turbidimetoric assay of CRP. Both Polystyrene-binding peptide (PS-tag) and Poly(methyl methacrylate)-binding peptide-(PMMA-tag) were introduced to its C-terminal region for site-specific attachment and orientation control of scFv, while scFv with D10-tag which was decapeptide only composed of Asp was used as a negative control. Three types of scFvs were over-expressed in an insoluble fraction of E. coli cell, and highly purified via IMAC under the reduced and denatured condition. Conformation of tagged scFvs were refolded by one-step dialysis using 50mM Tris-HCl, pH 8.5 for PMMA-tagged scFv (scFv-PM) and D10-tagged scFv (scFv-D10), and 50mM MES, pH 6.0 for PS-tagged scFv (scFv-PS). Consequently, according to the results of indirect ELISAs and sandwich ELISA, antigen-binding activities of all scFvs tested were fairly recovered by the refolding operation.

ScFv-immobilized PS latex beads were prepared by the passive adsorption method. Averaged diameter of scFv-PS were significantly increased probably because multi-point interaction of positively-charged PS-tag induced aggregation of PS latex beads which were negatively-charged. On the other hand, latex beads with scFv-PM and scFv-D10 were well-dispersed at the pH between 7.5 and 8.5, while scFv-PM-immobilized latex beads were aggregated at pH 6.5, suggesting that surface charge of scFv-PM-immobilized latex beads might be neutral at this pH which was correspondence to isoelectric point of scFv without PMMA-tag. These results expected that scFv molecules were site-specifically immobilized through the PMMA-tag. ScFv-D10-immobilized latex beads were not aggregated at the pH range between 6.5 and 8.5, suggesting that scFv-D10 molecules were randomly oriented and change in charge density of latex beads was not significant so much because the exposed D10-tag might be caught up on surface charge density which must be reduced

by immobilization of scFv.

Turbidimetoric assays were performed using both scFv-PM and scFv-D10-immobilized latex beads. Consequently, scFv-PM-immobilized latex beads had wider detection range of CRP, compared with scFv-D10-immobilized latex beads.

Thus, scFv-PM-immobilized latex beads developed in the present study will be useful for economic and sensitive detection of biomarkers using latex-turbidimetric assay.

466400 Enzymatic Conjugation Strategy for the Design of Artificial Biomolecular Assemblies

Abstract:

A variety of functional biomolecular assemblies are observed in natural biological systems. In particular, proteinaceous supramolecular assemblies, such as viral capsids and bacterial surface pili, are of great interest in bioscience and bionanotechnology because their well-organized structures associated with unique biological functions would give us insight in the design of artificial biomolecular assemblies in practical applications. To mimic natural biological assembly systems, we have established site-specific protein conjugation techniques by using two enzymes with different catalytic properties.

Microbial transglutaminase (MTG) that works in crosslinking reaction between the side chains of glutamine (Gln) and lysine (Lys) residues gives us opportunity to attach an artificial affinity moiety to the selective site of protein of interest (POI). We have designed artificial substrates with multiple biotin moieties, and conducted site-specific biotinylation of POI. As a proof-of-concept study, tetrabiotinylated endoglucanase was prepared, and mixed with streptavidin. Eventually, one-dimensional assembly comprised of these protein building blocks was obtained, and it exhibited the unique hydrolytic activity for cellulosic substrates.

Another enzyme candidate is horseradish peroxidase (HRP) that catalyzes radical coupling reaction between the phenolic side chains of tyrosine (Tyr) residue. We found that HRPâ€™s substrate specificity can be narrowed to a peptide tag fused genetically to POI. In line with the HRP-catalyzed polyphenol formation, we succeeded in creating proteinaceous polymeric materials via the site-specific radical coupling at a Tyr-containing tag (Y-tag). It is worth noting that both MTG and HRP provide us with site-specific and covalent conjugation strategies for a range of biomolecules, which is useful for the design of artificial biomolecular assemblies.

462831 Optimization and Scale-up of Microfiltration TFF for Reliable Clarification Processes

Abstract:

One of the most challenging steps in the manufacture of microbially produced therapeutics is the clarification of feed solutions containing solubilized inclusion bodies. In such applications the recovery and purity of the target protein is often limited. The high percentage of impurities results into low depth filter capacities applying the normal flow filtration (NFF) and insufficient separation efficiency in addition to the scale-up limits of the centrifugation. Tangential flow filtration (TFF) with a microfiltration membrane has been considered in such processes. Importantly, the control of permeate flow rate is essential to achieve filtration success criteria and limit fouling.
This study highlights the successes of a permeate-controlled TFF microfiltration process at various scales with different feed streams. The laboratory TFF system with a membrane area of 0.1 m2 was used for the initial evaluation. A pump controlled the permeate flux ensuring a stable and successful process operation. By using the model feed (BSA containing Whey feed stream), the protein sieving was investigated. The process simulation has confirmed the cake-mass resistance profile and protein sieving prediction. For the clarification process of solubilized IBs from Escherichia coli, the permeate-controlled TFF set-up was compared to the NFF process in regards to separation efficiency. Although the TFF membrane used has a smaller nominal pore size than the used depth filters, a Pellicon 1000 kDa PES membrane in V-screen cassette provided higher retention of impurities. The resistance curves were analyzed applying the cake-lift model for TFF and cake model for NFF. The overall fittings suggests that the advantage of the permeate-controlled TFF might be due to the control of the membrane pressure increase which results in low membrane fouling, effective separation and high protein sieving. The scale-up to the Phase I GMP runs demonstrated the robustness of this technology and justifiably used in a clarification process.

464171 Design Calculation Procedure for Flow-through Chromatography

Abstract:

Flow-through chromatography (FTC) is considered to be an efficient method for separating two components as the flow is continuous. It is also suited for a single-use chromatography operation. Although several papers have been published on the flow-through chromatography process design, it is still not clear especially how the mass transfer affects the separation performance.

In this study, we have demonstrated how to design FTC based on the distribution coefficient as a function of salt concentration $K(I)$ obtained from linear gradient elution experimental data along with the zone spreading (mass transfer) data. We have determined K-I curves of monomer and dimer from linear gradient elution (LGE) experiments. The zone spreading parameter values were evaluated based on the plate number N by using a simplified single-lumped zone spreading parameter model. Curves for monomer and dimer were numerically calculated in order to determine the sample feed volume VF that can allow complete separation. Then, VF values were calculated as a function of I. For low K values, the operation time was short whereas VF was quite small. By increasing K values (which corresponds to lowering I), VF increased. However, the separation time increased markedly. The separation time – the sample feed volume curves were found to be useful for designing FTC. It was also found that for difficult separations such as aggregate removal the mass transfer plays a significant role. Flow-velocity must be carefully tuned in order to increase the sample volume. For such cases monoliths or membranes are better choice.

464386 Purification of Therapeutic Proteins By Affinity Precipitation Using Peptide Conjugated Smart Biopolymers

Abstract:

While chromatographic bioprocessing is the dominant mode of separation for biological products, there is an ongoing search for more cost-effective and scalable processes to address the increasing diversity of therapeutic biomolecules and their increasing titers. Affinity precipitation using smart biopolymers has generated much interest recently for its potential for the simultaneous recovery and purification of biological products. The current work focuses on designing affinity peptides against therapeutic proteins followed by the recombinant engineering of elastin like polypeptide (ELP)-peptide conjugates to develop affinity precipitation processes for proteins. Peptides were designed using epitope mapping as well as high throughput approaches such as phage display. Promising leads were then examined using microarray screening and fluorescence polarization in order to establish peptides with sufficient binding affinity as well as elutability using appropriate modifications in the fluid phase conditions. ELP-peptide constructs against model proteins and proteins of therapeutic importance were recombinantly engineered and their efficacy for purifying the proteins by affinity precipitation were evaluated. Further, the effects of different factors on the precipitation process were studied, such as (i) peptide presentation on the ELP, (ii) peptide affinity towards target protein, (iii) absence and presence of amino acid linkers between the peptide and ELP and (iv) nature and length of these linkers. This study will help to develop design heuristics for ELP based constructs to facilitate effective protein purification by affinity precipitation.

461256 Adsorption and Separation of Native and Pegylated Proteins on Anion Exchange Resins with Varying Degrees of Polymer Grafting

Abstract:

Ion exchange resins containing charged polymer grafts have been shown to enhance protein binding capacity and mass transfer rates as a result of the dynamic nature of the polymer structure. Most of the previous work in this area has, however, been devoted to cation exchange resins. In this work, we have examined the adsorptive and chromatographic properties of native and PEGylated proteins on a range of anion exchange (AEX) resins that are based on the same macroporous backbone matrix, but that contain different amounts of grafted polymers. BSA, PEGylated BSA, and thyroglubulin are used as model proteins covering a broad range of molecular sizes and strength of interaction with the charged ligands on the resin. Adsorption isotherms and batch uptake kinetics were determined for each protein in order to quantify the macroscopic behavior, while confocal laser scanning microscopy (CLSM) was used to elucidate intraparticle transport for single and multicomponent adsorption. The results show that the performance of these resins depends on a balance of size-exclusion and electrostatic interactions. Polymer grafting increases the binding capacity and leads to an enhancement of adsorption rates for the smaller proteins. On the other hand, as molecular size increases a point is reached where size-exclusion becomes dominant resulting in poorer performance. An optimum grafted polymer content seems to exist that optimizes the balance of these two properties.

458491 Theoretical Analysis of Enhanced Mass Transfer in Modules with Novel Twisted Hollow Fiber Membrane

Abstract:

A number of approaches have been used to develop membrane modules with enhanced mass transfer. Rotating membrane systems generate Taylor vortices that can significantly increase flux and reduce fouling, but these modules have very low membrane packing densities (membrane area per unit volume) and are very difficult to scale-up. Helically wrapped hollow fibers generate Dean vortices in a much more scalable geometry, although these modules have not yet been commercialized for large-scale applications. The objective of this work was to examine the behavior of twisted hollow fiber modules in which Dean vortices are generated by the coiled flow that occurs within a twisted elliptical flow channel. Computational fluid dynamic simulations are used to examine incompressible flow through twisted elliptical and helically-coiled cylindrical fibers for typical flow Reynolds numbers employed in membrane filtration processes ($Re < 200$). The resulting flow fields are compared in terms of the distributions of vorticity and wall shear rate, the pressure drop, and the azimuthally-averaged normal velocity in the vicinity of the wall. The secondary flow in twisted elliptical fibers generates significant recirculation flow leading to near-wall velocities directed normal to the surface of the membrane (below figure). The azimuthally-averaged normal velocity near the wall of a twisted elliptical fiber is found to be two to four orders of magnitude larger than typical ultrafiltration velocities. For $Re < 75$, the average wall shear rate and normal velocity near the wall are found to be significantly higher in twisted elliptical fibers compared to their helical counterparts. Modules produced using twisted elliptical hollow fibers would be linearly scalable by changing the number of parallel fibers, without any alteration in the device hydrodynamics. These findings highlight the potential for mass transfer and flux enhancement using modules with novel twisted elliptical hollow fiber membranes.

473805 Advances and Challenges in Membrane Desalination and Osmotic Power Generation

Abstract:

Membrane based water treatment and desalination technologies have been gaining momentum in potable water production from seawater and brackish groundwater, in municipal and industrial water reuse, as well as for management and reuse of agricultural drainage water. Considerable progress has been made since the early days of Reverse Osmosis (RO) and more recent development in Forward Osmosis (FO) with respect to water development of high performance fouling and chlorine resistant membranes, improvements of in feed pretreatment and reduction in RO energy consumption. Given the increased level of water scarcity in various regions of the world there has been growing efforts to expand the use of otherwise unutilized inland groundwater resources, lower the cost of seawater desalination and even produce energy from salinity gradients (i.e., pressure retarded osmosis (PRO). Here a fundamental description and evaluation of RO, FO and PRO will be provided with a focus on assessment of the merits of various process alternatives to common RO process configurations, the viability of PRO and challenges in FO. In particular, recent results from CFD analysis of FO and PRO and fundamental theoretical analysis of RO energy consumption (for different configurations) will be presented to clarify overoptimistic expectations and address the remaining major challenges.

462750 Cactus Inspired Self-Humidifying Membranes for Energy Generation

Abstract:

Cactus inspired self-humidifying membranes for energy generation

Young Moo Lee

Department of Energy Engineering, College of Engineering, Hanyang University, Seoul 133-791, Korea

The regulation of water content in polymeric membranes is important in a number of applications, such as in reverse electrodialysis and in proton-exchange fuel-cell membranes. External thermal and water management systems add both mass and size to systems, and so intrinsic mechanisms of retaining water and maintaining ionic transport in such membranes are particularly important for applications where small system size and density is important. For example, in proton-exchange membrane fuel cells, where water retention in the membrane is crucial for efficient transport of hydrated ions, by operating the cells at higher temperatures without external humidification, the membrane is self-humidified with water generated by electrochemical reactions.

Here we report an alternative solution that does not rely on external regulation of water supply or high temperatures. Water content in hydrocarbon polymer membranes is regulated through nanometre-scale cracks in a hydrophobic surface coating. These cracks work as nanovalves to retard water desorption and to maintain ion conductivity in the membrane upon dehumidification. Hydrocarbon fuel-cell membranes with surface nanocrack coatings operated at intermediate temperatures show improved electrochemical performance, and coated reverse electrodialysis membranes

show enhanced ionic selectivity with low bulk resistance.

472029 Organic and Biofouling of Polymeric Membranes: Mitigation Using Surface Graft Polymerization and Assessment with Fouling Analytics

Abstract:

Membrane processes require both mitigation and management of fouling to ensure viable process economics, and performance that meets required standards in terms of selectivity or separation effectiveness.

In this talk we will discuss mitigation approaches that focus on surface modification of poly(ethersulfone) membranes using plasma or photo-initiated graft polymerization. We will show that selection of graft moieties is aided by employing a high throughput screening process, which is able to identify surfaces having antifouling properties for a range of feeds. Examples will be provided for feeds of interest in water purification, including humic substances, extracellular polymeric substances modeled by sodium alginate, and microbial fuel cell effluent. In addition, we will show that biofouling can be controlled using both amines and zwitterionic surfaces (betaines).

Membrane surface modification can be employed to fulfill multiple objectives of reducing flux decline and improving rejection. An example will be presented to show how surface charge can be manipulated to improve the rejection of oxyanions found in contaminated groundwater while preserving antifouling properties toward humic substances.

Fouling analytics â€“ including models that describe the rate of flux decline â€“ can be used to help manage the impact of fouling on membrane processes. In this talk we will show how such analysis can be used to interpret data from screening studies and aid in membrane selection, reduce energy consumption, and optimize cleaning schedules to control the temporal progression of fouling, especially as different fouling mechanisms evolve.

463629 Surface Modification of Ultrafiltration Membranes : Where Are We Now ?

Abstract:

The surface modification of ultrafiltration membranes to increase their resistance to fouling started at the beginning of the 90'. Lots of works have been published, with some of them turned into successful commercial products. If commercial membranes are nowadays more fouling resistant than ever, there is still room for improvement and we can try to understand how this could be achieved. Models borrowed from polymer physics can help understanding the behaviour of hydrophilic moieties attached by one end to a membrane surface. As for an example, a model known as the Alexander's model, further developed by de Gennes and co-workers, suggests that when the grafting density is is high enough, the polymer chains stretch towards the bulk solution and form a so called "brush". Then, the foulant molecules/particles can:
Diffuse through the brush, if their dimension is small compared to the average distance between the grafted polymers on a membrane surface
Compress the brush if they are too large to penetrate it.
This helps to define some criteria for the membrane modification to be efficient. In particular, the trade-off between length of the polymer chains and grafting density can be discussed. A comparison of the model predictions to data of adsorption of proteins on solid surfaces taken from the literature shows the relevance of this model and its limitation. The experimental minimum grafting density required to avoid the adsorption of proteins seems to be larger than the threshold density for which a "brush" of polymer" would form at the surface of the membrane. We analyse, by FTIR microspectrometry, the surface of modified membranes to reveal
a) the heterogeneity of the surface modification at the scale of the millimeter, showing that parts of the surface of the membranes are not covered with the copolymer, even after several hours of contacting the membrane to the solution,
b) that the dynamics of the adsorption of the polymer to the surface is slower than expected.
In cases where the surface modification is unevenly distributed over the surface of the membrane, the

existence of a protective brush at the surface of the membranes is also probably unevenly distributed. This may explain this discrepancy between the experimental threshold density and the one predicted by the theory.
In the presence of a filtration flux, this picture can be significantly changed. When comparing adsorption (no flux through) and fouling (in the presence of flux) experiments run with a membrane modified with a given co-polymer, we find that the optimum grafting density in terms of resistance to fouling is larger than the grafting density which gives a high resistance to adsorption.
This can be analysed by accounting for the contribution of the flux to the compression of the brush. By combining the transport equations at the surface of the membrane to the Alexander model, one can show that this contribution can be significant, especially for short polymer length. A higher flux would then require a higher grafting density.

Figure 1: Depending on the fouling molecule/particle radius and on the density of hydrophilic polymers grafted at the surface of a membrane, a molecule/particle can approach the surface either by insertion into the brush formed by the polymers or by compressing the brush. The "resistance" to the compression (V/kT) requires higher grafting densities for small molecules/particles. If the grafting density is lower than a threshold value (around 0.5% here) the grafted polymers are not expected to form a brush but may mask part of the surface of the membrane and reduce adsorption.

Some references:

[1] Surface modification of ultrafiltration membranes by low temperature plasma, II. Graft polymerization onto polyacrylonitrile and polysulfone, M. Ulbricht, G. Belfort, J Membr Sci, 111 (1996) 193-215

[2] S. Alexander, Adsorption of chain colloids with a polar head : a scaling description, J Physique 38(1977) 983-987

[3] S. I. Jeon, J. H. Lee, J. D. Andrade, P. G. de Gennes, Protein-Surface Interactions in the Presence of Polyethylene Oxide, J Coll Interface Sci 142(1991)149-158.

452700 Experimental Validation of Membrane-Vapor-Extraction Separation Process

Abstract:

Our group recently proposed a novel separation process called membrane vapor extraction (MVE), to separate dilute (bio)solutes from an aqueous fermentation broths [1]. In MVE, semi-volatile aqueous solutes vaporize at the upstream side of a membrane, diffuse in through the membrane pores, and subsequently dissolve into a nonpolar solvent highly favorable to the solutes but not to water. Since the two liquids are not in contact, MVE avoids formation of difficult to separate microemulsions.

Design analysis of a 1.5-m long, 30-m^2 membrane-area countercurrent MVE unit for processing 2-wt. % aqueous butanol by a prototype solvent (dodecane) at 40 °C demonstrated over 90 % recovery of the feel butanol with essentially no water loss [1]. The separation factor is well over 1000. Our initial design study, however, gave no experimental evidence for the calculated MVE-unit performance [1].

The present study experimentally validates the MVE process. A 6-cm wide by 10-cm long Teflon channel-flow cell with 0.1 mm gap thicknesses permits countercurrent (or concurrent) extraction studies. The flow cell is operated in the transient mode since axial concentration changes are minimal. 0.2-µm pore-diameter VersaporeR membranes (Pall Corporation, New York) are both hydrophobic (water contact angle = 128°) and oleophobic (dodecane contact angle = 105°). Accordingly, VersaporeR membranes are nonwetting to both feed and solvent, as required for MVE.

Figure 1 shows a typical result for extraction of 2-wt. % aqueous butanol into dodecane at 40 °C. Filled triangles illustrate loss of butanol from the aqueous feed stream, whereas filled circles demonstrate absorption of butanol into the countercurrent dodecane solvent stream. Solid lines represent *a priori* predictions using mass transfer coefficients from the Graetz-Lévêque analysis for laminar channel flow. The excellent agreemen confirms that vapor-phase diffusion resistance is negligible in the MVE-membrane pores. We present similar results for different operating conditions in MVE including feed concentration, temperature, and flow rate. In consort with the initial design analysis [1], our new experimental results confirm the viability of MVE to remove and recover dilute aqueous (bio)solutes with minimal energy expenditure.

1. D.E. Liu, C. Cerretani, R. Telles, A.P. Scheer, S. Sciamana, P.F. Bryan, C.J. Radke, and J.M. Prausnitz, "Analysis of Countercurrent Membrane Vapor Extraction of a Dilute Aqueous Biosolute," AIChE Journal, 61(9), 2795-2809 (2015).

Figure 1. Transient extraction of 2-wt % aqueous butanol (filled triangles) into dodecane (filled circles) using countercurrent MVE at 40 °C. Solid lines are predicted *a priori* with Graetz-Lévêque mass transfer resistances and a measured partition coefficient.

459986 How to Describe and Predict Plasticization of Glassy Polymeric Membranes in Gas Separations

Abstract:

<span">How to describe and predict plasticization of glassy polymeric membranes in gas separations

<span">

<span">Francesco M. Benedetti, Matteo Minelli and Giulio C. Sarti<span">

<span">Department of Civil, Chemical, Environmental and Materials Engineering (DICAM)

<span">Alma Mater Studiorum - University of Bologna, via Terracini 28, 40131, Bologna, Italy.

<span">

giulio.sarti@unibo.it

<span">

<span">Plasticization of glassy membranes due to gases or vapors is typically recognized through an increase in gas permeability as the upstream pressure increases. Qualitatively that is associated to an enhancement of polymer chain mobility with consequent increase in penetrant diffusivity, starting from a certain relatively high upstream pressure below which such effects are not present. The phenomenon is normally associated to the sorption of large quantities of low molecular weight species into the polymer matrix, leading to important volume dilation, reduced mechanical rigidity, depression of the glass transition. The increased mobility affects the transport properties of the plasticizing agent as well as of the other gases possibly present in the feed, thus reducing the membrane mixed gas selectivity. The quantitative description of gas permeability and, possibly, the prediction of its dependence on operating pressure is thus relevant for a better understanding of the phenomenon and for the selection of suitable operating conditions.

By considering several different glassy membranes, we have observed that the various different behaviors shown by gas permeability, including plasticization, can be described by considering only a solution-diffusion model in which penetrant molecular mobility varies with its concentration through an exponential law, with two adjustable parameters only. Diffusivity is taken as the product of molecular mobility and a thermodynamic factor, calculated by using the NELF model for thermodynamic properties of the glassy phase.

In particular, the fitting of the only two adjustable parameters of the model to the initially declining branch of the permeability-pressure curve allows the prediction of the plasticization pressure, if existing, by using only the solubility diffusivity model without invoking the onset of any additional physical phenomenon.

Interestingly, we have also inspected the permeability dependence on upstream pressure for CO_2in a commercial polyimide (Matrimid), at different downstream pressure values. The resulting behaviour is indeed unexpected and not fully consistent with the qualitative description recalled above. The model analysis has been thus focused on the experimental data obtained at different downstream pressures, offering a deeper insight on the so-called plasticization phenomenon.

The simultaneous consideration of permeability and solubility isotherms, together with the concentration and swelling profiles in the membrane, reveals that in some cases the plasticization pressure takes place after part of the membrane has turned into rubbery phase; while in other cases, e.g. for Matrimid, plasticization pressure takes place when the entire membrane is still glassy but with a polymer swelling sufficient for a permeability increase.

Finally, it is observed that all parameters used have a defined physical meaning which lead to good general correlations with properties of both polymer and penetrant, based on which permeability predictions can be obtained.

DIVIDING-WALL COLUMN SCREENING GUIDELINES AND APPLICATIONS

Frank (Xin X.) Zhu, Robert E. Tsai, Paul C. Steacy, Joe R. Haas, Honeywell UOP, Des Plaines, IL

Abstract

A dividing-wall column (DWC) is a distillation column with a vertical partition that can be implemented as a potentially attractive alternative to conventional multi-column or sidecut column arrangements; use of a DWC may enable capital cost and energy savings of around 30% or even higher to be achieved. Rules-of-thumb involving process conditions like feed composition and relative volatility are often cited for consideration when trying to determine if a DWC might make sense (or not) in a given flowscheme. The intent of this work is to present and review these guidelines in the context of refining and petrochemical processes and obtain an updated perspective that will allow for more detailed guidance to be provided for future screening evaluations.

Introduction

Distillation is the most common separation method used in the process industries. A reason for its prevalence is that it is a mature technology – one that is fairly well understood such that it can typically be designed and operated with high reliability. Distillation systems can furthermore be considered relatively simple units, generally consisting of just a column configured with overhead condensing and bottoms reboiling. Economies of scale are also favorable compared with other separation methods.

Distillation is nevertheless a fundamentally energy-intensive process, and distillation operations are consequently a major target of energy-saving efforts. Tradeoffs between operational savings and investment costs must be considered in any such assessment, however. For instance, thermodynamic losses might be reduced by outfitting a column with one or more side reboilers and side condensers, but there is a point where the cost of additional equipment is not economically justifiable. A configuration that is able to save both energy and capital expenditures versus other designs is clearly quite desirable and is why the concept of the dividing-wall column has come into prominence.

Theoretical Basis for Dividing-Wall Columns

The theoretical basis for a dividing-wall column can be understood by considering a multicomponent separation requiring three products: A, B, and C, where A is the lightest and C is the heaviest. Figure 1 shows how a direct sequence of two distillation columns can be used for the separation. A is separated in the first column, and a mixture of B and C exits the bottom for separation in the second column. For some mixtures, there is an inherent thermal efficiency with this process. In the first column, the concentration of the middle component (B) builds to a maximum value at an intermediate location below the feed but then is diluted past this point as the concentration of C continues to increase (Figure 2). Energy is wasted separating B to a maximum purity but not removing it.

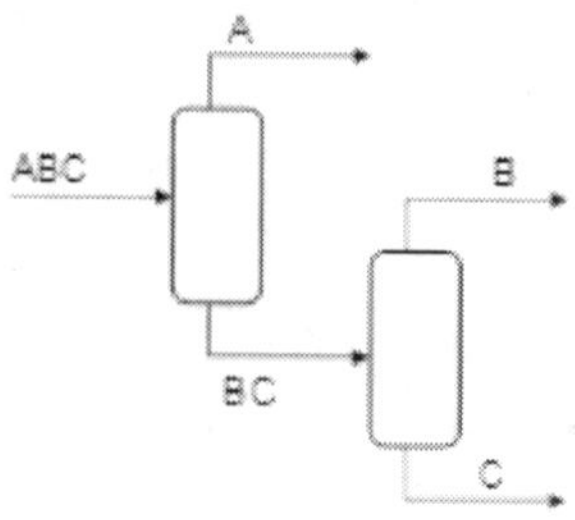

Figure 1. Direct Distillation Sequence

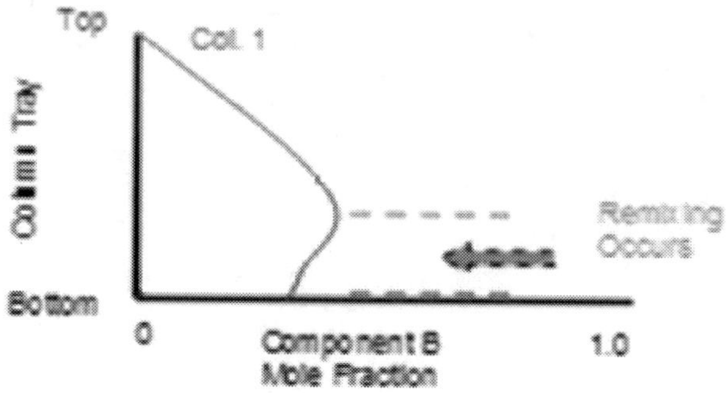

Figure 2. Profile of Component B in Column 1

Energy consumption may be reduced via use of a prefractionator arrangement (Figure 3, left). This configuration performs a sharp split between A and C in the first column (prefractionator), while allowing B to distribute between the overhead and bottoms streams. All of the A and some of the B is removed in the overhead of the prefractionator, while all of the C and the remaining B is removed in the bottoms of the prefractionator. The upper portion of the second (main) column then separates A from B, while the lower portion separates B and C. The fraction of B separated in the overhead of the prefractionator can be specified in order to minimize or eliminate remixing penalties.

The prefractionator arrangement can be thermally integrated to create a configuration commonly referred to as a Petlyuk Column (Figure 3, middle). Streams from the main column are used to provide vapor and liquid traffic in the prefractionator so that the system only has one condenser and reboiler.

The Petlyuk arrangement can furthermore be mechanically integrated, where the columns are combined in a single shell with a vertical partitioning wall (dividing wall) to distinguish the prefractionator and main sections (Figure 3, right). In summary, with this dividing-wall column (or DWC), utilities savings are achieved by addressing the thermal inefficiency issue, and capital savings are achieved by reducing equipment count from two condenser/reboilers/shells to just one. Also worth noting are indirect benefits such as reduced plot space as well as shorter piping/electrical runs. Capital cost and energy savings of around 30% are common when using a DWC vs. conventional two-column system. Even greater energy benefits of 50-60% have been cited for some unconventional designs.

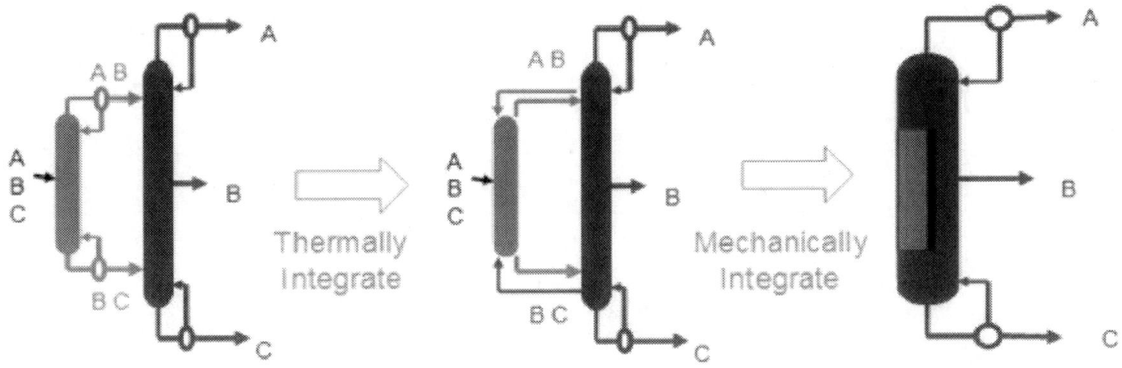

Figure 3. Evolution from Prefractionator Arrangement to Dividing-Wall Column

Review of Guidelines for Dividing-Wall Column Applications

The dividing-wall column concept and configuration for general fractionation purposes can be found in literature dating back to the mid-1900s, but first practice within the process industry did not occur until around 1985 (BASF). The delayed implementation of this technology may be attributed to factors such as design, operation, and control concerns, as well as less prior emphasis on process intensification. DWCs have gained somewhat more acceptance since their introduction, and the number of installed units has certainly grown within the past 15 years or so. UOP specifically can reference 20 completed designs in its portfolio, with close to half of those in current operation.

Overall though, it seems that the DWC is still treated with a bit of wariness when it comes to industrial practice. Our thought is that having a reliable set of basic guidelines, regarding when a typical DWC should or should not be considered, may help to direct future applications. It appears that mentality is shared with at least several others in both industry and academia as well. The sub-sections below first highlight existing recommendations, which have been extracted mostly from industrial-type publications such as *Chemical Engineering Progress* and contributed by companies like Linde, Sasol, Dow, and UOP.

Feed Composition

The first commonly cited rule-of-thumb indicates when a DWC should be considered with respect to feed composition:

- "...small amount of the middle component B, balanced presence of the two swing components A and C in the feed" [1]
- "Component B should be in excess, and components A and C should be present in fairly equal quantities" [2]
- "Consider a DWC when the middle or B component is at least 20% of the total stream flow" [3]

The compiled excerpts suggest that successful DWC applications tend to have balance between the light and heavy keys. However, the advice pertaining to the middle component is contradictory, ranging from "small amount" to "excess."

Several comparative scenarios (i.e., two-column versus DWC) were reviewed in an effort to gain further insight into this issue; these examples consist of designs from the UOP database as well as theoretical UOP or literature studies [1]. Table 1 shows the feed compositions associated with cases that were confirmed to benefit from implementation of a DWC, in terms of exhibiting operating cost savings of generally about 20-30% but ranging overall from 6% to as much as 40%.

Table 1. Feed Compositions of Successful DWC Applications

Case	Feed Composition (mol%)			
	A	B	C	Other
1	43	26	31	
2	38	10	38	14
3	16	49	29	6
4	20	60	20	
5 (Petyluk config.)	45	10	45	

The cases in Table 1 reflect a range of component B quantities (shown with gray shading), so each of the distinct assertions mentioned above appears to have a degree of validity; in other words, none of the statements is totally wrong, but that also means none is always right either. Rather than trying to pin down a value or range for the middle component, DWC applicability may be better blanketed under a specification that focuses on correspondence between the light and heavy keys, as this relation seems to be a more reliable indicator. While components A and C are not exactly equal in all of the cases, neither is dominant (i.e., more than 60% of the feed) such that there is semblance of balance between the two.

Relative Volatility

Relative volatility is presented as another heuristic. In this regard, DWCs are likely to lend themselves to situations with:

- "...balanced relative volatility ratio A/B to B/C" [1]
- "...split between A and B is at least as difficult as that between B and C" [2]

The common theme from the references is approximately equal relative volatility ratios between the light and medium ($\alpha_{A/B}$) and medium and heavy ($\alpha_{B/C}$) components. The second point suggests that deviation toward a larger B/C ratio ($\alpha_{A/B} \leq \alpha_{B/C}$) may also be acceptable.

Table 2 displays the relative volatility ratios between the A, B, and C components in the feed for the same DWC applications as in Table 1.

Table 2. Relative Volatilities of Successful DWC Applications

Case	Relative Volatility (Feed)	
	$\alpha_{A/B}$	$\alpha_{B/C}$
1	3.08	2.87
2	2.89	2.59
3	2.08	1.87
4	1.58	1.62
5 (Petyluk config.)	2.04	2.30

The A/B and B/C volatility ratios are quite close for each case, although some of the data technically support an opposite relation ($\alpha_{A/B} \geq \alpha_{B/C}$) to the one proposed in the second point above. Thus, comparability may be the key condition here, regardless of whether one ratio is slightly larger than the other. This makes sense when viewed from the opposite end of the spectrum. If one separation were far easier than the other, then it would not be too logical to force everything into a single column. The reflux rate would be governed by the more difficult separation, and energy would be wasted dealing with the entire A/B/C system. Hence, while it is fair to reason that a system exhibiting similar ratios might lend itself to a DWC process, there is probably no strong basis to broaden the guideline beyond that.

Pressure

Pressure is also often referenced as a critical parameter:

- "The flexibility of operating at distinctly different design pressures might outweigh the savings possible with a DWC" [2]
- "...may be more cost-effective to use two separate columns operating at different pressures" [4]
- "Don't use a DWC if the pressure difference between the conventional sequence is high" [3]

There is consistent indication that if a conventional scheme is comprised of columns at sufficiently distinct pressures, then application of a DWC is generally not a good idea.

This assertion appears to hold up well, although further clarification may be necessary. Pressure is clearly an important factor from a fundamental perspective, as combination of two totally different systems may first and foremost be troublesome while trying to maintain the same separation objectives (product specifications, etc.). Assuming that the desired targets can be achieved with a DWC, though, the main implication of pressure pertains to a typical DWC relying on only a single condenser and reboiler to provide liquid and vapor traffic. As a result, the condenser needs to operate at the coldest temperature for the multicomponent separation (and corresponding utility level), and the reboiler likewise needs to operate at the hottest temperature (and corresponding utility level). A two-column system, in contrast, has the flexibility to use intermediate utility levels.

As an example, one potential application that was examined involved a first pressurized column (165 psig) with a reboiler heated by very-high pressure steam (HHP steam) and a second atmospheric column with a reboiler heated by low pressure steam (LP steam), with much larger condensing and reboiling duties furthermore being associated with the latter. Combining this system into a single DWC was discovered to not only require a 25% greater total reboiling duty but also a shift of the entire load to HHP steam – certainly an unfavorable prospect. Not all cases may be so clear-cut, however, so quantification of how "different" two columns may be in pressure, while still being amenable to combination in a DWC, requires assessment of the particular situation, such as proximity to steam-level transition points and of course project economics.

Sidedraw Improvement

Consideration of a DWC versus a conventional sidedraw column is recommended:

- "...purity of the middle product is greater than can be achieved in a simple sidedraw column" [2]
- "...a DWC may be advantageous as it may accomplish the separation in a smaller column using less energy compared to the simple sidedraw setup" [2]

The DWC may offer advantages of either higher product quality for the same utility consumption or lower utilities for the same product quality.

This guideline is logical, because a sidedraw column is configurationally relatable to a DWC in terms of feed and product inputs and outputs; however, the use of the sidedraw always has the issue of contamination of either heavy material (when the sidedraw is located below the feed) or light material (when the sidedraw is located above the feed). The dividing wall can be interpreted as creating more stages of separation between the feed and sidedraw. (The light key contaminants boil up over the wall and condense back down, and the heavy key contaminants condense down below the wall and boil up over.) Opting for a DWC over a sidedraw column thus seems reasonable as a natural consideration.

Material and Geometric Construction

A final rule-of-thumb relates to physical limitations:

- "If either the superseded topping or tailing column would require a relatively expensive material of construction... may make the partitioned column so expensive that it is cheaper to build two smaller columns" [4]
- ...a DWC will likely be taller and have a larger diameter than either of the two conventional columns, and may surpass construction restrictions for a single tower" [2]

The DWC will most likely be bigger than either of the individual fractionation towers being combined as well as require the more-extreme metallurgy between the two, so a DWC will not be practical if issues like height or material-of-construction are already a limiting factor or concern in the conventional two-column system.

As with the previous heuristic, the statements here are thought to be fairly straightforward and do not warrant further discussion.

Proposed Guidelines for Dividing-Wall Column Applications

Table 3 provides a list of consolidated and/or revised rules-of-thumb pertaining to when a DWC design should perhaps be considered. Bear in mind that these all operate under presumption of a separation involving a multicomponent feed and at least three isolated products.

Table 3. Proposed Guidelines for DWC Applications

Category	Guideline
Feed Composition	Neither components A nor C should dominate (be more than 60% of the feed), although they do not necessarily have to be equal.
Relative Volatility	The relative volatility ratios of A/B and B/C should be comparable.
Pressure	The corresponding two-column system should be at sufficiently similar pressures as to not result in unfavorable shift in utilities.
Sidedraw Improvement	Greater product purity or reduced energy usage may be achieved versus a simple sidedraw column.
Material and Geometric Construction	The corresponding two-column system should not already be at mechanical design limits or require distinct metallurgies with vastly different costs.

References

1. Rév, E., Emtir, M., Szitkai, Z., Mizsey, P., and Fonyó, Z. (2001), "Energy savings of integrated and coupled distillation systems," *Computers and Chemical Engineering*, 25, pp. 119-140.
2. Schultz, M.A., Stewart, D.G., Harris, J.M., Rosenblum, S.P., Shakur, M.S., and O'Brien, D.E. (2002), "Reduce Costs with Dividing-Wall Columns," *Chemical Engineering Progress*, May, pp. 64-71.
3. Pendergast, J.G., Vickery, D., Au-Yeung, P., and Anderson, J. (2009), "Consider Dividing Wall Columns," *Chemical Processing*, January, pp. 23-27.
4. Becker, H., Godorr, S., Kreis, H., and Vaughan, J. (2001), "Partitioned Distillation Columns – Why, When & How," *Chemical Engineering*, January, pp. 68-74.

453718 Effect of Phase-Equilibrium Uncertainties on the Separation of Azeotropes Using Entrainers

Abstract:

Effect of Phase-Equilibrium Uncertainties on the Separation of Azeotropes Using Entrainers – Application to the Ethyl Acetate-Ethanol System System

Alejandro Cuevas,[1] Henry Z. Kister, Paul M. Mathias[2]

Fluor Corporation, Aliso Viejo, California 92698, USA

[1]*Presenting Author:* Alex.Cuevas@Fluor.com

[2]*Corresponding Author:* Paul.Mathias@Fluor.com

Many papers have been written highlighting the need to estimate process-design uncertainties that result from the uncertainties in correlated physical properties. The most significant source of property uncertainties on process design is from the correlations of phase equilibrium. However, uncertainty analysis is not a routine element of today's industrial process-design practice, mainly because education and awareness are lacking, and the proposed methods are difficult and cumbersome to apply.

In order to help rectify the situation, Mathias [1] developed an intuitive and easy-to-apply method based upon treating the mixture, for the purpose of activity-coefficient perturbation, as a set of pseudobinaries described by the Margules equation. Mathias [1] first applied his method to two case studies – (1) a propylene-propane superfractionator for which small changes in correlated relative volatilities have a large effect on the design of the distillation column; and (2) a dehexanizer column that separates a mixture containing many close-boiling hydrocarbon components – and demonstrated that the proposed approach provides quantitative insight into the effect of property uncertainties for both these diverse process designs, and helps to quantify the safety factors that need to be imposed upon the design. Mathias [2] also applied the methodology to a distillation train that separates the acetone-chloroform-benzene ternary mixture, which contains one maximum-boiling azeotrope, and showed that the approach quantifies the effect of property uncertainties on utility consumption and also identifies limits on operating variables (specifically recycle flow). In a third work, Mathias [3] applied the method to the separation of the water/1-butanol mixture, and demonstrated the use of the methodology to systems that exhibit liquid-liquid equilibrium, and for a case where the acivity-coefficient model cannot adequately describe the complex phase equilibrium.

In this presentation, the Margules perturbation method is applied to the separation of the ethyl acetate/ethanol binary mixture, which forms an azeotrope and hence requires an entrainer to break the azeotrope. Other papers have used process modeling to compare the performance of different entrainers. This publication demonstrates the added value of employing uncertainty analysis to the relative performance of azeotropic entrainers, and highlights the manner in which uncertainty analysis can improve distillation technology.

References

[1] Mathias, P. M., "Sensitivity of Process Design to Phase Equilibrium – A New Perturbation Method Based Upon the Margules Equation," *J. Chem. Eng. Data*, **2014**, 59, 1006-1015.

[2] Mathias, P. M., "Effect of VLE Uncertainties on the Design of Separation Sequences by Distillation – Study of the Benzene-Chloroform-Acetone System," *Fluid Phase Equilibria*, **2016**, 408, 265-272.

[3] Mathias, P. M., "Effect of Phase-Equilibrium Uncertainties on the Separation of Heterogeneous Azeotropes – Application to the Water/1-Butanol System," presented at PPEPPD-2016, Porto, Portugal, May 22-26, **2016**.

450996 Investigation of Fluid Dynamics in Structured and Random Packings Using X-Ray Computed Tomography

Abstract:

The flow behavior of one or more liquid phases in packed distillation columns is hardly understood, since with common optical measuring techniques local parameters in a packing are not achievable.

In this work the X-ray computed tomography (CT) is used as non-invasive measuring technique. The experiments are performed with a customized computer tomograph at our institute that enables measurements of flows in vertical objects. The detector and the X ray source are rotating around the column and the column is not moved, so that the flow inside is not influenced by centrifugal force. As a result of the high spatial (80 μm) and temporal (1000 projections per second) resolution, fluid dynamic parameters (like liquid hold-up, wetted packing surface, interfacial area and liquid film thickness) inside separation columns can be monitored with a high precision. All parameters are directly determined from the recorded CT images at different height positions along the column without a comparison to reference images of dry or wetted packing sections. For the evaluation several steps of image processing were applied to finally obtain segmented images with specific gray values for each component. The accuracy of the measuring technique including the developed segmentation algorithm was verified on a dry and wetted packing through a comparison of the mass of the (wetted) packing and the calculated values out of the segmented images.

The experiments were performed in a glass column with an inner diameter of 100 mm and a packing height of approximately 1000 mm. For the investigation of the fluid dynamics in structured packings SULZER Mellapak 500.Y and Raschig SuperPak 350.Y packings and as random packings Raschig SuperRing and Raflux and RSMR rings from RVT were used. For the investigation of the influence of the substance properties on the fluid dynamics different liquids (water and organic compounds) were used. The liquid load was varied between 2 and 50 $m^3 m^{-2} h^{-1}$ and the F-factor was increased to 2.5 $Pa^{0.5}$ or the flooding point.

For every packing type the integral liquid hold-up over the entire packing as well as the hold up profile along the packing height was determined. For an evaluation of the liquid distribution in radial direction the packing cross section was separated in three annuluses of equal area and the liquid hold-up was read out from the CT images for each annulus. For an analysis of the wetting efficiency the wetted packing surface area was determined and compared to the geometrical packing surface area. In a next step the interfacial area between liquid and gas phase was determined to obtain the real mass transfer area. The results show a not evenly wetted packing surface and a bigger liquid film thickness than expected through correlations from literature for each packing type.

Furthermore the flow morphology of the liquids was investigated using a three dimensional visualization of short packing sections along the column.

Vacuum Distillation
System Type

Paper 156d (Proceedings 448769)

Andrew W. Sloley, Principal Consultant, Advisian

Prepared for Presentation at the

American Institute of Chemical Engineers 2016 National Meeting

San Francisco, CA, November 13-18, 2016

www.**advisian**.com

Advisian
WorleyParsons Group

Vacuum Distillation – System Type

System Types

- Dry
- Damp
- Wet

Defined by

- How much steam
- Where it is
- How it's handled

Advisian / 2

System Type

Steam added to process drops partial pressure of oil

- Reduces temperature required to achieve a specific vaporization
- Increases vaporization at a temperature limit

Steam addition locations

- Dry, no steam
- Damp, velocity steam in heater coil only
- Damp, velocity steam in heater coil and stripping steam
- Wet, velocity steam in heater coil and stripping steam and vacuum system has precondensor
- Other: flash zone

Performance Possibilities

Some Numbers

Some numbers to get an idea of what's important

Basis

- 100,000 bpd unit for constant capacity units
- Crudes
  - Light (WTI – West Texas Intermediate 39.9°API)
  - Medium (TJL – Tia Juana Light 31.6°API)
  - Heavy (BCF24 – Bachaquero Field 23.3°API)
- 24' (7.32m) for constant diameter units

Advisian / 5

Typical Performance

Worldwide, typical cutpoint is probably below 1000°F (538°C)

Cutpoint often difficult to evaluate due to inaccuracies in TBP curve determination

What is cutpoint?

- Cutpoint is an estimate of yield
- Cutpoint: equivalent temperature on the TBP curve (true boiling point) that gives the same volume yield as found in the plant.
- Can be done on weight, but unusual.

Advisian / 6

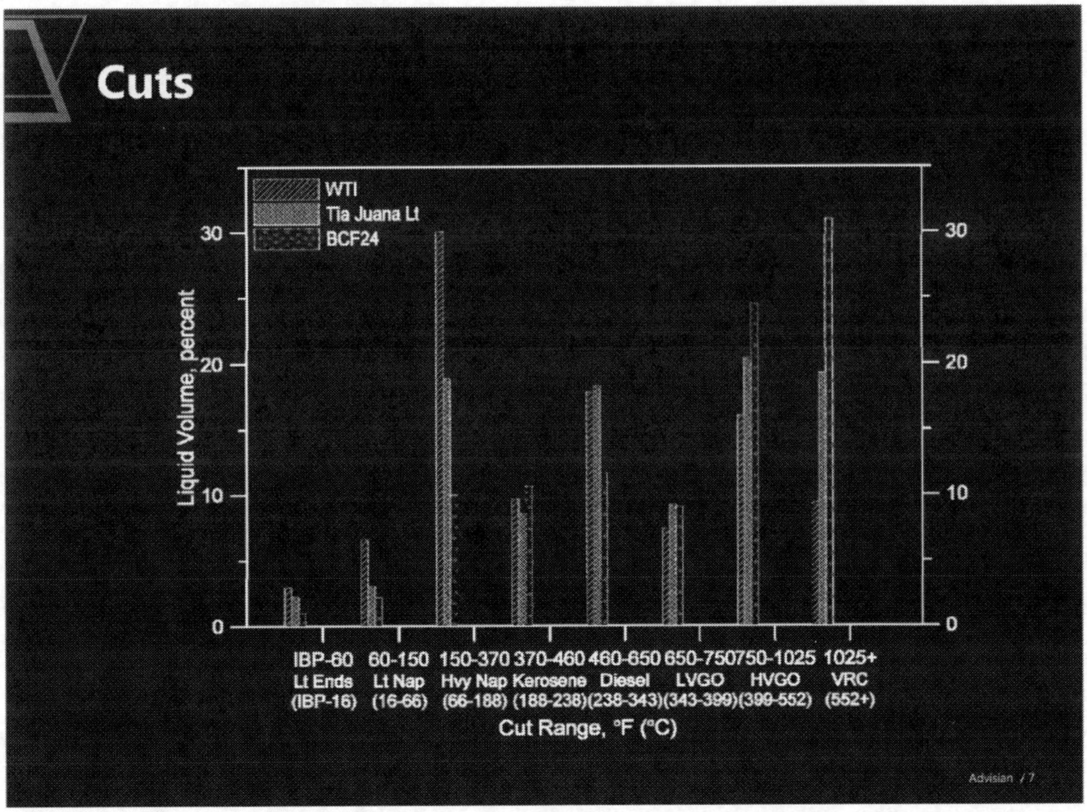

Cuts
WTI
Tia Juana Lt
BCF24
Liquid Volume, percent
IBP-60 Lt Ends (IBP-16)
60-150 Lt Nap (16-66)
150-370 Hvy Nap (66-188)
370-460 Kerosene (188-238)
460-650 Diesel (238-343)
650-750 LVGO (343-399)
750-1025 HVGO (399-552)
1025+ VRC (552+)
Cut Range, °F (°C)
Advisian / 7

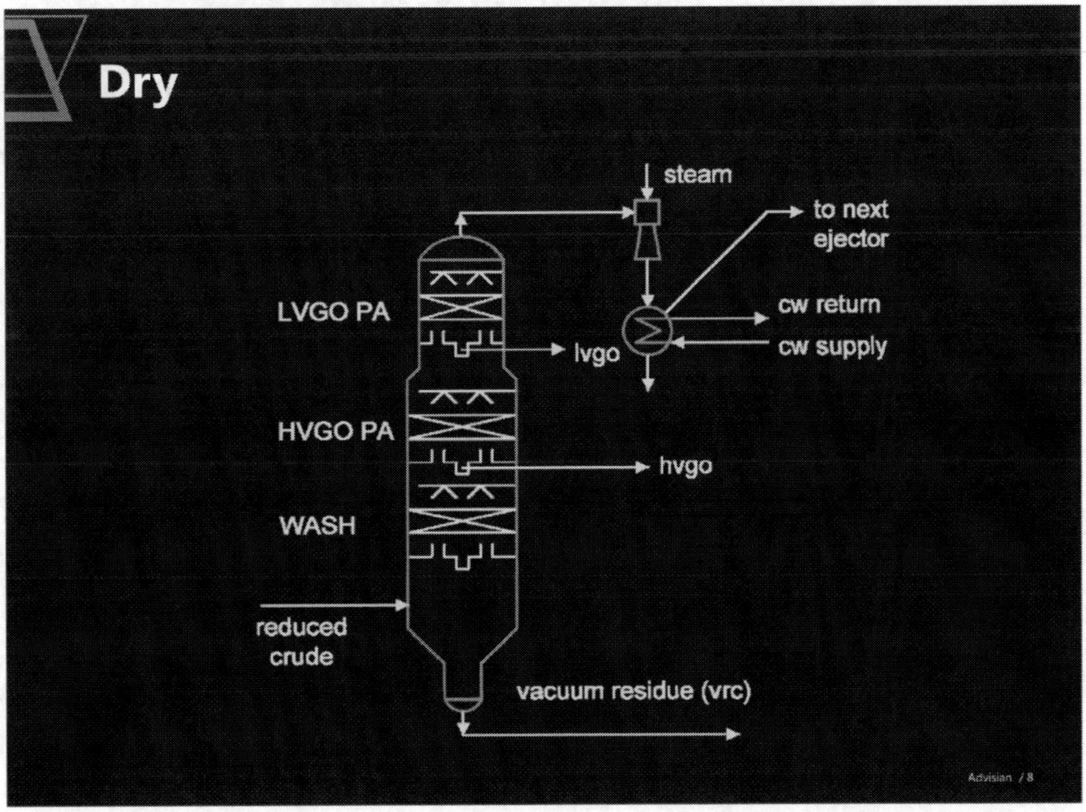

Dry
steam
to next ejector
LVGO PA
cw return
cw supply
lvgo
HVGO PA
hvgo
WASH
reduced crude
vacuum residue (vrc)
Advisian / 8

Dry

Where is the steam?

- Nowhere

How is the steam handled?

- None needed

Advisian / 10

Damp – 1

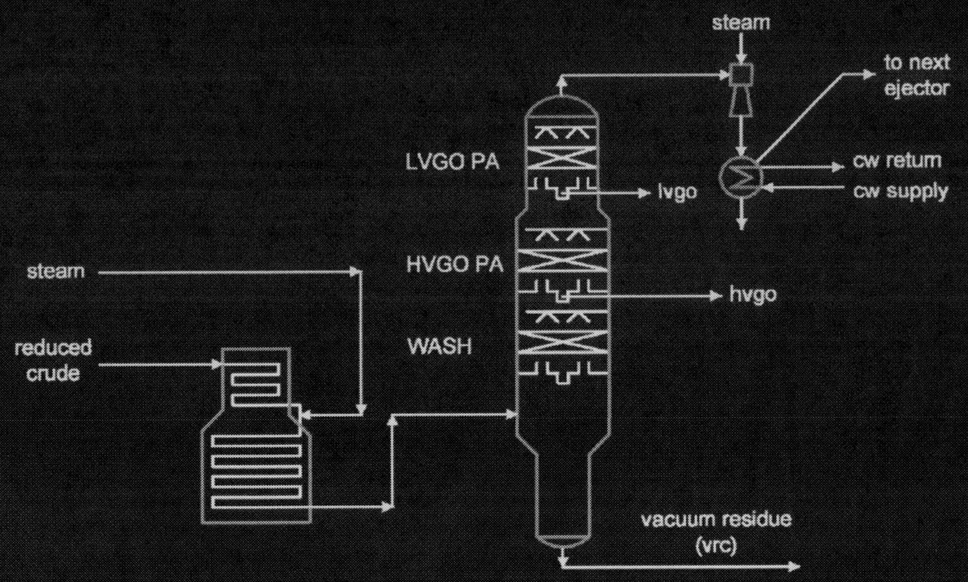

Advisian / 11

Damp – 1

Where is the steam?

- Heater
- Usually between convection section and radiant section
- Can be other places in heater

How is the steam handled?

- To first stage vacuum ejector

Advisian / 12

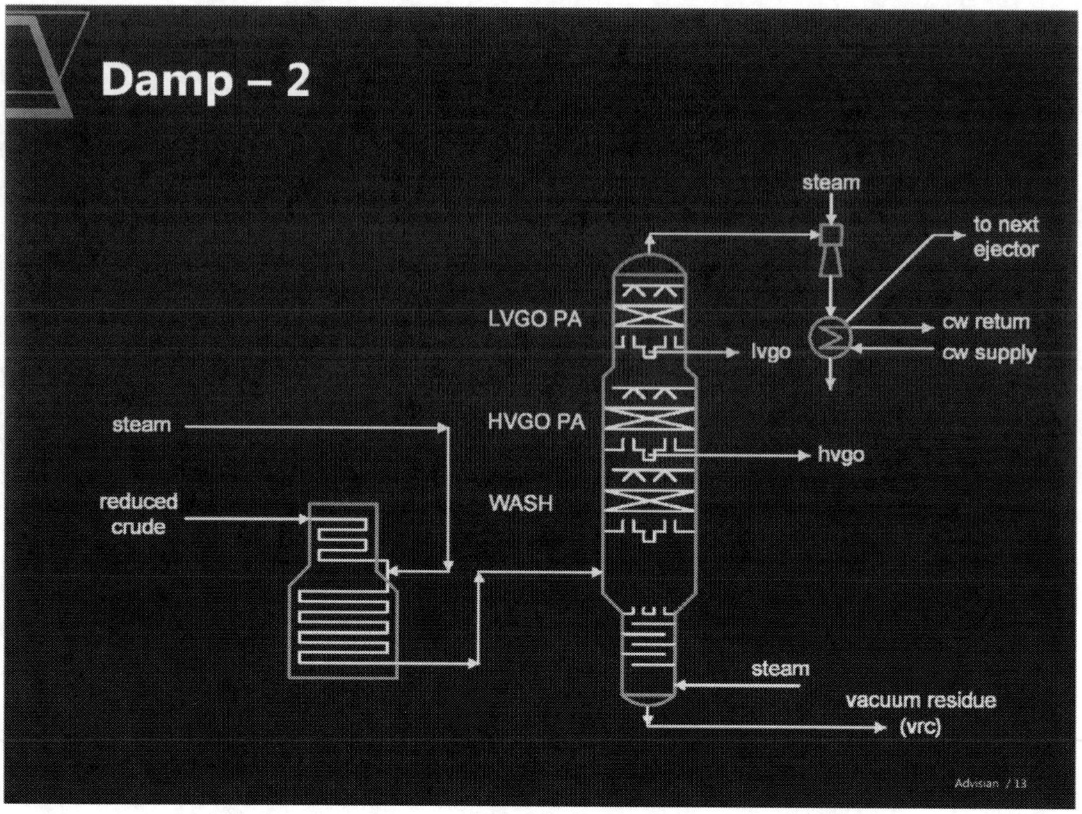
Damp – 2
steam
to next
ejector
LVGO PA
cw return
lvgo
cw supply
steam
HVGO PA
hvgo
reduced
crude
WASH
steam
vacuum residue
(vrc)
Advisian / 13

No Precondensor
Damp Vacuum Tower
Advisian / 14

Damp – 2

Where is the steam?

- Heater
- Usually between convection section and radiant section
- Can be other places in heater

Stripping section

How is the steam handled?

- To first stage vacuum ejector

Advisian / 15

Wet

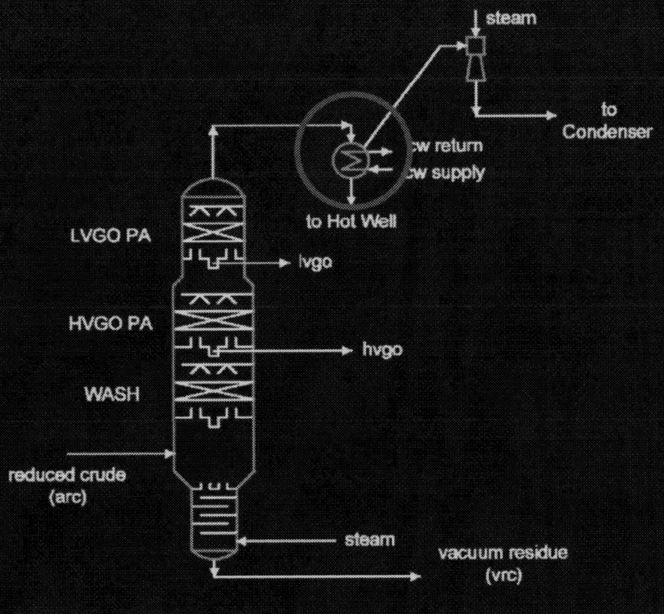

Advisian / 16

Wet

Where is the steam?

- Heater
- Usually between convection section and radiant section
- Can be other places in heater

Stripping section

How is the steam handled?

- To precondensor before first ejector

Advisian / 18

Wet Requirements, 1020°F+ Cutpoint

High-capacity stripping section

High-efficiency stripping section

High-capacity LVGO section

Advisian / 19

Some Questions

Reliability – what do equivalent reliability units look like?

- Operations

Capacity – how does capacity differ with type?

- Revamps

Investment – how does investment differ with yield and type?

- New

Advisian / 20

Reliability

Advisian / 21

Reliability Factors

Dry

- No stripping trays to lose
- High cracking rates in heater
  - Long residence time
  - High film temperatures

Damp

- Losing stripping trays loads up ejectors
- Lower cracking rates in heater
  - Short residence time
  - Heater outlet temperature can go up

Wet

- Pre-condenser prevents condensibles from loading ejectors if stripping trays lost upstream
- Lower cracking rates in heater
  - Short residence time
  - Heater temperature can go up

Advisian / 22

Unit Comparisons

Conditions varied for comparable run-length and reliability

Moderately pushed operation

Qualitative comparison

Medium Crude (TJL 31.6°API)

System	Atmos Cutpoint, °F	Vacuum Cutpoint, °F	Vac Heater Outlet, °F	Vac Heater Outlet, psia	Flash Zone, mmHg	Diam, ft
Dry	718	1,029	780	2	15	26.3
Damp-1	718	1,061	805	2	15	29.8
Damp-2	718	1,118	805	2	15	31.8
Wet	718	1,097	805	2.5	68	26.8

Capacity

Advisian / 25

Capacity Factors

Dry

- Low pressures required
- Low capacity for a given tower diameter

Damp

- Balance of steam to improve yield versus pressure
- Large ejector load

Wet

- Higher-pressure operation makes tower smaller
- More steam required to get yield up
- Small ejector load

Advisian / 26

Unit Comparisons

Stripping steam rate varied to get different yields

Capacity pushed to maximum

Pressure varied to see the effect of yields and capacity

Big question is damp versus wet for either

- Same capacity or
- Same yield

Damp units have both velocity steam and stripping steam (Damp-2)

Advisian / 27

Feed Rate versus Pressure

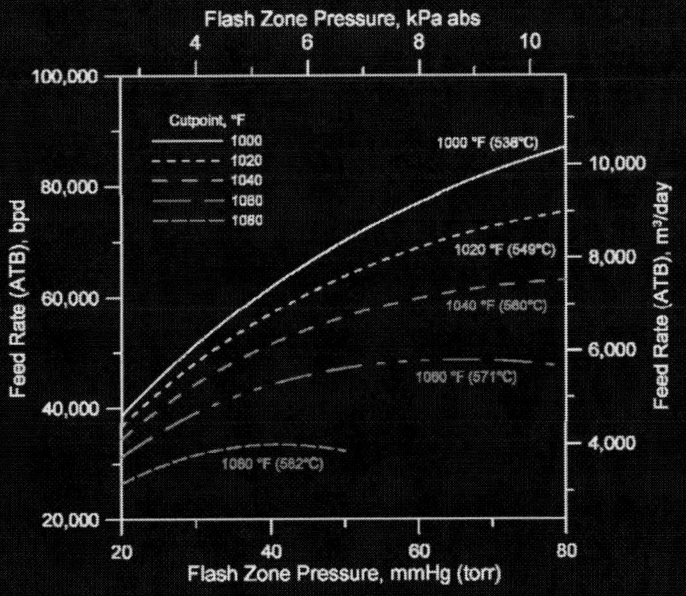

Advisian / 28

Product Rate versus Pressure

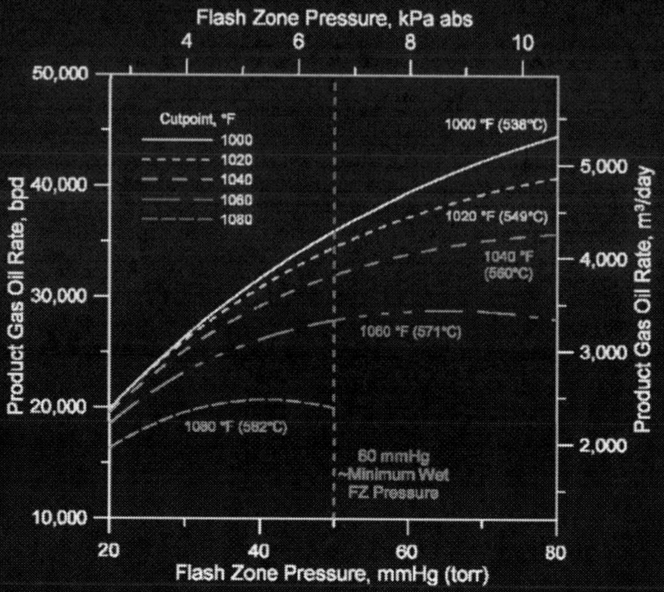

Capacity Versus Pressure

Higher pressure operation increases rates

- Extra stripping steam can keep yield up
- More difficult as yield (cutpoint) increases

Eventually, higher pressure forces capacity down

- Too much stripping steam required
- The higher the yield (cutpoint), the sooner this happens
- Wet operation not practical for cutpoints above 1070°F (577°C) unless very cold cooling water available

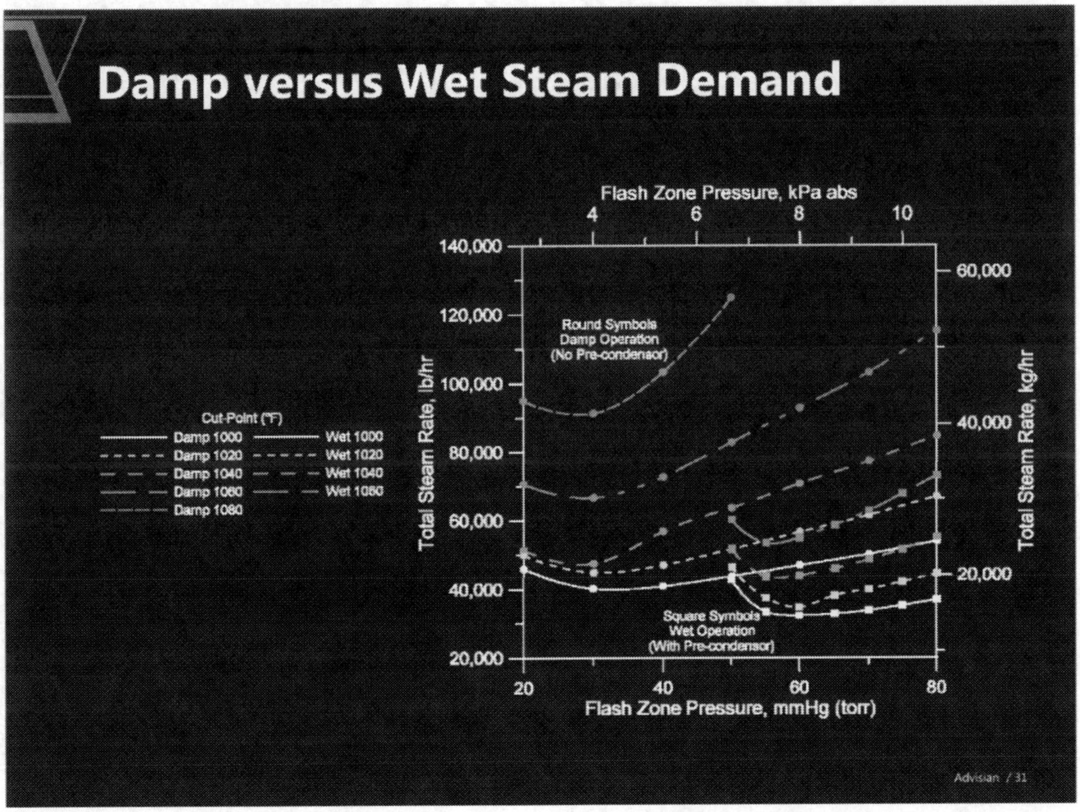
Damp versus Wet Steam Demand
Flash Zone Pressure, kPa abs
4
6
8
10
140,000
120,000
100,000
80,000
60,000
40,000
20,000
Total Steam Rate, lb/hr
60,000
40,000
20,000
Total Steam Rate, kg/hr
20
40
60
80
Flash Zone Pressure, mmHg (torr)
Round Symbols
Damp Operation
(No Pre-condensor)
Square Symbols
Wet Operation
(With Pre-condensor)
Cut-Point (°F)
Damp 1000
Damp 1020
Damp 1040
Damp 1060
Damp 1080
Wet 1000
Wet 1020
Wet 1040
Wet 1060
Advisian / 31

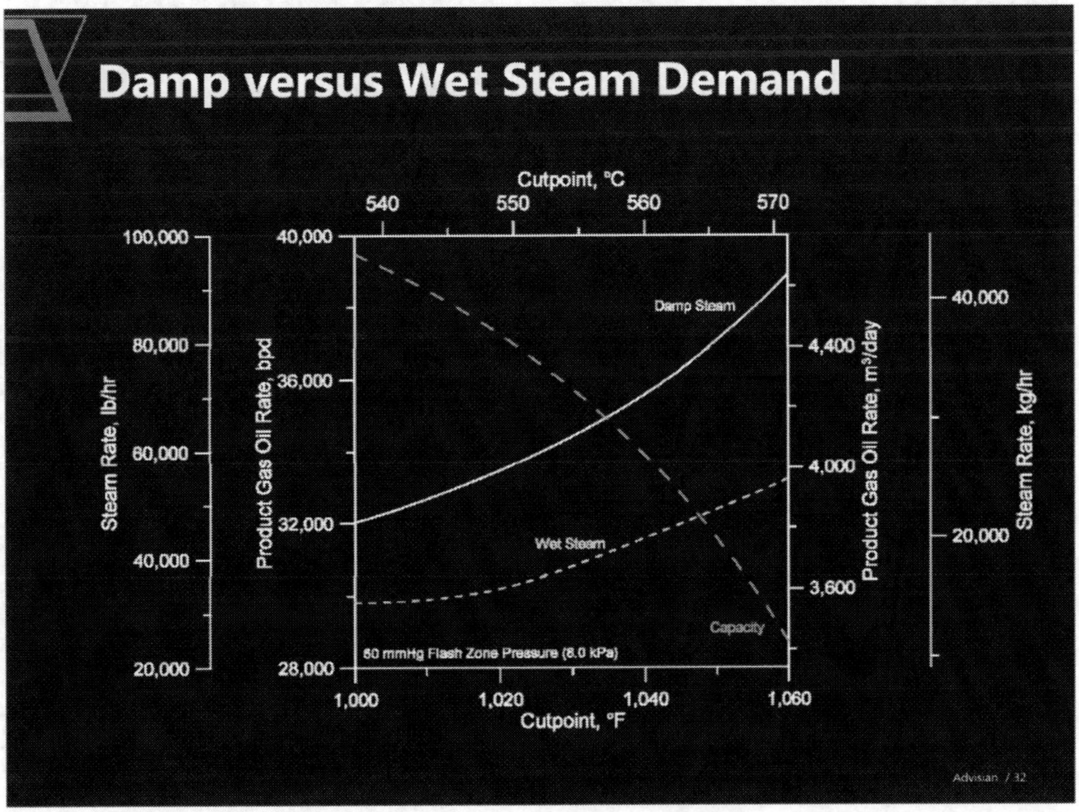
Damp versus Wet Steam Demand
Cutpoint, °C
540
550
560
570
100,000
80,000
60,000
40,000
20,000
Steam Rate, lb/hr
40,000
36,000
32,000
28,000
Product Gas Oil Rate, bpd
Damp Steam
Wet Steam
Capacity
60 mmHg Flash Zone Pressure (8.0 kPa)
4,400
4,000
3,600
Product Gas Oil Rate, m³/day
40,000
20,000
Steam Rate, kg/hr
1,000
1,020
1,040
1,060
Cutpoint, °F
Advisian / 32

Total Steam

Total steam = Velocity steam + Stripping Steam + Ejector Steam

At 1000°F and 60 mmHg stripping steam rate close to zero

Wet Operation Sweet Spot

Very favorable operation for wet units

- 1060°F or lower cutpoint (571°C)

Requires

- High-efficiency stripping section
- Normal stripping sections reduce the favorable operation point

Stripping Steam Rate versus Pressure

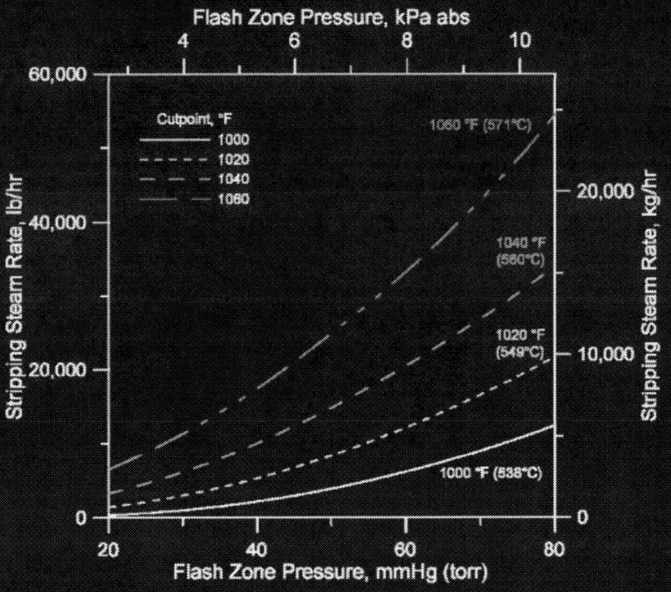

Constant Yield

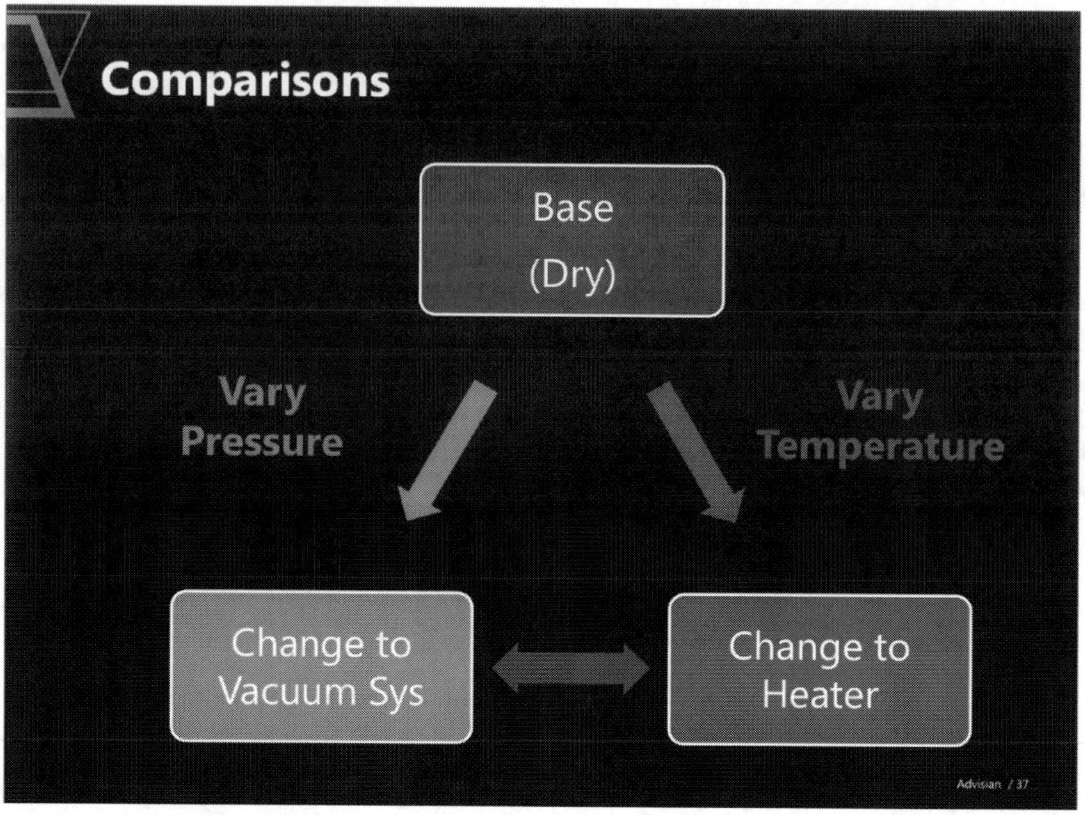
Comparisons
Base
(Dry)
Vary
Pressure
Vary
Temperature
Change to
Vacuum Sys
Change to
Heater
Advisian / 37

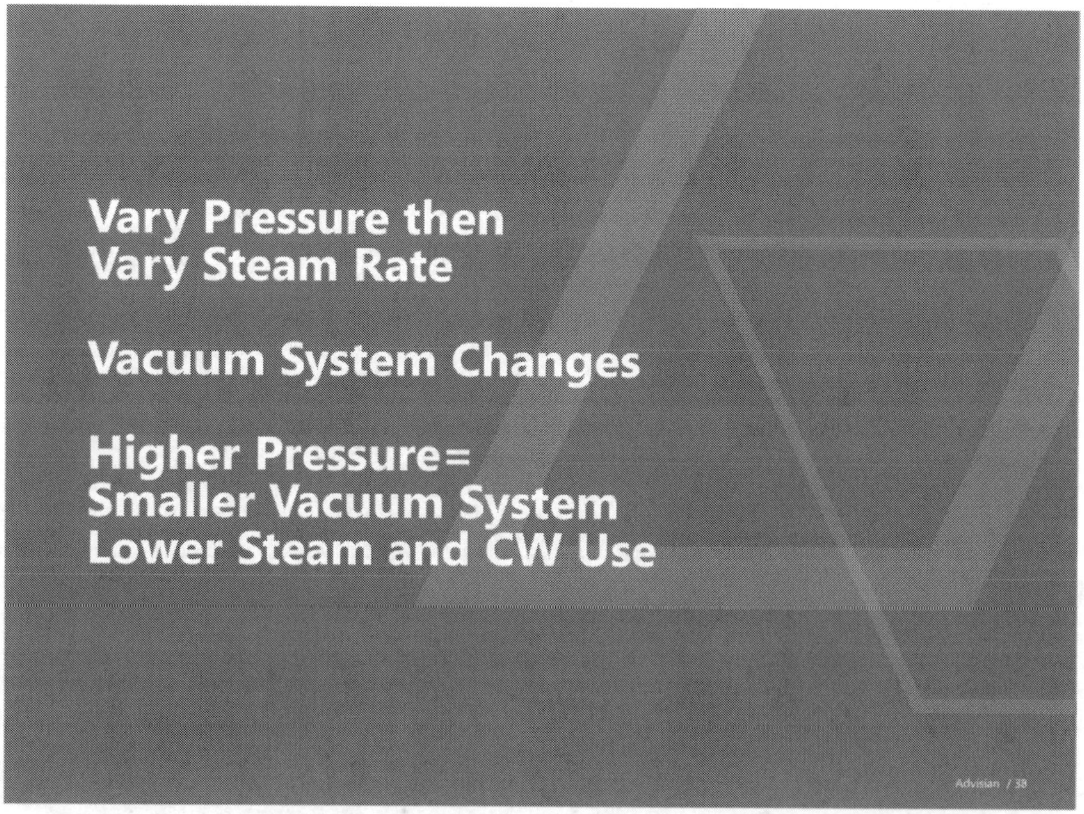
Vary Pressure then
Vary Steam Rate
Vacuum System Changes
Higher Pressure=
Smaller Vacuum System
Lower Steam and CW Use
Advisian / 38

Medium Crude (TJL 31.6°API)

System	Atmos Cutpoint, °F	Vacuum Cutpoint, °F	Vac Heater Outlet, °F	Vac Heater Outlet, psia	Flash Zone, mmHg	Diam, ft
Dry	718	1,030	781	2	15	26.4
Damp-1	718	1,030	805	2.4	53	18.6
Damp-2*						
Wet**	718	1,030	805	2.5	68	17.6

*Damp-2, same operating pressure as Wet.
**Wet, stripping steam rate reduced to match yields.

Advisian / 39

Vary Temperature

Heater Changes

Colder Heater =
Longer Run-Length
Lower Fuel Consumption

Advisian / 40

Medium Crude (TJL 31.6°API)

System	Atmos Cutpoint, °F	Vacuum Cutpoint, °F	Vac Heater Outlet,°F	Vac Heater Outlet, psia	Flash Zone, mmHg	Diam, ft
Dry	718	1,030	781	2	15	26.4
Damp-1	718	1,030	776	2	15	28.7
Damp-2	718	1,030	725	2	15	29.5
Wet	718	1,030	760	2.5	68	18.3

Advisian / 41

Conclusions

Advisian / 42

Damp versus Wet

Damp units required for high cutpoints

Current economics of capacity versus yield and quality favor wet units when operating at 1060°F (571°C) or lower

Revamps require very complex analysis

Advisian / 43

Fixed Yield Operation

Adding steam allows for higher pressure or colder heater outlet

Benefits complex – balance of

- Capital
- Operating costs
- Run-length

Adding velocity steam (Damp-1) very attractive

Next step more complex trade-offs

Advisian / 44

?
Advisian / 45

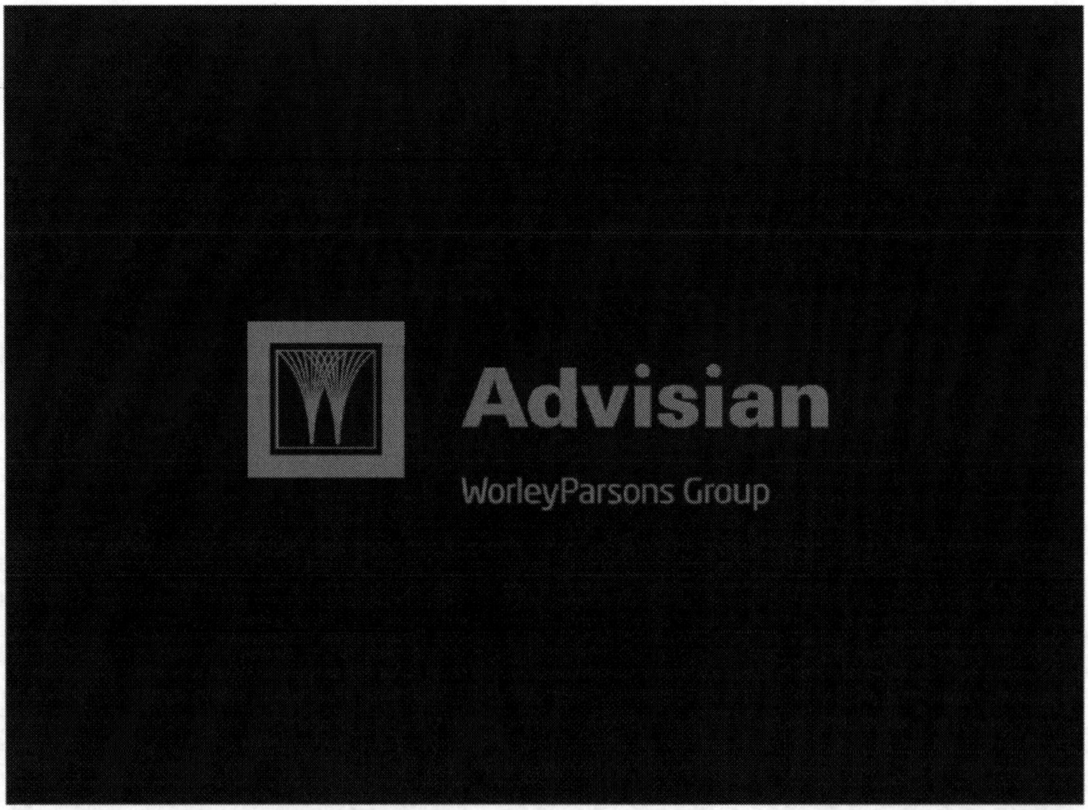
Advisian
WorleyParsons Group

469580 Highlights of the FRI Technical Activities for the Past 20 Years

Abstract:

Paul Steacy has been a guiding hand for Fractionation Research, Inc. (FRI) for over 20 years. Paul became a member of the Technical Committee in 1993, and has been the acting Chairman since 1995. FRI will provide an overview of technical accomplishments over his tenure during the past 20 years; including the evolution of the FRI test program, research projects and correlation development. Selected test results and video footage will be shown.

455209 Tuning Hybrid PIM-PI-OH/Z-MOF-Based Mixed-Matrix Membranes for Highly Efficient Propylene/Propane Separation

Abstract:

Energy-efficient separation of propylene (C_3H_6) from propane (C_3H_8
where C_3H_6 is used as feedstock for the production of a variety of important chemicals and plastics. A potentially significant marketplace exists for membrane technology in this separation if it can demonstrate the necessary performance. Currently, the required C_3H_6 purity is achieved by extremely energy-intensive cryogenic distillation. Recently, we reported the use of hydroxyl-functionalized polyimides of intrinsic microporosity (PIM-PIs) for
C
combined with a C_3H_6 permeability of 2.5 Barrer at a feed pressure of 4 bar. One potential route to improving the permeability *and* selectivity is development of high-performance polymer-nanoparticle hybrid materials, also known as mixed-matrix membranes (MMM). For the first time, we report the fabrication and exceptional pure- and mixed-gas C_3H_6/C_3H_8 gas permeation properties of a PIM-PI-based MMM. Using PIM-6FDA-OH as the polymer matrix, we integrated a zeolitic metal-organic framework (ZMOF), ZIF-8, at various loadings ranging from 20% up to even 65% by weight. SEM imaging demonstrated overall good adhesion between the polymer/nanoparticle interfaces, and establishment of percolative nanoparticle networks upon loadings greater than 45%. This nanoparticle clustering phenomena observed in sufficiently loaded membranes facilitates the formation of dual transport conduits necessary to defy conventional permeability/selectivity tradeoffs observed in solution processable polymers due to the existence of both continuous polymer *and* inorganic phases. The membranes were tested with a 50:50 C_3H_6/C_3H_8 feed mixture at pressures up to 7 bar (p/p_{sat} ~0.8). Not only did the membranes demonstrate exceptional pure-gas properties (a >35, $P(C_3H_6)$ >10 Barrer), but the high performance was also strongly preserved in mixed-gas conditions. Exceptional
C

observed minimal onset of plasticization, highlighting the strong potential of PIM-6FDA-OH/ZIF-8 mixed-matrix membranes for C_3H_6/C_3H_8 separations.

454808 Crosslinked Polyimide Hollow Fiber Membrane Performance Under Real Field Gas Conditions

Abstract:

Polymeric hollow fiber membranes have been used for CO_2 removal in natural gas processing industry for both off-shore and on-shore applications. While these membranes have attractive separation performance in terms of separation efficiency and productivity, the sensitivity of the polymer membrane to impurities and harsh operating conditions continues to be a drawback. These impurities, such as C3+ hydrocarbons, as well as a high partial pressure of CO_2, enhance the polymer chain's solubility in glassy polymers, lead to swelling (plasticization) of the polymer, resulting in a loss of separation efficiency.

Considerable laboratory research has been done on the novel membrane materials, to prevent the loss of hydrocarbons, validated under pure CO_2 or clean CO_2-Hydrocarbon gas mixtures. But very few have tested the membrane's durability against the impurities faced in a real field. Chevron, in collaboration with Georgia Tech, has developed a proprietary XLP (Cross-Linked Polyimide) hollow fiber membrane technology, tested with a real field gas slipstream (with high CO2 partial pressure and hydrocarbons beyond C3's). The presentation will cover the importance of membrane robustness under harsh feed conditions and the potential of XLP membranes to operate under such conditions.

451882 Mixed-Matrix Membranes Containing Engineered Nanoporous Materials for Biogas Purification

Abstract:

Biogas produced in anaerobic digestion process has been considered as a renewable energy source. For the recovery and the purification of biogas, various processes for CO_2 separation including amine-scrubbing and adsorption have been employed. Although membrane-based gas separation is an ideal platform for the biogas recovery/purification, the applications have been limited by the unsatisfactory performance of commercial polymeric membranes under the biogas separation conditions such as low CO_2 partial pressure. Thus, improving the performance of current polymeric membranes, which are inexpensive and mechanically stable, may facilitate the application of membrane technology in large-scale biogas upgrading processes. To this end, various porous materials that can selectively transport CO_2 can be incorporated into polymer matrices to design CO_2-selective mixed-matrix membranes.

Among various porous materials, metal-organic frameworks which comprise metal center and organic linkers to form three-dimensional porous structure have attracted vast research interests in the recent years as potential fillers for mixed-matrix membrane fabrication owing to their large pore volumes and tunable functionalities. In this work, a submicron-sized $Zn(pyrz)_2(SiF)_6$ metal-organic framework crystal possessing both an optimum pore size for selective CO_2 permeation and a strong affinity for CO_2 molecule was synthesized and incorporated into various polymer membranes to design mixed-matrix membranes for biogas upgrading. CO_2/CH_4 mixture gas permeation tests revealed that the separation properties of mixed-matrix membranes, especially selectivities, were significantly improved compared to those of pure polymeric membrane owing to the selective CO_2 uptake and transport in the metal-organic framework crystals.

Recently, our research group have developed hierarchical mesoporous Ca-A zeolites with amine-appended mesoporous domain for application in post-combustion CO_2 capture. Owing to the contributions of both the active sites in Ca-A zeolites and the amine groups in mesoporous surfaces, the material exhibited an excellent CO_2 uptake property even at low CO_2 partial pressure. Moving forward, amine-appended hierarchical Ca-A was employed as a filler for mixed-matrix membranes to improve CO_2 selectivity. Gas permeation testing revealed that the amine-appended hierarchical Ca-A is highly efficient in enhancing CO_2/CH_4 selectivity of polymeric membranes.

References

1. H. Gong, T. H. Nguyen, R. Wang, T. H. Bae, Separations of binary mixtures of CO_2/CH_4 and CO_2/N_2 with mixed matrix membranes containing $Zn(pyrz)_2(SiF)_6$ metal-organic framework, *J. Membr. Sci.* **2015**, *495*, 169-175

2. T. H. Nguyen, S. Kim, M. Yoon, T. H. Bae, Hierarchical LTA zeolites with amine-functionalized mesoporous domain for carbon dioxide capture, *ChemSusChem*, **2016**, *DOI:10.1002/cssc.201600004*

3. T. H. Nguyen, H. Gong, S. S. Lee, T. H. Bae, Amine-appended hierarchical Ca-A zeolite for enhancing CO_2/CH_4 selectivity of mixed-matrix membranes, *submitted*

460574 Novel Supramolecular Organic Framework (SOF) Based Mixed Matrix Membranes (MMMs) for CO_2/CH_4, CO_2/N_2, and N_2/CH_4, Separations

Abstract:

Novel supramolecular organic framework (SOF) based mixed matrix membranes (MMMs) for CO_2/CH_4, CO_2/N_2, and N_2/CH_4, separations.

Zoban V. Singh[1], LiLi Tan[2], Matthew G. Cowan[1,2], Wei Zhang[2], Douglas L. Gin[1,2], and Richard D. Noble[1]

(1) Department of Chemical and Biological Engineering, University of Colorado, Boulder, CO, (2) Department of Chemistry, University of Colorado, Boulder, CO

Abstract: Supramolecular organic frameworks (SOFs) represent a new class of microporous solids that display excellent CO_2-light-gas selectivity. Analogous to inorganic zeolites, SOFs conduct high-selectivity CO_2-light gas separation via a similar sorption based mechanism. However, forming large-scale membranes from pure microporous solids, such as zeolites, is challenging due to defect-spaces arising between the tight packing of zeolite crystals. We have previously presented a series of mixed-matrix membranes containing zeolites, which combine the excellent CO_2-light gas separation properties of the zeolite with the processability of a polymer.Â In contrast to zeolites, which are insoluble, SOFs can be readily dissolved into many conventional solvents, making solvent-based casting fabrication techniques for MMM formulation a convenience. In this talk, we present a new SOF based MMM platform. Upon crystallization and complete activation of the SOF at elevated temperatures, CO_2Â permeability and CO_2-light gas selectivity values skyrocket. Due to thermal crosslinks that form in the polymer matrix at higher temperaturesâ€"a

phenomenon pervasively reported in the literatureâ€"and due to the formation of larger crystalline surface areas in the SOF at elevated temperatures, sorption and diffusivity based permselectivity synergize. For these reasons, the separation performance of the new MMMs created with SOFs display incredibly high CO_2/CH_4, CO_2/N_2, and N_2/CH_4 selectivity. We believe, these are amongst the most selective MMMs ever created for the aforementioned gas pairs and represents a breakthrough in the development of high performance membranes for commercial applications given that all three separations hold major industrial and environmental significance. In addition to the excellent separation properties of the SOF, the ability to dissolve the dispersed phase into a solvent that can co-dissolve the polymer matrix, presents a significant advantage towards ease of processability and scale-up.

467667 Improved Carbon Molecular Sieve Membranes for Mixed Olefins/Paraffins Separations

Abstract:

Light olefins, such as ethylene (C_2H_4) and propylene (C_3H_6), are among the most important building blocks in the modern petrochemical industry. Olefins are primarily produced by thermal cracking of hydrocarbon feedstocks, and a series of separation processes is required to separate valuable olefins from paraffins. Because of the similar properties of the olefin and paraffin pairs, the separation is difficult and costly. For nearly seven decades, distillation has been commonly performed for recovery of olefins from mixtures with their respective paraffins. The distillation process requires energy-intensive operating conditions, including high pressure and low temperature, to perform the challenging separation. It is, therefore, appealing to develop olefins-selective membranes for use in a hybrid membrane-distillation process to retrofit the current distillation separation processes to lower the overall energy usage.

Carbon molecular sieve (CMS) membranes are able to exceed the limitation of the trade-off between productivity and efficiency of traditional polymeric membranes. In this work, starting materials and processing conditions were investigated in order to explore the impact on CMS membranes separation properties for mixed olefins/paraffins. CMS membranes were made having improved olefins/paraffins separation performance, which will be reported.

471839 Parallel-Aligned Graphene Oxide/Pebax Ultrathin Composite Membrane for Gas Separation

Abstract:

The application of graphene oxide (GO) in membrane offers a substantial opportunity to realize efficient separation at the molecular level. The separation mechanism for most of the existing GO membranes is based on the intergalleries between GO nanosheets. As a result, current research focus is to assemble GO into stacked laminates for membrane fabrication. However, the pure GO membranes are usually very brittle, and it is difficult to control the laminate structures (alignment, interlayer space) over a large surface area.

A promising alternative is to blend GO into a polymer matrix for gas separation membrane fabrication. Shen et al. [1] applied the favorable hydrogen bonding between Pebax and GO to assemble GO into laminate structures. The flat sheet membrane has molecular sieving spacing and straight diffusion pathways, allowing efficiency gas transport through the membrane. In another line of research, Berean et al. [2] blended GO into PDMS matrix. The formation of interfacial voids provides fast gas transport channels. However, most of the existing GO/polymer composite membrane are relatively thick, and the formation of thin GO/Polymer films are preferable for practical gas permeation. However, it is difficult to control the orientation of the GO laminates within the ultrathin layer, and the randomly orientated GO laminates would create non-selective defects.

In this work, we applied the dip coating technique to fabricate the GO/Pebax ultrathin coating layer on a porous PVDF membrane. Dip coating process is easy to control and scale-up, and the resultant membrane exhibited significantly improved mechanical properties. We then postulated the surface tension, capillary effect and gravity during the dip-coating process might facilitate the GO sheets alignment within the thin selective layer. The SEM and TEM images confirmed the presence of the shear-aligned GO laminate structures. Due to the confinement of Pebax polymer chains, GO nanosheets formed stacked laminates with intergallery distance of 0.35 nm, which is beneficial to the separation of CO_2 (0.33 nm) and N_2 (0.36 nm).

The gas separation results show the incorporation of GO within Pebax significantly improved the CO_2 permeance by 90 %, while the CO_2/N_2 selectivity was relatively unchanged. However, when the GO loading within the selective layer was too high, the aggregation of GO led to the loss of selectivity and permeance. Such composite membranes exhibited ideal stability over an extended testing period, indicating the aging effect was negligible. We further investigated the formation mechanism of the aligned structure, and discovered the thickness of the liquid layer was the key to the alignment: higher withdrawn speed led to thicker liquid film on a polymeric membrane, which required longer evaporation time. The constant convection force towards the liquid surface during the prolonged evaporation process has been discovered to lead to the heterogeneous GO structures [3]. Conclusively, the shear-aligned GO laminates can be achieved within a thin polymer layer by a facile dip coating process. This technique provides a facile approach to realize the large-scale application of GO/polymer membrane for gas separation.

[1] J. Shen, G. Liu, K. Huang, W. Jin, K. R. Lee, N. Xu, *Angew. Chem. Int. Ed.* **2015**, *54*, 578-582. *Angew. Chem.* 127, 588-592.

[2] K. J. Berean, J. Z. Ou, M. Nour, M. R. Field, M. M. Y. A. Alsaif, Y. Wang, R. Ramanathan, V. Bansal, S. Kentish, C. M. Doherty, A. J. Hill, C. McSweeney, R. B. Kaner, K. Kalantar-zadeh, *J. Phys. Chem. C*, **2015**, *119*, 13700-13712.

[3] C.-H. Tsou, Q.-F. An, S.-C. Lo, M. De Guzman, W.-S. Hung, C.-C. Hu, K.-R. Lee, J.-Y. Lai, *J. Membr. Sci.* **2015**, *477*, 93-100.

469301 Gas Separation Properties of a Porous Zn(II)-Based Metal Organic Framework (MOF)/Polymer Membrane for High Flux Gas Transport

Abstract:

Metal organic frameworks (MOFs) are crystalline porous materials consisting of polymeric inorganic networks with high surface areas, well-defined pore sizes and adjustable chemical functionality. MOFs have found applications in areas such as gas storage and separation, heterogeneous catalysis, sensors and drug delivery systems among others. MOF chemical properties make them ideal in the development of thin films and membranes for the adsorption and separation of gases, as their chemistry can be tailored to serve a particular application; recently, one of the main focuses of MOF membrane research is the separation and conversion of CO_2 from other small gas molecules as in the case of power plant emissions. This work focuses on the formation of microcrystalline Zn-based MOF membranes on porous supports and the study of their gas permeation properties and gas transport mechanisms. A previously reported Zn(II)-based MOF ($Zn_4(pydc)_4(DMF)_2$â€¢3DMF **(1)**)[1] with promising properties for CO_2 capture and separation, was studied for the formation of microcrystalline porous membranes. Porous alumina disks were used as the support material for membrane formation. Porous supports were modified by the formation of a functionalized silica layer followed by dip-coating in a dilute polyethylenimine (PEI) solution to enhance hydrogen-bond attachment and provide a smooth surface for membrane formation[2]. Seed crystals of **(1)** were synthesized as previously reported[1] and crushed into microcrystals, they were subsequently attached to the modified supports by a manual assembly procedure, XRD patterns of the seed crystals were studied. A MOF membrane was formed on the seeded supports using a secondary growth method from a synthesis solution of **(1)**[3], XRD patterns of the synthesized membrane confirmed the formation of **(1)**, while the surface and thickness of the membrane was studied by SEM imaging. As an additional step, a thin and low permeance polysulfone (PSF) layer (<500 nm) was used in order to seal the surface of the membrane and impart mechanical strength to the superficial MOF crystals, this thin film also corrects superficial defects of the synthesized films. Membranes containing PSF superficial layers were studied by SEM imaging to ensure that gas permeation through MOF crystals was not blocked by the polymeric layer. A constant volume-variable pressure system was used to measure the flux of gases through the **(1)** membrane; CO_2, N_2, H_2 and CH_4 gas fluxes were measured showing a high gas permeance compared to other reported MOFs and a strong relation between molecular weight and gas permeance, this suggested a Knudsen type transport mechanism through the pores of **(1)**. This was further confirmed by calculating the permselectivity of gases and comparing the obtained values to the Knudsen values. The effect of temperature on the permeance of gases was studied under conditions pertinent to the separation of CO_2 from flue gases. The transport mechanism of gases inside the pores of the membrane was confirmed by the formation of a thin MOF layer composed by only a few microcrystals in thickness, the permselectivity results calculated from the permeance of these films corroborated a Knudsen transport mechanism happening inside the pores of the MOF and not between the spaces among adjacent crystals. This is a method that can be potentially applied to the study of the diffusion mechanism of gases through the pores of different MOF crystals.

[1] Ahrenholtz, S. R.; Landaverde-Alvarado, C.; Whiting, M.; Lin, S.; Slebodnick, C.; Marand, E.; Morris, A. J. Inog. Chem. 2015, 54, 4328-4336.

[2] Ranjan, R.; Tsapatsis, M. *Chem. Mater.* **2009**, *21*, 4920-4924.

[3] Lai, Z.; Tsapatsis, M.; Nicholich, J. P. *Adv. Funct. Mater.* **2004**, *14*, 716-729.

459677 Sulfonated Network in Polymers of Intrinsic Microporosity for Gas Separation

Abstract:

Sulfonated polymer networks increase CO_2 uptake up to 13.5 % in sorption tests. In general, sulfonations of polymers were conducted in acid conditions such as sulfuric acid, chlorosulfonic acid or trimethylsilyl chlorosulfonate. However, polymer sulfonation may lead to further polymer cross-linking and degradation. We reported here the sulfonated network can significantly improve the gas separation performance of highly permeable polymers of intrinsic microporosity because the sulfonated groups tend to have strong molecular interactions with CO_2 and O_2. The presence of a higher degree of sulfonation in the polymer chains drives the membranes with better gas separation performance close to or above the Robeson upper bound lines for major gas-pairs separation. Moreover, they also displayed improved anti-plasticization properties up to 30 atm using a binary CO_2/CH_4 feed gas. The newly developed advanced polymeric membranes are potential candidates for air separation, natural gas separation and CO_2 capture.

465920 Intermolecular Structure of Binary Mixture in Confined Spaces

Abstract:

The multi-component adsorption which is the coexistence state of two or more kinds of molecules in the small space is used in many of the applications using porous materials such as molecular separation, purification, and electric devices. It is expected that the unusual molecular mixing state in confined spaces. We have been applying the X-ray diffraction (XRD), neutron diffraction(ND) to elucidate the structure of adsorbed molecules in confined carbon spaces. Furthermore, the Reverse Monte Carlo (RMC) method was applied to obtain the intermolecular structure of confined molecules. RMC can provide the intermolecular structure which reproduces the experimental scattering data using geometrical parameters of adsorbents and adsorbate molecules including the multi-component adsorption and solution systems. We try to determine the 3-dimensional structural information by combination analysis of the RMC method and experimental scattering data. We will describe the intermolecular structure of binary mixture including the water, and alcohols in the hydrophobic nanospaces.

462130 Dynamic Mean Field Theory with Hydrodynamic Interactions: Theoretical Development and Applications in the Study of Fluid Transport in Porous Adsorbents

Abstract:

Understanding the dynamics and equilibrium behavior of inhomogeneous fluids is of great importance in many fields of science and engineering, including gas and liquid separations in porous adsorbents and membranes, heterogeneous catalysis, enhanced oil recovery, microfluidics, etc. In these examples the structure of the fluids can become quite rich, especially if the fluid is transported through confined or complex geometries. In these examples, the thermodynamic view of a fluid under confinement can be quite different from what is observed in the bulk, while metastability as well as mass transfer limitations to equilibration can play an important role in this case [1]. Such challenging issues are tackled using molecular simulation methods and classical Density Functional Theory (DFT). The latter method can in principle describe accurately the thermodynamics of fluids confined in porous structures of arbitrary complexity. Furthermore, a dynamic version of the theory has been developed focusing primarily on the study of spinodal decomposition of a single fluid. Unfortunately, implementation of DFT in two or three dimensional systems is extremely expensive computationally even for equilibrium studies.

Coarse-grained methods based on lattice fluid models offer an attractive alternative for such cases, as has been demonstrated by several research groups over the last decade. Furthermore, a dynamic version the theory has been developed (Dynamic Mean Field Theory, DMFT) to describe the relaxation processes of fluids confined in porous materials [2]. The theory is based on a lattice fluid model and can be viewed as an approximation to Kawasaki dynamics, a process where the molecules hop between nearest neighbour sites under the action of transition probabilities such as those in the Metropolis Monte Carlo method, constrained by the restriction of single occupancy. DMFT reproduces the thermodynamic description from mean field DFT (MFT) in the appropriate limit so that it can describe both the dynamics and the thermodynamics of the system.

A limitation of DDFT and its lattice analogue, DMFT, is that when viewed within the context of the theory of transport phenomena it is essentially a continuity equation for the fluid density and produces overdamped dynamics, which characterize the interaction among colloidal particles immersed in a solvent acting as a heat bath. Thus the DDFT method is strictly valid for colloidal fluids, where there is friction between the colloidal particles and the solvent, so that one does not have to account for momentum currents and the velocity field is fully determined by the density field. For the case of molecular fluids on the other hand, there is no solvent present so the momentum of the particles is no longer damped by the friction with the solvent. In this case equilibration is realized primarily due to the action of viscosity and a momentum conservation equation must be included and be coupled with the continuity equation, so that the momentum current is treated on equal footing with the density field.

In this work we present a recently developed DMFT model that includes hydrodynamic interactions [3]. The new model satisfies simultaneously continuity and momentum equations under the assumptions of constant dynamic or kinematic viscosity and small velocities and/or density gradients. We first present applications of the theory to spinodal decomposition of subcritical temperatures for a lattice gas model in MFT. We find that the DMFT/HI model correctly describes the transition from diffusive phase separation at short times to hydrodynamic behaviour at long times. In addition, in common with DMFT, it provides a theory of transport that is fully consistent with the theory of the thermodynamics from density functional theory.

Accordingly, we apply the DMFT/HI model to study the transport of fluids in porous media. The model is extended to account for frictional effects due to the presence of the internal pore surface. We propose a simple formalism based on the concept of adding a local frictional force at the pore-solid interface. The frictional force is determined from the oscillator model employing the low density diffusion theory proposed by Bhatia and Nicholson [4]. The DMFT/HI model has been applied to study steady state and dynamic transport of a lattice fluid in a 2D slit pore at various degrees of wetting. Steady state simulations provide density and velocity profiles that are in qualitative agreement with results from rigorous Non-Equilibrium Molecular Dynamics simulations. Dynamic adsorption-desorption experiments reveal that hydrodynamic interactions result in faster uptake rates compared to the standard DMF model, and can have a major impact in the transport mechanism of the fluid at strong wetting conditions. In the latter case, the formation of the monolayer is slower than pore condensation during both adsorption and desorption. On the other hand, for the case of weak wetting conditions, the effect of hydrodynamic interactions does not alter the basic characteristics of the transport mechanism and the standard DMFT can still give a good qualitative picture of the transport process.

References

[1] P. A. Monson, Microporous and Mesoporous Materials 160, 47 (2012).

[2] P. A. Monson, Journal of Chemical Physics 128, 084701 (2008).

[3] E. S. Kikkinides and P. A. Monson, Journal of Chemical Physics 142 (2015).

[4] S. K. Bhatia and D. Nicholson, AICHE Journal 52, 29 (2006).

471544 Adsorption Stress in Nanopores

Abstract:

Adsorption Stress in Nanopores

Alexander V. Neimark

Department of Chemical and Biochemical Engineering,
Rutgers University, 98 Brett Road, Piscataway, New Jersey 08854-8058, USA

Email: aneimark@rutgers.edu; Web: http://sol.rutgers.edu/~aneimark/

Experimental demonstration of adsorption-induced deformation of microporous adsorbent has been known for a long time, starting from the pioneering work of Meehan in 1927 on charcoal swelling upon CO_2 adsorption. Later, it was shown with multiple examples that guest molecules adsorbed in pores exert a significant stress in the host solid matrix causing, depending on the conditions, not only swelling but also contraction, and sometimes matrix structural transformation and even collapse. With the advent of metal-organic frameworks, the adsorption deformation has attracted renewed attention due to the enigmatic phenomena of gate opening and breathing; the former discovered by Katsumi Kaneko in 2001. Adsorption deformation has important implications and potential for the development of high-performance adsorbents and drug delivery vehicles. It takes place in geological formations, like micro-mesoporous domains in coal and shale reservoirs, where the deformation induced by CO_2-methane displacement may cause significant reduction of permeability and even affect the mechanical integrity of reservoirs.

I will present a general thermodynamic approach to predicting adsorption stress and respective deformation in nanoporous materials based on the notion of the adsorption stress. The adsorption stress is defined as the derivative of the grand thermodynamic potential of the adsorbed phase with respect to the variation of the sample volume. It can be calculated based on adopted adsorbate-adsorbent models from the classical density functional theory or Monte Carlo atomistic simulations. The adsorption stress introduced into the linear elasticity theory provides a unified framework for extension of classical poromechanics to nanoporous materials. Special attention will be paid to the specifics of adsorption deformation in micropores of molecular dimensions and to the interpretation of the strain measurements in situ XRD and dilatometry experiments. I will show how the adsorption strain isotherms can be employed for evaluation of the micropore size distribution. Examples include micro- and mesoporous carbons and MOFs.

471194 Computer Simulations of Chemical Reactions and Proton Transport in Porous Materials and Functionalized Surfaces

Abstract:

Atomically-detailed computer simulations can be used to predict properties of materials that have not yet been synthesized, providing a way to screen hypothetical materials, identifying the most promising candidates for subsequent experimental exploration. We present electronic structure density functional theory (DFT) calculations for two classes of materials: the first a catalytic porous material for CO_2 hydrogenation and the second a functionalized graphene material having the ability to conduct protons anhydrously.

We show that catalytic moieties can be integrated into robust Zr-based metal organic frameworks (MOFs) that can be used to reduce CO_2 with H_2. We examine functionalized UiO-66 and UiO-67 and demonstrate that Lewis acid and base sites fixed within the MOFs can be very effective at heterolytically splitting H_2 into protic and hydridic species that can then be added simultaneously to CO_2 to produce formic acid. We investigate the effects of confinement on the reaction energies and barriers. We show that the reaction energies and barriers can be tuned by changing the functional groups and have identified BrÃ¸nstedâ€"Evansâ€"Polanyi (BEP) relationships that can be used to construct a Sabatier analysis. The optimum properties of the catalytic moieties can then be identified.

Intermediate temperature proton exchange membrane (PEM) fuel cells have many potential advantages over conventional (low temperature) PEM fuel cells, including faster kinetics, higher quality waste heat, and reduced poisoning of the electrodes. However, existing PEM materials are limited to operating at low temperatures (below about 80Â°C) because of the requirement that the membranes need to be well-hydrated to function. Thus, there is a need to develop new materials that can conduct protons at high rates under low-water conditions. We have performed DFT calculations to show that hydroxylated graphene is capable of conducting protons under anhydrous conditions.

We identify diffusion pathways and barrier heights for proton conduction along 1-dimenstional and 2-dimensional chains of hydroxyl groups on the surface of functionalized graphene. The barriers computed from DFT indicate that proton conduction should be faster than through hydrated Nafion membranes currently used in PEM fuel cells.

461025 Understanding Gate Adsorption of CO_2 on Elastic Layer-Structured Metal-Organic Framework-11

Abstract:

Metal-organic frameworks (MOFs) with a flexible framework show an adsorption-induced structural transition called gate adsorption. The gate adsorption leads to a steep rise in the adsorption isotherm at a threshold gas pressure depending on a gas molecule, and thus the flexible MOFs are promising for the development of new gas storage and separation processes. Understanding of the gate adsorption behavior is crucial to control gate adsorption pressure for these applications by tailored synthesis of flexible MOFs. We have therefore focused on elastic layer-structured metal-organic framework-11 ($[Cu(4,4'$-bipyridine$)_2(BF_4)_2]_n$)[1], which shows a typical gate adsorption behavior of CO_2, and determined precise structures of ELM-11 before and after the CO_2gate adsorption (closed and open state) by in situ synchrotron X-ray powder diffraction (XRPD) measurements. The determination of the both structures is essential to model the gate adsorption behavior with the aid of molecular simulations and free energy analysis.

The gate adsorption should occur at the pressure when the osmotic free energy change between the closed and open state, $\Delta\Omega_{os}$, becomes zero.[2] The osmotic free energy change is represented as a sum of the Helmholtz free energy change required to deform the host framework ($\Delta \boldsymbol{F}_{host}$), $\boldsymbol{P}\Delta\boldsymbol{V}$ term ($\boldsymbol{P}$ is pressure and $\Delta\boldsymbol{V}$ is volume change of host), and the grand potential of guest that is obtained from the integration of a fictitious adsorption isotherm of the guest on the open host framework, $\boldsymbol{N}_{open}$, as a function of chemical potential of the guest. The thermodynamic relation of $\Delta\boldsymbol{F}_{host}$ is then expressed as: $\Delta\boldsymbol{F}_{host} = \Delta\boldsymbol{U}_{host} - \boldsymbol{T}\Delta\boldsymbol{S}_{host}$, where $\Delta\boldsymbol{U}_{host}$ and $\Delta\boldsymbol{S}_{host}$ are differences in the internal energy and entropy between the closed and open structures.

The closed structure of ELM-11 at 273 K and the open structure of ELM-11 encapsulating CO_2 at 195–298 K were determined by our new structural refinement method[3,4] using the in situ synchrotron XRPD data. We obtained $\boldsymbol{N}_{open}$ at 195–298 K by grand canonical Monte Carlo (GCMC) simulations using the open framework structures from XRPD. We determined $\Delta\boldsymbol{F}_{host}$ such that the $\Delta\Omega_{os}$ value became zero at the experimental gate pressure, and found that there was a linear correlation between $\Delta\boldsymbol{F}_{host}$ and temperature. The least-squares fitting of $\Delta\boldsymbol{F}_{host} = \Delta\boldsymbol{U}_{host} - \boldsymbol{T}\Delta\boldsymbol{S}_{host}$ to the obtained $\Delta\boldsymbol{F}_{host}$ values provided $\Delta\boldsymbol{U}_{host} = 30.6$ kJ/mol and $\Delta\boldsymbol{S}_{host} = 65.9$ J/Kmol. We then confirmed that the $\Delta\boldsymbol{U}_{host}$ value obtained from the free energy analysis coincided with the value from DFT-D calculations using the closed and open structures (31.2 kJ/mol), which demonstrates that the gate adsorption behavior can be adequately described by the thermodynamic model proposed by Coudert et al.[2]

References:

1) A. Kondo ***et al., Nano Lett.* 6,**2581 (2006).

2) F.-X. Coudert ***et al., J. Am. Chem. Soc.* 130,**14295 (2008).

3) H. Tanaka ***et al., J. Phys. Chem. C* 119,**11533 (2015).

4) S. Hiraide ***et al., Dalton Trans.* 45,**4193 (2016).

459967 Confinement-Induced High Pressure Phases in Nanopores: Can the Pressure be in the Megabar Range?

Abstract:

There is an abundance of anecdotal evidence that nanophases adsorbed within nanoporous materials can exhibit high pressures as a result of the confinement[1,2]. For example, phase changes and chemical reactions that only occur at high pressures in the bulk phase occur in the confined phase at bulk phase pressures that are orders of magnitude lower. The pressure in the pore is a second order tensor, and for simple pore geometries has both a normal pressure component (normal to the walls) and one or more tangential components (parallel to the walls).

For simple fluids in pores that are up to a few nanometers in width, molecular simulations show that both the normal and tangential pressures can be locally very high (thousands or tens of thousands of bars) in the pore, even though the bulk phase in equilibrium with the pore is at a pressure of one bar or less. The cause of these high in-pore pressures will be discussed, and where possible comparison with experimental results will be made[3].

When the molecules in the confined nanophase react with each other chemically it may be possible to achieve even higher tangential pressures, in the megabar range. Evidence for this is provided by recent experiments on sulfur (an insulator at ambient conditions) in narrow single-walled carbon nanotubes, carried out by Kaneko and coworkers[4]. They find that the sulfur atoms within the pore covalently bond to form a one-dimensional phase that is metallic. In the bulk phase sulfur forms a metallic phase only at pressures above 95 GPa. In our recent molecular dynamics simulations of this system we find that the sulfur atoms are covalently bonded in the pore and that they experience tangential pressures in excess of 100 GPa as a result of the strong confinement[5].

1. Yun Long, Jeremy C. Palmer, Benoit Coasne, MaÅ‚gorzata Åšliwinska-Bartkowiak and Keith E. Gubbins,

"Pressure enhancement in carbon nanopores: a major confinement effect",Â Physical Chemistry Chemical Physics,Â 13, 17163-17170 (2011).

2. Yun Long, Jeremy C. Palmer, Benoit Coasne, Malgorzata Sliwinska-Bartkowiak, George Jackson, Erich A. Muller and Keith E. Gubbins, "On the Molecular Origin of High Pressure Effects in Nanoconfinement: Effects of Surface Chemistry and Roughness", Journal of Chemical Physics, 139, 144701 (2013).

3. H. Drozdowski, M. Kempinski, M. Sliwinska-Bartkowiak, M. Jazdzewska, Y. Long, J.C. Palmer and K.E. Gubbins, "Structural Analysis of Water Adsorbed in Activated Carbon Fibers", *Physical Chemistry Chemical Physics*, **14**, 7145-7153 (2012).

4. Y. Fujimori, A. Morelos-Gomez, Z. Zhu, et al., "Conducting Linear Chains of Sulphur Inside Carbon Nanotubes", *Nature Communications*, 4, DOI 10.1038/ncomms3162 (2013).

5. K.E. Gubbins, C.K. Addington and J.M. Mansell, to be published.

473522 On the Role of Surface Chemistry in Carbon Adsorption

Abstract:

To be added

463837 Periodic Cycled Separation Using Cops Trays

Abstract:

A realization of periodic cycled separation is presented which enables continuous vapor flow also during liquid drainage from the trays. With the proposed design, a new operation form is introduced where the trays are drained sequentially rather than simultaneously. Thus the vapor flow is not interrupted during the liquid drainage.

Previously we have compared separation factors and column efficiencies for a column stripping ammonia from water with air, using three different sets of internals: Conventional sieve trays, Sulzer BX gauze packings and periodically cycled trays. The periodic trays are advantageous compared to their predecessors. For different ratios of countercurrent vapor/liquid flow rates, column efficiencies are two times greater for periodically cycled columns than for columns with sieve trays. Depending upon the holdup and gas flow, the tray efficiency of a single periodic tray can be up to 300 % of an ideal stage.

In this communication we have explored operation at higher hold-up-to-gas-flow ratios, in order to obtain evenÂ higher column efficiencies. Also, the influence of carbon dioxide on pH measurement - central to the data logging procedure â€" are investigated. Also, aspects of design of period times are investigated.

*** Formatting Issue - Best Available Paper/Graphic ***

472793 Study of Mass Transfer Mechanism with Interfacial ConvectionÂ I: The Simulation Method and Its Validation

Abstract:

Investigation of interfacial mass transfer mechanism by means of computation mass transfer method is prospective with the development of computer technique. As the prerequisite, the mass transfer phenomenon must be correctly simulated to provide reliable data for analyzing, which requires a believable model for describing the overall scene of the mass transfer process. Firstly, the Rayleigh convection is simulated using random disturbance model based on Lattice Boltzmann method (LBM) and the results are compared to the schlieren observation for validation. The approach for simulation is proved believable by detailed analysis from six aspects, and it is recognized that the simulation method is superior both on controlling the numerical diffusion and capturing the characteristics of the absorption process. Furthermoreï¼Œanalogous to the turbulent small eddy model and surface divergence, two mechanism models to predict the interfacial mass transfer coefficientÂ k_LÂ are proposed with pseudo dissipation rate and surface divergence calculated from velocity field. Simulation results of 12 cases of CO_2 absorption process into different solvents fit the two proposed models well, which implies that eddies affect the interfacial mass transfer in the same manner with turbulent mass transfer, independent of the means how it is generated.

Key words:

Mass transfer; Interfacial convection; Lattice Boltzmann method; Mechanism Model

As

456930 Experimental Verification and Aspen Plus Simulation- Effect of Operating Conditions on CO_2 Removal

Abstract:

This investigation simulates unprocessed natural gas (NG) pretreatment system to remove more than 99% of CO_2 from NG using Aspen Plus. Data from an experimental bench-scale apparatus verify simulation results. NG compositions has changed to see its effect on CO_2 capture (%) and methane and CO_2 output vapor mole fraction. Effect of varying pressure on capture has also investigated. Costing parameters, exergy loss and heat exchangeâ€™s efficiency have been calculated. The results confirm low exergy losses of 39.57 kWh/treated NG for simulation and 44.8 kWh/treated NG for bench-scale apparatus. High heat exchangerâ€™s efficiency of +90% has been achieved. High pressure and low methane in NG composition will result in high CO_2 capture.

Efficient Pollution Abatement

Ed Cussler and Bill Arnold
Chemical Engineering and Materials Science; and
Civil, Environmental, and Geo-Engineering
University of Minnesota

Removing pollutants from mixtures seeks to minimize costs. Often, in public enterprises like water systems, such pollutant removal must be accomplished as efficiently as possible and at a cost less than some proscribed maximum. This means that the pollutant concentrations must be managed to respect both their concentrations and their toxicity. To accomplish this, we suggest a strategy for estimating separation costs for any pollutant. We then show how this can be applied to achieve a desired total toxicity for a product stream containing a spectrum of pollutants.

469351 Standing Wave Design of a Non-Isocratic Three-Zone Simulated Moving Bed for Continuous Purification of Therapeutic Proteins

Abstract:

Protein therapeutics is the fastest growing segment of new pharmaceutical products. Production of protein drugs usually involves many challenging purification steps, resulting in higher production costs than those of low molecular weight drugs. Conventional manufacturing of therapeutic proteins involves multiple batch purification processes, which can take many days or weeks to complete. For insulin production from recombinant E. coli, more than 10 weeks are required to produce high-purity insulin. Continuous capture and purification of therapeutic proteins can reduce manufacturing time, minimize proteolytic degradation of the target proteins, increase protein yield and purity, and reduce purification costs. A three-zone, open-loop, non-isocratic simulated moving bed with different adsorption conditions (pH, displacer concentration, ionic strength, solvent strength, or a combination of these factors) can achieve a much higher adsorbent productivity and a lower purification cost than the corresponding batch stepwise elution chromatography. Monoclonal antibody capture on Protein A is used as an illustrative example. The batch and continuous processes with the same adsorbent volume are compared for two cases: separation of a target protein from a non-adsorbing impurity (Case 1), and separation of a target protein from a competing impurity and a non-adsorbing impurity (Case 2). For both cases, the SMBs can achieve ~100% purity and 99+% yield. The SMB for Case 1 has an order of magnitude higher productivity and product concentration and the SMB for Case 2 has two orders of magnitude higher productivity and product concentration than the optimal batch stepwise elution process.

464777 Breaking through the Productivity Barrier: Integration of a Novel Modular Chromatography Scaffold and a New Resin Design to Achieve a Hyper-Productive Protein a Capture Process

Abstract:

Much attention has been made on the recent proliferation of protein A resins. Likewise there has been an increased focus on operating modes including SMB and multicolumn operations. All possess a variety of attributes that can affect Mab purification operations and costs. However very little innovation has been made on the chromatography bed itself and how that can enhance productivity or affect resin usage. The combination of two new developments, Chromasetteâ„¢ chromatography device with a supported bed design and Ampshere A3â„¢ resin is shown to be able to achieve 5-10 X productivity gains over current values. Process modeling was used to elicit the optimal design point for both bead properties as well as peak productivity operating velocities The novel Chromassetteâ„¢ bed design will be reviewed along with results for Amsphere A3 Mab capture. Future direction and possibilities of this new technology will also be discussed.

469924 Protein a Resin Evaluation in a Chromatography Process through Process Modeling and Doe Approach

Abstract:

Increased product yield at reduced cost and time has been the driving force of any manufacturing process. However, inevitable increase in future demand for biopharmaceutical drugs, along with intensified competition and stringent regulatory laws have exhibited an imperative need for established platform processes. In this regard, the implementation of continuous processing, in whole or in components, has demonstrated to increase the manufacturing productivity with key impact on speed, cost and facility implications. The Protein A chromatography process as a capture step has shown enormous potential in terms of high affinity to the mAb, translating to higher purity and considerable reduction in volume for downstream polishing steps. However, Protein A resin costs remains a high contributor to the overall economics of the mAb purification process. Protein A resin is considered to be highly sensitive and loses its performance efficiency after 100-120 sanitization cycles. We have approached the Protein A degradation as a leaching problem primarily caused by caustic in the sanitization phase. A Shrinking Core Model (SCM) was utilized to explain the Protein A leaching kinetics in the column. The model parameters viz rate constant, stoichiometry factor and effective diffusivity were further estimated and validated with Protein A affinity chromatography batch and breakthrough experiments. The binding capacity of Protein A resin in terms of residual Protein A ligands was appraised for process efficiency. A 3-level fractional factorial DOE study using MODDE was performed to describe the leaching mechanism by varying caustic concentration, with or without product and three different resins. A regression model was developed in SIMCA to ascertain the resin lifetime. The results and associated outcome confirmed that this model can be employed to predict the column conditions during 100-120 operating cycles. The model can also be applied as a decision tool for column switching in a continuous separation. The approach elucidates the conditions required to perform the affinity separation for mAb capture step, with indications of improving the resin performance and extending the resin lifetime. The outcomes of the evaluated approach supports the optimized utilization of the Protein A resin in continuous processing for improved biomanufacturing.

Key Words: Continuous processing, Process model, Protein A chromatography, mAbs

461990 Ultrafiltration of Ribonucleic Acids

Abstract:

There is a growing interest in the use of nucleic acids, and in particular RNA, as an alternative to conventional therapies in the treatment of AIDS, cancer, and a wide range of genetic disorders. Short interfering (siRNA) and micro (miRNA) RNA can provide unique ability to modulate gene expression. The successful commercialization of these RNA-based therapies requires the development of a cost-effective platform for the purification of the therapeutic RNA, including removal of host cell proteins, DNA, and other nucleic acid fragments. Several recent studies have demonstrated the feasibility of using small pore size ultrafiltration membranes for the purification of supercoiled plasmid DNA, but it is difficult to extrapolate these results to RNA due to differences in both the size and structure of the nucleic acids. The objective of this work was to evaluate the potential of using ultrafiltration for RNA purification.

Ribonucleic acid from Torula yeast was used as a model RNA. The RNA was suspended in Tris-EDTA buffer, with concentrations measured using the Quant-iT RiboGreen RNA Assay. RNA integrity was examined by Agarose Gel Electrophoresis with the gels visualized using GelStar Nucleic Acid Gel Stain. Ultrafiltration experiments were performed in a stirred cell apparatus using Ultracel composite regenerated cellulose membranes with 100 kD nominal molecular weight cutoff (pore size of approximately 10 nm).

RNA transmission increased with increasing filtrate flux, similar to behavior observed previously with plasmid DNA. This is likely due to the elongation of the RNA molecules in the converging flow field entering the membrane pores. However, the RNA transmission decreased with increasing ionic strength, which is exactly the opposite of the behavior observed with DNA. This effect was quite pronounced. For example, at a flux of 0.1 cm/s, the RNA transmission decreased from more than 60% in the TE buffer to less than 2% in the same buffer with 10 mM NaCl. This large effect of ionic strength on RNA transmission is likely due to changes in the secondary and / or tertiary structure of the RNA due to strong intramolecular interactions. These results provide important insights into the possible use of ultrafiltration for RNA purification, including the separation between DNA and RNA under different ionic conditions.

471281 Development of a Sustainable Tribo-Electrostatic Separation Approach for Dry Fractionation of Legume Flours

Abstract:

Conventional wet fractionation techniques for the production of plant protein powders require the use of solvents, concentrated acids and alkali that result in protein denaturation, thereby reducing the functionality and quality of the protein product. We have explored the potential of a dry or solvent free triboelectrification-based approach for producing protein-rich fractions from legume flours with the advantage of preserving their nutritional value and native functionalities. A vertical bench-scale tribo-electrostatic separator was developed and successfully applied in the classification of navy bean flour as a model system. The approach consisted of a fluidized bed flour reservoir, polytetrafluoroethylene (PTFE) tribo-charging tube, and a plate-type separation chamber. A tribo-electrostatic approach was used to charge proteins and carbohydrates in navy bean flour to different levels before fractionating them based on the magnitude and type of charge inside the separation chamber under the influence of an electric field. Optimization of a single-stage tribo-electrostatic separation of navy bean flour was done by conducting a mixed level full factorial experiment with the aim of investigating the influence of various process variables including air flow rate, tribo-charger length, plate voltage, and plate angle on the protein enrichment and fractionation efficiency. A multiple linear regression model was developed and used to analyze the experimental data. Using the optimum operating conditions identified, the protein content of the original navy bean flour was increased from 25.4% to ~38% accounting for 50% of the total protein. This novel solvent-free tribo-electrostatic separation approach has considerable commercial significance as it preserves the bio-functionality of the protein and averts the likelihood of toxic microbial contamination common in currently used wet processes.

470225 A Mathematical Model to Synthesize and Integrate Distillation Columns into Overall Processes

Abstract:

Distillation constitutes a widely used and energy intensive separation process with a high impact on energy cost and the sustainability of processes. For this purpose, there are several integration schemes to reduce distillation energy consumption and secure further integration of distillation systems with the overall process. Such schemes are thermally-coupled systems, multiple-effect distillation, heat pumping, feed preheating, vapors recompression and combinations of them. This paper focuses on the integration scheme of multiple-effect distillation, since it succeeds in high efficiencies from both stand-alone integration of distillation and integration with the rest heat exchangers of the process. Integration of the re-engineered systems with the overall process is possible to secure up to 100 % energy savings in distillation.

Multiple-effect distillation involves the use of parallel separation effects, which share the initial feed mixture and operate at different pressures. Under these conditions, available heat from the condensers of higher pressure effects can be used by the reboilers of lower pressure effects provided that the minimum temperature difference between heat sources (condensers) and sinks (reboilers) is secured. Energy consumption of the distillation system can be significantly reduced according to the number of effects, the operating pressures and the strategy of splitting the feed mixture among effects. Additional benefits can be succeeded by appropriately integrating the distillation effects with the heat exchangers involved in the rest process; the latter are also called background process needs.

Energy integration of hot/cold streams, either they belong to the multiple-effect distillation system or the background, naturally relates to the use of Pinch analysis. The procedure requires the appropriate placement of separation effects within the Grand Composite Curve of the integrated background needs over or under the pinch point [1]. The appropriate placement rule offers the guidelines for the selection of number, pressures and feed split fractions of effects. In particular, the number of effects defines the split of the overall distillation energy consumption among the multiple levels, while the feed split fraction defines the capacity of effects and, thus, the amount of energy that is required per effect. The operating pressures determine the temperature ranges of added effects and accordingly the temperature difference that is required for heat exchange among the effects and between the effects with the background. In a common multiple-effect distillation scheme, the higher pressure effect receives heat from hot utilities (or the background), while heat is cascaded to lower pressure effects until it is rejected to the background (or the cold utilities).

Both temperature ranges and duties determine the integration potential of the selected effects. At first, the temperature ranges of effects that need to get integrated should not overlap; moreover, the ranges should secure the minimum temperature difference for heat exchange. The temperature ranges also define whether the effects can get integrated with the background process over or below the pinch. Secondly, the number and split fractions define the energy consumption of each effect and, thus, the range of energy savings by integrating each other or with the background. The efficiency of multiple-effect integration schemes is graphically tested using the appropriate placement technique within the Grand Composite Curve as a trial an error procedure. In general the procedure involves the following steps: (1) add an effect, (2) change pressures of effects until effects are appropriately placed over/below the pinch of the background, (3) change the feed split fractions of effects until they are fully (or almost fully) integrated among them and with the background; otherwise go back to step (1). This procedure has been extensively used to integrate distillation columns [2,3], while other rigorous distillation models [4,5] have been also presented in literature to select operating pressures and split fractions; however, they assume fixed number of effects.

This paper presents a shortcut and linear optimization model to select the best synthesis plan of number, pressures and split fractions of distillation effects that minimize distillation energy cost securing the minimum capital cost of added effects. A transshipment model is used to debottleneck energy consumption of the overall process considering the needs of background hot/cold streams to be fixed and variable duties for the condensers/reboilers of effects.

The methodology assumes a pre-selected range of candidate pressures, where a distillation effect is capable to operate at each pressure level. Given the pressure of each candidate effect, the temperature ranges as well as the maximum reboilers and condensers duties can be estimated in advance assuming that each effect receives the whole feed mixture. Moreover, assuming that distillation duties are linearly depended on feed flowrate, the actual duties of effects can be linearly approximated as a fraction of their maximum duties; the fraction relates to the feed split fraction (variable) that is assigned to each candidate. Non-zero split fractions apparently denote the selection of an effect. Linear models are used for the estimation of capital costs of effects according to the selected fractions; the parameters of models were estimated by regressing distillation capital cost in the range of split fractions at each pressure.

The heat demands of candidate effects can be incorporated into the heat cascade of the overall process as additional variable hot/cold streams, whose contribution is managed through the associated split fractions. The hot/cold streams of the background are also incorporated in intervals of the cascade with fixed duties, since their energy contents are not affected by the choices made for effects. A transshipment model is used to debottleneck heat flows along the cascade using additional terms for the selection and contribution of effects to intervals. The model involves the equations of the transshipment model, constraints for selection of effects and the objective function, which is a cost function involving energy cost and distillation capital expenses; the latter is included to avoid solutions, which though they secure minimum energy cost, they aimlessly use more than the needed effects to succeed this target.

The proposed optimization model was used in the courses of a real-life biorefinery which fractionates lignocellulosic biomass into C5 Sugars, C6 Sugars and lignin. The Sugars are further fermented to produce bio-ethanol. The model was used to simultaneously integrate three different distillation columns – one for solvents recovery and two for ethanol purification – with the overall process resulting in high energy savings up to 82 % and zero distillation energy cost. Two effects have been selected for the solvents recovery distillation and one effect for each ethanol purification case. The operating pressures of all four distillations were appropriately changed to secure maximum integration with the background. The model not only detected promising multiple-effect structures for the solvents recovery by examining interactions among the synthesis specifications (number, pressures, split fractions) but also proceeded in re-engineering of operating pressures of ethanol distillations without aimlessly adding effects to secure fully integrated effects with the overall process.

References

1. B. Linnhoff, H. Dunford and R. Smith, 1983, Heat Integration Of Distillation Columns Into Overall Processes, Chemical Engineering Science, 38, 8, 1175–1188
2. M. Dias, M. Modesto, A. Ensinas, S. Nebra, R. Filho, C. Rossell, 2011, Improving bioethanol production from sugarcane: evaluation of distillation, thermal integration and cogeneration systems, Energy, 36, 6, 3691–3703
3. P. Kravanja, A. Modarresi, A. Friedl, 2013, Heat integration of biochemical ethanol production from straw – A case study, Applied Energy, 102, 32–43
4. M. Martín and I. Grossmann, 2011, Energy Optimization of Bioethanol Production via Gasification of Switchgrass, AIChE Journal, 57, 12, 3408–3428
5. M. Martín and I. Grossmann, 2013, On the Systematic Synthesis of Sustainable Biorefineries, Ind. Eng. Chem. Res., 52, 9, 3044–3064

458581 Separation of Complex Mixtures of Aromatic Isomers and Their Hydrogenated Derivates By Batch-Rectification

Abstract:

Liquid Organic Hydrogen Carriers (LOHCs) are aromatic compounds that allow storing renewable energy through a reversible catalytic hydrogenation reaction. This way it is possible to store and transport hydrogen safely with reasonable density and to overcome the limitations by its high volatility and low volumetric energy density.

Among others, benzyl toluene (H0-BT or Marlotherm LH®) is a promising LOHC with the ability to store up to 6.2 mass% hydrogen in its hydrogenated state (H12-BT) and to withstand even harsh conditions (>300 °C). However, BT is not a pure compound but a mixture of constitutional isomers dependent on the manufacturing process. Additionally, every isomer of H0-BT forms further stable intermediates (mostly isomers of H6-BT) and diastereomeres (of H12-BT) during the hydrogenation reaction. In total, more than 13 stable isomers could be detected in the reaction mixture, which makes the LOHC system a complex multicomponent mixture. The objective is to separate the multicomponent mixture to measure thermo-physical data and to perform further studies on the reaction mechanism.

In this contribution, a batch rectification processes, both in lab scale (2-250 ml) and pilot plant scale (5-20 l), are used to separate the isomeric mixture. Because of the isomers having almost equal vapor pressures, the distillation of this high boiling mixture is performed in a column with up to 40 separation stages and under vacuum conditions. First the H0-BT isomer mixture was distilled. To gain simplified mixtures of hydrogenated intermediates and diastereomeres a partial hydrogenation reaction is performed for each H0-BT isomer separately. The hydrogenated fractions are isolated as well. All separated fractions are analyzed by GC-MS.

The distillation routine presented in this contribution is part of a kinetic study defining a detailed hydrogenation reaction framework of the LOHC system. Pure isomers are used to calibrate a Raman Device to facilitate a non-invasive method for individual compound quantification during the LOHC hydrogenation reaction.

Main Fractionator Control
Regulatory Control

Paper 193c (Proceedings 448771)

Andrew W. Sloley, Principal Consultant, Advisian

Prepared for Presentation at the
American Institute of Chemical Engineers 2016 National Meeting
San Francisco, CA, November 13-18, 2016
Rev 0 – 2016 xx xx

www.**advisian**.com

Control

Regulatory
- Keeping system at the setpoint

Servo
- Changing the setpoint

Main fractionator control
- Mass balance control
- Mass in = mass out
- Changes in feed rate or composition
  - At fixed product composition
  - Must change product rates

Current Industrial Practice

Flow Rates on Products Controlled

Three basic configurations

1. **Direct** flow control on products
2. **Level control** on feed to sidestripper **followed by flow control** on stripper product
3. **Flow control** on feed to sidestripper **followed by level control** on stripper product

- All meet requirements
- Different dynamic response

Control Configurations

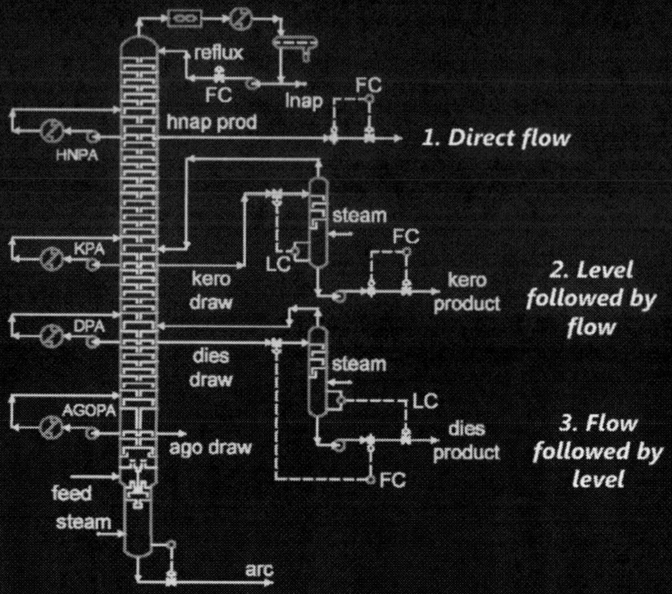

Control in Operation

Operator intervention to adjust

- Feed rate changes
- Feed composition changes
- Product composition changes
- Heat balance changes

Active area for advanced control systems

- Feed-forward elements
- Multiple inputs
- Problem area in model tuning

Basic Steps and Options

Advisian / 7

Control Basics

Basic system = SISO

- Single input, single-output

Setpoint moved by operator to reject disturbances

Cascaded in series to create more complex systems

Advisian / 8

Temperature, Pressure, Composition

Can temperature be used to infer composition?

Often done on simpler columns

Why not on main fractionators?

Temperature Cascaded to Rate

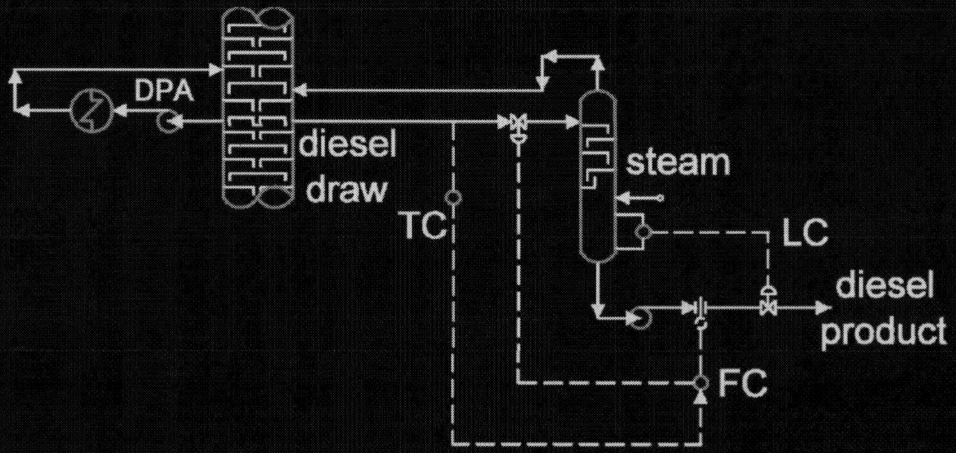

Temperature Advantages

Inexpensive measurement

Rapid response

Good accuracy and high precision of temperature reading

Inherent link between temperature and ASTM distillation specifications

So why not used more often?

Advisian / 11

Evaluation

Advisian / 12

Questions to Answer

Is temperature a reliable proxy for composition in a stable unit?

How well does temperature reject disturbances from

- Pressure disturbance?
- Heat balance disturbance?
- Feed composition disturbance?

Use atmospheric crude column as an example

Base crude, Tia Juana Light, 31.6° API, 0.8676 SG

Advisian / 13

Specifications

Stream	Light Specification	Heavy Specification
Light naphtha	None	Column overhead temperature
Heavy naphtha	None	ASTM D-86 Distillation
Kerosene	ASTM D-56 Flash Point	ASTM D-86 Distillation
Diesel	ASTM D-93 Flash Point or Minimum steam rate in sidestripper	ASTM D-86 Distillation
AGO	Minimum steam rate in sidestripper	Minimum liquid wetting rate in main fractionator (below draw)

2-sigma error band used on specification values to get expectation of 95% reproducibility on an independent test.

Advisian / 14

Is Temperature a Reliable Proxy for Composition in a Stable Unit?

Advisian / 15

Diesel

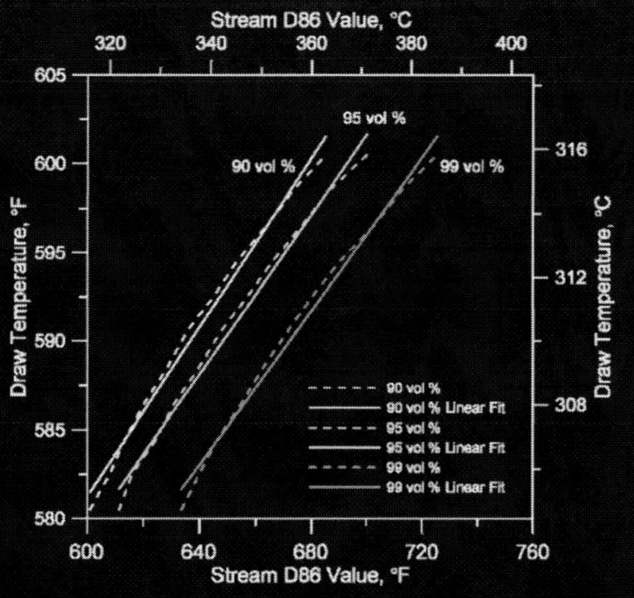

Advisian / 16

Stable Units

Temperature is reliable proxy for composition measurement

Example shown, diesel, was typical

Response close to linear, r^2 values from 0.973 to 1.00

Easy to use and program

Why not used more often?

Disturbance Consequences

Non-Symmetric Consequences

Specifications are not-to-exceed in one direction

- Failure on one side is serious – off-specification
- Failure on the other side, less serious, some product giveaway

~2-sigma error band defined in test method to give 95% reproducibility

- 2.5 % probability off-specification
- 2.5 % probability product give-away

How Well are Pressure Disturbances Rejected?

Pressure Variations

Changing operating conditions

- Feed vaporization
- Overhead condenser cooling water or air temperature changes
- Capacity

Advisian / 21

Pressure Variation Results

Stream	PSI	kPa
Heavy naphtha	2.05	14.1
Kerosene	3.39	27.1
Diesel	2.53	17.4

Operating pressure variation to shift reproducibility error on specification value expected from 2.5% to 50%.

Advisian / 22

Specification Value versus Pressure Change

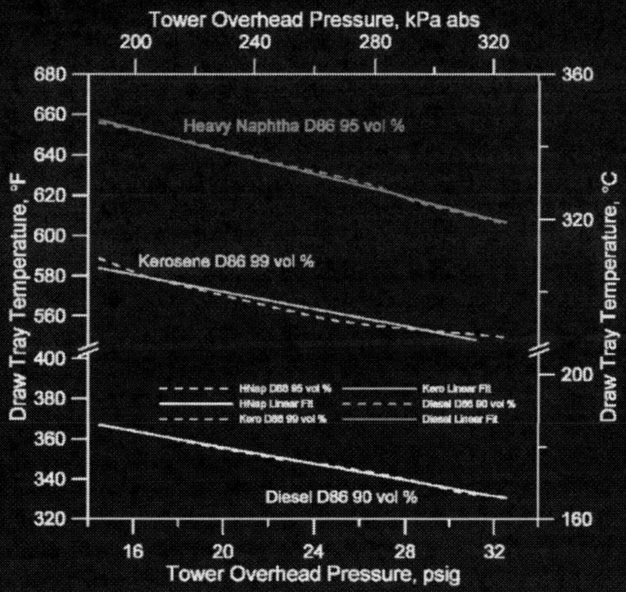

Pressure Variation

Significant source of error

Corrections possible – base on pressure at temperature measurement point

How Well are Heat Balance Disturbances Rejected?

Advisian / 25

Heat Balance Variations

Changing operating conditions

- Crude types
- Heat integration
- Internal reflux rates in main column
- Complex interactions

Advisian / 26

Heat Balance Shifts Examined – Diesel

Shift	M btu/hr	GJ/hr	% Probability of Off-Specification Result Independent Test
A. Base			2.5
B. Diesel pumparound to condenser	5.0	5.3	11.5
C. Diesel pumparound to condenser	10.0	10.6	36.3
D. Kerosene pumparound to condenser	5.0	5.3	10.4
E. Diesel pumparound to heavy naphtha pumparound	5.0	5.3	5.7

Advisian / 27

Specification Value versus Heat Balance Shift

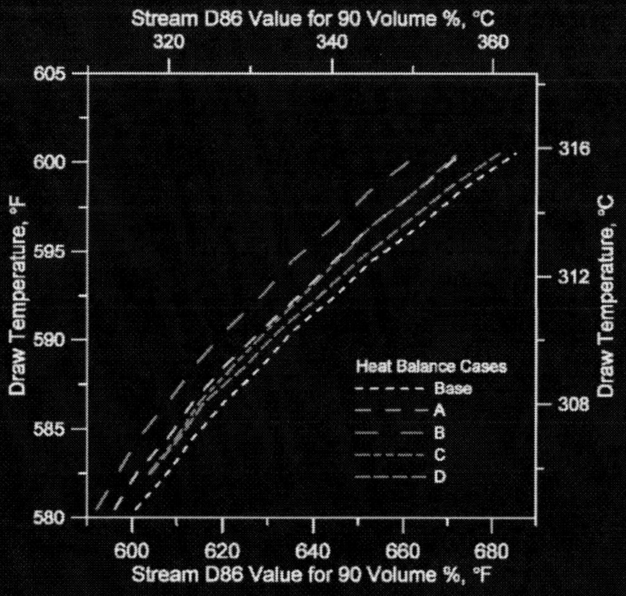

Diesel Product

Advisian / 28

Heat Balance Variation

Significant source of error

Correlations complex

- Not suitable for simple corrections
- Difficult to have simple rules for operators

Advisian / 29

How Well are Feed Composition Disturbances Rejected?

Advisian / 30

Feed Composition Variations

New crudes

Variation within crude

- Rail delivery (light-tight oils)
- Tanker truck delivery

Blending in tank farm

Recycle from slop

API° gravity most commonly monitored

Advisian / 31

Feed Composition Variations Examined

Crude	Abbreviation	API°	SG
Tia Juana Light	TJL	31.6	0.8676
Lagomedio	LM	31.5	0.8681
West Texas Sour	WTS	31.5	0.8681
Abu Al Bu Khoosh	BK	31.6	0.8676
Sepinggan Yakin Mix	SYM	31.7	0.8671

Crudes selected for wide variation in distillation curves with minimal change in API° gravity.

Advisian / 32

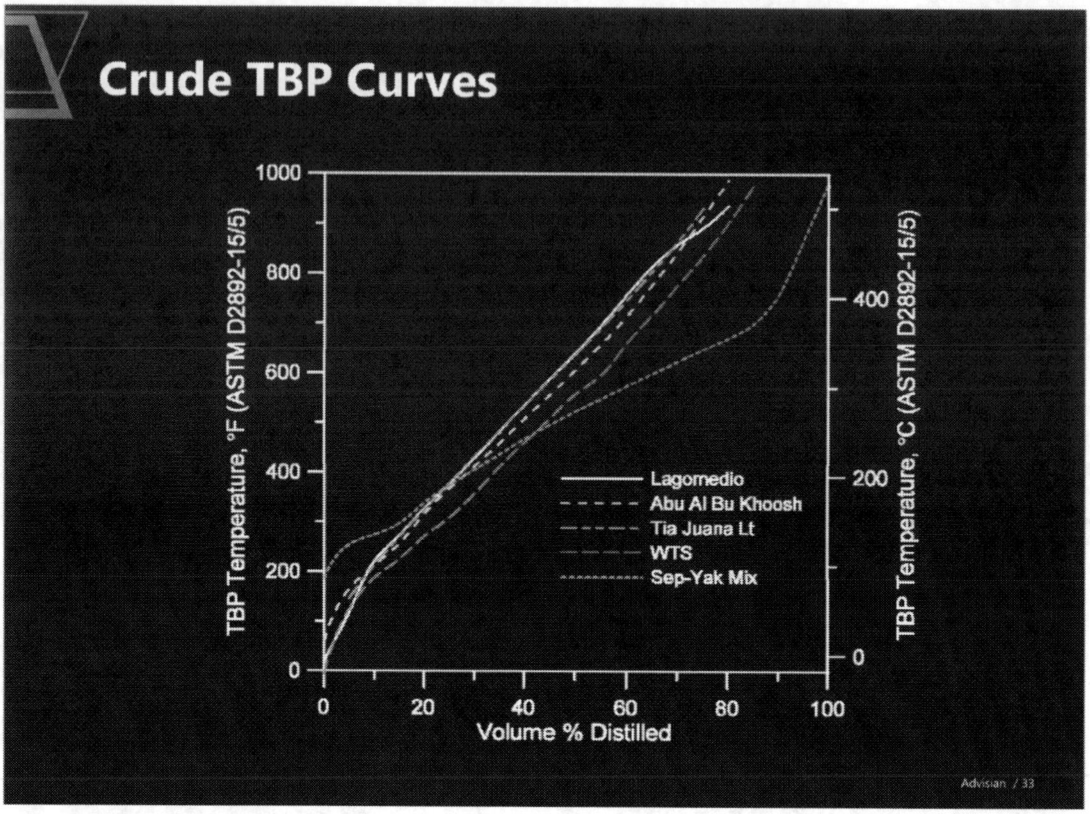

Crude TBP Curves
TBP Temperature, °F (ASTM D2892-15/5)
TBP Temperature, °C (ASTM D2892-15/5)
Volume % Distilled
Lagomedio
Abu Al Bu Khoosh
Tia Juana Lt
WTS
Sep-Yak Mix
Advisian / 33

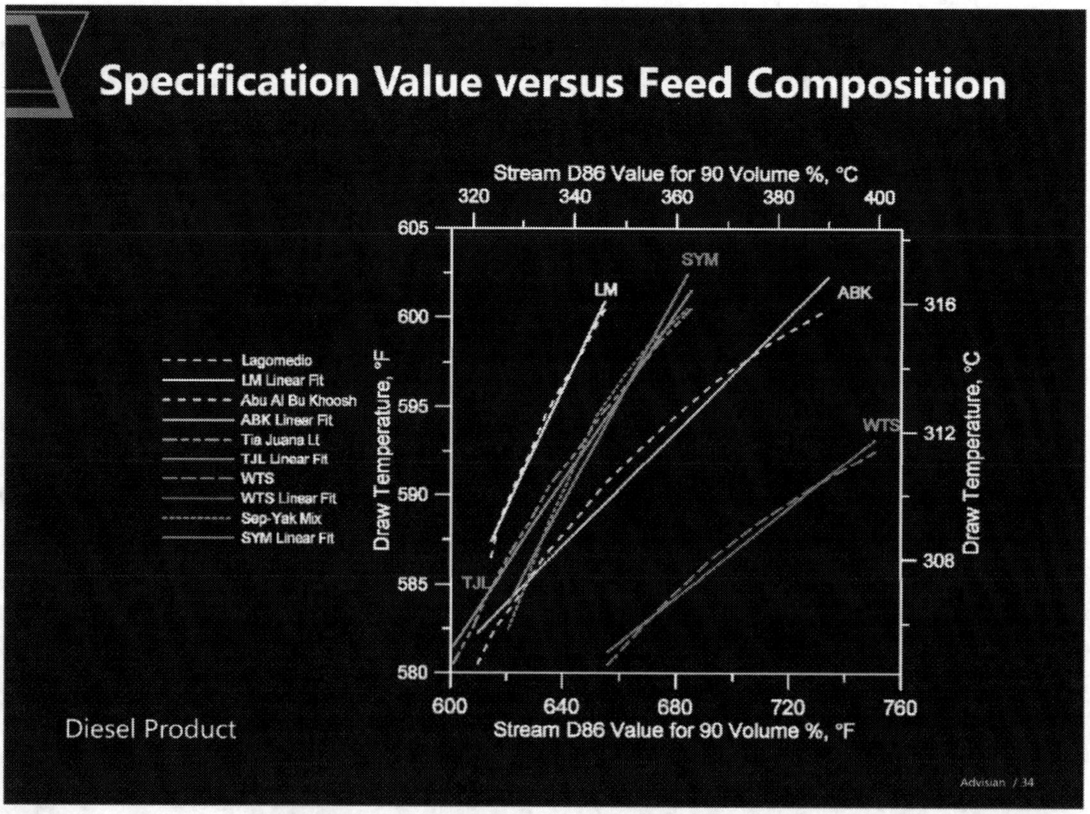

Specification Value versus Feed Composition
Stream D86 Value for 90 Volume %, °C
Stream D86 Value for 90 Volume %, °F
Draw Temperature, °F
Draw Temperature, °C
Lagomedio
LM Linear Fit
Abu Al Bu Khoosh
ABK Linear Fit
Tia Juana Lt
TJL Linear Fit
WTS
WTS Linear Fit
Sep-Yak Mix
SYM Linear Fit
LM
SYM
ABK
WTS
TJL
Diesel Product
Advisian / 34

Feed Composition Shifts Examined – Diesel

Crude	% Probability of Off-Specification Result Independent Test *If setpoint not changed*
Tia Juana Light – Base	2.5
Lagomedio	44.4
West Texas Sour	100.0
Abu Al Bu Khoosh	89.6
Sepinaggan Yakin Mix	71.2

Advisian / 35

Feed Composition Variation

Very large errors

Correlations very complex

Other performance characteristics worse than distillation temperatures

- Freeze
- Cloud

Temperature by itself not suitable

Advisian / 36

Conclusions

Advisian / 37

Today's Take-Away

Temperature suitable for stable units

Pressure correction required, straightforward

Heat balance correction required, complicated

Composition correction vital, extremely complicated

- Implies heat balance shifts
- Blending effects on product rates
- Internal reflux shifts due to heat balance and product rates

Simple SISO cascade from temperature to flow not suitable except for the most stable units

Advisian / 38

Advisian / 39

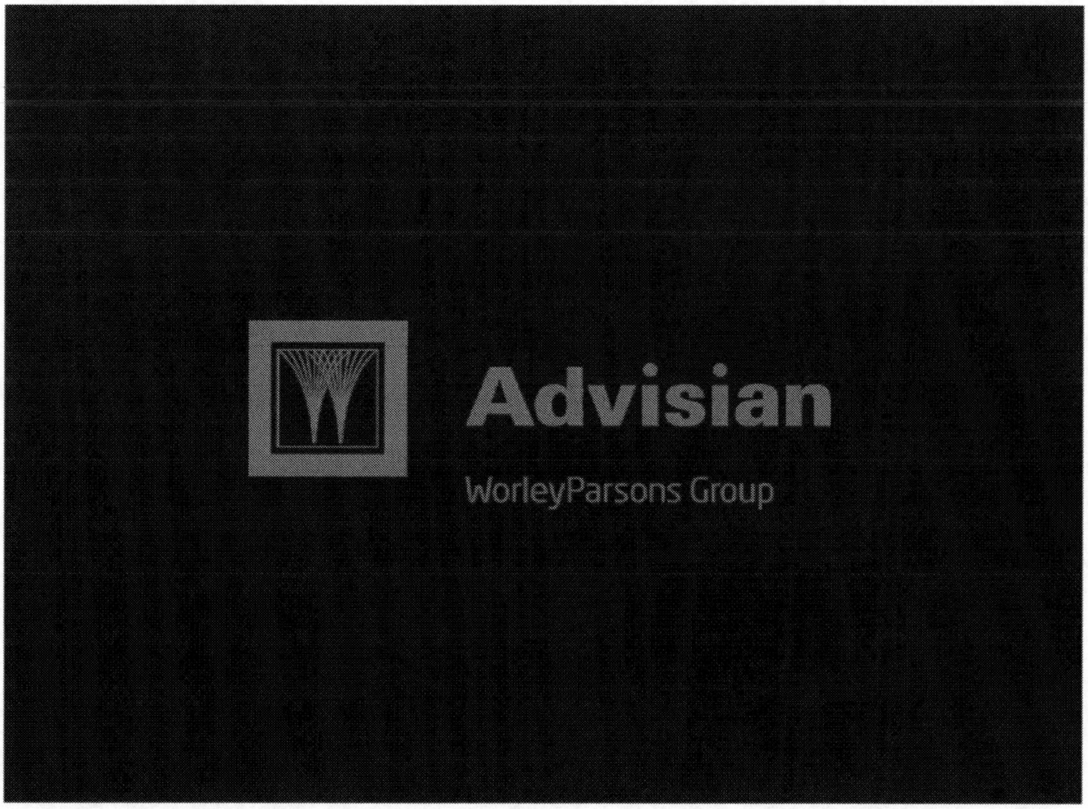
Advisian
WorleyParsons Group

460971 Tomographic Determination of Fluid Dynamic Parameters for Three Phase Systems in Packed Columns

Abstract:

The fluid dynamic behavior of two poorly miscible liquids and a gas phase in packed columns, as it occurs in three-phase distillation, is very complex, poorly understood and rarely described in literature. Therefore and due to the frequently occurring negative influence of the additional liquid phase on the mass transfer the objective of this work is to determine the fluid dynamics in a three-phase system.

The used X-ray computed tomography scanner (CT), which has been customized designed for process technology application, enables the determination of local fluid dynamic parameters under process conditions. X-ray source and detector rotate around the vertical axis and are height-adjustable what enables the investigation of the whole unmoved column without influencing the flow. The high spatial (80 µm) and temporal (1 fps) resolution of the generated CT images and the selection of suitable materials enable, in combination with subsequent image processing steps, a precise determination of the individual phases in the entire column.

All experiments were performed in a glass column with an inner diameter of 100 mm and a packing height of 1 m. Different surface-treated Mellapak 500.Y of the Sulzer Chemtech AG were used to investigate the influence of the surface modification on the results. Water/dichlorobenzene/air was used as three-phase system, because preliminary CT tests showed distinguishable gray values of all phases. The liquid load, both the ratio (1:3, 1:1, 3:1) and the total amount (2 to 40 $m^2\ m^{-3}\ h^{-1}$), and also the supply order of the liquids were varied. Besides the investigations over the entire column, detailed images in 1 mm and 150 µm intervals enable a closer look at local flow patterns. Moreover long term experiments were conducted to investigate temporal changes of the flow behavior. For a quantitative evaluation of the experimental data a new image processing routine was developed to receive reproducible and meaningful results. It is building on an evaluation method for two phase systems, developed at the same institute, and includes a new segmentation algorithm to assign every phase its own gray value. This allowed the quantification of amount and distribution of the phase interfaces and the hold-ups as a function of place, time and packing parameters.

As a result, for example, a significant reduction of the interfacial area between water and air was proven, because the water forms preferable rivulets in the packing corrugations and the 1,2-dichlorobenzene forms films, which overflow the packing and the water rivulets.

451353 Minimum Energy of Multicomponent Distillation Systems Using Minimum Additional Number of Heat and Mass Integration Sections

Abstract:

We propose a powerful and simple-to-use fact about heat and mass integration to consolidate distillation columns in a multicomponent distillation configuration. First, thermal couplings are introduced to all intermediate transfer streams in the distillation column. And then, distillation columns associated with a lighter final product reboiler and a heavier final product condenser are consolidated to produce a heat and mass integrated configuration. We call this novel set of configurations the HMP configurations. A systematic way of enumerating all HMP configurations for n-component separations is introduced, and the results for up to 7-component separations are presented for the first time. Also, we compare the energy savings of HMP configurations with the well-known fully thermally coupled (FTC) configuration via case studies. We demonstrate that HMP configurations can have very similar and sometimes even the same minimum total vapor duty requirement as the FTC configuration, while using far less number of column sections, intermediate transfer streams, and thermal couplings than the FTC configuration. These findings not only intensify the search space of distillation configurations, but also provide industrial practitioners with insights in designing energy efficient distillation configurations that are easy to operate and control.

473786 Heat Transfer Investigation and Dynamic Simulation of Internally Heat-Integrated Distillation Column

Abstract:

Distillation is the most widely applied process of separation in various chemical industries. But it is quite energy intensive and accounts for 3% of the total industrial energy consumption. The overall thermodynamic efficiency of this process is found to be as low as 5-20%. Therefore several methods have been developed to improve the thermal efficiency, such as heat pump-assisted distillation, multi-effect distillation, vapor recompression distillation, and dividing-wall distillation. Heat-integrated distillation column (HIDiC) is another attractive distillation scheme, which has received considerable attention due to the high energy saving potential by means of improving thermodynamic efficiency. Compared with the conventional distillation column, the energy consumption saving of HIDiC could be up to 30%-60%.

In a HIDiC, the rectifying section and the stripping section are set as two independent columns for a clear perspective, between which heat coupling could happen. Specifically a compressor is used to increase the pressure of the vapor from the top of the stripping section, which is then fed to the bottom of the rectifying section. The liquid leaving the bottom of the rectifying section is fed to the top of the stripping section after the pressure is reduced by a throttling valve. Thus the rectifying section is operated under higher pressure and temperature than the stripping section. A certain amount of heat is exchanged between the two sections, and it provides the necessary internal liquid flow for the rectifying section and internal vapor flow for the stripping section. As a consequence, the heat duties of the bottom reboiler and top condenser are reduced, and a considerable amount of energy is saved.

In this paper, the heat transfer between rectifying and stripping section of a pilot-plant concentric structured internal heat-integrated distillation column (HIDiC) was investigated experimentally. The equation of heat transfer coefficient (U) changing with the pressure ratio was regressed. The result shows that the values of U decreased with the increase of pressure ratio. In addition, a dynamic simulation was developed upon the HIDiC. The control scheme with larger gain and smaller period was chosen to operate the HIDiC equipment. It can be found that the HIDiC system can rapidly return to steady state when feed disturbances occurred. In the process of continuous operation, good stability and operability were observed.

- Abstract Text

460925 Structural Optimization of the Practical Tube-Typed Heat Integrated Distillation Column (HIDiC)

Abstract:

Heat integrated distillation column (HIDiC), since developed to substantially improve the thermal efficiency of distillation process, has always been a research focus in the field of advanced distillation. Amounts of studies on heat distribution, mathematics model establishment and physical model selection have been carried out for the progressive development of HIDiC. However, the industrialization process of HIDiC is nearly stagnant. There only two or three industrial-grade tube-typed HIDiCs were set up in Japan and Netherlands.

Certainly, considering the operability, economy and pressure capacity, the tube-typed HIDiC is the most practical one among the existing physical models. The production capacity of single tube in tube-typed HIDiC is the major design variable which deserves our devotion. When total production capacity of the HIDiC is decided, the production capacity of single tube have a great influence on both of the operating cost and capital cost. To determine the optimal production capacity of single tube, we initially investigated the relationship between production capacity of single tube and the thermal efficiency of tube-typed HIDiC by plenty of process simulations in Aspen Plus. Aiming to obtain accurate and guidance results, the heat transfer parameters, especially the heat transfer coefficient, were measured via the customized heat transfer testing equipment based on the principle of tube-typed HIDiC. For a specific separation target, we found that there is a maximum production capacity of single tube to guarantee that the tube-typed HIDiC could act as an energy saving distillation process. Also, varied with the production capacity of single tube, the highest thermal efficiencies are not all obtained at the state when either condenser or reboiler are not needed. Adopting the total annual cost (TAC) as the optimizing objective, a systematic optimization procedure, mainly based on the economic evaluation theory of distillation process by Douglas, was built to find out the optimal production capacity of single tube. Tested with different separation system, it was verified that the optimization procedure is qualified.

Further, we also felt anxious with the adaptability of the traditional HIDiC model, which separates the column into high pressure section and low pressure section where the feed stage is. There is no doubt that the initial HIDiC platform would more or less lose its advantage in energy saving when the feed stage is changed as the component of feed and separation requirement vary. To improve the adaptability, a novel HIDiC model was proposed. In this novel HIDiC, the high pressure section and low pressure section will share the same number of stages which is equal to half the number of all column stages, and the feed stage will be placed at either high or low pressure section as needed. The novel HIDiC is so called symmetrical heat transfer HIDiC (sym-HIDiC). Evaluated by process simulation in Aspen Plus, under their optimal structure and operating conditions, sym-HIDiCs had almost the same energy saving performance with comparison to tradition HIDiCs when they shared the same separation targets. However, to adapt the change of separation requirement, the tradition HIDiC will be anatomically rebuilt, but it only need to adjust the position of feed stage in sym-HIDiC as a conventional distillation column. Moreover, the pressure difference between two sections has decreased due to this specific heat transfer form. It enhances the safety of HIDiC, and also reduce the capital cost.

475310 Introduction to Altsep: Sustainable Separation Processes Roadmap Review

Abstract:

Creating a sustainable future requires sustainable manufacturing processes. This is especially true for the separation processes central to chemical manufacturing. Conventional fluid separations account for over a third of energy use and over half of the capital investment of chemical plants. To pave the way toward transforming the energy efficiency of fluid separations, the ACS Green Chemistry Institute (GCI) has partnered with AIChE to coordinate the development of a NIST-sponsored innovation roadmap. Based on a series of workshops, the draft AltSep roadmap has identified a number of research, development, and demonstration initiatives. These initiatives are critical for establishing a molecular property-based system for the industrial selection, simulation, and design of energy-efficient separation processes and accelerating the industrial application of such alternative separation processes.

A series of speakers will present on the following topics:

- Background and need for the roadmap,
- System to synthesize and design alternative separation processes, and
- Challenges to overcome barriers to adoption of sustainable separation solutions.

Following these presentations, feedback will be solicited from the audience to improve the quality of the roadmap findings.

457859 Effect of Membrane on the Performance of a Hydrogen-Vanadium Regenerative Fuel Cell

Abstract:

Hydrogen-vanadium regenerative fuel cells offer a feasible solution for storing electrical energy from the grid or directly from renewable energy sources such as wind and solar. Due to the relatively high cost of vanadium ($ per kg), the hydrogen-vanadium flow battery is attractive over the all-vanadium system due to the 50% reduction in the amount of vanadium solution required. Furthermore, by replacing the negative vanadium electrode with a hydrogen electrode the effect of self-discharge by the crossover of unwanted ions, mostly the smaller vanadium II and III ions, is expected to be greatly reduced. Also, by having a two-phase system, gaseous negative electrode and aqueous positive electrode, crossover species can be easily separated and recovered. In this presentation, the performance of electrospun perfluorosulfonic acid (PFSA)/polyphenylsulfone and PFSA/polyvinylidene fluoride (PVDF) composite membranes, relative to PFSA membranes such as Nafion, with regards to the self-discharge rate and electrical performance of a hydrogen-vanadium regenerative fuel cell will be discussed.

464994 Dynamics of Anhydrous Proton Transport on the Surface of Functionalized Graphene

Abstract:

Proton transport phenomena has been of tremendous importance for the application of proton transport membrane fuel cells for their efficiency and clean emissions. We are specifically focused with the applications of proton transport at interfaces for a better design of new materials for proton exchange membrane (PEM) fuel cells in anhydrous conditions. Study of transport of protons over the surface of hydroxylated graphene in anhydrous conditions is performed using density functional theory. The hypothesis is that a facile anhydrous proton transport will occur on a surface that has a continuous network of hydrogen bonded hydroxyl groups showing "grotthuss-like" mechanism. Simple configurations of 1-D single file and a 2-D hydrogen bonded network of hydroxyl groups on graphane membrane will sufficiently facilitate the transport of protons. Dynamical properties and energetics related to the transport of the excess proton is studied. In this work, facile transport of excess proton over the surface of functionalized graphene has been confirmed with computed energy barriers using nudged elastic band technique and their respective mechanism pathways will be discussed. Ab-initio molecular dynamics (AIMD) simulations are used to determine the dynamical properties of 1-D single file and 2-D network of fully hydroxylated graphene surface with one excess proton concentration. Correlations influencing the dynamical properties for the above patterns of hydroxylation are investigated. Stability of the functionalized graphene material for varying degree of hydroxylation, morphological defects and also excess proton concentrations will be reported. A monte-carlo based lattice model of 1-D hydrogen bonded network system mimicking the proton hopping mechanism observed from the AIMD simulations is developed. This is to support and verify the behavior of normal diffusion of excess proton transport for larger system size and longer time domain.

472540 Indirect and Direct Observation of Ionomer Colloidal Systems with Applications to Fuel-Cell Catalyst Layers

Abstract:

Manufacturing methods for fuel cell catalyst layers are critically important for achieving high material utilization and performance. Typically, electrode systems are cast from an 'ink' or 'colloidal' state, where the solid material is suspended in a solution containing a solvent and a polymer. For polymer-electrolyte fuel cells, a perfluorinated sulfonic-acid (PFSA) ionomer, typically Nafion, is used in the ink and catalyst layer, where its roles are to act as a binding agent and conduit for proton conduction to the reaction site.[1] Recently, it has been suggested that the ionomer film that surrounds the platinum reaction site may contribute to increased resistances that limit performance at low Pt loadings.[1,2] Thus, it is of great interest to understand the factors controlling the film formation in catalyst layers to optimize material proeprties and cell performance. The nature of the dispersion and casting process could have significant influence on the morphology and properties of the dispersion-cast membranes.[3-5] To isolate these effects, the morphology of PFSA dispersed in various solvents (glycerol, water/2-propanol, and NMP) was reported previously using small angle neutron scattering (SANS), where catalyst layers from the different solvents exhibit different performance.[6] The understanding of *local* 'ink' structures as a function of active material (polymer and solid) volume fraction and across solvent space and length scales can aid in understanding bulk properties and the impact of ink formulation in cast systems. In this talk, a direct observation of the colloidal suspension of the ionomer in different solvents will be described using confocal laser scanning microscopy (CLSM). CLSM has long been utilized as a tool for characterizing the structure of biphasic colloidal gels. We combine dynamic light scattering, ultra-small angle x-ray scattering, in-situ saxs (during printing) and confocal microscopy to understand PFSA size and structure in dilute (1,3, and 5 wt%) ionomer solutions. Moreover, a novel we explore how the structure evolves during printing processes using synchrotron based small angle x-ray scattering methods. Finally, we contrive a model 'ink' composed of attractive and repulsive silica spheres and PFSA suspended in different solvents (water, NMP, and isopropyl-alcohol) and use CLSM to

gain a greater understanding about solid|PFSA interactions in model inks.

1. A. Z. Weber and A. Kusoglu, *Journal of Materials Chemistry A*, **2**, 17207 (2014).

2. A. Kongkanand and M. F. Mathias, *The Journal of Physical Chemistry Letters*, 1127 (2016).

3. T. T. Ngo, T. L. Yu and H.-L. Lin, *Journal of Power Sources*, **225**, 293 (2013).

4. C.-H. Ma, T. L. Yu, H.-L. Lin, Y.-T. Huang, Y.-L. Chen, U. S. Jeng, Y.-H. Lai and Y.-S. Sun, *Polymer*, **50**, 1764 (2009).

5. Y. S. Kim, C. F. Welch, R. P. Hjelm, N. H. Mack, A. Labouriau and E. B. Orler, *Macromolecules*, **48**, 2161 (2015).

6. C. Welch, A. Labouriau, R. Hjelm, B. Orler, C. Johnston and Y. S. Kim, *Acs Macro Letters*, **1**, 1403 (2012).

466036 Multiscale Modeling of Polymer Electrolyte Membrane for Fuel Cell

Abstract:

In this study, we present a multiscale modeling of nanophase-segregation of the nanophase-segregated structures and transport properties of polymer electrolyte membranes for fuel cell. Molecular interactions between entities in full-atomistic molecular dynamics (MD) simulations are described by force field that is developed using quantum mechanical density functional theory (DFT).

The equilibrium structures are obtained from MD simulations, providing detailed information of local structures through pair correlation functions. The extents of nanophase-segregation in this full-atomistic structures are evaluated from the structure factor analysis. To scale up the simulations from full-atomistic modeling to meso-scale coarse-grained modeling, we employ Chi parameter as a coarse-grained energy parameter, which is used for coarse-grained simulations such as MesoDyn and dissipative particle dynamics (DPD) simulation. Using such multiscale modeling approach, we investigate various polymeric materials including Nafion, dendrion, PEEK, and so on.

453357 High Anion Conduction in Partially Fluorinated Multiblock Copolymers

Abstract:

High Anion Conduction in Partially Fluorinated Multiblock Copolymers

Lisha Liu[*†], John M. Ahlfield[‡], and Paul A. Kohl[‡]

[†]School of Materials Science and Engineering,

[‡]School of Chemical and Biomolecular Engineering

Georgia Institute of Technology

Atlanta, Georgia 30332-0100

Anion Exchange Membranes (AEM) are used in fuel cells, electrolyzers and electrodialysis devices. Alkaline conditions provide a route to overcoming fundamental issues with acid-based devices. The fuel cell and electrolyzer issues include the high cost of platinum catalysts, complex water management, and sluggish electrochemical reactions. However, the performance of AEM fuel cells is not as good as that of Proton Exchange Membrane (PEM) fuel cells partially because of the limitations of current anion exchange membranes, such as low ionic conductivity, poor stability at high pH, and high water uptake.

A series of partially fluorinated multiblock copoly(arylene ether)s with long head-group tether were synthesized for use in AEM fuel cells and electrolyzers. The multiblock copolymers were synthesized via polycondensation of hydroxyl-terminated oligomers and fluoro-terminated oligomers. The resulting multiblock structure has one hydrophilic block and one hydrophobic block. It was designed so that nanophase-separation occurs and efficient conductive channels are formed with low water uptake. Multiblock copolymers with different block lengths and ion exchange capacity (IEC) were synthesized to maximize ion conductivity and explore the relationship between chemical structures and membrane properties. Table 1 shows the membrane properties. A non-linear relationship was found between the number of head-groups on a hydrophilic block and the conductivity. Doubling the number of head-groups more than doubled the hydroxide conductivity. Hydroxide conductivity as high as 119 mS/cm at 80°C have been observed with a specific size block size: $X_{5.4}Y_7$, where X is the hydrophobic block and Y is the hydrophilic block. The hydrophobicity of the backbone has allowed synthesis of polymers with minimal water uptake. The ratio of conductivity-to-water uptake shows that less water is absorbed compared to conventional materials. The number of waters per ion within the polymer is as low as 4. This is reflected in the measurement of the amount of free-water and bound-water. No conductivity loss was observed after soaking the membrane in 1M NaOH solution at 60°C for over 600 hours.

Financial support from US Office of the Deputy Assistant Secretary of the Army for Defense Exports and Cooperation (DASA-DE&C) is gratefully acknowledged.

Table 1. Summary of properties of membranes

Block copolymer	**Molecular Weight (GPC)**	**IEC (Ion Exchange Capacity) (meq/g)**	**OH^- Conductivity (mS/cm)**				**Water Uptake**
			R.T.	40°C	60°C	80°C	(%)
Y_8-1	18K	1.18	13.12	17.91	24.34	36.05	35.85
$X_{3.1}Y_{3.6}$-1	88.6K	0.66	16.41	26.37	35.55	51.50	5.56
$X_{5.4}Y_7$-1	68.2K	0.73	14.17	21.08	27.94	34.72	8.00
$X_{5.4}Y_7$-2	68.2K	1.30	38.21	58.29	96.07	119.70	50.77
$X_{3.1}Y_8$-2	55.9K	1.56	23.13	45.41	66.41	94.03	26.67
$X_{3.1}Y_{3.6}$-2	66.0K	1.19	25.79	41.75	59.03	85.00	25.00
$X_{5.9}Y_5$-2	59.0K	1.10	22.14	33.56	50.29	66.70	19.57

1. X-hydrophobic block; Y-hydrophilic block; subscripted number-block lengths; 1 or 2-tether amount)
2. IEC (Ion Exchange Capacity) was calculated via ^1H NMR results.
3. OH^- conductivity was measured by four-probe conductivity cell.
4. Water uptake was measured at room temperature.

465994 Anion Exchange Membranes for Fuel Cell Applications Based on Advanced Cations

Abstract:

The potential of anion exchange membrane (AEM) fuel cells to provide inexpensive compact power from a wider variety of fuels than is possible with a proton exchange membrane (PEM) fuel cell, has continued to drive the research interest in this area. Alkaline catalysis in fuel cells has been demonstrated with non-precious metal catalysts, and with a variety of fuels beyond H_2 and methanol. Alkaline fuel cells (AFCs), based on aqueous solutions of KOH, have serious drawbacks associated with system complexity and carbonate formation. Anion exchange membrane (AEMs) fuel cells have a number of advantages over both PEM fuel cells and traditional AFCs; however, although anionic conductivity in AEMs can be comparable to PEMs the chemical stability of membrane attached cations in hydroxide is still not always sufficient for practical applications. Recently, it has been recognized that a number of advanced cations, may give AEMs the needed chemical stability. In some circumstances simple trimethyl benzyl ammonium cations are stable up to 60°C allowing us to being to study hydroxide and water transport in these systems. We use in-plane conductivity and multi-nuclear PFGSE to measure self-diffusion coefficients of the water and where possible, the ion, i.e. F^-, carbonate and bicarbonate. Together with temperature and RH dependent SAXS we couple this information with the morphological changes in the materails.

By their nature these organic cations form dipoles, which have a tendency to interact, see Figure below for a trimethylbenzyl cation with bromide for illustrative purposes. For maximum ion transport the cation should be distributed along the polymer chain, however, cation clustering is a common phenomena. It occurs in phase separated diblock polymers when an attempt is made to raise the IEC, no net increase is observed in ionic conductivity above a certain IEC. In many other less structured polymers it occurs in the fully humidified state when the distributed disordered cations order in agglomerates at a disorder/order T_d. This T_d is often observed in the operating temperature

range expected for a device, in DMA, DSC, and broadband electric spectroscopy. Interestingly, even though the SAXS clearly shows an agglomeration on the nano-scale, little difference is seen in the bulk conductivity with temperature. However, steric hindernce effects are expected to discourage aggregation, and may allow the cation to tune all properties of the AEM. As these phenomena are so prevalent in AEM materials the implications and science behind them are being investigated by us and will be discussed in this presentation. The implications of more complex cations proposed for stability in AEMs will be described.

462223 Sustanion™ Membranes for Electrolyzers and Fuel Cells

Abstract:

This paper will discuss the properties of Sustanion™ membranes for electrolyzers and fuel cells. The membranes are based on a styrene backbone that has been functionalized with Imidazolium groups and other additives. Different formulations show conductivities between 30 and 70 mS/cm in KOH and are unexpectedly stable. A CO_2 electrolyzer has been run for 6 months at room temperature with no significant loss in performance. We observe CO_2 reduction currents up to 600 mA/cm^2 at 3.25 V. The closest competitor needs to go to 6.5 V to obtain similar currents. An alkaline water electrolyzer has been run for 200 hrs at 60 C in 1 M KOH. We observe currents of 1 A/cm^2 at voltages less than 2V. Again, 3 time higher than the closest competitor. We are expecting to offer these membranes for beta-testing by various academic groups. Please let us know if you are interested.

460806 Design and Development of Electrolyte Membranes for Solid Alkaline Fuel Cells

Abstract:

Solid alkaline fuel cells (SAFCs) are another system that holds the potential to achieve high-energy conversion efficiency without Pt catalysts. Although most of metal catalysts can be used under alkaline environment, development of durable electrolyte membranes in alkaline media is the key for this technology. We are systematically designing and developing new materials 1,2) from the molecular level to the device level based on organic chemistry3) and quantum chemical simulation4). We have designed a new anionic polyelectrolyte membrane consisting of highly cross-linked and packed ion-exchange sites with aromatic backbone. Each aromatic monomer unit has three ion-exchange sites and the sites can cross-linked with other monomer units. The monomer was polymerized in pores of a porous membrane substrate and the membrane showed high ionic conductivity, exceptionally low fuel cross-over and good chemical stability. Those functionalities can satisfies the anticipated requirements for a viable solid state alkaline fuel cells.

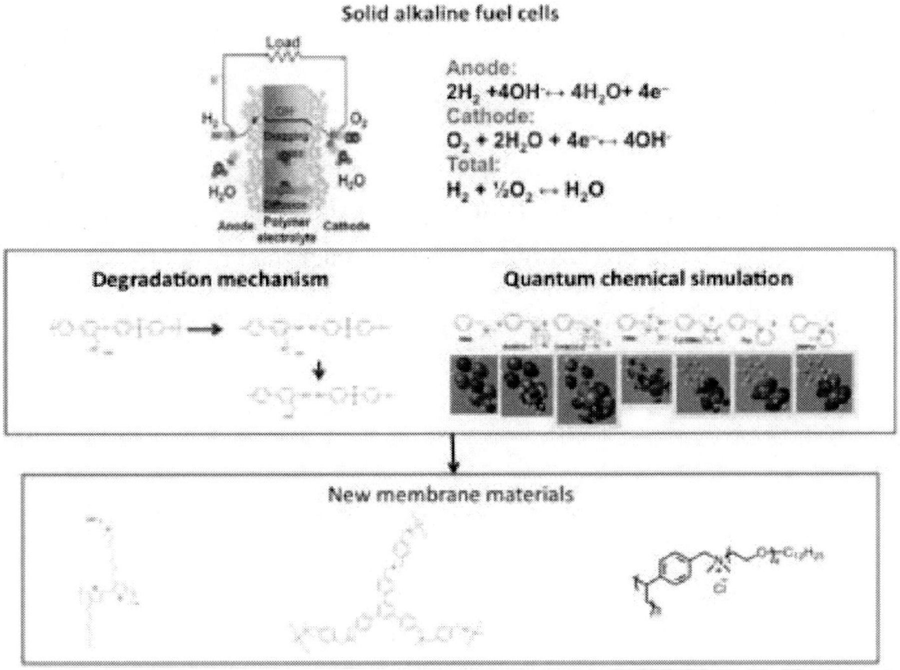

References

1.) S. Miyanishi, T. Fukushima, T. Yamaguchi, ***Macromolecules***, 48(8), 2576-2584 (2015)

2.) G. S. Sailaja, S. Miyanishi, T. Yamaguchi, ***Polym. Chem.***, 6, 7964-7973 (2015)

3.) S. Miyanishi, T. Yamaguchi, ***Phys. Chem. Chem. Phys.***, in press (2016)

4.) K. Matsuyama; H. Ohashi, S. Miyanishi, H. Ushiyama, T. Yamaguchi, ***RSC Advances***, 6, 36269-

36272 (2016)

466840 Membrane Electrolytes with Magnetic Field Aligned Graphene Oxide Nanosheets for Alkaline Fuel Cells

Abstract:

This study aims to elucidate the magnetic graphene oxide (GO) nanofiller alignment in polymer matrix on the resulting filmsâ€™ ionic conduction and methanol barrier properties. Magnetic iron oxide (Fe_3O_4) nanoparticles were fabricated and decorated onto the GO nanosheets. The magnetic GO-Fe_3O_4 nanofillers were mixed with quaternized polyvinyl alcohol (QPVA) to form polymer composite. By applying magnetic field on parallel sides along in-plane direction during drying and film formation step, the GO-Fe_3O_4 nanosheets were aligned in an orientation parallel to the direction of magnetic field and connected each other into elongated domains. The QPVA composite membranes were doped with alkali to become hydroxide-conducting electrolytes for direct methanol fuel cell application. The magnetic field-aligned QPVA/GO-Fe_3O_4 membrane showed higher ionic conductivity and lower methanol permeability compared to nonaligned QPVA/GO-Fe_3O_4 membrane. The aligned QPVA/GO-Fe_3O_4 composite exhibited a peak power density of 172.4 mW cm^{-2} at 60°C, 35.6% higher than that without magnetic field treatment.

468817 Development and Impacts of Intein-Based Bioseparations

Abstract:

Discovered in the early 1990â€™s, inteins are self-slicing proteins with the ability to excise themselves from a variety of host proteins. This self-splicing activity allows inteins to be moved into many contexts for a range of applications, and mutated inteins have now been demonstrated in a wide variety of protein engineering, labeling and purification approaches. One of the most impactful of the intein applications has been the development of self-cleaving affinity tags for recombinant protein purification. In this strategy, a conventional affinity or purification tag is appended to a self-cleaving mutant intein. The intein is further appended to a target protein, which can be easily purified via the tag. Once purified, the intein is induced to self-cleave, releasing the purified native target with minimal expense or additional processing. Early intein applications were plagued by issues with premature cleaving, required modifications to the target protein, or the need for high concentrations of thiol compounds to induce cleaving. These issues limited the applications where inteins could be used successfully, and prevented inteins from gaining widespread adoption in scientific research and industrial protein production. Recent work has solved many of these problems, however, most notably allowing inteins to be used for disulfide-bonded recombinant glycoproteins expressed in mammalian and other eukaryotic hosts. Here we will report a split intein system developed in our laboratory, where a small intein-segment tag on the target protein is captured by novel affinity resin. Controlled self-cleavage of the smaller segment releases the native protein, without the need for modification of its sequence or additional processing. This approach has been highly successful in solving the remaining problems with inteins, and has potential as a platform for the purification of virtually any recombinant protein expressed in any host system.

467599 Trends in Preparative Chromatography

Abstract:

Besides the widespread application of chromatography as an analytical technique, there is a large and growing interest in the pharmaceutical industry and in biotechnology to isolate and purify value added products using preparative liquid chromatography. This led to remarkable progress in the areas of a) developing and commercializing highly selective and efficient stationary phases, b) understanding and quantifying the consequences of overloading chromatographic columns and c) exploiting more efficiently the driving forces for the separation using advanced operating regimes [1]. In the lecture selected new process concepts of preparative chromatography will be explained. Hereby, in particular multi-column configurations and various ways of exploiting gradients will be illustrated [2-5]. These concepts offer potential for significant improvements of essential performance parameters as productivity, yield and solvent consumption.

[1] Guiochon G., Felinger A., Shirazi D.G., Katti A.M., Fundamentals of Preparative and Nonlinear Chromatography, 2nd Edition, Academic Press, New York, 2006 [2] Seidel-Morgenstern A., Keßler L. C., Kaspereit M., New Developments in Simulated Moving Bed Chromatography, Chem. Eng. & Techn. 31, 2008, 826 [3] Seidel-Morgenstern A., Preparative Gradient Chromatography, Chem. Eng. & Techn. 28, 2005, 1265 [4] Schmidt-Traub H., Schulte M., Seidel-Morgenstern A. (Eds.), Preparative Chromatography, 2nd Edition, Wiley-VCH Verlag, Weinheim, 2012Â [5] Kaspereit M., Seidel-Morgenstern A., Process Concepts in Preparative Chromatography, in: â€œLiquid Chromatographyâ€ (Eds. Fanali S., Haddad P., Poole C., Schoenmakers P., Lloyd D.), Elsevier, Oxford, 2013, Ch. 19, 427

466847 Molecular Insights into Protein Separations from Scattering Methods

Abstract:

Obtaining molecular-level insights into protein separations requires spatial resolution on the order of nanometers, which can be especially challenging for systems that include high concentrations, multiple phases and separation agents such as chromatographic media. Drawing on methods that are widely used in the soft matter community, we have used small-angle neutron scattering (SANS) and small-angle x-ray scattering (SAXS) to characterize the structure of systems arising in protein separations. Because both neutrons and x-rays have wavelengths on the order of Ångströms, these methods are able to access the size range of proteins, albeit results are obtained in reciprocal space and generally require model fitting to interpret. Examples that will be presented include protein precipitates and gels as well as ion-exchange adsorbents with and without adsorbed protein.

464362 Understanding and Enhancing Selective Fab Separations Using Multimodal Chromatography

Abstract:

This research seeks to determine what conditions are required to achieve selective separations of similar protein variants and to provide fundamental insight into the mechanisms underlying these separations. The retention of protein libraries on several multimodal cation-exchange systems demonstrated that the retention of many proteins proved to be sensitive to subtle changes in the ligand chemistry and geometrical presentation. All-atom explicit Molecular Dynamics (MD) simulations were then carried out to shed light on the multiple weak interactions that resulted in the unique selectivities achieved in these multimodal chromatographic systems. A range of biophysics techniques including NMR, ITC and AFM were also employed to study the energetics, kinetics and thermodynamics of protein binding to self-assembled monolayers (SAMs) of MM ligands. Finally, quantitative structure activity relationship models were developed using unique molecular descriptors inspired by the biophysics and simulation studies for predicting the chromatographic behavior of proteins on a many multimodal resin systems and to provide insight into the relative importance of various regions on Fabs for interacting with these ligands. This work provides fundamental understanding of the nature of these interactions at the molecular level and insight into the design of MM ligands with important implications for addressing challenging problems in downstream bioprocessing.

466693 Insights on Ligand-Protein Interactions in Affinity Membranes for Antibody Purification

Abstract:

Recently, we designed and synthesized a series of small molecules acting as synthetic ligands for IgG. These ligands are the results of several years of study of the nature of interaction of monoclonal antibodies, and in particular IgGs, with their natural ligand, namely protein A and with the available synthetic ligands. An integrated approach that combines molecular dynamics with experimental activity and crystallographic studies was used to find of a new binding site of IgG, which lies in proximity of the main binding site of protein A, known as the consensus binding site, but is not overlapped with it. These small ligands were thus specifically computationally designed to bind to the new binding site of IgG that is, according to our simulations, more easily accessible than the consensus binding site.

Preliminary experiments with commercial cellulosic membranes decorated with these ligands, in particular with one of them called HPTA, showed promising properties in terms of binding capacity, selectivity and overall recovery of IgG from complex mixtures. Indeed, there are several issues related to residual non-specific binding and to the antibody recovery with the elution step, that appear to be associated to the membrane surface and to the spacer arm and not to the ligand itself. This was demonstrated in fundamental studies in which the influence of the endcapping protocol and of the spacer arm on the new ligand selectivity was studied, using pure solutions of IgG, BSA and lysozyme, as well as mixtures of IgG and BSA and lysozyme.

Current work is focused on the modification of the properties of the membrane surface in proximity of the affinity ligand by functionalizing part of the surface with histidines or similar residues containing amines, that able to change their protonation state as a function of pH. A modification of the charge of the surface is indeed expected to change drastically the energy of interaction of the adsorbed antibody with the surface, thus improving selectivity.

459137 Polyelectrolyte Functionalized Membranes for Tunable Separations and Enzyme Reactions

Abstract:

The development of separation and greener chemical synthesis processes with reduced energy consumption and minimal environmental impact is critical for sustainable operation. Membrane processes are finding wide applications ranging from water treatment to reactors to advanced bio-separations. The development of bioinspired, tunable, nanocomposite membranes provide added opportunities in permeate flux and separation selectivity control. Depending on the types of functionalized groups (such as, chain length, charge of groups, biomolecule, etc.), these membranes could be used in applications ranging from metal (or oxyanions) separation to enzymatic catalysis. These membranes (with functionalized polypeptides or other polyelectrolytes) are also used for obtaining nanofiltration type separations stimuli responsive (pH and temperature). For example, we have demonstrated 90 % separations of arsenic oxyanions with high water flux at pressure of about 1 bar. By taking advantage of helix-coil transitions one can get tunable separations by altering pH. In addition, electrostatic self assembly in pores (layer-by-layer, LbL) can also be achieved through alternate adsorption of cationic and anionic poly-aminoacids or polyelectrolytes under convective flow conditions. Non-stoichiometric immobilization of charged multilayers within a confined pore geometry leads to an enhanced volume density of ionizable groups in the membrane phase leading highly tunable separations. Multilayer assemblies of polyelectrolytes provide excellent platform for protein/enzyme immobilization by providing reusability and high reactivity. Our results with Glucose oxidase enzyme with glucose as substrate showed very high (70 to 80% of homogeneous activity) activity, and continuous synthesis of hydrogen peroxide and gluconic acid. The presentation will include the role of nano-domain interactions for selective separations, polypeptide/polyelectrolyte assembly in membrane pores, and enzyme reactions for chemical synthesis and water applications. This research is supported by NSF KY EPSCoR program and by NIH-NIEHS-SRC.

470642 Acoustic Wave Separation: An Alternative Single-Use Technology for Harvest Operations

Abstract:

As mammalian cell culture has become more productive, there has been a drive to scale down manufacturing processes into a smaller footprint and shift to single-use technologies. The high cell densities that drive this productivity create significant harvest challenges that must be overcome in order to enable a scaled-down, disposable process. Beyond direct depth filtration (DF) and early prototype single-use centrifugation systems, there are few options for single-use harvest technologies that are commercially available. Acoustic wave separation (AWS) is a novel technology that clarifies cell culture in a continuous manner utilizing 3D standing acoustic waves. Acoustic forces generated by the standing waves trap cells and cell debris, which agglomerate to form large cell particles that settle out of solution. As the primary harvest step, this technology has the potential to reduce high cell density cultures to levels that can be managed by the secondary DF step using significantly less filter area.

In this presentation, the results of an evaluation of AWS technology for a range of cell culture processes will be reviewed. Parameters affecting AWS performance were explored, including: feed packed cell mass, feed and solids stream flow rate, voltage, and number of stages. Optimized process conditions delivered greater than 95% reduction in normalized turbidity, 92% in total cell density and 80% in packed cell mass, while maintaining a step yield of greater than 85%. In comparison to a platform direct DF harvest, pre-processing with AWS was able to reduce the DF area requirements by approximately 6x, providing an estimated 6.7x cost savings. Based on these results, AWS provides a promising alternative to current single-use harvest technologies.

475621 A Refined Model for Adsorption of Simple Gases on Graphite

Abstract:

In this presentation, we extend the work of a number of investigators [1] to develop a new model for graphite that accounts for two important factors: (1) an energetic corrugation of the surface that is stronger than that given by the usual Steele model for the carbon in a graphene layer, and (2) the distinction between the top graphene layer and the lower layers because the former is in contact with only one neighbouring graphene layer. This model gives rise to a distribution of the interlayer spacing with the bulk interlayer spacing (for second and higher graphene layers) consistent with experimental XRD measurements and improves the modelling of experimental isotherms and isosteric heat. We illustrate the potential of this new model for graphite with a systematic and comprehensive study of adsorption of noble gases over a range of temperatures, from below the triple point to the boiling point. This study highlights the differences between the new model and the Steele model, especially at temperatures below the triple point where the corrugation is more strongly manifested, and the improved agreement between simulation results using the new model and experimental data. Anisotropy of the graphene layer also accounts for a shift in the 2D-transition for the first two layers and the sub-steps in the isotherm, associated with the spike in the plot of the isosteric heat versus loading. By decomposing the isotherm and the isosteric heat into contributions from the first and second layers, we can determine the mechanisms underlying these transitions. Adsorption of linear molecules, such as nitrogen, carbon dioxide and ethylene, and polar molecules, such as methanol and ammonia, will also be discussed in the presentation.

References:

[1] M. Cole and J. Klein, "***The interaction between noble gases and the basal plane surface of graphite***", Surf. Sci., 124 (1983) 547; R. J. Gooding, B. Joos and B. Bergersen, "***Krypton on graphite: Microstructure at zero temperature***", Phys. Rev. B, 27 (1983) 7669; M. L. Klein, S. F. O'Shea and Y.

Ozaki, "***Interaction potentials and the properties of xenon overlayers physisorbed on the graphite basal plane***", J. Phys. Chem. B, 88 (1984) 1420; G. Vidali and M. Cole, "***Lateral variation of the physisorption potential for noble gases on graphite***", Phys. Rev. B, 29 (1984) 6736; M. Schobinger and F. Abraham, "***Energetics of the incommensurate phase of krypton on graphite: a computer simulation study***", Phys. Rev. B, 31 (1985) 4590; E. Ustinov, "***Effect of crystallization and surface potential on the nitrogen adsorption isotherm on graphite: A refined Monte Carlo simulation***", Carbon, 100 (2016) 52.

Acknowledgement: This work is supported by the Australian Research Council, DP160103540, CF acknowledges the support from Decra, DE160100959.

460784 Adsorption and Transport Properties of Alkanes in Kerogen-Containing Shale-Rocks

Abstract:

Oil and natural gas from deep shale formations are expected to greatly impact the entire world economy as these are spread out everywhere around the globe. The abundance of shale gas resources worldwide --and the fact that burning natural gas emits less CO_2 than other fossil fuels-- has created the expectation of a golden age of natural gas in a global energy system. This revolution relies on the large-scale deployment of new technologies allowing the production of hydrocarbons (oil, and especially natural gas), from source rock formations that were considered unproductive until very recently. The modeling of the hydrocarbons flow in nanoporous rocks such as shale has become an important new area of fluid mechanics. Gas/oil shale are sedimentary rocks with ultralow permeability. Estimates of long-term production and technically recoverable resources remain highly uncertain. The fundamental mechanisms controlling shale gas extraction remain poorly understood, and the classic theories and simulation techniques used by the oil and gas industry have proven inadequate for shale source rocks. Flow through shale poses a distinctive challenge that is new to the oil and gas industry: a large part of pores in shale have typical widths in the order of a few angstroëms, and are within an organic porous material (kerogen) containing adsorbed hydrocarbons. At these scales, the pore size is on the order of the mean free path of the hydrocarbon molecules, and the Navier-Stokes equations with no-slip boundary condition cannot adequately represent the flow. In the present talk, a generic model describing the flow in a multi-scale porous medium (such as shale) and fully taking into account the thermodynamics of confined fluids in nano and sub-nanopores, (including adsorption processes) is proposed without postulating any transport or diffusion mechanism. The model assumes that the rock pore void consist of different types of domains with different of pore sizes starting at the sub-nanometer level with a realistic atomistic description of the kerogen porosity (close to that of common porous carbons). This is a key improvement compared to current attempts to model flow (and production) in shale which all assume that the flow has to comply to Darcy's equation.

470282 Mapping the Adsorption and Phase Behavior of Fluids in Micro-Mesoporous Adsorbents with Hierarchical Pore Structure: *Towards an Advanced Pore Structural Characterization*

Abstract:

Recently, much effort has been devoted to the development of materials which exhibit hierarchical pore structure with the appropriate balance of micropores and mesopores, as well as advantageous pore connectivity. The network of interconnected micro- and mesopores facilitates efficient transfer of fluids to and from catalytically active sites, leading to benefits in numerous catalytic applications. With the development of these materials, a large amount of work has been dedicated to their pore structural characterization. Detailed insights about the pore architecture (pore size distribution, pore volume, and pore interconnectivity) are important because they control transport phenomena, diffusional rates, and govern selectivity.

Gas adsorption is well suited for this characterization because it assesses a wide range of pore sizes, spanning the entire micro- and mesopore range [1]. Argon (87 K) adsorption coupled with methods such as non-local density functional theory (NLDFT) allow one to obtain an accurate pore size distribution over the entire micro- and mesopore size range using a single method [1,2]. However, for an in-depth characterization of the pore network, a detailed understanding of the adsorption and phase behavior of fluids confined in such complex pore structures is needed. Within this context, we focus on a systematic experimental study of the adsorption and phase behavior of argon, nitrogen, and carbon dioxide in a series of mesoporous Y-zeolites with hierarchical pore structure and tailorable mesopore size [3] over a wide range of temperature (from below the bulk triple point to above the boiling point). An advanced interpretation of the hysteresis behavior, also as a function of temperature, allows one to obtain a detailed picture of the pore network. For this we utilized hysteresis scanning experiments, performed at various temperatures.

Our results not only lead to a better understanding of the fundamental aspects of the adsorption and phase behavior of fluids in such hierarchically structured micro-mesopore networks, but also contribute to the development of new methodologies for an accurate and detailed pore structural characterization, which is crucial for advancing the application of nanoporous materials such as mesoporous zeolites in catalysis, separations, and other industrial processes.

[1] Thommes, M., Cychosz, K.A. *Adsorption* 20, 233 (2014).

[2] Landers, J., Gor, G.Y., Neimark, A.V. *Colloids Surfaces A* 437, 3 (2013).

[3] Garcia-Martinez, J., Xiao, C., Cychosz, K.A., Li, K., Wan, W., Zou, X., Thommes, M. *ChemCatChem* 6, 3110 (2014).

471467 Speedy Standing Wave Design and Splitting Strategies for the Separation of Mixtures with Three or More Components Using Simulated Moving Beds

Abstract:

Simulated Moving Beds (SMBs) consume orders of magnitude less solvent, require an order of magnitude less sorbent, and take up less space than batch chromatography operations. As a result, SMBs are economical for large-scale separations. Although SMBs have many advantages, they have not been widely used for large-scale production. One key barrier is the complexity of the design and optimization of SMBs. A four-zone SMB for a separation of four components with linear isotherms has 31 variables, which include 15 material properties and 16 design parameters (four yields, five operating parameters, and seven equipment parameters).

Furthermore, for a mixture of three or more components, more than one SMB may be needed in tandem to recover one or more target products. If there is only one target product in the feed and it happens to be the most (or least) retained component, then only one SMB is needed to recover the pure product. If the target product is an intermediately retained species, then at least two SMBs in tandem are required to recover the pure product. When the target component is not the most (or the least) retained species in a mixture of four or more components, there can be many splitting options (or choice of which components are recovered in which product streams). There are no general rules for finding the most economical splitting strategy in the literature.

The Standing Wave concept was used for size-exclusion systems with three components to develop analytical expressions for solvent consumption and sorbent productivity as a function of dimensionless groups. The resulting Speedy Standing Wave Design (SSWD) equations have been incorporated into a computer program to quickly find the optimal design parameters for maximum productivity, minimum solvent consumption, or minimum cost.

In this work, the SSWD is extended to systems with three or more components with linear adsorption isotherms. The SSWD equations can be simplified for diffusion-controlled systems. The computer program based on the SSWD equations can quickly find the design and operating parameters which can achieve the desired product yields and are optimal for productivity, solvent consumption, or separation costs. In addition, SSWD program can find the optimal SMB designs for each of the many possible splitting options in order to compare the performance of each splitting option. The program can also evaluate different adsorbents for the various SMBs in the splitting cascade.

The number of possible splitting options (SO) is related to the number of components (n) by the following equation: $SO = (n^2-1)(n^2+2n)/24$. To recover an intermediate product, the first split should be between the target solute and the adjacent solute with the largest difference in retention factor. In other words, the easiest separation should be performed first. Solvent consumption can be significantly reduced by allowing certain components to distribute between the product ports. For the same yield, purity, and productivity, choosing the correct splitting strategy can save solvent and increase product concentration by an order of magnitude.

466033 Adsorptive Recovery of Biobutanol in Vapor Phase

Abstract:

The depletion of oil reserves and the increasing costs for crude oil have attracted the attention of sustainable and more environmentally-friendly alternatives for petroleum based fuels and chemicals. Biobutanol, produced by ABE (acetone, butanol and ethanol) fermentation of renewable feedstocks (biomass), is a promising candidate as renewable platform molecule. A typical ABE fermentation generates a mixture of acetone, butanol and ethanol with a total concentration of 2 wt% (3:6:1 ratio, respectively) diluted in water [1]. However, high separation costs due to the presence of other co-products and low final concentration prevents biobutanol to become a viable competitor to other biofuels [2,3]. Adsorption has been identified as an energy-efficient technique. In previous work, we have already proposed ZIF-8 and SAPO-34 as shape selective materials for the recovery of butanol in liquid phase conditions [4,5]. No attention has been paid yet to the recovery of acetone, ethanol and butanol from the head space of the fermenter, in spite of a significant amount of these relatively volatile organics in the vapor phase. Â In this work, a selection of relevant zeolites and MOFs (with hydrophobic/ hydrophilic properties, shape/non-shape selective, potential kinetic selectivityâ €¦), was applied for the vapor phase separation and recovery of butanol in the headspace of the fermentation chamber. Working in vapor phase offers several advantages; ABE concentrations relative to water vapor are higher, adsorbents donâ€™t suffer from stability issues (aqueous conditions and low pH) and no clogging or fouling problems due to microbial cells and inorganic salts. Separation performance was tested in dynamic conditions by performing vapor phase breakthrough experiments, where the effect of composition, humidity of the vapor stream, contact time and total pressure on selectivity was investigated. The effect of water on the adsorption of the organics could be minimized by using very hydrophobic materials. Adsorbents with complementary selectivity were identified; the combination of different materials in sequential columns with alternating adsorption and desorption steps allows to obtain butanol in very high purity and recovery. Dynamic process simulations have been performed for this multicolumn recovery process, accounting for the complex multicomponent equilibria of the different adsorbents (including favorable, unfavorable and sigmoidal isotherms). Such simulations allow to identify the optimal combination of materials to maximize product selectivity and recovery. **References** [1] D. Antoni, V.V. Zverlov, W.H. Schwarz, Appl. Microbiol. Biotechnol. 77 (2007) 23-25. [2] T.C. Ezeji, N. Qureshi, H.P. Blaschek, Chem. Rec., 4 (2004) 305-314 [3] C. Dellomonaco, F. Fava, R. Gonzalez, Microb. Cell Fact. 9 (2010) 3 [4] J. Cousin Saint Remi, T. Rémy, V. Van Hunskerken, S. Van der Perre, T. Duerinck, M. Maes, D. De Vos, E. Gobechiya, C.E.A. Kirschhock, G.V. Baron, and J.F.M. Denayer, ChemSusChem 4 (2011) 1074-1077. [5] J. Cousin Saint Remi, G. Baron, J. Denayer, Adsorption 18 (2012) 367-373.

463395 Molecular Insight into Water Effects in a Stable Metal-Organic Framework: An *in-situ* Experimental and Molecular Modeling Analysis

Abstract:

Competitive water adsorption can have an important impact on metal-organic framework (MOF) performance properties. This can range from occupying active sites in catalytic reactions to co-adsorbing at the most favorable adsorption sites for gas separation and storage applications. The novel question addressed in this work is: for a MOF that is stable after moisture exposure, what are the reversible, loading-dependent structural changes that occur during water adsorption?

To explore this question, a combination of *in situ* synchrotron powder diffraction analysis and molecular simulation techniques are applied to provide insight into the important role of loading-dependent water effects in the stable Zn-DMOF-TM structure. Through this analysis, insight into changes to crystallographic lattice parameters, water siting information, and water-induced defect incorporation are obtained. Overall, this work shows that even for stable MOFs that maintain their porosity and crystallinity characteristics after water exposure, important molecular-level structural changes can still occur during water adsorption.

464335 A Continuous Crystallization Process to Separate Mixtures of Enantiomers Exploiting Two Coupled Fluidized Bed Crystallizers

Abstract:

Crystallization is one of the most effective ways to separate mixtures of enantiomers. So far, primarily batch processes are used. In particular, Preferential Crystallization (PC) has proven to be applicable in batch processes to crystallize selectively pure enantiomers from conglomerate forming racemic mixtures [1]. This contribution presents results of applying a fluidized bed crystallizer configuration for continuous PC of pure enantiomers from a racemic mother liquor [2]. To crystallize both enantiomers simultaneously two fluidized bed crystallizers are coupled via their liquid phases. Saturated racemic mother liquor is supplied from a feed tank with a specific flow rate into the bottom sections of the two double jacketed fluidized bed crystallizers. To initiate the crystallization process in both crystallizers seeds of the opposite enantiomers are added. The preferential growth of the seed crystals reduces the supersaturation in both vessels. In order to allow for continuous production of crystals with a defined size, the lower sections of both crystallizers are conical to offer the opportunity of classified product removal at a certain height. The depleted solutions leaving both crystallizers are returned into the feed tank. In this way always racemic solution is supplied to both crystallizers. Nuclei of the counter enantiomer and other small particles, which have a negative effect on the purity of the product, are taken out with the fluid phases at the tops of the two crystallizers and subsequently dissolved in the heated recycle lines. To assure that the process can be continued continuously, larger particles and formed agglomerates are withdrawn at the bottom, crushed and reinjected to provide permanently new seeds. As a proof of feasibility we will describe the set up developed [3] and present experimental results for the separation of the enantiomers of D/L-asparagine monohydrate [4]. The status of theoretically describing the process using detailed [5] and reduced models [6] will be discussed as well as challenges and opportunities for the separation of other chiral substances. The financial support of German Research Foundation (DFG) within the priority program "Dynamic simulation of interconnected solids processes DYNSIM-FP (SPP 1679)" is gratefully acknowledged.

1. H. Lorenz, A. Seidel-Morgenstern, Angewandte Chemie- International Edition, 53, 2014, 1218.

2. H.-H. Tung, E. L. Paul, M. Midler, J. A. McCauley, Crystallization of Organic Compounds: An Industrial Perspective, Wiley, Hoboken, New Jersey, 2009.

3. D.. Binev, H. Lorenz, A. Seidel-Morgenstern, Chemical Engineering Science 133, 2015, 116.

4. D. Binev, H. Lorenz, A. Seidel-Morgenstern, Crystal Growth & Design, 16, 2016, 1409.

5. K. Kerst, L. Medeiros de Souza, A. Bartz, A. Seidel-Morgenstern, G. Janiga, Computer Aided Chemical Engineering , 2015, 263.

6. M. Mangold, L. Feng, D. Khlopov, S, Palis, P. Benner, D. Binev., A. Seidel-Morgenstern, Journal of Computational and Applied Mathematics 289, 2015, 253.

466054 Tuning the Properties of Crystalline Particles Temperature Cycling in a Continuous Tubular Crystallizer

Abstract:

Tuning the properties of crystalline particles –Temperature Cycling in a Continuous Tubular Crystallizer

P. **Neugebauer**[1], M. **Besenhard**[2], J.G. **Khinast**[1,2]

1 Graz University of Technology, Institute for Process and Particle Engineering, 8010 Graz, Austria

2 Research Center Pharmaceutical Engineering, 8010 Graz, Austria

Correspondence:

e-mail: peter.neugebauer@tugraz.at; tel: +43 (316) 873 - 30424

e-mail: khinast@tugraz.at; tel: +43 (316) 873 – 30400; fax: +43 (316) 873 - 1030400

The utility of tubular reactors for the continuous production of crystalline particles by now has been demonstrated for a wide range of different substrates. Mostly, these systems profit by the ease of supersaturation control by the installation of sections at different temperatures.

Experimental designs employing water baths or jackets have shown to be adequate to benefit from the high surface:volume ratio offered by tubular crystallizers, guaranteeing efficient heat flow from or to the crystallization slurry[1–3].

In the present study we established an easy-to-use platform which allows the modification of crystals from ingredients of pharmaceutical interest under controlled conditions. Our installation facilitates easy changing of process settings such as flow rate, crystal loading and temperature trajectory. Simultaneously, it provides immediate results enabling us to experimentally test a broad range of settings in a short period of time.

Within our setup (Figure 1) a tubular crystallizer with hot and cold zones was shown to be adequate to oscillate dissolution and growth efficiently via temperature cycling. By this, the feasibility of crystal engineering could be shown for pharmaceutical ingredients. By establishing a segmented flow mode interfering effects such as aggregation and breakage could be minimized and by thorough characterization of the temperature cycler unwanted nucleation could be overcome.

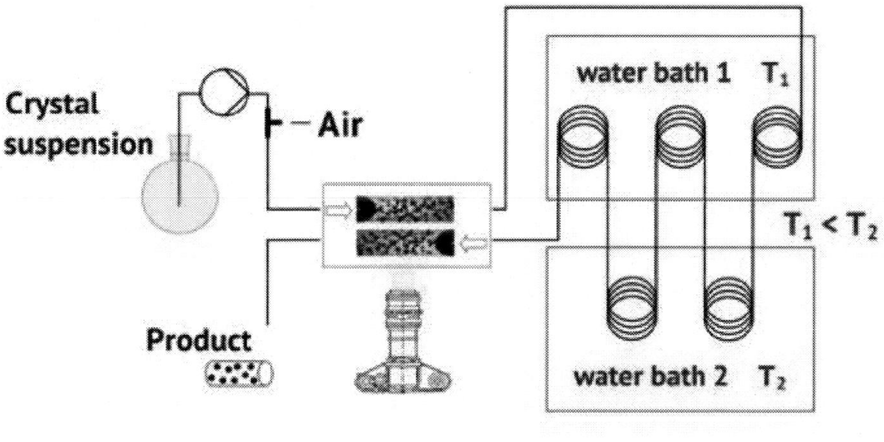

Figure 1: Schematic drawing of the experimental setting

(1) Neugebauer, P.; Khinast, J. G. Continuous Crystallization of Proteins in a Tubular Plug-Flow Crystallizer. *Cryst. Growth Des.* **2015**, *15*, 1089–1095.

(2) Besenhard, M. O.; Neugebauer, P.; Da-Ho, C.; Khinast, J. G. Crystal Size Control in a Continuous Tubular Crystallizer. *Cryst. Growth Des.* **2015**, *15*, 1683–1691.

(3) Jiang, M.; Zhu, Z.; Jimenez, E.; Papageorgiou, C. D.; Waetzig, J.; Hardy, A.; Langston, M.; Braatz, R. D. Continuous-Flow Tubular Crystallization in Slugs Spontaneously Induced by Hydrodynamics. *Cryst. Growth Des.* **2014**, *14*, 851–860.

453235 Continuous Crystallization Using a Sonicated Tubular System for Controlling Particle Size in an API Manufacturing Process

Abstract:

A pH swing crystallization that was very sensitive towards scale-up and required specific equipment due to the non-Newtonian characteristics of the fluid, has been investigated using continuous crystallization (CC) as a means to improve particle size control, scalability and potentially reduce manufacturing cost. A sonicated tubular crystallizer was designed and shown to allow the pH swing crystallization to be operated robustly without blockage, to give a wide range of desired particle size. The results were compared with crystallization in a premixing semi-batch system and a simple addition semi-batch system under similar conditions. Interestingly, it was found that the sonicated tubular system was the only one of the three systems that gave the desired control for small particle size (1~7 Î¼m). Even though the productivity of the sonicated tubular crystallizer still needs to be increased to allow CC and numbering-up to become cost competitive with batch crystallization, we have demonstrated the potential of continuous tubular crystallization as a useful alternative method to conventional batch operation.

463983 Continuous Reactive Crystallisation for Control of Salt Formation Processes

Abstract:

Salt crystallisation is of great importance to the pharmaceutical industry (majority of pharmaceutical products are marketed as salts) but it is often not a well-controlled process. This is because salt formation via a pH shift can be highly sensitive to mixing, and under batch or semi-batch conditions this can lead to severe issues with reproducibility and consistency of quantitative particle attributes, including solid form, crystal size distribution and crystal habit. The salt crystallisation process typically takes the form of a batch titration where the basic drug is dissolved in a solvent before the acid is slowly added. The issue with this procedure is that when the two solutions first come into contact the local supersaturation will build up very quickly and uncontrolled nucleation may occur leading to an undesired salt stoichiometry or crystal size distribution. For this reason, greater control over mixing to achieve greater control over the supersaturation build up is required to improve the consistency of the process and the quality of the particulate product. The aim of this research is to develop a quantitative reactive crystallisation model including a solution speciation model, a mixing model and a population balance model to guide the operation of continuous reactive crystallisation processes. We show the development of this quantitative reactive crystallisation model and how it is used to select operating conditions for the continuous crystallisation of the example 2:1 salt of 3,5-dinitrobenzoic acid and ethylenediamine. By doing this we can control which of the two polymorphic forms (monoclinic & triclinic) is crystallised, the crystal habit and the crystal size distribution.

456813 Process Intensification for Continuous Slug-Flow Crystallization

Abstract:

In the pharmaceutical and chemical industries, the continuous generation of crystals of target size distribution has the potential to improve efficiency for post-crystallization operations [1]. The control of crystallization processes can be challenging when undesirable phenomena such as particle attrition and breakage occur. Slug-flow crystallization has been demonstrated to allow enhanced control of crystal properties such as size and shape [2][3]. The slurry flow is combined with an air flow and fed to a tube to induce a multiphase hydrodynamic instability that spontaneously generates well-mixed slugs where the crystals continue to grow. These slugs are well-mixed without having the mixing blades as in traditional crystallizer designs that induce undesirable particle attrition.

This presentation will elaborate: (1) the process design and strategies for continuous slug-flow crystallization that controls the crystallization phenomena; (2) nucleation systems based on multiphase hydrodynamics from joining two streams in dual impinging jet, coaxial, or radial micromixers, which has been demonstrated to produce seed crystals of uniform crystal size distribution (CSD) [4][5], with a nucleation rate that depends on the liquid flow rate [4][5][6]; and (3) nucleation facilitated by advanced ultrasonication (e.g., focused and indirect) which provides extra degrees of freedom for controlling the CSD by decoupling the nucleation rate from liquid flow rate [3]. Process intensification strategy to improve the crystal growth control is also discussed which makes the whole process more robust. Experimental validation confirms that the proposed crystallizer designs reduce production time and equipment cost by orders of magnitude while suppressing secondary nucleation, attrition, and aggregation—dominant but undesired phenomena that worsen the ability to control the crystal properties.

References

[1] Z. Sun, N. Ya, R. C. Adams, and F. S. Fang, "Particle size specifications for solid oral dosage forms: A regulatory perspective," *Am. Pharm. Rev.*, vol. 13, no. 4, pp. 68–73, 2010.

[2] R. J. P. Eder, S. Schrank, M. O. Besenhard, E. Roblegg, H. Gruber-Woelfler, and J. G. Khinast, "Continuous sonocrystallization of acetylsalicylic acid (ASA): Control of crystal size," *Cryst. Growth Des.*, vol. 12, no. 10, pp. 4733–4738, 2012.

[3] M. Jiang, C. D. Papageorgiou, J. Waetzig, A. Hardy, M. Langston, and R. D. Braatz, "Indirect ultrasonication in continuous slug-flow crystallization," *Cryst. Growth Des.*, vol. 15, no. 5, pp. 2486–2492, 2015.

[4] M. Jiang, Z. Zhu, E. Jimenez, C. D. Papageorgiou, J. Waetzig, A. Hardy, M. Langston, and R. D. Braatz, "Continuous-flow tubular crystallization in slugs spontaneously induced by hydrodynamics," *Cryst. Growth Des.*, vol. 14, no. 2, pp. 851–860, 2014.

[5] M. Jiang, M. H. Wong, Z. Zhu, J. Zhang, L. Zhou, K. Wang, A. N. Ford Versypt, T. Si, L. M. Hasenberg, Y. E. Li, and R. D. Braatz, "Towards achieving a flattop crystal size distribution by continuous seeding and controlled growth," *Chem. Eng. Sci.*, vol. 77, pp. 2–9, 2012.

[6] A. J. Mahajan and D. J. Kirwan, "Micromixing effects in a two-impinging-jets precipitator," *AIChE J.*, vol. 42, no. 7, pp. 1801–1814, 1996.

455203 Continuous Cooling and Membrane-Assisted Crystallization with Airlift Crystallizer

Abstract:

Crystallization plays a substantial role in separation and purification of chemical and pharmaceutical compounds. The common problems in batch crystallisation processes often include large variations in the product quality per batch, inflexibility of the process and complex process controllability. The reason could be difficult handling of wet or semidry crystalline materials and high necessity of operating labour force in batch processes [1]. Furthermore, a common problem in conventional stirred crystallizers is the entanglement of multiple phenomena such as primary nucleation, growth and secondary nucleation.

Initially, batch experiments with airlift crystallizer confirmed that it can suppress secondary nucleation to a higher level than conventional stirred crystallizer [2].

Subsequently, experiments with membrane unit showed its feasibility in assisting crystallization processes by providing sufficient rate and level of supersaturation.

Based on results from membrane unit experiments effective parameters are known as feed temperature and air flow rate. The rate and level of supersaturation can be directly controlled and manipulated by changing these parameters for different process requirements.

In this study the performance of a continuous operated airlift crystallizer is discussed and secondly it is shown that membrane distillation can be applied as an alternative technology for the generation of supersaturation in combination with an airlift crystallizer.

Next, membrane unit and airlift crystallizer were combined. The setup is capable of operating in batch and continuous mode, cooling and membrane-assisted operation. Membrane-assisted Crystallization (MaC), firstly meet the requirements for decoupling the crystallization phenomena toward a more flexible process optimization and secondly it has the potential to operate in continuous mode and to significantly reduce energy consumption and operating labor force.

In this contribution feasibility of performing cooling and membrane-assisted crystallization using airlift crystallizer is presented. It is shown that a continuous membrane-assisted airlift crystallization is believed to be more controllable, flexible and energy efficient.

A model is developed and validated which is capable of predicting the membrane flux and level of supersaturation generation and the crystallization kinetics. The model is used to design and optimize the process.

1. Myerson, A., *Handbook of industrial crystallization*. 2002: Butterworth-Heinemann.

2. Lakerveld, R., J.J. Van Krochten, and H.J. Kramer, *An air-lift crystallizer can suppress secondary nucleation at a higher supersaturation compared to a stirred crystallizer.* Crystal growth & design, 2014. **14**(7): p. 3264-3275.

472727 Continuous Heterogeneous Crystallization of Active Pharmaceutical Ingredients on Excipients Surfaces

Abstract:

Novel continuous heterogeneous crystallization processes were developed, which the active pharmaceutical ingredient is crystallized directly on the surface of an excipient within the crystallizer. The product is then dried, and formed into tablets without the need for complex downstream processing steps such as milling, sieving, granulation, and blending. The APIs and excipients systems selected by study of heteroepitaxial mechanisms, and surface functionality of substrate. The effect of various process parameters, such as temperature, residence time, mode of operation, and excipient loading were studied. Three different process types in two categories of direct crystallization on surface of crystalline excipient and on surface of a carefully selected polymer (with surface modification) were studied. The excipient selection, and process parameters design have significant impact on drug loading, avoiding bulk nucleation and crystallization, control of API crystals shape and size, and control of process. In general, a decrease in temperature as well as the increase in residence time, both increased the drug loading of excipient with API. Within the range of our analysis, proportional drug loading can be easily tailored for a wide range depending on required drug dosage per tablet. The fully continuous processes, and eliminating the downstream steps result in producing crystalline compounds and final form (tablets) in significantly faster, more efficient, with lower footprint, and more economical.

455255 Design of Taylor-Couette Disc Contactors

Abstract:

Many biorefinery processes suffer from poor economics. Target constituents do not show up in appropriate concentration for simple separation technologies. They rather have to be separated from side streams, low quality broths or multicomponent mixtures with low concentration but bad matrix properties. The gap between expectations and needs may be solved by combining simple solvent extraction with reactive separations. Liquid-liquid extraction is a leading technology to raise the economic feasibility of separation processes and provides access to combination of chemical reaction and separation in one reactor type. For application in reactive bioseparations the Taylor-Couette Disc Contactor (TCDC), a hydrodynamic hybrid of a Taylor-Couette Reactor and Rotating Disc Contactor (RDC), may offer advantageous operation features. This type of liquid-liquid contactor does not contain stator rings which provide dead zones for crud accumulation. Furthermore an increased shaft diameter improves the classical RDC design and tie-in a similar flow pattern compared to banded two phase flow of a Taylor-Couette Reactor. Increased rotor discs diameters form compartments during operation, stabilize the flow pattern and inhibit high axial dispersion. Compared with Taylor-Couette Reactors the active reactor volume is significantly increased which results in more stable banded Taylor-Couette vortexes even at higher flow rates.

In order to ensure economical and efficient operating points of any kind of reactor, they have to be designed appropriate for their application. For the design of continuously operating reactors, like the TCDC, the ideal plug flow reactor (PFR) and the continuously stirred tank reactor cascade (CSTR) model are applied. With these basic design concepts either the separation efficiency for a given column height, or the column height for continuous separation processes can be obtained. However, the initial points for these calculations is given by information about the hydrodynamic characteristics of the reactor. With given phase ratio and rotational speed the mean sauter diameter and dispersed-phase hold-up can be determined by empirical correlations. Hence the specific mass-transfer area in the column can be predicted and the reactor behavior is characterized.

For hydrodynamic characterization of the TCDC empirical correlations for drop size estimation and dispersed-phase hold-up, originally developed for RDC design are applied to predict drop size and hold-up in the TCDC. Based on experimental data of the mean sauter diameter and dispersed-phase hold-up at varying rotational speeds, the design rules are adjusted for the needs of TCDC design. Thus, it is possible to predict mass-transfer area in the TCDC with given flow rates and rotational speed.

In terms of modelling mass transfer, experiments confirm complete backmixing in each compartment, best approaching apparatus design with CSTR cascade model.

452109 Influence of Ions on Coalescence in Liquid/Liquid Dispersions - a Systematic Experimental Analysis

Abstract:

In several industrial processes liquid/liquid dispersions play an important role. The efficiency of these processes is mainly determined by the drop size distribution, which is the result of the competing phenomena drop breakage and coalescence. Although existing models for the coalescence rate include physical properties and fluid dynamic characteristics like energy dissipation rate and relative velocity, other parameters such as the influence of ions are either not or merely insufficiently implemented. In a first step, the system is reduced to the fundamental behaviour of single droplet coalescence to quantify the impact of these parameters separately and in particular the influence of ion species and concentration. With a fully automated test cell - based on the principle to observe the collision of a pendant drop and a second rising drop using high speed imaging (see Fig. 1, a) - serial examinations of binary droplet collisions under variable system conditions are possible.

With this setup the influence of different ion species in various concentrations on the coalescence probability for the system toluene/water was investigated systematically. It was observed that coalescence is inhibited with increasing ion concentration and that coalescence times are significantly influenced by the ion species. The experiments were repeated with n-heptane/water to test the reproducibility of results for different test systems.

To validate the single drop results, experiments in a lab-scale stirred tank (DN 150) were conducted (see Fig. 1, b). The drop size distributions were measured with an endoscope technique for selected ion species and concentrations. Additionally, further important parameters like dispersed phase fraction and specific energy dissipation rate were varied and analysed. The comprehensive systematic parameter analysis is presented and single drop and stirred tank results for the investigated test systems are compared and discussed.

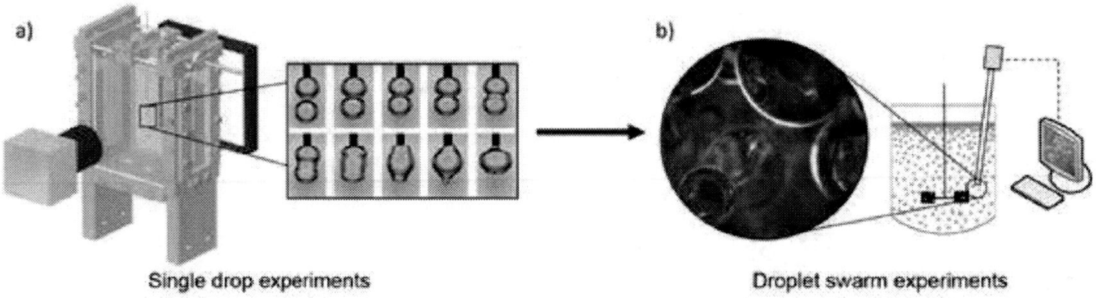

Fig. 1: Systematic coalescence analysis: from single drops (a) to droplet swarms (b)

Financial support within the DFG project "Coalescence efficiency in binary systems" is gratefully acknowledged.

457449 Prediction of Liquid-Liquid and Solid-Liquid Equilibria of Isomers

Abstract:

In chemical industry there are several reactions which produce a linear product and a number of branched side products simultaneously. One example is the hydroesterification of long-chain unsaturated esters of fatty acids where linear and branched diesters are produced. For application only the linear ones are of interest resulting in a demand to separate these isomers. This separation is a challenging task since distillation is very elaborative because of small separation factors and extraction often does not meet the high purity demands. Solvent crystallization is a promising alternative because linear and branched molecules show different solubilities in a solvent. The challenge is the appearance of a superposition of a liquid-liquid equilibrium (LLE) and a solid-liquid equilibrium (SLE) during the separation process. Since this superposition can affect the product properties, it has to be understood in order to build up suitable crystallization processes.

The aim of this contribution is the experimental and theoretical investigation of the mentioned superposition of LLE and SLE for linear and branched molecules. To be certain that the final crystallization product shows the desired properties, one has to avoid a liquid-liquid demixing during the crystallization process. Therefore, it has to be known in advance whether a demixing will take place or not under the defined process conditions. To reduce the experimental effort thermodynamic modelling is commonly applied. In the course of modelling one has to fit pure component parameters to experimental data. For branched molecules, it is often not possible to gain experimental data, because they are barely commercially available in a high purity. Therefore, phase equilibria of systems containing branched molecules can often not be modelled applying the common procedure. To overcome this limitation a model-based experimental design, which allows for the prediction of phase equilibria of branched molecules, was developed. Since in the above mentioned hydroesterification a lot of different isomers are produced that differ only in the position of the side group, it is crucial that our methodology is able to cover all configurations of isomers. For the validation of the developed approach different binary and ternary mixtures of linear and branched alkanes dissolved in an alcohol were

investigated.

The basic idea of the developed methodology is to perform only experiments with linear molecules and use the gained results for the prediction of phase equilibria of branched molecules. For this purpose, it is mandatory to use a thermodynamic model that considers the architecture of the branched molecules. The lattice cluster theory (LCT) was developed to consider the architecture of the molecules without any additional adjustable parameters [1]. The LCT, therefore, allows for the prediction of thermodynamic properties of branched molecules based on pure component parameters of linear molecules. To consider the self-association of the alcohol the LCT is combined with the chemical association lattice model (CALM) [2]. The combination of the LCT with a chemical association model was already successfully applied for the calculation of liquid-liquid equilibria of hyperbranched polymers [3] and for the calculation of the superposition of liquid-liquid and solid-liquid equilibria of hyperbranched polymers dissolved in an alcohol [4].

The developed methodology can be divided into three steps. First of all experiments with linear molecules have to be performed. In this contribution LLE measurements of binary systems of a linear alkane and an alcohol were performed. These experimental data were then used to fit the required model parameters. In the framework of the LCT there is only one adjustable parameter for a binary system, namely the interaction energy Î"Îµ. Additionally, there are two adjustable parameters to consider the self-association of the alcohol. In the second step the fitted parameters are used for the prediction of the LLE of binary systems including a branched alkane and an alcohol. To validate the predicted phase equilibria LLE measurements for these systems were performed. In the third and last step LLE and SLE of ternary systems containing a linear alkane, a branched alkane and an alcohol were predicted based on the binary subsystems. The predicted phase equilibria were again validated by experiments.

Using the developed model-based approach one is able to predict the superposition of LLE and SLE for binary and ternary systems containing branched molecules by only knowing the binary LLE of the corresponding linear molecule. The predicted phase equilibria are in good agreement to the experimental data for all investigated systems.

References

[1] J. Dudowicz, K. F. Freed, W. G. Madden, Macromolecules 23 (1990) 4803 â€“ 4819

[2] D. Browarzik, J. Mol. Liq. 146 (2009) 95 - 104

[3] T. Zeiner, C. Browarzik, D. Browarzik, S. Enders, J. Chem. Thermodyn. 43 (2011) 1969 - 1976

[4] T. Goetsch, P. Zimmermann, S. Enders, T. Zeiner, Chem. Eng. Process. 99 (2016) 175 - 182

460979 Application of X-Ray Computed Tomography for Flow Visualization in Packed Extraction Columns

Abstract:

Hydrodynamic parameters such as drop size distributions and dispersed phase holdup have been studied extensively with various optical measuring techniques. Non-invasive optical measuring techniques however are limited to non-opaque systems mostly at low disperse holdup. Industrial extraction columns however are often operated close to the flooding point at high disperse hold-up. Until now there is no optical measuring technique to evaluate flow conditions in random or structured packings used for extraction columns. Computed tomography allows non-invasive observations of fluid dynamics in such systems and is not limited to low disperse hold-up conditions. The measuring technique is established for investigations on hydrodynamics in opaque packed distillation columns.

In this work X-ray computed tomography as a non-invasive measuring technique is applied to investigations in packed extraction columns. The tomographic setup allows high spatial (80 µm) and temporal resolution (1 fps) tomographic measurements of the packed column. It has been designed for process technology applications and is mounted on a gantry system. Therefore the packed column is not exerted to any centrifugal forces during the measurement.

The technique allows for the determination of drop shapes, drop size distributions and local holdup measurements inside the packed section, which are hardly accessible with common optical measuring techniques such as particle video microscopy.

This work presents a suitable test-system for tomographic studies in packed extraction columns and methods for the investigation of bubble flow inside the packed section. The experiments are conducted in a glass column with internal diameter of 50 mm and packing length of 600 mm.

The local dispersed phase hold-up is measured non-invasively over packing height and at the intersection between two packing segments for varying liquid loads. The tomographic images enable qualitative measurements of the evolving drop-size distribution over the full packing height. An algorithm was developed for the qualitative processing of the tomographic images. It includes automatic image segmentation, bubble detection and bubble size measurements.

Beside quantitative measurements of drop size distributions and local holdup, we are able to qualitatively visualize drop deformation and breakage phenomena inside the channels of the structured packing. Multiple measurements at different packing heights reveal preferred bubble paths inside the structured packing.

455466 An Inline Phase Separator Using the T-Connector with a Modified Needle for Microfluidic Mass Transfer Studies

Abstract:

A phase separator made from a T-connector and a modified needle was developed for implementing rapid inline liquid-liquid phase separation during microfluidic extraction studies. The phase separator was based on the selective wettability of fluids in microchannels. All components of the phase separator were commercially available and easy to assemble, connect, clean, and repair in the laboratory. Classical extraction systems with interfacial tensions ranging from 49.6 to 1.68 mN/m worked well in this phase separator over flow rates ranging from 150 and 1500 Î¼L/min. The advantage of the new phase separator was the needle could promote droplet coalescence at the pinpoint. Microfluidic mass transfer experiments with four different extraction systems were investigated by the T-connector-based phase separator. Contrast to the gravity phase separator, the novel separator can provide more accurate mass transfer data.

475286 Predicting emulsion coalescence: Multi-scale simulation techniques

Abstract:

In optimizing industrial solvent extraction performance, a throughput-limiting process is the efficient separation of raffinate and loaded organic phases via coalescence of dispersed droplets. New research findings are needed in order to quantify the physical and chemical regimes that favour coalescence, which will aid in the design and optimization of hydrometallurgical processing equipment; with both economic and environmental benefits. Coalescence remains difficult to model due to the disparate length scales involved: a successful model must simultaneously capture both droplet and film-scale dynamics. While multiphase flows have been well simulated by interface-capturing methods, the detail of thin-film drainage is difficult to simultaneously resolve using these techniques alone.

This study demonstrates the predictive power of a multi-scale simulation method, in which droplet-scale interface-capturing techniques are coupled to a lubrication analysis of thin-film drainage. The developed multi-scale interface-capturing (MSIC) method was applied to binary droplet interactions in emulsion systems. Depending on the droplet diameter ratio used, the MSIC method can discriminate between bouncing and coalescence outcomes. The present results support the development of physically based coalescence relations for application in unit-scale population balance models.

461505 Upgrading Waste for Food, Fuel and Chemicals Production Via Metabolic Engineering

Abstract:

Waste is rapidly becoming a major problem of the modern world. Disposing of waste in an environmentally acceptable manner presents major issues and in many cases it incurs major costs to communities all over the world. On the other hand, waste, in its various forms, (gases, liquids or solids) could be a valuable feedstock if it could be converted to biofuels and other products. This is possible for some types of waste that is available in aggregated form and at low or negative cost. Various waste-processing technologies have been developed in recent years including industrial biotechnology. In this task, the latter is aided by critical advancements in Metabolic Engineering that transforms microbes into little chemical factories capable of converting feedstocks to a variety of products. In this talk, I will present a technological platform based on native or engineered microbes that convert waste gases and municipal wastes to lipids for biofuel production, specialty products, food and bioingredients. The platform is centered around volatile fatty acids that can be generated from waste streams by various biological and thermochemical methods.

470941 Scaling up and Industrializing the Production of Viral Vectors and Cells for Therapeutic Use

Abstract:

Cell and Gene Therapies have the potential of revolutionizing medicine in the 21st Century. Bioprocessing the therapeutic agents for these therapies still face many challenges during scale up. The utilization of scale down unit operations from bioreactors to downstream purification is very useful in defining the design space and preventing costly scale up errors. Case studies will be presented to demonstrate the challenges and solutions associated with process development of such therapeutics. Optimization for adherent cells cultures in two different platforms â€“ microcarriers and flatware â€“ is presented. The cultures are scaled up to industrial scale for both platforms. Optimal conditions for the production of viral vectors in fixed bed bioreactors are determined through DOEâ€™s and scaled up according to the scaling rules associated with these type of reactors. Purification development and scale up for these vectors are presented.

459696 Snapshot of Sequential SNARE Assembling States Between Membranes Shows That N-Terminal Transient Assembly Initializes Fusion

Abstract:

Many prominent biological processes are driven by protein assembling between membranes. Understanding the mechanisms then entails determining the assembling pathway of the involved proteins. Because the intermediates are by nature transient and located in the intermembrane space, this determination is generally a very difficult, not to say intractable, problem. In this presentation, I will present a new setup with sphere/plane geometry. Using this setup, we have been able to freeze one transient state in which the N-terminal domains of SNARE proteins are assembled. A single camera frame is sufficient to obtain the complete probability of this state with the transmembrane distance. I will show that it forms when membranes are 20 nm apart and stabilizes by further assembling of the SNAREs at 8 nm. This setup that fixes the intermembrane distance, and thereby the transient states, while optically probing the level of molecular assembly by Förster resonance energy transfer (FRET) can be used to characterize any other transient transmembrane complexes.

454158 Modulating Amyloid and Virus Particle Interactions Using Osmolytes

Abstract:

Osmolytes, compounds that modulate cell volume and fluid balance, play an important role in modulating the hydrophobic interaction of proteins and larger biomolecular nanoparticles. It is likely that this modulation is caused by the water caging ability of osmolyte molecules. While in the lab of George Belfort, we were working on methods to modulate the interaction of hydrophobic amyloid proteins and peptides. His work, and the work of others, has shown that amyloid kinetics can be modulated with osmolytes. In my independent lab, we have utilized the ability of osmolytes to change the interaction of hydrophobic molecules by developing osmolyte flocculation technology (OFT) for virus particle purification. OFT has the potential to greatly reduce both development time and cost for future vaccine products. While osmolytes have many uses in protein stabilization, the ability to modulate hydrophobic interactions may open a new avenue of use for these versatile compounds.

467720 Building in Vitro Models of Human Barriers Tissues

Abstract:

We have focused our recent efforts on building a â€œBody-on-a-Chipâ€ as a human surrogate for both drug testing and as model for testing the potential human toxicity of chemicals or chemical mixtures. Such a model consists of barrier tissues (such as skin, GI tract, blood brain barrier, kidney, etc.) and integral body organs (such as liver, heart, fat, etc). These barrier tissues control the uptake and presentation of drugs and chemicals to the rest of the body. Typically cells representing the barrier tissue are supported on a membrane, but ideally a very open porous membrane so that transport limitations are controlled by the cell barrier rather than the supporting membrane. In this talk I will focus on examples and issues in terms of building barrier tissue models and their integration with the whole body model.

457504 Structural Insights into Protein Splicing Inhibition By Platinum Therapeutics As Potential Anti-Microbials

Abstract:

Abstract

Inteins, self-splicing protein elements, interrupt genes and proteins in many microbes, including the human pathogen *Mycobacterium tuberculosis*. The intein in the mycobacterial recombinase, RecA, is specifically inhibited by cisplatin, *cis*-[Pt(NH_3)$_2$$Cl_2$], widely used to treat a variety of cancers. Cisplatin also inhibits mycobacterial growth by targeting inteins. Structural and kinetic analyses of cisplatin binding to the RecA intein were performed to investigate the inhibition of protein splicing by this platinum(II) compound. Mass spectrometric and crystallographic studies of cisplatin binding revealed a complex in which two platinum atoms bind to the intein at N- and C-terminal catalytic cysteine residues in the presence of the reducing agent tris(2-carboxyethyl)phosphine (TCEP). Kinetic analyses of NMR spectroscopic data support a two-step mechanism for cisplatin binding, in which a Pt(II) unit first rapidly interacts at the N-terminus, followed by a slower, first-order binding event involving both the N-terminus (Pt) and C-terminus (Pt+TCEP). Notably, the ammine ligands of cisplatin, which remain bound to platinum in DNA adducts formed as an early step in the mechanism responsible for its anticancer activity, are no longer bound to the metal atom in the intein adduct, as revealed by crystallography. We also explored two additional platinum-based compounds with and without ammine ligands for their ability to inhibit intein splicing. A platinum 4,4,4-trifluorobenzoylacetonate (tfbz) complex, [Pt(tfbz)(NH_3)$_2$](NO_3), and K[Pt($Cl_{3Â}$

to cisplatin, in the low micromolar range. These findings should inform the design of additional platinum-based compounds for targeting inteins in microbial pathogens.

455930 Experimental and Simulation Study of Ethanol Steam Reforming Reaction for Hydrogen Production in Large Scale Catalytic Membrane Reactor

Abstract:

Experimental and simulation study of ethanol steam reforming reaction for hydrogen production in large scale catalytic membrane reactor

Rui Ma, Bernardo Castro-Dominguez, Ivan P. Mardilovich, Anthony G. Dixon, Yi Hua Ma

Center for Inorganic Membrane Studies, Department of Chemical Engineering, Worcester Polytechnic Institute, 100 Institute Road, Worcester, MA 01609, USA.

Production of hydrogen from ethanol steam reforming (ESR) has attracted more interest due to the massive application of hydrogen and carbon neutral property of bio-ethanol. In hydrogen production, the process intensification concept has been applied to ESR by conducting the reaction in a catalytic membrane reactor (CMR) in order to generate and separate pure hydrogen in the same unit simultaneously. Since the product is being removed during the reaction, the conversion is improved.

In the present study, ESR was carried out in a large scale CMR with a 150 cm^2 membrane surface area. The process was investigated both experimentally and with Computational Fluid Dynamics (CFD) simulation. A defect-free membrane was prepared as Pd/Au/Pd/Au and placed in the center of a double screen annular cage while a nickel-based commercial catalyst (HiFUEL R110) was loaded between the double screens surrounding the membrane as shown in Figure 1, thus the catalyst particles were not in direct contact with the membrane surface. The reaction was performed

under different operating conditions: liquid hourly space velocity (LHSV), operating pressure and temperature as well as steam to ethanol (S/E) ratio. Under 300 hours of operation, 100% conversion of ethanol was achieved under all conditions; H_2 with 99.9% purity was produced with the rate of 0.38 g/h at LHSV= 3.77 h^{-1}, P=5 bar, T= 500 ¡C and S/E= 5. It was shown that the process is enhanced by high pressure, high S/E ratio and high temperature. The effect of each condition is discussed in detail in this work.

Simulation models were developed in 1-D and 2-D utilizing Polymath and COMSOL Multiphysics respectively. The kinetics used in the model were validated against experimental outcomes from previous publications using the 1-D model. In the 2-D model, the hydrogen permeation process was simulated as a flux term on both shell side (retentate) and tube side (permeate) applying SievertsÕ law:

-n·Ni=PH2[*PH2shell-PH2tube*] $-n \cdot N_i = \bar{P}_{H_2}\left[\sqrt{p_{H_2}^{shell}} - \sqrt{p_{H_2}^{tube}}\right]$ (1)

in which *PH2shell* $p_{H_2}^{shell}$ and *PH2tube* $p_{H_2}^{tube}$ represent H_2 partial pressure at the retentate side and permeate side respectively.

For the 2-D simulation, the equation of motion and the species continuity equations were solved simultaneously using a finite element method. The effect of the catalytic bed was considered by implementing the Darcy-Forchheimer law to describe the extra resistance to the flow, in which Ergun equation was used to describe the pressure drop in the porous medium. The simulation results were compared with experimental data and showed an accuracy of 84% for 1-D simulation and 91% for 2-D simulation.

Applying the simulation model, the benefits of process intensification were observed by comparing the H_2 production rate from ESR in a traditional packed bed reactor (PBR) and in the CMR. While operating the reaction at higher temperatures and higher pressures, an improvement of up to 122% of H_2 generation in CMR was shown.

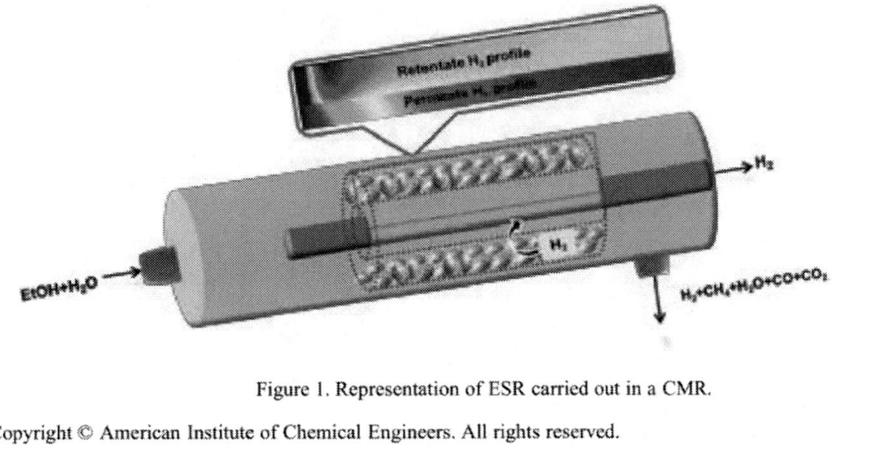

Figure 1. Representation of ESR carried out in a CMR.

463670 Membrane Reactors for Equilibrium-Limited Alcohol and Light Hydrocarbon Synthesis

Abstract:

Reaction and separation are two of the most crucial steps in chemical plants. Performance of each of these two steps can strongly affect the profitability of a given process by affecting its raw materials and energy demand. To improve the overall performance of the process, reaction and separation can be combined into a multifunctional Membrane Reactor (MR). Such reactors typically consist of a catalyst bed in contact with a high-temperature membrane and increase process synergy, reduce by-product formation as well as amount of reactants consumed. Additionally, a considerable reduction in the overall energy costs is to be expected. While there are several reports on the applications of MR for both gas and liquid phase reactions, further studies are still required to move towards commercialization of these MRs.

In this talk, we will present two different examples of high-temperature MRs. The first is an extractive MR that is used for methanol (MeOH) synthesis, and the other a contactor MR that is used for carbon (CO_2) capture and utilization (CCU). Synthesis of MeOH from syngas is a severely equilibrium-limited reaction, meaning that a large fraction of the unreacted syngas feed, must be recycled to make this equilibrium-limited process economically feasible. The ability of MR to increase the per-pass yield makes them an optimal choice for process intensification, particulalry for coal-derived and biomass-derived syngas from air-blown gasifiers that contains considerable amounts of inert N_2.

Power plants are responsible for a large fraction of the man-made carbon-dioxide emissions, and are major contributors to the green house effect. The key challenge here is the dilute CO_2 concentration in the flue-gas, which make the conventional approaches such as direct sequentration technically as well as economically inefficient. Coupling of reaction and separation in a contactor MR, provides the driving force for the separation which moderates the challenge, and can also produce valuable products (e.g., synthetic natural gas, alcohols, etc.) which can be used as fuels to reduce the overall energy demand of the plant, and to improve its efficiency.

During our talk, we will present two MR configurations, and will discuss the influence of reaction conditions and separation efficiency on the final yield of the reaction. The different types of high-temperature membranes that have been investigated will be described, and the influence of membrane characteristcs on overall system performance will be described. Efforts to optimize MR performance both experimentally and numerically will be furher outlined.

464110 Effect of Ni-doping on the Hydrogen Permeation Flux in La5.5WO111.25 Mixed Protonic-Electric Conductor

Abstract:

Ethylene is used as a raw material for producing various polymers. Currently, thermal steam cracking is the conventional process technology for ethylene production. However, the dehydrogenation of ethane dehydrogenation through thermal steam cracking is an expensive and energy intensive process, because the dehydrogenation of ethane is endothermic and is limited by the thermodynamic equilibrium. According to the characteristic of the reaction, the membrane reactor combines the reaction and separation process in a single step. Consequently, the limited-equilibrium reaction is able to shift to the product side. In this study, the performance of a dense ceramic hydrogen permeable membrane reactor for the ethane dehydrogenation, according to the equilibrium reaction C_2H6 $C_2H_4 + H_2$ with a $Pt\text{-}Sn/Al_2O_3$ catalyst was investigated. A dual-phase ceramic membrane with extremely high hydrogen permeation flux flow has been used for the in situ removal of H2 to overcome thermodynamic constraints. The conversion of ethane in membrane reactor could be increased by ~15% at 775 , in comparion with the fixed-bed reactor. Meanwhile, about 75% ethylene selectivity was obtained.

472991 Multi-Scale (Pellet-Reactor Scale) Membrane Reactor Modeling and Simulation: High Temperature and High Pressure Water-Gas Shift Reaction

Abstract:

The objective of this work is to develop a mathematical model and simulation of the Membrane Reactor (MR) in order to carry out the water gas shift (WGS) reaction. MR subsystems integrate reaction and separation in a single unit, with the aim of attaining increased process efficiency, and process compactness. In the MR, a membrane that is selective to hydrogen is used to enhance the WGS reaction's kinetic rate, and to possibly overcome equilibrium conversion limitations imposed by thermodynamics.

The MR system is composed of a reaction zone packed with catalyst pellets, and a permeation zone, where the reaction products permeate. For the reaction zone (classic packed bed reactor), after completing a single-pellet isothermal/non-isothermal steady-state, stand-alone simulations, we couple our model with an isothermal/non-isothermal steady-state packed-bed reactor model to form a hybrid multi-scale reactor model. The catalyst pellet simulation is repeatedly carried out along the reactor bed length (yielding the effectiveness factor along the reactor length for the locally prevailing reaction conditions), and is coupled with a 1-D (axial) reactor model that captures species transport/reaction along the reactor length. Finally, classic packed bed reactor is coupled with a permeation zone to create full membrane reactor (MR) system.

The velocity and species' concentration profiles along the reactor length are captured by momentum/species transport models accounting for convection/reaction /diffusion mechanisms. In the derivation of the model's equations, the Reynolds Transport Theorem was applied separately to each of the domains; the pellet's domain, the reactor's domain and the permeation's domain. The rigorous Maxwell Stefan and dusty gas models are applied to describe mass diffusion fluxes. The effectiveness factors are calculated along the membrane reactor. Finally, performances of the classic packed bed reactor and the membrane reactor are

compared.

469589 Pervaporation-Assisted Furfural Production from Xylose

Abstract:

Furfural produced from the biomass-derived xylose may serve as a platform molecule for the sustainable production of fuels, lubricants, and other specialty chemicals. The acid-catalyzed dehydration of xylose to furfural is burdened by side reactions that form degradation products, known as humins, which reduce the yield of furfural. The formation of humins can be minimized by removal of furfural during its production. In aqueous-phase reactions, this is conventionally achieved by steam stripping, but may also be accomplished by pervaporation, a membrane process. The goal of this study was to demonstrate improvements in pervaporation-assisted furfural production. We previously described a pervaporation-assisted reactor which improved furfural production compared to a membrane-free reactor, but it was limited by two main factors: (1) permeation-induced depletion of reaction-volume which restricted experiments to moderate xylose conversion, and (2) interstage cooling and heating required to accommodate the relatively low membrane temperature. We improved upon the system by incorporating an inlet flow to balance the permeation of material from the reactor. This permitted the study of complete-conversion batch reactions and continuous pervaporation-assisted furfural production. Furthermore, we used cross-linked polymer membranes, which allowed us to eliminate the interstage cooling and heating and to operate the pervaporation-assisted reactor isothermally. This resulted in improved membrane performance as a consequence of increased flux. We then explored the use of metal halides as homogeneous catalysts for pervaporation-assisted reactions, which have combined Lewis- and Brønsted-acid functionality and have been shown previously to produce furfural selectively. While homogeneous catalysts are traditionally thought to be cumbersome because of the need for product/catalyst separation, we show that such a separation is simplified in a pervaporation-assisted reactor because of the lack of permeation of the metal halide catalysts.

472244 Zeolite Membrane Permselective to Water Vapor and Its Application to Dehydration Reactions

Abstract:

Several types of zeolite composite membranes have been developing for separation between iso- and n-isomer, olefin and paraffin, inorganic gas and hydrocarbon, and water and alcohol. Particularly, in the water removal for alcohol-water mixture it has been shown that the composite membranes composing of several kind zeolites such as LTA, ERI OFF (T), CHA, FAU can be synthesized hydrothermally on porous ceramic support. Those have higher water permselectivity, whereby being commercialized in the dehydration process for alcohol-water mixtures.

As an application of zeolite membranes to chemical reactions, various esterification reactions have already been examined mostly in liquid-phase perfectly mixed reactor to aim an equilibrium shift by selective water separation. Differently from such a batch operation, a vapor-phase one, making efficient and continuous production possible, is considered to be advantageous. Dehydration reaction of alcohols, limited thermodynamically, is also industrially important and operated at higher temperatures, producing olefins together with water. This study focuses on vapor-phase esterification and dehydration reaction of alcohol. A membrane reactor incorporated with chabazite (CHA) membrane and solid-acid catalyst is employed for removing water vapor selectively from the reacting mixture. The results in detail will be discussed.

466775 Esterification and Extraction with Supported Liquid Membrane Permeation

Abstract:

Wood pulping processes often generate aqueous effluents containing carboxylic acids; particularly acetic acid. Biorefinery concepts increasingly target integrated material loops including side streams and side stream constituents, and value added products, such as esters, are desired approaches. Esterification of carboxylic acids from pulping effluents may enable production of added value products from pulping effluents. In the present project liquid supported membrane permeation is used to combine esterification reaction with subsequent extraction of the reaction product. Teerachaiyapat et al [1] showed that a hollow fiber membrane module can be used as esterification reactor in principle. The set up used in the present work, originially developed behind the background of heavy metal recycling, has been adapted for reactive membrane separations. The membrane unit offers challenging technological features in isolating value-added products from dilute effluents with a complex matrix. Since cross flow operation is applied, fouling as well as scaling is not a serious issue of process stability limiting significance [2]. The flexibility in membrane composition gives access to the transfer of non-ionogenic as well as ionogenic species. Even highly water soluble solvents, reactants and catalysts may be applied without need of raffinate treatment, since loss of solvent is limited by the physical barrier of the membrane.

The results confirm applicability of the membrane process for the combination of separation by membrane permeation with rigid membranes and esterification of acetic acid with octanol. Via combination of membrane permeation and esterification, the reaction equilibrium related to the aqueous phase may be overcome because of limited solubility of water in the membrane phase. Acetic acid is transferred from the aqueous bulk phase to the rigid interface of the membrane where it converts to the nearly water insoluble ester by releasing water to the aqueous phase while the ester diffuses into the membrane bulk phase. Esterification in the aqueous phase would need the catalyst sulfuric acid which cannot be recycled. In phase transfer catalysis diethylhexylphosphoric acid (DEHPA) or sulfonates (e.g. Dodecylbenzenesulfonic acid) may be used as catalyst in the solvent phase. DEHPA is not catalyzing this reaction sufficiently while dodecylbenzenesulphonic acid (DBSA) is catalyzing the esterification reaction to a sufficient extent. Emulsion formation of DBSA in the solvent phase is avoided by separation from the aqueous phase with the rigid membrane.

[1] T. Teerachaiyapat and P. Ramakul, "Application of hollow fiber supported liquid membrane as a chemical reactor for esterification of lactic acid and ethanol to ethyl lactate," *Korean J. Chem. Eng.*, vol. 33, no. 1, pp. 8–13, 2016.

[2] M. Siebenhofer, H. Noll, and M. Kienberger, "Liquid Membrane Permeation with Supported Liquid Membranes and their Application in Li-ion Battery Recycling," *Sep. Sci. Technol.*, vol. 50, no. 18, pp. 2937 – 2947, 2015.

473103 Investigation of the Degradation of Endocrine Disrupting Compounds in a Membrane Bioreactor

Abstract:

The presence of endocrine disrupting compounds (EDCs) and pharmaceutically active compounds (PHAC) in sewage, industrial, and domestic waters is becoming a major health and environmental concern. The traditional activated sludge process is designed to eliminate solids, organic carbon and nitrogen loadings from wastewaters, but not trace contaminants such as EDCs. Membrane bioreactors (MBR) have become a preferred method for treating municipal and other industrials wastewaters. The objective of this study is to use advanced membrane technologies to develop wastewater treatment processes for removal of EDCs in order to recover and reuse wastewater to augment drinking water supplies. A set of model EDCs including acetaminophen, amoxicillin, atrazine, estrone, and triclosan were selected to study the removal by membrane bioreactor to represent different classes of compounds. Lab-scale MBR operated with real wastewater was tested under various operating conditions, such as retention time and volatile suspended solids concentrations to remove the spiked EDCs. The MBR was operated in two configurations batch and semi-continuous. In both operating modes, the module MBR can reach desired chemical oxygen demand (COD < 30 ppm), total ammonia nitrogen (TAN <5 ppm), and nitrate nitrogen (NO_3-N< 5 ppm) in different retention times. Concentrations of the targeted EDCs and PHAC in the influent, effluent, and mixed liquor suspended solids of the two reactors at different retention time were determined in the period of operation. Different modes of removal mechanisms for these EDCs were identified.

457704 Size-Based Separation of Supercoiled Plasmid DNA Using Ultrafiltration

Abstract:

Most previous studies of membrane-based separations have shown no effect of DNA size on plasmid transmission through small pore size ultrafiltration membranes, consistent with the predicted behavior for the elongation of highly flexible polymer chains. However, supercoiled plasmids are known to have a â€œbranchedâ€ structure with the number of branches dependent on the DNA length. This difference in branching could lead to a significant dependence of the transmission on the plasmid size, providing opportunities for size-based separations using ultrafiltration.

Data were obtained with 3.0, 9.8, and 16.8 kbp plasmids using both cellulosic and polyethersulfone ultrafiltration membranes with different nominal molecular weight cutoffs. Initial experiments were performed with purified samples of the supercoiled or linear isoforms, with the results used to identify appropriate conditions for plasmid separation.

Plasmid transmission increased with increasing filtrate flux due to the elongation of the plasmids in the converging flow field. However, the flux dependence was different for each plasmid due to differences in the extent of branching of the twisted supercoiled DNA. This behavior provided a significant selectivity that could be used to separate the 3.0 and 16.8 kbp supercoiled plasmids using small pore size ultrafiltration membranes. In contrast, transmission of the linear (un-branched) plasmids was independent of the plasmid size. These results provide important insights into the use of ultrafiltration for the purification of plasmid DNA.

Keywords: Ultrafiltration, DNA, supercoiled, plasmid, membrane separation

461413 High Capacity Protein a Membranes for Mab Capture: an Alternative to Column Chromatography

Abstract:

Bind and elute membrane chromatography is receiving increased attention in the downstream purification of therapeutic proteins. The success of flowthrough AEX membrane chromatography and the development of next-generation high capacity materials have enabled a membrane-based affinity mAb capture step in a single-use format. In this talk, several novel Protein A hydrogel prototypes for mAb capture were tested in bind and elute mode. Different formulations of pore size and permeability were screened to find an optimal membrane with desired flow characteristics and binding capacity. A faster Protein A process with improved impurity clearance was demonstrated for 2 mAbs when compared to a traditional column process. High yield and impurity clearance is maintained over 50 cycles while operated at residence times as low as 2 second and loadings up to 30 g/L. Further, continuous operation of the developed capture step is demonstrated by integrating multiple membranes into a multi-column chromatography skid (BioSMB) to facilitate rapid membrane cycling. This integration capability into continuous downstream operation along with the improved impurity clearance and productivity demonstrates these membranes are a promising option for future downstream purification.

461762 Parvovirus-Retentive Filter Performance and Removal Mechanisms during Process Interruptions

Abstract:

Parvovirus retention by some commercially available membrane filters can significantly decline when flow is stopped and restarted during processing. These process interruptions can be a normal part of the filtration process or they can be unanticipated and their impact may not have been evaluated during viral clearance validation. Mitigation of the risk of decreased viral removal includes careful selection of the filter, operating parameters, process controls, and validation strategy. Each of these approaches necessitate an improved understanding of the underlying physical phenomena. However, due to differences in membrane structure and chemistry, the results and underlying governing phenomena can vary between different filters.

In this study, experimental data were generated with a commercial parvovirus-retentive filter which previously had not shown a significant decrease in virus retention following process interruptions. A wide range of parameters related to the operating conditions, the nature of the flow stoppage, and the membrane filter were evaluated. Custom single-layer devices enabled quantification of the impact of process interruptions which had been limited in previous work with standard two-layer devices. The results were compared to a new, simplified mechanistic model of virus retention during flow interruptions, which is based on the principle of size exclusion and accounts for properties of the membrane structure. The model results were consistent with the experimental data, predicting that the decreased clearance associated with flow stoppages is largely independent of the membrane pore size distribution, but may be sensitive to the membrane morphology (i.e. the effective number of separation layers). Both the model and experimental data confirmed that for the single-layer, asymmetric membrane studied, only small decreases (<0.5 logs) occur in virus retention following a process interruption for a wide range of operating parameters. This increased understanding and the experimental data are important tools that can support the development and justification of risk mitigation strategies.

467766 Use of Fluorescent Nanospheres to Study Capture Behavior in Virus Filtration Membranes

Abstract:

Virus clearance is an essential part of the downstream process to remove the viral contamination from biotherapeutics. Several recent studies have demonstrated the complex behavior of commercial virus filters, including the observed decline in virus retention with increasing volumetric throughput and the transient loss in virus retention in response to a process disruption observed with some (but not all) virus filters. Although fluorescently labeled viruses can be used to directly visualize virus capture in different virus filters, the preparation of these fluorescently-labeled viruses is challenging and the results can be influenced by the presence of both proteinaceous and cellular debris. The objective of this work was to use fluorescently labeled nanospheres with confocal microscopy to probe the capture behavior of several commercial virus filters.

Carboxylate-modified fluorescently labeled polystyrene nanospheres with 20, 40, and 100 nm size were obtained from ThermoFisher. Filtration experiments were performed with Ultipor® DV20, Viresolve® Pro, and Viresolve® NFP membranes. The membranes were examined by laser scanning confocal microscopy after challenge with one or more of the nanospheres. Images obtained after filtration of the 20 nm nanospheres show very similar capture profiles to those obtained with fluorescently labeled X174 bacteriophage (25 nm in size), demonstrating the validity of using the nanospheres as models for small viruses. Results obtained with mixtures of the 20, 40, and 100 nm nanospheres show that the different size nanoparticles are captured at slightly different depths within the Viresolve® Pro filter, consistent with the graded pore size within this highly asymmetric membrane. Experiments performed in which the virus filtration was "disrupted" part way through the experiment show multiple capture zones within the Ultipor® DV20 membrane, consistent with the migration of previously captured particles within the filter when the pressure is released. This type of migration was not seen with the Viresolve® Pro, which also provided highly robust virus retention even in response to a pressure disruption. The results from this study clearly demonstrate the potential of using small fluorescently labeled nanospheres as a powerful tool to study the capture behavior in virus filtration membranes.

471666 Using Ultrafiltration to Concentrate Detergent-Solubilized Membrane Proteins without Concentrating Detergents

Abstract:

Membrane proteins (MPs) carry out various physiological functions in living organisms and as such are important targets for pharmaceutical products and for the design of novel sensors and synthetic membranes (Kim et al. 2012; Kowalczyk et al.; Shen et al. 2014). The challenge with using MPs for such applications lies in purifying high concentrations of functional MPs. The hydrophobic exterior of MPs necessitates their stabilization in detergent micelles. Due to low yields obtained, MPs often need to be concentrated for subsequent biophysical characterization or use in applications. The widely-used ultrafiltration-based concentration of MPs in commercial filters often leads to a concomitant increase in concentration of empty detergent micelles. The resulting high detergent concentrations can destabilize MPs, interfere with subsequent structural and biophysical characterization of MPs, and present challenges in applications. We present a detailed analysis of the hydrodynamic conditions that selectively promote detergent passage during MP ultrafiltration. We studied two different classes of transporter MPs, the bacterial water channel protein, Aquaporin Z (AqpZ) and the archaeal light-driven chloride pump, Halorhodopsin (pHR) in nonionic detergents, octyl-β-D glucoside (OG) and decyl-β-D maltoside (DM) respectively. The observed sieving coefficient (S_o)- the ratio of detergent concentration in permeate to that in retentate, was evaluated in the stirred cell and centrifugal systems. The sieving coefficient for detergent for both pure detergent and MP-detergent systems was much greater in the stirred cell than in the commercial filters (**Figure 1A**). Thus, almost complete transmission of detergent can be achieved by operating under conditions that encourage concentration polarization (CP). This contradicts current approaches in soluble protein concentration systems, including centrifugal filters, which are designed to maximize flux by minimizing CP (**Figure 1B**). Thus a rethinking of the design of ultrafiltration-based systems of MP concentration to promote CP may be required for downstream processing of MPs.

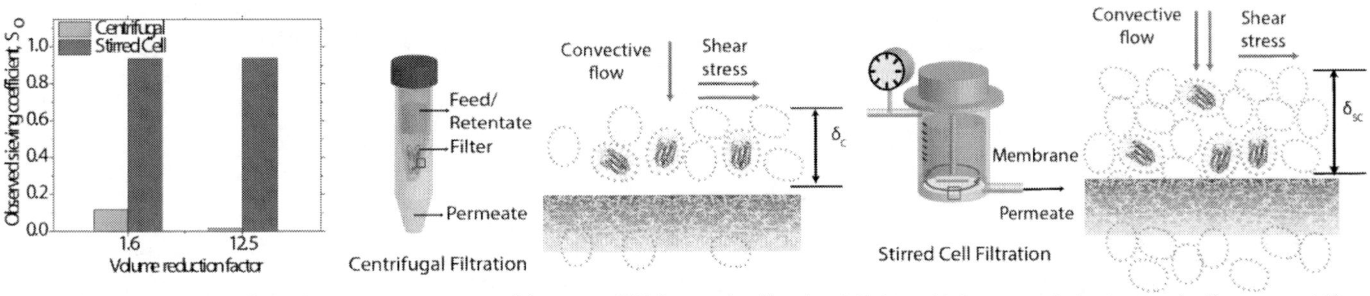

Figure 1. A. Concentration polarization promotes passage of detergent, DM during ultrafiltration. DM shows higher transmission in stirred cell versus centrifugal filter under equivalent conditions of transmembrane pressure of 11 psi (75.9 kPa) and 14 psi (96.6 kPa) and concentration factors of 1.6 and 12.5 respectively. B. Protein concentration using centrifugal filtration is designed to minimize concentration polarization by maximizing shear stress. This is achieved by inducing circulatory flow by using a specific geometry of the filter. In pressure- driven filtration using a stirred cell, the shear rates can be controlled and thus the extent of concentration polarization can be tuned to study the effect of concentration polarization on detergent passage.

References

Kim Y-R, Jung S, Ryu H, Yoo Y-E, Kim SM, Jeon T-J. 2012. Synthetic biomimetic membranes and their sensor applications. Sensors 12(7):9530-9550.

Kowalczyk SW, Blosser TR, Dekker C. Biomimetic nanopores: learning from and about nature. Trends in Biotechnology 29(12):607-614.

Shen Y, Saboe PO, Sines IT, Erbakan M, Kumar M. 2014. Biomimetic membranes: a review. Journal of Membrane Science 454:359-381.

464074 Novel Polytriazole Ion Exchange Membranes for Bioseparations

Abstract:

There is a need for processes of membrane manufacture using solvents, which are not associated with hazards, which would be safer and greener than those used so far in industrial processes. Ionic liquids (IL) and dimethyl carbonate (DMC) are considered environmentally friendly solvents, because are non-toxic, relatively inert and safe under ambient conditions.

This work presents a new green process for manufacture of polytriazole ion exchange membranes for protein separation. This involves the solubilization at room temperature of a polytriazole with free OH groups in ionic liquids such as, 1-ethyl-3-methylimidazolium acetate (IL) and dimethyl carbonate (DMC). The interesting fact is that the polytriazole polymer is not soluble in pure IL or DMC, but it is soluble in different ratios mixture of these two solvents. We experimentally evaluated the solvent ratios leading to one phase solution, delimitating the phase diagram for this system. The membranes were obtained by phase inversion, using water as precipitation bath or mixture of water â€“ methanol and water â€“ ethanol. We studied by SEM how the ratio of the solvents and how different precipitation baths influence the membrane morphology. The membranes were activated for lysozyme adsorption by grafting sulfonic groups. Two sultone structures were used as grafting agents: 1,3 propane sultone and 1,4-butane sultone. The modification of the membranes took place by ring opening mechanism. The activation was done by immersion in 5%wt solution of grafting agent in water, containing small amount of NaOH, at 65 Â°C.Â The grafting sulfone groups on the membrane were confirmed by FTIR, NMR and TGA, and the membranes morphology was investigated by SEM.

The ion-exchange membranes were preliminary characterized in bind and elute mode using pure lysozyme solutions. The Tris buffer, chosen for equilibration and adsorption, was optimized in terms of molarity and pH. The membranes modified with 1,4-butane sultone showed higher adsorption than the membranes modified with 1,3-propane sultone.

Indeed, the values of static binding capacity obtained are promising and comparable to that of commercial ion-exchange membrane adsorbers used for polishing steps in antibody manufacturing.

Acknowledgement

This research was supported by King Abdullah University of Science and Technology (KAUST)

453434 Non-Coulombic Structure of Ionic Liquid Confined in Carbon Micropores with HR MC-Aided X-Ray Scattering

Abstract:

We applied the HRMC simulation-aided X-ray scattering for elucidation of the 3D structure of ionic liquid of 1-ethyl-3-methylimidazolium bis(trifluorometanesulfonyl) imide (EMI-TFSI) confined in nanopores of carbide-derived carbons of 0.7 nm and 1.0 nm in width. We analyzed the plausible anion-anion, cation-cation, and anion-cation structures using the corresponding electron radial distribution functions obtained from the HRMC. We found the unprecedented accumulation of co-ions in the first coordination shell of EMI-TFSI in the nanopores, although the co-ions in the first coordination shell should be highly repulsive electrostatically. The occupation percents of anions around an anion in 0.7 nm-pores and in bulk liquid are 23 % and 7 %, respectively. The unusual accumulation of co-ions cannot be understand by Coulomb's law. The high accumulation of co-ions at the first coordination shell can be ascribed to the creation of image charges in the electrical conductive carbon walls.

465325 Virtual Synthesis of Ultra Permeable and Selective Microporous Polymer Materials for Gas Separation: A Molecular Simulation Study

Abstract:

Microporous organic polymers (MOPs) are a promising class of materials that are attracting increasing interest due to their potential applications in areas such as heterogeneous catalysis, separations, sensing, and energy applications. Among MOPs, polymers of intrinsic microporosity (PIMs) and conjugated microporous polymers (CMPs) are highly advantageous due to the tunability of the porosity and functionality of their framework. In this study, we present an atomistic simulation study of PIMs that are ultra-permeable due to enhanced porosity arising from the packing and CMPs that are highly selective due to functionalizations of the of conjugated linkers. For PIMs, gas permeability and selectivity were evaluated, and it was found that permeability to be almost 7 times larger compared to PIM-1 at the same time obtaining relatively higher selectivities for He/N_2, H_2/N_2, O_2/N_2, CO_2/N_2, and CH_4/N_2. For functionalized CMPs, CO_2/N_2 mixed-gas separation performance was evaluated for vacuum-swing adsorption applications, and it was found that functionalizations increase the selection performance by a factor of 11.

460819 Pressure Tensor of Fluids in Nanopores of Various Geometries: Molecular Simulation Studies By Mechanical Route and Thermodynamics Route

Abstract:

The pressure tensor components of LJ argon adsorbed in carbon pores of different simple geometries, i.e., ideal slit, sphere and cylinder, are studied and compared. We find that the pore geometry has large effects on the adsorption isotherm, the structure of the confined phase and the in-pore pressure of the adsorbate. Depending on the wall curvature, the attraction from the carbon wall acting on the contact layer of adsorbate is strengthened, and thus the confined phase is more compressed, which results in a more enhanced tangential pressure, especially at the contact layer. For the same reason, pore fillings occur at lower pressure in a spherical pore than in a slit or cylindrical pore of the same size. Moreover, our new framework to calculate pressure tensor components based on the thermodynamic route (or volume perturbation method) works well for calculating the pressure in pores of these three geometries.

465197 A Rigorous Thermodynamic Framework for Isosteric Heat of Adsorption

Abstract:

Isosteric heat of adsorption is an important thermodynamic quantity from both experimental and theoretical point of view. It is essential not only for understanding thermal effects and heat management related to gas adsorption, but also for developing and validating adsorption models and materials force fields. Despite a long history of theoretical studies and extensive literature, controversies often arise on the analysis of isosteric heat due to empirical assumptions to describe gas adsorption, and due to the inconsistency between calorimetrically measured isosteric heat and the indirectly correlated isosteric heat from adsorption isotherms. In this talk, we present a rigorous thermodynamic framework to analyze isosteric heat effects without any unjustified assumptions on the pore geometry of the porous adsorbents or operating conditions. The new thermodynamic framework avoids problematic division of the thermodynamic system into a bulk gas phase and an adsorbed phase or empirical estimation of the pore volume. It enables a rigorous prediction of the isosteric heat of adsorption from adsorption isotherms in a thermodynamically consistent manner. We demonstrate applications of the new theoretical framework with simulation results and experimental data for gas adsorption in model adsorbents and in realistic nanoporous materials, respectively, and compare the results with conventional thermodynamic models. We will show that theoretically predicted isosteric heats of adsorption for nanoporous materials including MOF and zeolites are in good agreement with experimental calorimetry data.

465519 Molecular Simulation for Adsorption-Based Separation of Bio-Renewable Chemicals

Abstract:

Renewable chemicals produced from biomass are typically obtained in dilute solutions and possess low volatilities and thermal instabilities. As a result, separation of these compounds from solution by adsorption is an attractive, low energy alternative to distillation. Unfortunately, these separations involve complex sorbate molecules with multiple functional groups from solution (often water) that lead to failure of models designed for adsorption of gases. Herein, we show how molecular simulation fills the necessary gap in understanding multicomponent adsorption of bio-renewable chemicals on all-silica zeolites. Through high-throughput screening simulations and rigorous Gibbs ensemble simulations from solution phase, we identify advantageous all-silica zeolites for several separations.

468796 Exploring the Link Between the Properties of Flexible, MOF-like Adsorbent Materials and Adsorption Behavior Via Flat-Histogram Monte Carlo

Abstract:

Recent years have seen the discovery and development of adsorbent materials that deform or flex under stress, including gas pressure due to exposure to an adsorbate and external mechanical stress. Materials that deform due to gas pressure may prove to be extremely useful in practical engineering applications. Beneficial properties of deformable sorbents include enhanced gas selectivity due to expansion or contraction of pore space and relatively large operating capacity due to large changes in pore volume [1]. Simulation modeling of such materials is advancing, but it is typically difficult to separate the individual effects of sorbent flexibility and the gas-adsorbent interaction from each other; only the net result is available. Recently, Shen and Siderius introduced a technique for simulating flexible porous materials based on flat-histogram Monte Carlo methods for studying adsorption in deformable adsorbents [2]. The method is able to extract adsorption isotherms, free energies, and limits of stability. In particular, this method allows for independent specification of the adsorbent deformation characteristics and the gas-adsorbent interaction properties. Furthermore, for a given gas-adsorption interaction, it allows for rapid evaluation the effect of different deformation characteristics on the net adsorption characteristics of the adsorbent-adsorbate system. In the present work, we apply this simulation method to study flexible adsorbents that resemble metal-organic frameworks (i.e., materials composed of coordinated metal centers and connecting organic ligands) and construct the adsorption isotherms and free-energy diagrams for adsorption of a single-component gas. We then identify certain adsorption characteristics that follow from specific material properties, including the relative strength of adsorbing regions, the nature of our model's â€œligandsâ€, and the material's flexibility, that together give rise to the observed behavior. One can envision the use of this method for designing new sorbent materials to achieve certain performance objectives, such as high operating capacity via a pore-opening response.

[1] J. A. Mason et al., Nature, 527:357 (2015).
[2] V. K. Shen and D. W. Siderius, J. Chem. Phys., 140:244106, (2014).

462894 Molecular Design of Zirconium Metal-Organic Frameworks for CO_2 Separation

Abstract:

Current investigations on metal-organic frameworks (MOFs) for CO_2 capture have developed considerably large amount of promising candidates, and have suggested several design guidelines for such applications. However, systematical evaluation of certain combination of metal cluster and organic linker remain elusive. In this work, a series of MOFs are molecularly designed from bottom up for CO_2/N_2 and CO_2/CH_4 separations. Specifically, containing zirconium (Zr) clusters and N-rich tetrazolate linkers, 120 hypothetical MOFs are computationally constructed using reversed topological approach based on 40 topological nets. The simulation results match well with experimental data. And the Zr-MOFs not only outperform over their carboxylate counterparts in term of CO_2/N_2 and CO_2/CH_4 separation, but also offer higher CO_2 capacity than many existing MOFs and representative adsorbents, attributed to the rich nitrogen donors from tetrazolate linkers. Among the 120 candidates, the MOF with eft topology (MOF-eft-P) possesses the highest CO_2 uptake and selectivity for CO_2/N_2 and CO_2/CH_4 mixtures, with performance comparable to the best MOFs experimentally reported. The breakthrough simulation also suggests its excellent separation power for both gas mixtures. Furthermore, the structural properties are characterized and correlated with adsorption and separation performance parameters. In order to reach the highest separation performances, moderate void fraction and high isosteric heat of adsorption are desirable. This study not only elucidates the design strategy for new MOFs, but also reveals the molecular insight into CO_2 separation in Zr-MOFs.

467793 Applicability of Consistency Criteria to Calculate BET Areas in Metal-Organic Frameworks (MOFs) with Simultaneous Micro- and Meso-Porosity

Abstract:

Metal–organic frameworks (MOFs) are extremely attractive porous materials for application such as gas storage, chemical separations and catalysis in large part due to the exceptionally high surface areas that they can provide. Indeed, the performance of a MOF for certain applications can be somewhat anticipated by the value of this property. MOF surface areas are typically estimated in experiments by applying the BET theory to measured nitrogen isotherms, where an estimated Brunauer, Emmett, and Teller (BET) nitrogen monolayer loading is converted to a "BET area." However, the meaning of MOF BET areas remains a subject of debate, especially in regard to the relation between the BET area and the geometrically calculated surface area, which can only be obtained computationally.

Since there is sometimes ambiguity in terms of selecting the so-called "BET range," most recent emphasis has been placed on the utilization of four so-called "BET consistency criteria" of Rouquerol during the BET calculation. To elucidate the general applicability of these four criteria without the caveat of imperfect (or improperly activated) MOF samples or measurement uncertainties, we simulated nitrogen adsorption in "perfect crystals" of graphene and 25 MOFs having different pore-size distributions and evaluated the effectiveness of these criteria in directing the calculation of BET areas from the simulated isotherms. Additionally, the BET areas thus obtained were compared with the corresponding geometrically calculated, nitrogen-accessible surface areas (NASAs).

We also analyzed molecular configurations from the simulations to elucidate the contributions of "pore filling" and "monolayer formation" to the nitrogen adsorption loadings in different MOF pores, revealing the origin of inaccuracies in BET-calculated monolayer loadings that largely explain discrepancies between BET areas and NASAs. We find that the BET calculation can significantly overestimate the true monolayer loading even if all consistency criteria are satisfied. This scenario is especially common in MOFs combining mesopores ($d \geq 20$ Å) and large micropores ($d = 10$–20 Å), due to the overlap of pore-filling and monolayer-formation regimes of these two kinds of pores.

It is important to note, however, that while it is not always possible to satisfy all consistency criteria, we find that it is critical to minimize the deviation from these criteria during the BET range selection for a consistent comparison of BET areas of different MOFs, as well as for comparing simulated and experimental BET areas of a particular MOF. Furthermore, to accurately assess the quality of a MOF sample, it is best to compare experimental BET areas with simulated BET areas rather than with calculated NASAs.

References:

[**1**] D.A. Gomez-Gualdron, P.Z. Moghadam, J.T. Hupp, O.K. Farha, R.Q. Snurr, J. Am. Chem. Soc. 2016, 138, pp. 215-224.

466692 Using Adsorption Calorimetry to Evaluate the Adsorption Properties in Porous Materials

Abstract:

The enthalpy change of the adsorbent-adsorptive system during physical adsorption is a basic characteristic which can be used both to characterize the solid and as an input for adsorption processes. Direct measurement of adsorption energies can be performed coupling microcalorimetry with a manometric/volumetric device. We have the possibility to obtain measurements not only at room temperatures but also at 77 K. Calorimetry gives a direct measurement of the probe-surface interactions and at 77 K our set-up allows a high resolution representation of both the isotherm and adsorption energies. From a fundamental standpoint, the comparison between a probe molecule with a significant quadrupole moment and one without any permanent moment can be beneficial for the characterization of adsorbents. From a more applied point of view, comparing the heats of adsorption measured at room temperature could be used to predict separation properties. Moreover, the heat release during adsorption gives an idea as to the ease of bed regeneration, and has a strong impact on the performance of an adsorbent bed since, in physisorption, an increase in temperature will lead to a decrease in amount adsorbed.

The present study aims to highlight the continuing interest of adsorption microcalorimetry for the characterization of adsorbate-adsorbent interactions, such as Metal Organic Frameworks materials with various gases. There is a growing interest on â€œMetal Organic Frameworksâ€ (MOFs) in many areas dealing with porous materials. These hybrid solids are formed of metallic centres linked to each other through organic chains. Along with their flexibility, one of the most interesting structural features in some MOFs is their open metal centres (or coordinatively unsaturated sites). These open metal centres play a strong role as specific sites for gas adsorption. The ability to modulate the strength of the gas-MOF interaction by changing the metal centre is of interest for possible applications in the separation of mixtures where specific interactions may well be necessary. These include propane/propene and CO_2/N_2.

Two well-known MOF structures with open metal sites are considered: MIL-100(M) and M_2(dobdc). In the case of MIL-100, the Al, Fe, Cr and V forms are considered whereas for the M_2(dobdc) the Fe, Mn, Zn, Co and Ni forms are studied. The adsorption enthalpies are measured at 77 K, for H_2, N_2, Ar, CO, CH_4, and at 303 K, for N_2, CO, CO_2, CH_4, C_2H_6, C_3H_6, C_3H_8 and C_4H_{10}. All of this data allows a number of trends to be made including the role of gas polarizability and that of the metal electronegativity.

466016 Gravimetric Analysis Under Process-Near Conditions at Nanogramme Resolution: New Sorption and Solvation Apparatus for Very Small Sample Quantities

Abstract:

With the development and commercialization of the Magnetic Suspension Balance (MSB) at Ruhr-University and Rubotherm GmbH at the end of the 1980s [1], the gates were opened for the development of a wide range of apparatuses for the analysis of weight changes in fluids at process-near conditions. Applications are well known and vary from generation of reference density data of pure fluids and mixtures [2], thermogravimetric analysis [3] to adsorption and absorption [4]. The main limiting factor in the conventional MSB has often been the configuration, which has been limited to a bottom-loading device, as the sample has necessarily been suspended by an electromagnetic coupling from a laboratory microbalance. Recently, based on a patented technology developed at ETH Zürich, Rubotherm GmbH has developed a new Nano Magnetic Suspension Balance (NMSB) where the magnetic levitation coils have the dual function of suspending and weighing the suspended body enclosed under high pressure. The electric suspension current is directly proportional to the weight of the suspended body, and is measured at high resolution, thus dispensing with the hitherto required laboratory microbalance. The new balance is able to measure weight changes down to 10^{-8} g, where conventional MSBs are limited by the laboratory microbalance to a highest mass resolution of 10^{-6} g. Additionally, the new NMSB permits free configuration of the associated apparatus, either as a top or bottom loading balance.

A new apparatus has been developed within the RESOLV cluster of excellence (funded by the German Research Foundation) which takes advantage of the new NMSB, regarding the higher resolution as well as the free choice of measuring configuration. To realize this, a flexible interface has been devised to permit conversion of the same balance between a top and bottom loading configuration, depending on the measurement to be performed. The top loading configuration is particularly advantageous for thermogravimetric analysis with elevated temperatures, as the high temperature low density area is situated above the NMSB itself which is in the low temperature, high density area of the pressure cell. This configuration is used for thermogravimetric analysis and adsorption quantification on small sample quantities of porous materials; here we have obtained high quality results of N_2 and CO_2 adsorption on metal organic framework materials with sample quantities in the range of 5 mg (many 100s of factors smaller than i.e. volumetric methods). The bottom loading configuration is somewhat simplified and is used for experiments with high dead weights but which require the high resolution of the new NMSB. Here we are investigating weight changes in moistening and drying processes on two dimensional model surfaces simulating new coatings against biofouling [5]. These samples combine high weight (of the wafer substrate) with very low specific surface area, resulting in a small measurable weight difference. The flexible new apparatus is connected to a gas dosing system comprising a component for static dosing of dry gases and a component for dosing binary gas mixtures under flow conditions mixed with saturated vapours. Results of these dynamic mixture measurements will also be presented.

[1] H. W. Lösch, PhD Thesis, Ruhr-University Bochum (1987)

[2] W. Wagner, R. Kleinrahm, Densimeters for very accurate density measurments of fluids of large ranges of temperature, pressure and density, Metrologia, 41 (2004) 24-39

[3] C. Seibel, T. M. Fieback, New sample carrier systems for thermogravimetric analysis under forced flow conditions and their influence on microkinetic results, Rev. Sci. Instr., 86 (2015) 95104

[4] H. W. Lösch, R. Kleinrahm, W. Wagner, Neue Magnetschwebewaagen für gravimetrische Messungen in der Verfahrenstechnik, Chem. Ing. Tech., 66 (1994), 1055-1058

[5] A. Rosenhahn, S. Schilp, H. J. Kreuzer, M. Grunze, The role of „inert" chemistry in marine biofouling prevention, Phys. Chem. Chem. Phys. (2010) 4275-4286

467101 Can Compressing MOFs Lead to Increased Separation Properties ? Gas Adsorption on Mechanically Constrained Soft Porous Crystals

Abstract:

Gas separation processes involving adsorption may have advantages over other separation methods (simplicity, environmental, cost) although one drawback still remains which is the need for increased selectivity. A simple way to improve the selectivity of a separation process may be to increase the pore confinement by applying an external pressure to compress the porous solid to induce molecular sieving. Indeed, some MOFs can be considered as soft porous crystals whose structure transits when they are exposed to stimuli such as adsorbed molecules[1], pressure[2] or temperature. This structural flexibility, leading to a change in the pore size, strongly influences the selectivity of the adsorbent for some gas mixtures. For example, the MIL-53 solid in its narrow pore form exhibits a very good CO_2/CH_4 selectivity, which is lost when the structure switches to the large pore form with the increase of gas pressure[3]. By maintaining the narrow pore form via mechanical compression, it may be possible to control its selectivity over a wider range of pressure. Releasing the mechanical pressure will lead to the large pore form which could be easier to regenerate.

For this purpose, we have developed a novel methodology to tune the adsorption behavior of mechanically responsive materials by coupling the effects of internal gas pressure and external mechanical pressure. In order to pilot the structural flexibility of the adsorbent during the gas adsorption, we built an experimental device to apply a mechanical pressure (up to 25 tons) on the porous material via a uniaxial press system which equally allows gas adsorption up to 15 bars.

Promising applications of this concept will be explored in this communication including mechanically induced desorption, thus avoiding thermal treatment, and pressure induced gate opening.

[1] S. Bourrelly, P.L. Llewellyn, C. Serre, F. Millange, T.Loiseau, G. Ferey, J. Amer. Chem. Soc., 2005, 127(39) 13519-13521.

[2] I. Beurroies, M. Boulhout, P.L. Llewellyn, B. Kuchta, G. Ferey, C. Serre, R. Denoyel, Angew. Chem. Int. Ed., 2010, 49(41), 2010, 7526-7529.

[3] L. Hamon, P.L. Llewellyn, T. Devic, A. Ghoufi, G. Clet, V. Guillerm, G.D. Pirngruber, G. Maurin, C. Serre, G. Driver, W. Van Beek, E. Jolimaitre, A. Vimont. M. Daturi, G. Ferey, J. Amer. Chem. Soc., 2009, 131(47) 17490-17499.

465259 Adsorption-Induced Deformation of Hierarchical Mesoporous Structures: Stresses Normal to the Pore Walls and Along the Pore Walls

Abstract:

All porous materials deform upon adsorption of fluids. Similarly to adsorption isotherms, experimentally measured strain isotherms during adsorption (strain as a function of gas pressure) contain a lot of information about the porous material, e.g. its pore size distribution [1] and its elastic properties [2]. There are two approaches to measure the strain isotherms experimentally. The first is in-situ dilatometry, i.e. measurement of the macroscopic elongation of a sample, which works only for monolithic samples [3]. The second is based on in-situ small angle X-ray scattering (SAXS) on ordered porous materials, allowing measurements of the lattice constant as a function of gas pressure [4]. This method only works for sufficiently ordered pore lattices such as ordered mesoporous materials.

Hierarchical silica monoliths with disordered macropores and ordered cylindrical mesopores [5] provide a unique medium, where the adsorption-induced deformation can be studied by means of both aforementioned experimental approaches [6]. While the strain isotherm measured by in-situ SAXS is related to the radial stress normal to the pore walls, the strain isotherm measured by in-situ dilatometry is determined mostly by the axial stress component. The theory of normal adsorption-induced stress in mesoporous materials and corresponding strain has been developed in our earlier work [7]; here we present the extension of this theory for the axial deformation and illustrate it with the experimental data on hierarchical silica-based monoliths.

1. Kowalczyk, P.; Balzer, C.; Reichenauer, G.; Terzyk, A. P.; Gauden, P. A. & Neimark, A. V. Using in-situ adsorption dilatometry for assessment of micropore size distribution in monolithic carbons *Carbon,* **2016**, *103*, 263
2. Gor, G. Y.; Bertinetti, L.; Bernstein, N.; Hofmann, T.; Fratzl, P. & Huber, P. Elastic Response of Mesoporous Silicon to Capillary Pressures in the Pores *Appl. Phys. Lett.*, **2015**, *106*, 261901
3. Balzer, C.; Wildhage, T.; Braxmeier, S.; Reichenauer, G. & Olivier, J. P. Deformation of porous carbons upon adsorption *Langmuir,* **2011**, *27*, 2553
4. Günther, G.; Prass, J.; Paris, O. & Schoen, M. Novel insights into nanopore deformation caused by capillary condensation *Phys. Rev. Lett.*, **2008**, *101*, 08610
5. Hartmann, S.; Brandhuber, D.; Hüsing, N. Glycol-modified silanes: Novel possibilities for the synthesis of hierarchically organized (hybrid) porous materials *Acc. Chem. Res.*, **2007**, *40*, 885
6. Balzer, C.; Morak, R.; Erko, M.; Triantafillidis, C.; Hüsing, N.; Reichenauer, G. & Paris, O. Relationship Between Pore Structure and Sorption-Induced Deformation in Hierarchical Silica-Based Monoliths *Z. Phys. Chem.*, **2015**, *229*, 1189
7. Gor, G. Y. & Neimark, A. V. Adsorption-induced deformation of mesoporous solids *Langmuir,* **2010**, *26*, 13021

470750 Monte Carlo Simulation of High Pressure CO_2 Adsorption Isotherms for Characterization of Micro-Mesoporous Carbons

Abstract:

CO_2 has been proposed as probe molecule to characterize micro-mesoporous carbons due to its ability to adsorb at elevated temperatures, which allows for higher diffusion rates and penetration into narrow micropores. Recently, a quenched solid density functional theory (QSDFT) kernel of isotherms was developed [1] for the adsorption of CO_2 at elevated temperature and pressure, for the first time enabling the characterization of both the micro- and mesoporosity of a carbon material in a single measurement. While this effort represents a significant improvement in the capabilities of computational methods for carbon characterization using CO_2 adsorption, DFT cannot completely capture the subtler properties of CO_2 that are a product of its non-spherical molecular geometry, nor the effects of heterogeneous pore structure present in amorphous carbons.

This work presents a novel adsorption model for CO_2 adsorption in heterogeneous (rough) carbon nanopores at 273 K, calculated using Hybrid Reverse (HR) and Grand Canonical (GC) Monte Carlo. The models of slit-shaped and cylindrical pores are developed using heterogeneous pore walls, which molecular structure is fitted with HRMC to the radial distribution function of a reference amorphous carbon. Validation of model parameters is done by: (1) vapor-liquid equilibrium calculations in a bulk system; (2) prediction of adsorption isotherm on graphite surface; (3) comparison of the isotherms with experimental data and the QSDFT isotherms. The calculated isotherms are combined into adsorption kernels and used to generate the surface area, pore volume, and pore size distribution of several characteristic carbon materials.

1. Cimino R., C.K., Thommes M. and Neimark A. V., *Characterization of Hierarchically Structured Porous Carbons*

by High Pressure CO2 Adsorption.

461101 API Particle Formation: Technological Approaches to Forming Unique API Characteristics

Abstract:

The formation of highly differentiated APIs to enable specific downstream filtration, drying and drug product performance, in terms of flowability, blending, content uniformity and in vitro disintegration and dissolution is becoming an ever important feature in pharmaceutical process design.

Within this presentation the application of technological platforms, such as continuous processing, hybrid batch/continuous wet milling enabled system, will be discussed in detail. A series of case studies will show the work flow and design details associated to engineering and population balance models, reactor design, supersaturation monitoring / control to enable the delivery of highly differentiated API particle characteristics. The interface between the drug substance and drug product will be discussed at length to demonstrate the ability of engineering strategies to enable predictable downstream process performance.

466385 Seeding Policy and Optimization of Batch Crystallization Processes

Abstract:

Two important components of a recipe for the operation of a seeded batch crystallization process are the seed properties and the temperature (or supersaturation) profile. When seeds are used in batch crystallization, in most cases the objective is to relieve as much of the generated supersaturation as possible by the growth of the seeds, which corresponds to minimizing the nucleated mass.

Doki et al. (2002) showed that at least for some systems, nucleation could be effectively suppressed by using a sufficient seed mass, called the critical seed loading, and that the required seed mass depends on seed size. Later, Tseng and Ward (2014) showed that the critical seed loading could be determined knowing only crystal nucleation kinetics, which permitted the determination of the critical seed loading ratio for a large number of chemical systems for which nucleation kinetics have been measured. In this contribution we show that similar analysis can be used to determine a feasible range of seed sizes and corresponding seed loadings that will generate product crystals of a desired size (or size distribution) with negligible nucleation.

Furthermore, extending previous work by Hofmann et al. (2010), we show that nearly-analytical solutions for the optimal supersaturation trajectory (and Praeto fronts for multi-objective cases) can be determined for a variety of different objective functions. The nearly-analytical nature of the results allows greater understanding of the effect of the objective function on the shape of the optimal trajectory. This method also requires only knowledge of the nucleation kinetics, and therefore can be applied to determine the optimal supersaturation trajectory for a large number of systems.

Combining the two methods together, we develop a systematic procedure for rapidly determining operating recipes for seeded batch crystallization with a small number of experiments.

References

Doki, N., N. Kubota, M. Yokota and A. Chianese (2002). Determination of critical seed loading ratio for the production of crystals of uni-modal size distribution in batch cooling crystallization of potassium alum. Journal of Chemical Engineering of Japan 35(7): 670-676.

Hofmann, S., and J Raisch (2010). Application of optimal control theory to a batch crystallizer using orbital flatness, 16th Nordic Process Control Workshop, Lund, Sweden, 25â€“27 August 2010; Nordic Working Group on Process Control.

Tseng, Y. T. and J. D. Ward (2014). Critical seed loading from nucleation kinetics. AIChE Journal 60(5): 1645-1653.

466256 Defining a Design Methodology for Conversion of a Batch API Crystallization to a Continuous Crystallization

Abstract:

Crystallization is an essential unit operation for small molecule active pharmaceutical ingredient (API) manufacturing. In recent years there has been an increasing focus on continuous crystallization of such APIs. Advantages of continuous processing include increased control of product attributes, reduced equipment foot print, potential elimination of the requirement for scale-up activities, greater flexibility of equipment infrastructure and reduced expenditure.

This study focuses on creating a defined methodology for converting an existing batch crystallization process to a continuous system. Many previous studies have focused on characterizing continuous crystallization equipment and addressing processing challenges arising from continuous operation such as fouling and attaining steady state operation. The equipment used for this study was a paired CSTR system operated as an MSMPR. The use of this type of system for continuous crystallization has been previously demonstrated in literature [1-4].

The main goal of the design methodology was to transition an existing batch cooling crystallization process to a continuous format, while (at least) maintaining the yield and desired PSD of the existing batch crystallization. The API chosen as the case study for application of the design methodology was produced by a cooling crystallization following a salt formation in solution.

The initial stages of the design methodology consisted of a deeper investigation of the existing batch process, to include solubility screening, solvent selection optimisation, metastable zone width investigation and solid form characterisation. Concurrently, investigations of the existing crystallization process were performed with online monitoring of the process using FBRM and FTIR at 1L scale. From this work an improved understanding of induction times, nucleation temperature and the effect of the cooling rates and hold times on PSD as well as the crystal form was generated.

Temperature selection for continuous cooling crystallization is crucial, and a number of temperature set points were evaluated in batch format to improve confidence for continuous operation. A temperature strategy was selected for first implementation in continuous format, to be sequentially modified as understanding grew.

Following this work a two-step continuous cooling crystallization process was implemented using 2 glass jacketed 1 L reactors. Hot solution was fed to the first crystallizer using a peristaltic pump. Pressure transfer zones were used for intermittent transfer of slurry from the first crystallizer to the second crystallizer and on to a holding vessel for isolation. The effect of start-up strategy on particle size was investigated and was optimized to give the fastest time to steady state for the desired particle size. Various start up strategies were evaluated, including start up with a saturated solution, start up with a seed bed (with various seed loadings), as well as start up with crystal slurry from a previous batch respectively in the first tank. Through systematic trials the initial temperature strategy was sequentially modified until the desired operation and particle size production was achieved. Particle size was measured offline using a Malvern Mastersizer. Crystal form of the product was measured using XRD. The effect of various process parameters such a residence time, agitation and slurry transfer regime on the final PSD were investigated. The product yield was found to compare well with the existing batch process.

This project was made possible by funding from Enterprise Ireland Innovation Partnership Project IP-2014-0356

1. Hou, G., et al., *Development and Characterization of a Single Stage Mixed-Suspension, Mixed-Product-Removal Crystallization Process with a Novel Transfer Unit.* Crystal Growth & Design, 2014. **14**(4): p. 1782-1793.

2. Morris, G., et al., *Estimation of Nucleation and Growth Kinetics of Benzoic Acid by Population Balance Modeling of a Continuous Cooling Mixed Suspension, Mixed Product Removal Crystallizer.* Organic Process Research & Development, 2015. **19**(12): p. 1891-1902.

3. Power, G., et al., *Design and optimization of a multistage continuous cooling mixed suspension, mixed product removal crystallizer.* Chemical Engineering Science, 2015. **133**: p. 125-139.

4. Polster, C.S., et al., *Pilot-Scale Continuous Production of LY2886721: Amide Formation and Reactive Crystallization.* Organic Process Research & Development, 2014. **18**(11): p. 1295-1309.

467662 Tailoring Crystal Size Distribution with Sonocrystallization

Abstract:

Control over the product crystal size distribution (CSD) is very important in industrial crystallization for efficient downstream processing and improved formulation properties[1] Sonocrystallization is a prominent technique to reduce the crystal size and to generate uniform crystal size distributions[2, 3]. Current state of the art includes the sonocrystallization assisting PAT(Process Analytical Technology) to better understand the sonocrystallization process to better control the crystal size distribution to meet the specific requirements of an active pharmaceutical ingredient (API). Various approaches to use ultrasound for size distribution control in a batch cooling crystallization have been investigated. Solute concentration is monitored by In-situ React IC-IR and FBRM used to identify the metastable zone width and crystal chord length distributions. Crystallization of Piracetam in Isopropyl alcohol (IPA) was studied in a batch crystallizer with a working volume of 500ml. A calibration based approach was used for the solute concentration measurement. Crystal size measurement using light microscopy and Image-processing is a promising technique to determine the actual CSD rather than measuring with Laser diffraction technique. Width of the crystal size distribution characterized as Span = (D(90)-D(10))/D(50) determines the final product size distribution. It was observed that, conventional cooling crystallization with operating conditions of agitation rate and cooling rate resulted in bigger and broader crystal size distribution. However, reduced Meta stable zone width (MSZW), lower crystal size and lower Span were observed in case of continuously sonicated crystallization. Application of ultrasound in MSZW at low power of ultrasound and lower cooling rate leads to creation of lower nucleation rate and bigger crystals with narrow distribution. The application of ultrasound prior to nucleation leads to higher nucleation rate (which was observed from FBRM count rate) and smaller crystals with a more narrow distribution. Careful supersaturation control after the nucleation step can lead to the formation of uniform product crystals. The only drawback of application of ultrasound in crystallization process is that the heat liberation due to high power ultrasound, which will be difficult to control by cooling systems. The temperature rise because of liberation of heat of crystallization will also lead to partial dissolution of nuclei which also helps in serve as a feedback control system of the nucleation. Total supersaturation was consumed due to ultrasound induced nucleation and concentration followed the solubility curve, where applied intermittent sonication serve as heating cycles rather than creating more nucleation. For optimization, the influence of time and power of ultrasound, and cooling profile were investigated. Crystal size distribution and span were compared between different control strategies including uninsonated, continuous sonicated, sonicated prior to nucleation and sonicated in MSZW followed by intermittent sonication. Additionally continuous crystallization with ultrasound in a two reactor system of Sono-Nucleator (50ml) and subsequent growth of crystals in a 3liters of reactor was studied.

The results of this research provides better understanding of the sonocrystallization process because of the inclusion of in-situ monitoring systems along with ultrasonic probe in a batch crystallizer. An insight is obtained about how to choose process trajectories to produce smaller crystals with narrow CSD and bigger crystals with narrow CSD, and about implementation in a larger scale batch and continuous crystallization.

References:

1. S. Kim, B. Lotz, M. Lindrud, K. Girard, T. Moore, K. Nagarajan, M. Alvarez, T. Lee, F. Nikfar, M. Davidovich, S. Srivastava and S. Kiang, *Organic Process Research Development*9 (2005), 894–901.

2. Luque de Castro, M. D.; Priego-Capote, F. Ultrasonics sonochemistry 14 (2007), 717–24

3. Jieqiong Li,Ying Bao, Jingkang Wang, Chemical engineering technology 36 (2013), 1341–1346

This work has been supported by Science Foundation Ireland, Grant number: 12/RC/2275.

469553 Process Parameters in the Purification of Curcumin By Cooling Crystallization

Abstract:

Structurally related impurities (often by-products) are of great importance in the pharmaceutical industry. Such impurities are hard to remove completely by crystallization, because of their structural similarity to the solute molecule. Impurities may become incorporated into structures and influence nucleation[1,2] and growth[2], and thus polymorphic outcome, crystal shape and size.[2-4] Often multi-step successive crystallizations are needed to reach a desired purity, significantly lowering the product yield.

Curcumin, a major active component of the Indian spice Turmeric (Curcuma Longa), has a wide range of bioactivities (antimicrobial, anti-inflammatory and anticancerogenic properties). Extract obtained from Turmeric comprises a mixture of curcumin, demethoxycurcumin (DMC) and bisdemethoxycurcumin (BDMC), known as curcuminoids. Commercially available crude curcumin contains significant amounts of the other two curcuminoids, typically about 17% DMC and 3% BDMC. Separation and purification of curcumin is quite challenging, since curcuminoids are structurally very similar, with differences only in the number of methoxy groups.

In this work, purification of crude curcumin by successive cooling crystallizations has been investigated.[5] Statistical experimental design techniques were used to identify the effect of cooling rate, seeding, seed polymorph and agitation on the yield, crystal purity, particle size, shape and structure as well as determining the minimum number of crystallization steps needed to arrive at a particular product specification. Isopropanol was selected as the solvent for the curcumin crystallization based on determined solubilities of the individual curcuminoids. The composition of solid curcumin samples and concentration of individual curcuminoids dissolved in the mother liquor were quantified by HPLC. The polymorphic composition of the crystals was determined by Raman spectroscopy and by X-ray powder diffraction.

Particle size and shape distribution was determined by light microscopy and image-processing analysis, using Feret's diameter and aspect ratio as particle characteristics. Molecular dynamics simulations were performed to calculate curcumin lattice energy as a function of curcuminoid impurity concentration (GAFF 600 molecules, Gromacs).

The results showed that cooling crystallization of curcumin is sensitive to process parameters including seed polymorph, stirring method, cooling rate and the number of crystallization steps, the latter being closely related to impurity concentrations. Manipulation of these process parameters allows for polymorph selection, particle habit and size control, purity and yield control. It was found that removal efficiency of DMC, the major impurity present in crude curcumin, exponentially decreases as a function of the number of successive crystallizations. Crystallization operated in seeding mode and slow cooling rate increases the purity of the obtained curcumin crystals. In particular, using seeds of the metastable Form II improves the purification of curcumin significantly. The computational results showed that incorporated impurity decreases the lattice energy, and that this decrease is more pronounced for the metastable Form II, indicating a weaker affinity of the impurity molecules for the metastable Form II curcumin. At substantially optimal crystallization conditions high purity curcumin (>99.1%) can be obtained by a reduced number of successive crystallization steps. Such a decrease in the total number of successive crystallizations increase the overall curcumin yield from 28% to 50%.

1. Pino-Garcia, O.; Rasmuson, A. Cryst. Growth Des. **2004**, 4, 1025–1037.
2. Shtukenberg, A.; Lee, S.; Kahr, B.; Ward, M. Annu. Rev.Chem. Biomol. Eng. **2014**, 5, 77–96.
3. Cashell, C.; Corcoran, D.; Hodnett, B. K. Cryst. Growth Des. **2005**, 5, 593–597.
4. Liu, J.; Svärd, M.; Hippen, P.; Rasmuson, A. C. J. Pharm. Sci. **2015**, 104, 2183–2189.
5. Ukrainczyk, M.; Hodnett, K. B.; Rasmuson, A. C. submitted to Org. Process Res. Dev. (2016).

This work has been supported by Science Foundation Ireland, Grant number: 12/RC/2275.

460409 Influence of the Composition of Water/Methanol Mixtures and Temperature on the Crystal Form and Solubility of Sodium Dehydroacetate

Abstract:

The slurry transformation experiments for the monohydrate of sodium dehydroacetate with different compositions of water and methanol at varying temperatures were investigated in this study. It was found that the temperature and water content affected not only the solubility but also the crystal form of the precipitated material. The sodium dehydroacetate anhydrate was obtained with a low water content; however, the monohydrate of sodium dehydroacetate was the more stable form with a high water content. At the same time, the solubility of anhydrate and monohydrate of sodium dehydroacetate were measured with different compositions and temperatures. From powder X-ray diffractionï¼ˆPXRDï¼‰patterns, thermal gravimetric analysis (TGA) data and scanning electron microscopy (SEM), a new needle hydrate form of sodium dehydroacetate was shown on the surface of anhydrous phase when the water content was between 2 % and the critical transformation values. In situ Raman spectra and microscope images demonstrated that this needle hydrate was not the metastable intermediate nor was it the result of concomitant crystallization during the transformation. Pure needle hydrate of sodium dehydroacetate was obtained using evaporation crystallization. Moreover, the PXRD patterns and TGA results revealed that this needle hydrate was another new form of sodium dehydroacetate monohydrate. A possible explanation for these observations was found in the relationship between the change of water content, the interaction between solute and solvent and the influence of supersaturation.

463058 Reaching Optimal Productivity in Batch Viedma Ripening Processes Using Minimal Data

Abstract:

In 2004 Viedma reported that nearly racemic mixtures of L-sodium chlorate and D-sodium chlorate undergo a deracemization process in the presence of an attrition source, i.e., the initially racemic mixture reaches chiral purity within several days of simply stirring the suspension[1]. This discovery, subsequently termed Viedma ripening (VR), was subsequently shown to also be applicable to organic molecules that exhibit a conglomerate phase diagram. It was successfully applied to a host of different molecules, among them amino acids[2,3] and pharmaceuticals[4]. VR thus represents a new and promising way to arrive at enantiomerically pure products.

One of the key observations made in early studies on VR was that the rate towards enantiomeric purity proceeds in an autocatalytic fashion, i.e., starting from a low enantiomeric excess there is considerable lag time before the conversion rate speeds up. The rate is finally slowing down again towards the end of the process.

The overall Viedma ripening process was later identified to consist of a complex interplay between racemization in solution, enantioselective agglomeration, attrition/breakage, and size-dependent solubility/Ostwald ripening.[5,6] Being a rather slow process, attempts at optimizing deracemization processes focused on shortening the time needed to reach enantiomeric purity and hence to increase productivity.[7,8] However, no study so far has attempted a thorough optimization of process conditions with the goal of maximizing productivity.

The incentive to optimize a process towards optimal productivity and shortened batch times is evident, but the flexibility to exchange equipment on existing plants is limited. Consequently, the focus of process optimization should be shifted towards the remaining process conditions, such as amounts of seeds (and enantiomeric purity of seeds), seed distribution, and suspension density. Hence, the key objective of this contribution is to introduce a methodology that achieves optimal productivity (maximizes mass of pure enantiomer produced per time) for batch Viedma ripening processes with the use of as little experimental data as possible.

The methodology presented here seeks to optimize a batch-after-batch production of enantiopure crystals. Within such a processing strategy, the current batch can be conveniently run by re-using the mother liquor from the previous batch (minus appropriate purging to avoid impurity build up), adding a defined amount of a racemic mixture of solids and by seeding the process with some enantiomerically pure seeds (part of the product from the previous batch). Due to VR, such a batch will reach enantiopurity after some time. The goal of this methodology is to keep this time as short as possible. It can be shown that such a processing strategy is mathematically equivalent to an ideal continuous plug flow reactor with a recycle stream consisting of part of its effluent stream. Since Viedma ripening exhibits autocatalytic features, this analogy can be used to find an optimal seeding policy for batch Viedma ripening processes. It will be shown that the only data needed for this strategy is a curve of enantiomeric excess versus time, obtained from a single experiment by standard techniques (e.g. HPLC). It will further be shown that an optimal suspension density can be established without additional data.

The strategy is first shown to work based on extensive model results employing population balance equation models and various kinetic expressions (to ensure a wide range of applicability) and then exemplified using an experimental study on Naproxenethylester crystallized from ethanol, which is racemized in solution using a base (sodium ethoxide in this case).

References:

[1] Viedma, "Chiral symmetry breaking during crystallization: complete chiral purity induced by nonlinear autocatalysis and recycling", Phys. Rev. Lett., 2005, 94, 065504

[2] Viedma et al., "Evolution of Solid Phase Homochirality for a Proteinogenic Amino Acid", J. Am. Chem. Soc., 2008, 130, 15274–15275.

[3] Noorduin et al., "Emergence of a Single Solid Chiral State from a Nearly Racemic Amino Acid Derivative", J. Am. Chem. Soc., 2008, 130, 1158–1159.

[4] Noorduin et al., "Fast attrition enhanced deracemization of naproxen by a gradual in situ feed", Angew. Chem. Int. Ed., 2009, 48, 4581–4583.

[5] Iggland and Mazzotti, "A population balance model for chiral resolution via Viedma ripening", Cryst. Growth Des., 2011, 11, 4611–4622.

[6] Gherase et al., "Experimental and Theoretical Study of the Emergence of Single Chirality in Attrition-Enhanced Deracemization", Cryst. Growth Des., 2014, 14, 928–937.

[7] Iggland et al., "Complete solid state deracemization by High Pressure Homogenization", Chem. Eng. Sci., 2014, 111, 106-111.

[8] Noorduin et al., "Scaling Up Attrition-Enhanced Deracemization by Use of an Industrial Bead Mill in a Route to Clopidogrel (Plavix)", Org. Process Res. Devel., 2010, 14, 908-911.

455479 Turbidity Control Cleavage of Stable Emulsions

Abstract:

Emulsions are major issues in different processes. It is for example a common task in biobased processes to recover valuable components from the fermentation broth. This is for example done by liquid-liquid extraction. Despite the advantages of this unit operation like the processing at low temperature there are a few drawbacks which have to be overcome. The ability of emulsification is a specific problem which is often encountered in biobased industry since surface active components may be side products in the broth. Both the extract and the raffinate phase can be contaminated with unwanted turbidity. However the results are very different. While a contamination in the extract phase may reduce the product quality, stable solvent fines in the raffinate phase limit the wastewater quality and the economic feasibility of the process due to solvent loss.

Turbidity control of the raffinate phase can be performed using different techniques like thermal, chemical, mechanical or electrical splitting. While the first two are quite expensive in terms of energy and material demand the latter two can be provided in an efficient way. In this context admixture of flocculants as example for chemical splitting shall be mentioned. This operation may help to control turbidity, but a second separation step is necessary to recover the flocculating agent. Additionally a negative impact on product quality has to be considered.

This project focuses on turbidity control of aqueous effluents using flotation as example for mechanical splitting and electro coalescence, which can be seen as a representative of electrical splitting technology. A dodecane spiked aqueous feed emulsified with surfactants and a high shear mixer is prepared as effluent. To compare the results with the cleavage of an industrial effluent the settling behavior is investigated in the flotation equipment as well as in the electric field coalescence equipment. For this experiments the aqueous phase consisted of different mixtures of alcohols, carboxylic acids and water and the organic phase of undecane and 4â€‘dodecylbenzenesulfonic acid (4DBSA). Phase separation experiments are monitored using a camera to see even slightest changes in the system.

Flotation technique as well as electro coalescence as well as technology combinations are applicable for successful turbidity control.

465796 Novel Crown Ether with Bulky and Rigid Subunits: Experimental and Theoretical Alkali Metal Selectivity Studies

Abstract:

Crown ethers (CEs) are unique macrocylic compounds capable of forming strong complexes with metal ions which fit well into the cavity of the CE. Because of this size-match selectivity, CEs had been studied in the fields of phase transfer catalysis, metal sensing and metal ion extraction. Among the CE class, twelve to 14 membered CE rings are known to form stable complexes with lithium (Li^+) in both organic and aqueous solutions even in the presence of other alkali metal ions. Specifically, dibenzo-14-crown-4 ether (DB14C4) and its derivatives are known as Li^+ complexant because of its ideal cavity dimension (1.2-1.5 Å) for complexation with Li^+ (1.2 Å). The two benzo groups provide rigidity that lessens its affinity with bigger alkali metal ions such as sodium (Na^+, 2.0 Å) potassium (K^+, 2.7 Å) and cesium (Cs^+, 3.34 Å).DB14C4 has only been applied as carriers in electrodes with moderate sensitivity towards Li^+ or as a carrier in liquid-liquid extraction (LLE) systems due to its difficult synthesis involving multiple reaction steps and low over-all yields. Although efforts have been made to develop new efficient methods for DB14C4 synthesis, complexation with larger metal ions is still inevitable due to the formation of"sandwich"-type complex lowering its selectivity towards Li^+.

To address this issue, a new class of 14-crown-4 ether containing two bulky subunits has been synthesized to prevent the formation of sandwich type CE-metal ion complex formation. The bulky subunits are designed to act as blocking moieties, which hinder the complexation with bigger metal ions. As formation of 1:1 CE-metal ion complex is promoted, enhanced Li^+ selectivity was achieved compared to its rigid type counterpart. However, decomplexation of metal ions is difficult due to steric hindrance and limiting interaction of water or other solvents with the metal ion once it is inside the cavity of the bulky CE. This was alleviated by designing CEs with one bulky moiety but the selectivity was greatly compromised due to flexibility of the CE backbone. Moreover, the overall yield is very poor because of difficulty in the ring closure reactions when using bulky tertiary diols as starting substrate.

Herein, intermolecular cyclization of bulky bis-epoxide with 1,2-dihydroxybenzene was performed to synthesize novel lithium selective 14 membered CE having both rigid and bulky subunits. The rigid aromatic groups enhanced the rigidity of 14-crown-4 ether backbone while the bulky subunits provided a blocking mechanism to prevent bigger metal ions to from a complex as well to alleviate the formation of higher ordered sandwich type complexes. The reaction route was optimized by changing the reaction solvents, temperature, and catalyst using commercially available bis-epoxides, neopentyl diglycidyl ether (2a) and 1,4-butanediol diglycidyl ether (2b). The synthesized CEs with different bulky structures were assessed by experimental LLE and theoretical density functional theory calculations (DFT) to understand the effect of the bulky structures. Parameters obtained from the DFT calculations such as cavity size, oxygen to metal ion (O-M^+) distances as well as metal ion binding energies (ΔE_{M+}) were investigated to correlate with the metal ion selectivity obtained from the experimental extraction results.

Efficient synthesis of the di-hydroxy CEs (3a-h) via intermolecular cyclization of bulky bis-epoxide with 1,2-dihydroxybenzene was achieved using *t*-BuOH as solvent with the appropriate metal hydroxide catalyst that fits the cavity size of the target CE. The bulky bis-epoxide intermediate was synthesized via etherification of the bulky diols (2,2-diethyl1,3-propanediol, pinacol, bicyclopentyldiol, cis-1,2-cyclohexanediol, cis-cyclopentanediol and 1,2-dihydroxybenzene) with allyl bromide to obtain di-alkene intermediates (1c-h). Subsequent oxidation of the di-alkene intermediate with a peroxy-acid (m-CPBA) gave bis-epoxide intermediates (2c-h).The structures of the synthesized CEs and intermediates were confirmed by ^{1}H and ^{13}C NMR, FTIR, and HRMS. The alkali metal (Li^+, Na^+, K^+, Cs^+) complexing abilities of CEs 3a-h was assessed by liquid-liquid extraction of alkali metal perchlorates in dichloromethane-water system and reported as metal distribution (D_{M+}) and selectivity ($\alpha_{Li/M+}$). All the CEs synthesized showed high Li^+ distribution but dihydroxy-benzotetramethyl-14-crown-4 ether (3d) and dihydroxy-benzo-bicylopentyl-benzo-14-crown-4 ether (3e) having bulky tetramethyl and bicyclopentyl respectively, showed highest Li^+ selectivity in the presence of other alkali metals. Furthermore, no D_{K+} and D_{Cs+} were analyzed for 3d and 3e. Selectivity of $\alpha_{Li/Na}$=3344 and $\alpha_{Li/Na}$=2512 achieved for 3d and 3f, respectively is among the highest in the literature. In comparison to DB14C4 having both rigid moieties, showed distribution not only towards Li^+ but also with bigger alkali metals. The complexation of DB14C4 towards bigger metals can be attributed to the formation of 2:1 sandwich type complex.It is noteworthy to mention that CE dihydroxy-benzo-neopentyl-15-crown-4 ether (3a) and dihydroxy-benzo-16-crown-4 ether (3b) with a 15 and 16-crown-4 membered ring respectively, showed preferences towards Na^+due to a bigger cavity size.

To explain the metal ion preference behavior of the synthesized CEs with bulky and rigid structures as well as the tendency of the CEs to form sandwich type complexes, DFT calculations was employed. The lowest energy structures of the synthesized free CE as well as the binding energies of the 1:1 or the 2:1 CE-metal ion sandwich type complex were determined using the Vienna ab initio simulation package (VASP) described by the Perdew-Burke-Ernzerhof (PBE) generalized gradient functional and the projector augmented wave (PAW) method.

The cavity size of the synthesized CEs is defined as the empty space at the center of the CE hetero atoms and was estimated as the average distance of the opposite heteroatoms subtracted by the Van der Waals radius of oxygen atoms (1.4 Å). Among the CEs synthesized, 3d and 3e have ideal cavity sizes (1.2 Å and 1.21 Å respectively) for Li^+ having an ionic diameter of 1.2 Å. The results complement the experimental LLE which could be the main reason for the high Li^+ selectivity. Bigger cavity sizes were obtained for 3a, 3b and 3c (1.36, 1.63, and 1.34 Å respectively) which also possess high distribution with the bigger Na^+due to the bigger 15-16 membered ring formed.

The average O-M^+ distances were also determined to assess the stability of the complex. The results showed larger O-M^+ distances on the bulky side of the CE compared to the rigid side especially for 3d and 3e with K^+ and Cs^+ complexes. On the other hand, a typical CE-metal ion complex (DB14C4-Li^+) showed equal O-M^+ distances in both sides. Similar trend was also observed in DB14C4-Li^+ crystal structures available in the literature. Thus, the bulky groups present in the synthesized CEs provide steric hindrance and hence destabilizing complexes with bigger metal ions as revealed by the varying O-M^+distances in the bulky and rigid side. In addition, the bulky structures could also prevent formation of sandwich type complexes.

To confirm the tendency of the synthesized CEs to form 1:1 or 2:1 CE complex, a simple enthalpy exchange reaction mechanism (ΔH_{M+}) towards other alkali metals (Na^+, K^+ and Cs^+) with respect to Li^+ in the presence of water was calculated based on the DFT binding energies. The results showed high preference towards Li^+ due to a positive ΔH_{Na+}, ΔH_{K+}, and ΔH_{Cs+} calculated for the 1:1 CE-metal ion complex. However, for the 2:1 complex, only 3d and 3e showed positive ΔH_{M+}suggesting that 3d and 3e prevented the formation of sandwich type complexes.

In summary, novel lithium selective 14-16 membered crown-4 ethers having both bulky and rigid subunits were synthesized in high yields via intermolecular cyclization of the bulky bis-epoxide intermediate with 1,2-dihydroxybenzene under basic conditions. Experimental LLE showed that 3d and 3e had the highest Li^+ selectivity which can be attributed to their comparable cavity size with Li^+ diameter estimated by DFT calculations. DFT calculations also showed instability of complexes towards bigger metal ions (Na^+, K^+, Cs^+) due to the effect of the bulky structure providing bigger O-M^+ distances. The bulky subunits not only prevented the formation of higher ordered complexes but also fine-tuned the cavity size of the CE. The potential of these types of CEs having both rigid and bulky functionality as selective chelates for selective sensing or extraction of Li^+in dilute solution such as seawater is currently under investigation.

This research was supported by the Basic Science Research Program through the National Research Foundation of Korea (NRF) funded by the Ministry of Science, ICT and future Planning (2015R1A2A1A15055407) and by the Ministry of Education (No. 2009-0093816).

463939 Extractive Desulfurization of Fuels Using Ionic Liquids

Abstract:

Conventional hydro desulfurization (HDS) process is considered as an efficient sulfur removal process in oil refining industry. However, in the case of aromatic sulfur compounds (thiophenes, dibenzothiophenes and their alkylated derivatives), HDS process has some weak points. High level of hydrogen consumption and operating at high temperature and pressure which make it a hazardous process are the main challenges and impose high level of fixed and commercial costs. In this regard, research for finding alternative deep desulfurization techniques have been continued during last years and still is going on. Among various methods, extractive desulfurization using ionic liquids (ILs) as green solvents attracted more attention during last decades. This process can be run at ambient temperature and pressure. Meanwhile ILs have great capacity because of their special properties compared to common organic solvents. They are usually thermally stable and have negligible vapor pressure which makes them environmental friendly solvent. On the other hand, ILs formed by an anion and cation that can be easily designed for specific tasks by an appropriate combination of their ions. Previous works in this field show that imidazolium based ILs have greater selectivity while pyridinium based ILs are stronger extractive solvents. Some anions are not stable and decompose to harmful and toxic byproducts during extraction. Besides solubility and selectivity concerns, density and viscosity are important in extractive desulfurization as well. Usually, ILs with lower density show higher desulfurization efficiency.

Considering above facts, in this work various ILs have been preliminary screened based on their physical and chemical properties. Finally, extraction of thiophene from normal alkanes using ILs including nitrate $[NO3]^-$ or thiocyanate $[SCN]^-$ as anion and imidazolium or pyridinium as cation have been studied. Several ionic liquids such as 1-Octyl-3-methyl imidazolium thiocyanate [Omim][SCN], 1-hexyl-3-methyl imidazolium thiocyanate [Hmim][SCN], 1-butyl-3-methyl imidazolium thiocyanate [Bmim][SCN], 1-octyl-3-methyl imidazolium nitrate [Omim][NO3], 1-buyl-3-methyl imidazolium nitrate [Bmim][NO3], 1-hexyl-4-pyridinium nitrate [HPy][NO3] and 1-butyl-4-pyridinium nitrate [BPy][NO3] were synthesized. The structure of the synthesized ILs was checked using nuclear magnetic resonance (NMR) spectroscopy. The purified IL was dried and degassed for 24 hr. at 343.15 K under vacuum, then were kept in bottle under argon gas. The mass fractions of water in the dried products, measured by Karl Fischer coulometer, were less than 1×10^{-3}. These ILs were proved to be stable and inexpensive. Meanwhile they have low viscosity and density while are environmental friendly.

Liquid-Liquid Equilibrium (LLE) data for ternary mixtures containing sulfur compound, fuel model and IL were experimentally measured. Selectivity and distribution coefficient were calculated using experimental data. The experimental LLE were determined by preparation of immiscible mixtures of three components of IL+ *n*-alkane+ thiophene with total mass of 4gr. Each component was weighed by a laboratory balance with the precision of $\pm10^{-4}$ g. The mixture in the glass vial was closed firmly with a cap and septum and stirred by a shaker for 120 min at 300 *strokes per minute***.** The glass vials were placed for at least 12 hr into a water bath with a precision of ±0.1° K. The two phases in the glass vials were separated and analyzed after reaching the equilibrium condition. The composition of each phase was determined using gas chromatography (GC). All analyses were repeated three times and the experimental uncertainties for the top and bottom phase compositions were less than $\pm1.10^{-3}$ and $\pm2.10^{-3}$in mole fraction, respectively.

The experimental LLE data for ternary mixtures showed that cation chain length of IL has a remarkable effect on selectivity of aromatic sulfur compounds. It was found that the IL with lower molecular weight of cation has higher selectivity and lower solute distribution coefficients. Besides, the experimental data demonstrated that the selectivity increases as alkane chain length of fuel model increases while the solubility of ionic liquids in the fuel models decreases, hence the costs of desulfurization process will be decreased. Moreover, the Results showed that for all systems when thiophene concentration decreases, the selectivity and distribution coefficient values increases. It means that ILs are good solvent at deep extractive desulfurization process. Meanwhile The selectivity remarkably increases as chain length of IL's cation decreases for all systems while the solute distribution coefficient decreases slightly.

In order to check, the temperature on selectivity, extraction of thiophene from n-decane by BMIM[NO_3] and OMIM[NO_3] have been studied three different temperatures (288, 298 and 313 K). Temperature showed no great effect in desulfurization and the study showed that the selectivity increases as the solute mole fractions in the hydrocarbon-rich phase decreases. Also the selectivity and distribution ratio of thiophene was reduced with increasing the temperature.

Our results proved that [Bmim][SCN] has the highest selectivity and lowest alkane solubility, in the system containing thiophene and *n*-decane. In other words [Bmim][SCN] is a more suitable ionic liquid solvent than others for extractive desulfurization of higher alkane.

Finally measured experimental data were correlated using non- randomness two liquid (NRTL) theory and the model parameters were obtained. Ternary phase equilibrium diagrams for mentioned ternary mixtures including the experimental and calculated tie-lines have been reported. It can be seen that the model can correlate the experimental data efficiently.

472256 Diffusion of Ions in Charged and Uncharged Polymers (Invited Talk)

Abstract:

Charged polymer membranes are widely used for water purification applications involving control of water and ion transport, such as reverse osmosis and electrodialysis. Efforts are also underway worldwide to harness separation properties of such materials for energy generation in related applications such as reverse electrodialysis and pressure retarded osmosis. Additional applications, such as energy recovery ventilation and membrane-assisted capacitive deionization, rely on polymer membranes to control transport rates of water, ions, or both. Improving membranes for such processes would benefit from more complete fundamental understanding of the relation between membrane structure and ion sorption, diffusion and transport properties in both cation and anion exchange membrane materials. Ion-exchange membranes often contain strongly acidic or basic functional groups that render the materials hydrophilic, but the presence of such charged groups also has a substantial impact on ion (and water) transport properties through the polymer.

We are exploring the influence of polymer backbone structure, charge density, and water content on ion diffusion properties. Results from some of these studies will be presented, focusing on diffusion of salt, primarily NaCl, through various neutral, positively charged and negatively charged membranes via concentration gradient driven transport (i.e., ion permeability) and electric field driven transport (i.e., ionic conductivity). One long-term goal is to develop and validate a common framework to interpret data from both electrically driven and concentration gradient driven mass transport in such polymers and to use it to establish structure/property relations leading to rational design of membranes with improved performance.

461520 A Predictive Model for the Determination of Mixed Gas Transport and Solubility in Glassy Polymers

Abstract:

The description of the mixed gas transport and sorption in polymers is of fundamental importance for the development gas separation membranes, and for the evaluation of their performances under real conditions [1,2]. However, very few approaches are suitable to that aim, especially for glassy phases, for which the Dual Model Sorption model (DMS) is still largely used, in spite of its limited, if any, predictive ability [3].

The transport model recently proposed by Minelli and Sarti has already proved its ability in the description of the permeability and solubility of pure gases and vapors in glassy polymers, on the basis of a simple fundamental approach [4]. The model considers the diffusion coefficient as the product of the molecular mobility and the thermodynamic factor, related to the concentration dependence of the penetrant chemical potential. The model relies on the thermodynamic description provided by the nonequilibrium thermodynamics for glassy polymers, NET-GP (and the nonequilibrium lattice fluid, NELF, in particular), which provides the thermodynamic behavior of the polymer/penetrant system, in a predictive fashion [5]. In parallel, a simple exponential function of penetrant concentration is used to describe the mobility coefficient behavior, which contains the only two adjustable parameters of the model: the penetrant mobility at infinite dilution and the plasticization factor.

In this work, the transport model has been extended to describe the transport and the solubility of gaseous mixtures in glassy polymeric membranes. In particular, the experimental permeability data of CO_2/CH_4 mixtures in conventional as well as in innovative glassy membranes have been examined, and described by the present model, with no need of any additional parameters with respect to the pure gas case. The NET-GP approach can be properly employed to represent the multicomponent solubility in glassy polymers, as already proved for various systems [6], and the transport model accounts for the presence of a second component through its concentration dependence of the diffusion coefficient. The model is able to describe well the experimental data for all the systems investigated, and the permeability behaviors of both penetrant species are accurately represented in a wide feed pressure ranges.

Interestingly, the model is able to highlight complelive sorption phenomena among the two penetrants, and the effects produced on both solubility and diffusivity coefficients. Furthermore, polymer dilation induced by CO_2 (or other swelling agents) can be simply identified, and its plasticization effect in case of real mixtures is quantitatively identified.

References:

[1] B. Freeman, Y. Yampolkii, Membrane Gas Separation John Wiley and Sons, 2010

[2] P. Bernardo, E. Drioli, Ind. Eng. Chem. Res. 48 (2009) 4638.

[3] W.J. Koros et al., J. Poly. Sci. B: Polym. Phys. 19 (1981) 1513.

[4] M. Minelli, G.C. Sarti, J. Membr. Sci. 435 (2013) 176.

[5] F. Doghieri, G.C. Sarti, Macromolecules 29 (1996) 7885.

[6] M. Minelli et al., Macromolecules 44 (2011) 4852.

462446 *In Situ* Pressure-Contact Time-Resolved Fourier Transform Infrared Attenuated Total Reflectance Spectroscopy: A New Method to Measure Liquid Diffusion in Free-Standing Polymer Films

Abstract:

A conventional method (*ex situ* gravimetry) for measuring liquid diffusion in free-standing polymer films typically consists immersing a film into a liquid, manually removing the film at a given time interval, removing the film and blotting the excess liquid from the surface of the film, weighing the film, re-immersing the film in the liquid, and then repeating this process many times to construct weight gain *versus* time data. However, this *ex situ* technique can introduce significant error when the sample size is small (typical of polymers synthesized at the academic laboratory scale) or take significant experimental time when the sample is thick, and this technique does not provide any molecular level detail. In this study, a new technique was developed to accurately measure liquid diffusion in free-standing polymer films with molecular level detail: *in situ* pressure-contact time-resolved Fourier transform infrared attenuated total reflectance (FTIR-ATR) spectroscopy. Previously, many studies have demonstrated the accurate measurement of liquid diffusion in polymer films using FTIR-ATR spectroscopy; however, polymers typically require solution casting on to an ATR element (or crystal) to ensure adequate adhesion or contact between the polymer film or crystal, where the infrared signal is measured inside the polymer close to the polymer/crystal interface. The new method developed in this study allows for an accurate measurement of liquid transport in polymer films for the case where the films cannot be easily solution cast or solution casting can alter the properties of polymer films that are processed using alternative methods (*e.g.*, extrusion). A custom-made cell was designed to introduce the liquid above the polymer film while also applying an adequate pressure between the polymer film and the ATR crystal to ensure adequate adhesion for accurate FTIR-ATR measurements. Liquid transport was investigated in free-standing polymer films with this new technique and compared with liquid transport measurements in free-standing polymer films using conventional gravimetry and liquid transport measurements in solution-cast polymer films using the conventional FTIR-ATR technique. Experiments were performed on both rubbery and glassy polymer films and also as a function of temperature and film thickness. Non-Fickian diffusion was observed in glass polymer films, specifically diffusion-relaxation phenomena was observed, where the relaxation behavior differed for free-standing films compared to solution-cast films. The results from this new technique will be discussed in detail and in comparison to conventional gravimetry and conventional FTIR-ATR results.

457315 Penetrant and Ion Dynamics in Model Microphase Separated Copolymers

Abstract:

Nanostructured AB block copolymers have been of great interest for transport applications such as lithium ion battery electrolytes and CO_2 separation membranes. Various nanostructures (lamellae, cylinder, gyroid, sphere) form as functions of the fraction of A monomer, A-B interaction strength, and molecular weight, and one can choose the A monomer to allow transport of the target molecules while the B monomer provides mechanical robustness. Here, we study both diblock and tapered AB block copolymers (TBCs) where a "tapered" mid-block is inserted between two pure blocks of A and B monomers. The taper has a gradient in composition from pure A to pure B (or from pure B to pure A for an inverse taper) and taper length can be used an additional tuning parameter to control the nanostructure. A recent experiment shows a certain length of taper can improve both polymer and ion dynamics for battery electrolyte systems [1]. To provide physical insight into such results and predict which systems may improve transport, we perform molecular dynamics simulations using a simple coarse-grained bead-spring model. Monomer-sized penetrants with favorable interactions with one type of monomer are added to lamellar structures of normal and inverse TBCs at a range of taper length from 0% (diblocks) to 100% (gradient copolymer). We observe both polymers' and penetrants' dynamics increase as taper length increases for normal TBCs but non-monotonically change for inverse TBCs. For the inverse TBCs, nonintuitive trends in the dynamics arise from competing effects of local A-B mixing and chain conformations. We have also considered salt-doped systems (with full long-ranged Coulomb interactions) to explore how ion transport is different than that of uncharged small penetrant molecules.

[1] W.-F. Kuan, R. Remy, M. E. Mackay, and T. H. Epps, III, RSC Adv. **5**, 12597 (2015).

451322 The Kinetics of Diffusion in Keratin Fibres Using Mass Spectrometry

Abstract:

The Kinetics of Diffusion in Keratin Fibres Using Mass Spectrometry

N. Ali[1], D.R. Williams[1], J. Marsh [2] and S. Godfrey[3]

[1] Surface Particles Engineering Laboratory, Department of Chemical Engineering

Imperial College, Kensington, London SW7 2BY

[2] The Procter & Gamble Company, 11810 East Miami River Road, Cincinnati, Ohio 45252

[3] Researchand Development, Procter & Gamble, Darmstadt Germany.

Abstract

Over the years there has been considerable progress in understanding the structure and properties of hair. However, we still do not fully understand about the structure of hair and the penetration chemistry of actives in the hair.

Hydration of hair, for example, is a fundamental process. It governs swelling behaviour, structure, diffusion rates and mechanical properties of hair. Despite its importance, keratin water interactions (hydration) are poorly understood. This is particularly true for hydration along single hair fibres and across hairs from different body sites, genders, ethnicities and ages. Addressing these key knowledge gaps will allow us to unlock critical insights to enable innovation.

Hydration of hair, for example, is a fundamental process. It governs swelling behaviour, structure, diffusion rates and mechanical properties of hair. Despite its importance, keratin water interactions (hydration) are poorly understood. This is particularly true for hydration along single hair fibres and across hairs from different body sites, genders, ethnicities and ages. Addressing these key knowledge gaps will allow us to unlock critical insights to enable innovation.

In this study, the time dependent mass uptake of solutes has been demonstrated using Gas Chromatography Mass Spectrometry (GC-MS). The effect of solute concentration and temperature on the kinetic uptake of monoethanolamine by hair fibres has been investigated. Diffusion coefficient has been determined using Crankâ€™s equation for an infinite cylinder.

462941 A Combined Experimental and Modeling Approach to Study the Sorption and Diffusion Phenomena in Materials

Abstract:

A combined experimental and modeling approach to study the sorption and diffusion phenomena in materials

Hom N. Sharma, Stephen J. Harley, Yunwei Sun, and Elizabeth A. Glascoe

Lawrence Livermore National Laboratory

L288, 7000 East Ave., Livermore, CA 94550

925-423-5368

sharma@llnl.gov

Moisture sorption and diffusion in materials are of great interest in wide range of applications from fuel cells to microfluidics and pharmaceuticals.(1-3) The uptake and outgassing of moisture is also associated with aging and compatibility issues in a system or sub-system, which is important to establishing the lifetimes and viability of current assemblies and screening new materials for future designs. The process is dynamic and consists of different sorption modes and varies dramatically in different materials.(2, 4) Therefore, an in-depth understating of the moisture uptake and outgassing is essential.

In this study, we investigate the moisture sorption and diffusion using a combined experimental and modeling approach. Various materials including polymers are investigated over a wide range of temperature (30 – 70 °C) and relative humidities (RH) (0 – 90%) to quantify the moisture transport mechanism as shown Fig. 1. Further, a reactive transport model is developed, which includes (a) a triple-mode sorption model that includes absorption, adsorption, and pooling of species, (b) molecular diffusion, and (c) chemical reaction kinetics. Using a 1D or 3D model, we can simultaneously simulate the transport and chemical reactions of mobile species through materials. The equation of diffusion coupled with Langmuir adsorption and pooling sorption can be given as:

$$\frac{\partial C}{\partial t} = \frac{\partial (C_H + C_L + C_P)}{\partial t} = \frac{\partial}{\partial x}\left(Dx\frac{\partial C_H}{\partial x}\right) + \frac{\partial}{\partial y}\left(Dy\frac{\partial C_H}{\partial y}\right) + \frac{\partial}{\partial z}\left(Dz\frac{\partial C_H}{\partial z}\right)$$

where C, C_H, C_L, C_P, D are the total concentration of sample bulk mass, mass concentration in Henry mode, concentration in Langmuir mode, pooling sorption mode concentration, and effective diffusion coefficient, respectively.

Our results show that the particular mode could become dominant at certain RH regions indicating the material specific diffusion-sorption behavior. For example, Langmuir mode is dominant in Zircar RS1200 at the beginning while Henry mode gradually increases and reaches to its maximum. Above 50% RH, pooling (clustering) starts and becomes dominant sorption mode while Langmuir attains its equilibrium as shown in the figure below. In this presentation, we will discuss our combined experimental and modeling approach using diffusion-sorption mechanisms along with the parameter estimation and uncertainty quantification techniques in various polymeric and non-polymeric materials.

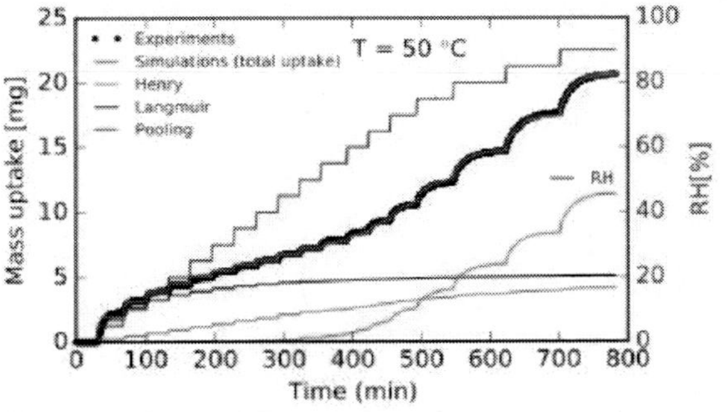

Fig. 1: Experiments and multi-mode diffusion-sorption modeling results of dynamic moisture uptake by Zircar RS1200 at a range of relative humidities at 50 °C.

This work performed under the auspices of the U.S. Department of Energy by Lawrence Livermore National Laboratory under Contract DE-AC52-07NA27344.

Reference:

1. Harley SJ, Glascoe EA, Maxwell RS. Thermodynamic Study on Dynamic Water Vapor Sorption in Sylgard-184. J Phys Chem B. 2012; 116(48):14183-90.

2. Harley SJ, Glascoe EA, Lewicki JP, Maxwell RS. Advances in Modeling Sorption and Diffusion of Moisture in Porous Reactive Materials. Chemphyschem. 2014; 15(9):1809-20.

3. Davis EM, Elabd YA. Water Clustering in Glassy Polymers. J Phys Chem B. 2013; 117(36):10629-40.

4. Sun YW, Harley SJ, Glascoe EA. Modeling and Uncertainty Quantification of Vapor Sorption and Diffusion in Heterogeneous Polymers. Chemphyschem. 2015; 16(14):3072-83.

458148 Triptycene-Containing Polybenzoxazole (PBO)-Based Membranes: New Synthesis and Gas Transport Properties

Abstract:

Triptycene-containing Polybenzoxazole (PBO)-based Membranes: New Synthesis and Gas Transport Properties

[1]Shuangjiang Luo, [1]Ashish Kushwaha, [2]Junyi Liu, [2]Haiqing Lin, [1]Ruilan Guo* (rguo@nd.edu)

[1]*Department of Chemical and Biomolecular Engineering, University of Notre Dame, Notre Dame, IN 46556-5637*

[2]*Department of Chemical and Biological Engineering, University at Buffalo, State University of New York, Buffalo, NY 14260-4200*

Polybenzoxazole (PBO)-based membranes, such as thermally rearranged (TR) polymers, hold great promise for gas separation that unite high gas permeability and high selectivity. It has been shown that gas separation performance of these thermally converted PBO membranes are largely determined by the precursor structures and the thermal treatment protocols. Herein we report the preparation and transport properties of two new series of PBO-based polymer membranes that were thermally derived from triptycene-containing *o*-hydroxy polyimide and polyamide precursors via thermal rearrangement (TR) process and thermal cyclodehydration (TC) process, respectively. Incorporation of triptycene units into poly(hydroxyimide) precursor structures led to significant increase of fractional free volume and created ultrafine microporosity in the converted PBO-based TR membranes, which enabled both high gas permeabilities and high selectivities. Although TC process of poly(hydroxyamide) precursor led to moderate improvement in the separation performance of the resulting triptycene-containing PBO membranes as compared to TR process, the PBO films converted via TC process exhibited excellent mechanical properties that are superior to many other TR membranes previously reported in the literature as well as the triptycene-containing TR membranes in this study. In particular, the PBO film thermally rearranged at 450 °C showed H_2 pure gas permeability of 807 barrer, CO_2 permeability of 269 barrer, and CO_2/CH_4 and H_2/CH_4 selectivities of 67 and 202, respectively, at 35 °C and 11 atm, which are far beyond the upper bound limits. In this talk, synthesis and characterization of these new PBO membranes will be presented. Fundamental transport properties including pure gas permeation data (with H_2, N_2, CO_2, and CH_4) will be discussed to illustrate the fundamental relationship between microscopic structures with macroscopic transport properties for these new triptycene-containing PBO membranes.

468297 Water Sorption and Diffusion in Nanofibrillated Cellulose Films and Composites

Abstract:

Nanocelluose is an innovative nano-sized material, which is attracting significant interest in view of its possible applications in various fields, such as, among the others, gas barrier materials or gas separation membranes. Indeed, nanocellulose films and nanocomposites are strongly barrier towards gases in dry conditions, while the permeabilities increase up to 2 or more orders of magnitude in humid environments, with significantly high selectivities toward CO_2. Water vapor can thus tune the nanocellulose properties from a barrier materials to a very promising membrane and the study of water sorption in these materials is crucial to completely understand the potentialities and current limitation of such new material for this kind of applications.

This work investigates the solubility and transport of water vapor in different types of Nanofibrillated Cellulose (NFC) as well as in polyvinylamine (PVAm)-NFC nanocomposites, by means of direct sorption and diffusion experiments in different experimental conditions.

In particular, NFC obtained by different cellulose pretreatments (carboxymethylation or enzymatic) prior to mechanical processing, were tested spanning over wide ranges of water activity and temperatures, between 16 to 65°C, while PVAm-NFC composite are investigated at various NFC loadings at 35°C.

Interestingly, the water diffusivity increased both with temperature and water activity, approaching a maximum value, independent on the operative condition considered for the tests. In addition, non-Fickian behaviors are observed especially at higher temperatures and R.H. inspected, clearly indicating the existence of different processes controlling the kinetics water sorption in the NFC film.

Similar behavior was also observed for in the case of nanocomposites, which also presented analogous water diffusivity, even if they showed significantly higher water uptake (larger than 50 % wt.) with respect to the pure NFC films that never exceed 10%wt.

The experimental data are then analyzed by appropriate modeling descriptions, seeking for the clear understanding of the process of moisture diffusion in nanocellulose based materails. Different approaches have been considered for solubility, accounting for both absorption and adsorption processes, and diffusivity, covering single or multiple Fickian diffusion, diffusion relaxation models, as well as the so-called parallel exponential kinetics model (PEK), often employed to represent the water sorption in cellulose fiber.

464834 Diffusion in Pebax Block Copolymers: Effects of Diffusant Mass and Polymer Composition

Abstract:

The rate at which small molecules diffuse through polyether-b-polyamide (PEBAX) copolymers is crucially important to their applications as controlled release media and gas separation membranes. PEBAX may also be blended with additives and used in medical devices, in which case accurate predictions of how fast those additives are released into the human body will serve the needs of public health. These uses have motivated us to experimentally measure a set of representative diffusion (D) and partition (K) coefficients over a wide range of diffusant masses (MW, ca. 50 â€" 550 Da) and polyether contents (0 â€" 84 mol. %) via a variety of gravimetric and spectroscopic sorption techniques.Â Here we will show that log(D) vs. log(MW) is linearly proportional with a slope of -4.7 within PEBAX 4033. The effect of polyether content on D and K can be well-accounted for by the effective medium theory of Sax and Ottino for transport in heterogeneous materials. These results broadly define upper limits on patient exposure to device additives, which will be useful in subsequently gauging the potential risks of being exposed to additives.

471436 Modeling Gas and Vapor Transport Mechanisms in Mesoporous Membranes Using Dynamic Mean Field Theory

Abstract:

Mesoporous membranes with technologically relevant pore sizes provide an energy efficient means for separations such as CO_2 from effluents and volatile organic compounds from air. Continuous progress in material synthesis over the years has led to techniques which allow reasonable fine tuning of geometry and surface chemistry of mesopores within the membranes. However, absence of design rules linking mesopore properties with confinement and flow mechanisms which determine permeation and separation efficiency have proven a hindrance towards designing better materials. Towards development of such design rules through the study of underlying transport processes, we have investigated the problem of permeation of pure gas and condensable vapor as well as their mixtures under pressure gradients, through the application of dynamic mean field theory (DMFT), a lattice-based theory. We initially modeled permporometry, an experimental technique where light gas permeates in presence of a condensable vapor in the pore. Interestingly, the maximum flux of light gas was found adjacent to the strongly adsorbed surface layer. We then studied separation of a condensable vapor from its mixture with light gas under significant pressure gradient. We discovered capillary condensation confined to the high pressure (feed side) of the pore. Experimentally relevant model parameters were used to investigate effects of such a localization on separation efficiency. Effects of modification in geometry and orientation were studied through ink-bottle like structures while effects of surface chemistry were explored through modification in interactions. To gain a perspective on the results, experimental studies from the literature, and computationally expensive dual control volume â€“ grand canonical molecular dynamics (DCV-GCMD) atomistic simulations, were used for comparison.

472868 Understanding Protein Adsorption on Polymeric Ligands from Molecular Dynamics Simulations

Abstract:

Poly (N-vinylcaprolactam) (PVCL) is a prototype thermo- and salt-responsive polymer. PVCL is a widely investigated biomaterial that has potential applications in drug delivery and protein purifications as responsive hydrophobic ligands for protein adsorption and elution during down stream processing. Here classical molecular dynamics simulations were conducted to investigate the energetics and interaction mechanisms between the PVCL polymer chains and bovine serum albumin (BSA) in aqueous solutions. Water structures surrounding the polymer chains and BSA as well as their interactions including hydrogen bonding, electrostatic and van der Waals interactions were determined. Significant insights were obtained on the effects of polymer hydration state (i.e. collapsed or hydrated), polymer chain length as well as the presence of salt ions on the protein-ligand binding interactions.

452112 Molecular Simulations of Ion Permeation in Graphene Oxide Membranes

Abstract:

Molecular Simulations of Ion Permeation in Graphene Oxide Membranes

The scarcity of clean water for human consumption is a serious problem facing many communities around the world. Improving the efficiency of desalination and decontamination processes could alleviate this problem. To enable the full potential of desalination and decontamination to be harnessed, membrane materials with improved properties must be found. Graphene oxide (GO) membranes may present a breakthrough in this field because they display fast water transport and selective ion permeation, as well as being amenable to large-scale, low cost production by chemical means. Once careful control of the spacing between layers has been realised, GO is potentially an excellent candidate material for aqueous separation applications, such as desalination or decontamination.

In this work, molecular dynamics simulations are used to investigate the permeation of ions into graphene capillaries and improve understanding of the experimentally observed selectivity. By generating single ion potentials of mean force for a range of ions and pore widths, we show how the selectivity is dependent on the hydrated structures and hydration free energies of the ions.

The selectivity of anions appears to be dominated by the energetic cost of dehydration upon confinement. We predict how this property could be exploited for the use of GO to clean up weakly hydrating and problematic anionic contaminants (e.g. radioactive $^{99}TcO_4^-$), with a high selectivity over competing species. For cations, the observed ion selectivity is dependent on the effect of dehydration, relative to the strength of cation – pi interactions with the sp^2 electrons in graphene. These interactions have been accounted for in our simulations. The results have improved

understanding of ion selectivity in GO and enabled us to make predictions about how to optimise the design of these membranes for a given application.

452990 Mass Transfer Enhancement of Non-Dispersive Solvent Extraction By Dean Vortices: Modeling, Experiments and Optimization

Abstract:

Dean vortices induced in helical hollow fiber membranes (HHFM) are simplest and most effective to increase the mass transfer rate of the membrane processes. As such, it is important to understand the underlying mechanisms of the mass transfer behavior of HHFM processes by modeling and simulation. This work focuses on the mass transfer enhancement of non-dispersive solvent extraction with Dean vortices in purified terephthalic acid (PTA) wastewater treatment by modeling, experiment and optimization. Experiments showed that the mass transfer rate of the HHFM extraction was doubly enhanced compared with the straight hollow fiber membrane extraction. A new mathematical model of the HHFM extraction was developed in an orthogonal helical coordinate system with a perturbation solution to describe the velocity profile. The key parameters affecting the intensity of Dean vortices of HHFM extraction are the Graetz number of the lumen side and the dimensionless curvature. Numerical analysis of the mass transfer showed that the highest concentration area exists in the area of the radial velocity vortices, where the velocity of Dean vortices is zero. An optimal group of operating parameter for PTA wastewater treatment is suggested.

455772 Simulated Permeation and Characterization of Pegylated Gold Nanoparticles in a Lipid Bilayer System

Abstract:

PEGylated gold nanoparticles are considered suitable nanocarriers for use in biomedical applications and targeted drug delivery systems. In our previous investigation with the alkanethiol-functionalized gold nanoparticle, we found that permeation across a protein-free phospholipid membrane resulted in damaging effects of lipid displacement and water and ion leakage. In the present study, we carry out a series of coarse-grained molecular simulations to explore permeation of lipid bilayer systems by a PEGylated gold nanoparticle. Initially, we examine molecular level details of a PEGylated gold nanoparticle (constructed from cycled annealing) in water and find a distribution of ligand configurations (from mushroom to brush states) present in nanoparticles with medium to high surface coverage. We also find that the characteristic properties of the PEGylated gold nanoparticle do not change when it is placed in a salt solution. In our permeation studies, we investigate events of water and ion penetration as well as lipid translocation while varying the ligand length, nanoparticle surface coverage and ion concentration gradient of our system. Results from our studies show (1) The number of water molecules in the interior of the membrane during ligand-coated nanoparticle permeation increases with PEGn-SH surface coverage, ligand length, and permeation velocity but is not sensitive to the ion concentration gradient. (2) Lipid molecules do not leave the membrane; instead they complete trans-bilayer lipid flip-flop with longer ligands and higher surface coverages. (3) The lack of formation of stable water pores prevents ion translocation. (4) The PEGylated nanoparticle causes less damage to the membrane overall due to favorable interactions with the lipid head groups which may explain why experimentalists observe endocytosis of PEGylated nanocarriers in-vivo.

462462 An Integrated Multiscale Modeling and Experimental Approach to Design Fouling-Resistant Membranes

Abstract:

An integrated multiscale modeling and experimental approach to design fouling-resistant membranes

[1]Linkel Boateng, [1]Ryan Monk, [2]Peng Xie, [1]Anna Malakian, [1]Steven Weinman, [2]David Ladner, [3]Ilenia Battiato, [1]Scott Husson, [1]Sapna Sarupria

[1]Department of Chemical & Biomolecular Engineering, Clemson University, Clemson, SC 29634

[2]Department of Environmental Engineering & Earth Science, Clemson University, Anderson, SC 29625

[3]Department of Mechanical Engineering , San Diego State University, San Diego, CA 92182

Abstract

Membrane fouling is one of the largest costs associated with membrane processes in water treatment. The fouling propensity of a membrane depends greatly on its surface properties such as chemistry and roughness. An in-depth understanding of membrane fouling behavior under different conditions requires extensive investigations. In our research efforts, we use a combination of modeling and experimental studies to identify mechanistic strategies for fouling control in membranes. The overall goal of the study is to engineer new nanofiltration and reverse osmosis (RO) membranes that combine chemical modifications with physical patterning to reduce fouling.

Molecular dynamics (MD) simulations are performed to elucidate membrane fouling mechanisms and to estimate membrane properties at the molecular scale. Specifically, we calculate the potential of mean force between the foulants and the membrane surface to quantify the strength of foulant-membrane interaction. This information obtained from the MD simulations is incorporated into a computational fluid dynamics (CFD) model to predict the fouling behavior on the membrane at the continuum scale. We will discuss the details of this multiscale modeling approach and report results on the fouling behavior of the membrane at the molecular and continuum scales. The modeling results will be compared to experiments to assess the effectiveness of the current multiscale approach.

467402 Optimizing the Operation of a Direct-Flow Filtration Device

Abstract:

Direct-flow filtration is an ultrafiltration membrane process in which hydrostatic pressure forces a liquid through the semipermeable walls of a hollow fibre module in the absence of a cross flow; the far end of the feed channel is capped and therefore all of the feed fluid is forward to pass the membrane walls. This filtration mode offers an efficient combination of the benefits of classical dead-end and crossflow filtration providing there is periodic back-flush. In this study, mathematical models for the flow in a 2D filtration channel and in a 3D hollow fibre have been developed to investigate the flow behaviour within such direct-flow filtration devices. These models were originally developed by applying lubrication equations to systems with walls assumed to be zero wall thickness. This constraint was lifted in the second part of the work. The space taken by the walls of hollow fibres is not negligible and the influence of wall thickness was important.

The pressure profile in the permeate channel influences the pressure profile in the feed channels. We study the influence of permeate spacing and other parameters, and found that there was a strong axial dependence of the filtration performance. Figure 1 shows the transmembrane pressure difference (TMP) profiles versus axial distance at different module spacing parameter values in (a) a 2D channel and (b) a 3D module. The module productivity was evaluated for various input and operating conditions in order to find the operating regimes of the device that maximize the spatial uniformity in the TMP. Figure 2 shows the cumulative fluid volume transported through the fibre walls as function of axial position for (a) a 2D channel and (b) a 3D module for various values of module spacing parameters. Ideally one should optimize the use of the entire membrane area. The membrane area is not evenly used especially when the module are tightly packed. Behaviour of 2D channels is a rough guide to the behaviour of 3D modules. Additional studies, which will be part of the presentation, indicate there exists an optimum packing range, below which reduction in packing density will have very little effect on process performance, however productivity will be reduced as packing density is reduced.

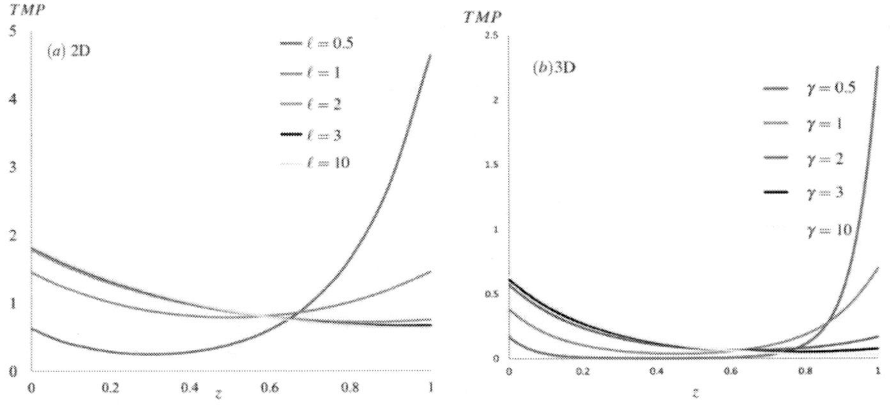

Figure 1. The transmembrane pressure difference (TMP) profiles versus axial distance, at different module spacing parameter values in (a) a 2D channel and (b) a 3D module.

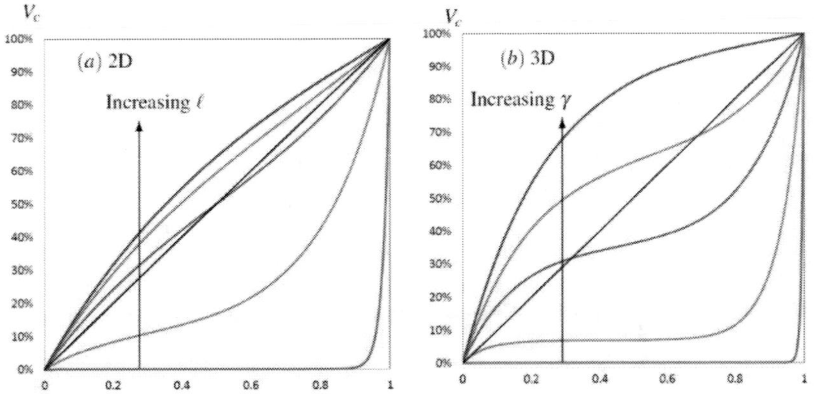

Figure 2. The cumulative fluid volume transported through the fibre walls as function of axial position for (a) a 2D channel and (b) a 3D module for various values of module spacing parameters.

468081 The Effect of Pore Penetration on Composite Membranes

Abstract:

Composite membranes typically consist of at least one nonporous layer on top of a porous structure. The porous support restricts diffusion in the top layer because molecules can exit the layer only where a pore is present. Computational fluid dynamics (CFD) has been used to simulate the geometric restriction and an empirical model was developed to quantify the effect of the porous support on the permeance of composite membranes.

In this work, modeling results are compared to experimental data for composite membranes consisting of a dense polydimethylsiloxane (PDMS) layer on a polyetherimide (PEI) ultrafiltration support. Significant discrepancies are observed between the experimental and modeling results. Pore penetration is believed to be a likely reason for the difference between the theoretical and measured permeances. A resistance-in-series model and CFD were employed to calculate the effect of pore penetration on the permeance of composite membranes. Results from these two approaches are in good agreement, and provide a good fit to experimental composite membrane data if pore penetration depth is allowed to vary. An analysis of the PDMS/PEI composite membrane data, suggests that the depth of PDMS penetration into the PEI pores increases as the thickness of the PDMS selective layer increases.

465985 Water Desalination through Zeolitic-Imidazolate Framework Membranes: A Molecular Simulation Study

Abstract:

Fresh water scarcity has become a major global concern due to the rapid growth in population, energy demand and industrialization. As over 95% of water on the Earth is seawater, there is considerable interest to desalinate seawater to supply freshwater. A handful of desalination techniques have been developed including reverse osmosis (RO) and multistage flash distillation (MSFD). Compared to thermal-based MSFD, RO is more energy efficient. In the present scenario, 42 million kg of potable water is produced daily by RO, which accounts for 60% of world desalination capacity. The capital cost of current RO technology is high and it needs to be improved prior to propagation. Aiming to develop efficient RO membranes, a large number of studies have been conducted. In the past decade, zeolitic-imidazolate frameworks (ZIFs) have emerged as a new class of nanoporous materials which possess exceptional chemical and thermal stability. In addition, their pore size and affinity are readily tunable, thus there is a large opportunity to explore ZIFs with improved desalination performance.

In this work, water desalination through ZIF-25, -71, -93, -96 and -97 membranes is investigated by molecular simulation. The ZIFs possess identical rho-topology, but differ in functional groups. Due to the presence of various functional groups, the aperture size and polarity differ in the five ZIFs. The rejection of salt (NaCl) is found to be around 97% in ZIF-25, and 100% in the other four ZIFs. The permeance ranges from 27 to 710 kg/(m^2.hr.bar), about one ~ two orders of magnitude higher compared to commercial RO membranes. Because of a larger aperture size d_a, ZIF-25, -71 and -96 exhibit much higher water flux than ZIF-93 and -97. The flux in ZIF-25, -71 and -96 is governed by the polarity of functional group rather than d_a. With hydrophobic â€"CH3 group, ZIF-25 has the highest flux despite the smallest d_a among ZIF-25, -71 and -96. With the highest flux, the life time of hydrogen bonding in ZIF-25 is shorter than in ZIF-71 and -96. Furthermore, water molecules undergo fast flushing motion in ZIF-25, but frequent jumping in ZIF-96 and particularly in ZIF-97. An Arrhenius-type relationship is found between water flux in ZIF-25 and temperature, and the activation energy is predicted to be 6.5 kJ/mol. This simulation study provides a microscopic insight into water desalination in a series of ZIFs, reveals the key factors governing water flux, and suggests that ZIF-25 might be an interesting reverse osmosis membrane for high-performance water desalination.

468342 Molecular Dynamics Simulations of Permeability in Gel Phase Mixed Lipid Bilayers

Abstract:

Lipid membranes are vital to all complex forms of life and can control the transport of material into cells and the body. While membrane composition and structure clearly play a role in determining if and how fast a solute can permeate across a membrane barrier, a better understanding of the relationship between membrane composition, structure, and permeability is needed to be able to improve, replicate, and control the barrier function of biological membranes. Although NMR spectroscopy and neutron scattering can measure global bilayer permeabilities, they cannot provide key molecular-level insight. Molecular simulations can uncover the physical mechanisms of permeation and provide a better understanding of the dependence of bilayer permeability on the local environment. Here, using atomistic molecular dynamics simulations, we investigate how chemical architecture of the bilayer molecules affects the structural properties, finding a complex balance between the size and hydrophilicity of the headgroups, relative length of lipid tails, and the attraction between tails. Furthermore, we focus on how these structural differences influence the passive permeation of water molecules through the bilayer structures, finding a strong dependence on the size and hydrophilicity of the headgroups, chain length, and side-branching.

452566 Impact of the Surface Energy of Particulate Foulants on Membrane Fouling

Abstract:

Foulant-foulant and foulant-membrane interfacial interactions play an important role in dictating the extent of fouling. In order to understand the impact of the surface energy of particulate foulants on the fouling extent, the direct observation through the membrane (DOTM) technique was used to characterize critical flux and also assess the initial evolution of foulant deposition. Polystyrene and glass with diameters approximating 10 μm were used as foulants because (i) the similar particle diameter eliminated the differences due to particle back-transport, and (ii) the Gibbs free energies (ΔG) of foulant-membrane and foulant-foulant interactions were both negative (i.e., attractive) and both positive (i.e., repulsive) respectively for polystyrene and glass. Results indicate that: (1) because of the attractive and repulsive *foulant-membrane* interactions of the polystyrene and glass, respectively, (i) critical flux was lower for polystyrene than glass, (ii) the phenomenon of a flowing particle layer was observed at a lower cross-flow velocity (CFV) for glass than polystyrene, and (iii) relaxation was more effective for glass than polystyrene; (2) because of the attractive and repulsive *foulant-foulant* interactions of the polystyrene and glass, respectively, clustering was observed for the former but not latter; (3) out of the four particle deposition mechanisms (namely, convective, entrapment, dead zone and clustering) distinguished, the first three occurred for both attractive and repulsive interfacial interactions, but the forth (i.e., clustering) only occurred when foulant-foulant interactions were attractive.

452791 Operating Strategy for the Energy-Efficient Reverse Osmosis (EERO) Desalination Process

Abstract:

Operating Strategy for the Energy-Efficient Reverse Osmosis (EERO) Desalination Process

William B. Krantz[a,c], Tzyy Haur Chong[b,c], Siew-Leng Loo[b,c]

[a] Dept. of Chemical & Biological Engineering, University of Colorado, Boulder CO 80309-0424

[b] School of Civil & Environmental Engineering, Nanyang Technological University, Singapore 639798

[c] Singapore Membrane Technology Center, Nanyang Technological University, Singapore 637141

The energy-efficient reverse osmosis (EERO) process, patented in 2014 by the Singapore Membrane Technology Center, reduces the osmotic pressure differential (OPD) and the cost for potable water production from a saline water feed by combining Single-Stage Reverse Osmosis (SSRO) with a countercurrent membrane cascade with recycle (CMCR). The retentate from the SSRO stage is the feed to the CMCR and its permeate is blended with that from the CMCR to produce the potable water product. Key features of the EERO process are: (1) optimum injection of the SSRO retentate into the CMCR; (2) operating all stages at the same OPD; (3) countercurrent retentate and permeate flow; (4) permeate recycle to the retentate side in the CMCR; and (5) retentate recycle to the permeate side by employing one or more NF stages in the CMCR. These features reduce the concentration difference across the membrane in each stage, thereby reducing the OPD and net specific energy consumption (SEC_{net}).
Since the permeate from an NF stage has a reduced divalent salt concentration and is the feed to the terminal CMCR stage, the latter can be operated at a higher recovery. Operating the EERO stages at the same OPD requires that the recoveries of adjacent CMCR stages add to one. Hence, fouling in the NF stage whose feed has a high divalent salt concentration can be mitigated by operating it at a lower recovery (higher safety factor).

Optimal operation of the EERO process can require using NF membranes with very low salt rejections in one or more of the CMCR stages. If a membrane with a higher than optimal rejection is used, near optimum EERO performance can be achieved at the cost of a small increase in the SEC_{net} by boosting the pressure in the NF stage.

EERO process performance for different operating strategies is compared to that for SSRO and two SSROs in series based on the OPD and SEC_{net}. The annualized total cost of water production for EERO operation at 75% recovery is shown to be less than that for SSRO operation at 50% recovery.

At 75% recovery EERO results in a two-fold increase in the retentate concentration that not only reduces the brine disposal, but also translates to a four-fold increase in the energy density possible via a hybrid EERO-PRO (pressure-retarded osmosis) process for recovering the osmotic potential energy in the brine. As such, this hybrid process would cross the 10 W/m^2 threshold for economic PRO operation.

Acknowledgment

This research was supported by the Singapore National Research Foundation under its Environmental & Water Technologies Strategic Research Program and administered by the Environment & Water Industry Program office (EWI) of the PUB. The Singapore Membrane Technology Center, Nanyang Environment and Water Research Institute. Nanyang Technological University is supported by the Economic Development Board of Singapore.

456121 Separation and Modeling of Oil-in-Water Emulsions Stabilized By Different Types of Surfactants Using Electrospun Fiber Membranes

Abstract:

In this work, we examine the influence of surfactants on the microfiltration of oil-in-water emulsions using electrospun fibrous membranes of poly(trimethyl hexamethylene terephthalamide) (PA6(3)T) with fiber diameters of [121 ± 22 nm]. Four types of surfactants (anionic, cationic, non-ionic and zwitterionic) were used to create monodisperse emulsions of dodecane in water, with average droplet sizes of [475 ± 14 nm]. Both dead-end and cross-flow filtration configurations were employed to study the interaction (i.e. fouling and cleaning) between the emulsion and membrane.

In both dead-end and cross-flow configurations, the emulsions prepared with anionic and non-ionic surfactants exhibited the highest fluxes, around five-fold higher than that for the emulsion stabilized by cationic surfactant. The emulsion with zwitterionic surfactant led to intermediate flux. The oil rejections were generally as high as 90%, with the exception of the emulsion stabilized by non-ionic surfactant, whose rejection was closer to 60%. The observed fluxes using the cross-flow system were about one order of magnitude higher than those observed with the dead-end system, due to a mechanism for foulant removal (cleaning) attributed to hydrodynamic forces. The differences in the fluxes and rejections for emulstions stabilized by different surfactants is traced to electrostatic interactions between the emulsified droplets and the membrane, based on zeta potential measurements and flux recovery after backflushing.

To study the fouling and cleaning mechanism, several models were also applied in this work. Blocking filtration models suggest that the membranes fouled initially in the regime of complete blocking, but then transitioned to cake filtration as the flux continued to decline. Foulant resistivity modeling supported a series resistance model, consistent with cake filtration. The emulsion stabilized by anionic surfactant showed the lowest resistance, while that stabilized by cationic surfactant showed the highest resistance. Using the surface renewal model, the emulsion stabilized by anionic surfactant was found to be most readily removed by hydrodynamic cleaning during cross-flow filtration.

Lastly, the decay of flux was well described by the conformally coated fiber model, wherein the fibrous structure of the membrane is considered explicitly, and the flow through the membrane is modified as a foulant layer builds up on the fibers.

461223 Development of Cordierite Honeycomb Membrane Filters for Produced Water Treatment

Abstract:

Oil and gas exploration and production processes produce large volume of waste as produced water that needs to be treated before reuse or discharge due to fresh water scarcity and upcoming tighter regulation. In this talk, we will present Corningâ€™s effort to develop chemically durable cordierite honeycomb microfiltration membranes for oil-water filtration. Two types of cordierite membrane have been developed with the pore size of 0.2 and 0.5 micron, respectively. The clean water permeance is in the range of 1500-2000 L/m2/h/bar. The chemical durability test demonstrated that they are durable in the solutions from pH3 to pH13 at the temperature up to 80 degree C. The oil-water filtration performance will be discussed.

464875 Influence of Graphene Quantum Dot Surface and Pore Modification on Membrane Fouling

Abstract:

Influence of Graphene Quantum Dot Surface and Pore Modification on Membrane Fouling

Membranes have been widely utilized for many applications including water softening, pharmaceutical purification, and dairy processing. In systems with organic foulants or sparingly soluble salts present, fouling of the membrane surface leads to reduced performance and shortened membrane lifespans. Graphene oxide has been of great interest as an antifouling material for membrane modification due itâ€™s inherit hydrophilicity and the presence of negatively charged carboxyl groups. Graphene oxide quantum dots are highly hydrophilic particles that maximize functionality in a size (1-3 nm) that makes them a particularly interesting candidate for membrane functionalization.

In this work, the benefits of GQD modification toward surface fouling behavior and pore blockage by organic foulants will be studied. Through EDC/NHS coupling, GQDs were covalently bonded to the surface of nanofiltration membranes.Â Zeta potential analysis shows higher negative surface charge properties after GQD coupling and increased rejection of Na_2SO_4 (97% compared to 76% for the standard membrane). The GQD coupled membrane exhibits double the normalized flux after 2 Â½ hours of recovery of solution containing 400 ppm BSA when compared to the standard membrane.

GQDâ€™s were integrated throughout the pore structure of cellulose ultrafiltration membranes by means of phase inversion. A common ionic liquid solvent between GQDs and cellulose membranes was used to deter aggregation and promote the selective entrapment of GQD in the membrane pore network. The effect of water flux, selectivity and antifouling behavior for BSA will be studied.

Acknowledgements

The author acknowledges the support of the Southern Company in funding and collaboration, and additional funding from the National Science Foundation EPSCoR for funding. Nanostone Membrane of Oceanside, CA has provided support and collaboration in the development of membranes and spiral wound membrane modules.

465388 Water Desalination through Molecular Layer-By-Layer Membranes: Insights from Molecular Dynamics Simulations

Abstract:

Reverse osmosis (RO) membranes based on polyamide (PA) membranes synthesized via interfacial polymerization (IP) of trimesoyl chloride (TMC) and m-phenylenediamine (MPD) represent one of the most common membranes used for producing potable water from seawater. However, typical RO membranes synthesized via IP can have several drawbacks which can impact the efficiency of the membrane. In particular, IP membranes lack control over (1) film thickness, (2) surface roughness, (3) cross-linkage, and (4) local chemical composition, which influence the membrane performance and efficiency. As an alternative to these challenges, ultraâ€'thin PA membranes synthesized via molecular layer-by-layer (mLbL) deposition have been introduced.

Although PA-mLbL have been studied in previous work, there are still challenges in both experimental characterization and understanding the relationship between membrane performance and membrane variables (grafting density and monomer ratio). In this work, molecular dynamics simulations were employed to characterize and to investigate membrane performance with respect to these membrane variables. We first applied a general simulated polymerization algorithm, Polymatic, developed in our group [Theor. Chem. Acc. 2013, 132, 1334] to virtually synthesize RO membranes via mLbL deposition on silicon substrates. The resulting simulated structures for twenty cycles are characterized by analyzing densities, membrane thickness, free volume, swelling, molecular orientations and local chemical compositions. Membrane performance was evaluated by analyzing water and salt dynamics in mLbL membranes with respect to grafting density and monomer ratio. This simulation study provided molecular level insight into water desalination using PA-mLbL membranes and illustrates that molecular simulations can play a crucial role in the understanding of next generation RO membranes synthesized via mLbL deposition.

468883 Calculation of the Isosteric Heat of Adsorption Using Quenched Solid Density Functional Theory

Abstract:

Calculation of the Isosteric Heat of Adsorption using Quenched Solid Density Functional Theory

Richard T. Cimino[1], Piotr Kowalczyk[2], Alexander V. Neimark[1] [1]Department of Chemical and Biochemical Engineering, Rutgers University, Piscataway, NJ 08854

[2]School of Engineering and Information Technology, Murdoch University, Perth, WA, 6150

When a fluid adsorbs on a surface, it releases energy in the form of heat. At equilibrium, the change in energy (enthalpy) of the adsorbate upon adsorption is characterized by the isosteric heat, which obeys the Clausius-Clapeyron (C-C) Equation: $q_{iso} = RT1T2/(T2-T1)(lnp2-lnp1)_N$. There are several methods for measuring the heat of adsorption, which along with the adsorption isotherm provide valuable information about the state of the adsorbed fluid in a material and the materialâ€™s pore structure. One of the direct methods is to calculate the isosteric heat from the adsorption isotherms measured at different temperatures using the C-C equation. Theoretically, the heat of adsorption can be estimated in the course of Monte Carlo simulation by calculating respective fluctuations of potential energy and number of particles in the Grand Canonical ensemble. Alternatively, the classical density functional theory can be applied either to directly calculate the adsorption heat from the free energy functional or by calculating the adsorption isotherms at different temperatures (at least two) and applying the C-C equation in the same way as it is done experimentally. Here, we employ the latter approach based on the quenched solid density functional theory (QSDFT). We explore the isosteric heat of adsorption of several probe molecules (N_2, Ar, and CO_2) on reference mesoporous silica materials and several classes of microporous carbons with hierarchical pore structures. The calculated values of

the isosteric heat are compared with experimental measurements and the results of direct MC simulations.

465868 Molecular Insight into Seawater Pervaporation through Zeolitic Imidazolate Framework Membranes

Abstract:

The scarcity of freshwater has escalated as a global concern due to increasing population, energy demand and industrialization. Currently, there is considerable interest to desalinate seawater, which constitutes over 95% of water on the Earth. Among several desalination techniques, multistage flash distillation (MSFD) and reverse osmosis (RO) have been majorly utilized. However, phase transition occurs in energy intensive thermal-based MSFD; and RO needs pressurization and is associated with easy fouling. Recently, pervaporation (PV) has emerged as a relatively new economically viable separation technique. As a unique integration of membrane permeation and evaporation, PV offers several advantages such as low energy consumption, high separation capability and easy scaling-up. To date, most PV membranes still suffer from low mass transfer (flux or permeability) compared to graphene and carbon nanotube membranes. Therefore, increasing interest is to explore new PV membranes that exhibit superior performance for water desalination. With large surface areas and high porosities, zeolitic imidazolate frameworks (ZIFs) have emerged as a new class of porous materials that could be used as membranes for pervaporation.

In this study, an atomistic simulation study is reported for seawater pervaporation through five zeolitic imidazolate framework (ZIF) membranes including ZIF-8, -93, -95, -97 and -100. Salt rejection in the five ZIFs is predicted to be 100% due to the sieving of small apertures. The hierarchy of water flux is as ZIF-100 >> -8 > -95 > -93 > -97. With the largest aperture, ZIF-100 possesses the highest water permeability of 5 ´ 10^{-4} kg×m/(m^2×hr×bar), which is substantially higher compared to commercial reverse osmosis membranes, as well as zeolite and graphene oxide pervaporation membranes. In ZIF-8, -93, -95 and -97 with similar aperture size, water flux is governed by framework hydrophobicity/hydrophilicity; in hydrophobic ZIF-8 and -95, water flux is higher than in hydrophilic ZIF-93 and -97. Furthermore, water molecules in ZIF-93 move slowly and remain in the membrane for a long time, but undergo to-and-fro motion in ZIF-100. The lifetime of hydrogen bonds in ZIF-93 is found to be longer than in ZIF-100. This simulation study quantitatively elucidates the dynamic and structural properties of water in ZIFs, identifies the key governing factors and suggests ZIF-100 is an intriguing membrane for seawater pervaporation.

449695 Effects of Adsorbent Material Properties on the Selective Separation of Supercritical Multicomponent Fluid Mixtures

Abstract:

A great deal of effort has, thus far, been put into screening existing porous materials, including zeolites and metal-organic frameworks (MOFs), for their ability to selectively adsorb desired molecules from a fluid mixture on the basis of molecular size and shape. [1,2] However, when different molecules have similar sizes or lack characteristic structures whose complement can be readily engineered in a porous material, selective separation becomes difficult. Here we use computationally efficient flat-histogram Monte Carlo (MC) simulation methods [3] to study the adsorption of prototypical coarse-grained binary supercritical fluid mixtures in flexible porous media. We demonstrate an extensible methodology for studying multicomponent adsorption which yields the complete free energy surface of the confined mixture. We focus on a canonical binary mixture which is particularly cumbersome to separate via traditional distillation techniques, and thus is a representative candidate for alternative approaches such as adsorbents or other mass-separating agents. Interrogation of the adsorption behavior at different overall bulk compositions reveals regimes in which the material's selectivity for a single component may be a highly non-monotonic function of applied pressure. We illustrate how this may be controlled by systematically tuning the adsorbent's chemical and mechanical properties. By controlling the relative strength and range of interaction each species has with the adsorbent we identify a regime in which the selective capture of dilute components may be enhanced by increasing the bulk pressure, even if the species is less entropically favored inside the material. At high pressure, the adsorbent's flexibility becomes increasingly important and can also lead to local extrema in the selectivity of that material. We elucidate the effects of a material's flexibility on its separation efficiency and discuss its relationship to the resulting heterogeneous fluid structure inside the porous adsorbent. This allows us to propose general heuristics for the design of adsorbent materials to achieve selective separation for certain classes of fluid mixtures.

References

[1] Wilmer et al., Nature Chemistry 4, 83-89 (2012).
[2] Lin et al., Nature Materials 11, 633-641 (2012).
[3] Errington and Shen, Journal of Chemical Physics 123, 164103 (2005).

472612 Screening of Copper Open Metal Site MOFs for Olefin/Paraffin Separations Using DFT Derived Force Fields

Abstract:

Industrial production of ethylene and propylene involves a catalytic cracking process that is followed by high-pressure cryogenic distillation of light olefin/ paraffin mixtures. Due to their importance to the chemical and polymer industry, efficient adsorption processes for olefin/paraffin separations are likely to result in significant energy savings. Although a small number of Metal Organic Frameworks (MOFs) have been experimentally evaluated, large-scale computational screening studies are not available due to the difficulty in describing the complex interactions of the olefin with the open metal site (OMS).

In this work, we develop new Density Functional Theory derived force fields (FFs) that accurately reproduce the experimental adsorption isotherms C_2 and C_3 hydrocarbons in CuBTC. Further, by using extensive DFT calculations, we demonstrated that the FF transferable to other MOFs containing open Cu sites. Using these transferrable force fields, we evaluate the performance of 94 distinct Cu-OMS MOFs for the industrially important propylene/propane separation, and identify 20 MOFs that are predicted to have attractive properties as adsorbents. Our approach is easily applicable to related olefin/paraffin separations in Cu OMS MOFs and to MOFs with other under-coordinated metal sites.

456242 Applicability of Polarizable Force Fields for M-MOF-74

Abstract:

Climate change caused by the increasing carbon dioxide concentration in the atmosphere is one of the major challenges of todays society. Carbon capture and sequestration could play an important role in quickly mitigating this development. Metal-Organic Frameworks (MOFs) haven been shown to have potential for post combustion carbon capture. Especially, the family of M-MOF-74, with M = Co, Cr, Cu, Fe, Mg, Mn, Ni, Ti, V, and Zn has attracted attention due to promising properties for various gas separation applications. Depending on the metal ion, M-MOF-74 exhibits a distinct adsorption behavior. Experimentally determined adsorption isotherms of carbon dioxide show more or less pronounced inflections at a loading of approximately one CO2 molecule per metal ion. This behavior is a result of a strong affinity between some of the metal ions and adsorbed CO2 molecules. To accurately predict the interaction strength of different metal ions in molecular simulations is challenging. Traditionally, molecular simulation studies focus on re-parameterizing force field parameters to account for the enhanced interactions. However, this procedure is cumbersome and it is required for every structure with framework specific interactions. We feel that a polarizable force field has the potential to overcome the limitations of this procedure and to result in improved force fields with better transferability and physical justification. Several quantum mechanical studies suggest that polarization plays an important role in the strong affinity between the open metal sites in M-MOF-74 and CO2. Here, we apply the procedure of Lachet et al. to include explicit polarization in the conducted Monte Carlo simulations. The procedure uses the induced point dipole method to account for polarization. To achieve reasonable simulation times, polarization is considered exclusively between guest molecules and the framework and back-polarization is neglected. Besides polarization, repulsion, and dispersion interactions are considered via a standard Lennard-Jones potential and static charge distributions are modeled via point charges. We developed a simple procedure, to investigate if polarizable force fields are able to capture the distinct difference in adsorption behavior between the varying metal ions. First, the force field is adjusted using two global scaling factors to reproduce the experimental adsorption isotherm for CO2 in Mg-MOF-74. Second, the force field with the same scaling factors is applied to simulate the adsorption of CO2 in M-MOF-74 with the remaining 9 types of metal ions. For the majority of metal ions, the simulation results are in reasonable agreement with experimental measurements. The affinity for CO2 varies depending on the metal ion integrated in the framework. Hence, the potential of polarizable force fields to describe MOFs incorporating open metals sites is shown. The next step will be to derive a consistent set of force field parameters from quantum mechanical calculations to develop a predictive model for CO2 adsorption.

465261 Adsorptive Separation of Ethane and Ethylene Using All-Silica Zeolites

Abstract:

Ethane and ethylene form a significant portion of the global sales of the petrochemical industry. Their separation is very challenging because these two compounds have very similar sizes, shapes, and self-interaction strengths. Industrially, the mixture is separated using cryogenic distillation, but separation factors range only from 1.2 to 2.3. Small separation factors necessitate very accurate molecular models in order to be able to computationally study potential separation processes. In this work, we develop improved force fields for these C2 species and use them to screen adsorption performance in all-silica zeolites.

460763 Hydrogen Adsorption on Metal Hexaboride Surfaces: An Ab Initio Study

Abstract:

Alkaline earth hexaborides are thermoelectric materials with unique thermophysical properties that have a broad variety of applications with great potential for new uses in fields such as lightweight armor development, gas storage, and n-type thermoelectrics. Considering their potential as possible hydrogen storage media, the adsorption properties of these materials have not yet been extensively studied. In this work, density functional theory (DFT) along with Car-Parrinello molecular dynamics (CPMD) is used to study the effects of hydrogen loading and surface terminations on hydrogen adsorption properties within these materials. Idealized (100) surfaces are determined using the slab method with various surface metal:boron ratios and geometries, and preferred geometries are found to depend on the particular metal. Hydrogen effects are studied with CPMD and self-consistent field calculations and show a theoretical maximum loading on the surface. These results provide very valuable insights on the physical and chemical mechanisms of hydrogen adsorption in metal hexaboride materials.

473100 Simulation of Gas Phase Supersaturation and Particulate Matter Formation in Post Combustion CO_2 capture Plants Using Aspen Plus

Abstract:

Over the past years CO_2 emission reducing technologies were widely investigated due to alarming climate change, particularly, global warming. One of the most promising and perspective methods for capturing CO_2 is applying Carbon Capture and Storage (CCS) of flue gases using reactive absorption based Post Combustion CO_2 Plants. One of the most popular solvents for absorbing CO_2 is a 30 wt.% aqueous solution of the monoethanolamine (MEA). Apart from mechanical losses, solvent loss due to evaporation and particulate matter (PM) phase emissions result in the increase in total operational cost and environmental impact. Prediction of PM formation in PCCC columns is a complex task requiring modelling of several steps including gas-phase supersaturation, PM formation (nucleation) and PM dynamics (growth and loss). Pre-existing PM into PCCC columns facilitates formation of new particles as a result of gas phase supersaturation. Hence, developing a simulation tool to understand the key parameters impacting the supersaturation of the gas phase in PCCC columns is of paramount importance. Few attempts were made to setup a simulation tool using Aspen Plus simulator to predict supersaturation of gas streams in PCCC columns in the absence of particulate phase, and in the presence of the flue gas PM. However, such simulation models still need to be improved to better understand the interactions between the particulate phase and other phases inside the column. In a recent study, equilibrium between the gas phase and the PM phase along the absorption column was assumed. However, the true mass and heat transfer between the particulate phase and gas phase need to be determined using a nonequilibrium (rate-based) model. The objective of this study is to firstly implement a rate-based model in Aspen plus to simulate cocurrent interactions between gas phase and particulate phase. The simulation results are validated using experimental data available in the literature. Then using the developed model, sensitivity analysis is perfomred to understand the effect of different operating conditions on MEA loss through the PM phase. Our preliminary data show

that the gas and particulate phases reach equailibrium along the given absorption column, and also MEA losses through PM phase is comparable to the MEA loss through the gas phase.

470137 Development of Functional Membranes for Use in Nuclear Forensics Applications

Abstract:

Clandestine activities involving the separation or manipulation of special nuclear material (SNM) for the purpose of developing a weapon of mass destruction is likely to result in the contamination of environmental waters. The capability to conduct isotopic analyses for waterborne SNM would be a powerful nuclear forensics tool. Despite widespread interest, there currently is no fieldable, separation technique capable of selectively concentrating uranium directly from groundwater for isotopic analysis.

This contribution describes our efforts to develop a high throughput analytical technique for waterborne isotopic analysis by using reactive, functional membranes as alpha spectroscopy substrates. Alpha spectroscopy is an analytical technique that allows for the discrimination of isotopes of uranium based on differences in characteristic energies for emitted alpha particles. In this work, alpha spectroscopy substrates were prepared by two methods: (1) physical deposition of a uranium-selective, water-soluble polymer film on ultrafiltration membranes and (2) grafting uranium-selective ligands from the surface of ultrafiltration membranes. Uranium was loaded onto the substrates by filtering uranium-contaminated water through the ultrafiltration membranes. The uranium-selective, water-soluble polymer was prepared by the copolymerization of 2-hydroxyethyl methacrylate and ethylene glycol methacrylate phosphate. The ligand-grafted membrane was prepared by the UV-polymerization of ethylene glycol methacrylate phosphate with N,N-methylene bisacrylimide. Membranes were characterized by Fourier-transform infrared spectroscopy before and after modification to support the deposition or grafting of the polymer on the membrane surface. The capacity for uranium, 1.9 mmol U/ g, was determined from equilibrium binding experiments.

The effect of membrane preparation method and membrane pore size on peak resolution in the alpha spectrum was investigated for pure uranium containing solutions at pH = 6. To mimic more realistic conditions, the selectivity of the membrane was tested using uranium-233 in simulated groundwater. Both uranium-coated membranes prepared from distilled water and groundwater showed resolutions of 40-60 keV in the alpha spectrum and detection efficiencies of 12% for uranium-233. The membranes showed both high resolution and fast preparation time. The permeability of the polymer-coated, ultrafiltration membranes (MWCO 100 kDa) was determined to be 3.74 LMH/kPa which means 2 L of contaminated groundwater can be processed per hour at 345 kPa (50 psi) in a 45 mm ultrafiltration cell. The results of this research are laying the foundation for the development of a novel rapid, field portable method for the separation and isotopic identification of waterborne SNM.

448583 Crystallization Separation of Na_2so_4 $\cdot 10H_2o$ and NaCl from Coal Gasification Water Based on Pseudo-Ternary System (Na_2SO_4+ NaCl +H_2O) Phase Equilibria

Abstract:

Due to high contents of Na_2SO_4 and NaCl, coal gasification wastewater (CGW) is of poor biodegradability and great difficulty to treat. At present, CGW is often delivered to evaporation ponds, producing huge amounts of mixed salts with impurities. With high water solubility, mixed salts can lead to secondary pollution and thus are identified as hazardous wastes which need further treatment. Under this background, a fractional crystallization method for obtaining high-purity $Na_2SO_4 \cdot 10H_2O$ and NaCl from CGW was developed based on phase equilibria.

The phase equilibria of pseudo-ternary system (Na_2SO_4 +NaCl+H_2O) in CGW were investigated at T= (268.15 to 373.15) K. The CGW was provided by a gas-making plant in the northwest of China. The solubility of $Na_2SO_4$and NaCl together with the density of the equilibrium liquid phase was determined by the isothermal solution saturation method. The equilibrium solids were investigated by the Schreinemaker's method of wet residues and X-ray powder diffraction. According to the experimental data, the phase diagrams and the diagrams of density versus composition were plotted and turned out to exhibit diverse characteristics in different temperature ranges, including (313.15 to 373.15) K, (293.15 to 303.15) K, (273.15 to 288.15) K and at 268.15K.

On the base of phase equilibria, a fractional crystallization method was developed to separate pure $Na_2SO_4 \cdot 10H_2O$ and pure NaCl. Purity analysis of the obtained products was carried out. It was found that both $Na_2SO_4 \cdot 10H_2O$ product and NaCl product were of high purity (≥98.5% by mass fraction) and their heavy metals contents were far

below identification standards for hazardous wastes. The obtained products could be used as raw materials by downstream industries. The fractional crystallization method showed great promise for the treatment of CGW and provided an important approach for solving the pollution problems caused by coal industries in China.

458840 Use of Real-Time Analytics and Apoptosis Assays for Enhanced Understanding of Cell Growth and Viability in a CHO-Based Process for Monoclonal Antibody Production

Abstract:

In a cell culture process for biopharmaceutical manufacturing, monitoring cell growth and viability is crucial for understanding cell behavior, as well as for determining cell culture duration and timing of harvest. Standard industry practice typically includes daily sampling of bioreactors for cell count and metabolite profiles. However, daily measurements are often too infrequent to detect rapid changes in metabolite concentrations and viability, and may not capture early indications of decline in cell health. We have evaluated the use of apoptosis assays as well as in-line monitoring techniques to gain a more complete understanding of cell growth, health and viability throughout the cell culture process.

In this study, we have applied markers of early, middle and late apoptosis to evaluate changes in cell growth, health and viability during a fed-batch process in 5L bioreactors. The markers that were monitored included 7-AAD, Caspases, and TUNEL assay for DNA nicking. Multiple early and mid-stage apoptotic markers were compared for their effectiveness in evaluating cell health and predicting cell growth trends in a bioreactor. Additionally, to provide more frequent cell growth measurements and metabolite readings to align with apoptotic assays, in-line probes were incorporated for real-time data analysis. Through calibration and method development, a capacitance-based Incyte probe was used for real-time monitoring of viable cell density (VCD), and an in-line Raman spectroscopy probe was used for real-time monitoring of glucose and lactate.

The results demonstrated that apoptotic markers can provide additional insight into cell health beyond what is learned from daily cell counts, thereby greatly deepening our understanding of cell culture conditions and process parameters during development. Markers of early apoptosis were observed several days prior to a drop in viability as measured by a commercial cell counter, while late-stage apoptotic markers were observed hours in advance or concurrently with the VCD and viability drop. Further insight into the health and productivity of the culture may be obtained by correlation of in-line metabolite and cell growth profiles with cell apoptotic progress.

In summary, optimization and application of real-time analytics and apoptosis assays in a CHO cell process can provide a rich and complete data set for better visualization of cell growth and health throughout the culture process, thereby enabling process improvement to maximize product titer and quality.

468928 Providing a Molecular Basis to Modeling Protein Adsorption to Multimodal Carbohydrate-Based Ligands

Abstract:

Recognition and binding of proteins to carbohydrate-based multimodal ligands plays a critical, but poorly understood, role in both fundamental cellular (e.g., cell glycocalyx and viral protein binding) and industrial processes (e.g., antibody drug bioseparations). Manufacturing costs of biopharmaceutical drugs can be drastically reduced through development of inexpensive multimodal (or mixed-mode) chromatographic bioprocesses for protein isolation and purification. However, unlike classical affinity or ion exchange based bioseparations that have been well studied, multimodal chromatographic (MMC) systems are still in their developmental stages [1]. Inexpensive and biocompatible MMC matrices/ligands derived from carbohydrates have tremendous potential for industrial bioseparations. However, we currently have a poor understanding of the molecular mechanisms driving protein binding to traditional carbohydrate (e.g., sepharose) and/or glycoprotein (e.g., concanavalin A) derived chromatographic matrices. This has been further stymied by the lack of suitable experimental tools and predictive adsorption models that explain the heterogeneous nature of protein-carbohydrate interactions. One of the challenges with traditional methods (e.g., solution depletion) used for monitoring protein adsorption is that bulk ensemble measurements provide limited insight into the underlying molecular mechanisms driving protein-ligand binding. Studying binding interactions of proteins and carbohydrate-based ligands at the single-molecule level is critical to explaining observed macroscale measurements, which will ultimately lead to the development of mechanistically-relevant protein macroscale adsorption models.

Here we study the adsorption of carbohydrate-binding proteins to two distinct model polysaccharide substrates with identical molecular composition but varying surface hydrophilicity using complementary bulk and single-molecule based adsorption techniques. We have modified a recently developed optical tweezers force spectroscopy technique [2] to characterize the single molecule protein binding-unbinding force dynamics to polysaccharide surfaces and probed the protein-ligand bond lifetimes at the single molecule level. Analysis of our records was used to identify trends in the force-binding lifetime datasets to find that the unbinding force and total binding lifetimes reflects the underlying heterogeneity of the available binding sites for carbohydrate-derived ligands. This distribution reflects the spectrum of varying affinity sites present on the polysaccharide surface and ultimately provides a molecular basis for using various multi-site Langmuir-type adsorption models to characterize protein adsorption to polysaccharide surfaces. Interestingly, simple regression analysis of bulk adsorption data using multi-site Langmuir-type models reveals that traditional protein bulk adsorption measurements are unable to provide data with sufficiently high resolution to discriminate between the low and high affinity protein binding sites for polysaccharides if the difference in the protein-carbohydrate ligand unbinding force is relatively small. In summary, we now have the necessary tools to systematically characterize protein adsorption to multimodal carbohydrate-derived ligands for optimizing industrial bioseparations using a first-principles approach.

References

[1] Pinto, I. F., Aires-Barros, M. R. & Azevedo, A. M. Multimodal chromatography: debottlenecking the downstream processing of monoclonal antibodies. Pharm. Bioprocess. 3, 263â€"279 (2015).

[2] Brady, S. K., Sreelatha, S., Feng, Y., Chundawat, S. P. S. & Lang, M. J. Cellobiohydrolase 1 from Trichoderma reesei degrades cellulose in single cellobiose steps. Nature Communications. 6, 10149 (2015).

472914 Operational Control Strategy Development Via First Principles Mapping

Abstract:

Product and Operational control strategies are essential elements in efficient and effective manufacturing of quality medicines. Many forms of guidance exist (e.g., PQLI) to help define and structure documented control strategies. Most control strategies are good at identifying the controlled variables and critical parameters that are essential to ensure product quality. What is often missed, however, is identifying the underlying first principle transformations that are the core of every deliberate step that is taken in making medicine. Each process step has an objective to be met. In fact, when we look at each process step, there are multiple objectives to be met. We must ensure the product quality is delivered, but we must also ensure that we do it cost effectively, protect the operator, protect the community, protect the environment, meet demand, etc. To manufacture the Product, we must design a Process Map as a collection and sequence of Process Stages and Process Operations that combine to deliver certain features and requirements of the product. The process operation is simply stated as an action (or transformation) on an object with a qualifier, e.g. *Disperse powder to less than 1% variation.* The mechanisms for transformation can always be traced to first principles of science. Thus, the Control Strategy is based upon key first principles and associated attributes and variables, at the most fundamental level. To satisfy the criteria of the transformation we must choose and implement effective Control Methods. For the example of "disperse powder", we might choose a vertical agitated tank that will deliver an adequate environment for the first principles attributes of shear and fluid velocity. For each process operation, we must consider objectives and control methods for all Features, such as quality, safety, cost, etc. In the case of Safety, for example, we can provide control methods for preventing dust explosions based upon the first principles of ignition source control.

This presentation will define and demonstrate the means to construct and implement a structured product control strategy via process mapping using first principles-based transformations and chosen control methods.

460708 Nanoextrusion: A Platform Enabling Comparative Assessment of Nanocomposites Vs. Amorphous Solid Dispersions for Drug Dissolution Enhancement

Abstract:

A relatively novel nanoextrusion process was used to produce extrudates of griseofulvin (GF), a poorly water-soluble drug, in a polymeric matrix with the goal of enhancing the dissolution rate and thus the bioavailability. Unlike a traditional hot melt extrusion (HME) process, nanoextrusion process (Baumgartner et al., 2014; Khinast et al., 2013) uses a wet-milled suspension of the drug as a feed along with an extrusion polymer and disperses the drug nanoparticles in the polymeric matrix while evaporating water, thus yielding extrudates in the form of nanocomposites. Simply, nanoextrusion process acts as a continuous drier unlike the traditional HME process. Being a continuous process and having the capability to handle viscous drug suspensions, it could be advantageous over various pharmaceutical drying operations in the manufacturing of solid nanoparticle-based formulations. In this study, we propose that as another potential advantage of the nanoextrusion, extrudates with different forms of the drug can be prepared by the same process via use of different extrusion polymers: nanocrystalline drug dispersed in the polymeric matrix (nanocomposites) or amorphous drug molecularly dispersed with the polymer (amorphous solid dispersion, ASD). It is hypothesized that the drug nanocrystals in the nanocomposites dissolve faster than the amorphous drug in the ASD when the drug dose is so low not to cause supersaturation in the dissolution medium (low drug dose). Specifically, hydroxypropyl cellulose (HPC) and Soluplus were used to stabilize wet-milled suspensions and to form the matrix of the extrudates. Feed suspensions of GF were prepared by wet stirred media milling using HPC and Soluplus in the presence of sodium dodecyl sulfate (SDS), an anionic surfactant. The milled suspensions along with additional polymer (HPC/Soluplus) were fed to a co-rotating twin-screw extruder, which dried the suspensions and formed various extrudates. The extrudates were milled into powders via mortar–pestle for further analysis. The effects of drug particle size in the feed suspensions, stability of drug suspensions, and polymeric matrix type–size on the drug dissolution from the milled extrudates were examined. The suspensions were characterized via laser diffraction, while the extrudates and their milled powders were characterized by laser diffraction, SEM, and XRD. Digital microscopy was used to visualize the changes of different polymeric matrices when exposed to water. Statistical analysis with Korsmeyer-Peppas model helped us to identify the drug release mechanisms. By using polymers with different interactions/miscibility with the drug, two different forms of the drug were formed: extrudates with nano/micro-crystalline drug particles dispersed in the HPC matrix as a secondary phase (nanocomposites or microcomposites) and extrudates with amorphous drug molecularly dispersed within the Soluplus matrix (amorphous solid dispersion). In the composites, drug particle size was the dominant factor that affects the dissolution performance, whereas the matrix size appears to play a significant role in the drug dissolution performance in the amorphous solid dispersion. We have demonstrated that for low-dose drugs, e.g. highly potent poorly soluble drugs, drug nanoparticles in the nanocomposites may actually dissolve faster than amorphous form of the drug in an amorphous solid dispersion, depending on the size of the polymeric matrix.

References

Baumgartner, R., Eitzlmayr, A., Matsko, N., Tetyczka, C., Khinast, J., Roblegg, E., 2014. Nano-extrusion: A promising tool for continuous manufacturing of solid nano-formulations. Int. J. Pharm. 477, 1 11.

Khinast, J., Baumgartner, R., Roblegg, E., 2013. Nano-extrusion: a one-step process for manufacturing of solid nanoparticle formulations directly from the liquid phase. AAPS PharmSciTech 14, 601 604.

468257 Evaporation of Binary Mixtures and Shell Formation in Spray Dried Droplets

Abstract:

Evaporation of binary mixtures and shell formation in spray dried droplets

P.C. Valente[1], Í. Duarte[1], T. Porfirio[1], X. Liu[2], F. Zhang[2] & M.N. Temtem[1]

[1] R&D Drug Product Development, Hovione Farmaciência S.A, Sete Casas, 2674-506 Loures, Portugal;

[2] College of Pharmacy, University of Texas at Austin, U.S.A;

Amorphous solid dispersions (ASDs) of poorly water soluble active pharmaceutical ingredients (APIs) in hydrophilic polymeric matrices are a commonly used strategy to increase the solubility and dissolution rate of oral-dosage forms [1]. However, since the polymer is hydrophilic and the API is poorly water soluble, it is often difficult to find a common solvent for both and a mixture of co-solvents is used instead. Solvent mixtures also offer the possibility of influencing particle morphology and drying rate [2] and have been reported to offer advantages in the solid state drug-polymer miscibility and on the stability of the ASDs [3,4].

To fully explore the potential of spray dried ASDs from co-solvent drug-polymer solutions, a fundamental understanding on the solvent chemistry, droplet composition history throughout drying and its impact on the drug-polymer interaction, stability and phase behavior are required. These mechanisms are, in turn, strongly influenced by the process conditions, such as thermodynamic layout, atomizing characteristics and spray drier geometry.

The focus of the present work lies on the evaporative dynamics of binary solvent droplets and how they are influenced by process conditions. Unlike the single component counterpart, modelling the evaporation of binary mixtures depends not only on the process parameters, but also on the mass (and momentum) transfer within the droplet [5]. For example, if one considers solely the molecular diffusion for solvent mixtures with starkly different vaporization enthalpies, the evaporation rate of each solvent will be proportional to their initial mass fraction and, counter-intuitively, will be independent of the vaporization enthalpies [6]. However, whenever there is a non-negligible relative velocity between the droplet and the surrounding drying gas, the internal circulation within the droplet can lead to a steep increase in the effective mass diffusion rate. The evaporation rate of each solvent will also strongly depend on their vaporization enthalpy (see Figure 1). This leads to the more intuitive case where the more volatile solvent tends to evaporate first.

We show that these two limits lead to dramatically different drying kinetics and can strongly affect the onset of drug/polymer precipitation and shell formation. After shell formation the drying kinetics tend to be strongly modified, highlighting the importance in predicting the onset of shell formation to bound the applicability of the drying models. These two limiting cases of infinitely slow or infinitely fast droplet component diffusion may be used to formulate scalar (0D) evaporation models [5,6]. In practical spray drying situations, both limits can be attained depending on the binary mass diffusion coefficients, atomization characteristics, temperature profile or drying gas flow velocity, to name a few parameters. These evaporation models are particularly useful when coupled to a numerical solver to predict stability and phase separation of amorphous solid dispersions [7,8] or a numerical solver of the Navier-Stokes equations, commonly denoted as computer fluid dynamics (CFD) [9,10].

We extend the previously developed platform for phase separation prediction of drug-polymer-solvent systems to systems with binary solvent blends in an effort to generalize a screening methodology for ASDs and compare the data against experimental results.

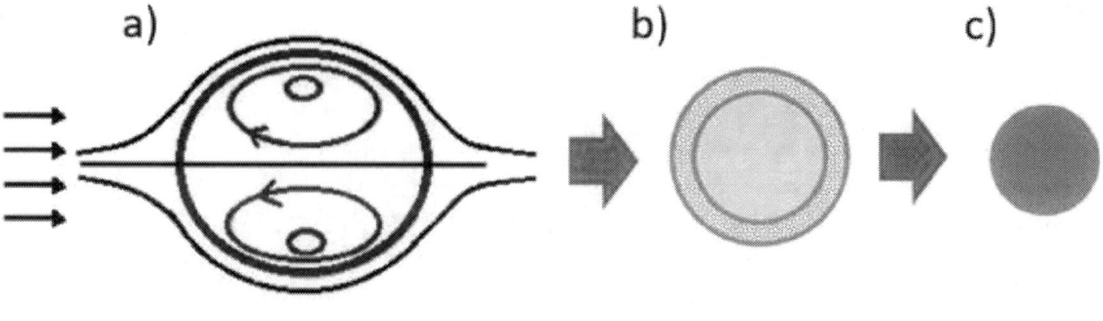

Figure 1: Sketch of various stages throughout the droplet drying history: a) droplet with internal circulation caused by the relative velocity between the droplet and the surrounding gas, b) shell formation and c) the final particle.

References

[1] Perrie Y., Rades T. "Pharmaceutics - Drug Delivery and Targeting", Pharmaceutical Press, 2010.

[2] Wulsten, E., Kiekens, F, van Dycke, F., Voorspoels, J. & Lee, G., "Levitated single-droplet drying: Case study with itraconazole dried in binary organic solvent mixtures", Int. J. Pharmaceutics, 2009, 378(1-2) pp. 116-121

[3] Janssens, S, Anné, M., Rombaut, P. & Van den Mooter, G., "Spray drying from complex solvent systems broadens the applicability of Kollicoat IR as a carrier in the formulation of solid dispersions", Eur. J. Pharm. Sci. 2009, 237, pp. 241-248.

[4] Paudel, A. & Van den Mooter, G., "Influence of Solvent Composition on the Miscibility and Physical

Stability of Naproxen/PVP K 25 Solid Dispersions Prepared by Cosolvent Spray-Drying", Pharm. Res. 2012, 29, pp. 251-270.

[5] Sirignano, W. A. "Fluid dynamics and transport of droplets and sprays", Cambridge University Press (2nd Ed.), 2010.

[6] Law, C. K. "Recent advances in droplet vaporization and combustion", Prog. Energy Combust. Sci, 1982, 8 pp. 171 – 201, 1982.

[7] Duarte, Í., Santos, J. L, Pinto, J. G & Temtem, M., "Screening methodologies for the development of spray-dried amorphous solid dispersions", Pharm. Res., 2015, 32 pp. 222 – 237, 2015.

[8] Prudic, A., Ji, Y. & Sadowski, G., "Thermodynamic Phase Behavior of API/Polymer Solid Dispersions", Mol. Pharmaceutics, 2014, 11(7), pp. 2294-2304

[9] Sazhin, S.S et al., "A simplified model for bi-component droplet heating and evaporation", Int. J. Heat Mass Transfer, 2010, 21 pp. 4495-4505

[10] Maqua, C., Castanet, G. & Lemoine, F., "Bicomponent droplets evaporation: Temperature measurements and modelling", Fuel 2008, 87(13-14), pp. 2932-2942.

455382 Impact of Tooling Design during Development and Risk Evaluation of Tri-Layer Tablet Formulations

Abstract:

Multi-layer tablet is one of the formulation strategies utilized to mitigate chemical and/or physical incompatibility between different drug substances when delivering fixed dose combination drugs. Compressing multi-layer tablets has significant challenges because the interface(s) between neighboring layers are subject to defects such as delamination. It is therefore important to understand the factors that contribute to the strength of the interfaces. During early stages of drug development, feasibility studies are often conducted by making multi-layer compacts with round flat-faced tooling as the exact size and shape for the commercial tablets may not be finalized at that point. However, the link between different tooling designs and multi-layer tablet performance is not well established. This study examines the major engineering considerations when using mini-piloting tools to gauge the risk of interfacial defects in a tri-layer tablet. The first layer (layer A) of the tri-layer tablet was a wet granulation formulation of 56% drug load. The middle layer (layer M) was the inert layer consisting of a placebo blend. The third layer (layer B) was a dry granulation formulation of 33% drug load. The impact of tablet weight and dimensions (diameter and thickness) were evaluated to gain understanding during scale up/down as needed in early feasibility assessment and in troubleshooting. The factors in tooling selection, including tablet shape (round vs. oval), cup depth (0 vs. 0.054â€), and embossing (small vs. large), were evaluated to gain insight on the impact of tooling design on the interface strength of the tri-layer tablet. Tablets were compressed under different main compression forces while keeping the first and second layer tamping forces constant. Axial test was performed to quantify the interface strength under tensile stress. Â The thickness, diameter, and aspect ratio of the tablet were found to have a profound impact on the interface strength at a given compression pressure and on the compression pressure under which the interface start to fail, causing tablet delamination. These were the results of combined factors of the compressibility of the individual layers, the force transmission through the layers during compression (effect of tablet thickness), and the ejection force (effect of aspect ratio). The presence of a 0.054â€ cup depth was found to lower the interface strength to less than 50% of that of a flat tablet. This cup depth effect is consistent with the observations made in the monolithic tablet regime and can be explained by the non-uniform density distribution caused by the curvature of the tablet interface. The oval shaped tablets were found to reduce the interface strength by 30 N or more compared to the round shaped tablets, suggesting that compacts made with round flat faced toolings may provide false positive results when used to assess the interface risks of a multi-layer tablet. Large embossing on the tooling appeared to help enhancing the interface strength, probably by providing more surface areas for the neighboring layers to interlock. This study elucidates the effects of scale-related factors like tablet weight and dimensions on the interface strength of a tri-layer tablet. These effects, not previously well studied or even realized during development, can significantly impact the feasibility evaluation of a multi-layer tablet. It also highlights the importance of tooling design selection on the performance of a tri-layer tablet that must be accounted for when using minipiloting tools for accelerated development of fixed dose combination products.

471306 End-Point Determination and Scale-up/Scale-Down of High Shear Wet Granulation Process Using an in-Line Di-Electric Probe

Abstract:

The main goal of this study is to demonstrate anovel in-line process analytical technology (PAT) for detecting end-point of granulation in a high shear wet granulation (HSWG) process. This technique enabled us to monitor liquid distribution and granule densification across different scales. While many reports in the literature have focused on indirect techniques such as power consumption or torque (which are scale and geometry dependent), in this method a physical property (dielectric permittivity of medium) dependent on the liquid amount and granule porosity (or density) is monitored directly during the course of granulation.

A di-electric probe was modified for recording complex dielectric permittivity of wet mass. The output of the probe was monitored and recorded during liquid addition and wet massing, and samples were taken at corresponding time intervals to correlate the final property of granules to the instrument output. Dependence of the real part of the effective dielectric permittivity on solid fraction of moist powder was suggested by Louge et. al. (1997) for non-spherical particles (snow). We adopted this model for high shear wet granulation process and monitored (in-line)Â density of wet mass during granulation.

Use of this in-line characterization tool enabled us to successfully transfer a 50% drug load API to a 600 L top-driven granulator without the need for any development batch by monitoring granulation end-point based on physical properties of wet mass. Specifically, we could understand the impact of changing process parameters (impeller speed, spray time and wet massing) and API properties (particle size /density) on final granule attributes and tablet CQAâ€™s which was consistent during scale-up/down to geometrically similar or non-similar granulators.

M.Y. Louge, R.Â Steiner, S.C. Keast, R. Decker, J. Dent, and M. Schneebeli, "Application of capacitance

instrumentation to the measurement of density and velocity of flowing snow", *Cold Regions Sci. & Tech.*, 25, 47-63 (1997).

466340 Modeling of Taylor-Couette Disc Contactors

Abstract:

Modeling of Taylor-Couette Disc Contactors (TCDC)

Annika Grafschafter and Matthaeus Siebenhofer

The Taylor-Couette Disc Contactor (TCDC) is a liquid-liquid extraction column [1]. From the hydrodynamic point of view, it is a hybrid of a classical Rotating Disc Contactor (RDC) and a Taylor-Couette Reactor. The TCDC design is based on RDC design, but without stator rings and an increased shaft diameter. The flow pattern in the single compartment is induced by centrifugal force, counteracted by friction of the continuous phase. It compares well to banded two phase flow of a Taylor-Couette reactor. The modeling of two-phase flow in chemical process applications is not an extraordinary task when it is possible to predict the specific mass transfer area and the hydraulic operation range. Two-phase contactors can be appropriately designed by accurate prediction of the hydrodynamic characteristics, e.g. the mean sauter diameter and dispersed-phase hold-up, giving access to the basic design of the plug flow reactor (PFR) and the continuously stirred tank reactor (CSTR) cascade. For the design of stirred liquid-liquid extractions columns literature recommends application of the PFR model. To consider axial back mixing, the classical HTU-NTU concept as deduced from the PFR model, is extended with a height of dispersion unit (HDU-value) with the aid of axial dispersion coefficients [2]. However, the axial dispersion coefficient of both phases is difficult to determine.

Due to the formation of toroidal vortexes (banded two phase flow), each compartment of the TCDC rather corresponds to two ideal stirred tank reactors in series with ideal mixing, suggesting CSTR cascade design.

Modeling of hydraulics, including the prediction of the stirring speed, the drop size distribution and mean sauter diameter as well as hold up, flooding and the specific power demand as well as the residence time distribution of the continuous phase [1] and the disperse phase [3] in the single compartment and the column has been validated with

experiments. Mass transfer experiments with the systems n-butanol-undecane-water and acetic acid-Cyanex 923-water have confirmed that the concentration profile along the column height and the separation efficiency can be precisely modeled with CSTR cascade design

[1] Aksamija E., (2015), Der Taylor-Couette Disc Contactor; ein vereinfachtes und optimiertes Design von Drehscheibenextraktoren, PhD Thesis, University of Technology Graz

[2] Mersmann A., Kind M., Stichlmaier J., (2005), Thermische Verfahrenstechnik, Springer-Verlag Berlin Heidelberg, p.382 â€“ 383

[3] Sarzio R., (2015), Experimentelle Validierung der CFD-Simulationen mehrphasigen StrÃ¶mung

in der RDC Kolonne, Master Thesis, University of Technology Graz

475160 Effects of Physical Properties and Phase Dispersion on Sieve Tray Extractor Performance

Abstract:

Sieve trays are very often used in commercial-scale liquid extraction applications in the chemical and petrochemical industries because of their reliable scale-up, absence of moving parts and ability to handle high throughputs. Properly designed sieve trays eliminate backmixing and provide for the reformation of drops. The hydraulic and mass transfer characteristics depend on many parameters such as the tray geometry, physical properties and which phase is dispersed. In particular, liquid viscosity and interfacial tension can have a significant effect on mass transfer and drop coalescence. Published sieve tray extraction data are quite limited regarding the variation of interfacial tension, solvent viscosity and density. The overwhelming majority of available data involve an aqueous phase and a hydrocarbon phase with very little information published on two immiscible hydrocarbon phases.

The paper expands the publically available database on sieve tray extraction performance with the study of the sulfolane/n-heptane test system. Toluene was chosen as the transferring solute. In this study, following effects were studied:

- flow velocities
- addition of transferring solute
- mass transfer direction
- which phase is dispersed
- coalescer material of construction
- density difference, interfacial tension and sulfolane viscosity

In this work, the hydraulic and mass transfer results are compared with the SRP predictive models for sieve tray

extraction design.

461743 Liquid-Liquid Extraction in Intensified Small Scale Units

Abstract:

Liquid-Liquid Extraction in Intensified Small Scale Units

Garciadiego Ortega Eduardo, Tsaoulidis Dimitrios, Angeli Panagiota*
Department of Chemical Engineering, University College London, Torrington Place, WC1E 7JE, London, UK. * Corresponding author; e-mail: p.angeli@ucl.ac.uk

ABSTRACT

Intensification of solvent extraction processes leads to reduction of costs and solvent inventories, and improves the safety and sustainability of the operations in a wide range of industries. The use of small channels as contactors improves mixing and increases the surface area available for mass transfer as compared to conventional contactors. Extractions performed in millimetric channels operating in slug flow regime, show high volumetric mass transfer coefficients (k_La) and extraction efficiencies (%E) [1,2]. These studies have shown the links between hydrodynamic characteristics like plug lengths (and thus interfacial area) or internal circulation patterns in each phase with the mass transfer performance of the contactors.

The throughput achieved in one channel, however, is small compared to industrial rates. To increase throughput in small scale devices, scale out (instead of scale up) is used; this can however, require a prohibitively large number of small channels. To overcome this problem we followed two strategies. In the first strategy, the size of the channels is increased enough to allow large throughputs while still preserving the benefits of small scale operation, such as laminar flows and well defined, regular patterns. Scale out to desired throughputs would then involve a smaller number of channels. In the second one, the velocities of the two liquid streams are increased when they are brought into contact (impinging-jets mixer). It has been found that when jets of the same phase collide very high energy dissipation rates are achieved, which can be almost two orders of magnitude higher than in conventional contactors [3]. Impinging-jet mixers have not been studied for jets of immiscible liquids, but the high energy dissipation rates are expected to produce very small drops and increased interfacial areas.

For the scale-out strategy, the main challenge is to develop manifolds which produce uniform flow distribution in the many parallel channels. The two single phases can be distributed in separate manifolds and fed into the contacting channels (*split-combine*), or alternatively, the two phases can be first mixed before the two-phase mixture is distributed into several channels (*combine-split*). Both these configurations have been studied for gas-liquid flows operating in Taylor flow regime [4, 5]. Al-Rawashdeh et al. (2012) found for the split-combine approach that the manifold was successful for only one flow rate condition [4]. For the combine-split approach good flow distribution was again achieved only at limited conditions [5]. Neither of these approaches has been tested for liquid-liquid flows.

The scale up of liquid-liquid flows in small channels was investigated in this work following both strategies. Water and kerosene were used as the immiscible liquids. Flow patterns and drop sizes were studied with high speed imaging. For the manifold studies both the split-combine and combine-split concepts were used to construct set-ups with 2 and 4 outlet channels with 1 and 2 mm internal diameter. For the design of the impinging-jet mixer (Figure 1) two different nozzles have been used (0.25 and 0.5 mm internal diameter) and two different mixing chambers (2 and 3 mm). Drop size measurements showed that interfacial area in the case of the impinging-jet mixer is increased up to 8 times compared to T-junction capillaries for the same operating conditions. In addition, it was found that smaller droplets formed with increasing energy dissipations. The goal of the current work is to identify the different flow configurations that produce small drops with narrow size distribution.

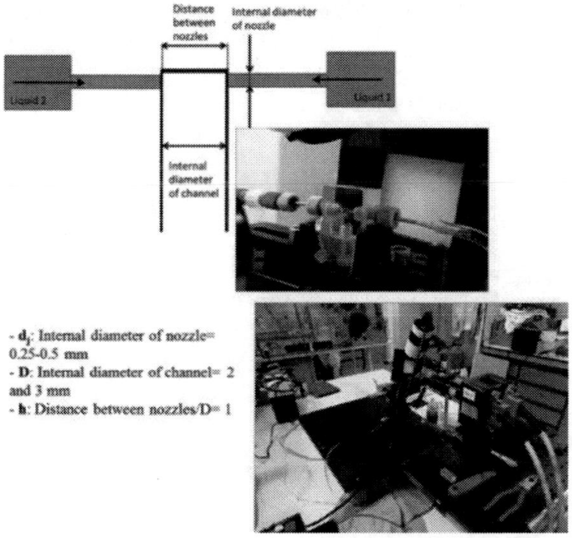

Figure 1. Schematic and photograph of the experimental setup of the confined impinging-jet mixer.

References:

[1] Tsaoulidis D, Dore V, Angeli P, Plechkova NV, Seddon KR. Dioxouranium(VI) extraction in microchannels using ionic liquids. *Chem Eng J*. 2013;227:151-157. doi:10.1016/j.cej.2012.08.064

[2] Li Q, Angeli P. Intensified Eu(III)extraction using ionic liquids in small channels. *Chem Eng Sci*. 2016;143:276-286. doi:10.1016/j.ces.2016.01.004

[3] Saien J, Zonouzian AE, Dehkordi AM. Investigation of a two impinging-jets contacting device for liquid-liquid extraction processes. *Chem. Eng. Sci*. 2006;61:3942-3950. doi:10.1016/j.ces.2006.01.034

[4] Al-Rawashdeh M, Fluitsma LJM, Nijhuis TA, Rebrov EV, Hessel V, Schouten JC. Design criteria for a barrier-based gas–liquid flow distributor for parallel microchannels. *Chem Eng J*. 2012;181-182, 549-556. doi:10.1016/j.cej.2011.11.086

[5] Hoang DA, Haringa C, Portela LM, Kreutzer MT, Kleijn CR, Van Steijn V. Design and characterization of bubble-splitting distributor for scaled-out multiphase microreactors. *Chem Eng J*. 2014;236:545-554. doi:10.1016/j.cej.2013.08.066

470170 Hydrodynamics of Two-Phase Ionic Liquid Solvent Systems in Countercurrent Chromatography

Abstract:

In industry, current approaches for liquid-liquid extraction separations often involve the use of organic solvents which have low flash points and are inherently unsafe. The long residence times, and the large volumes in the mixer-settler or pulsed column extraction units that are most commonly used, further increase the high hazard ratings for the process. A promising intensified separation technology is high-performance countercurrent chromatography (CCC). It is a form of liquid-liquid chromatography that takes places along a continuous length of tubing, which is wound around a drum (called a bobbin). Within the CCC column, i.e. the coiled tubing, one of the liquid phases is held stationary by centrifugal forces generated as a result of rotating the bobbin in a planetary motion, while the second (mobile) phase is then continuously pumped through it in such a way that there is good retention of the stationary phase. Rotation of the column results in a variable centrifugal field. This creates a series of simultaneous mixing and settling zones along the length of tubing, promoting solute transfer between the phases, and therefore, separation of species with different partition coefficients. Due to the high-speed rotation of the column, a large number of stages can be generated in a single machine. For example, a sample injected into the mobile phase can experience 800 repetitive mixing and settling steps in a minute when the motor rotor speed is 800 rpm, thus promoting highly efficient separations.

In this work, the application of ionic liquid biphasic solvent systems in a custom-designed CCC device has been investigated for the potential separation and recovery of valuable and intrinsically important metals, such as uranium, thorium, lanthanides and noble metals. Due to their dual nature and unique physiochemical properties, ionic liquids have gained increasing attention in recent years as potentially suitable candidates for replacement of volatile organic solvents in liquid-liquid extractions. Some of their most favourable features include their negligible vapour pressure, chemical and thermal stability, nonflammability and their adjustable miscibility and polarity. The unique combination of the two technologies, ionic liquids and CCC, represents a new and exciting philosophy to liquid-liquid separations. It overcomes the problems associated with the extractive process related to the use of organic solvents (flammability, volatility) and to the performance of the equipment (large inventories, long residence times) and will result in safer, more environmentally friendly approaches.

The viscosity of ionic liquids is often higher than that of common molecular solvents and can pose significant problems for traditional CCC devices, which are mostly low pressure. As a result, to date, their application to CCC has been very limited and greater insight into the hydyrodynamics that govern or control the extraction process is required. To investigate the hydrodynamic behaviour of the ionic liquid biphasic solvent system, visual observations have been carried out on a transparent 2-D spiral CCC column at different angles around the central axis of rotation using a high-speed camera. The images acquired have revealed the phase distribution and the mixing and settling areas within the column undergoing planetary motion, as well as the interfacial characteristics that can promote mass transfer (such as waves). A necessary condition to separate solutes using the CCC technique is that the column retains some of the stationary phase. Therefore, the influence of several operational parameters on the volume fraction of stationary phase retained was also studied including mobile phase flow rate, mobile phase pumping direction (Head --> Tail or Tail --> Head), rotational speed and direction of rotation. The settling and mixing characteristics of the two-phase mixture, and phase retention within the coils have been related to the densities, viscosities and interfacial tension of the ionic liquid solvent systems employed.

Acknowledgements

The project was funded by the UK Engineering and Physical Science Research Council (EPSRC) (Project title: Application of ionic liquid-liquid chromatography (ILLC) to extraction of metals), in collaboration with QUILL Research Centre at The Queen's University of Belfast, Northern Ireland, U.K.

469059 Microfluidic and Millifluidic Separations for Recovery of Plutonium from Residues

Abstract:

Operations within the plutonium facility at Los Alamos National Laboratory routinely generate residues that are too rich in plutonium to discard, but not pure enough to be used as feed for the electrorefining purification process. These residues are currently purified for electrorefining using either a nitric acid-based anion exchange process or a hydrochloric acid-based solvent extraction process. While historically successful, these processes were sized for higher throughputs rather than inherent criticality safety, and the flammable solvents and resins require risk management.

A multi-disciplinary team at Los Alamos is testing milli- and microfluidic contactors for use in the chemical separation of actinides. These highly efficient separations at short length-scales have the potential to purify plutonium residues with a smaller footprint than conventional aqueous processing. The slow flowrates required for millifluidic processing allows smaller volumes of solution to be used, which makes it easier to design a process that is inherently safe from a nuclear criticality accident. This in turn makes automation feasible. If a need for increased capacity returns, the process can be scaled up by adding parallel systems. By contacting the aqueous phase with a liquid anion exchanger (Aliquat 336) in an ionic liquid carrier, the flammable solvents and resins can be eliminated. Additionally, higher efficiency separations can result in a decreased volume of liquid transuranic waste requiring disposal.

Microfluidic contactors using co-current (slug flow) and counter-current (membrane separated) configurations were fabricated and tested. A higher throughput, commercially available milliliter-scale unit has been purchased for production scale testing. (LA-UR-16-23287)

459330 Future Production of Americium-241 Oxide at Los Alamos National Laboratory

Abstract:

Future Production of Americium-241 Oxide at Los Alamos National Laboratory

Mary Ann Stroud, Louis Schulte, David Kimball and Brad Skidmore

Actinide Engineering and Science Group, Los Alamos National Laboratory, Los Alamos, NM

LA-UR-16-23214

Americium-241 oxide has been produced at United States Department of Energy (DOE) sites in the past with specification requirements of chemical purity of > 95 wt% AmO_2 and ^{241} Am isotopic purity > 99 wt%. This stockpile has been depleted and there is currently no domestic supplier of $^{241}AmO_2$. An ongoing need exists in industry for $^{241}AmO_2$, to be used primarily as Am-Be neutron sources for oil and gas well logging. The DOE Office of Science Isotope Program has funded Los Alamos National Laboratory (LANL) to re- establish a capability for ^{241} AmO_2 production at the LANL Plutonium Facility (LANL-PF).

Am rich residues are a readily available by-product of LANL-PF Pu metal purification operations. LANL-PF aqueous chloride operations efficiently separate the bulk of the Pu from these Am rich residues using solvent extraction. LANL-PF is in the process of starting up the CLEAR (Chloride Extraction and Actinide Recovery) glovebox line to produce purified $^{241}AmO_2$ from the resulting Am rich effluent streams. The CLEAR line will use extraction chromatography (EC) to purify and recovery the ^{241}Am from the effluent stream. The EC operation uses di-(4-*t*-butylphenyl)-N,N-diisobutylcarbamoylmethylphosphine oxide $(4\text{-}t\text{-}BuC_6H_4)_2P(O)CH_2C(O)N(i\text{-}Bu)_2$ as an extractant, along with diamyl amylphosphonate [DAAP, $(C_5H_{11})_2P(O)OC_5H_{11}$] as a diluent, on an inert resin bead. The EC columns provide a simple, reusable method of concentrating actinides from HCl waste effluent streams. The new CLEAR line will also add ion exchange (IX) columns using Reillex-HPQ resin to recover and purify Pu. Incorporation of EC and IX requires minor additional equipment, and the processes are relatively straightforward to run. Large scale EC runs on HCl process effluents have demonstrated efficient recoveries (~99%) of Am and Pu from process effluent streams.

464920 Evaluation of the Mass Transfer Rate in Columns of Small Diameter Zeolite Particles

Abstract:

The mass transfer rate in a zeolite adsorbent column is generally assumed to be controlled by intraparticle diffusion, which is a reasonable assumption for large diameter particles (> 1 mm). Under this assumption, breakthrough curves can be used to determine a mass transfer coefficient (MTC) with a linear driving force (LDF), often assumed constant in column dynamics models. A similar assumption is often applied to the mass transfer rate of smaller diameter zeolite particles (< 1 mm). However, since the diffusion path inside the particle is significantly reduced, the assumption of intraparticle diffusion controlling the mass transfer rate is no longer necessarily valid. By reducing the internal diffusion resistance of the particle, mass transfer external to the particle becomes more prominent in determining the overall adsorption rate.

In this presentation, we will evaluate if the assumption of intraparticle diffusional resistance as rate limiting is appropriate for small diameter X-zeolite particles. Breakthrough curves were measured over a range of gas velocities and pressures for a commercial Li-X zeolite (~0.5 mm in diameter). The mass transfer zone (MTZ) lengths calculated from the breakthrough curves provide insight into the mechanism controlling the mass transfer rate. Additionally, the breakthrough curves were matched to simulated breakthrough curves to determine if a constant MTC is suitable for the zeolite particle.

468484 A Simple Graphical Check of Consistency for Zero Length Column Desorption Curves

Abstract:

The mass balance for a zero length column (ZLC) system leads to a very simple check which can be used to ensure internal consistency of the experimental curves. This is based on reporting experiments run at different flowrates on a c/c_0 vs Ft plot, where F is the volumetric flowrate. The general theory of the ZLC will be used to identify what characteristics should be shown by a valid set of experimental runs at different flowrates. The proposed method is applied to two different sets of results for carbon dioxide on 13X zeolite reported in both the c/c_0 vs t and the c/c_0 vs Ft plots. The consistency check is used to identify by simple inspection of the experimental curves if the data are reliable. The results lead to an important recommendation: any future contribution reporting ZLC curves should include the c/c_0 vs Ft plot which can confirm the internal consistency of the results and indicate if the experiments are carried out under kinetic control and hence a mass transfer coefficient can be measured reliably.

470966 Understanding the Transport of Gases in Carbon Molecular Sieve Using the Volumetric Frequency Response Technique

Abstract:

Frequency response methods have been developed as an alternative technique for the determination of mass transfer mechanisms controlling the transport of gases into and out of microporous adsorbent materials. Recent results in the literature are showing that this technique seems to be the best for discriminating the controlling mechanism. It is also possible to reveal two or more different controlling mechanisms that may dominate in different frequency ranges. The present talk will focus on the use of the volumetric frequency response (VFR) approach for the identification on the main mechanism governing the transport of gases such as O_2, N_2 and CO_2 in pelletized and ground molecular sieve carbon (MSC 3K 172 from Japan Enviro Chemicals, Osaka, Japan). In a recently commissioned VFR apparatus, experiments were carried out with pure gases at pressures of 200, 400 and 750 Torr, temperatures of 20, 30, 40 and 50 °C and frequencies between 0.7×10^{-5} and 10 Hz using pelletized CMS. Additional experiments were conducted at 750 Torr and 30 °C using a ground sample of this CMS with particle sizes between 30-50 Tyler mesh. The results are showing that the transport mechanism can be unequivocally identified when fitting the frequency response curves to various transport models. The results are also revealing that conducting FR experiments at different temperatures and pressures, as well as using different particle sizes helps to ensure that the correct transport mechanism has been identified. During this presentation the latest results obtained from the system will be presented and interpreted.

452608 Fixed Bed Modeling Revisited; Generalized Solutions for S-Shaped Isotherms

Abstract:

Both theoretical models and experiments that study the effects of the shape of the equilibrium isotherm on the fixed bed dynamics are limited to systems which obey rectangular (irreversible), linear and favourable equilibrium isotherms rather than unfavourable or more complicated equilibrium isotherms, such as S-shaped (Type IV or V). Furthermore, most modeling studies focus on the application of models on particular systems, hence, the derived conclusions are either system-specific or they are based on phenomenological models which exhibit no physical significance to the experimental setup. Moreover, the theoretical models are further restricted as they are solved under certain assumptions; mostly constant pattern and ideal plug flow. This paper presents results on rather unusual fixed bed dynamics, i.e. for systems that follow S-shaped isotherms, and fixed beds that operate in non-asymptotic and non-ideal flow conditions. The solutions are generalized by considering models that are based on classical dimensionless parameters, and the derived conclusions are applicable to any system that obeys the underlying assumptions.

473077 Study of Adsorptive Reactor (AR): Dynamic Multi-Scale (Catalyst /Adsorbent/Reactor Scale) Modeling and Simulation

Abstract:

Hybrid configurations of coupling reaction systems with some form of separation systems substantially improve reactant conversion or product selectivity and, for reversible reactions, establish a more favorable reaction equilibrium than that which could be achieved under conventional reactor operation.

Catalyst-adsorbent configurations with fixed-bed columns can be employed as an Adsorptive Reactor (AR) for simultaneous reaction and separation. This study presents Dynamic Multi-Scale (Catalyst /Adsorbent/Reactor Scale) Adsorptive Reactor Modeling and Simulation. First, a single-catalyst pellet isothermal/non-isothermal dynamic and a single-adsorbent pellet isothermal/non -isothermal dynamic simulations are performed. Then, these pellets are coupled with an isothermal/non-isothermal dynamic packed-bed reactor model to form a hybrid multi-scale Adsorptive Reactor (AR) model. The catalyst/adsorbent pellets simulation is repeatedly carried out along the reactor bed length and is coupled with a reactor model that captures species transport/reaction along the reactor length. The velocity and species' concentration profiles along the Adsorptive Reactor length are captured by momentum/species transport models accounting for convection/reaction /diffusion mechanisms. In the derivation of the model's equations, the Reynolds Transport Theorem was applied separately to each of the domains; the catalyst pellet's domain, the adsorbent pellet's domain and the adsorptive reactor's domain.

466779 Simulation and Optimization of Vacuum Pressure Swing Adsorption Processes for Coalbed Methane Purification

Abstract:

This study focuses on the dynamic simulation and optimization of vacuum pressure swing adsorption (VPSA) processes which allow more efficient recovery of methane from coalbed methane (CBM) gas than pressure swing adsorption (PSA) processes. In other words, this work presents 1) the newly developed simulation model (Ko, 2016) which can calculate an accurate interstitial gas velocity within adsorption beds, 2) the successful scale up design (up to 10,000 times) results by using the optimization approach (Ko et al., 2005), and 3) the effect of different operating conditions on the performances such as the purity and recovery of methane. The optimal design conditions of VPSA pilot plants (10Nm^3/h of feed flow rate) to purify coalbed methane (CBM) gas are also obtained according to different performance specifications by using the developed simulation and optimization approach. As a conclusion, the developed model for the simulation and optimization is very efficient in the scale up design as well as in predicting the performances accurately.

470821 On the Limitations of Breakthrough Curve Analysis in Fixed-Bed Adsorption

Abstract:

This work examined in detail the *a priori* prediction of the axial dispersion coefficient from available correlations versus obtaining it and also mass transfer information from experimental breakthrough data and the consequences that may arise when doing so based on using a 1-D axially dispersed plug flow model and its associated Danckwerts outlet boundary condition. These consequences mainly included determining the potential for erroneous extraction of the axial dispersion coefficient and/or the LDF mass transfer coefficient from experimental data, especially when non-plug flow conditions prevailed in the bed. Two adsorbent/adsorbate cases were considered, i.e., CO_2 and H_2O vapor in zeolite 5A, because they both experimentally exhibited significant non-plug flow behavior, and the H_2O-zeolite 5A system exhibited unusual concentration front sharpening that destroyed the expected constant pattern behavior (CPB) when modeled with the 1-D axially dispersed plug flow model. Overall, this work showed that it was possible to extract accurate mass transfer and dispersion information from experimental breakthrough curves using a 1-D axial dispersed plug flow model when they were measured both inside and outside the bed. To ensure the extracted information was accurate, the inside the bed breakthrough curves and their derivatives from the model were plotted to confirm whether or not the adsorbate/adsorbent system was exhibiting CPB or any concentration front sharpening near the bed exit. Even when concentration front sharpening was occurring with the H_2O-zeolite 5A system, it was still possible to use the experimental inside and outside the bed breakthrough curves to extract fundamental mass transfer and dispersion information from the 1-D axial dispersed plug flow model based on the systematic methodology developed in this work.

460836 Using Three NIST Reference Zeolite Materials to Study the Effect of Pore Metrics on the Measurement of High-Pressure CO_2 adsorption Isotherm

Abstract:

Many commercial and industrial applications, such as gas separation, catalysis, environmental remediation, and gas capture and storage relies on solid adsorbents with suitable pore metrics. Ongoing efforts to cost-efficiently curtail carbon dioxide (CO_2) emissions from stationary sources (such fossil fuel-fired power plants and natural gas purification stations) have increased interest in developing CO_2 adsorbents that require little energy to recover the adsorbed CO_2. Although new and improved CO_2 adsorbents continue to emerge, the pace of innovation is hindered by the lack of reliable criteria to optimize experimental parameters and the lack of robust methods to characterize the CO_2 adsorption behavior of materials, especially at high-pressures. In an effort to establish such criteria, we have selected three NIST reference zeolite materials: RM-8850 (zeolite Y), RM-8851 (zeolite A) and RM-8852 (ZSM-5 zeolite) to measure high-pressure CO_2 adsorption isotherms at 20 °C and up to 45 bar. Based on their distinct pore characteristics, these reference materials were used to test methods to optimize experimental parameters in adsorption isotherm measurements. The preliminary results and details of the optimized measurement criteria will be presented.

469048 Measurement of the Density of Microporous Materials Needed to Convert Net Adsorption to Absolute Adsorption

Abstract:

In high pressure adsorption measurements it is possible to convert net adsorption to absolute adsorption by determining the density of the solids which includes the volume of the micropores [1]. The most direct way to measure this density is by means of mercury intrusion porosimetry. In this contribution we develop an alternative methodology which allows to determine the same density using standard pycnometry.

Results for both helium and water pycnometry are presented for different hydrophilic and hydrophobic commercial pelletized materials which confirm that the density of microporous materials needed to convert net adsorption to absolute adsorption can be measured without the use of mercury.

The general conclusions that can be drawn are that helium pycnometry provides reliable results while water pycnometry requires careful pre-treatment of the samples in order to yield the correct densities.

EU IRSES Grant 295156 - Offshore Gas Separations (OFFGAS) is gratefully acknowledged.

[1] S. Brandani, E. Mangano, and L. Sarkisov. Net, excess and absolute adsorption and adsorption of helium. Adsorption, 22(2), 261-276.

456069 Generalized Perturbation Method to Determine Multicomponent Isotherm for Saturated Systems through a Modeling-Design Concurrent Approach

Abstract:

Determining adsorption isotherms for multi-component systems requires substantial experimental effort. Single component adsorption as well as binary/ternary interactions need to be studied at many different compositions. An example is xylenes isomers on molecular sieve. Industrial xylene adsorption process interests in the competitive adsorption of five components (p-xylene, m-xylene, o-xylene, ethyl benzene and para-diethyl benzene) at saturation. The adsorption properties are usually determined by studying single component adsorption experiments and fitting the data to competitive adsorption models. These adsorption can be characterized by empirical models [1] or by Ideal Adsorbed Solution Theory (IAST) [2]. When the number of components is large, a number of experiments must be performed at various compositions, which requires substantial effort.

The perturbation method is an experimental technique that an equilibrium of an adsorbent with the liquid phase is perturbed by a small injection of a sample with different concentration [3]. This method, where only the retention time of the sample is measured, can reduce analysis effort for multi-component systems. While this method has been reported for a dilute system which consists of two solutes, applications for a larger number of components, or for saturated systems where interactions with the mobile phase cannot be ignored, have not been reported.

We extended the perturbation method to study concentrated (non-dilute) system to model the adsorption isotherm near or at the saturation. Furthermore, we also adopted a concurrent approach, where isotherm modeling and process design are carried out iteratively. In this proposed approach, a process model with preliminary isotherm parameters is used to design an adsorption process, and this model also predicts some representative compositions in the process. At these predicted compositions, isotherm measurement tests are carried out. The isotherm determined from the experimental data refines the process model, which leads to better prediction of representative compositions. These two steps, isotherm determination and process design, are carried out concurrently and iteratively.

We present a simulated moving bed (SMB) process for xylene isomer separation as an example. A rigorous dynamic model for the SMB process is implemented and optimized by a deterministic method. The internal concentration profiles are used to determine the compositions where experiments for isotherm determination is carried out. By this iterative approach, experimental effort is reduced to minimal, and model reliability can be increased.

1. Tournier, H., et al., *Adsorption equilibrium of xylene isomers and p-diethylbenzene on a prehydrated BaX zeolite.* Industrial & Engineering Chemistry Research, 2001. **40**(25): p. 5983-5990.

2. Rota, R., et al., *Generalized Statistical-Model for Multicomponent Adsorption Equilibria on Zeolites.* Industrial & Engineering Chemistry Research, 1988. **27**(5): p. 848-851.

3. Heuer, C., et al., *Design of the simulated moving bed process based on adsorption isotherm measurements using a perturbation method.* Journal of chromatography A, 1998. **827**(2): p. 175-191.

473232 Multi-Layer Multi-Component Adsorption Isotherm Model:Theoretical and Experimental Validation

Abstract:

Multi-layer multi-component adsorption isotherm model: Theoretical and experimental validation

Ju Weon Lee[1] Andreas Seidel-Morgenstern[1, 2]

[1] Max-Planck-Institute for Dynamics of Complex Technical Systems, Magdeburg, Germany

[2] Institute of Process Engineering, Otto von Guericke University, Magdeburg, Germany

In liquid chromatography, one of many isotherm models introduced based on theoretical and heuristic backgrounds is selected to analyze and describe the retention behaviors of adsorbates [1, 2]. However, most conventional isotherm models are not suitable in cases that adsorbate mixture shows ununiformed behaviors, e.g. in a certain range isotherm shape is changed from Langmuir-like to anti-Langmuir and vice versa, or some components shows different behaviors. Recently, several researchers introduced multi-layer isotherm models for a single component to express theoretical clues for the mixed isotherm behaviors [3-5].

In this work, a novel multi-layer multi-component (MLMC) adsorption isotherm model developed on a theoretical basis was introduced to explain the retention behaviors that the conventional theoretical isotherm models cannot explain. The following equation shows the structure of the MLMC model.

$$q_i = \sum_{j=1}^{N} q_{i,j} = \sum_{j=1}^{N} \left[q_{Max} \left(\prod_{k=0}^{j-1} \frac{\psi_k}{1+\psi_k} \right) \frac{K_{i,j} c_i}{1+\psi_j} \right], \psi_k = \sum_{i=1}^{M} K_{i,k} c_i$$

where M and N are the number of components and adsorption layers, respectively.

Case Study: Mixed binary system ( )

$$n = \infty, K_{2,i} = \cdots = K_{\infty,i}$$

For mixed binary systems (one component obeys a Langmuir-like isotherm, and the other obeys an anti-Langmuir isotherm), a generalized Langmuir isotherm model (concentration terms in the denominator can be assigned to positive or negative) was introduced in a heuristic manner [6]. Figure 1 shows the simulated elution profiles with the MLMC model.

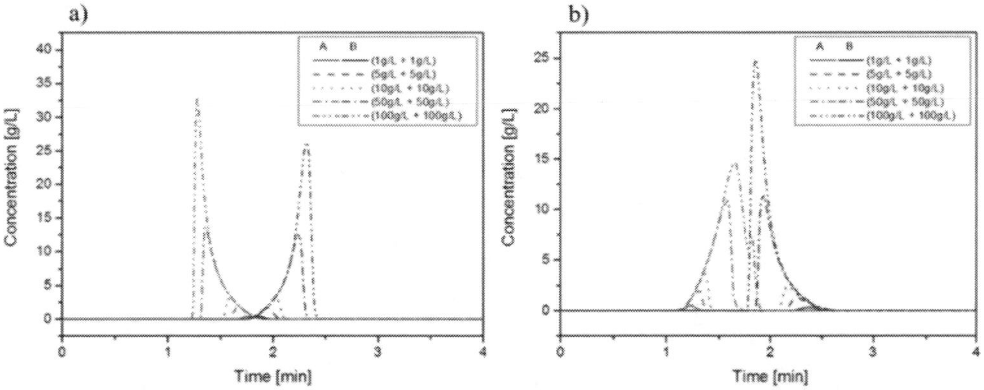

Figure 1. Simulated concentration profiles of the binary mixtures with the MLMC model.

$K_{2,A} < K_{1,A} = K_{1,B} < K_{2,B}$ a) , b)

$K_{1,A} < K_{2,A} = K_{2,B} < K_{1,B}$

References

[1] T.L. Hill, "An Introduction to Statistical Thermodynamics" (1960) Dover Publications Inc.

[2] G. Guiochon, S.G. Shirazi, A.M. Katti, "Fundamentals of Preparative and Nonlinear Chromatography" (1994) Academic Press.

[3] C.-H. Wang, B.J. Hwang, Chem. Eng. Sci. 55 (2000) 4311.

[4] F. Gritti, W. Piatkowski, G. Guiochon, J. Chromatogr. A 978 (2002) 81.

[5] P. Vajda, A. Felinger, J. Chromatogr. A 1324 (2014) 121.

[6] M. Mazzotti, J. Chromatogr. A 1126 (2006) 311.

KINETIC STUDIES OF N_2 ADSORPTION ON POLYMERS OF INTRINSIC MICROPOROSITY (PIM-1) TO STUDY THE EFFECT OF THE C-N FUNCTIONALITIES

Paramita Ray[†], Cheng-Yu Wang[£], Angela D.Lueking[†]

[†]Pennsylvania State University, University Park, PA, United States
[£]Feng Chia University Taichung, Taiwan

ABSTRACT

Polymers of Intrinsic Microporosity (PIM-1) (Figure 1) are a unique class of organic polymers which are porous because of the presence of a spiro centre that prevents rotation of polymer chain which causes them to possess internal free volume (IFV)[1]. PIM-1 are unique compared to other porous carbon nanostructures because of the presence of regular C-N surface functionalities. The permeability and selectivity of PIM-1 for gas separation can be tuned via chemical modification of the nitrile group or photochemical modification of the spirocentre[2, 3, 4]. Because of such structural attributes these polymers possess a high surface area and can be probed for N_2 adsorption and entrapment. Here, we explore the role of the CN surface functionalities of PIM-1 in inhibiting gas diffusivity, by providing a comparative study of N_2 and Ar diffusion rates.

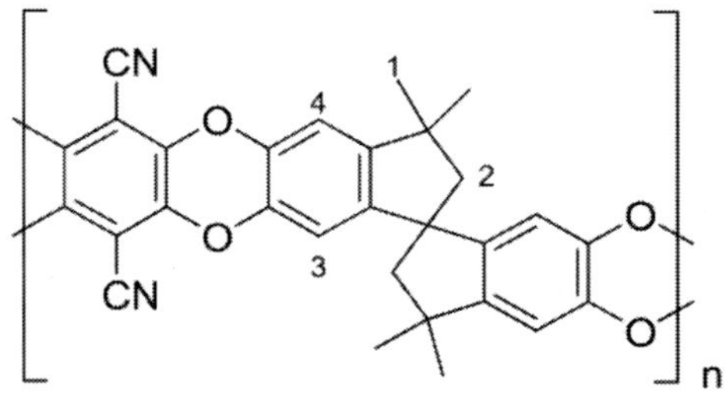

Figure 1: Polymer of Intrinsic Microporosity (PIM-1)

INTRODUCTION

Nitrogen confinement in porous carbon materials has been of interest because of the design of structures used for purification of air, high-density gas storage containers [5], new optical devices [6], chemical sensors [7] etc. The capability of porous materials for confinement of gases depends on several factors such as the surface functionalities, pore dimensions, diffusion rate of the gas molecules. The motivation of the adsorption studies on PIM-1 with N_2 was to ascertain if a π-π stacking interaction inhibited the diffusion rate of N_2 within PIM-1. The rate of N_2 and Ar diffusivity into PIM-1 are compared, due to their similar kinetic diameters (0.36 and 0.34nm, respectively), as it is anticipated that Ar will not be subject to π-π stacking with the CN group.

METHODS AND MATERIALS

PIM-1 was synthesised by the nucleophilic substitution reaction of tetrafluoroterephthalonitrile and 5,5',6,6'-tetrahydroxy-3,3,3',3'-tetramethyl- 1,1'-spirobisindane (> 96%). The resulting product was a yellow powder which was characterised by H^1NMR, FTIR, and Raman spectroscopy to confirm successful synthesis and the presence of the CN stretch. Prior to adsorption measurements, PIM-1 samples were evacuated at 120°C under high vacuum for 24hrs. A Micromeritics ASAP 2020 volumetric adsorption unit was used to collect Nitrogen adsorption isotherms at 77 K. The resulting N_2 adsorption was a type IV isotherm, indicative of the presence of both micropores and mesopores. The BET surface area, determined in the relative pressure region P/P_0 of 2.6 x 10^{-6} to 0.032, was calculated to be 666m^2/g. The pore size distribution calculated from the NL-DFT method indicated a bimodal pore size distribution with an average pore size of 0.8nm.

N_2 and Ar adsorption rates were collected on a Hiden Intelligent Gravimetric Analyzer (IGA-001) at both near-ambient temperature and at 77 K and pressures ranging between 50-180 mbar. To correct for buoyancy, the density of PIM-1 was measured in He up to 20 bar, and determined to be 1.4g/cc, which is in agreement with previously reported values for PIM-1[8,9]. Additional high-pressure (20-80 bar) ambient temperature data collected on a differential Sieverts apparatus[10] was too rapid to obtain reliable rate data. At 77K, the saturation pressure of Ar of 230 torr[11] was utilised for determination of relative pressure (i.e. P/Po). The pressure ramp rate of 100 mbar/min led to a 1-2 minutes for pressure stabilization. In analysing the kinetic data, the pressure step was treated as if it were instantaneous, which was previously shown to have a negligible effect on the resulting rate parameters

Adsorption rate data was fit to the Linear Driving Force (LDF), Double Exponential (DE), and Stretched Exponential (SE) models, with theory described elsewhere [12, 13]. Data were fit both via linearization of the corresponding kinetic equation, as well as a numerical fit using least squares analysis provided in commercial software. The DE model provided the best fits for the majority of the rate data, and is thus the focus in this paper. Consistent with a bimodal pore size distribution, the DE model is based on the linear combination of two exponential driving forces for adsorption, which are generally associated with (a) a slow diffusion into a pore with high energy barrier and (b) a fast diffusion into pores of low activation barrier, viz.:

$$A_1(1-e^{-k_1 t})+A_2(1-e^{-k_2 t})\ldots\ldots\ldots\ldots\ldots\ldots (1)$$

Where k_1, k_2 are mass transfer coefficients, corresponding to each diffusion regime, and A_1, A_2 are weighting factors (which sum to 1) and t is the time. The mass transfer coefficient is directly proportional to the diffusion coefficient, normalized by the characteristic length squared ($k=D/L^2$). We assumed the characteristic length to be the particle size of PIM-1, which was determined with a Zeta analyser to be 18.7 nm (Figure 2). The particle size was determined within one hour of solvation in dichloromethane, prior to an observed gel formation (after 20 hours). All adsorption rate measurements were collected sequentially on the same sample, such that the characteristic length of diffusion is not expected to vary.

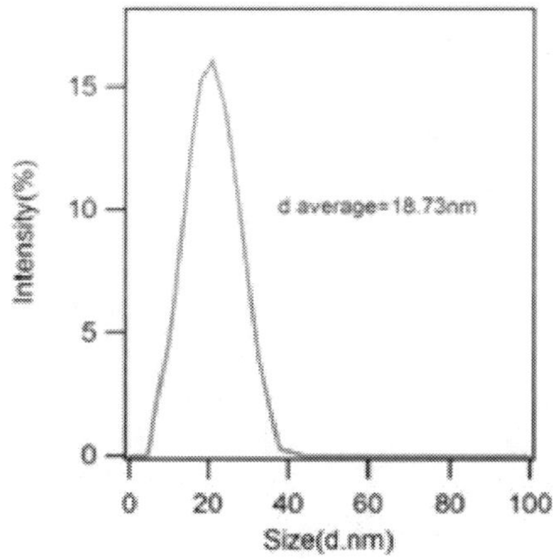

Figure 2: Particle size distribution of PIM-1 measured in a solution of dichloromethane

THEORY

A unified diffusion theory has been proposed to describe gas diffusivity, Knudsen diffusivity, liquid diffusivity, and configurational diffusion within a solid, either in the gas-translational (GT) or solid vibration (SV) regime [14]. In the configurational diffusion regime, gas molecules must overcome an activation barrier to "hop" to the next site. In the Knudsen diffusion regime, the pore dimension is less than the mean free path, and thus diffusion is inhibited by collision with the pores. The unified equation is as follows:

$$D = guL\ exp\left[-\frac{E}{kT}\right] = gL\sqrt{\frac{8kT}{\pi m}}\ exp\left[-\frac{E}{kT}\right] \quad \text{.............} (2)$$

Where u is the velocity at which the molecule travels, E is the activation barrier for diffusion (non-zero only for the case of configurational diffusion), k is the Boltzmann constant, and T is temperature; g is a geometric factor to account for directionality, and typically taken as 1/3 for diffusion that is fully three-dimensional, i.e. when the diffusing species is not "channelled" in a particular direction, such as the case for gaseous, liquid, or Knudsen diffusion. For configurational diffusion within a solid, g is treated empirically. L is the diffusional length (the average distance between collisions or jumps), taken as the mean free path for free gaseous diffusion, the pore diameter for Knudsen diffusion, and the distance between adjacent sites in configurational diffusion. The latter expression in (3) substitutes the molecular velocity from statistical mechanics, where m is the molecular mass.

In a porous material, presuming that the characteristic length and temperature are constant, a comparison between two gaseous species in the Knudsen or configurational regime leads to:

$$\frac{D_1}{D_2} = \sqrt{\frac{m_2}{m_1}}\, exp\left[-\frac{\Delta E}{kT}\right] \quad \text{.............. (3)}$$

Where ΔE has been introduced to denote the difference in activation energy. Rearranging:

$$\Delta E = E_1 - E_2 = kT\, ln\left[\frac{D_2}{D_1}\sqrt{\frac{m_2}{m_1}}\right] \quad \text{...........(4)}$$

This analysis presumes low coverage, such that diffusion is not influenced by other gaseous molecules. One will note that differences in diffusivity of two gases will be more pronounced at lower temperature, i.e. T is in the denominator of Eq. (3). Eq. 4 was used to compare the rates of N_2 adsorption to that of Ar. Ar is expected to have a lower "free" diffusivity due to a higher molecular mass (Eq 2), independent of size. However, in the configurational diffusion regime, a gas may become impeded such that ΔE is nonzero. Adsorbate-adsorbate interactions at high gas loading may also affect the diffusivity[15].

According to previous reports[14] the configurational diffusion regime is reached when the mean free path of the gas molecule is comparable to that of the pore size.

$$\frac{\lambda}{P} \rightarrow 1(Configurational) \text{ .... (5a)}$$

$$\frac{\lambda}{P} < 1(Knudsen) \text{.................. (5b)}$$

Pressures were chosen to correspond (approximately) to both the configurational and Knudsen diffusion regime based on the considerations in Eq. 5.

RESULTS AND DISCUSSION

At near-ambient temperature, the heat of adsorption of N_2 on PIM-1 is 16kJ/mol (=0.15eV), at an uptake of 0.18wt% as determined from supercrical adsorption isotherms at 272.6 K and 298 K from the Clausius-Clayperon equation (Figure 3).

(a) (b)

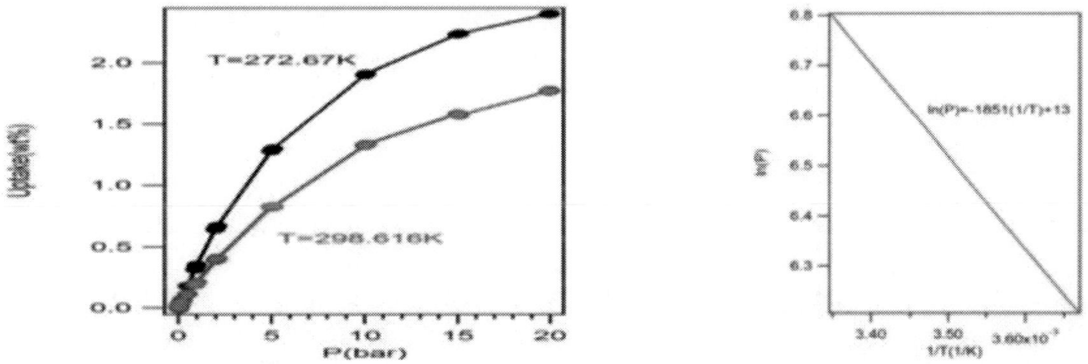

Figure 3: (a) Nitrogen adsorption isotherm on PIM-1 upto 20bar collected at 298K(bottom red curve) and 272K(top black curve), (b) Plot of $\ln(P_2/P_1)$ vs 1/T to obtain the heat of adsorption of N_2 on PIM-1.

The rates of N_2 and Ar adsorption to PIM-1 were studied at 77K at similar reduced pressure (P/Po). The rates of both N_2 and Ar data were well-described by the DE model at all pressures studied. Select rate data is shown in Fig 4a, with a compilation of all rate constants in Fig. 4b.

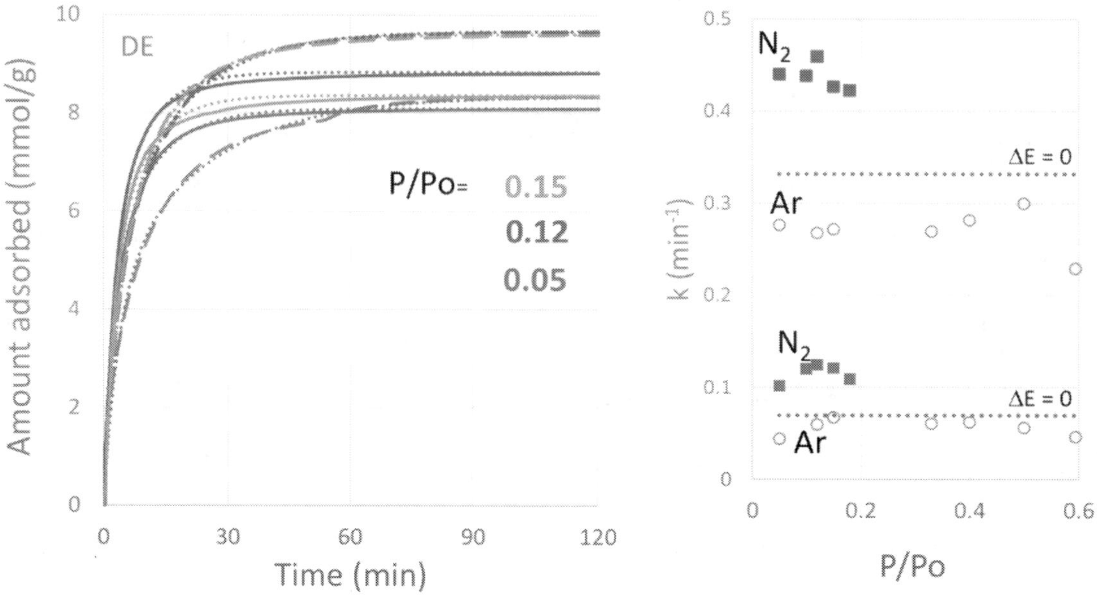

Figure 4: (a) Experimental N_2 (solid lines) and Ar (long dashed lines) adsorption rate data with the corresponding DE model fits (dotted lines) for diffusion into PIM-1 at 77K and three select relative pressures.P/Po=0.12(top solid line), (b) Corresponding double exponential fits for all rate data for N_2 (squares) and Ar (circles); the dotted line is the expected diffusivity of N_2 if the barrier for diffusion of N_2 and Ar were equal, according to Equation (3).

In all cases, the diffusivity of N_2 exceeded that of Ar, as is expected based on the molecular mass for non-activated diffusion (Eq. (2)). The diffusivity was not a function of pressure for the small pressure range studied, suggesting adsorbate-adsorbate interactions were negligible. Using average values of diffusivity, Eq. (4) suggests that the activation barrier for diffusion that is 3.5 meV higher for Ar in regime 1 of the DE model and 2 meV higher in regime 2 of the DE model. This difference is fairly insignificant (for reference, the heat of adsorption was 150 meV, as noted above, Fig 3). Thus, no real impediment of N_2 diffusivity is seen due to the interactions with the C-N functional groups for PIM-1. The next step is to reduce the pore size of PIM-1 via UV irradiation, and return to these comparative diffusivity studies. In contrast, a strong barrier for N_2 diffusivity is needed to explain its slow uptake in carbon molecular sieves, relative to the heavier O_2 and Ar [16].

CONCLUSIONS:

The kinetics of N_2 diffusion has been studied and results were fit to a double exponential decay model. The fitted mass transfer coefficient showed no real pressure dependence over the limited pressure range studied (i.e. 50 -180 mbar). In all cases, the diffusivity of N_2 exceeded that of Ar as expected based on differences in the molecular mass. Thus, no effect of impediment of N_2 diffusivity due to strong π-π interactions is noted in as-synthesized PIM-1 and the CN groups do not impede the diffusivity of N_2 in PIM-1 at 77K relative to Ar.

REFERENCES

1. Nicholas T. Tsui Alex J. Paraskos Lokman Torun Timothy M. Swager Edwin L. Thomas Minimization of Internal Molecular Free Volume: A Mechanism for the Simultaneous Enhancement of Polymer Stiffness, Strength, and Ductility *Macromolecules*, **2006**, *39* (9), pp 3350–3358

2. Li, F. Y.; Xiao, Y.; Chung, T.-S.; Kawi, S. High-Performance Thermally Self-Cross-Linked Polymer of Intrinsic Microporosity (PIM-1) Membranes for Energy Development. *Macromolecules* **2012**, *45*, 1427-1437

3. Satilmis, B.; Budd, P. M. Base-catalysed hydrolysis of PIM-1: amide versus carboxylate formation. *RSC Adv.* **2014**, *4*, 52189-52198.

4. Song, Q.; Cao, S.; Zavala-Rivera, P.; Lu, L. P.; Li, W.; Ji, Y.; Al-Muhtaseb, S. A.; Cheetham, A. K.; Sivaniah, E. Photo-oxidative enhancement of polymeric molecular sieve membranes. *Nat Commun* **2013**, *4*, 1918.

5. DeCoste, J. B.; Peterson, G. W. Metal-organic frameworks for air purification of toxic chemicals. *Chem Rev* **2014**, *114*, 5695-5727.

6. Durgun, E.; Ciraci, S.; Yildirim, T. Functionalization of carbon-based nanostructures with light transition-metal atoms for hydrogen storage. *Physical Review B* **2008**, *77*.

7. Lim, S. H.; Elim, H. I.; Gao, X. Y.; Wee, A. T. S.; Ji, W.; Lee, J. Y.; Lin, J. Electronic and optical properties of nitrogen-doped multiwalled carbon nanotubes. *Physical Review B* **2006**, *73*.

8. Larsen, G. S.; Lin, P.; Siperstein, F. R.; Colina, C. M. Methane adsorption in PIM-1. *Adsorption* **2010**, *17*, 21-26.

9. Frentrup, H.; Hart, K. E.; Colina, C. M.; Muller, E. A. In Silico Determination of Gas Permeabilities by Non-Equilibrium Molecular Dynamics: CO2 and He through PIM-1. *Membranes (Basel)* **2015**, *5*, 99-119.

10. Sircar, S.; Wang, C.-Y.; Lueking, A. D. Design of high pressure differential volumetric adsorption measurements with increased accuracy. *Adsorption* **2013**, *19*, 1211-1234.

11. Lueking, A. D.; Kim, H.-Y.; Jagiello, J.; Bancroft, K.; Johnson, J. K.; Cole, M. W. Tests of Pore-Size Distributions Deduced from Inversion of Simulated and Real Adsorption Data. *Journal of Low Temperature Physics* **2009**, *157*, 410-428.

12. Klafter, J. and Shlesinger, M. F. On the relationship among three theories of relaxation in disordered systems. *PNAS* **1986,** *83,* 848.

13. Chen, B. *et al.* Surface Interactions and Quantum Kinetic Molecular Sieving for H_2 and D_2 Adsorption on a Mixed Metal - Organic Framework Material.*JACS.* **2008,**6411–6423.

14. Xiao, J. Diffusion of Hydrocarbons Theory. *Chemical Engineering Science* **1992,** *47,*1123-1141.

15. Li, L.; Bell, J. G.; Tang, S.; Lv, X.; Wang, C.; Xing, Y.; Zhao, X.; Thomas, K. M. Gas Storage and Diffusion through Nanocages and Windows in Porous Metal–Organic Framework Cu2(2,3,5,6-tetramethylbenzene-1,4-diisophthalate)(H2O)2. *Chemistry of Materials* **2014**, *26*, 4679-4695.

16. Rynders, R. M., Rao, M. B. & Sircar, S. Isotope Exchange Technique for Measurement of Gas Adsorption Equilibria and Kinetics. *AIChE Journal* **1997,***43,* 2456.

463207 Organic Solvent Nanofiltration with Novel Polymeric Membranes

Abstract:

A very large variety of membranes polymeric or otherwise have been investigated over the last two decades for organic solvent nanofiltration (OSN). The materials and structures used to make OSN membranes include among others the following: poly (dimethylsiloxane) (PDMS); mixed matrix membranes (MMMs) of PDMS with zeolites and other fillers; asymmetric integrally skinned polyimide (PI) membrane crosslinked with aromatic or aliphatic diamines; polyaniline; polypyrrole; interfacially polymerized polyamide with or without functionalized carbon nanotubes; carbon-based membranes including graphene. Most of the studies involved polymeric membranes. One of the weaknesses of most polymeric membranes is their varying tendencies for swelling with demanding organic solvents aprotic or otherwise; this gets reflected in the solute rejection behaviors of such polymers. Diamine-crosslinked PI membranes show excellent resistance to organic solvents but have some swelling in the presence of water. To this end we have started studying membranes from particular classes of fluoropolymers which are extremely inert. Initially these membranes were studied for pervaporation-based selective removal from organic-organic mixtures and aqueous-organic mixtures. Next they were studied with various solvents for OSN over a considerable pressure range. We report preliminary results of solvent flux and solute rejection for such membranes. These studies were made with dense flat membranes of different thicknesses supported on appropriate solvent-resistant porous supports.

457259 Negligible Aging in Poly(ether-ether-ketone) Membranes Widens Application Range for Solvent Processing

Abstract:

Negligible aging in poly(ether-ether-ketone) membranes widens application range for solvent processing

João da Silva Burgal, Ludmila Peeva, Andrew Livingston*

Department of Chemical Engineering and Chemical Technology, Imperial College, Exhibition Road, London SW7 2AZ, UK

E-mail: a.livingston@imperial.ac.uk

Presenting author: Ludmila Peeva (l.peeva@imperial.ac.uk)

Keywords: ageing, organic solvent nanofiltration, high-temperature operation

Conventional molecular separation processes such as evaporation and distillation exert substantial energy demands due to the latent heat of vaporization of liquids. Membrane technology has lower energy consumption than conventional separation processes, requiring only one-tenth of the energy to process an equivalent volume of liquid. In spite of this advantage, membranes have yet to make a significant impact in industrial organic solvent based systems, due to unresolved issues including their stability over time. Most commercially available nanofiltration and reverse osmosis membranes are either integrally skinned asymmetric (ISA) membranes, or thin film composite (TFC) membranes. Both types are made from glassy polymeric materials which age over time leading to adverse changes in membrane performance; many OSN membranes display more than 30 % permeance decline over the first few days of operation.

This ageing phenomenon is caused by re-arrangement of the polymer chains in the membranes from the non-equilibrium excess state created during the membrane formation process, towards a stabilized equilibrium state, in a time-dependent manner via structural relaxation as represented in **Figure 1**. Usually, ageing is manifested through a gradual decrease in membrane volume (densification), which leads to increased brittleness, decreased permeability, decreased enthalpy, and alterations in other properties. For example, the permeance of polyimide ISA Organic Solvent Nanofiltration (OSN) membranes created by phase inversion decreases from around 1-2 L m^{-2} h^{-1} bar^{-1} to zero after annealing at 100 °C-200 °C. This is consistent with the negligible intrinsic solvent permeance of solvent-cast polyimide films, in which polymer chains reach their highest density.

Organic solvent nanofiltration membranes have been prepared from poly(ether-ether-ketone) (PEEK) by phase inversion. Remarkably, these membranes undergo negligible ageing, even under extreme conditions of high temperature air annealing, and high temperature solute filtration with agressive solvents. This negligible ageing of PEEK membranes is contrasted with substantial ageing of crosslinked polybenzimidazole (PBI) and polyimide (PI) membranes. After air annealing at 120 °C, PBI and PI membranes become brittle and lose all permeance, whereas PEEK membranes remain flexible and retain a constant permeance of ~ 0.2 $L.h^{-1}.m^{-2}.bar^{-1}$ for tetrahydrofuran (THF). The structural change in PBI and PI membranes is attributed to polymer transition from a non-equilibrium glassy state towards an equilibrium state at which chain packing precludes permeation of solvent. In contrast, fabrication of PEEK membranes results in polymer in a quasi-equilibrium glassy state with intrinsic permeance, which is not adversely affected by accelerated-ageing environments. High temperature filtrations in DMF up to 140 °C for the three polymeric membranes showed PEEK performance to be stable and consistent with negligible ageing, whereas PBI and PI rapidly loose stability.

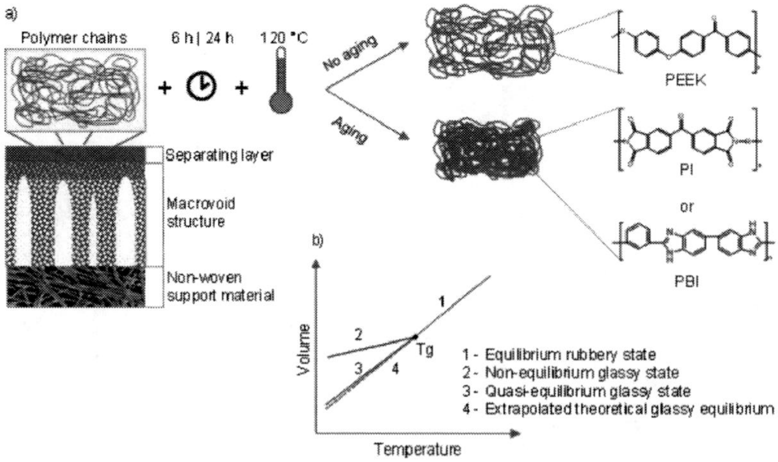

Figure SEQ Figure * ARABIC 1 – a) Schematic of the effect of time and temperature on the polymer chains of ISA membranes. PEEK polymer chains show no signs of ageing densification, in contrast to PI and PBI which undergo ageing. b) Qualitative volume vs. temperature diagram for a glass-forming polymer depicting the equilibrium rubbery (1) and equilibrium glassy (4) states as well as the quasi (3) and non-equilibrium glassy (2) states.

461384 Active Layer Brush Membranes for Pervaporation: A New Class of Synthetic Membranes

Abstract:

Performance and characterization of a new class of synthetic membranes are presented. These membranes are based on a brush-like active layer and were created through graft polymerization of vinyl monomers from light active porous poly(ether sulfone) supports. The grafted material or active layer is the rate limiting barrier to species transport and endows the membrane with selectivity, and are collectively known as polymer brushes. Various grafted polymer brushes were tested with an emphasis on those synthesized from hydrophobic monomers. Irradiation and reduction-oxidation methods were used to graft the polymer brush layers. The membranes were tested for their transport properties (selectivity and permeability) and characterized for polymer morphology, grafting density and solvent effects on mean brush height. Several techniques, including Attenuated Total Reflectance â€“ Fourier Transform Infrared Spectroscopy (ATR-FTIR), X-Ray Photoelectron Spectroscopy (XPS), Time of Flight â€“ Secondary Ion Mass Spectroscopy (TOF-SIMS), Atomic Force Microscopy (AFM), and Scanning Electron Microscopy (SEM) were used. The membranes were tested for recovery of small organics (e.g. acetic acid, isobutanol) from aqueous mixtures with pervaporation. An attractive feature of these brush skinned membranes are their ease and low cost of manufacture.

460010 A Versatile Polytetrafluoroethylene Membrane Separator for Microreactor Applications

Abstract:

Here we report the development of a simple, low-cost capillary-based membrane separation module for liquid-liquid phase separation and microextraction.[1,2] Designed specifically for microfluidic and microreactor applications, the separator comprises a length of microporous polytetrafluoroethylene (PTFE) capillary bonded to rigid PTFE capillary for direct coupling into a flow reactor. The separator is shown to provide highly efficient separation of aqueous-fluorous, aqueous-organic, and organic-fluorous segmented flows between 100 and 2,000 $\mu L\ min^{-1}$, and under a range of flow rate ratios, making it ideal for post-processing chemical syntheses performed in droplet and segmented flow microreactors.

We demonstrate that complete separation can be achieved for a range of liquid-liquid combinations by tuning the flow restriction at the separator outlet to account for differences in fluid viscosity and surface tension. Using the separator, we demonstrate the following flow processing operations: (i) inline spectroscopic analysis of the solvent phase in a solvent:carrier segmented flow process â€" where removal of the carrier phase circumvents the need for high speed detectors synchronized to the droplet/segmented flow; (ii) inline aqueous-organic product extraction; (iii) continuous recovery of a carrier phase from a solvent:carrier segmented flow process; and (iv) the ability to regenerate a stable segmented flow post-separation, and thereby facilitating inline processes that require a combination of continuous and segmented flow regimes, such as multistage purification.

(1) Bannock, J. H.; Phillips, T. W.; Nightingale, A. M.; de Mello, J. C. *Anal. Methods* **2013**, *5*, 4991â€"4998.

(2) Phillips, T. W.; Bannock, J. H.; de Mello, J. C. *Lab Chip* **2015**, *15*, 2960â€"2967.

478210 Organic Solvent Reverse Osmosis using Carbon Molecular Sieve Membranes

Abstract:

High-performance membranes derived from carbon materials have shown excellent chemical resistance, high molecular selectivity and fast mass transport across the membrane. It is well-known that carbon molecular sieve (CMS) membranes are able to surpass the "polymer upper bound trade-off line" for many molecular pairs. CMS membranes with tailored ultramicropore and micropore dimensions show both high processability of polymeric membrane and the high selectivity of inorganic membranes. Moreover, slit-like transport pathways through the CMS materials translate into higher productivity (i.e., higher mass transport across a diffusion barrier) than that of cage-like zeolite structures due to the longer jump lengths between pore windows. CMS membranes have been proven to be effective in a variety of gas separation processes such as olefin/paraffin separation, natural gas separation and air separations. However, the low fluxes observed in CMS hollow fibers (due to porous substructure collapse during pyrolysis) hinder scale-up of CMS membranes for industrial separation applications. The performance of carbon molecular sieve separation membranes, which exploit the effect of mass transport across a selective diffusion barrier in order to separate molecules, can be improved by reducing the thickness of the membrane. In this study, a novel post-spinning cross-linking treatment is introduced to maintain the asymmetric porous morphology in CMS hollow fiber membranes even after high temperature pyrolysis processes. A dual-layer dry-wet spinning process was used to prepare asymmetric poly(vinylidene fluoride) (PVDF) hollow fibers with thin selective skin layers. Cross-linking of as-spun hollow fibers "locked-in" the porous substructure and the fiber morphology remained unchanged after the cross-linking process. Moreover, this technique effectively prevented substructure collapse in asymmetric CMS hollow fiber membranes. These asymmetric CMS hollow fiber membranes have significant potential for utilization in organic solvent separations, as they exhibit essentially zero solvent-induced swelling and possess high selectivity for similarly sized molecules.

470838 Modeling and Characterization of Organic Solvent Nanofiltration PDMS Membranes for the Separation of Organic Mixtures

Abstract:

Permeation rates and solute rejection are key factors in organic solvent nanofiltration processes and were investigated for two different PDMS-based membranes, both through an extensive experimental campaign and a model-based analysis. The characterization was done using different n-alkane solutes with variable chain length, corresponding to molecular weight in the range from 170 to 310 kg/kmol and several solvents with different properties (toluene, n-hexane, ethanol and butanol). In addition, the corresponding alkane/alcohol mixtures were investigated; all the tests were conducted at 25°C and 35 bar as transmembrane pressure. Significant differences in fluxes and rejection behavior for the OSN processes were measured for the case of different solvent composition, in both membranes. The results are interpreted in terms of the solution-diffusion model, also utilizing the results from sorption experiments of pure solvents performed ad hoc at 25°C for the investigated membranes. The model was developed to calculate the organic-solvent fluxed through a dense OSN membrane in ternary and quaternary systems. To quantify the solubility of the organic solvents/solutes in the membrane in the feed and permeate side, the "Perturbed Chain-Statistical Associating Fluid Theory" (PC-SAFT) equation of state was used modified through a term that binds the elastic contribution to the free energy , induced by the extension of the polymer chains between cross-linking points by "affine-network" model. The thermodynamics properties were explained for both membrane using the PC-SAFT parameters of uncrosslinked PDMS and diversifying the contribution of elastic energy , through the degree of crosslinking values. The binary interaction parameters were evaluated using the sorption experiments results. The thermodynamic approach used allows to represent the permeability and rejection as a function of solvent composition. The diffusion-coefficients (membrane-solvent/solute) are in agreement with the literature data. The solution-diffusion theory is able to describe the flux and most of the time the rejection of solute in the mixtures.

469441 Recovery of Organic Solvents from Vegetable Oil Streams

Abstract:

Membrane separation is increasingly used for the processing of organic solvents systems. Apart from targeting at very specific separations, such membranes are suitable for upgrading of solvent recovery. In current industry, a combination of membrane technology and distillation/evaporation is the most relevant.

Currently used distillation/evaporation is expensive (energy) and sometimes introduces degradation of products (lowering quality). Less heat input in the products by using membranes should improve continuous quality at lower costs. But when solvent is recovered with membranes the throughput of the membrane is not constant-due to concentration polarization and fouling. This leads to a point where membranes are less effective. In a constellation of membrane and evaporation this means that there are optimal conditions –for membranes in technical sense and for the process for the overall economics.

We have modeled and experimented a few systems to explore the optimal regimes. In this area membrane permeation and separation as well as ease of evaporation and energy costs play a role. Solsep has also experimented with systems that can handle high viscosity feeds and promise relatively high and well-defined mass-transfer properties. This could shift the optimal point to less evaporation. For eg acetone-veggy oil streams it has been found that membranes can be extremely attractive giving well-defined product quality with decreased costs. Enviroment-friendliness and low space needed is also an attractive feature.

Advantage (over current membrane systems)

-robust in organic solvents (here proposed: methylene chloride, acetone, toluene)

-also designed for high viscosity

Advantage over currently used technology (evaporation):

-low temperature, low(er) energy costs

-better product quality

-adaptable to current technology (evaporation, "flashing")-hybrid

-less or other fouling that may be better controllable (also evaporation has fouling)

We will discuss work on the removal of solvents and possible use in processing of oils and fats industry. A special pilot is on acetone recovery from seed oil fractionation plan has been built. On this system extensive piloting is done and still in progress.

463923 Transport Behaviour of Alkanes and Aromatic through Polysulfone/Date Pit Mixed-Matrix Membrane – an Investigation for Separation Potential

Abstract:

Naturally occurring composites are a new set of materials that are currently generating interest in varieties of applications. Date pit is a member of this set of green materials. The presence of cellulose, lignin and hemicelluloses in date pit constitutes varieties of advantages in its application in separation processes. Thus, the objective of this study is to investigate the potential of date pit based mixed-matrix membrane for separation of alkanes and aromatic mixtures. Polysulfone based mixed-matrix membranes containing date pits of different loading (0 - 5%) were prepared using the dry-wet phase inversion technique. The resulting membranes were characterized using SEM, XRD and transport properties tests. The transport properties were evaluated in terms of permeability, solubility, diffusivity and porosity using n-hexane, C-15, C-30 and Toluene. Porosity was found to increase by about 11% for membrane with 5 wt% date pit as compared to pure polysulfone membrane. Permeability of n-hexane, C-15, and C-30 through the prepared membranes are in the range of 0.021 – 0.18, 0.85 –2.36 and 0.34– 1.61 x $10^{-7}cm^2/s$, respectively. All the composite membranes were soluble in toluene and hence cannot be used to separate mixtures containing toluene. However, the mixed-matrix membrane was observed to possess some separation potential for mixtures of alkanes.

471609 Challenging Separations in the Power Generation Industry

Abstract:

The power generation industry has diverse challenging physical and chemical separations, including many in which membranes are important. For coal-fired steam generators, physical separations of ash (particulate matter) from the flue gas using electrostatic precipitation, wet scrubbing, or fabric filters pose challenges on various fuel types (interference from other constituents). Chemical separations include adsorption of contaminants such as mercury, reactive separations targeting sulur dioxide or sulfuric acid, and removal of carbon dioxide. Environmental contaminants (such as mercury, selenium, and nitrates) must also be removed from water used in the flue gas cleanup and ash handling.

However, other power generators include challenging separations. Energy storage technologies such as batteries and flow batteries rely on membranes to maintain charge separation and respond quickly when electrical demand is high. Advanced nuclear power generators need process contaminants removed to maintain high efficiency.

Poly-generation (such as hydrogen and power production) relies on separation technologies. Produced hydrogen must be separated from co-products using hydrogen separation membranes, and utilized in fuel cells that requires different membranes. Advanced hydrogen generation technologies such as electrolysis and photolysis require advanced membranes, and thermochemically produced hydrogen has challenging separations in each of the unit operations.

This talk will survey the key progress and remaining challenges in each area of power generation: coal-fired, storage, nuclear, and hydrogen.

473791 Self-Adaptive Water Treatment and Desalination of Source Water of Variable Feed Quality

Abstract:

Membrane based water treatment and desalination technologies have been gaining momentum in potable water production from seawater and brackish groundwater, municipal and industrial water reuse, as well as for management and reuse of agricultural drainage water. Considerable progress has been made since the early days of Reverse Osmosis (RO) water treatment/desalination regarding reduction in energy consumption, development of high performance fouling and chlorine resistant membranes, as well as improvements of in RO feed pretreatment. Given the increased level of water scarcity in various regions of the world and in particular in the Western US there has been growing efforts to expand the use of otherwise unutilized inland groundwater resources. A major challenge, however, is to implement a strategy that produces treated water of the right quality for the right use at the desired location. Temporal variability of source water quality, however, is a major challenge. Also, many of the various locations/communities, where water purification/desalination is needed, are remote and are not part of a centralized water infrastructure. In such locations, it is critical that water purification is accomplished at high recovery. Addressing the above challenges requires the development of a robust strategy for self-adaptive operation of membrane based filtration and desalination operations that will enable the deployment of distributed water systems. Such water systems have to readily adapt to changing water capacity demand and water quality fluctuations. Accordingly, an approach will be described whereby the concept of self-adaptive distributed water systems is developed, following a rapid process field evaluation approach that then serves to optimize process configuration. The above is enabled via direct real-time membrane monitoring and considerations of high recovery in a small plant footprint with salt harvesting via novel accelerated process of chemical demineralization. Examples of recent advances in the above areas will be discussed along with presentation of successful field demonstrations of seawater, brackish and industrial water treatment and reuse, in addition to purification of impaired groundwater.

473771 Desalination Performances of Large Hollow Fiber-Based DCMD Devices

Abstract:

Membrane distillation (MD) can be utilized to desalt highly saline waters that have considerable osmotic pressure. Current reverse osmosis desalination technology is not designed for saline waters having high osmotic pressure. Realization of this goal requires MD membranes to be highly fouling resistant. Fouling is a major obstacle that MD technique should potentially eliminate or control by selecting appropriate membrane material/surface, flow manipulation, additional force field and cleaning procedure. In direct contact membrane distillation (DCMD), Song et al. (2008) have shown that a rectangular cross-flow hollow fiber membrane configuration with a specific type of plasma polymerized superhydrophobic coating on the membrane outside surface and a novel method of operation ensured no precipitation based fouling from $CaCO_3$ and $CaSO_4$. We dwell on this novel aspect compared to conventional methods of fouling control. The simulation model of Song et al. (2008) assumed the membrane mass transfer coefficient, k_m, as an adjustable parameter for large hollow fiber modules with crossflow in a given range guided by estimated values and Zukauskas equation for predicting heat transfer coefficients in the two boundary layers. Here k_m was predicted from an enhanced model from Li and Sirkar (2016) without any adjustable parameters. To develop efficient designs, the model further simulated the performance of hollow fiber modules where the fiber ID was varied. Effects of fiber length on water vapor flux and water production rate were also modeled. This analysis will help design optimum hollow fiber membrane modules for desalination by DCMD for brines with low as well as high salinity.

472716 Development of a Multi-Stage Membrane Distillation System for Treatment of Hydraulic Fracturing Flow Back Waters

Abstract:

The reuse of wastewater for beneficial uses has become increasingly important in recent years. There is an urgent need to develop innovative and more effective technologies for treatment of wastewaters. Many of these wastewaters such as hydraulic fracturing flow back waters, contain very high dissolved solids (TDS). For example, treatment of hydraulic fracturing flowback waters can be very challenging as they often contain very high TDS, in excess of 200,000 ppm as well as dissolved surfactants and small organic compounds. Treatment of these highly impaired wastewaters is therefore very challenging. Pressure driven membrane processes such as reverse osmosis are impractical for treating very high salinity wastewaters due to the high osmotic back pressure that must be overcome. Membrane distillation has been proposed as a new unit operation for treatment of very high TDS wastewaters. One of the advantages of membrane distillation is the fact that low grade waste heat may be used.

Here we have developed a multi-staged membrane distillation system for maximizing water recovery from actual hydraulic fracturing flowback waters. The multi-stage system has been designed to different on-site applications for wastewaters with specific TDS levels. We show that ideal membrane properties vary as the TDS of the feed increases above 300,000. At lower TDS values membranes should be selected that maximize flux. However at TDS values above 300,000, membrane surface properties such as roughness becomes important. At theses higher TDS values leakage of the feed through the membrane pores and into the permeate side is a major concern. In addition appropriate pretreatment of the feed to remove dissolved organic compounds is essential prior to membrane distillation.

Here different membranes have been used in the different stages and performance of the system in terms of water recovery potential and specific energy consumption has been evaluated. Significant concentration of the feed has been conducted and over 60% water recovery has been achieved for actual hydraulic fracturing flowback waters from

Marcellus shale containing over 100,000 ppm TDS.

465789 Role of Membrane in Iemb Treatment of Groundwater Highly Contaminated with Oxyanions

Abstract:

Role of Membrane in IEMB treatment of groundwater highly contaminated with oxyanions

Perchlorate contamination of ground water is a worldwide concern. Several sites in Israel's coastal aquifer are contaminated with perchlorate, with hundreds of ppm found in one area accompanied with significant concentrations of nitrate and chlorate as well. This has prevented water production from wells in the area.

The ion-exchange membrane bioreactor (IEMB) [1] is a hybrid process for safe treatment of groundwater highly contaminated with oxyanions (perchlorate, nitrate, and chlorate). By Donnan dialysis anionic contaminants migrate across an anion exchange membrane (AEM) from a feed-water compartment to a bio-compartment. Once in the bio-compartment, the anions undergo microbial reduction to safer species such as chloride ions and nitrogen. The AEM acts as a barrier and keeps both compartments completely separate. Glycerol is used as an exogenous carbon and electron source for the biodegradation process [2]. This arrangement keeps the carbon source, reaction byproducts and bacteria confined in the bio-reactor thus preventing secondary contamination of the treated water. This contrasts with standard anaerobic bioreactors for treating such contaminations, for which the Ministry of Health requires additional barriers including MF, UV and GAC.

The present study examined the role of membrane type in performance of the IEMB in removing perchlorate and other anions (nitrate and chlorate) at levels of hundreds mg L^{-1} from ground water with a mixture of high levels of oxyanions. This was based on Donnan dialysis studies and was studied initially for synthetic and actual ground water fed to the water side while feeding 0.1 N of NaCl to the bio-compartment. To understand the results, conductivity and selectivity measurements were carried out on the different membranes. The anion exchange membranes examined included ACS (Tokuyama Soda), PCA-100 (PC-Cell GmbH), and AMI-7001 (Membrane International).

For ACS membrane and all experiment conditions studied here, perchlorate dominates the flux across the AEM and fluxes deviated from linear dependence on the driving force at less than 0.5 mM. Even though perchlorate concentration is considerably lower than nitrate and chlorate, its flux is greater than the sum of the other anion fluxes. At an effective perchlorate driving force (EDF) value above 0.7 [mM] perchlorate had a negative effect on the other anions resulting in a decrease in the flux for both nitrate and chlorate while its own flux only increased.

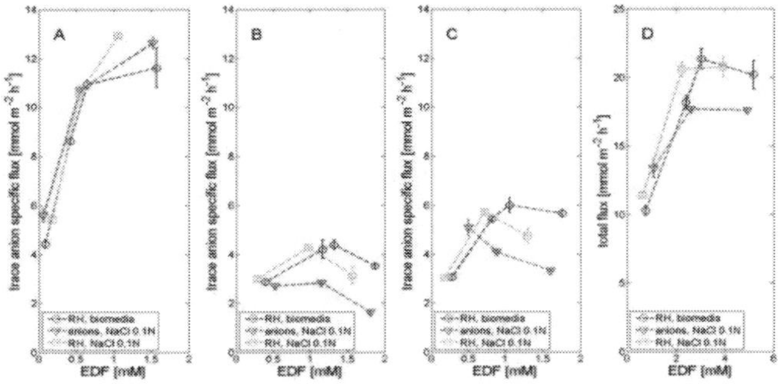

Figure 1 – Results of Donnan dialysis on ACS membrane. Oxyanions specific flux vs. the Effective Driving Force (EDF) of (A) perchlorate, (B) chlorate, (C) nitrate and (D) anions total flux vs. the total EDF, for IEMB Donnan dialysis experiments fed with (○) RHGW and bio media, (▼) synthetic anions solution and NaCl 0.1N and RHGW and NaCl 0.1N (■) into the water compartment and bio-compartment, respectively.

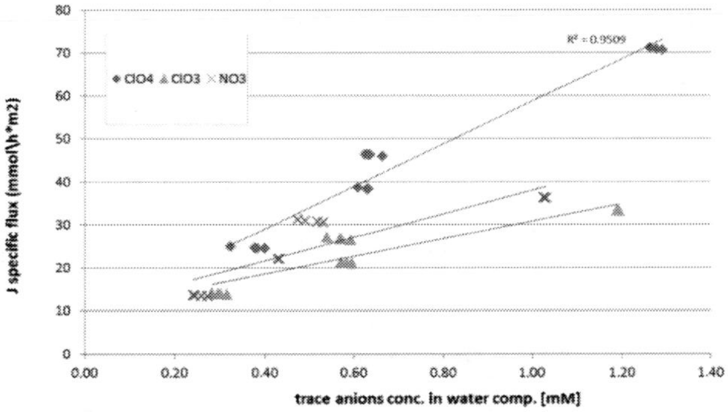

Figure 2: Specific flux in Donnan dialysis on synthetic mixture of oxyanions using PCA-100 anions exchange membrane. Bio-side contained 0.1 N NaCl.

On the other hand, experiments with PCA-100 membranes showed a linear dependence of flux on driving force to much higher levels and much less intereference of the perchlorate on the transport of nitrate and chlorate (Figure 2). Tests of membrane conductivity and regenerability supported the improved results with the PCA-100 membranes. Another type (AMI-7001) also showed promise based on conducitivity but may be problematic in terms of polymer structure. The membrane selectivity for perchlorate and the other oxyanions over chloride was clearly lower for the PCA-100 membranes than ACS. A qualitative model will be proposed to explain these findings and relate it to the differences in ion exchange membrane structure.

If time allows, the implications of these findings for operating the IEMB as a plug flow reactor will be outlined.

[1] A.D. Fonseca, J.G. Crespo, J.S. Almeida, M.A. Reis, Drinking water denitrification using a novel ion-exchange membrane bioreactor, Environ. Sci. Technol. 34 (2000) 1557–1562.

[2] S. Fox, Y. Oren, Z. Ronen, J. Gilron, Ion exchange membrane bioreactor for treating groundwater contaminated with high perchlorate concentrations., J. Hazard. Mater. 264 (2014) 552–9.

470031 Economic Comparison of Pressure Driven Membrane Processes to Electrically Driven Processes for Use in Hydraulic Fracturing

Abstract:

Hydraulic fracturing has provided many new resources for the United States in oil and gas. These resources continue to be an economic driver going forward in many areas of the country. However, significant challenges remain with the environmental sustainability of this technology. One of the areas that is of particular concern is the re-use of the water that is used in fracking operations. The current method of dealing with this water, deep well injenction, takes water out of the eco-system and has been shown to cause earthquakes. This water has high levels of suspeneded and disolved organics as well as high levels of dissolved salts. Because of this the osmotic pressures of these streams are significantly more than seawater reverse osmosis in several instanced. Our group has looked at several possible alternatives of dealing with this water including reverse osmosis, nanofiltration, electrodialysis, and electrodeionization. All of these technologies have different advantages depending on the concentration of the water and ultimate re-use objective. This study has shown that electrodeionization is the most flexible technology, due to selective ion removal and high water recoveries, with the potential to work for both re-use and drinking production as well as acceptable economics at high and low ion concentrations. Experimental results as well as economic modeling will be presented in this talk.

470715 Development and Testing of Amine-Type Scavengers for the Removal of H_2s from Liquid Sour Crudes

Abstract:

As the worldwide transportation sector grows due to global population expansion as well as growing economies, particularly in China and India, fossil fuel hydrocarbon based energy will continue to be the primary energy feedstock over the next few decades. Consequently, meeting the demand for fossil fuels has been met with many environmental and processing challenges, especially for crudes with higher H_2S content. Crude oils that have sulfur concentrations more than 0.5 wt% are considered sour crudes, since they are characterized by a foul, odorous smell. Sour crudes are of lower quality and present serious health and environmental concerns. Therefore, sustainable measures to lower the sulfur content (i.e. crude oil sweetening) are of significant importance, financially and environmentally. Organic sulfur compounds, such as di-benzothiophene, are usually converted using hydro-desulfurization reactions and are not seen has problematic for transportation of crude oils from the wellhead to the refinery. Removing of H_2S at the wellhead before transporting *via* pipeline or railcar increases the value of crude oil and in some cases is necessary to conform to legal transport laws. Hydrogen sulfide, however, is normally removed using amine based absorbing materials, known as scavengers. Hydrocarbon companies are exploring the prospect of greener and more cost-efficient methods for the scavenging (i.e. chemical absorption) of H_2S from crude oils. The most used non-regenerative absorbent for H_2S scavenging is a class of compounds known as triazines. Triazines are typically formed from a condensation reaction between an amine and a carbonyl compound. In-line scavenging of H_2S is the favored technique for producing subsea wells containing crude oils with minute H_2S concentrations especially when the well is coupled to a host facility for which there is no H_2S scavenging or where it is too costly to implement. This work explores the fundamental characteristics of seven common scavengers, namely through FTIR and Raman spectroscopies, to identify the molecular facets for top-performing scavengers. In particular, this work identifies which scavenger-type is best for a particular crude oil type (i.e. acidity, aqueous content, H_2S content, etc.). Each of the scavengers were tested in H_2S-contaminated crude oils at various conditions, including temperature and reactor stirring rates. These results will be shown along with a proper delivery system to add scavenger compounds during the filling of a rail tanker or even injection into a pipeline.

475200 Compact Rotating Packed Bed Contactor for Process Intensification in Offshore and Onshore Gas Separations

Abstract:

Conventional tray or packed towers have been traditionally used for gas separations. The towers perform satisfactorily, but they have disadvantageous characteristics for sites where space and weight are at a premium such as offshore marine environments or onshore retrofits: footprint, height (need to be vertical, especially for tray towers), weight, foaming issues, large liquid inventory, relatively low turn-down ratio, high cost of installation and difficulty in modularization.

Rotating Packed Bed technology (also known as "Higee") is one of the recognized Process Intensification technologies in industry. It can shrink the process unit size significantly as compared with conventional technologies. The RPB internals consist of a high surface area donut shaped rotating packed bed mounted on a shaft in a pressure vessel. Gas enters from the outer radius surface of the packed bed, flows radially inwards, and passes through the whole packing before exiting through the center eye. Liquid is sprayed through a series of nozzles in the center eye into the inner radius surface of the packing and flows radially outwards through the packing due to the centrifugal force. The liquid, moving counter-currently to the gas, exits at the outer surface of the packing. Internal seals are used to ensure that there is no gas or liquid bypass or short-circuiting around the packed bed.

The high centrifugal force generated by rotation in the RPB allows the formation of thin liquid films, small droplets and high liquid-gas interfacial areas in the packing leading to significantly higher mass transfer coefficient than conventional (tray or packed tower) technology. The significantly higher mass transfer coefficient of the RPB leads to reduced footprint, weight, height and CAPEX. In addition, RPB offers other benefits such as high turn-down ratio, low liquid hold-up, elimination of anti-foaming agent.

The use of RPB as an alternative compact technology for gas processing will help to monetize Chevron's

equity gas both onshore and offshore through the use of reduced size and weight equipment.

This talk will review the latest developments in the technology and will discuss the efforts within Chevron to develop the technology for the gas processing application.

458255 Application of Computational Fluid Dynamics Simulation in Study of Amine Absorber at Off-Shore Operation

Abstract:

Crude natural gas contains acid gases such as CO_2 and H_2S. The acid gases must be removed to achieve clean natural gas for the wide usage. One of the most popular processes used to remove the acid gases is the amine absorber. In offshore operations, the gas processing is commonly carried out on the FPSO (Floating, production, storage and off-loading) unit located near oil wells. The absorber efficiency on FPSO may be deteriorated because of ocean wave interaction.

This work aims to study the effect of ship tilting and motion on the acid gas removal efficiency in an amine absorber with structured packing, by using CFD (Computational fluid dynamics) simulation. First, a porous media CFD model was developed in the framework of Eulerian two-fluid flows to mimic fluid behaviours inside structured packing. Parameters of the CFD model were adjusted to fit experimental data in two cases of vertical standing and static tilting. Second, CFD simulation was used to improve the design of a liquid redistributor subjected to the motion. The results in the effective interfacial area, CO_2 concentration profile and uniformity index were compared for three cases of vertical standing, ship tilting and ship motion.

467987 Dynamic Simulation of CO_2 Absorption Using Mixtures of Solvents

Abstract:

Dynamic Simulation of CO_2 Absorption using Mixtures of Solvents

Jorge Gabitto and Costas Tsouris

The removal of CO_2 from gaseous streams is a very important industrial technology with applications in (i) reduction of greenhouse gas emissions from flue gases in coal-fired power plants and (ii) removal of radioactive C_{14} in the nuclear industry, among others. One of the well-known technological alternatives for CO_2 capture is absorption/stripping with aqueous solvents, such as alkanolamines and their mixtures. A dynamic model of the absorption process for dilute carbon dioxide mixtures in mixed solvents solutions is proposed. A new reaction scheme that allows the use of different solvents has been added to increase the versatility of the proposed model. Chemical reactions between CO_2 and the high pH solution have been included in the model through the use of an enhancement factor for chemical absorption. Interfacial equilibrium constants have been calculated based upon Henryâ€™s constant values for the different gas-phase species calculated using fugacity ratios obtained by the Peng-Robinson equation of state. Mass and energy balances have been derived for all the species in the gas and liquid phases. Numerical techniques have been used to calculate the axial profiles of temperature and concentrations of the resulting species in both phases. The model has been validated using experimental data available in the literature for different columns, packing materials, and operating conditions. Simulations under different operating variables and dynamic inputs have been conducted. Simulation results for different solvents have been compared and analyzed. These results show the potential of the proposed model to optimize the operation and control of these separation processes.

463922 Hydrodynamics and Mass Transfer of Amine Absorber According to Gravity Center Position

Abstract:

Flow maldistribution affects CO_2 removal efficiency of amine absorber for offshore operations such as LNG-FPSO (Floating Production Storage Off-loading). Under ship motion, flow maldistribution is caused by acceleration forces depending on the distance from the center of gravity (COG). Therefore, the position of COG is a significant factor in the offshore amine absorber.

The purpose of this study is to investigate the relationship between the position of COG and CO_2 removal efficiency of the amine absorber. CFD (Computational Fluid Dynamics) simulations were performed with the three dimensional (3D) geometry of a pilot-scale amine absorber having Mellapak 250.X (M250X) structured packing. A porous media Eulerian CFD model which includes three momentum forces (porous resistance, gas-liquid drag force, dispersion force), mass transfer and chemical reaction was applied to the CFD simulation in order to express characteristics of the M250X packing. The CFD model was validated with experimental data of a vertical standing case. Finally, three cases with different positions of COG were examined to find the effect of COG on the CO_2 removal efficiency of the amine absorber.

469359 Mass Separation By Metamaterials

Abstract:

Mass Separation by Metamaterials

In this talk, we introduce a novel physical approach to perform mass separation based on mass-diffusion metamaterials that can cloak one compound while concentrating the other, being this the first metamaterial that can sort chemical and biomolecular species. Such novel metamaterial provides the basis for a new method to manipulate mass diffusion and achieve separations in chemical and biomolecular engineering. We employ coordinate transformations to the Fickâ€™s law equation to simultaneously manipulate the diffusion paths of different species in both transient and steady state regimes. The design of an ideal, non-homogeneous, anisotropic mass-separator metamaterial device is first introduced. We also present a design strategy that allows for experimental realization of such metamaterial using homogeneous isotropic materials. A practical proof-of-concept is introduced by separating a binary mixture of oxygen and nitrogen diffusing through a polymeric matrix. The simultaneous metamaterial manipulation of two different chemical species has not been reported. This work paves the way to achieve mass separations using metamaterials devices in chemical and biomolecular science and technology.

Rational design of metamaterial structures enables unprecedented manipulation of physical properties in research areas that include optics, acoustics, mechanics, and thermal transport. Initially, metamaterial cloaking shells designed by coordinate transformations and conformal mapping were shown to render objects invisible to light. In recent years, metamaterial theory has expanded beyond the concept of invisibility, making possible a variety of novel optical devices such as light concentrators, splitters, and perfect lenses. The principle behind optical metamaterials is the invariance of Maxwellâ€™s equations under coordinate transformations. Following the remarkable success in optics, metamaterials have been further extended to other research areas by recognizing that the fundamental equations of sound, elastic vibrations, matter-wave phenomena, as well as those of heat (and mass transfer under steady-state conditions), are also invariant under coordinate transformations. This extension of metamaterial theory and concepts has led to a large number of counter-intuitive methods to control optical, acoustical, mechanical, and thermal properties.

Mass flow manipulation is of utmost importance in many physical processes in chemical and biomolecular science, since separation and catalysis require precise control of the diffusion of relevant species. Separation systems are critically important in several different research areas ranging from chemical manufacturing, recovery of biological solutes from wastes, to semiconductor crystal growth, and generation of artificial kidneys. Towards this end, a systematic method to achieve precise manipulation of molecular diffusion paths is presented by coordinate transformations and metamaterials. We introduce a novel approach to obtain mass separation based on metamaterials that can sort chemical and biomolecular species by cloaking one compound while concentrating the other. We present a practical case where a mixture of oxygen and nitrogen is manipulated using a metamaterial that cloaks nitrogen and concentrates oxygen, laying the foundation for molecular mass separation in biophysical and chemical systems through metamaterial devices.

466429 Nanocellulose Based Facilitated Transport Membranes for CO2 Separation

Abstract:

A new type of facilitated transport membranes, based on Microfibrillated Cellulose (MFC), have been investigated for CO_2 separation applications. In particular, the films were obtained by mixing an aqueous suspension of MFC with Lupamin (a Polyvinylamine produced by BASF), exploring different solid contents and concentrations.

The permeability of CO_2, CH_4 and N_2 was measured at 35 °C as a function of relative humidity. Water sorption experiments were also performed, allowing a better understanding of the permeability results with respect to the actual water content within the membrane. As a reference, pure MFC films have been also prepared and their gas permeability tested in the same conditions.

The overall results suggest that both MFC and MFC/PVAm films have really interesting performances for both the CO_2/CH_4 and CO_2/N_2 separations, showing very high selectivity values (higher than 400) and placing most materials well above the trade-off curve of 2008 Robesonâ€™s plot.

MFC films showed higher selectivity but lower CO_2 permeability with respect to the MFC-Lupamin blends, likely because of the different level of water absorbed by the two polymeric matrixes. Indeed, pure MFC never exceeded 10 %wt water uptake, while the Polyvinylamine blends showed a clear upturn in the water sorption curve for RH larger than 60%, reaching a mass uptake higher than 50%wt at the maximum water activity inspected.

451730 CMS Membranes for Nitrogen and Sour Gases Purification from Natural Gas

Abstract:

Nitrogen and sour gas removal from natural gas is traditionally an energy intensive process, due to similarities between the thermodynamic properties of the two species. Membrane separation shed new light for this application . In particular, Carbon Molecular Sieve Membranes (CMSM) have shown higher values in both the selectivity and fluxes.

This research prepared CMSMs from a commercially available polymer Kapton® Polyimide for the separation of N_2/CH_4 and CO_2/CH_4 gas pairs. The membrane were pyrolyzed at 3 different temperatures (600, 700, and 800°C) and tested for pure gas of N_2, CO_2, CH_4, and Helium.

The permeation data showed that the membrane with good separation performance for CH_4/N_2 gas pair as well as for CO_2/CH_4 gas pair with interesting results in gas separation. By comparing the permeation fluxes at different temperatures and membrane of different thicknesses, the effect of surface diffusion/surface flux were found to be significant for sorptive species like CO_2 and CH_4, but less prominent on N_2 and helium.

ID# 469304

Reverse-Selectivity of the Supported Ionic Liquid Membranes

Amir Khakpay, Paul Scovazzo

University of Mississippi, Department of Chemical Engineering, 134 Anderson Hall University, MS 38677, United States

Abstract

This experimental study explores the reverse-selective potential of RTIL based membranes by lowering the solubility of the non-condensing gases like N_2 and CH_4. Particular attention was given to the reverse-selective behavior in the propane/methane and propane/nitrogen separations. For this purpose, a series of experiments were conducted in a batch gas permeance system. The permeance of the above-mentioned gases was obtained using polyvinylidene fluoride (PVDF) as the support and [emim][Tf_2N], [emim][BF_4], [emim][TfO], [emim][DCA], [emim][SCN], and [bmim][NO_3] as immobilized room temperature ionic liquid (RTIL). Furthermore, the permeance of CH_4 and N_2 were investigated in terms of hydrogen bond accepting ability of the RTIL, as an indication of the gas/ionic liquid solution nonideality. The results showed that an increase in the nonideality of the solution leads to a decrease in the permeance of CH_4 and N_2. Moreover, reverse-selectivity decreases with an increase in the viscosity of the RTILs.

Keywords: RTIL, SILM, permeance, selectivity, reverse-selectivity, solubility, diffusivity.

1. Introduction

Membranes have emerged as viable gas-separation systems due to their lower energy consumption, higher reliability, and lower initial cost requirements [1,2]. Membrane technology for separation of gas mixtures has progressed considerably over the past decades [3]. Gas separation using membranes for industrial applications are mainly for the separation of simple gases; such as, H_2, O_2, N_2, CH_4, CO_2, or vapor recovery from gas mixtures [4]. Efforts to improve membrane performance have been directed at the membrane material synthesis and fabrication methods. Although most polymeric membranes are low-priced, easy to make, and easy to scale up, their gas separation performance does not always meet the required separation and efficiency [3].

Room temperature ionic liquids (RTIL) are salts that are liquid at room temperature [5]. RTILs have drawn great attention in recent years for the fabrication of novel and effective materials for special tasks such as carbon dioxide capture [6]. RTILs are environmental friendly solvents with special properties such as negligible vapor pressure, chemical and thermal stability, recyclability, and non-flammability [7]. The integration of ionic liquids and membranes is predicted to play an important role in providing cost-effective and energy efficient CO_2 separation technologies to replace traditional methods [8]. Supported ionic liquid membranes (SILM) are porous membranes in which ionic liquid is immobilized inside the pores of a polymeric or inorganic support [9,10].

The solubility of the CH_4 and its deviation from the ideal solution was studied by Carvalho et al. [11]. The nonideal solubility behavior of the CH_4 in the RTILs results from a delicate balance between the solute-solute, solvent-solvent and solute-solvent interactions [11]. According to Carvalho et al. [11], the nonideal solution behavior of CH_4 solution in RTILs is due to the RTILs' polarity. The higher the polarity, the larger the deviation from the ideal solution behavior [11]. Therefore, in order to decrease the solubility of CH_4 in the RTILs, the polarity of the solution should be maximized and as a result, the selectivity of the CO_2/CH_4 separation will be maximized [11].

Separation of higher hydrocarbons such as propane and butane is a typical stage in the natural gas treatment. Natural gas is treated to eliminate water, CO_2, sulfur components, mercury vapors, and higher hydrocarbons [12]. The separation of higher hydrocarbons using membranes is an interesting alternative and economic analyses show feasibility with short payback periods [12,13]. Glassy polymers such as polyacetylenes, functionalized polysulfones, and thermally-rearranged cardocopolybenzoxazoles show excellent gas transport properties and reverse selectivity for the separation of condensable hydrocarbons or CO_2 from permanent gases [14].

The purpose herein is to explore the reverse-selective potential of RTIL based membranes by lowering the solubility of the non-condensing gases like N_2 and CH_4. This solubility is one characteristic of reverse-selective membranes affecting membrane selectivity. To our best knowledge, the reverse-selective behavior of the RTIL based membranes has not been explored previously.

The C_3H_8/N_2 separation is a prime indicator for the reverse selectivity [15]. Therefore, the main objective of this work is to examine the reverse-selective potential of RTIL based membranes for C_xH_y/CH_4 separations as well as C_3H_8/N_2 separation in order to prove, or disprove, reverse-selectivity. One of the potential applications of the reverse-selective membranes discussed by Freeman and Pinnau [16] is the separation of larger molecules like C_3H_8 from CH_4.

2. Experimental

2.1. Materials

The following RTILs were purchased from IOLITEC Inc. (Tuscaloosa, Alabama): 1-Ethyl-3-methylimidazolium bis[(trifluoromethyl)sulfonyl]imide ([emim][Tf_2N]), 1-ethyl-3-methylimidazolium thiocyanate ([emim][SCN]), 1-ethyl-3-methylimidazolium trifluoromethanesulfonate ([emim][TfO]), 1-ethyl-3-methylimidazolium dicyanamide ([emim][DCA]), and 1-ethyl-3-methylimidazolium tetrafluoroborate ([emim][BF_4]). Moreover, 1-butyl-3-methylimidazolium nitrate ([bmim][NO_3]) was prepared in the lab. Table 1 contains the relevant properties of the RTILs [17–35].

Table 1. Physical properties of the used RTILs

RTIL	Chemical formula	Melting point °C	Molar volume (m^3/mol)×10^4	Viscosity at 30°C (cP)	Hydrogen bond accepting ability (β)	Reference
[emim][Tf_2N]	$C_8H_{11}F_6N_3O_4S_2$	-16	2.59	26.28	0.42	[18–20,26,35]
[bmim][NO_3]	$C_8H_{15}N_3O_3$	-	1.75	144.00	0.74	[17,21,35]
[emim][SCN]	$C_7H_{11}N_3S$	-	1.52	20.79	0.71	[22–25,35]
[emim][TfO]	$C_7H_{11}F_3N_2O_3S$	-10.92	1.79	35.98	0.57	[22,27,28,35]
[emim][DCA]	$C_8H_{11}N_5$	-4.85	1.61	14.78	0.64	[29–31,35]
[emim][BF_4]	$C_6H_{11}BF_4N_2$	14.85	1.55	23.35	0.55	[32–35]

Ultrahigh purity nitrogen and methane were purchased from NexAir (Memphis, Tennessee). Also, propane was obtained from Conley Gas (LA Porte, Texas). The porous hydrophilic polyvinylidene fluoride (PVDF) (obtained from Millipore Corporation) and porous hydrophobic polytetrafluoroethylene (PTFE) (obtained from Pall Corporation) were used for membrane fabrications as stabilizing supports. The characteristics of the supports used in this work are given in Table 2.

Table 2. Characteristics of the PVDF and PTFE supports used in this study

Polymer support	Diameter (mm)	Thickness (µm)	Porosity	Tortuosity	Nominal pore size (µm)
PVDF	47	125	0.70	1	0.1
PTFE	47	178	-	-	0.5

2.2. Apparatus

A detailed description of the experimental apparatus appears in Morgan et al. [19], the following is a brief summary. The apparatus consists of two stainless steel chambers as feed and permeate chambers, a vacuum pump (Fisher Scientific Maxima C Plus, Model M8C with an ultimate pressure rating of 10^{-4} Torr), a 0-5-psia pressure transducer (Omega PX811-005AV), and a PC for recording data. Furthermore, the unit is in an insulated box with a temperature controller. There is a septum port to inject test gases to the feed chamber using a syringe. In this work, two Viton O-rings were used to hold vacuum.

2.3. Methods

2.3.1. SILM fabrication

The SILMs were prepared by impregnating the polymer supports with 1 mL of RTIL. First 0.5 ml of the RTIL was spread on a watch glass using a syringe. Then, the polymer support was placed on it to absorb the RTIL. This process minimized the amount of trapped air during immobilization of the RTIL. In addition to minimize the trapped air in the back of the active side of the support, the active side was placed on top of the liquid. Upon completion of the membrane wetting, the remaining 0.5 mL of the RTIL was spread over the membrane until it was completely soaked with the RTIL. The prepared membrane was degassed and dehydrated by a vacuum desiccator overnight, after which, the excess of the RTIL was removed from the surface of the membrane using a filter paper before mounting in the apparatus.

2.3.2. Mounting and testing the prepared membrane

Before mounting the newly prepared membrane, the old one was removed from the apparatus plus the feed and permeate chambers cleaned using ethanol and Kimwipes. To prevent the displacement of liquid from pores under pressure, a layer of the hydrophobic Polytetrafluoroethylene (PTFE) was placed beneath the prepared membrane before mounting it inside the cell. The PTFE prevents the RTIL from being displaced from the support under pressure. The hydrophobic nature of the PTFE support prevents its wetting with the RTIL. After mounting the membrane inside the cell, the system was put under vacuum (< 4 Pa) overnight for degassing. A gas volume of 10 ml was injected into the cell. The pressure at the permeate chamber was measured continuously for 5 hours using a pressure transducer. After each experiment, vacuum was pulled to remove the remaining gases and to degas the RTIL for 45 minutes. Next, a new sample was injected into the cell. All experiments were performed at least three times to test the reproducibility and repeatability of tests. Furthermore, experiments were carried out using a new support for each RTIL to ensure that variability due to support was accounted for in the tests.

3. Results and Discussion

3.1. Permeance results

In this study, the experimental permeances were determined based on the assumption that gas transport is completely through the RTIL and the effect of the PVDF support on the gas permeance is negligible due to very low PVDF permeance for the test gases.

The following RTILs were selected according to their hydrogen bond accepting ability: [emim][Tf_2N], [emim][BF_4], [emim][TfO], [emim][DCA], [emim][SCN], and [emim][NO_3]. The permeances were determined using the steady state region of data. The pure gas permeances for N_2, CH_4, and C_3H_8 were determined in the initial pressure range of 0.76 to 1.51 psi at 30°C and are presented in

Table 3. Based on the reported data in literature, the effect of the cation on the gas permeance is negligible compared to that of the anion [10,36,37]. Therefore we did not vary the cation but kept it as an experimental constant. In this work, all of the cations are [emim] except for [bmim][NO_3], since [emim][NO_3] is solid at room temperature. The permeances are the lowest for [bmim][NO_3], while the highest permeance values are for [emim][Tf_2N]. The permeance of gases in the RTILs is mostly affected by the gas solubility in the RTILs.

Table 3. Experimental gas permeances at 30°C

Gas	Permeance (GPUs) with standard deviations					
	[emim][Tf_2N]	[emim][BF_4]	[emim][TfO]	[emim][DCA]	[emim][SCN]	[bmim][NO_3]
N_2	0.142±0.006	0.042 ±0.002	0.074±0.004	0.028±0.004	0.048±0.008	0.023±0.006
CH_4	0.26±0.01	0.078±0.003	0.17±0.02	0.092±0.004	0.096±0.009	0.070±0.006
C_3H_8	0.67±0.03	0.14±0.01	0.34±0.02	0.26±0.01	0.64±0.08	0.130±0.008

3.2. CH_4 and N_2 permeances based on the hydrogen bond acceptor property

The nonideal solubility of CH_4, studied by Carvalho et al. [11], can be attributed to the polarity of the RTILs, which can be determined using Kamlet-Taft parameters. The hydrogen bond accepting ability (β) was determined for the tested RTILs by Lungwitz's work [35]. The values of β are reported in Table 1.

Figure 1 shows the CH_4 and N_2 permeances versus β values. The plotted data are consistent with β values except for [emim][BF_4], in which it seems that the interactions of CH_4 with the RTIL is different from the other RTILs. [emim][BF_4] shows different behavior and it breaks down at room temperature in the presence of water. The R-squared values for the fitted lines when neglecting [emim][BF_4] are 0.99 and 0.92 for CH_4 and N_2, respectively. Moreover, when [emim][BF_4] are considered, the R-squared values were decreased to 0.66 and 0.74 CH_4 and N_2, respectively. The nonideal behavior of CH_4 in the RTILs, unlike CO_2, is due to the strong solute-solvent interactions, which leads to a positive deviation from the ideal solutions. The permeance of CH_4, as it was proposed by Carvalho and Coutinho [11], decreases with an increase in the hydrogen bond accepting ability of the RTILs; higher β values leads to more deviation from the ideal solution. The nonideal behavior of the CH_4 solubility in the RTILs confirms Carvalho's hypothesis [11]. In Figure 3, N_2 shows similar behavior with respect to the β values except for [emim][BF_4]. Therefore, as before, the N_2 solubility can be attributed to the interactions between N_2 and the RTILs. Furthermore, the permeance and nonideal solubility of N_2 in the RTILs can be correlated by hydrogen bond accepting ability of the tested RTILs.

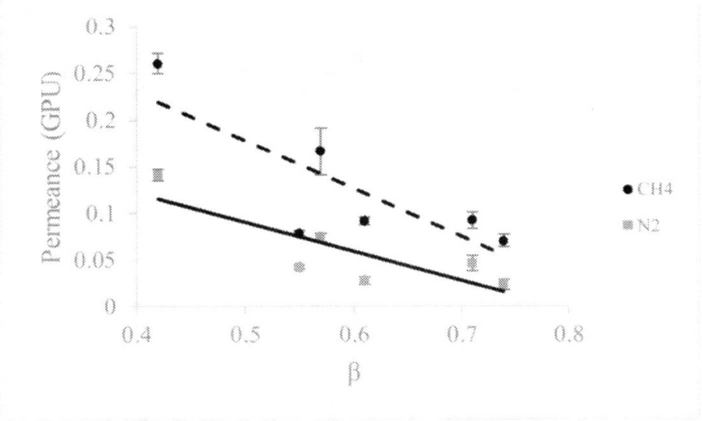

Figure 1. The experimentally determined N_2 and CH_4 permeances versus β values at 30°C. The hydrogen bond accepting ability of the RTIL has negative impact on the N_2 and CH_4 permeances and an increase in the hydrogen bond accepting ability of the RTIL leads to a decrease in the N_2 and CH_4 permeances.

3.3. Selectivity results

The ideal selectivity, α_{ij}, was determined by dividing the permeance of the faster permeating gas i with the permeance of the slower permeating gas j. Table 4 shows the selectivities for C_3H_8/CH_4 and C_3H_8/N_2 separations at 30°C. Moreover, perm-selectivity ranges for different gas pairs separations are:

- C_3H_8/CH_4: 2 to 7
- C_3H_8/N_2: 3 to 14

For the C_3H_8/CH_4 separation, the highest selectivity is for [emim][SCN] with a value of 7, while the lowest selectivity is for [bmim][NO_3] with a value of 2. Furthermore, The highest and the lowest selectivity values for the C_3H_8/N_2 separation are 3 and 14 for [emim][BF_4] and [emim][SCN], respectively.

The highest reported selectivities for the C_3H_8/CH_4 and C_3H_8/N_2 separations in literature for polymer membranes are 5.7 and 7.5, respectively [1,12,13]; compared to respective highest in this study of 7 and 14 (for [emim][SCN]) The comparison of the RTIL-membranes and polymeric membranes shows that RTIL-membranes are better than the polymeric membranes for the separation of C_3H_8/CH_4 and C_3H_8/N_2. Moreover, RTIL-membranes showed better performance when compared to the polymeric membranes.

Table 4. Experimental gas selectivities at 30°C

Gas	Selectivity					
	[emim][Tf_2N]	[emim][BF_4]	[emim][TfO]	[emim][DCA]	[emim][SCN]	[bmim][NO_3]
C_3H_8/CH_4	2.6±0.2	1.8±0.2	2.1±0.4	2.8±0.2	7±2	1.8±0.3
C_3H_8/N_2	4.8±0.4	3.3±0.4	4.7±0.5	9±2	14±4	6±2

3.6. Reverse-selective behavior of the C_3H_8/CH_4 and C_3H_8/N_2 separations

Reverse-selective membranes permeate larger molecules faster than the smaller ones; as shown by our reported selectivities of C_3H_8 over CH_4 and N_2. These membranes are getting more attention due to their efficiency in the separation of volatile hydrocarbons from permanent gases [14]. The favorable separation of condensable gases or vapors over permanent gases can be achieved by the competitive gas sorption or solubility. The first attempt to investigate the reverse-selective behavior of SILMs was done by Yokozeki et al. [14]. They tried to separate CO_2 from hydrogen [14]. However, to our best knowledge, the reverse-selective ability of SILMs for the separation of volatile hydrocarbon gas pairs has not been investigated. In this section, the reverse-selective behavior of C_3H_8/CH_4 and C_3H_8/N_2 separations using SILMs is investigated. Table 7 shows the kinetic diameters for the test gases.

Table 5. Kinetic diameters for the test gases

	CO_2	N_2	CH_4	C_3H_6	C_3H_8	C_4H_8	C_4H_{10}
Kinetic diameter (Å)	3.30	3.64	3.80	4.50	4.30	4.50	4.30

The kinetic diameter of C_3H_8 is larger than those of CH_4 and N_2; therefore, for the diffusion mechanism based membranes, CH_4 and N_2 should permeate faster than C_3H_8. However, for the reverse-selective membranes, C_3H_8 should have higher permeance. Figure 2 shows the permeance of C_3H_8, CH_4, and N_2 against kinetic diameter of the gases. The selected RTILs are [emim][Tf$_2$N], [emim][DCA], [emim][SCN], and [bmim][NO$_3$]. As seen in this figure, the permeance difference for the C_3H_8, CH_4, and N_2 varies considerably for the less viscous (viscosity <60 cP) RTILs confirming reverse-selective behavior of the C_3H_8/CH_4 and C_3H_8/N_2 separations. The difference between gas permeances decreases with an increase in the viscosity of the RTILs. As the viscosity increases, the permeance differences for C_3H_8, CH_4, and N_2 decrease, which may indicate a change from solubility dominated transport to the dominance of a gas diffusion mechanism for the gas transmission. The higher solubility of C_3H_8 compared to CH_4 and N_2 in the tested RTILs is responsible for the reverse-selectivity behavior.

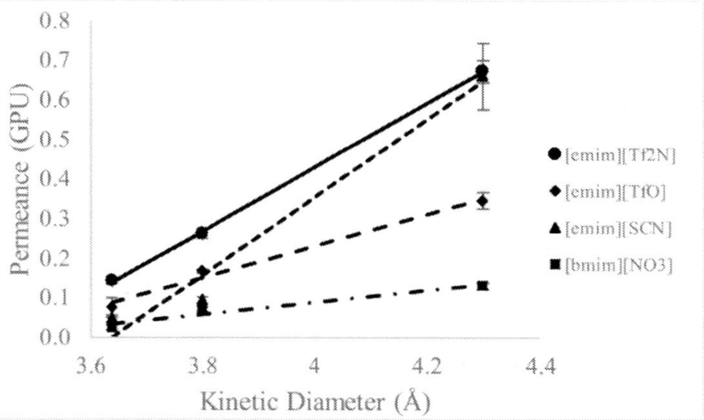

Figure 2. The permeances of C_3H_8, CH_4, and N_2 versus kinetic diameter at 30°C. The lines are provided to show trends and behavior of RTILs. The lines are drawn to show the trend. The permeance difference decreases with an increase in the RTIL viscosity.

The effect of the hydrogen bond accepting ability of the RTILs on the selectivity of C_3H_8/CH_4 and C_3H_8/N_2 in Figure 3. The β parameter has a positive effect on the selectivity of the C_3H_8/CH_4 and C_3H_8/N_2 separations and an increase in the β value leads to an increase in the selectivities. However, RTILs with large β values have larger viscosity. Therefore, there is a tradeoff between choosing a large β and a high selectivity because of a change in the transport mechanism. In Figure 3, [bmim][NO$_3$] is not considered because its viscosity is one order of magnitude higher than the other tested RTILs and the permeances are not quantfiable. CH_4 and N_2 are slow permeable gases and according to the Carvalho's hypothesis [11], to decrease the CH_4 solubility in the RTILs, the nonideality of the solution should increase. Our data and results confirm the Carvalho's theory [11] with great accuracy. N_2 showed a similar behavior to CH_4 and its permeance change can be explained by a change in the β value.

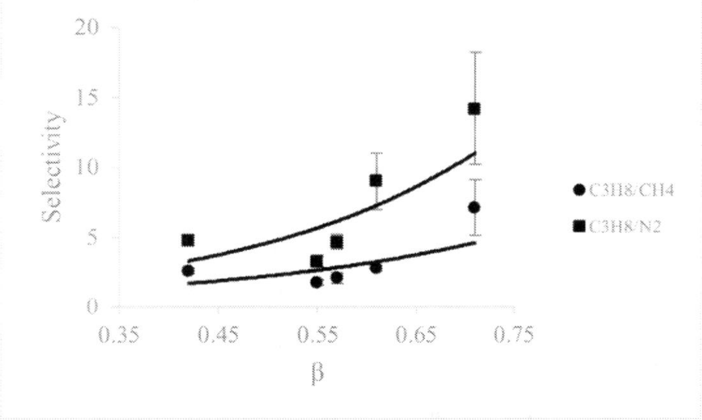

Figure 3. The effect of the hydrogen bond accepting ability of the tested RTILs on the selectivities of C_3H_8/CH_4 and C_3H_8/N_2 separations at 30°C. The lines are determined using proposed correlations. The selectivity of the gas pairs are well correlated using the hydrogen bond accepting ability of RTIL.

4. Conclusions

This work measured the permeance of selected gases in RTIL based membranes using PVDF support. Following RTILs were examined: [emim][Tf_2N], [emim][BF_4], [emim][TfO], [emim][DCA], [emim][SCN] and [emim][NO_3]. The permeance of the N_2 and CH_4 is the lowest for [bmim][NO_3]. The permeances of the N_2 and CH_4 are decreasing according to the hydrogen bond accepting ability confirming Carvalho's hypothesis [11]. Furthermore, selectivities of the selected gas pairs were obtained. In addition, the reverse-selective behavior of the C_3H_8/CH_4 and C_3H_8/N_2 separations were examined. The higher permeance of C_3H_8 in comparison with those of CH_4 and N_2 is due to solution mechanism of permeance in the RTILs. However, the selectivities were decreased for the viscous RTILs which may be due to an increasing dominance of diffusion mechanism for the permeance.

5. References

[1] I. Tirouni, M. Sadeghi, M. Pakizeh, Separation of C3H8 and C2H6 from CH4 in polyurethane–zeolite 4Å and ZSM-5 mixed matrix membranes, Sep. Purif. Technol. 141 (2015) 394–402. doi:10.1016/j.seppur.2014.12.012.

[2] A. Khosravi, M. Sadeghi, Separation performance of poly(urethane-urea) membranes in the separation of C2 and C3 hydrocarbons from methane, J. Memb. Sci. 434 (2013) 171–183. doi:10.1016/j.memsci.2013.01.025.

[3] M. Askari, T.-S. Chung, Natural gas purification and olefin/paraffin separation using thermal cross-linkable co-polyimide/ZIF-8 mixed matrix membranes, J. Memb. Sci. 444 (2013) 173–183. doi:10.1016/j.memsci.2013.05.016.

[4] P. Li, M.R. Coleman, Synthesis of room temperature ionic liquids based random copolyimides for gas separation applications, Eur. Polym. J. 49 (2013) 482–491. doi:10.1016/j.eurpolymj.2012.11.016.

[5] P. Scovazzo, J. Kieft, D. Finan, C. Koval, D. Dubois, R. Noble, Gas separations using non-hexafluorophosphate [PF6]− anion supported ionic liquid membranes, J. Memb. Sci. 238 (2004) 57–63. doi:10.1016/j.memsci.2004.02.033.

[6] J.E. Bara, E.S. Hatakeyama, C.J. Gabriel, X. Zeng, S. Lessmann, D.L. Gin, R.D. Noble, Synthesis and light gas separations in cross-linked gemini room temperature ionic liquid polymer membranes, J. Memb. Sci. 316 (2008) 186–191. doi:10.1016/j.memsci.2007.08.052.

[7] D.Y. Xing, N. Peng, T.S. Chung, Formation of cellulose acetate membranes via phase inversion using ionic liquid, [BMIM]SCN, As the solvent, Ind. Eng. Chem. Res. 49 (2010) 8761–8769. doi:10.1021/ie1007085.

[8] J. Albo, T. Yoshioka, T. Tsuru, Porous Al2O3/TiO2 tubes in combination with 1-ethyl-3-methylimidazolium acetate ionic liquid for CO2/N 2 separation, Sep. Purif. Technol. 122 (2014) 440–448. doi:10.1016/j.seppur.2013.11.024.

[9] A. Dahi, K. Fatyeyeva, D. Langevin, C. Chappey, S.P. Rogalsky, O.P. Tarasyuk, A. Benamor, S. Marais, Supported ionic liquid membranes for water and volatile organic compounds separation: Sorption and permeation properties, J. Memb. Sci. 458 (2014) 164–178. doi:10.1016/j.memsci.2014.01.031.

[10] E. Santos, J. Albo, a. Irabien, Acetate based Supported Ionic Liquid Membranes (SILMs) for CO2 separation: Influence of the temperature, J. Memb. Sci. 452 (2014) 277–283. doi:10.1016/j.memsci.2013.10.024.

[11] P.J. Carvalho, J. a. P. Coutinho, The polarity effect upon the methane solubility in ionic liquids: a contribution for the design of ionic liquids for enhanced CO2/CH4 and H2S/CH4 selectivities, Energy Environ. Sci. 4 (2011) 4614. doi:10.1039/c1ee01599k.

[12] M. Arruebo, J. Coronas, M. Menéndez, J. Santamaría, Separation of hydrocarbons from natural gas using silicalite membranes, Sep. Purif. Technol. 25 (2001) 275–286. doi:10.1016/S1383-5866(01)00054-5.

[13] A. Khosravi, M. Sadeghi, Separation performance of poly(urethane–urea) membranes in the separation of C2 and C3 hydrocarbons from methane, J. Memb. Sci. 434 (2013) 171–183. doi:10.1016/j.memsci.2013.01.025.

[14] C.H. Lau, P. Li, F. Li, T.S. Chung, D.R. Paul, Reverse-selective polymeric membranes for gas separations, Prog. Polym. Sci. 38 (2013) 740–766. doi:10.1016/j.progpolymsci.2012.09.006.

[15] A. Javaid, D. a. Krapchetov, D.M. Ford, Solubility-based gas separation with oligomer-modified inorganic membranes: Part III. Effects of synthesis conditions, J. Memb. Sci. 246 (2005) 181–191. doi:10.1016/j.memsci.2004.08.017.

[16] B. Freeman, I. Pinnau, Separation of Gases Using Solubility-Selective Polymers, Trends Polym. Sci. 5 (1997) 167–173.

[17] B. Mokhtarani, A. Sharifi, H.R. Mortaheb, M. Mirzaei, M. Mafi, F. Sadeghian, Density and viscosity of 1-butyl-3-methylimidazolium nitrate with ethanol , 1-propanol , or 1-butanol at several temperatures, J. Chem. Thermodyn. 41 (2009) 1432–1438. doi:10.1016/j.jct.2009.06.023.

[18] A.A. Miran Beigi, M. Abdouss, M. Yousefi, S.M. Pourmortazavi, A. Vahid, Investigation on physical and electrochemical properties of three imidazolium based ionic liquids (1-hexyl-3-

methylimidazolium tetrafluoroborate, 1-ethyl-3-methylimidazolium bis(trifluoromethylsulfonyl) imide and 1-butyl-3-methylimidazolium methylsulfate), J. Mol. Liq. 177 (2013) 361–368. doi:10.1016/j.molliq.2012.10.025.

[19] D. Morgan, L. Ferguson, P. Scovazzo, Diffusivities of Gases in Room-Temperature Ionic Liquids: Data and Correlations Obtained Using a Lag-Time Technique, Ind. Eng. Chem. Res. 44 (2005) 4815–4823. doi:10.1021/ie048825v.

[20] C. Lourenço, C.I. Melo, R. Bogel-Łukasik, E. Bogel-Łukasik, Solubility advantage of pyrazine-2-carboxamide: Application of alternative solvents on the way to the future pharmaceutical development, J. Chem. Eng. Data. 57 (2012) 1525–1533. doi:10.1021/je300044x.

[21] K. Seddon, A. Stark, M.-J. Torres, 627221.pdf, in: Clean Solvents, ACS Symposium Series, 2002: pp. 34–49.

[22] M.G. Freire, A.R.R. Teles, M. a a Rocha, B. Schröder, C.M.S.S. Neves, P.J. Carvalho, D. V. Evtuguin, L.M.N.B.F. Santos, J. a P. Coutinho, Thermophysical characterization of ionic liquids able to dissolve biomass, J. Chem. Eng. Data. 56 (2011) 4813–4822. doi:10.1021/je200790q.

[23] S. Seki, S. Tsuzuki, K. Hayamizu, Y. Umebayashi, N. Serizawa, K. Takei, H. Miyashiro, Comprehensive refractive index property for room-temperature ionic liquids, J. Chem. Eng. Data. 57 (2012) 2211–2216. doi:10.1021/je201289w.

[24] M. Królikowska, T. Hofman, Densities, isobaric expansivities and isothermal compressibilities of the thiocyanate-based ionic liquids at temperatures (298.15-338.15 K) and pressures up to 10 MPa, Thermochim. Acta. 530 (2012) 1–6. doi:10.1016/j.tca.2011.11.009.

[25] G. Mchale, C. Hardacre, R. Ge, N. Doy, R.W.K. Allen, J.M. Macinnes, M.R. Bown, M.I. Newton, Density - Viscosity Product of Small-Volume Ionic Liquid Samples Using Quartz Crystal Impedance Analysis, 80 (2008) 5806–5811.

[26] R.L. Gardas, M.G. Freire, P.J. Carvalho, I.M. Marrucho, I.M. a Fonseca, A.G.M. Ferreira, P G T Measurements of Imidazolium-Based Ionic Liquids, (2007) 1881–1888.

[27] R.L. Gardas, H.F. Costa, M.G. Freire, P.J. Carvalho, I.M. Marrucho, I.M. a. Fonseca, A.G.M. Ferreira, J. a. P. Coutinho, Densities and Derived Thermodynamic Properties of Imidazolium-, Pyridinium-, Pyrrolidinium-, and Piperidinium-Based Ionic Liquids, J. Chem. Eng. Data. 53 (2008) 805–811. doi:10.1021/je700670k.

[28] P. Wachter, H.G. Schweiger, F. Wudy, H.J. Gores, Efficient determination of crystallisation and melting points at low cooling and heating rates with novel computer controlled equipment, J. Chem. Thermodyn. 40 (2008) 1542–1547. doi:10.1016/j.jct.2008.05.012.

[29] P. Wachter, C. Schreiner, H.-G. Schweiger, H.J. Gores, Determination of phase transition points of ionic liquids by combination of thermal analysis and conductivity measurements at very low heating and cooling rates, J. Chem. Thermodyn. 42 (2010) 900–903. doi:10.1016/j.jct.2010.03.003.

[30] P.D. a. Bastos, F.S. Oliveira, L.P.N. Rebelo, A.B. Pereiro, I.M. Marrucho, Separation of azeotropic mixtures using high ionicity ionic liquids based on 1-ethyl-3-methylimidazolium thiocyanate, Fluid Phase Equilib. 389 (2015) 48–54. doi:10.1016/j.fluid.2015.01.011.

[31] E. Quijada-Maldonado, S. Van Der Boogaart, J.H. Lijbers, G.W. Meindersma, a. B. De Haan, Experimental densities, dynamic viscosities and surface tensions of the ionic liquids series 1-ethyl-

3-methylimidazolium acetate and dicyanamide and their binary and ternary mixtures with water and ethanol at T = (298.15 to 343.15 K), J. Chem. Thermodyn. 51 (2012) 51–58. doi:10.1016/j.jct.2012.02.027.

[32] M.R. Currás, P. Husson, A.A.H. Pa, M.F.C. Gomes, J. Garc, D.D.F. Aplicada, D. Ciencias, U. De Vigo, E.- Vigo, C. Universite, U.B. Pascal, I.D.C. De Clermont-ferrand, High-Pressure Densities of 2 , 2 , 2-Trifluoroethanol + Ionic Liquid Mixtures Useful for Possible Applications in Absorption Cycles, (2014).

[33] Y. a. Sanmamed, D. González-Salgado, J. Troncoso, L. Romani, a. Baylaucq, C. Boned, Experimental methodology for precise determination of density of RTILs as a function of temperature and pressure using vibrating tube densimeters, J. Chem. Thermodyn. 42 (2010) 553–563. doi:10.1016/j.jct.2009.11.014.

[34] J. Vila, B. Fernández-Castro, E. Rilo, J. Carrete, M. Domínguez-Pérez, J.R. Rodríguez, M. García, L.M. Varela, O. Cabeza, Liquid-solid-liquid phase transition hysteresis loops in the ionic conductivity of ten imidazolium-based ionic liquids, Fluid Phase Equilib. 320 (2012) 1–10. doi:10.1016/j.fluid.2012.02.006.

[35] S. Spange, R. Lungwitz, A. Schade, Correlation of molecular structure and polarity of ionic liquids, J. Mol. Liq. 192 (2014) 137–143. doi:10.1016/j.molliq.2013.06.016.

[36] J.E. Bara, R.D. Noble, D.L. Gin, Effect of " Free " Cation Substituent on Gas Separation Performance of Polymer - Room-Temperature Ionic Liquid Composite Membranes, (2009) 4607–4610.

[37] M. Hasib-ur-Rahman, M. Siaj, F. Larachi, Ionic liquids for CO2 capture—Development and progress, Chem. Eng. Process. Process Intensif. 49 (2010) 313–322. doi:10.1016/j.cep.2010.03.008.

470980 Activated Carbon Nano-Adsorbents, a Great Opportunity to Enhance the PDMS Pervaporation Membrane Performance

Abstract:

Biobutanol produced from Acetone-Butanol-Ethanol (ABE) fermentation is considered to be a reasonable replacement for fossil fuels. However, the need for inexpensive feed-stocks, the reduction of product inhibition via in-situ recovery to achieve higher final butanol concentration, higher yield, efficient method for solvent recovery and water recycle are the obstacles which prevented making this bio-alcohol economically viable.

Pervaporation, a membrane based-process, has a great potential for in-situ butanol recovery among different solvent recovery methods such as adsorption, liquid-liquid extraction, gas stripping, etc. Based on the studies which have been done on the pervaporation separation of the butanol, polydimethylsiloxane (PDMS) have been selected to be a reliable membrane as a result of its high hydrophobicity, ease of preparation, and their good thermal, chemical and mechanical properties. Moreover, Activated carbon adsorbent showed to have outstanding characteristics in the adsorption separation of the butanol from the ABE model solution and the binary solution of butanol. This work is the first attempt to embed the nano activated carbon particles in the matrix of the PDMS membrane to improve the performance of the butanol separation by pervaporation by increasing the flux and the selectivity of the membrane.

According to the results, permeate flux for a PDMS nano composite membrane of 6 wt% increased by 42.6% at 37°C compared to the neat membrane. Furthermore, 52% increase in the separation factor was achieved at the same concentration of the nano adsorbent.

470212 Mmif-Based Mixed-Matrix Membranes for CO_2 Separation

Abstract:

Metal organic frameworks (MOFs) are promising candidates for separation and purification applications due to their tunable chemistry. Hence, MOFs are considered promising candidates as dispersive phase in mixed matrix membranes (MMMs). The organic linkers in MOFs have affinity to polymer chains so the control of MOF/polymer interface is easier than zeolite/polymer interface. Zeolite based MMMs have been studied for the last two decades and much effort has been spent to ensure good adhesion between organic and inorganic phases.

The molecular simulation studies in the literature revealed several possible MOF structures as outstanding candidates for MOF-based MMMs. One of these MOFs, which became prominent for CO_2/CH_4separation, is microporous metal–imidazolate framework (MMIF) [1]. However, no experimental data is available in the literature to confirm the predicted performances of these MOFs, in particular no MMM studies using these MOFs are reported.

In this study, first, MMIF has been synthesized and characterized in order to experimentally verify its capability of CO_2 capture. Then, it is incorporated into polyimides to prepare MMMs. Membranes are characterized with SEM, XRD, TGA, DSC and pure and mixed-gas permeability tests for CO_2/CH_4 and CO_2/N_2separations.

Contrary to simulation results and Maxwell model predictions reported in the literature, pure gas permeability measurements at 35°C and 4 bar showed that Matrimid based MMMs containing 10% MMIF improved the permeability of Matrimid by 15% and no improvement in selectivity was observed. The permeability and selectivity remained stable with an increase in MMIF loading to 20%. On the other hand,

mixed gas selectivity for CO_2/CH_4improved significantly to 88 for 10% loading.

This work is supported by TUBITAK (Turkish Scientific and Technological Research Council) through Project No. 113M776.

Reference:

1. Erucar, I, Keskin, S., "Screening Metal Organic Framework-Based Mixed-Matrix Membranes for CO_2/CH_4 Separations", *Ind. Eng. Chem. Res.*, 50 (22), pp. 12606–12616 (2011).

464695 Stability of Highly Propylene-Selective Zeolitic-Imidazolate Framework, ZIF-8, Membranes

Abstract:

Currently propylene/propane separation is performed with a highly energy-intensive cryodistillation process mainly due to their close boiling points. Although a variety of membranes with different materials have been studied for membrane-based propylene/propane separation, there are currently no membranes and membrane processes that are commercially available. This is mainly due to the fact that the majority of new membrane materials proposed failed to meet the separation performance requirements with respect to the propylene permeability and propylene/propane separation factor.[1,2,3]

Zeolitic imidazolate frameworks (ZIFs), a subclass of metal organic framework with zeolite topology, are composed of divalent metal nodes (typically Zn or Co) interconnected with imidazolate-based ligands, forming three-dimensional crystalline structures with pores and cavities in the scale of molecules.[4] ZIFs are considered very promising gas separation membrane materials because of their chemical and thermal stability in combination with their ultra-micropores (less than 0.5 nm). For example, membranes of ZIF-8, made of Zn and 2-methyl-imidazolate with a SOD structure, were reported having excellent propylene/propane separation due to the fact that its effective aperture size lies between propylene and propane.[5] Highly propylene-selective ZIF-8 membranes were synthesized using several methods including *in situ* counter diffusion and microwave-assisted secondary growth methods reported by Kwon et al.[6,7]

For industrial applications of ZIF-8 membranes, their long term stability is critically important. There are, however, only a few such studies have been reported.[8] In our long-term stability studies, we observed that time-dependent propylene/propane permeation and separation behaviors depend on synthesis methods.

Here, we would like to discuss the relationship between the membrane stability and inherent surface defects resulting from different synthesis conditions and to show a post-synthetic treatment as an effective means to stabilize membranes.

References and Notes

1 Koros, William J. and Fleming, G.K. (1993), "Membrane-based gas separation", *Journal of Membrane Science*, 83, pp 1-39

2 Baker RW., (2002) "Future Directions of Membrane Gas Separation Technology" *Industrial & Engineering Chemistry Research.*, 41 (6), pp 1393-1411

3 Tanaka, K., Taguchi, A., Hao, JQ, Kita, H., and Okamoto, K., (1996), "Permeation and separation properties of polyimide membranes to olefins and paraffins", *Journal of Membrane Science*, 121, pp 197-207

4 Park KS, Ni Z, Côté AP, et al., (2006), "Exceptional chemical and thermal stability of zeolitic imidazolate frameworks", *Proceedings of the National Academy of Sciences*, 103 (27), pp 10186-10191

5 Zhang C, Lively RP, Zhang K, Johnson JR, Karvan O, Koros WJ., (2012), "Unexpected Molecular Sieving Properties of Zeolitic Imidazolate Framework-8", *The Journal of Physical Chemistry Letters.*, 3 (16), pp 2130-2134.

6 Kwon HT, Jeong H-K., (2013), "In Situ Synthesis of Thin Zeolitic–Imidazolate Framework ZIF-8 Membranes Exhibiting Exceptionally High Propylene/Propane Separation", *Journal of the American Chemical Society*, 135 (29), pp 10763-10768.

7 Kwon HT, Jeong H-K., (2013), "Highly propylene-selective supported zeolite-imidazolate framework (ZIF-8) membranes synthesized by rapid microwave-assisted seeding and secondary growth", *Chemical Communications*, 49, pp 3854-3856

8 Liu, D., Ma, X., Xi, H., Lin, YS., (2014), "Gas transport properties and propylene/propane separation characteristics of ZIF-8 membranes", *Journal of Membrane Science*, 451, pp 85–93

470036 Predicting Mosced Parameters of Nonelectrolyte Solids from Chemical Structure for Extraction and Crystallization Processes

Abstract:

Accurate and efficient models to predict the phase-behavior of nonelectrolyte solids in a wide range of solvents are central to the design of novel separation processes. A promising design tool is the MOSCED limiting activity coefficient model which is parameterized for 133 solvents. However, before predictions may be made for a solute of interest, solute MOSCED parameters are required. For novel compounds of interest, the necessary MOSCED parameters are unavailable and sufficient reference data is likely unavailable to regress the necessary parameters.

In this study we explore the use of electronic structure calculations to generate the reference data necessary to obtain solute MOSCED parameters. Specifically, we use solvation free energies computed using the popular software packages QChem and Gaussian, which employ the SM8 and SMD solvation models, respectively. Results will be shown wherein the method is successfully applied to a wide range of compounds of biological, environmental, and industrial interests. We additionally demonstrate how the limiting activity coefficients provided by MOSCED may be used to parameterize an excess Gibbs free energy model and used to make equilibrium solubility predictions.

SIMULATION-BASED ANALYSIS FOR HIGHLY SOUR NATURAL GAS SWEETENING USING MEMBRANES/AMINES HYBRID SYSTEMS

Paper ID #: 450972

Mohammed Alkatheri[a], *Ricardo Grandas*[a,b], *Alberto Betancourt-Torcat*[a], *Ali Almansoori*[a]*

[a]*Department of Chemical Engineering, The Petroleum Institute, Sas Al Nakhl Campus, Abu Dhabi, P.O. Box 2533, United Arab Emirates.*

[b]*Department of Chemical Engineering, Universidad Industrial de Santander, Carrera 27 calle 9, Bucaramanga, Santander, Post Code 678, Colombia.*

*Corresponding Author's e-mail: aalmansoori@pi.ac.ae

Keywords: Highly sour, natural gas processing, membranes, amines, hybrid system, economics, gas sweetening.

Abstract

The most widely employed sweetening gas scheme considers chemical absorption by solvents for acid gases removal. While this scheme is both well-known and proven, its application for tapping highly sour gas resources can be operationally challenging, costly, and wasteful. Nevertheless, new polymeric membranes have been applied for bulk H_2S removal from natural gas, including very high H_2S concentrations and operating pressures. Recent developments on this area may contribute to the development of unconventionally high sour gas resources or potentially retrofitting existing plants. For instance, in a primary stage the membrane system could be used to reduce the bulk concentration of H_2S and CO_2 in the feed gas. Sequentially, the final sweet gas product specifications could be met using an amine-based system. Accordingly, this type of hybrid scheme could potentially help reducing capital and operating expenditures. The present study examines the sweetening of highly sour gas with 15% H_2S (i.e., over 20% of H_2S and CO_2 combined) applying a simulation-based analysis. The proposed hybrid process was simulated using ProMax® v3.2. The simulation results show that the operating/utility expenditures of the sweetening process could be reduced using a hybrid system approach (instead of a stand-alone amine system).

Introduction

Recently, a growing number of countries have expressed their worries over meeting their future natural gas requirements and export commitments. This trend is particularly noticeable in the Middle East region mainly due to growing power and desalinated water demands, higher living standards, and gas injection for enhanced oil recovery (EOR) operations to span the life time of existing oil fields [1]. To meet the growing natural gas requirements producers have been forced to tap highly sour gas fields, once considered economically unattractive. The presence of high sulfur content in natural gas represents a serious techno-economic challenge [2]. Gas absorption using amines is the most common sour gases removal process for natural gas sweetening. In this process, the sour gas is contacted by an aqueous solution of alkanolamines (e.g., MEA, DEA, DGA and MDEA) in a tray packed tower. However, the process is highly energy intensive due to the high thermal requirements involved in the regeneration of the solvent. The corresponding energy requirements are proportional to the acid gases concentration and natural gas feed flowrate. Hence, the implementation of new methods such as hybrid systems can be a promising solution to overcoming the technical difficulties and important operational costs involved in highly sour gas processing. In a first or 'pre-separation' stage a membrane module could help reducing the acid concentration in the feed gas; while a sequential stage composed of a gas absorption system could meet the final gas product purity specifications. Furthermore, highly sour gas processing represents an opportunity to assess new technologies. These technologies may enable cost reductions along with technical advantages. For instance, in the membrane module the partial pressure difference between permeate and feed streams is the separation driving force; while in gas absorption the chemical reactions between solvents and acid gases promote the separation. As a result, membrane separation includes: low capital costs, no additional facilities (e.g., solvents storage), easy installation, space efficiency, and low operability costs compared with absorption units. Accordingly, the hybrid process proposed in this work takes advantage of the high H_2S concentrations that favor partial pressure difference-based processes, and H_2S selective materials to reduce operating costs.

Methodology and Economic Analysis Model

Figure 1 displays the hybrid system proposed in the present study to treat highly sour natural gas. The ProMax® simulation software was used to model the hybrid system. The hybrid system is composed of a single membrane stage (pre-treatment), followed by an absorption stage. The membrane module is used for bulk acid gas removal (particularly H_2S) from the crude natural gas feed. Then, the final pipeline specifications (i.e., 1 mole% CO_2, <4 ppm H_2S) are to be met by the gas absorption unit. The membrane module separates the inlet feed gas in two streams: 1) retentate stream (generally rich in methane and other hydrocarbons), 2) permeate (acid gas rich). The membrane's permeate gas can preferably be reinjected underground, and used in EOR operations. Consequently, the proposed hybrid approach can potentially help saving capital and operating costs.

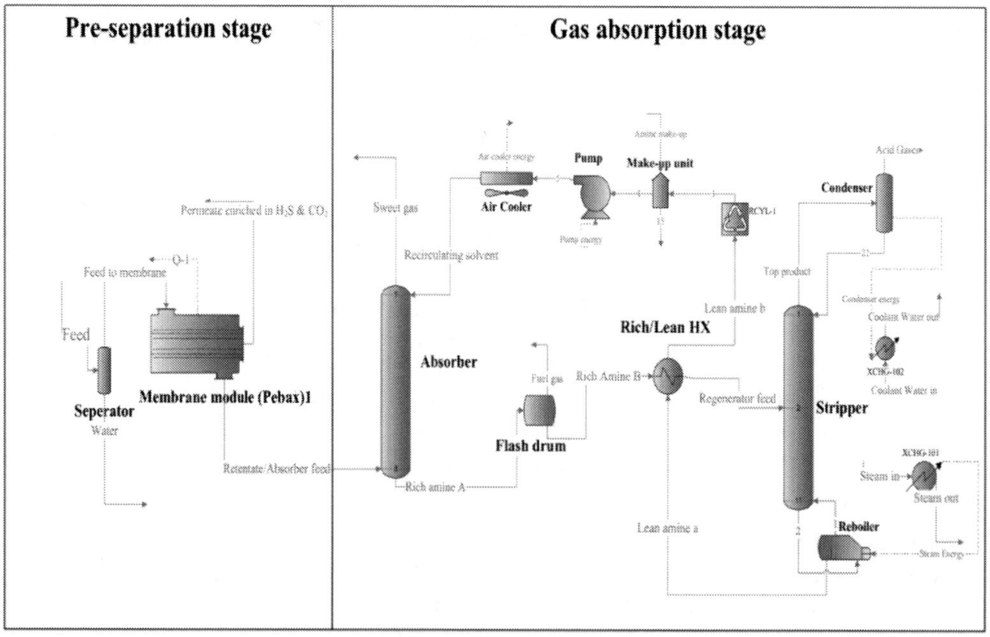

Figure 1. Hybrid process flow diagram.

Commercial Pebax™ [3-5] was chosen as the membrane material in the pre-treatment stage. Pebax grades are considered to be useful polymers for acid gas removal given their high H_2S/CH_4 selectivity, and good permeability. The membrane's permeability and selectivities are listed in Table 1 [6]. Moreover, H_2S, CO_2 and CH_4 are the only three gases considered to permeate through the membrane module. This because methane is the most valuable component of the natural gas mixture; whereas H_2S and CO_2 are the acid gases that need to be separated.

Table 1. Pebax membrane properties [6].

Property	Value
Permeability of CH_4 (stdcm^3/cm^2.s.cm Hg)	1.8
Components	**Selectivity with respect to CH_4**
H_2S	77.78
CO_2	17.22

The membrane module used in the present study represents a selective separation method. The transport of the gas through the dense (nonporous) polymer membrane occurs by a solution-diffusion mechanism. In particular, the gas dissolves in the membrane high-pressure side of the polymer, then diffuses through the polymer phase, and desorbs or evaporates at the low pressure side. The membrane performance can be estimated using the flux equation for each component as follows:

$$J_i = Q_i(P_1 x_i - P_2 y_i) \qquad (1)$$

where P_1 and P_2 are the feed and permeate side pressures, respectively.

More detailed information on the membrane performance calculation can be found in Macab [7]. The ProMax® embedded membrane tool was used to perform the membrane module calculations. Since the acid gas selectivity is not optimum, methane losses occur in the membrane stage. The methane losses were calculated because they represent a loss opportunity cost in the sweetening process. The operation of the absorber is modeled using the "TSWEET Kinetics" model. This model accounts for the varying absorption rates of the different acid gases; particularly the relative slow absorption of CO_2 that is kinetically limited.

The economics of the gas sweetening process was evaluated based on the operating costs of the hybrid system. This, because adding the membrane module to the traditional stand-alone absorption process can potentially help saving costs. Accordingly, the stand-alone and hybrid systems are compared to evaluate the techno-economic feasibility of adding the extra pre-separation stage to the traditional gas absorption method. Moreover, since the membrane elements must be periodically replaced, its cost was considered as an operating expense. Additionally, the opportunity cost of the permeate's CH_4 losses is accounted as part of the hybrid system operating cost [8]. The separation process economic evaluation heavily depends on the chosen analysis method, and economic parameter values. The method applied in the present analysis can be found with more details in [8, 9].

The membrane module (consisting of pressure vessel, membrane element, associated piping, valving, etc.) was assumed to be \$180/m^2 [10]; while the membrane elements replacement costs was assumed to be half the membrane module cost (i.e. \$90/m^2) [10, 11]. Additionally, as the membrane life time is typically short (e.g. three years), the membrane capital recovery cost was approximated to be equal to the membrane capital cost divided by the membrane life time. In the current study, the operating labor costs were neglected because all process configurations analyzed are identical, and labor costs depend on the number of unit operations. Thus, this particular cost factor can be neglected as a distinct standard for comparison purposes. Moreover, for the present analysis only the membrane's module costs are considered as capital investments because they are a one-time spending. On the other hand, the capital cost of the absorption unit was not considered for several reasons:

1. The life time of the amine sweetening unit is much larger than that of the membrane module (e.g., the membrane life time is approximately 3 years while the amine sweetening unit can reach up to 35 years).
2. The present analysis aims to compare processes with similar equipment and almost same sizes. Therefore, disregarding the amine sweetening unit will not have significant effects on the final results.
3. One of the main objectives of the present study is to assess the viability of adding a membrane module to existing amine sweeting plants.

The main objective of the present study consists of identifying the optimum values of key process variables. Particularly, the H_2S concentration in the membrane's retentate stream (i.e., which enters the gas absorber) is a key independent variable that significantly influences the overall cost of the hybrid system. A summary of the economic analysis procedure, cost equations, and techno-economic assumptions is shown in Table 2.

Table 2. Economic parameters and assumptions.

<table>
<tr><td rowspan="5">Cost analysis for membrane module.</td><td>Membrane Capital Cost, MCC ($/yr) [8]: $180/m^2 of membrane.</td></tr>
<tr><td>Membrane Replacement Cost, MRC ($/yr) [8]: $90/m^2 of membrane.</td></tr>
<tr><td>Membrane Hydrocarbon Losses, MHL ($/yr) [12]: $ 19.19/MWh.</td></tr>
<tr><td>Membrane Capital Recovery Cost, CRC ($/yr):
$CRC = \frac{MCC}{Membrane\ life\ time}$ (2)</td></tr>
<tr><td>Membrane Operating Cost, MOC ($/yr):
$MOC = MRC + MHL + CRC$ (3)</td></tr>
<tr><td rowspan="4">Cost analysis of gas absorption unit.</td><td>Utility Cost ($/yr).
Electricity Cost, EC [12]: 0.06 $/kWh.
Total power load = Pump power + Air Cooler power.
Steam Cost, SC [12]: $ 29.29 /1,000 kg.
Coolant Water Cost, CC [12]: $ 14.8/1000 m^3.
Total Utility Cost,
$TUC = EC + SC + CC$ (4)</td></tr>
<tr><td>Absorption stage Hydrocarbon Losses Cost, AHL [12]: $ 19.19/MWh.</td></tr>
<tr><td>Amine Losses Cost, ALC ($/yr).
Solvents Prices [13].<table><tr><td>Solvent type</td><td>($/kg)</td></tr><tr><td>MEA</td><td>0.5</td></tr><tr><td>DEA</td><td>3.8</td></tr><tr><td>MDEA</td><td>2.6</td></tr><tr><td>DGA</td><td>4.1</td></tr></table></td></tr>
<tr><td>Absorption Stage operating cost, AOC ($/yr).
$AOC = ALC + AHL + TUC$ (5)</td></tr>
<tr><td>Total Separation Cost.</td><td>Total Separation Cost, TSC ($/yr).
$TSC = AOC + MOC$ (6)</td></tr>
<tr><td>Other Assumptions.</td><td>Stream Factor: number of working days per year = 300 day.
Membrane life time = 3 yr.</td></tr>
</table>

Hybrid Process Analysis

In the present study, the stand-alone absorption system is considered to be the baseline to benchmark the proposed hybrid system approach, and also choose the top performing amine in terms of operating costs. The stand-alone absorption system was simulated using different alkanolamines (i.e., MEA, DEA, DGA, and MDEA) to assess each solvent's individual performance for treating highly sour gas. The rich loading and solvent concentration values used in the simulations were chosen within the range that enables carbon steel to be safely used as the plant's equipment material [14]. Moreover, the feed composition, flow rate, pressure, and temperature are strong functions of the natural gas reservoir. Accordingly, in the present analysis the highly sour gas feed conditions were obtained from data available in the literature (see details in Table 3) reflecting local Gulf country's conditions [15].

Table 3. Highly sour gas feed assumed operating conditions [15].

Operating Conditions	**Value**
Feed pressure	70 bar
Feed temperature	40 °C
Feed standard flow	250 MMSCFD
Sweet gas specification	4 ppm H_2S; $CO_2 < 1\%$ mole
Main components	**Mole %**
CO_2	5
H_2S	15
CH_4	64
Ethane	10
C_3+	4
Methyl Mercaptan	1
Ethyl Mercaptan	1

The performance of the stand-alone absorption systems using the four alkanolamines was assessed based on their corresponding total operating costs. Accordingly, the total operating cost of single absorption systems using the four different types of alkanolamines (MEA, DEA, DGA, and MDEA) for treating highly sour gas can be found listed in Table 4.

Table 4. Total operation cost for single absorption systems using different types of alkanolamines.

Variable	**Type of amine**			
	MEA	**DEA**	**MDEA**	**DGA**
Total operating cost (MM $/yr)	164	96	72	86

According to the results shown in Table 4, MDEA and DGA are the top performing solvents to be used in the absorption unit for the current specified highly sour feed gas conditions. Accordingly, both amines were selected for performing the techno-economic analysis on the hybrid system. Additionally, a pre-separation stage was introduced in the system before the absorption unit as first treatment stage (see Figure 1). The pre-separation stage consists of a membrane module, where the membrane elements are made of Pebax material. The resulting hybrid system was modeled in ProMax® to assess its performance when treating highly sour gas compared with conventional stand-alone absorption systems. For the analysis, different scenarios were evaluated. Each scenario considers a H_2S removal level in the membrane module combined with an absorption unit using one of the selected amines.

The single and total annual operating costs of the pre-separation stage as a function of the percentage of H_2S removed by the membrane module is shown in Figure 2. According to the figure, the main module cost is the methane losses through the membrane. Methane losses will inevitably occur as a result of its permeability through the membrane (see Table 1). Methane losses account for approximately 85% of the total operating cost in the membrane module; while the capital and replacement costs represent around 15%. Furthermore, the membrane module costs are directly proportional to the percentage of acid gases removed by the membrane module. For instance, the higher the acid gases removal rate the higher the membrane area requirements; and therefore, higher methane losses.

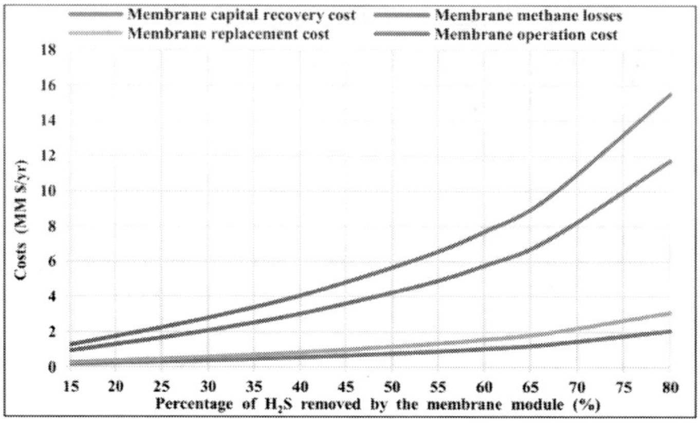

Figure 2. Total and single costs for the membrane stage as a function of the H_2S removed.

Regarding the absorption stage, Figure 3 shows the single and total annual operating costs for the absorption stage. From the figure it is worth pointing out that the H_2S content in the absorber's feed has no significant effects on MDEA and DGA losses. On the other hand, the hydrocarbon losses in the absorption stage are inversely proportional to the H_2S fraction removed by the membrane module. Furthermore, over 95% of the absorber's operating costs are utility expenses.

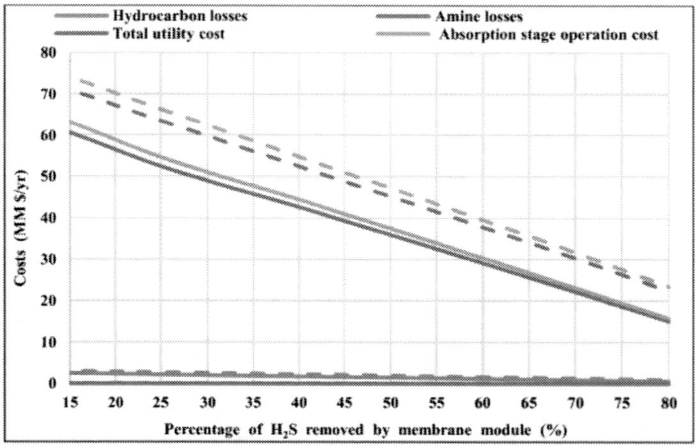

Figure 3. Total and single operating costs for the absorption stage as a function of the H_2S removed for MDEA (continuous line) and DGA (dashed line).

Figure 4 shows the single and total utility costs in the absorption unit. It can be observed that as the amount of H_2S removed in the membrane stage increases, the absorber utilities demand decrease. This because less acid gases must be removed by the absorber. Accordingly, the recirculating solvent flow rate in the hybrid system drops for either solvents MDEA or DGA. Consequently, the power consumed to cool and pump the solvent back into the absorber, as well as the steam used to regenerate the alkanolamines are reduced. Both energy forms (i.e. steam and electricity) account for the majority of the utilities cost in the absorption stage.

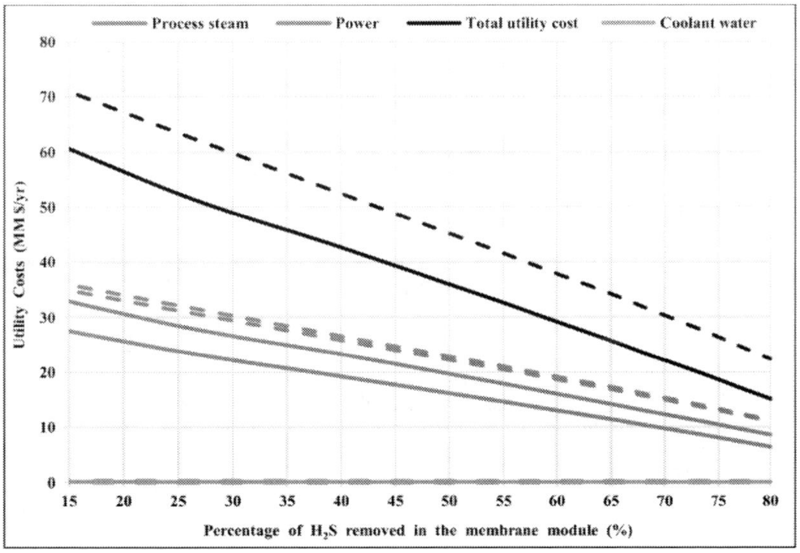

Figure 4. Total and single utility costs for the absorption stage as a function of the H_2S removed in the membrane stage using MDEA (continuous line) and DGA (dashed line) as solvents.

Figure 5 depicts the total costs for the hybrid systems as well as the stages costs (i.e., pre-separation and absorption stage) as a function of the H_2S removal level in the membrane module. From the figure it can be observed that the higher the H_2S removed by the membrane module, the lower the total hybrid system's cost. Conversely, the membrane module operating costs increase as the H_2S removal rates increase. This because higher H_2S removal rates demand larger membrane areas; which also imply higher methane losses (see Figure 2). However, there is a physical limit on how much H_2S can be removed by membranes. This since reducing the H_2S content in the natural gas to ppm levels would require massive membrane areas and imply very significant methane losses (e.g., making the process unprofitable). Moreover, comparison between the stand-alone and hybrid systems show that the latter involves lower operating costs (see Figure 5). Therefore, the hybrid system can potentially perform better than the conventional stand-alone absorption unit for treating highly sour natural gas.

It is well known that the composition and operating conditions of the gas feed may vary with time for existing amine units. Therefore, major adjustments to adapt the existing amine units might become necessary to overcome such changes. However, adding a membrane separation stage before an existing amine absorption unit enables the amine unit to operate under higher feed range capacities and acid gas concentrations. Moreover, hybrid processes can also provide investment flexibility because the membrane capacity can be easily expanded by incorporating additional elements to existing membrane modules or by installing incremental membrane modules. On the other hand, the economics of such retrofitted hybrid plant might differ from those presented in this study given that for the former many additional factors need to be taken into account. Furthermore, in this study the permeability of higher hydrocarbons through the membrane module has not been considered. Higher hydrocarbons may condense to various extents during the membrane separation process. The condensation is undesirable and can be prevented by heating the feed stream. Nonetheless, this would translate into additional equipment; thus, increasing the hybrid system's capital and operating costs.

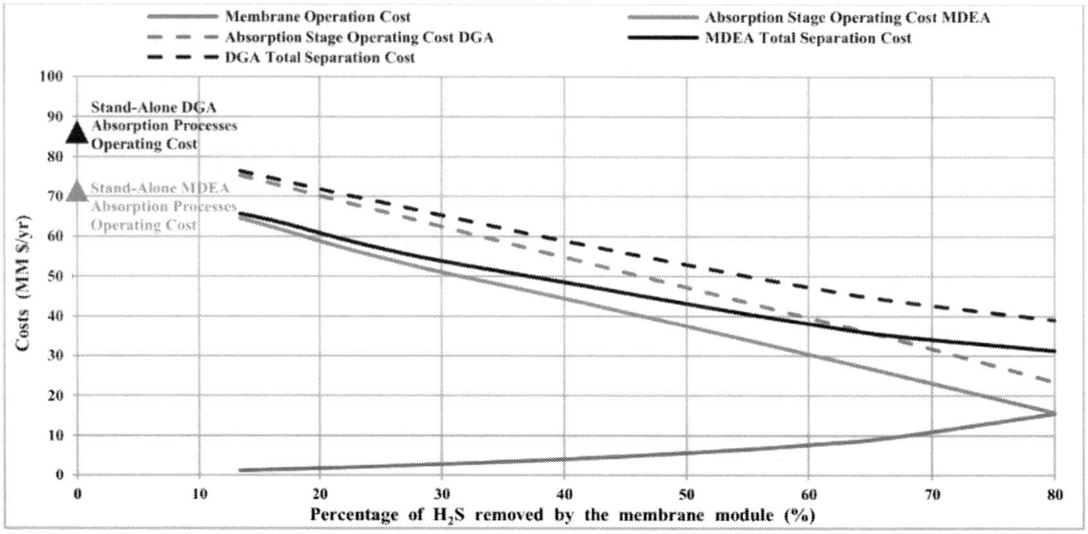

Figure 5. Total and stage costs for the hybrid system (using MDEA and DGA) as a function of the H_2S removed by the membrane module.

Conclusion

In this study, a hybrid system consisting of a single membrane module followed by a conventional amine-based absorption unit is proposed. Based on a preliminary simulation study and information available in the literature, stand-alone absorption units using DGA and MDEA as solvents were found to be the most energy efficient processes for sweetening highly sour gas. Accordingly, both solvents (DGA and MDEA) were selected to be used in the proposed hybrid system analysis.

The economics of the hybrid system showed that it can be more cost and energy-effective than the conventional stand-alone absorption process. Therefore, the implementation of hybrid system concepts can help reducing the consumption of steam and electricity utilities resulting in important cost savings. Moreover, hybrid systems can potentially be more environmentally friendly due to higher energy efficiency rates.

Hybrid processes are more flexible in terms of operability. For instance, retrofitting existing absorption-based gas plants with membrane modules for bulk H_2S removal may potentially reduce sweetening costs; while providing a higher operating flexibility for treating feed flow rates and composition changes. However, there is a limit on the H_2S removal level that can be achieved using membrane modules; thus, relying entirely on membranes to reduced acid gases content into ppm levels is infeasible due to methane losses an vast membrane area requirements. Additionally, methane losses occupy a large fraction of the membrane module operating cost. This represents one of the major concerns/challenges for employing membrane modules due to reductions in the methane recovery rates.

References

1. Gierman M., et al., (2014), Sour-To-Acid concept for sour gas fields, in Sulphur, BCInsight Ltd:

London, UK. p. 62-65.
2. U.S. Energy Information Administration, United Arab Emirates in International energy data and analysis, (2015).
3. Vaughn J.T. and Koros W.J., (2014), Analysis of feed stream acid gas concentration effects on the transport properties and separation performance of polymeric membranes for natural gas sweetening: A comparison between a glassy and rubbery polymer. Journal of Membrane Science, 465: p. 107-116.
4. Kraftschik B., et al., (2013), Dense film polyimide membranes for aggressive sour gas feed separations. Journal of Membrane Science, 428: p. 608-619.
5. George G., et al., (2016), Polymer membranes for acid gas removal from natural gas. Separation and Purification Technology, 158: p. 333-356.
6. Bryan Research & Engineering, Inc., ProMax software, Built-in values (2013).
7. Mccabe W.L., Smith J.C., Harriott P., (1993), Unit operations of chemical engineering. 5th ed.
8. Bhide B.D., Voskericyan A. and Stern S.A., (1998), Hybrid processes for the removal of acid gases from natural gas. Journal of Membrane Science, 140(1): p. 27-49.
9. Hao J., Rice P.A. and Stern S.A., (2008), Upgrading low-quality natural gas with H_2S- and CO_2-selective polymer membranes: Part II. Process design, economics, and sensitivity study of membrane stages with recycle streams. Journal of Membrane Science, 320(1–2): p. 108-122.
10. Niu M.W. and Rangaiah G.P., (2014), Retrofitting amine absorption process for natural gas sweetening via hybridization with membrane separation, International Journal of Greenhouse Gas Control, 29: p. 221-230.
11. Hao J., Rice P.A. and Stern S.A., (2002), Upgrading low-quality natural gas with H_2S- and CO_2-selective polymer membranes: Part I. Process design and economics of membrane stages without recycle streams. Journal of Membrane Science, 209(1): p. 177-206.
12.Turton R, Bailie R., Wallace B. W., Joseph A. S., Debangsu B., (2009), Analysis, Synthesis and Design of Chemical Process. 3rd ed.
13. Ochienga R., Berrouka A. S., Petersa C. J., Slagleb. J., (2013), Amine-based gas-sweetening processes prove economically more viable than the Benfield HiPure process, in Bryan Research & Engineering, https://www.bre.com/Support-Technical-Articles.
14. Bryan Research & Engineering, Inc., (2015), Level 2 Training: Sour Gas Processing Removing Undesirable Components from Gases and Liquids.
15. Slagle, J.C., (2013), Highly Sour Amine Sweetening Process Study for Conditions Common in the Middle East, in The Abu Dhabi International Petroleum Exhibition and Conference 2013: Abu Dhabi, UAE.

455938 A Computational Fluid Dynamics Study of a Pilot-Scale Multitube Membrane Module for Hydrogen Purification

Abstract:

A Computational Fluid Dynamics Study of a Pilot-scale Multitube Membrane Module for Hydrogen Purification

Rui Ma, Bernardo Castro-Dominguez, Ivan P. Mardilovich, Nikolaos K. Kazantzis, Anthony G. Dixon, Yi Hua Ma

Center for Inorganic Membrane Studies, Department of Chemical Engineering, Worcester Polytechnic Institute, 100 Institute Road, Worcester, MA 01609, USA.

Palladium alloy membranes play an important role in the separation of hydrogen, which is an essential material in industrial manufacturing and also a clean energy carrier with an expanding market. A pilot-scale multitube membrane module has been built and tested previously; however, a thorough understanding of the flux distribution, species profiles and mass transfer limitations is needed to further optimize the model performance. For this purpose, a 3-D computational fluid dynamics (CFD) simulation model was built utilizing COMSOL Multiphysics and validated with the experimental data.

The module contains seven Pd/Au/Pd membrane tubes with one at the central axis and six placed symmetrically around the axis, giving a total permeable area of 1050 cm^2. Coal-derived syngas was fed into the module from the shell side (retentate side) and hydrogen permeated through the membrane was collected from the tube side (permeate side) while the unwanted gas was depleted from the shell side. The species continuity equations and the equation of motion were solved simultaneously by treating the fluids on both sides of the membrane as concentrated mixtures. In this model no species is considered as solvent or solute, thus, while hydrogen is being removed from the retentate side,

the gas mixture properties such as density varied along with the composition change. A flux term was defined at the tube surface to describe the process of hydrogen transport through the membrane using SievertsÕ law (Eq. 1):

-n·Ni=PH2[*PH2shell-PH2tube*] $$-\mathbf{n}\cdot\mathbf{N}_i = \mathcal{P}_{H_2}\left[\sqrt{p_{H_2}^{shell}} - \sqrt{p_{H_2}^{tube}}\right] \quad (1)$$

in which PH2 $\mathcal{P}_{H_2}$ is hydrogen permeance [$Nm^3h^{-1}m^{-2}bar^{-0.5}$] and *PH2shell, PH2tube* $p_{H_2}^{shell}, p_{H_2}^{tube}$ are the hydrogen partial pressure on the shell side and tube side [bar], respectively.

A fine mesh with 2,000,000 elements and a stabilization method with a tuning parameter of d=*0.25* $\delta = 0.25$ were applied in this model in order to achieve better convergence. The simulation showed a mass balance error of 2%, which could be further reduced by applying a finer mesh. The module showed a high accuracy compared with experimental data (permeate flow rate [L/min], retentate H_2 percentage [%], and H_2 recovery [%]). From the simulation results of the module, concentration polarization can be visualized and characterized, and the influence of operating conditions on membrane performance including hydrogen recovery and membrane usage can be analyzed to gain insight.

It was also observed that the performance of the seven membrane tubes is differentiated based on their relative position due to the following reasons: 1) The central tube displays a competing role when its performance is comparatively assessed against the 6 surrounding tubes which leads to a lower H_2 flux of the central tube in comparison to the other six. 2) The retentate outlet is on the reactor side as shown in Figure 1, which makes the flow pattern and H_2 profile in the reactor asymmetric and causes the driving force of each tube to vary. Finally, within the context of the present study, it is demonstrated that the results derived and the data generated could reliably inform efforts for further model scale-up.

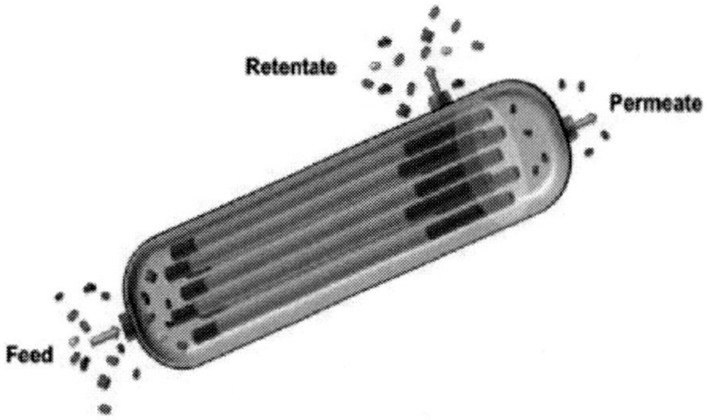

Figure 1. Representative sketch of the module

458949 Microscopic Study of Acid Gas Transport in Zeolitic Imidazolate Frameworks By Diffusion NMR

Abstract:

Zeolitic imidazolate frameworks (ZIFs) are a subset of metals organic frameworks and exhibit high micropore volume and surface area as well as exceptional thermal and chemical stabilities. These properties have made ZIFs an attractive candidate for a wide range of energy efficient industrial separations. One such application deals with the removal hydrogen sulfide (H_2S) and carbon dioxide (CO_2) from methane (CH_4) in natural gas streams. To successfully use ZIFs in such a separations, fundamental knowledge of microscopic diffusion of H_2S, CO_2 and CH_4 inside pore networks of ZIFs is essential. Furthermore, understanding of possible changes in transport properties of ZIFs caused by exposure to H_2S is also needed to characterize their potential in H_2S separations. In this study we use pulsed-field gradient nuclear magnetic resonance (PFG NMR) to measure microscopic diffusion of H_2S and CO_2 in a variety of ZIF types including ZIF-11, ZIF-8, ZIF-7, and ZIF-90.

The combination of high magnetic fields and large gradients in PFG NMR allows us to probe diffusion for displacements in the range of micrometers with great accuracy. As a result, it becomes possible to resolve intra- and inter-particle diffusion by studying diffusion behavior over a broad range of displacements and the corresponding diffusion times. The PFG NMR experimental data will be discussed in the context of the corresponding results of molecular simulations performed by the collaborators.

469382 Gas Separation Using Novel Carbon/Carbon Composite Membranes

Abstract:

Carbon membranes are porous inorganic membranes materials with promising applications in a large variety of industrial applications such as the separation of CO_2 and water vapor from gas mixtures. Research into carbon membranes has shown that they do have advantages over polymer membranes in terms of their superior separation performance, thermal resistance, and chemical inertness. In this study, carbon/carbon composite membranes (CCCM) were made for gas separation. The CCCMs were prepared by infusing a polyimide resin in Bucky Paper and carbon nanofiber mats to increase the mechanical strength of the membrane. Pyrolysis in a nitrogen environment of the polymer-infused carbon-based supports (at different temperatures for various times) produced nano-porous membranes. The structures of the CCCMs were tested for their gas separation properties.

455248 Development of Activated Carbons for Adsorption Heat Pumps

Abstract:

Adsorption heat pumps (AHPs) system have been receiving a lot of attentions due to the ability of heating and cooling using relatively low-level heat sources. Activated carbon (AC)-ethanol pair is considered to be promising for AHP because of high ethanol adsorption capacity of ACs and a relatively high vapor pressure of ethanol even at low temperature levels. To further improve performances of ACs as an adsorbent for AHP (*e.g.* effective adsorption amount of ethanol, adsorption/desorption kinetics, and cycle-ability during the AHP operations), there are various modification factors such as surface area, pore size and the distribution, surface functional groups, pore depth and packing density, to be considered and optimized. Among them, the surface functionalities and pore structures are considered to strongly influence on adsorption behavior of polar ethanol molecules. To clarify the influence of these factors, model ACs having controlled pore structures and surface functionalities have to be developed.

In this study, ACs having different content of surface functionalities but similar pore structure or similar content of surface functionalities but different pore structure were prepared by heat treatments in H_2 and/or careful adjustments of carbonization and KOH activation conditions. Then, the ethanol adsorption performances of the prepared ACs were investigated by using a gravimetric adsorption apparatus.

It was found that abundant surface functionalities decreased the adsorption amount of ethanol and shortened the apparent adsorption equilibrium time due to the diffusional hindrance of ethanol molecules by strong interaction between the surface functional groups and ethanol molecules. On the other hand, as was predicted by our GCMC simulation, the AC having micropores of 1.6 nm in pore width was found to show the highest effective adsorption amount in the AHP operation range (P/P_0 = 0.1-0.3), which reached to 1.7 times higher than ultrahigh surface area commercial AC, Maxsorb III. Furthermore, the superior ethanol adsorbability of the prepared AC was maintained upon adsorption/desorption cycling without remarkable pore structural and morphological degradations at least up to 10 cycles.

We believe that the findings confirmed a potential of the AC-ethanol pair for AHPs.

455874 Superior Ion Adsorption of Polyelectrolyte Threaded Metal Organic Frameworks

Abstract:

The high porosity, high surface area, uniform and ordered structure of metal organic frameworks (MOFs) are explored for selective ion adsorption. We report new composites of polyelectrolytes synthesized within the MOF porous MOF by extending the concept of â€œ*in situ* polymerization in MOFâ€ developed by Kitagawa and co-workers.[1]

Sodium poly(4-styrene sulfonate) threaded in MIL-101 denoted as NaPSS~MIL-101 is synthesized directly with polymerization in situ of the MOF. The polyelectrolyte threaded NaPSS~MIL-101 exhibits superior exchange kinetics, high selectivity with co-ion rejection, reversibility, and durability. The ion adsorption kinetics is compared to commercial ion-exchange resin IR-120. The NaPSS~MIL-101 polyelectrolyte threaded in MOF has high surface area of 1850 m^2/g and a large specific volume 0.85 mL/g. The fixed charges of the polyelectrolyte are exposed for full interaction with solvated ions and solvent, without the need of swelling as in conventional ion-exchange materials IR-120. Excellent selectivity based on charge is demonstrated when NaPSS~MIL-101 is immersed into a solution of two organic dyes. The anionic Acid Blue 9 is excluded though it has significant van der Waals affinity to high surface porous materials.

Another example of polyelectrolyte synthesized in MOF is demonstrated by anionic polyvinyl benzyl trimethylammonium hydroxide (PVBTAH) threaded in ZIF-8 (PVBTAH~ZIF-8). The new composite materials is synthesized in steps of chloro-monomer impregnation, *in situ* polymerization, amination, and alkaline ion exchange.[3] The synthesized non-cross-linked PVBTAH~ZIF-8 material also exhibits superior ion-exchange kinetics compared to conventional ion-exchange resins.[3]

References

[1] T.Uemura, K.Kitagawa, S. Horike, T. Kawamura, S. Kitagawa, M. Mizuno, and K. Endo, *Chem. Commun.* 48 (2005) 5968-70.

[2] Liang Gao, Chi-Ying Vanessa Li, and Kwong-Yu Chan, *Chem. Mater.* 27(10) (2015) 3601-3608.

[3] L. Gao, C.Y. V. Li, K.Y. Chan, and Z.N. Chen, *J. Am. Chem. Soc.* 136 (2014) 7209-7212.

458263 Treated Laterite As Potential Adsorbent for Removal of Heavy Metals from Drinking Water

Abstract:

Treated laterite as potential adsorbent for removal of heavy metals from drinking water

Somak Chatterjee, Sirshendu De.

Department of Chemical Engineering, Indian Institute of Technology, Kharagpur, Kharagpur â€" 721302, India.

Contamination of groundwater by heavy metals is serious threat to environmental safety. These heavy metals undergo chemical transformation and pollute the water, which causes harmful diseases in the human body. Some of the inorganic form of the material is not metabolized and accumulates in the tissues and blood stream of human body. According to world health organization, maximum permissible limit of these heavy metals in drinking water should not be greater than 10 Âµg/L. Processes like advanced oxidation, precipitation ion exchange and membrane separation are some of the water treatment techniques and can be used in the removal of heavy metals. However, each of these processes has inherent limitations. Adsorption is an age old famous process which has been continuously used for these problems. Adsorbing agents are treated either physically (roasting, calcination) or by use of common chemicals (like zinc activation of charcoal made from coconut fibers). In most of the cases, the chemical agent is leached to the treated stream and becomes a potential threat in addition to the persistent danger. Also, cost of starting material and treatment becomes important parameters in judging the sustainability of these treatment procedures. Therefore a cost effective, environment friendly and high performance yielding material is required to address this issue. Laterite is present in Rarh region of the eastern part of India (Kharagpur, 22.32° N, 87.32° E) and can be chemically treated in bulk scale for arsenic removal from ground water [20]. Preparation of treated laterite occurs in three steps, i.e., acid dissolution followed by hydrolysis of the dissolved mass and finally washing [20]. Treated laterite (TL- prepared using

hydrochloric acid and sodium hydroxide) was successfully utilized for this purpose. However, change in the treatment procedure has been done to make the material viable for cationic metal removal (lead, cadmium, zinc, nickel and copper). The adsorbent was characterized using scanning electron microscopy (SEM), Transmission electron microscopy (TEM), X-ray diffraction (XRD), energy dispersive X-ray (EDX), and Fourier Transform Infrared Spectroscopy (FTIR) to highlight its physical and chemical properties. Optimized equilibrium conditions were 1 g/L adsorbent concentration, 0.26 mm size and a pH of 7±0.2. Adsorption capacity of heavy metals was evaluated and was obtained to be 16 mg/g (Lead), 4.5 mg/g (cadmium), 3.5 mg/g (zinc), 2 mg/g (nickel) and 3 mg/g (copper). The adsorption was observed to be exothermic for lead whereas it was endothermic for other heavy metals. Equilibrium time of 24 to 30 hours is required to attain a 99 % adsorption of heavy metals by TL. Adsorption kinetics is explained by second order kinetics.

Keywords: heavy metals; laterite; adsorption; equilibrium; kinetics

459413 Influence of Phase Transition Temperature on CO_2 Adsorption/Desorption Behavior of Thermo-Responsive Gel Slurry

Abstract:

Reduction of CO_2 emissions from fossil fuel power plants and relating industrial factories is becoming increasingly important for addressing climate change concern. Development of efficient technology to capture and recover CO_2 from flue gas is an important issue. Major commercial method for CO_2 recovery from flue gas is chemisorption using amine-based sorbent. This method is energy consuming in desorption and regeneration process of the sorbent. Thus, cost effective process in the regeneration of spent sorbent and an eco-recovery of chemisorbed CO_2should be required. Therefore, researches for energy saving chemisorption process have been attracting much attention for decade.

In this study, thermo-responsive hydrogel copolymerized with amine monomer was examined as an energy saving CO_2 adsorbent. The thermo-responsive hydrogel reveals volume phase transition from a swollen "hydrophilic" state to a collapsed "hydrophobic" state at its lower critical solution temperature (LCST). Preceding study performed by other research group indicated the potential of this recovery concept. Using the thermo-reversible change in adsorption capacity of the gel for CO_2, selective separation and recovery of CO_2 is available. The potential advantage of the thermos-responsive hydrogel in the chemisorbed CO_2 recovery is large reduction of desorption/regeneration temperature due to pKa shift of amine group accompanied with the volume change of the gel. Poly(N-isopropylacrylamide)-polyamine copolymer hydrogel slurry was synthesized as the thermo-responsive adsorbent for CO_2 in this study. Temperature swing adsorption and desorption experiments of CO_2 were performed using CSTR (Continuous Stirred Tank Reactor) type gas flow adsorber. The LCST of the gel was controlled based on the additives in the system

solution. Influence of the LCST shift on the CO_2adsorption/desorption properties was investigated.

References

[1] Yu Hoshino, Kazushi Imammura, Megchen Yue, Gen Inoue, and Yoshiko Miura,, J. Am Chem. Soc., 2012, 18177-18180

[2] Valerij Y. Grinberg,Tatiana V. Burova, Natalia V. Grinberg, Alexander S. Dubovik, Angel Concheiro, and Carmen Alvarez-Lorenzo,,Langmuir, 2014, 30, 4165–4171

460774 Experimental Evaluation and Computer Simulation of the Adsorption Separation Process of Oxidative Coupling of Methane Effluent Gases over 5A Zeolite Molecular Sieves

Abstract:

Ethylene is a major industrial olefin that is used as building block for the synthesis of a large variety of chemicals. Generally, it is obtained in oil refineries by steam or thermal cracking of light oil fractions. In last few decades, an alternative process for the synthesis of ethylene has been proposed by using methane from natural gas (NG). Because of the large NG reserves worldwide, and the possibility to produce methane from bio-digestion processes, methane turns out to be a feedstock with huge potential for commodities production in the near future.

Production of ethylene from NG involves an oxidative coupling of methane (OCM) where it reacts with oxygen in an exothermic process at 750-850°C using a selective catalyst. Major products of reactions are ethylene, ethane, carbon dioxide and carbon monoxide. The process is carried out in a fluidized bed reactor reaching conversions of around 30%. As the complete oxidation of methane is thermodynamically more favorable, the product stream contains large amounts of CO_2, sometimes larger than 25%. Separation of ethylene from this complex mixture involves energy intensive operations. Initially, $CO_2$is removed by absorption with amine solutions (MEA and MDEA). Afterwards, the paraffin/olefin mixture separation is accomplished by cryogenic distillation.

Among others, pressure swing adsorption (PSA) process appears as a promising alternative for the cost-effective separation of OCM effluent gases. In this process, an adsorbent material is used for the selective separation by using cycling operation of high and low pressures. Among different materials, zeolites have been used as industrial adsorbents in different applications (ethanol dehydration, natural gas sweetening, N_2â€"O_2separation, isomers separation, etc.) and also in the selective separation of olefin-paraffin mixtures.

In this work, the experimental evaluation of a zeolite 5A molecular sieve as adsorbent material for the separation of OCM effluent gases was carried out, and the computer simulation of the process was accomplished. Synthesized 5A molecular sieve was characterized by measuring shape, size, crush strength, water adsorption capacity, adsorption isotherms at -196, 25, 35 and 50°C, surface area, and pore size distribution.

Subsequently, the molecular sieve was packed in the adsorption unit of an OCM Miniplant (TU Berlin), and the breakthrough curves using OCM effluent gases were experimentally evaluated. Nitrogen was used as carrier gas due to its negligible adsorptive capacity. Experiments were carried out with single or mixed gases with molar percentage up to 20% at 30, 60 and 90°C and 2 and 5bar. Additionally, two model multicomponent mixtures were prepared and processed taking into account the conditions of OCM effluent gases. In the first model, a mixture similar to the OCM reactor effluent was used. The second corresponded to a mixture close to the obtained after carbon dioxide removal. 5A molecular sieves demonstrated to be selective for OCM gases separation, mainly for CO_2 free effluent gases.

Adsorption modelling was carried out in the Aspen Adsorption® software, using convection and dispersion models for mass transfer, and the Ergun equation for the momentum balance. A linear driving force was employed to model the mass transfer, an the equilibrium condition was modeled with a Langmuir type isotherm. Energy balance considered gas conduction and heat transfer to environment. 20 nodes were set to solve the differential partial equations of the mass and heat balance using the numerical Method of Lines (MOL). For each simulation, the model was validated by comparing the experimental molar fraction in outlet stream and the gas temperature inside the column. Mass and heat transfer coefficients were used as the adjustable parameters. The results show good agreement of the model with experimental observations, so it can be used for process up-scaling and further economic evaluation.

462490 CO_2 Capture Performance of CaO-Based Pellets Stabilized By Metal Oxide Ceramics and Shelled By Mesoporous Glassy Silica, Titania and Zirconia

Abstract:

Calcium oxides (CaO) have been considered as a promising materials for carbon capture at high temperatures due to their high carbon dioxide (CO_2) capacity, low cost, and environmental benignity. Commercialized applications of these materials have been hindered by their severe acitivity loss during the long term performance due to the sintering of the vulnerable CaO particles exposed to high temperatures. In an attempt to overcome these problems, the CaO-based pellet as a core has been synthesized using co-precipitation method doped by zirconium-based ceramic dopants with the range of ratio 5-10% wt. The synthesized CaO-based pellets were then coated by the high thermal resistant mesoporous zirconium dioxide, titanium dioxide or silicon dioxide to provide a reliable protection against the sintering process. The coreshell pellet sorbents were fabricated via the modified sol-gel techniques. The performance and structures of the sorbents have been investigated using TGA, BET, SEM and XRD. The highly crystalline ZrO_2-coated sorbents exhibited a significantly lower capacity decay compared to the uncoated sorbents and natural Cadomin, which could be attributed to the formation of a layered high-temperature-resistant ceramic through the solid-state reaction of the compact and highly crystalline ZrO_2 layer with CaO. The sorbent doped by $CaZrO_3$ and coated by zirconium dioxide, UF CaO-$CaZrO_3$@Zr, showed the highest stability among the synthesized sorbents. The activity loss of this sorbent was 20% during 20 cycles of carbonation/oxidation and CO_2 uptake at the first cycle was 10.23 Mole CO_2 per kilogram of sorbent, whereas an activity loss of 68.98 % and 51.2% was obtained for the natural Cadomin and un-coated CaO cores, respectively. The sorbent doped by $Al_2Zr_3O_9$ and coated by ZrO_2, CaO-$Al_2Zr_3O_9$@Zr, revealed activity loss of 29.3% during 20 cycles and CO_2 uptake of 10.27 mole per kilogram of sorbent at the first cycle.

Moreover, sorbents coated by Silicon dioxide and Titanium dioxides exhibited same performance with uncoated sorbents, which proved that SiO_2 and TiO_2 did not decrease the sintering of CaO particles.

462082 Evaluation of Separation Techniques for Ddsq Based Nanostructured Chemicals

Abstract:

Double Decker Polyhedral Oligomeric Silsesquioxanes or DDSQâ€™s are building blocks that have been used with success in several applications. The synthesis of DDSQâ€™s results in unavoidable production of mixtures containing *cis* and *trans* isomers when two different moieties (Y and X) are located in each deck. In this work DDSQ cages were synthesized making the Y moiety methyl, isobutyl, or cyclochexyl and the X moiety p-phenylamine or m-phenylamine. The cis and trans isomers were isolated by preparatory chromatography aid by TLC. A common absorption wavelength of 254nm was obtained for all the evaluated isomers by UV-vis spectroscopy. However, significant differences were obtained in the absorption peak height between para and meta moieties in the Y position at the same concentration. The individual isomers as well as the cis/trans mixtures were analyzed by adsorption normal phase HPLC and by partition HPLC. From these experiments was found that for cis/trans mixtures the trans peak showed lower retention times than cis peak. Also, an inverse relation was observed between the retention time and the size of the Y group.

462634 The Effects of Reactor Design on Partially Etched Titanium Carbide-Derived Carbon

Abstract:

Carbide-derived carbons are a class of amorphous carbon adsorbents with a narrow pore size distribution. Titanium carbide-derived carbon (TiC-CDC) with residual metal was synthesized by partial chlorination at 500°C. This partial metal removal in the carbide creates vacancies, about which the carbon reorganizes to form an amorphous, porous carbon structure. To understand the titanium removal process on a bulk scale, three reactor designs were tested: (1) a flow-over horizontal-bed reactor, (2) a vertical flow-through packed-bed reactor, and (3) a fluidized-bed reactor. These reactors were chosen to investigate how various Cl_2 flow patterns impact the etching uniformity on individual TiC-CDC particles. The horizontal- and packed-bed reactors non-uniformly etch TiC-CDC particles, while the fluidized-bed reactor produces samples with uniformly etched particles that follow the core-shell model of Ti extraction.

Purification of Olive Leaf Extract Via Polycarboxyl Ester Resin

*Selin Şahin**

Istanbul University, Engineering Faculty, Department of Chemical Engineering, 34320 Avcilar, Istanbul, Turkey

*Corresponding Author E-mail: selins@istanbul.edu.tr
Phone: +90-212-4737070 (17679), Fax: +90-212-4737180

Introduction

In this study, olive tree (*Olea europaea*) leaf was selected as the potential antioxidant source rich in phenolic antioxidants such as oleuropeosides including oleuropein and verbascoside, flavonoids including luteolin, luteolin-7-*O*-glucoside, apigenin-7-*O*-glucoside, diosmetin-7-*O*-glucoside, diosmetin, rutin and catechin, and simple phenolics including tyrosol, hydroxytyrosol, vanillin, vanillic acid and caffeic acid. Oleuropein is the chief constituent of the olive leaf extract, which is mostly responsible for the antioxidative, antimicrobial, antiviral, antiatherogenic, and cardioprotective, antihypertensive and anti-inflammatory properties. For this reason, it is a great value of seperating these plant extracts rich in phenolic compounds and using them as antioxidants in food, pharmaceutic and cosmetic industries. Purification after extracting the plant material might be essential to obtain extracts rich in phenolic content due to the many other compounds such as sugars, protein or metals in the plant extracts.

Generally, phenolic contents of natural extracts are purified by adsorption-desorption processes by polymeric adsorbents. Polymeric macroporous resins are commonly applied as purification matrices in food industry owing to their high efficiency, low pollution and procedural simplicity. This study aims to investigate the static adsorption of phenolics in olive leaf on several macroporous resins. Adsorption kinetics and thermodynamics of the resin XAD 7HP were investigated by means of static adsorption and desorption tests. At every step of the open batch procedure, the efficacy of the purification was assessed by the determination of total phenolic content and the major phenolic compound, oleuropein. Taking into account of ecological and dietary restraints, generally recognized as safe (GRAS), ethanol solution was used as solvent agent in the extraction process.

Results and Discussions

Evaluation of the polycarboxyl ester resins

Figs.1a and b indicate the comparative results of adsorption and desorption ratios of the resins of total phenolic material (TPM) and oleuropein during recycling, respectively. Four cycles including adsorption and desorption steps were applied to the resins. Taking into adsorption capacity of oleuropein, XAD 7HP showed better performance than XAD 4 (91% versus 74%). On the other hand, XAD 16 were found as the most successful resin with the highest ratio of both TFM and oleuropein (82% and 97%) regardless of desorption ratio. When a comparison is made between XAD 16 and 7HP, desorption of the target materials have to be considered. Therefore, XAD 7HP was found superior to XAD 16 with higher desopriton ratios of TFM and oleuropein. Regarding recycling, XAD

7HP showed recovery of around 90% in each cycle. However, desorption performance was 96% only in first cycle. So, the best resin has been proved to be Amberlite XAD 7HP without being recommended for recycling.

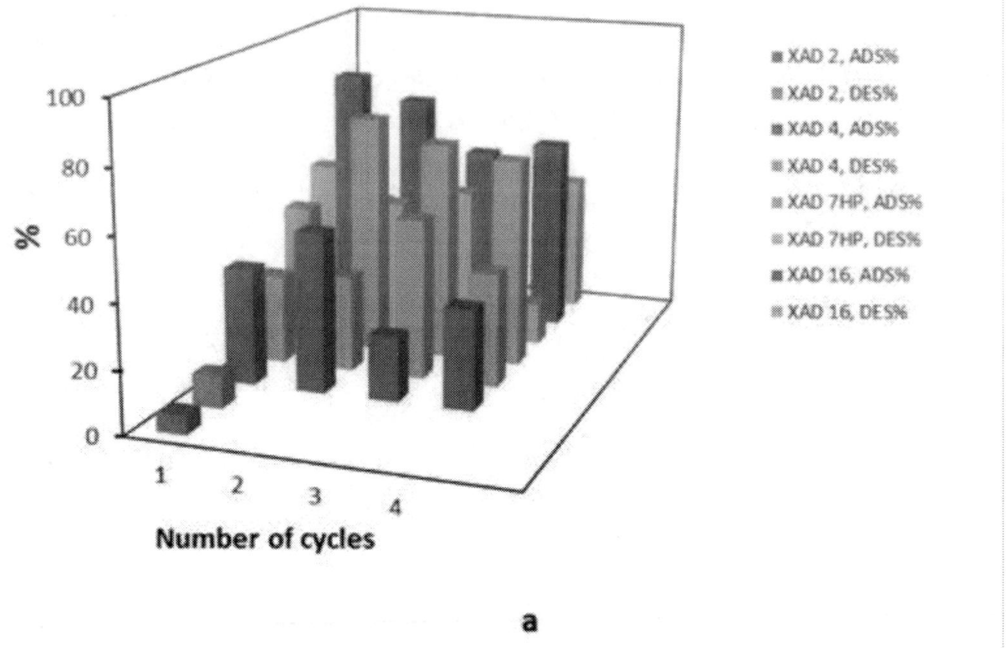

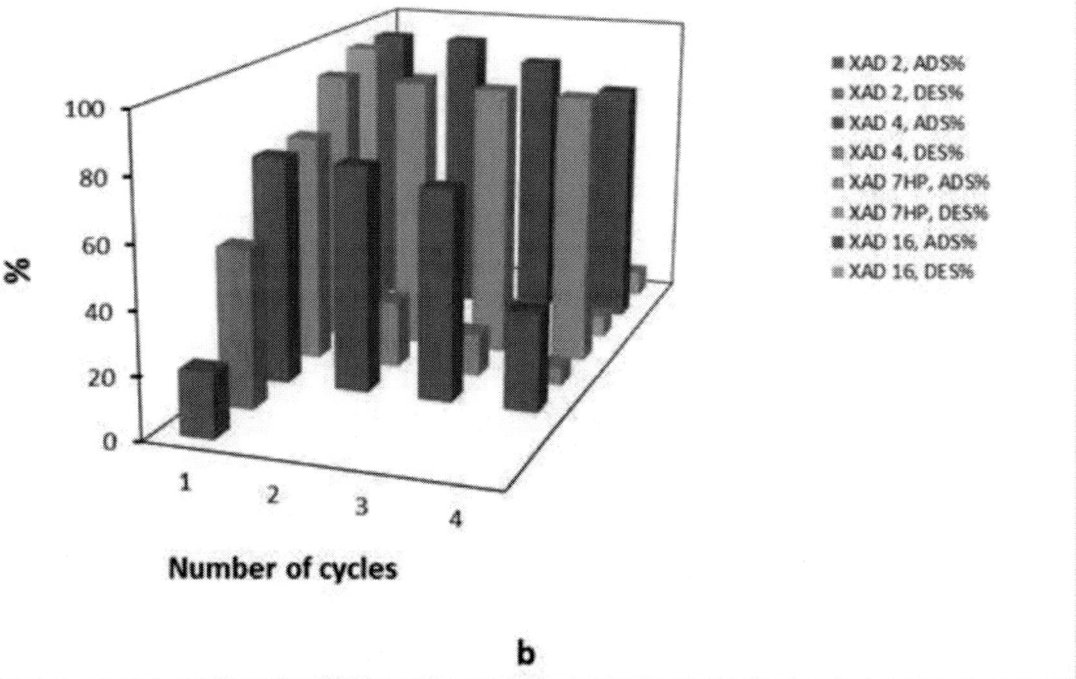

Fig.1. Recovery of TFM (**a**) and oleuropein (**b**) with the macroporous resins from an adsorption to a desorption step in each cycle

Adsorption kinetics on resin XAD 7HP

To identify the rate of adsorption, the experimental kinetic data were investigated exhaustively by the pseudo-first-order, pseudo-second-order and Elovich kinetic models and the lineer forms of these models were given below.

The pseudo-first-order kinetic model of Lagergren (S., 1898):

$$\ln(q_e - q_t) = \ln q_e - (k_1 t) \quad (1)$$

Pseudo-second-order kinetic model of Ho (Ho, 2006):

$$t / q_t = 1/(k_2 q_e^2) + t / q_e \quad (2)$$

$$h_0 = k_2 q_e^2 \quad (3)$$

Elovich model (Ho, 2006):

$$q_t = \alpha \ln(a\ \alpha) + \alpha \ln t \quad (4)$$

where q_e and q_t are the adsorption capacity at equilibrium, and at time t (mg g^{-1}) and k_1 (min^{-1}) is the rate constant of first-order adsorption model, k_2 (g $mg^{-1}min^{-1}$) is the rate constant and h_0 (mg g^{-1} min^{-1}) is the intial sorption rate of second-order adsorption model, α (mg g^{-1} min^{-1}) is the initial sorption rate for Elovich model, and the parameter a is the desorption constant during any one experiment.
The values of relevant parameters and the corresponding linear regression correlation coefficient R^2 were given in Table 1. The correlation coefficient for the kinetic models were found to be extremely high. On the other hand, calculated theoretical q_e is highly in agreement with experimental q_e indicating the adsorption kinetic on XAD 7HP can be defined by the second-order kinetic model and chemical sorption being the rate-controlling step (Ho and McKay, 1999; Emik, 2014). Moreover, the experimental reults were also fitted to the Elovich equation as seen in Table 1. Although the Elovich equation was firstly used in the kinetics of chemisorption of gases on solids, it has been successfully applied for the adsorption of solutes from a liquid solution. According to Elovich model, the surface active sites show different activation energies for chemisorptions owing to their heterogeneous structure (Chowdhury *et al.*, 2011). This finding also is strengthens the assunption that chemical sorption is the rate-controlling step.

Table 1. Kinetic parameters for the adsorption of TFM and oleuropein on XAD 7HP at 25 °C

	Pseudo-first-order		Pseudo-second-order			Elovich		
	R	k_1	R	k_2	h_0	R	α	a
TPM	0.9947	0.027	0.9989	130.21	26.042	0.9961	5.731	0.029
Oleuropein	0.9927	0.020	0.9961	$8x10^{-5}$	0.468	0.9942	6.609	0.032

Adsorption isotherms on resin XAD 7HP

The equilibrium adsorption isotherms were used in order to give information about the interaction between the adsorbates and the resin XAD 7HP. Langmuir and Freundlich isotherms were calculated at different temperatures (20, 25, 30 and 35 °C). The Langmuir model can be described as the following equation:

$$Q_e = \frac{Q_m \times C_e}{K_L + C_e} \qquad (5)$$

where Q_e is the adsorption capacity of the adsorbent (mg/g), C_e is the final concentration of the solute in the solution (mg/mL), Q_m is the maximum adsorption capacity (mg/g) and K_L is the Langmuir constant (mL/mg).

The Freundlich model can be expressed by the following equation:

$$Q_e = K_F \times C_e^{1/n} \qquad (6)$$

where K_F is the Freundlich constant and 1/n is the emprical constant depending on the system and the temperature.

Table 2 summarizes the parameters of Langmuir and Freundlich models for the adsorption of TPM and oleuropein on the resin XAD 7HP at different temperature values, respectively. Generally, the experimental results of adsorption on the resin XAD 7HP were fitted to both of the models at the same temperature values. Maximum adsorption capacity decreased by temperature. *1/n* value of the Freundlich isotherm changed between 0.5 and 0.8, which shows that adsorption can occur favourably, except for the adsorption of oleuropein at 20 °C (Ribeiro *et al.*, 2002). The values of 1/*n* less than 1 represent a favorable sorption (Chowdhury *et al.*, 2011). Q_m values of both TPM and oleuropein decreased by increase in temperature. This result indicates that the given adsorption system is exothermic.

Table 2. Adsorption isotherm parameters of TFM and oleuropein on resin XAD 7HP

Temperature (°C)	Adsorbate	Langmuir model			Freundlich model		
		R^2	Q_m	K_L	R^2	1/n	K_F
20	TPM	0.8550	25.58	44.51	0.8836	0.59	1.02
	Oleuropein	0.9426	19.42	56.33	0.8899	1.19	0.34
25	TPM	0.9822	19.46	19.10	0.9538	0.49	1.78
	Oleuropein	0.9019	2.24	1.67	0.9274	0.64	0.73
30	TPM	0.9911	12.39	18.19	0.9864	0.56	0.99
	Oleuropein	0.9886	9.03	11.99	0.9980	0.75	0.74
35	TPM	0.9670	11.86	12.41	0.9985	0.52	1.30
	Oleuropein	0.9954	4.80	20.51	0.9948	0.81	0.24

Conclusions

Selective adsorption of TFM and oleuropein with polycarboxyl ester resins was fulfilled successfully. Of the four resins tested, amberlite XAD 7HP showed the best performance regarding adsorption (91%) and desorption ratio (97%) for oleuropein without being recommended for recycling. Pseudo-first and second-order and Elovich kinetic models were efficient to represent the experimental data for the adsorption of TPM and oleuropein with high correlation coefficients. Additionally, the equilibrium data were fitted to Langmuir and Freundlich isotherms at four different temperature values. The parameters of the Lagmuir demonstrated that the proposed adsorption system is exothermic.

Acknowledgement

The author wishs to thank to The Scientific & Technological research Council of Turkey (TÜBİTAK) for the support of this Project. Project number is 114M728.

References

Chowdhury, S., Mishra, R., Saha, P. & Kushwaha, P. (2011). Adsorption thermodynamics, kinetics and isosteric heat of adsorption of malachite green onto chemically modified rice husk. *Desalination*, **265**, 159–168.

Emik, S. (2014). Preparation and characterization of an IPN type chelating resin containing amino and carboxyl groups for removal of Cu(II) from aqueous solutions. *Reactive and Functional Polymers*, **75**, 63–74.

Ho, Y.. & McKay, G. (1999). Pseudo-second order model for sorption processes. *Process Biochemistry*, **34**, 451–465.

Ho, Y.-S. (2006). Review of second-order models for adsorption systems. *Journal of Hazardous Materials*, **136**, 681–689.

Ribeiro, M., Silveira, D. & Ferreira-Dias, S. (2002). Selective adsorption of limonin and naringin from orange juice to natural and synthetic adsorbents. *European Food Research and Technology*, **215**, 462–471.

S., L. (1898). About The Theory Of So-Called Adsorption Of Soluble Substances.

464262 Process for Removing Heavy Metals and Disinfecting Wastewater Using Immobilized Biochar

Abstract:

An innovative process has been developed that removes heavy metals, particularly lead, from wastewater and disinfects the water at the same time using a combination of immobilized biochar and a bactericide in either polysulfone or alginate beads. The advantages of immobilization in a polymeric bead is that the adsorbents can be used as a fine powder, increasing the surface area/mass ratio, and by incorporating these powders into an open polymeric matrix the exposed biochar is more accessible to flow and, for packed bed operations, enables higher flow rates to be used with less pressure drop. While bactericides have obviously been used to treat drinking water, they have not been used in this particular configuration in conjunction with immobilized biochar. The technical significance of developing this new process is important because it could help make potable drinking water and remediation of waters contaminated with toxic metals less expensive. Preliminary batch and packed-bed experiments in our lab indicate that a generally spherical polymeric bead can be formed in large quantities that has improved properties over the non-immobilized biochar while maintaining a significant concentration of bactericide. Experimental data will be presented showing the effectiveness of the concept and plans for further development.

464988 KOH Activation of Petroleum Coke for CO_2 Capture: Effect of Carbonization Temperature

Abstract:

The CO_2 concentration in the atmosphere is over 400 ppm currently, which is about 40% higher than in the mid-1800s. Since CO_2 is a major greenhouse gas, increasing CO_2 concentration has caused global warming and climate change problems such as droughts and floods. To reduce CO_2 emissions, many researches have been carried out for developing CO_2 capture and storage (CCS) technology. Among CCS stages, CO_2 capture accounts for two thirds of the total cost, indicating importance of developing efficient CO_2 capture technologies. Adsorption using solid materials is one of promising CO_2 capture technologies and it has advantages of low energy consumption and easy regeneration. Among various adsorbents, CO_2 adsorption on activated carbons is based on high surface area and it can be obtained by activation of carbon-containing precursors. Petroleum coke, which is a byproduct of heavy oil upgrading processes, contains a high carbon content more than 80%, and therefore it is considered as a promising precursor for low-cost activated carbon. In this study, activated carbons having high surface area were synthesized through KOH activation of petroleum coke for CO_2 capture. Prior to KOH activation, carbonization process was carried out and the effect of carbonization temperature on the pore development and CO_2 adsorption capacity of the activated carbons was investigated.

464118 High-Throughput Computational Screening of Metal-Organic Frameworks for the Capture of Thiols

Abstract:

The removal of undesirable odors from foul-smelling fruits and vegetables (e.g. durian and onion) is an interesting issue especially in the Southeast Asian. In this study, we report high-throughput computational screening of 137953 hypothetical metal-organic frameworks (MOFs) and 4763 computation-ready, experimental MOFs (CoRE-MOFs) for the capture of volatile thiols (e.g. methanethiol and ethanethiol are the major components of odors from durian). Quantitative structure-property relationships are established between adsorption capacity and the geometric and energetic characteristics of MOFs (largest cavity diameter, isosteric heat of adsorption, density, void fraction and surface area). The peaks of adsorption capability are about 700 and 1000 mg/g for methanethiol and ethanethiol, respectively, at isosteric heat $Q^0{}_{st}$ of (30, 42) and (35, 48) kJ/mol. Several analytical limits are proposed and they can be used to determine the performance of thiol adsorption in MOFs. This simulation study identifies potential MOFs for the capture of thiols and reveals that the understanding of structure-property relationships is useful toward the development of new remarkable MOF for specific applications.

465008 Synthesis of Sodium Promoted Hydrotalcite and Its CO_2 Adsorption Properties at High Temperature

Abstract:

In the past centuries, global warming has been seriously accelerated since carbon dioxide (CO_2) emission caused by fossil fuel-based energy system has been gradually increasing. In this respect, CO_2 is classified into a main anthropogenic greenhouse gas so that the increase of CO_2 concentration in the atmosphere must be inhibited. To diminish CO_2 emission, lots of researches have been carried out. Among various CO_2 capture technologies, CO_2 can be directly captured using solid sorbent materials. Layered double hydroxide, also called as hydrotalcite, has been considered as a promising for CO_2 capture at high temperature range (200-500 °C). Hydrotalcite typically has fast adsorption and desorption kinetics, thermal stability, and ease of regeneration by pressure or temperature swing adsorption method. However, its low CO_2 adsorption capacity should be enhanced for practical sorption processes. Alkali metal precursors, usually potassium carbonate (K_2CO_3), are used for enhancing the CO_2 adsorption capacity of the hydrotalcite since alkali metal increases the basicity of hydrotalcite. However, as an alkali metal precursor, sodium nitrate ($NaNO_3$) has not been studied in previous studies. Therefore, in this study, $NaNO_3$-promoted hydrotalcite was synthesized and the effect of $NaNO_3$ on CO_2 adsorption properties of hydrotalcite was investigated using several analysis methods. The structural and textural properties were analyzed using X-ray diffraction (XRD) and N_2 adsorption analysis, and the CO_2 adsorption ability at high-temperature (200-400 °C) was measured using a thermogravimetric analyzer (TGA).

465125 Electrospun Poly(Glycidyl Methacrylate)-Based Crown Ether Nanofibers As Lithium Ion Adsorbent

Abstract:

Crown ethers (CE), especifically 12 to 14-membered CE rings, are most selective towards Li^+ due to their matching cavity size (1.2-1.5 Å) with the ionic diameter of Li^+(1.36 Å) [1]. However, CEs are often integrated in liquid–liquid extraction or liquid membrane systems which are tedious to perform and have low to moderate efficiency [2]. Thus, a more effective and convenient immobilization technique is by incorporating the CEs in solid supports.

Herein, electrospun poly(glycidyl methacrylate)-based crown ether (PGMA-CE) nanofibers (NF) were prepared, characterized and tested for lithium ion (Li^+) adsorption. The CEs were covalently attached on the PGMA NFs via "click" chemistry. Poly(glycidyl methacrylate) (PGMA) is an excellent candidate as CE support since it possesses the pendent epoxide groups, which are excellent sites for CE attachment. Glycidyl methacrylate (GMA) monomers were polymerized via radical polymerization to synthesize the PGMA. The PGMA polymer was then electrospun to produce the PGMA NFs. After which, PGMA NFs was azidated to introduce the azide moieties in the PGMA NFs via the epoxide ring-opening mechanism (PGMA-N_3). Meanwhile, the CE hydroxydibenzo-14-crown-4 ether (HDB14C4) was modified to contain a terminal alkyne group. HDB14C4 was esterified with propiolic acid (PA) using dicyclohexylcarbodiimide (DCC) and 4-dimethylaminopyridine (DMAP) as coupling reagent and catalyst, respectively, to synthesize dibenzo-14-crown-4 propiolate (DB14C4P). Finally, the CE-"clicked" PGMA NFs were produced via alkyne-azide cycloaddition (CuAAC), involving the DB14C4P and PGMA-N_3. The success of each synthesis step was confirmed through Fourier transform infrared (FTIR) spectroscopy, thermogravimetric and elemental analyses. Meanwhile, the successful modification of CE was determined using ^1H and ^{13}C nuclear magnetic resonance (NMR) spectroscopy. Different types of prepared nanofibers were characterized using scanning electron microscopy (SEM), ^{13}C and ^{15}N solid-state nuclear magnetic resonance (SS-NMR) spectroscopy and universal testing machine (UTM). The performance of PGMA-CE NFs as Li^+ adsorbent was investigated for selective Li^+ capture. Batch adsorption studies at varied Li^+ concentrations (7~70 mg/L), pH=11 and solid/liquid ratio of 0.625 mL/mg reveal its Langmuir-type behavior and pseudo-second-order rate of Li^+ uptake. The maximum Li^+ adsorption capacity (q_m) was determined to be 12.3 mg g^{-1}, which is the highest among other organic-based Li^+ adsorbents reported in literature. Adsorption loss after five adsorption-desorption cycles was negligible, which suggests the long-term stability of PGMA-CE NFs. In the presence of competing cations, the PGMA-CE NF was highly selective towards Li^+ following the uptake sequence: $Li^+ > Na^+ > K^+ > Mg^{2+} > Ca^{2+} > Sr^{2+}$. Overall results indicate the potential use and reusability of PGMA-CE NFs as a highly effective and high capacity Li^+adsorbent.

This research was supported by Basic Science Research Program through the National Research Foundation of Korea (NRF) funded by the Ministry of Education (2009-0093816) and Ministry of Science, ICT and future Planning (2015R1A2A1A15055407).

References:

1. C.J. Pedersen, Cyclic polyethers and their complexes with metal salts, *J. Am. Chem. Soc.*, 89 (1967) 7017–7036

2. J.G. Huddleston, H.D. Willauer, R.P. Swatloski, A.E. Visser, R.D. Rogers, Room temperature ionic liquids as novel media for 'clean' liquid–liquid extraction, *Chem. Commun.*, 16 (1998) 1765–1766

465176 Cycle Type Comparison Under Rapid Pressure Swing Adsorption Conditions

Abstract:

Pressure swing adsorption is a process used for a variety of industrial gas separations and purifications including air separation and hydrogen purification. Rapid pressure swing adsorption (rPSA) is an advancement of this technology for the purpose of process size reduction. This is advantageous for small scale processes such as portable oxygen concentrators where the main objective is a lightweight device. While the process at an industrial scale is well known, rPSA is a relatively new technology at a minuscule scale; much less is known about its operation. RPSA is operated at speeds 1-to-2 orders of magnitude faster than larger scale processes. Small adsorbent particles for mass transfer and high gas velocities to sustain large flowrates allow for high speed operation.

The main objective of this study was to compare rPSA cycle performance under different cyclic operations. Cycle types include vacuum swing adsorption (VSA), VPSA and PSA. Cycles operating with an equalization step are also compared to Skarstrom cycles. A commercial Li-X zeolite (~0.5 mm diameter) was used as the adsorbent in a 2-column air separation process. The system is capable of cycles under 3 seconds with an oxygen product purity of 96% (argon free air).

465390 Performance Evaluation of PEI-Silica and KOH-Activated Carbon for Carbon Dioxide Capture in Power Plants

Abstract:

Many decision makers of energy supply such as governments have turned attention to CCS technology to deal with global climate change mainly caused by CO_2 emission. Post-combustion CO_2 capture process using dry-sorbents, which separates CO_2 from flue gas of industrial faculties, is one of the most promising capture technologies. It is because proper sorbents can be chosen by conditions of flue gas easily, and sorption process can be eco-friendly operated when compared to capture processes using wet-sorbent. In this study, two types of dry sorbents (PEI-silica and KOH-AC) were evaluated for CO_2 sorption from power plant emission gas. Due to the PEI impregnation, PEI-silica had relatively small surface area and large pores. On the other hand, KOH-AC had high surface area and micro-pore was well developed. The adsorption amount of both sorbents was similar at 30°C and 1 bar CO_2. Due to chemisorption, the sorption capacity of PEI-silica was almost independent of temperature, showing very strong sorption affinity. However, the sorption capacity of KOH-AC was significantly affected by sorption temperature, showing weak sorption affinity. With respect to sorption rate, both sorbents could reach approximately 80 % of sorption capacity within 1 min. According to the results of TGA and DTG analysis, the regeneration temperature for PEI-silica was limited at 130°C while the regeneration of KOH-AC could be applied by gases up to 150°C. Under hot N_2 regeneration condition, the working capacity in a cyclic test was 10 wt.% for PEI-silica and 8.5 wt.% for KOH-AC. However, PEI-silica was a little unstable in the cyclic operation. Furthermore, due to the sorption affinity of water vapor on PEI-silica, the working capacity in a wet flue gas condition decreased up to 6 wt.% in a cyclic test.

465302 Separation of RDX and TNT from Expired Energetic Compounds

Abstract:

At present, the decayed and expired explosives/ammunitions are treated by destruction, incineration, or explosion methods. However, these methods can lead to pollution, resource waste, and safety issues. Due to tightening demand for environmental protection and regulations, the concept of green technology to cope with the social needs and environmental regulations is highly required to treat decayed and expired explosives/ammunitions. Therefore, eco-friendly technologies for recycle or transform methods are needed to replace the destruction and incineration for decayed and expired explosives.

In the study, two-step method was developed to recover energetic materials (TNT and TDX) from decayed and expired explosive composites: solvent extraction and simulated moving bed chromatography (SMB). Since the energetic composite used in the study contained TNT, RDX, and paraffin wax as a desensitizer, RDX and TNT were extracted from the composite by using the solubility difference in particular solvents. The composition of the extracted solution and the amount of undissolved fragment-free TNT were determined by varying dissolution time and a ratio of composition to solvent. The obtained TNT-rich solution was applied to further purification method, which was a SMB process for a continuous production of TNT and RDX with high purity. The characteristics of each column were obtained from the moment method for HPLC column and the operating condition for SMB process using 2-2-2-2 configuration was determined by a triangle theory. Then, the variation of purity, recovery, solvent consumption and productivity was evaluated at each operating condition of SMB. The developed technology combined with solvent extraction and SMB process can contribute to reusing decayed and/or expired energetic composites as a green technology.

465924 Design of Metal Based Nanofiber Scaffolds for the Removal of Heavy Metal Ions in Harsh Effluents

Abstract:

Wastewater containing heavy metals may be generated from fertilizers with the use of arsenical pesticides, from dumping products containing heavy metals with household waste such as Cd-Li batteries, from lead soil contamination resulting from airborne deposition due to lead-based petrol additives, or from mining effluents such as the release of mercury (estimated to 1,000 ton per year) during the gold extraction process [1, 2]. Membrane filtration and especially nano-filtration has been suggested as a preferable treatment to remove heavy metals from contaminated streams as it is capable of achieving strict discharge criteria while providing high efficiency, and has higher fluxes at lower pressures compared to reverse osmosis, for a relative process cost-efficiency and minimized environmental impact compared to electro-chemical techniques [3, 4]. However, it is confronted to some limitations including the reduction degree, the usually required de-complexation, and the material or membrane stability towards abrasion and pH variation [3]. Thus, new techniques able to stand the harsh water conditions are required.

Adsorption is an outstanding technique as it can potentially achieve high yields of heavy metal removal even for low concentration effluents. It can be a more cost-effective process to remove and recover heavy metals, based on the choice of the adsorbent material [4, 5]. Among the investigated adsorbents for heavy metal ions, metal oxide nanoparticles showed high adsorption capacities towards a wide range of heavy metal contaminants, however their application is hindered by their tendency to agglomerate in liquid streams due to attraction forces such as Van der Waals between the oxide layers [5].

Metal-based nanofiber scaffolds are a class of materials combining the enhanced specific surface area and controlled flow pattern of nanofiber mats with metal unique surface properties [6]. While single metals such as Cu, Ti, Ni, Zn and Fe have been investigated, metal alloys such as Cr and Cu can offer new opportunities with the combination of several metals useful properties (such as corrosion resistance and electrical conductivity) in one membrane.

Electro-spinning has been chosen as a fabrication route by using a precursor composed of metal salt or metal salts mixture to form a sol-gel network with a polymer template, followed by an annealing post-treatment to remove the polymer and induce the nucleation and coalescence of metal and metal oxide grains [7, 8]. The particle coarsening process has been imaged in situ by TEM and an Ostwald ripening type phenomenon was observed. The materials were further treated under a 15% H_2 in N_2 reductive atmosphere to lower the oxidation degree of the structure in order to improve the adsorption capacity with the increase of oxygen vacancies at the surface of the nanofiber web. A reduction of 6.2 At % of oxygen in the global composition of a Cu/CuO nanofiber web has been achieved after one hour treatment at 400 C, according to x-ray photo-spectroscopy measurements.

Compared to the as-spun fiber mat size, a shrinkage from 55% and up to 85% has been obtained during thermal treatment due to the polymer removal and the fiber crystallization. However, the homogeneity of fiber diameter and pore size distributions resulting from the electro-spinning process has been conserved.

For an electro-spun copper salt in polyvinyl alcohol, it has been observed from TEMs that when the temperature is higher than 700°C, the fiber morphology is lost as the coalescence of bigger grains forms necklaces. This resulted in a loss of mechanical resistance and excessive brittleness. In-situ SAXS annealing and TEM diffraction were used to observe the reorganization of microstructure and it was noticed that the monoclinic structure of cupric oxide crystals were formed from 120 C. The diffraction patterns showed that the reactivity of the exposed crystallographic plans of copper crystals obtained after reduction increased with the temperature, resulting in the need of a compromise between sufficient mechanical strength (2 to 5 bar in compression for adsorptive micro-filters) and higher surface reactivity. The nano-scale texture of metal nanofibers - measured by AFM - with high angle grain boundaries enhance the surface reactivity even for low specific surface areas (of 7.6 m^2/g and 11.8 m^2/g for an as-spun and annealed copper in PVA fiber mat), while the nanofiber mat porosity and sub-micrometer pore size provides microfiltration efficiency with low liquid flow resistance for ease of operation with liquid streams. In this study the synergy between the fiber microstructure and the nanofiber web surface reactivity will be explored to improve the heavy metal ion adsorption performance of single metal and metal alloy nanofiber mats.

1. Järup, L., *Hazards of heavy metal contamination.* British Medical Bulletin, 2003. **68**(1): p. 167-182.

2. Programme, U.N.W.W.A., *The World Water Development Report 1: Water for People, Water for Life.* 2003, UNESCO: Paris.

3. Mullett, M., R. Fornarelli, and D. Ralph, *Nanofiltration of Mine Water: Impact of Feed pH and Membrane Charge on Resource Recovery and Water Discharge.* Membranes, 2014. **4**(2): p. 163-180.

4. Fu, F. and Q. Wang, *Removal of heavy metal ions from wastewaters: A review.* Journal of Environmental Management, 2011. **92**(3): p. 407-418.

5. Hua, M., et al., *Heavy metal removal from water/wastewater by nanosized metal oxides: A review.* Journal of Hazardous Materials, 2012. **211–212**(0): p. 317-331.

6. Ramaseshan, R., et al., *Nanostructured ceramics by electrospinning.* Journal of Applied Physics, 2007. **102**(11): p. 111101.

7. Pomogailo, K., *Metallopolymer Nanocomposites.* Materials Science, ed. Springer. Vol. 81. 2005.

8. Barakat, N.A.M., et al., *Spider-net within the N6, PVA and PU electrospun nanofiber mats using salt addition: Novel strategy in the electrospinning process.* Polymer, 2009. **50**(18): p. 4389-4396.

467595 CO2, N2 and CH4 Adsorption into Novel MOF/Graphene-like Composites

Abstract:

In this work we measured, with a pressure decay technique, the adsorption in static conditions of different gases (N_2, CH_4, CO_2) in novel types of metal organic frameworks (MOFs) selective to CO_2, modified with different strategies, for CO_2 capture processes in post and precombustion as well as natural gas upgrading.

The starting MOF material is HKUST-1 or $Cu_3(BTC)_2$, which shows high adsorption and selectivity for CO_2. Starting from this structure, we evaluated the effect of intercalation, in the assembling stage, of carbonaceous graphene-like layers (GL) that induce modifications to the structure and porosity distribution, generating the so-called hybrid MOF-GL materials (MGLs).[1] We also investigated the effect of the partial replacement of Copper centers with Zinc centers (ZnMOFs), and of the simultaneous presence of Zinc centers and of graphene like layers (ZnMGLs). The different materials show rather different gas adsoprtion properties, that depend on the chemical structure and on the material porosity. In particular, the samples with the highest adsorption capacity in comparison to HKUST-1 are the MGL ones, while the samples ZnMOF and ZnMGL have a lower CO_2 uptake. The majority of adsorption isotherms shows a type I behaviour.

References

[1] M. Alfè, V. Gargiulo, L. Lisi, R. Di Capua, Synthesis and characterization of conductive copper metal-organic framework/graphene-like composites, *Materials Chemistry and Physics* 147 (2014) 744-750

458568 Measurement of Oxygen Solubility in Toluene and Benzyl Alcohol Using a Tube-in-Tube Membrane Contactor

Abstract:

Oxidation of alcohols is an important process in chemical industry, and the high demand for green chemistry has driven researchers to employ molecular oxygen rather than traditional, toxic chemical oxidants.[1] However, the solubility of oxygen in the substrate could potentially limit the extent of the reaction. Hence, data on the solubility of oxygen under reaction pressures and temperatures is essential information for reactor design and process optimization. In this study, a novel method using a Teflon AF-2400 tube-in-tube membrane contactor is proposed for the measurement of oxygen solubility. The semi-permeable membrane contactor allows the liquid substance to be saturated with oxygen at specific temperatures and pressures. When the pressure reduces to atmospheric pressure, the volumetric flow rate of oxygen released can be accurately measured and used for the solubility calculation. This method was initially employed to investigate the oxygen solubility in toluene, and the results agreed well with the literature[2,3](where the data were obtained with traditional equilibrium apparatus). The solubility of oxygen in benzyl alcohol was then experimentally measured with pressures up to 1.0 MPa at temperatures from 293K to 393 K. An empirical equation of the Henryâ€™s law constants as a function of temperature was also determined.

References

1. Davis SE, Ide MS, Davis RJ. Selective Oxidation of Alcohols and Aldehydes over Supported Metal Nanoparticles. *Green Chem.* 2013;15(1):17-45.

2. Wu X, Deng Z, Yan J, Zhang Z, Zhang F, Zhang Z. Experimental Investigation on the Solubility of Oxygen in Toluene and Acetic Acid. *Ind. Eng. Chem. Res.* 2014;53(23):9932-9937.

3. Li A, Tang S, Tan P, Liu C, Liang B. Measurement and Prediction of Oxygen Solubility in Toluene at Temperatures from 298.45 K to 393.15 K and Pressures up to 1.0 MPa. *J. Chem. Eng. Data.* 2007;52(6):2339-2344.

451922 Unconventional Strategies in Industrial Melt Crystallization

Abstract:

Unconventional strategies in industrial melt crystallization

GUILLOU L.

Fives Proabd, 20C rue de Chemnitz – F-68100 Mulhouse

The melt static crystallization (MSC) is usually described through an arrangement of steps consisting in:

- Controlled cooling and crystallization
- Draining of mother liquor containing impurities
- Controlled partial melting of the crystals to remove remaining wetting impurities a.k.a. "sweating"
- Melting of remaining purified crystals

This arrangement of steps defines one purification stage which can be multiplied in order to develop fractional crystallization processes that allow improvement of recovery yield or purity.

It is quite often considered that these features limit the scope of application of the MSC to very specific cases such as the treatment of concentrated feedstock and applicability to limited capacities only.

Through this contribution different cases of unconventional strategies for crystallization from the melt will be

examined:

- Supercharging and in situ recrystallization for lean concentrations
- Substitution of the sweating step by a washing step
- Extractive crystallization

Supercharging MSC

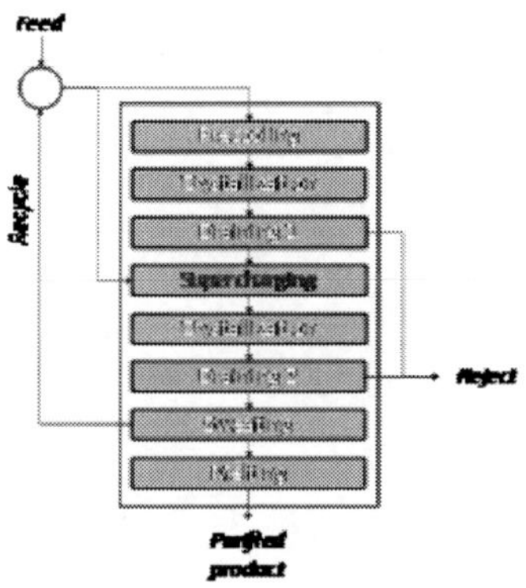

MSC with combined washing

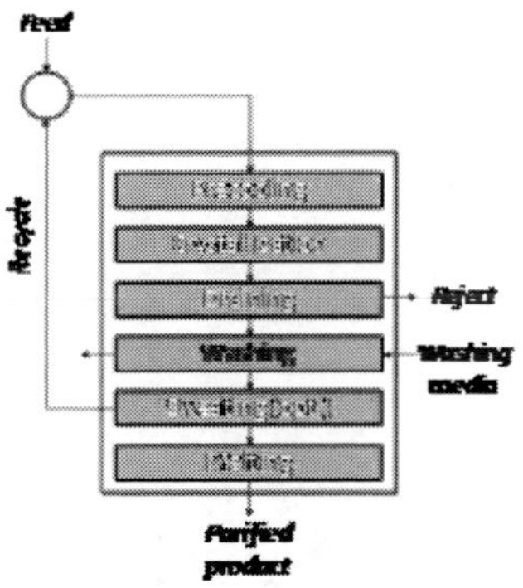

Extractive MSC

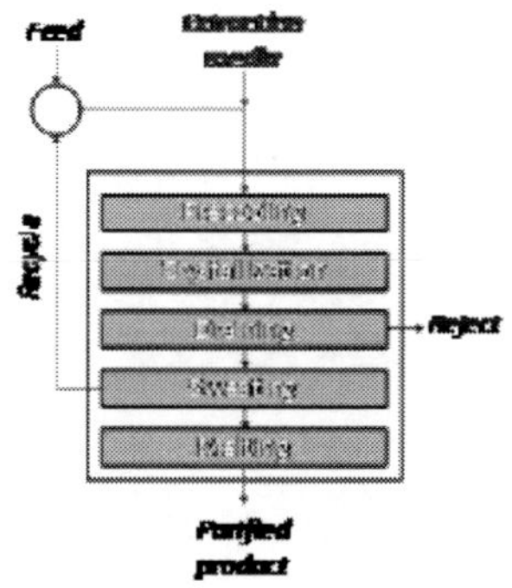

We will show how supercharging is used to treat feedstock containing less than 20% of crystallizable compound and how it contributes to increase the recovery yield per pass vs. suspension crystallization.

We will present and discuss the advantages of washing vs. sweating in terms of improvement of impurity removal for very high capacity plants (beyond 100 ktpy).

Finally we will show how extractive crystallization can be advantageously used for separating a three crystallizable compounds blend and allow achieving better recovery than an alternate based distillation approach.

Through selected industrial examples, the declination will be introduced. It will be shown that they are not necessary meant to be considered as specific tools used for solving specific issues. They can in fact be considered as generic strategies For sure they are not as much widespread than the standard MSC step arrangement which remains the most common. But they are neither to be considered as being that rare: more than 20% of the industrial purification plants using a Proabd® MSC process nowadays involve one or more of such strategies.

Keywords: Crystallization, purification, energy optimization.

452033 Amine Loaded Silica Nanomaterials for Advanced Carbon Dioxide Capture

Abstract:

Works dedicated to carbon dioxide capture and conversion on silica sorbents have shown that while these materials can achieve moderate capture capacities and fast diffusion speeds, they cannot outperform the current liquid amine systems employed industrially. In order to overcome these challenges our group and others have investigated the use of amine loaded silica sorbents as carbon dioxide capture systems, where the loaded amines act as chemisorbents to allow for increased selectivity and capacity for CO_2. However, these materials show a significant loss of diffusion speed and amine efficiency at the highest amine loadings, often because the effect that pore-amine interactions have on capture performance have not been investigated. In our work we have used multi-porous silica materials loaded with both tetraethylenpentamine (TEPA) and polyethylenimine (PEI) amine groups, in order to see what affect these amines will have when loaded onto multi-porous sorbents. We have shown that on MCM-36, which contains multi porous systems that are inaccessible from one another, that amine loading leads to decreases in capture performance due to pore blocking effects. However, when amine is loaded into a 3-D porous sorbent (in this case 3-Dimensional Disordered Zeolite Beta spheres) amine loading can be preferentially placed into the mesoporous channels and allow for quick diffusion to the loaded amine sites through the microporous spheres.

460955 Energy Efficient Solvent Recovery Via Reactive Distillation

Abstract:

Reactive separations are powerful technologies for separation of azeotropic multicomponent carboxylic acid mixtures. By reducing the boiling point and the enthalpy of evaporation via chemical reaction energy savings can be achieved and separation into pure constituents is possible.

Waste from the pulping industry may be a perfect feedstock material for biobased chemicals. The black liquor thickening condensate for example contains acetic acid and formic acid. Liquid-liquid extraction is perfectly suited for the wastewater treatment of this condensate to remove the carboxylic acids with low energy demand. But the solvent recovery is normally done by energy intensive distillation and additionally, the crude distillate is an azeotropic multicomponent mixture. By extending the solvent recovery to reactive distillation with an appropriate reactive agent, the energy demand can be reduced by lowering the boiling points and the enthalpy of evaporation of the reaction products. Esterification with methanol as example produces the low boiling methyl esters methyl formate and methyl acetate with an enthalpy of evaporation four times smaller compared to water. The combination of distillation and chemical reaction may lead to complete conversion by shifting the equilibrium of the esterification reaction to the product side via continuous removal of the products. The elevated temperature in the reboiler enhances the reaction rates whereby the application of a catalyst is not mandatory. Additionally the reaction products have higher market values compared to the carboxylic acids. Reactive distillation experiments in batch and continuous mode showed complete conversion of formic acid. Acetic acid is converted to a lesser extent. Methyl formate can be isolated in high grade quality in a first step. For acetic acid removal different routes are applicable. Exemplarily acetic acid conversion may be completed and methyl acetate removed from the solvent in a second step.

464601 Microscopic Diffusion Studies of Pure and Mixed Methane and Carbon Dioxide in ZIF-11 By NMR

Abstract:

Zeolitic imidazole frameworks (ZIFs), a subclass of metal organic frameworks, have shown promise in challenging gas separations. Specifically, the high surface area and pore volume, along with great chemical and thermal stability and great control over the pore sizes has already made ZIFs an excellent choice material for both adsorption-based and diffusion-based separations. The tunable ZIF pore aperture, which is in the range of about 0.2 â€" 2 nm, can result in molecular sieving phenomena, allowing for great diffusion-based selectivity.

Fundamental understanding of gas diffusion in ZIFs is of crucial importance for their potential applications in separations. Here we report microscopic studies of diffusion of CO_2, CH_4, and a CO_2/CH_4 mixture inside ZIF-11 particles. These gases are commonly found in natural and coal bed gases, where CO_2 is an impurity. Diffusion measurements were performed using ^{1}H and ^{13}C pulsed field gradient (PFG) NMR utilizing a high field (17.6 T) and high field gradients (up to 30 T/m) to obtain diffusivities for the length scales of displacements in the range of about 1-20 mm. The diffusivities for one component gases are correlated with those in the corresponding gas mixture. The role of particle external surface in the measured transport properties is discussed.

464619 Modeling MOF/Polyimide Interfaces in Mixed Matrix Membranes for CO2 Separation

Abstract:

The separation of CO2 from industrially important gas mixtures like natural gas, syngas and flue gas is an enormously crucial issue in chemical industry. An attractive class of material is mixed-matrix membranes (MMMs) where metal organic frameworks (MOFs) are incorporated into polymer matrices. The principle issue in these applications is the plasticization of the polymers at high partial pressures of CO2, which can lead to reduced membrane selectivity and unpredictable transport properties at the polymer/MOF interfaces.

In this molecular simulation study, we focused on the interfacial interactions between the polymer chains and MOF particles. For this purpose, the ZIF-8-Matrimid® 5218 MMM was modeled and the interfacial interactions of polymer-MOF were investigated via Molecular Dynamics (MD) and Monte Carlo (MC) simulations. First, Matrimid® 5218 was modeled alone and the model was validated by reproducing experimental CO_2 sorption data. CO_2 sorption was simulated by a series of Grand Canonical MC (GCMC) runs and subsequent MD runs to simulate chain relaxation and swelling of the polymer during gas uptake. Generalized AMBER Force Field (GAFF) was used to model the polyimide while atomistic charges were calculated using Semi-Empirical methods via B3LYP functional. The MMM model was constructed with the smallest possible crystal structure of ZIF-8 and 2 polymer chains. CO_2sorption runs were performed up to 30 bar in order to investigate the plasticization of MMM's and subsequent changes at the polymer/ZIF-8 interfaces. Simulation results indicated that the interactions between the functional groups of the polymer and ligands at the MOF surface may have significant contribution on the gas transport through the interface.

Acknowledgment: This work was supported by TUBITAK trough grant no. 113M776.

465534 Performance Evaluation of Silica-PEI Adsorbents in a Two-Bubbling-Bed CO_2 Capture System

Abstract:

In this study, the performance of silica-PEI adsorbents which is supplied by the University of Nottingham (UNOTT) have been tested in a continuous CO_2 capture system. It consists of a bubbling fluidized-bed sorption reactor and a bubbling fluidized-bed regeneration reactor. We used a simulated flue gas which contains 15 vol.% of CO_2 w/o moisture (N_2-balance). In the regeneration reactor, N_2 itself has been used as a fluidizing medium in order to check the material balance of the capture system. Based on the continuous experiments, the dynamic sorption capacity and the CO_2 removal performance have been evaluated according to the operating temperature, solid circulation rate, and bed height. And also, the amount of total carry over of both reactors has been measured in order to evaluate the mechanical strength of the adsorbents.

465997 MOF-Based Mixed Matrix Membranes for CO_2/N_2 Separations

Abstract:

Development of new membrane materials for separation of CO_2 from N_2 is crucial for the mitigation of CO_2 emissions. In this study, CO_2/N_2 separation potential of metal organic framework (MOF)-based mixed matrix membranes (MMMs) were investigated. We first performed atomically-detailed simulations to obtain adsorption data of CO_2 and N_2 gases in MOFs using Grand Canonical Monte Carlo (GCMC) simulations. Then Molecular Dynamics (MD) simulations were performed in order to predict diffusion properties of gases in MOFs. Simulation results were used to obtain CO_2 and N_2 permeabilities of MOFs.Â A large number of MOFs was found to be located above the upper bound established for polymeric membranesâ€™ CO_2/N_2 separation performances. We then combined the results of our simulations for MOFs with the experimentally available data of gas permeability of polymers to predict permeability MOF-based MMMs. Gas permeability and selectivity of 700 new MMMs composed of 70 different MOFs and 10 different polymers were calculated for CO_2/N_2 separation. Several MOF-based MMMs were able to exceed the upper bound established for CO_2/N_2 separation. Results showed that for polymers with low CO_2 permeability but high CO_2 selectivity, the identity of the MOF filler is not important while for polymers that have high CO_2 permeabilities but low CO_2 selectivities, selection of the MOF strongly affects the separation properties of the MMMs. We believe that our results will guide both experimental and computational studies in order to find the best performing MOF/polymer MMMs for CO_2/N_2 separations.

Acknowledgement: Financial support provided by The Scientific and Technological Research Council of Turkey (TUBITAK) Grant MAG-213M401 is gratefully acknowledged.

463940 The Mechanism Study of Water Curtain Diluting Ammonia Release Dispersion

Abstract:

The mass transfer mechanism between water curtain and ammonia cloud, and a function of ammonia absorption was built based on two-film theory. Then, the function was programmed using UDF and embedded into CRD to build a numerical model of ammonia forced dispersion and the model was validated by Dandrieux experiment. Then a forced ammonia dispersion concentration distribution model was built based on RED model and the calculated results were validated. The results show that the model can accurate calculate ammonia antural and forced concentrations, and provided a method for calculating ammonia release accident consequences and water dilution effectiveness. The disaster reduction under various water curtain parameters was calculated by the gas release probability function method. The results show that the increase of water flow rate is beneficial to reduce the ammonia concentrations, but when the water flow is too bigger, the increasing of dilution effectiveness is not obvious. The water curtain should be set in the passive dispersion region.

472338 Surface Energy Enhancement of Poly(butylene terephthalate) By Addition of TiO_2 and Polyhedral Oligomeric Silsesquioxane (POSS) Nanoparticles

Abstract:

The presence of water droplets in fuels is a major factor affecting the performance of diesel engines, contributing to fuel system corrosion, fouling and performance degradation. Coalescing filters are efficient and effectual for the removal of water drops from fuel. One disadvantage of the filtering approaches is that the hydrophobicity of fibers can cause increased liquid holdup in the filter which in turn reduces the filterâ€™s permeability. Due to its excellent chemical and mechanical properties, poly(butylene terephthalate) (PBT) largely used to produce fibers for coalescing filters, for the removal of water drops from fuel. One disadvantage of the filtering approach is the hydrophobicity of PBT fibers which causes liquid holdup in the filter, which in turn reduces the permeability. Surface energy enhancement of PBT by addition of hydrophilic fillers could decrease the hydrophobicity and modify the filters performance. In this research, poly(butylene terephthalate) (PBT) based nanocomposites containing nano- Titanium dioxide and Polyhedral Oligomeric Silsesquioxane (POSS) nanomaterial were prepared by melt compounding. The influence of these nano materials on surface energy of the resultant nanocomposites was investigated by the use of contact angle measurement (CAM). Contact angle measurements of the resultant PBT nanocomposites showed that the wettability of such composites depends on the particular nanoparticles used. PBT/TiO_2 (5 wt.%) and PBT/Trisilanolethyl POSS (5 wt.%) showed the lowest water contact angle and the highest wettability.

Liquid side mass transfer coefficient measurements in tall structured packing beds by reactive absorption with dilute MDEA solutions

John Roesler, A. Royon-Lebeaud, P. Alix

IFP Energies nouvelles – Rond-Point de l'Echangeur de Solaize, BP3, 69360 Solaize, France

Introduction

Mass transfer parameters are required for rate based models used in the design of absorption columns such as for natural gas treatment. In these gas treatment columns, depending on the nature of the sour feed gas and on the specifications of the sweet gas, the rate of absorption of one of the constituents to be removed can become locally limited by the liquid side mass transfer coefficient, k_La_e, which then becomes an important parameter. Various methods have been proposed in the literature using physical desorption or chemical absorption.

Physical absorption methods use low solubility gases, typically O_2 or CO_2 (Mackowiak 1999, Benadda *et al* 2000, Moucha 2005). The accuracy of the methods tends to rapidly decrease as the extent of absorption increases, due to growing uncertainties of liquid phase analysis relative to the equilibrium driving force. This often limits bed heights to one or two HTU's to avoid high concentrations of dissolved gases in the liquid phase.

Physical desorption also uses low solubility gases such as O_2, CO_2 to determine k_La_e (Rejl *et al* 2015, Kunze *et al* 2015). More recently toluene has been successfully used (Wang *et al* 2014). Desorption is more accurate than physical absorption as the accuracy of liquid phase analysis then depends more on the detection limit. Bed heights of up to 3 to 4 HTU's can be used.

The chemical absorption measurements have generally used the Danckwert's plot method which allows to determine simultaneously k_L and a_e (Benadda *et al* 1996). This method is labor intensive as it requires test with varying amounts of chemical reaction catalysts to the solvent and gives rather large uncertainties on k_L. Slow reactions with carbonates (Weiland *et al* 1993) have also been used. These methods require well established thermo-chemistry of the reaction system in order to properly assess its accelerating effect.

The above methods have various drawbacks among which the difficulty to operate at large bed heights. We have developed an alternate chemical absorption method based on diluted CO_2 in air into a dilute amine solution consisting of MDEA that provides a means of measuring k_L from about 1 HTU to beyond 5 HTUs of packing. The choice of MDEA was motivated among other reasons by the fact that the thermo chemistry is well established.

Materials and methods

Experiments were performed in a 146 mm diameter column using various structured packing blocks (Mellapak 250.X, Mellapak 250.Y, and IFPACC 2X^{TM}) with bed heights up to 4m. The column was operated in a counter-current flow mode with an ascending gas and descending liquid which is recycled in a 500 L tank. The column is operated at 1.5 bars absolute, with liquid temperatures between 15 and 20°C and gas temperatures between 5 and 25 °C. The gas is dry air or N_2 with various concentrations of added

CO_2 up to 1 %vol. The liquid is water in the case of oxygen desorption tests or a solution with between 2.5 and 3.5 %wt MDEA in the case of CO_2 chemical absorption tests.

Desorption of O_2

The reference method was the desorption of O_2 from water in a flow of N_2 . The liquid in the tank is first loaded with O_2 by flowing air and water through the column. The air is then replaced with nitrogen for the desorption measurements. The liquid side mass transfer coefficient, $k_L a_e$, is obtained by :

$$k_L a_e = \frac{V_L}{Z} \ln\left(\frac{C^0_{O2,in}}{C^0_{O2,out}}\right) \qquad (1)$$

where V_L is the superficial liquid velocity, Z the packing height and $C_{O2,in}$, $C_{O2,out}$ the concentrations at the inlet and outlet of the packing respectively. The method was used for bed heights between 0.2 and 1.5m, beyond which uncertainties became significant.

Chemical absorption with MDEA

The method uses the absorption of diluted CO_2 in air up to 1%vol in a diluted solution of MDEA (2.5 to 3.5%wt) restricted to CO_2 loadings between 0.1 and 0.2 mol/mol. The kinetic rates are otherwise too fast at the low end due to the presence of OH^- while at the high end the sensitivity to equilibrium uncertainties is of concern. Gas samples are analyzed for CO_2 with an on-line NDIR analyzers. The amine solution is collected at the column extremities and analyzed by titration in terms of amine content and CO_2 loading.

The solute concentration measurements yield:

$$E.k_L a_e = \frac{He \cdot V_G}{ZRT} \ln\left(\frac{y^*_{CO_2} - y^o_{CO_2 in}}{y^*_{CO_2} - y^o_{CO_2 out}}\right) \qquad (2)$$

where E is the acceleration factor, He is the Henry constant, V_G the gas superficial velocity, y^o_{CO2in} and y^o_{CO2out} the solute concentrations at the inlet and outlet of the packing and y^*_{CO2} is the solute gas side equilibrium concentration at the liquid-gas interface which is constant (nearly constant solvent loading along the column). The thermo chemical properties are evaluated from a well-established in-house model.

The method was used with bed heights of 0.44m (two packing layers) and up to 4 m, with increasing uncertainties at low bed heights due to small absorption extents.

Results and Discussion

Both the reference O_2 desorption and the MDEA methods were found to provide the same results with beds of seven packing blocks (Figure 1). In order to account for the different solute liquid side diffusivities all $k_L a_e$ values are back corrected to CO_2 which, on the basis of the double film theory, is done as follows for a solute i:

$$k_{L,CO2}a_e = \frac{k_{L,j}a_e}{\sqrt{D_{L,O_2}}}\sqrt{D_{L,j}} \quad (3)$$

The values presented are normalized relative to the one obtained with the MDEA method at a liquid load of 60 $m^3/h/m^2$. The discrepancies are larger at low liquid loads due to the low extents of CO_2 absorption (< 5% under 30 $m^3/h/m^2$) that lead to higher measurement uncertainties.

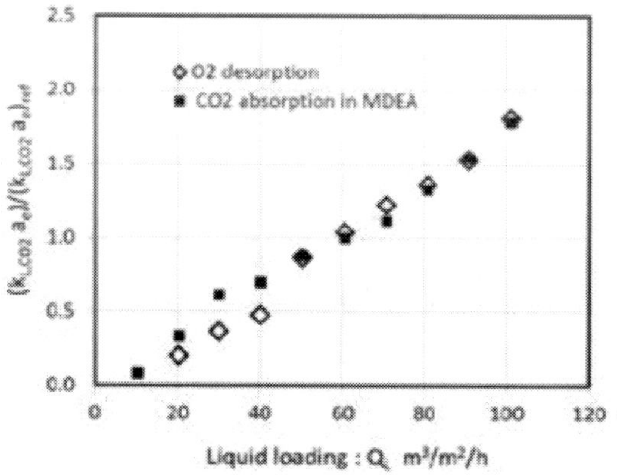

Figure 1 : Comparison liquid side mass transfer coefficients obtained with 7 layers of IFPACC-2X™ packing at $F_s = 0.65$ $Pa^{0.5}$. Reference k_La_e value for 60 $m^3/h/m^2$ with MDEA method.

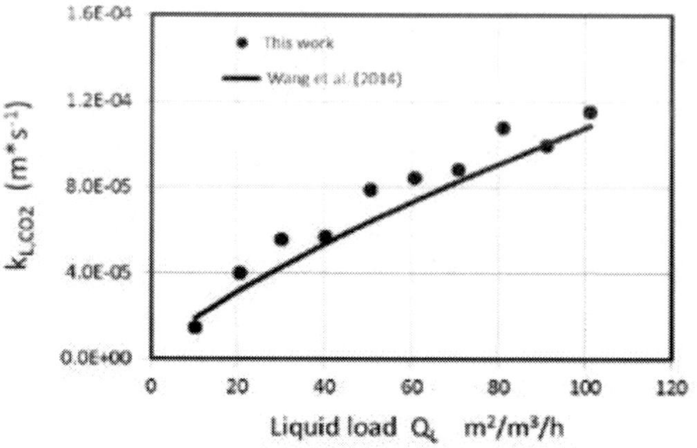

Figure 2 : Liquid side mass transfer coefficients obtained with the MDEA method over a 4m bed of Mellapak 250.X. The solid line is the value given by the correlation of Wang et al. (2014) after correction for the diffusivity differences of toluene relative to CO_2.

Figure 2 compares Mellapak 250.X results obtained with the MDEA method to the correlation of Wang et al. (2014) obtained with toluene desorption. The agreement is good and falls within the measurement uncertainties that combine those of a_e and k_L. The acceleration factor remains below 1.1 for k_L values above approximately 5×10^{-5} m/s, reaching a factor 2.75 at the lowest value of 2×10^{-5} m/s . The accuracy of the thermo chemistry thus becomes important only for the smaller values. The 4m bed height for the MDEA case corresponds to about 5 HTU's with a maximum extent of CO_2 absorption of 20% leaving ample margin applying the method to larger bed heights.

The MDEA method is currently under further investigation to apply it with CO_2 partial pressures below 100 Pa in order to implement it at reduced operating costs in terms of solvent and solute consumption in a 1 m diameter column.

References

Maćkowiak, J., 1999, *Chemie Ingenieur Technik*, 1+2(71), pp.100–104.
Benadda, B. et al., 2000. *Chemical Engineering Science*, 55, pp.6251–6257
Moucha, T. et al., 2005. *Chemical Engineering Research and Design*, 83, pp.88–92.
Rejl, F.J. et al., 2015. *Chemical Engineering Research and Design*, 99, pp.20–27.
Kunze, A. et al., 2015. *Chemical Engineering Research and Design*, 104, pp.440–452.
Wang, C. et al., 2014. *Energy Procedia*, 63, pp.1727–1744.
Benadda, B. et al., 1996. *Chemical Engineering and Processing*, 35, pp.247–253.
Weiland, R.H et al., 1993. *Industrial & Engineering Chemistry Research*, 32, pp.1411–1418.

473705 Prodias: Processing Diluted Aqueous Systems

Abstract:

A main challenge the process industry is facing today in introducing renewable raw material into their value chains, is the development of cost- and energy-efficient water removal and product-recovery techniques. In order to unlock the potential of the renewable-based product market for the European process industry, a re-thinking of downstream process development and the development of suitable methodologies for fast-track development of tailored downstream processes as well as the optimization of separation technologies are urgently needed.

PRODIAS (Processing Diluted Aqueous Systems) is a project that aims at providing new cost- and energy efficient downstream-processes to facilitate a competitive use of renewable feedstocks in the process industry. In order to realize the overall goal the following key deliverables have been defined:

1. Optimization and further development of single technologies.

2. Tuning of up-stream process in order to facilitate the downstream processing.

3. Development of hybrid combinations of complementary single technologies incl. apparatus in order to achieve cost-effectiveness and improved performance.

4. Demonstration of technologies in industrially appropriate environments of the three application areas represented in the PRODIAS project.

5. Process characterization to identify the key physical property data required for the design and operation of the technologies.

6. Development and validation of an integrated design approach.

This research receives funding from the European Community's Framework Programme for Research and Innovation Horizon2020 and involves 9 partners: BASF (Germany), Cargill Haubourdin (France), University of Kaiserslautern (Germany), Imperial College London (Great Britain), Alfa Laval (Sweden), GEA Messo PT (The Netherlands), Xendo (The Netherlands), UPM (Finland) and Enviplan (Germany).

This poster will present an overview of PRODIAS, the challenges identified and the structure of the project.

455679 Lab and Pilot Studies of Nanoporous Polyethersulfone Membranes As Support for Composite Membrane in CO_2 Separation

Abstract:

Nanoporous polyethersulfone (PES) membranes were prepared from PES/N-methyl-2-pyrrolidone (NMP)/2-methoxyethanol (2-ME) casting solutions with water as coagulant by both vapor- and non-solvent-induced phase inversion steps successively under various preparation parameters. 2-ME was incorporated into the casting solution as a pore-forming additive because of its high affinity with water, leading to more open membrane morphology. A detailed study of the effects of different parameters, including polymer concentration, 2-ME/NMP ratio, relative humidity, water vapor exposure time, and water coagulation bath temperature, was conducted in lab-scale experiments to determine the operation guidelines for pilot-scale fabrication. 14-inch wide PES membranes were fabricated successfully by using a continuous casting machine. A parametric study in the pilot-scale fabrication was also carried out. The surface morphologies of the PES membranes were characterized by scanning electron microscopy (SEM). The morphological differences between both scales by the same preparation conditions were compared. Based on the guidelines established from the lab-scale experiments, the casting solution composition and coagulation bath temperature were optimized in the pilot-scale continuous casting machine to fabricate the 14-inch wide PES membrane with a desirable morphology suitable for use as the substrate of composite membrane in CO_2 separation. The developed fabrication process also showed the potential for the commercial production of the PES membrane in a larger scale by the vapor- and non-solvent-induced phase inversion technique. In addition, a pilot-scale PES membrane with a pore size of ~ 69 nm and surface porosity of ~ 17% was employed in the continuous Zeolite-Y (ZY) nanoparticle deposition. A thin ZY layer with a controlled thickness and a uniform coverage was prepared. PES membranes deposited with ZY particles were also utilized as the substrate of composite membrane for CO_2 capture from flue gas in coal-fired power plants. From the gas transport measurement by using a simulated flue gas, the prepared composite membrane with an amine-containing selective cover layer showed a CO_2 permeance of > 800

GPU (1 GPU = 10^{-6} cm^3 (STP)$\cdot cm^{-2}\cdot s^{-1}$) and a CO_2/N_2 selectivity > 140, indicating the great potential for CO_2 capture from flue gas.

455687 Effects of SO_2 on Amine-Containing Facilitated Transport Membrane Performance for CO_2 Capture from Flue Gas

Abstract:

Post-combustion carbon capture is a key approach to control CO_2 emissions from power plants using fossil fuels. Treating real flue gas may be problematic as it contains not simply just CO_2 and N_2. The minor contaminants, such as SO_2 and O_2, may be the major factors that can limit the long-term stability of a membrane. The typical flue gas after flue gas desulfurization (FGD) treatment is at its discharge temperature of 57°C with 40 – 50 ppm SO_2. The SO_2 level can be further reduced to even 1 ppm by a caustic scrubber, i.e., 20 wt.% NaOH. With the presence of SO_2 in the feed gas, SO_2 could preferably react with amine carriers and hinder the permeation of CO_2 in amine-containing facilitated transport membranes. Possible amine degradation and competitive reactions raise the concerns on the membranes for use in real flue gas. Amine-containing facilitated transport membranes were tested at 102 and 57°C, respectively. SO_2 concentrations of 0.7 – 5 ppm were employed in our transport study. Unstable membrane performance was observed for the membrane tested at 102°C in the presence of SO_2. It was found that amines reacted with SO_2 to form bisulfite and sulfite products irreversibly at 102°C. On the contrary, the membrane showed a stable separation performance in the presence of 1 – 3 ppm SO_2 at 57°C. Moreover, an O_2 concentration of 3% was also added in the feed gas to simulate the real flue gas conditions. The membrane also showed a stable separation performance upon exposure to 3 ppm SO_2 and 3% O_2 for > 100 hours. In addition, infrared spectra were collected with respect to time by using an in-situ FTIR, and the growth of sulfite characteristic peaks was identified. The spectra results were consistent with the observed membrane separation performance, which is an affirmation of the SO_2 interference of the amines in the membrane.

The stable membrane performance at 57°C with SO_2 at ppm levels and O_2 at 3% indicates the applicability of the developed amine-containing membrane for use in CO_2 capture from flue gas.

455982 Enrichment of CH_4 from CO_2-CH_4 Mixtures By Selective CO_2 Separation with Ion-Exchanged Zeolite Membrane

Abstract:

Recovery of methane from CO_2-CH_4 mixtures in the land-fill gas as well as the low-grade natural gas field is increasingly important from the viewpoint of efficient utilization of natural and fossil fuel resources. Although there are several existing methods to recover methane such as cryogenic distillation and adsorption, membrane separation, having a reputation as an energy-efficient process, can be worthy of consideration. In this study, zeolite membrane is concerned in terms of its higher permselectivity for CO_2.

A high silica Na-type chabazite (CHA) zeolite membrane was hydrothermally synthesized on the outer surface of a porous a-alumina tube. The synthesized membranes were characterized by scanning electron microscopy (SEM), X-ray diffraction (XRD), and Energy-dispersive X-ray spectroscopy (EDX). The results of this study show that it is useful for CO_2/CH_4 separation. Single-gas permeation for CO_2 and CH_4 was tested over the temperature range of 313~373 K. Then the Na-CHA zeolite membrane was ion-exchanged with K or Cs, or both cations. The cation composition was determined with the EDX measurement. K-, Cs-, and KCs-type CHA membranes showed higher CO_2 permselectivity compared with the Na-type CHA membrane before ion-exchange. This was because with increasing cation diameter in the order Na, K and Cs the effective zeolite pore size became small, whereby decreasing the CH_4 permeance. Especially, KCs-CHA zeolite membrane had the highest CO_2 permeance and selectivity, which were corresponded to the world highest level.

456123 Enantioselective Separations Using Supported Liquid Crystalline Membranes

Abstract:

The production of optically pure enantiomers is a critical issue in fields including pharmaceuticals, foods, and agrochemicals. In the case of pharmaceuticals, often only one drug enantiomer is effective while the other enantiomer is ineffective or even toxic. Small molecule thermotropic liquid crystals (LCs) exhibit self-assembled orientations such as smectic, nematic, and isotropic phase depending on temperature changes. With an addition of chiral dopants, LC molecules in nematic phase oriented along one direction are reoriented in a helical way known as cholesteric liquid crystal (ChLC) phase. Herein we report a novel ChLC embedded polymer membranes, which features transport phenomena of membrane chromatography through the membrane pores. The ChLC phase induced by adding cholesterol-based chiral dopants in nematic LC phase exhibits left-handed chirality· at low temperature below 40°C and left-handed enantiomers transport slower than right-handed enantiomers due to the interactions between left-handed enantiomer and left-handed chiral phase in the membrane. The LC membranes showed selectivity up to 1.17 for the porous membranes by the interactions between the lefthanded S-1-phenylethanol and the lefthanded LC surface.

456138 Seeding Effect on the Purification of Enantiomers By a New Technique Combining Distillation and Crystallization

Abstract:

Stripping crystallization (SC) is a new separation technique which combines distillation and crystallization. Basically, SC is operated at a triple-point condition, in which the liquid is simultaneously vaporized and crystallized due to the three-phase equilibrium. In essence, the process is continued until the liquid phase is completely eliminated and only the pure solid crystals remain in the feed. Thus, crystal washing is not required since only pure solid crystals remain in the feed and no impurities are adhered on the crystal surfaces at the end of SC. In the previous research, SC has been successfully applied to separate p-xylene from the mixed xylenes by Shiau et al. (2005, 2008, 2013).

Ibuprofen, also called 2-(isobutylphenyl)-propionic acid, is a non-steroidal anti-inflammatory drug widely used to treat headaches and minor pains. Although it is known that S-ibuprofen is responsible for the anti-inflammatory effects while R-ibuprofen might cause some long-term side effects, ibuprofen is still sold as a racemic mixture in the market due to the complexity involved in the separation of enantiomers. The experiments show that subcooling of the liquid ibuprofen mixture is observed during the temperature-decreasing process. Ultrasound mixing or magnetic stirring only slightly reduces subcooling. However, once crystallization is induced in the subcooled liquid mixture, crystals are formed abruptly, leading to a product with a lower purity. Due to the structure similarity between S-ibuprofen and R-ibuprofen, R-ibuprofen might be incorporated into the S-ibuprofen crystal lattice when S-ibuprofen crystals are formed abruptly. Therefore, the ibuprofen mixture can not be further purified by SC without seeding.

Subcooling can be nearly eliminated by seeding with ultrasound mixing or magnetic stirring during the temperature-decreasing process. Subsequently, crystals are gradually formed preferentially on the seeds, leading to a product with a higher purity. Thus, seeding with ultrasound mixing or magnetic stirring can be applied to further purify the ibuprofen mixture. As magnetic stirring helps to uniformly disperse the seeds, seeding with magnetic stirring leads to a product with a higher purity compared to seeding with ultrasound mixing.

References:

1. Shiau, L. D.*, Wen, C. C. and Lin, B. S. "Separation and purification of p-xylene from the mixture of m-xylene and p-xylene by distillative freezing", *Ind. Eng. Chem. Res*, **44(7)**, p2258-2265, 2005.

2. Shiau, L. D.*, Wen, C. C. and Lin, B. S. "Separation of p-xylene from the multicomponent xylene system by stripping crystallization", *AIChE Journal*, **54(1)**, p337-342, 2008.

3. Shiau, L. D.*; Liu, K. F. "Investigations into the effects of the cooling rate on stripping crystallization", *Ind. Eng. Chem. Res*, **52**, p1716-1722, 2013.

Keywords: crystallization; vaporization; purification; thermodynamics process; ibuprofen

456305 Separation of Antioxidants from Strawberry By Supercritical Extraction with Carbon Dioxide and Entrainers

Abstract:

The effective extraction and purification process of antioxidants from strawberry is expected as a manufacturing method of new food source. Extraction with supercritical carbon dioxide has advantages that organic solvent is not necessary and property of extracts is controllable by changing entrainer, temperature and pressure. In this study, the receptacle of strawberry was extracted with supercritical carbon dioxide and entrainer. The effect of entrainer and temperature on the properties of extract was evaluated.

The extraction of the receptacle of strawberry was conducted by semi-batch system with carbon dioxide and mixtures of carbon dioxide - ethanol, acetone and water. The antioxidant activities in the presence of entrainer were higher than that in the absence of entrainer at 313 K and 20 MPa, which indicates that entrainers are effective for the extraction of antioxidants. From the HPLC analysis, the cinnamic acid was found as one of the products and quantified. In the presence of entrainer, acetone and water gave the highest yield of cinnamic acid and total saccharide, respectively. The polarity of entrainer strongly affected the properties of extract. In the extraction with supercritical carbon dioxide â€“ ethanol mixture, the values of antioxidant activity, the yield of total saccharide and cinnamic acid increased with increasing temperature. The high temperature region was preferred to the extraction of antioxidants from strawberry.

456307 CO_2 Removal from CO_2-CH_4 Mixtures Using a Shell-and-Tube Separator and Its CFD Analysis

Abstract:

For more precise analysis for various types of membrane separators, modelling using computational fluid dynamics (CFD) is very useful because of taking into account all the distributions with respect to flow rate, concentration, pressure and temperature. Even for a shell-and-tube separator, being the simplest configuration among the flow types, to more accurately determine the concentration dependence of zeolite membrane permeance during separation of CH_4-CO_2 mixture those phenomena should be considered.

The performance of the gas separator for binary CH_4-CO_2 mixture with zeolite membrane is experimentally known to be influenced by the composition, which can change along the length of separator: this makes it more difficult to formulate the concentration dependency. Then, it was assumed that the permeances of CO_2 and CH_4 were as follows; $P_{CO2}=P_{CO2}{}^{0}(1+a(1\text{-}x))$ and $P_{CH4}=P_{CH4}{}^{0}(1+bx)$, where x was the mol fraction of CO_2. The CFD code used in this study is STAR-CCM+. The constant *a* and *b* is determined so as to agree with the experimental result. It was made clear that the permeances of two gases for the chabazite zeolite membrane show the concentration dependences assumed. In addition, the concentration distribution and flow rate within the separator were visualized, helping understand the separation manner inside the separator employed.

457999 Liquid Methanol Sorption, Diffusion and Permeation in Water Swollen Polymers

Abstract:

Liquid methanol sorption, diffusion and permeation in cross-linked PEGDA (XLPEGDA) and ion exchange polymers based on sulfonated polystyrene has been investigated and compared to sodium chloride transport. Methanol was selected as an uncharged model molecule because of its complete solubility in water.

The mechanism of solutes diffusion in water swollen polymers was experimentally and theoretically analyzed, and the role of the polymer fixed charges was elucidated. In particular, methanol sorption and diffusion in charged and uncharged polymers are ruled by the membrane degree of swelling, with more swollen membranes exhibiting faster transport rates and greater sorption levels. The electrostatic phenomena affecting ion transport in swollen polymers, i.e. ion-ion interactions and Donnan exclusion, are not involved in methanol transport. The extent to which methanol transport rate in the membrane is reduced relative to that in pure water is discussed and related to the membrane water content using the Mackie-Meares model [1]. The issue of convective transport and its effect on the diffusion coefficient were quantitatively addressed.

Finally, Flory-Rehner theory for ternary systems [2] was used to quantitatively interpret methanol sorption isotherms in swollen polymers.

[1] J.S. Mackie et al., Proc. Royal Soc. 1955 232 498-509

[2] P.J. Flory et al., J. Chem. Phys. 1943 11 521-526

458206 Study of Copper-Based Commercial Sorbent for Desulfurization of Syngas at Warm Temperature

Abstract:

Gasification is a reaction process that converts carbonaceous solid fuel, such as coal, biomass and mixture of them, into gaseous fuel which is called synthesis gas or syngas. The temperature of syngas is in the range of 800 °C to 1,800 °C, and it is decreased to room temperature for removing the pollutants with traditional gas cleaning processes. Then, syngas is needed to be heated up to the operating temperature, based on the requirement from applications. In order to increase the overall efficiency of syngas application, the desulfurization process development trends to higher temperature. However, because of material limitations in dry desulfurization technologies, the optimum temperature for copper-based sorbent has been estimated to be in warm-temperature (300-600°C) range that is currently of industrial interest. The characteristics of selecting appropriate sorbents according to material's physical and chemical properties, such as sulfur capacity, attainable sulfur concentration in gas, possibility of regeneration, mechanical stability over multiple sorption-regeneration cycles, etc., are the parameters to design the reactors and to affect the energy consumption in the processes. The sulfur capacity is influenced by the reacting temperature, space velocity, gas composition and cycles of sorption-regeneration.

In this study, a commercial sorbent with particle size of 30~50 mesh is used for removing hydrogen sulfide from syngas (containing 1% vol. H_2S) down to ppm levels in a fixed-bed reactor. The effect of regeneration temperature and cycles of sorption-regeneration on the sulfur capacity were studied as reference to build an operation strategy. The XRD analysis indicates that sulfidation process occurs in partial reduction under reducing atmospheres, i.e., $CuO \rightarrow Cu/Cu_2O \rightarrow Cu_xS$ ($x \leq 2$). The copper-based sorbent maintains excellent sulfur removal efficiency (above 90% of initial content) during ten adsorption-regeneration cycles at higher regeneration temperature. The results show that the amount of sulfate decreases as regeneration temperature increases. Considering the trend that energy consumption increases as regeneration temperature increases, a big gap between sulfidation and regeneration temperature results in efficiency decrease. The study shows that the performances of the copper-based commercial sorbent seem to be promising for application in syngas desulfurization. It is necessary to decrease regeneration temperature and minimize sulfate formation by introducing a suitable additive in future work.

458215 Steady-State Behavior and Control of Four-Product Divided-Wall Columns for Alcohols Separation

Abstract:

Kaibel divided-wall column (KDWC) and extended Petlyuk divided-wall column (PDWC) are capable of separating mixtures of four components into high purity products. This work studies the separation of a four-component mixture of methanol, ethanol, 1-propanol and 1-butanol (MEPB) in KDWC and PDWC, respectively. The rigorous calculation models of KDWC and PDWC were set up on the basis of short-cut calculation at first. Simulation showed that KDWC could save 36.46% energy consumption compared to the conventional direct separation configuration, while PDWC could save up to 45.95% energy. The prefractionator of KDWC performs a difficult ME/PB split, and then the main column splits M/E and P/B in the different parts of the column. The prefractionator of PDWC splits M/B, and then the upper and lower parts of middle column in PDWC splits M/P and E/B, respectively. After that, the main column of PDWC splits M/E and P/B at the same time. The intermediate compounds donâ€™t remix in the main column of KDWC and PDWC, but remix in the other parts of the two DWCs. The composition control structure were proposed in Aspen Dynamic, investigating the effect of feed flow rate and composition disturbances. Dynamic control results show that KDWC can handle ±10% disturbances of flow rate and feed composition. PDWC can handle ±10% and ±20% disturbances of flow rate and feed composition. However, a larger deviations of ethanol product can be found when the disturbances of P and M disturbances happened. Combined with the control results and V_{min} diagrams for composition disturbances, it is found that the concentration of ethanol product in upper side stream will decrease when the disturbance make M/E split harder. Otherwise, an easier M/E split will increase ethanol purity.

459524 The Determination Method of Intraparticle Diffusivity By Using Circulating Type Shallow Bed Adsorber

Abstract:

Adsorption is used in various fields owing to high selectivity to a certain substance in gas and liquid phase, and the technique is important for separation in chemical engineering. For design of adsorbers, kinetic and equilibrium parameters are required. The kinetic parameters consist of fluid film mass transfer coefficient (k_F) and intraparticle diffusivity (D_S). To determine D_S, Fixed-bed breakthrough curve method, shallow bed method, the completely mixed batch reactor (CMBR) method and the circulating type fixed bed adsorber (CFBA) method are employed. Fujiki et al. proposed a determination method of D_S and k_F from liquid phase fixed bed breakthrough curve. But it is difficult to analyse experimental concentration profile by use of the fixed bed breakthrough curve. CMBR method is a convenient and environmental friendly experimental technique, because 1~2 L volume of test solution is required and the values of DS and kF can be obtained from one concentration decay curve. However, since fixed bed is generally used in chemical industrial separation, the determination of D_S by using fixed bed is preferred.

It was found by numerical calculation that the concentration history obtained by CFBA method was similar to that by CMBR method, when contact time was 5.00~28.0 sec. So the circulating type shallow bed adsorber (CSBA) method is focused on in this study. This method has the advantage of simple and low costs procedure. Furthermore, the value of D_S and k_F can be determined from one concentration history. Both values were determined by comparing concentration experimentally obtained with theoretical curve. As a result, the values of D_S and k_F were similar to those reported previously using CMBR and CFBA. The results indicate that the accurate values of D_S and k_F are simply determined by the circulating type shallow bed adsorber.

458732 Surface-Enhanced Membrane Separations for Biofuel Processing

Abstract:

In thermochemical or biochemical pathways for biomass/bio-oil conversion to biofuels or chemical products, separations play a significant role to improve efficiency or yield in various processing steps, each of which represent a unique challenge on its own separation requirements. Membrane separations are being explored to understand their technology and economic impacts in several processing applications: for example, improving carbon recovery and energy efficiency, preserving catalysts (higher reaction yields and longer lifetimes), enhancing the purification of desirable intermediates or products, and enabling reactive separations and biomass/bio-oil fractionation. This presentation will focus on a new class of surface-functionalized porous membranes that could fully take advantage of extreme surface material properties (e.g., superhydrophilicity or superhydrophobicity) to enhance selective separation mechanisms. Membranes with tailored pore structure and surface coating nanomaterials have demonstrated initial success in high-flux selective permeations for both liquid- and vapor-phase processing applications. Two case studies are on achieving molecular separations of oxygenates/water/hydrocarbon mixtures (vapor or liquid) and microemulsion separations relevant to thermochemical conversion pathways. The underlying separation mechanisms and material/process challenges corresponding to different cases will be discussed.

460586 CO_2-Selective Membranes Containing Sterically Hindered Polyvinylamine

Abstract:

CO_2 capture from flue gas (main component: N_2) is essential for eliminating greenhouse gas emission, and CO_2 separation from synthesis gas is effective for hydrogen purification. Due to the inherent advantages of membrane separation, membrane technology is receiving more and more attention for CO_2 separation. For most conventional polymer membranes, CO_2/N_2 separation follows the solutiondiffusion mechanism. The key challenge of solutiondiffusion membranes is the tradeoff between permeability and selectivity, i.e., the Robeson's Upper Bound. However, for facilitated transport membranes, the nonreactive gas component, e.g., N_2, diffuses through the membrane via the solutiondiffusion mechanism, whereas CO_2 follows the facilitated transport mechanism via its reversible chemical reaction with carriers (often amines). Therefore, CO_2 transport is facilitated, and both high CO_2 permeability and high CO_2/N_2 selectivity can be achieved.

Amine carriers are the key for the CO_2 separation performance of a facilitated transport membrane. Amine carriers exhibiting a high CO_2 loading capacity and reaction rate are always desirable. It has been reported that the CO_2 loading capacity of sterically hindered amine can be twice of unhindered amine in aqueous solution. In this work, highmolecularweight polyvinylamine (PVAm) was successfully synthesized and modified into sterically hindered polyamines with different degrees of steric hindrance via alkylation reactions. The membranes were synthesized by coating the sterically hindered polyamine solutions with suitable viscosity onto a nanoporous polyethersulfone substrate. Under the typical conditions for CO_2 capture from flue gas (57 °C, 1 atm, 100% relative humidity, 20% CO_2 in the feed gas balanced with N_2 on a dry basis), polyvinylamine membranes showed an average CO_2 permeability of 214 Barrers and CO_2/N_2 selectivity of 48.5. Although the content of amino groups reduced by incorporation of alkyl groups, CO_2 permeability was improved by 24% and CO_2/N_2 selectivity was increased by 14% by modifying polyvinylamine into poly(NmethylNvinylamine). These results have demonstrated the effect of steric hindrance with the incorporation of the methyl group for enhancing CO_2 separation performance in the solid membrane phase. This is one of very few pieces of work demonstrating the steric hindrance in the solid membrane phase. Poly(NisopropylNvinylamine) and poly(NtertbutylNvinylamine) membranes showed 15% and 11% improvements on CO_2 permeability, respectively, with no significant changes in CO_2/N_2 selectivity. Comparing the CO_2 separation performances of the above polyamine membranes and their corresponding steric hindrance degrees, we found that the CO_2 permeability increased first significantly with increasing the steric hindrance degree and then less pronouncedly with further increasing the steric hindrance degree. This could be explained by the reduced amino group content in the membranes and the lower reaction rate due to the incorporation of bulky groups onto amino groups. Based on the gas transport results, poly(N-methyl-N-vinylamine) has demonstrated the largest improvement on membrane performance, making it a promising candidate as the next-generation fixedsite carrier for CO_2 facilitated transport membranes.

460757 Adsorption of Water Contaminants in Continuous Flow Systems Using Carbon Cryogels with a Microhoneycomb Structure

Abstract:

A microhoneycomb monolith, which is a miniature version of a honeycomb-shaped monolith, has fairly straight macro channels, a few tens of μm in diameter which are formed by thin channel walls, a few μm in thickness. This monolith shows a minimized hydraulic resistance because of its straight channels and the lengths of the diffusion paths within its walls are extremely short[1]. Such a monolith is suitable for high-throughput separation of diluted water contaminants, e.g. phenolic compounds and various dyes, because a low hydraulic resistance and fast response due to short diffusion path lengths are compatible. This cannot be achieved in a column packed with particles.

In this work, to demonstrate the potential of such microhoneycomb monoliths for the separation of typical contaminants, phenol and methylene blue adsorption were conducted in batch systems and continuous flow systems using carbon cryogels with a microhoneycomb structure (carbon microhoneycombs, CMHs). The obtained monolithic CMHs had fairly straight channels, 25-45 μm in diameter, and the thickness of the walls which form the channels was 5-10 μm. The pressure drop which occurred when water was passed through the CMHs was 90-370 times lower than that of a column packed with particles having similar diffusion path lengths. This indicates that the CMHs can process fluids without causing a severe hydraulic resistance when compared columns packed with particles. From nitrogen adsorption/desorption experiments, it was found that the CMHs have a hierarchical micro-meso porous structure giving BET surface areas in the range of 513-1070 $m^2 \cdot g^{-1}$. The phenol and methylene blue adsorption isotherms of the CMHs are typical Langmuir-type which indicates monolayer adsorption. The monolayer uptake increases with the increase in carbonization temperature of the CMHs which represents that hydrophobic surfaces are preferable for the adsorption of phenol and methylene blue. The obtained phenol and methylene blue breakthrough curves show typical sigmoidal-shapes and the calculated dynamic capacities are equal to the values calculated from the isotherms. In addition, the length of unused bed, which represents the length of the zone in which adsorption takes place, is only a few mm. These results indicate that the adsorbates can quickly gain access to the adsorption sites in the CMHs' channel walls even in flow systems. The simulated breakthrough curves of a column packed with particles which shows the same total pressure drop and has the same material properties are a few-times broader when compared with those of the CMHs measured under the same flow conditions. This result indicates that the CMHs can effectively separate contaminants in flow systems when compared with typical columns packed with particles. The results indicate that CMHs have a high potential to be used for effective separation of water contaminants such as phenol and methylene blue.

[1] Mukai, S. R.; Nishihara, H.; Tamon, H., Chem. Commun. 2004, (7), 874-875.

462883 Modeling and Experimental Studies on the Start-up of a Divided Wall Column for High-Purity Separation of Aromatic Reformate

Abstract:

The increasing application of divided wall columns in chemical industry is driven by the fact that it can save up to 30% energy and capital cost than conventional separation sequence. Only few works have been reported on the start-up of divided wall columns, which is inherently a strongly nonlinear process. Here, for the first time the start-up of a divided wall column pilot plant is explored for separating an aromatic reformate, which contains trace water, eight aromatic components and two paraffins. Very high product purities above 99.95wt.% are specified. A four step start-up method is proposed and testified in Aspen Dynamics simulation and real operation of the pilot plant. The proposed start-up method is consisted of pre-feeding stage, benzene withdrawing stage, toluene withdrawing stage and continuous feeding stage. In the first three stages, the column pressure drop is selected as the key parameter for control the reboiler duty. In the last stage, the reboiler temperature is the key control parameter for the divided wall column. Besides, controlling the temperature on sensitive plate (Set point equals to steady-state value) by withdraw flowrate in the same section can reduce up to 40% start-up time than withdrawing at a constant flowrate in the second and third stages. The four-loop temperature control structure has been testified effectively on the pilot plant when facing 10% feed flow rate and composition fluctuations. The energy cost is calibrated by a steam vortex flowmeter, and results showed that the energy consumption of the divided wall column can reduce up to 30% than that of the conventional separation sequence, which is also consistent with the simulation results.

463497 High-Throughput Screening of Metal-Organic Frameworks for CO_2 Capture at High Humidity

Abstract:

One of the main challenges for the development of technology to capture CO_2 from power plant flue gas is to find or create materials with high selectivity for CO_2 at high humidity. Metal-organic frameworks (MOFs) have gained significant attention due to their excellent performance in gas adsorption and separation. However, water adsorption in MOFs is generally detrimental to CO_2 separation, and the removal of water content from flue gas prior to CO_2 capture would add to the overall cost. In this study, the CoRE (Computation-Ready, Experimental) MOF database of existing MOF structures has been computationally screened to find MOF candidates that are highly selective toward CO_2 over H_2O using a hierarchical computational screening approach. MOF candidates were selected based on the ratio of Henryâ€™s constants of CO_2 over H_2O computed via Widom insertion method, corresponding to the selectivity of CO_2 over H_2O, and the adsorption isotherms of selected MOFs with high selectivities were assessed using more accurate partial charges from density function theory (DFT) calculations by grand canonical Monte Carlo (GCMC) simulation. Finally, the top MOF candidates with high CO_2/H_2O selectivity were identified. Decomposition of interaction energy between MOF and adsorbate show that the electrostatic interaction contributes to the majority of the adsorption energy of H2O in these selected MOFs. This work shows a path for the discovery of adsorbent materials for CO_2 capture at high humidity by large-scale computational screening.

464067 In silico Screening of Metal-Organic Frameworks for CO_2/CH_4 and CO_2/N_2 Separation

Abstract:

We report a molecular simulation study to screen 4764 computation-ready, experimental metal-organic frameworks (CoRE-MOFs) for CO_2 separation from flue gas (CO_2/N_2) and natural gas (CO_2/CH_4). Quantitative relationships are established, for the first time, between metal type and adsorbent evaluation criteria (adsorption selectivity and capacity, working capacity and regenerability). It is found that alkalis exist in 75% of alkali-MOFs as nonframework ions or open metal sites, and 75% of alkaline-MOFs contain alkalines as open metal sites; thus alkali and alkaline-MOFs exhibit high adsorption selectivity and large capacity. Combining selectivity, working capacity and regenerability, however, alkali and alkaline-MOFs possess the lowest performance for CO_2 separation. Among ~1000 lanthanide-based CoRE-MOFs, 50% contain lanthanides as open metal sites and have the highest performance. Best 30 CoRE-MOFs are identified for CO_2/N_2 and CO_2/CH_4 separation, and they mostly contain lanthanides. Furthermore, we predict the breakthrough curves in two identified CoRE-MOFs and demonstrate their superior separation performance. This modelling study highlights the central importance of adsorbent evaluation by holistic criteria, and suggests that lanthanides could be the first choice in the design of new MOFs for CO_2 separation.

464260 Water Vapor Adsorption Characteristics of Aerogels for Desalination

Abstract:

Adsorption-Desalination (AD) system has been suggested and has proven to be a low-energy desalination method employing a low-temperature heat-driven adsorption/desorption cycle for desalination. Only commercialized silica gel is used to produce fresh water by desalinating seawater vapor in this process, but the poresâ€™ limitations restrict the amount of water uptake and harsh operating conditions. The development of novel porous adsorbents able to achieve high water vapor uptake at low temperature and relative pressure is a high priority to be incorporated into the AD system. Aerogels are has emerged as a promising candidate as an adsorbent because of their high pore and thermal characteristics. In this study, hydrophilic or hydrophobic aerogels were prepared by sol-gel method using different silica sources. Raman spectroscopy was applied to investigate structural changes of aerogels during water vapor adsorption. Enhancement of adsorption/desorption rate of aerogels were examined by dynamic vapor adsorption analyzer. Our results offer the potential in application of AD system utilizing aerogels with high water vapor uptake.

464407 Potentials of Significant Energy-Saving Via Hybrid Extraction Distillation Separation System: *N*-Propanol Dehydration

Abstract:

Potentials of Significant Energy-Saving via Hybrid Extraction–Distillation Separation System: *n*-Propanol Dehydration

Wei-Lun Chang, Bor-Yih Yu and I-Lung Chien*

Department of Chemical Engineering, National Taiwan University, Taipei 10617, Taiwan

*Corresponding Author¡¦s E-mail: ilungchien@ntu.edu.tw

Keywords: *n*-Propanol; Dehydration; Hybrid Separation Process; Extraction; Azeotropic Separation; Design and Control

Abstract

n-propanol is one of the common solvents widely used in chemical processes. However, it forms a minimum-boiling azeotrope with water that makes the dehydration impossible with only one distillation column. There are two recent literatures to tackle this separation task via extractive distillation, one using NMP [An *et al.*, 2015] and the other one using ethylene glycol [Pla-Franco *et al.*, 2015] as entrainer. Unfortunately, the enhancement on relative volatility of *n*-propanol over water is not that large compared to other extractive distillation systems, resulting in a large entrainer-to-feed ratio and also high energy consumption.

Hybrid extraction–distillation process was reported to be an energy-saving design for pyridine dehydration [Chen *et al.*, 2015]. In this work, this improved separation method is studied for the purpose of *n*-propanol dehydration. This hybrid process can be viewed as a derivative from heterogeneous azeotropic distillation method, because they both require adding an appropriate third component to form a new minimum-boiling heterogeneous azeotrope and large liquid-liquid envelope. The difference is that the needing of a pre-concentration column in heterogeneous azeotropic distillation is replaced in the hybrid process with an extraction column. Note that the solvent-rich stream into the extraction column comes from the organic reflux of the heterogeneous overhead from the azeotropic column. Most importantly, the solvent flow rate is an additional degree-of-freedom (DOF) that can significantly affects the energy consumption in the hybrid process, while this DOF is not available for heterogeneous azeotropic distillation. In this proposed hybrid process, DIPE is selected as the extraction solvent considering its separation performance on extraction, heat of vaporization, density, and less toxicity. Compared to the extractive distillation with NMP, steady-state results show that this improved design saves at least 64.1% reboiler duty using also much less column stages.

As for the dynamic control of this kind of hybrid separation process, Chen et al. (2015) made a trade-off between optimal steady-state design and dynamic controllability to reject the feed flow rate or composition disturbances with a fixed solvent-to-feed ratio. In this work, a novel control strategy is proposed based on closed-loop and open-loop sensitivity tests. Here, an adjustable solvent-to-feed ratio during dynamic control allows the operation of the optimal steady-state design instead of a trade-off design. Dynamic simulation results show that both *n*-propanol and water products can still be maintained at high-purities despite large variations in feed flow rate and feed composition changes.

References:

Y. An, W. Li, S. Huang, J. Ma, C. Shen, C. Xu; Design/optimization of energy-saving extractive distillation process by combining preconcentration column and extractive distillation column; *Chem. Eng. Sci.*, **135**, 166–178, 2015.

J. Pla-Franco, E. Lladosa, S. Loras, J. B. Monton; Approach to the 1-propanol dehydration using an extractive distillation process with ethylene glycol; *Chem. Eng. and Process.*, **91**, 121–129, 2015.

Y. C. Chen, K. L. Li, C. L. Chen, I. L. Chien; Design and Control of a Hybrid Extraction–Distillation System for the Separation of Pyridine and Water; *Ind. Eng. Chem. Res.*, **54**, 7715–7727, 2015.

465073 H_2TiO_3 Composite Nanofibers for Lithium Recovery from Seawater and Aqueous Resource

Abstract:

Lithium shortage is inevitable given the massive growth in its demand for use in transportation industry and various technologies. Thus, recovery of lithium from seawater and brine through ion exchange has become an attractive option to potentially alleviate the impending crisis on lithium shortage. Adsorption is considered as one of the simplest and most promising methods to mine lithium from aqueous resources. Manganese-based lithium ion sieves (LIS) have been the most popular and well-studied lithium adsorbents.

Recently, titanium-based inorganic materials (i.e. Li_2TiO_3) have demonstrated a strong potential as LIS for the selective capture of lithium ions. As a precursor, Li_2TiO_3 is transformed as the final LIS material (H_2TiO_3) via acid delithiation. H_2TiO_3 (HTO) has high theoretical capacity (126.47 mg/g) and a chemically stable structure as indicated by its low titanium dissolution upon acid pickling. However, HTO application in actual systems is limited due to the difficulty in its handling. The HTO powders must be incorporated in a suitable solid support for its efficient utilization. Herein, electrospun nanofibers (NFs) with unique dimensional properties were employed to serve as three-dimensional frameworks for HTO.

Different polymeric materials such as polyacrylonitrile (PAN), polysulfone (PSf), polyvinyl chloride (PVC), and polyvinylidene fluoride (PVDF) were screened to identify the most suitable matrix for the HTO powder. NFs with varied HTO loading were prepared by electrospinning method. The composite nanofibers were characterized by SEM-EDS, UTM mechanical test, TGA and then applied for selective lithium ion (Li^+) adsorption. Selection of the most suitable composite was conducted through a surface response methodology (RSM) using central composite design (CCD) of experiments.

From the developed optimization models and experimental results, it was determined that PAN composite NFs with 100 wt% HTO loading exhibited the highest adsorption capacity (31.4 mg g^{-1}) and superior mechanical property. It retained up to 93.6% of the adsorption capacity of the pure HTO powder (33.6 mg g^{-1}). The captured Li^+ were easily recovered by mild acid treatment (0.25M HCl) of the spent PAN-HTO NF, which was also regenerated for reuse. Cycled experiments showed the reusability (at least 5 times) and stability of the composite NF, which also effectively recovered Li^+ from an actual source (seawater). Overall, the results of the study showed the feasibility of using PAN-HTO NF for continuous and efficient recovery of lithium ions from aqueous resources.

This research was supported by Basic Science Research Program through the National Research Foundation of Korea (NRF) funded by the Ministry of Education (2009-0093816) and Ministry of Science, ICT and future Planning (2015R1A2A1A15055407).

465611 Sulfur-Based Copolymer Nanofiber Prepared Via Inverse Vulcanization for Heavy Metal Sequestration

Abstract:

Recent progress on inverse vulcanization has opened various opportunities to develop advanced materials for environmental remediation and energy storage applications. Herein, the development of a nanofiber adsorbent with extremely high capacity for various heavy metals, especially mercury (Hg^{2+}), is presented.

Processable sulfur copolymer was synthesized via facile inverse vulcanization of elemental sulfur with 2-carboxyethyl acrylate (CEA). Diradical chain ends of elemental sulfur produced from ring opening polymerization was stabilized by direct addition of CEA copolymer. Rheology of resulting sulfur-CEA copolymer (poly(S-r-CEA)) dope solution was manipulated by adjusting the molar ratio of the components. Poly(S-r-CEA) products were viscoelastic (i.e. 0< phase angle (Î´) <90Â°). Temperature-dependence of viscosity was analyzed via oscillatory rheology in order to determine appropriate temperature for the dope solution which can be processed into a nanofiber via electrospinning.

The prepared poly(S-r-CEA) nanofiber was tested for removal of heavy metals and was found out to have very high affinity towards Hg^{2+}. Extended X-ray Absorption Fine Structure (EXAFS) spectroscopy was employed in order to elucidate poly(S-r-CEA)-Hg^{2+} complex. Adsorption isotherm data revealed that Hg^{2+} sequestration of poly(S-r-CEA) follows monolayer Langmuir-type and that the calculated maximum adsorption capacity could reach up to 1000 mg g^{-1}. This adsorption capacity of poly(S-r-CEA) nanofiber is comparable to current Hg^{2+} benchmark adsorbent but features significant advantages such as cost and availability of main raw material, ease of preparation, and effectiveness of treatment process. This work therefore presents a new promising materials for the removal of $Hg^{2+}$and other heavy metal ions from contaminated water for environmental remediation.

This research was supported by Basic Science Research Program through the National Research Foundation of Korea (NRF) funded by the Ministry of Science, ICT and future Planning (2015R1A2A1A15055407) and 2015 Research Fund of Myongji University.

465542 Preparation of Amine Modified Bimodal Mesoporous Silica Particles for CO_2 Separation

Abstract:

Mesoporous silica has been extensively studied for many applications such as CO_2 adsorbents, drug delivery, catalysis, and so on. Among them, bimodal mesoporous silica (BMS), which had two different sizes of mesopores, have drawn much attention recently because it exhibited unique diffusion properties in the pores. For this reason, when BMS was modified to have affinity with CO_2, it was expected that both of high CO_2 adsorption capacity and fast adsorption rate would be achieved by small and large pores, respectively. Therefore, we prepared various kinds of amine modified BMS, and the effects of pore structure of BMS samples and the type of introduced amine on CO_2 adsorption/desorption properties were examined in this study. In the experiments, BMS were prepared by co-condensation of two different silica precursors. Tetraethylorthosilicate (TEOS) and triblock copolymer Purulonic 123 (P123) were used as silica precursors and templating agent, respectively. Polymethylhydrosiloxane (PMHS) was not only used as silica precursors but also as template for large pores. First, various BMS samples were prepared with different PMHS concentration, and the pore structures of the resulting samples were investigated. Bimodal mesoporous silicas were successfully synthesized and it was found that the pore diameter and the pore volume could be controlled by changing PMHS concentrations. After that, amines such as polyethyleneimine (PEI) or aminopropyl residue were introduced in the pores of BMS samples and CO_2 adsorption/desorption properties were measured. In addition, MCM-48 was also prepared as samples of monomodal mesoporous silica and then amine modification was carried out in the same way for comparison. The results showed that by optimizing the pore structures, the PEI modified BMS showed improved CO_2 adsorption/desorption properties.

466076 Improving Energetic Potential of a Batch Distillation through Bottom Flashing Route

Abstract:

Increasing consumption of fossil fuels to meet the global energy demand has led to an *exponential* rise in carbon emissions. This has stimulated an intensive research in finding a renewable energy vector and in improving the energy efficiency performance of the long serving process units. In this light, distillation, a century old chemical process unit that accounts for 95% of all fluid separations, is a potential candidate for boosting its heat reversibility. It is surprising to note that a conventional distillation column (CDC) can secure a maximum thermodynamic efficiency of about 20%.

This work aims at developing a thermal integration approach for a batch distillation under the framework of heat pumping. The heat pump systems, namely vapor recompression and bottom flashing, are being used in continuous flow distillation column for a long time with showing a substantial improvement in energy savings and cost. Because of increasing use of batch separation in biochemical, fine chemicals, pharmaceutical and food processing industries, more research attention needs to pay to improve the thermodynamic efficiency of batch distillation. In this direction, we attempt to explore the feasibility of bottom flashing arrangement for a batch column that separates a binary mixture of cyclohexane and toluene.

The bottom flashing loop is connected with the conventional batch distillation through receiving the reboiler liquid and giving back a boil-up vapor. For this phase conversion, a throttling valve and a compression system are additionally involved. Actually, the reboiler content is proposed to use in this batch distillation with bottom flashing (BDBF) column as a heat sink against the overhead vapor employed as a heat source. To make it feasible, the temperature of heat sink needs to decrease so that there exists a reasonable driving force (here 15^0C) between the heat source and heat sink. For this, we employ a throttling valve that acts as a flasher for pressure reduction. Subsequently, a compressor is used for the recovery of that lost pressure. In between, the flashed liquid is used to replace the cooling medium in the top condenser, and this, in turn, gets vaporized and mixed with flashed vapor before subjected to compression. The compressed vapor then acts as boil-up vapor and it leads to reduce the hot utility supply to the reboiler. By this way, the BDBF scheme improves the energetic and economic potential over a conventional batch distillation (CBD).

Like the continuous batch column, the BDBF configuration operates at a transient state. Hence the constituent elements of the bottom flashing arrangement and its associates, namely flasher, top condenser, compressor and reboiler, need to be operated at a transient condition so that one can run the BDBF at a close dynamics, if not same, with its conventional counterpart. In fact, this is a prerequisite required to make a fair comparison between them for performance evaluation in terms of energy consumption.

To illustrate the BDBF scheme, we model the CBD fitted with the bottom flashing loop and simulate the resulting differential algebraic equation system. The differential equations are obtained by the application of the conservation principle and the algebraic equations represent the tray hydraulics, phase equilibrium, phase enthalpy and so on. In the present study, we assume ideal vapor phase and the Wilson activity coefficient model is employed to predict the liquid phase nonideality. The column operates at atmospheric pressure with a stage pressure drop of 0.3 kPa and a 80% tray efficiency.

It is obvious from the qualitative analysis that the BDBF column requires no external cooling medium and a reduced amount of hot utility, indicating a reduction of operating cost. At the same time, however, this thermally integrated column involves an additional operating cost for the compressor that runs with electricity, which is much more expensive than the thermal utility. Keeping these points in mind, we perform a quantitative analysis in terms of energy savings. For this, one needs to estimate the total energy consumption (Q_{Cons}) made in both the startup and production phase.

For the representative binary system, the CBD consumes a total of 4.35 E6 kJ/cycle, whereas the BDBF column involves a reboiler duty (Q_R) of 1.37 E5 kJ/cycle and compressor duty (Q_{Comp}) of 3.89 E5 kJ/cycle. So the total amount of energy consumed (Q_{Cons}) by the BDBF column is 13.04 E5 kJ/cycle that is obtained from:

$$Q_{Cons} = Q_R + 3\, Q_{Comp}$$

Note that the multiplication factor 3 is adopted to convert the electrical energy to thermal energy. Accordingly, it is computed that the BDBF column secures an about 70% savings in utility consumption with reference to the conventional batch distillation. The example system shows that the bottom flashing mechanism is capable of providing a substantial improvement in energy savings for batch processing.

466666 Improving the Performance of Commercial Polyethylene Based Anion Exchange Membranes in an Electrodialysis Process Using Functionalized Iron Oxide Nanoparticles

Abstract:

Introduction

Desalination is becoming essential in the context of an increasing water demand and a growing world population. The desalination capacity on 2015 was around 86.5 million $m^3 \cdot day^{-1}$ (http://idadesal.org/). Due to economic reasons, brackish water tends to be the preferable water source when available. Electrodialysis is one of the desalination technologies used in the production of freshwater from brackish water (Fernandez-Gonzalez et al., 2015). This option has been recently pointed out as a cost-effective technology for the desalination of brackish water (McGovern et al., 2014).

Fouling is an issue that affects membrane desalination in general and electrodialysis in particular. Since most of organic foulants are negatively charged, this phenomenon is more pronounced in anion exchange membranes. At the same time, in order to avoid scaling, the anion exchange membranes must be impermeable to sulphate ions (Mulyati et al., 2013). Thus, to prevent fouling and scaling in electrodialysis, the desired characteristics of anion exchange membranes are high monovalent ion selectivity and improved anti-fouling properties.

In the present work, nanocomposite monovalent anion exchange membranes, by coating a standard commercial available anion exchange membrane with a sulfonated poly (2,6-dimethyl-1,4-phenylene oxide) (sPPO) and different loads of sulfonated iron oxide nanoparticles (0.2% $g \cdot g^{-1}$ - 0.6% $g \cdot g^{-1}$), have been developed, characterised and tested for anti-fouling and anti-scaling properties .

According to previous studies (Güler, 2014) this coating is expected to provide a negative charged surface to the anion exchange membrane that will also change the hydrophilicity of the surface thus the anti-fouling properties of the membrane. Additionally, due to electrostatic repulsions between divalent anion and the negative charge of the membrane surface, much more important than the repulsion to monovalent ions, this modification will confer monovalent ion selectivity to the nanocomposite anion exchange membrane. Thus the modification of the membrane will boost its performance against fouling and scaling.

Material and Methods

The commercial membranes modified in this work were the heterogeneous anion exchange membranes RALEX AM-PP by a physical coating (Yin & Deng, 2015) using sPPO and Fe_2O_3-SO_4^{2-}. There are no chemical bonds but physical interactions between the membrane surface and the film of sPPO and Fe_2O_3-SO_4^{2-}. The surface modification does not affect the bulk membrane structure.

Several analytical techniques were used in order to characterize the surface properties of the membranes. The surface roughness was determined by means of Atomic Force Microscopy (AFM). Scanning Electron Microscopy (SEM) was used to observe the morphology of the surface. Fourier Transform Infrared spectroscopy (FITR) was used to determine the chemical structure of the membrane and the composition of the negatively charged layer. The modifications in the hydrophilicity of the surface were characterised by measuring the water contact angle. The monovalent ion selectivity of membranes was investigated by measuring the ion flux of Cl^- and SO_4^{2-} through the anion exchange membrane in a four-compartment electrodialysis cell. Concentrations of Cl^-, SO_4^{2-} and Na^+ were determined by Ion Chromatography. The antifouling potential of the membranes was evaluated by measuring the potential difference between both sides of the membranes over time in the presence of a reference foulant (sodium dodecyl sulphate).

Results and Conclusions

FTIR results confirm the successful coating of the membrane with sPPO and iron oxide nanoparticles. Both results examined by SEM and AFM showed an improvement in the surface homogeneity of the modified membrane. Water contact angle measurements of the commercial (100º) and the modified membranes (up to 59.2º) showed an increase in surface hydrophilicity. The best results obtained from the modified membrane can improve the anti-fouling performance up to 24% when the surface homogeneity and hydrophilicity increase. An improvement in monoselectivity of membranes up to 34% was also observed. The evolution of the concentration of Na^+ in the dilute compartment confirmed no changes in the selectivity Cl^-/Na^+ of the membranes.

In conclusion, the modification of a commercial anion exchange membranes with a coating of sPPO and iron oxide nanoparticles proved the simultaneous improvement of the monovalent ion selectivity of the membrane and its anti-fouling and anti-scaling properties without losing Cl^-/Na^+ permselectivity.

Acknowledgements

Financial support from MICINN under project 2014-57833-R and CTQ2013-48280-C3-1-R-D is gratefully acknowledged. The author C. Fernandez-Gonzalez thanks the Ministry of Education for a FPI grant BES-2012-053461.

References

Fernandez-Gonzalez, C., Dominguez-Ramos, A., Ibañez, R. and Irabien, A. (2015), Sustainability assessment of electrodialysis powered by photovoltaic solar energy for freshwater production. Renewable and Sustainable Energy Reviews, 47, 604-615

McGovern, R.K., Zubair, S.M.andLienhard V, J.H. (2014), The cost effectiveness of electrodialysis for diverse salinity applications. Desalination, 348, 57-65

Mulyati, S., Takagi, R., Fujii, A., Ohmukai, Y. and Matsuyama, H. (2013), Simultaneous improvement of the monovalent anion selectivity and antifouling properties of an anion exchange membrane in an electrodialysis process, using polyelectrolyte multilayer deposition. J. Membr. Sci, 431, 113-120

Yin, J. and Deng. B. (2015), Polymer-matrix nanocomposite membranes for water treatment. J. Membr. Sci., 479, 256-275

Güler, E., van Baak, W., Saakes, M. and Nijmeijer, K. (2014), Monovalent-ion-selective membranes for reverse electrodialysis. J. Membr. Sci., 455 254-270

Rapid Evaluation of Distillation Process Alternatives Involving Complex Column Configurations for the Separation of Multicomponent Azeotropic Mixtures

Chafika Adiche, Technische Universität Darmstadt, Thermal Process Engineering, Darmstadt, Germany, Belaid Ait Aissa, Technische Universität Clausthal, Thermal Process Engineering, Germany, present address: Evonik Industries, Germany.

Abstract

We present in this work the application of a recently developed generalized design approach [1] for rapid and efficient evaluation of several distillation process alternatives to separating multicomponent highly nonideal azeotropic mixtures into three products, with respect to the minimum energy requirement and the thermodynamic efficiency of the process.

1 Introduction

The synthesis of the optimal distillation process design for the separation of multicomponent azeotropic mixtures is until today a challenging task, particularly if complex column configurations are involved.

In this work, a generalized design approach of distillation columns with complex configurations developed recently [1] is proposed for the synthesis of optimal distillation column design to separating multicomponent highly nonideal azeotropic mixtures into three products. The investigated distillation process designs in this work comprise direct sequence, indirect sequence, single-feed columns with a side stream above or below the feed section, columns with prefractionator and thermally coupled columns [Figure 1].

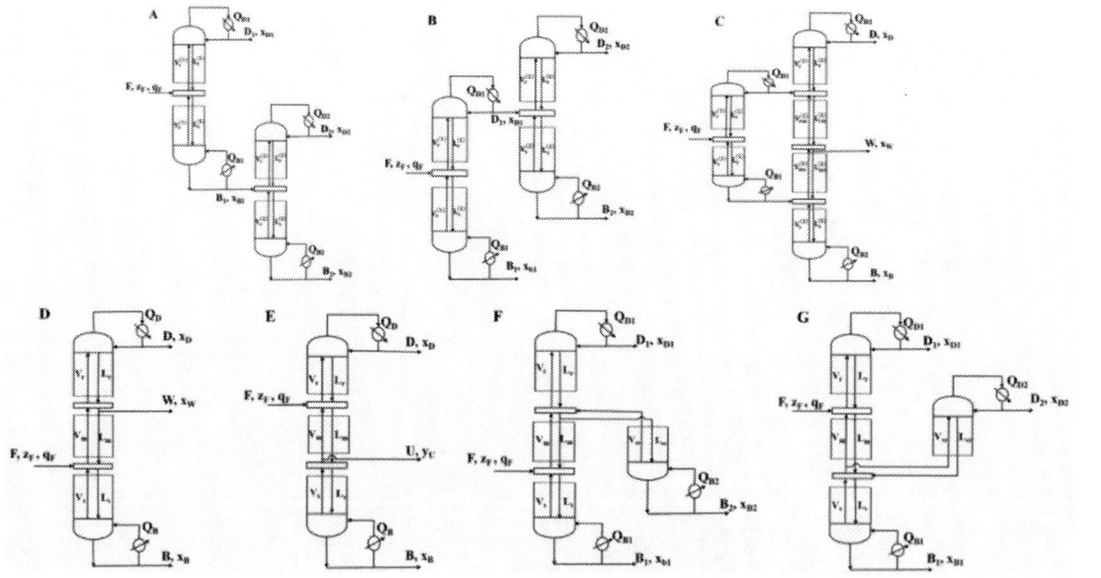

Figure 1. Basic distillation column designs for the separation of a multicomponent mixture into three products.

2 Design approach

The design approach applied in this work is based on the distillation theory of ideal behaving mixtures and on the decomposition of the whole distillation process into all constituting mass transfer sections as shown in Figures (2.a) and (2.b) for a column with a side stripper. At pinch conditions, all possible constant composition zones in the process including those around the feed sections and those around the side stream sections are identified, so that no conceptual splitting of a thermally coupled distillation column into a sequence of simple columns is needed. Furthermore, each set of constant composition zones is characterized by the relevant mass transfer sections along with the respective Eigenfunctions and split, as illustrated in Figure (2.c) for a column with a side stripper separating a ternary ideal behaving mixture.

Accordingly, simple short-cut methods valid for arbitrary feed thermal condition, for any number of components and for different types of splits have been developed for the determination of all relevant design parameters at pinch conditions of the previously mentioned column configurations separating multicomponent mixtures [1].

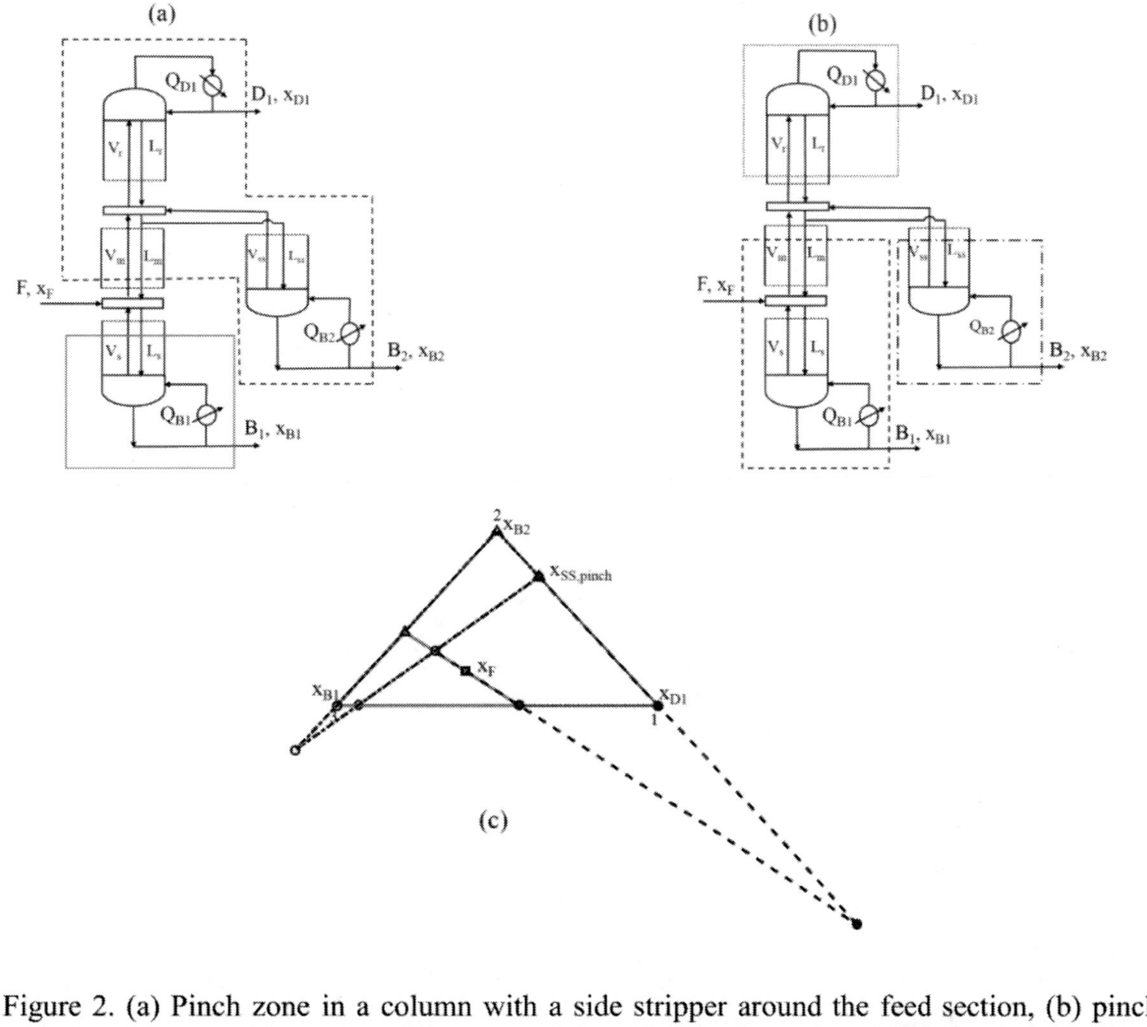

Figure 2. (a) Pinch zone in a column with a side stripper around the feed section, (b) pinch zone in a column with a side stripper around the side stream section, (c) separation geometry in a column with a side stripper at pinch conditions. **O**: unstable node pinch, Δ: saddle pinch, ●: stable node pinch.

The design approach presented above was applied to the determination of optimal distillation systems separating multicomponent azeotropic mixtures. For this purpose, azeotropic mixtures were approximated by appropriate pseudo-ideal mixtures [2]. This approximation approach requires (i) the knowledge of the thermodynamic structure of the azeotropic mixture along with the corresponding residue curve map, and (ii) the analytical characterization of the relevant distillation region where the separation is taking place. Accordingly, an appropriate set of relative volatilities of the singular points (pure component (s) and azeotrope (s)) forming the corresponding pseudo-ideal mixture may be determined in a transformed system of compositions by means of a rigorous thermodynamic model applied at feed composition [2].

In this work, separation examples of the ternary highly non-ideal azeotropic mixture of methanol, isopropanol and water were investigated. This mixture exhibits a minimum boiling temperature azeotrope constituted of isopropanol and water; giving rise to two distillation regions separated by a simple distillation boundary line [Figure (3.a)].

For all considered column configurations, the separation of the mixture of methanol, isopropanol and water is taking place inside the distillation region delimited by methanol, azeotrope and water as illustrated in Figure (3.b) for a column with a side stripper.

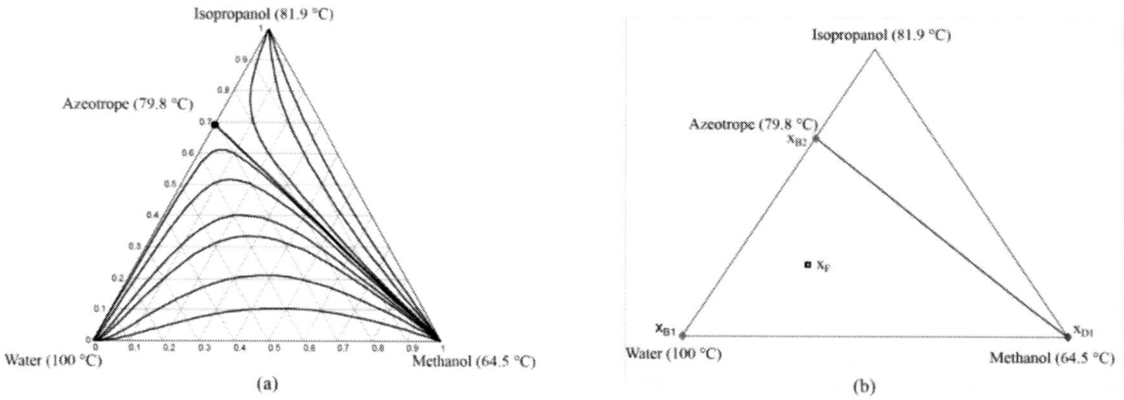

Figure 3. (a) Residue curve map of the mixture of methanol, isopropanol and water, (b) the equivalent pseudo-ideal mixture to separate the azeotropic mixture with a composition of x_F = [0.20 0.25 0.55] (here in a column with a side stripper) is constituted of methanol, azeotrope and water.

3 Results and discussion

For all investigated distillation column designs in this work [Figure 1], the developed short-cut methods were successfully validated using rigorous simulation (*Radfrac* models).

The design parameters determined in this work for each column design, constitute good initial values for rigorous simulation as it is shown in Table 1 for a column with a side stripper. They allow therefore the determination of more accurate design parameters leading to the desired product specifications [Figure 4].

Table 1. Design parameters of a column with a side stripper

Design parameters	This work	Rigorous simulation (Radfrac)
D_1/F	0.2	0.19
B_2/F	0.36	0.36
$V_{s,min}/F$	1.28	1.1
$L_{ss,min}/F$	0.95	0.96

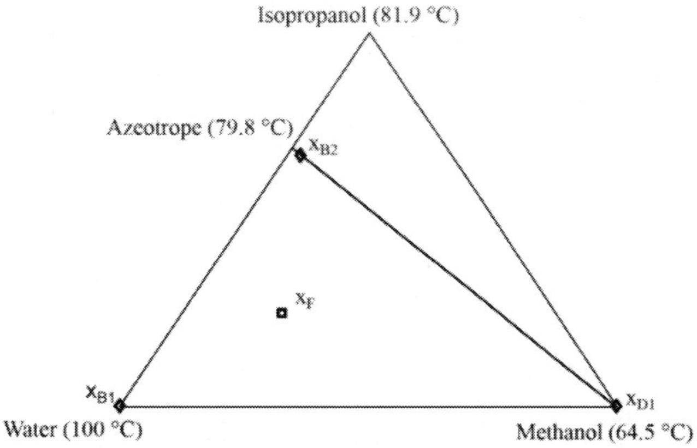

Figure 4. Product compositions in a column with a side stripper (obtained using rigorous simulation).

Furthermore, the developed design approach provides useful and simple design tools for systematic and rapid evaluation of multicomponent azeotropic distillation systems involving complex column designs. Figure 5 shows that for the studied separation examples, the column with prefractionator (C), and the single-feed column with a side stream above the feed section (D) were found to be the optimal column designs in terms of both thermodynamic efficiency η [3] and energy requirement $\sum_i Q_{B_i}$.

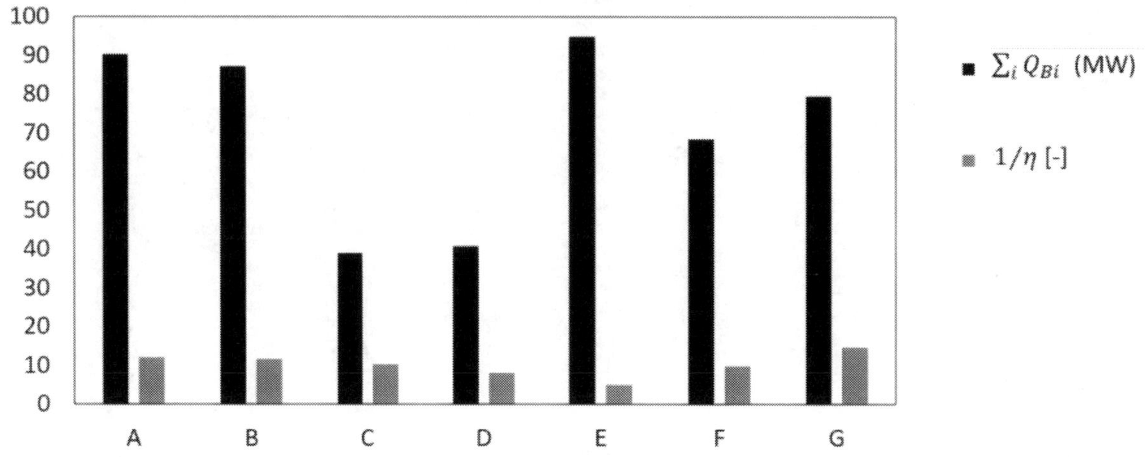

Figure 5. Screening of the investigated distillation column designs [Figure 1] for the separation of a mixture of methanol, isopropanol and water with $x_F = [0.20\ 0.25\ 0.55]$, with respect to energy requirement and thermodynamic efficiency, ambient temperature T_0 = 298.15 K, temperature of the hot source T_{steam} = 377.6 K [3].

4 Conclusion

For all investigated complex column configurations, the design tools presented in this work to separating a ternary highly non-ideal azeotropic mixture into three products, have been successfully validated using appropriate RadFrac (Aspen Plus ®) models. They provide thereby good design estimates for fast and reliable screening of all considered process alternatives by means of rigorous simulation, thus leading to the determination of more reliable design parameter and finally to the establishment of optimal distillation column designs.

References

1. Adiche, C. and Ait Aissa, B. (2016), ''A Generalized Approach for the Conceptual Design of Distillation Columns with Complex Configurations'', *Chem. Eng. Res. Des., Part A*, 109, pp. 150-170.
2. Adiche, C. and Vogelpohl, A. (2011), ''Short-cut Methods for the Optimal Design of Simple and Complex Distillation Columns'', *Chem. Eng. Res. Des. Part A.*, 89, pp. 1321-1332.
3. Seider Warren D., Seader J.D., Lewin Daniel R. (1999), ''Process Design Principles: Synthesis, Analysis, And Evaluation'', John Wiley & Sons, Inc.

467419 Micro-Computed Tomography of Swollen Heterogeneous Ion-Exchange Membranes

Abstract:

Heterogeneous ion-exchange membranes are widely used in electro-separation processes such as e. g. electrodialysis or electrodeionization. Their optimal electrochemical characteristics are thus of utmost importance for ideal performance of large-scale industrial units. Two of the parameters that strongly affect their characteristics are their composition and the surface and volume structure. Heterogeneous ion-exchange membranes consist of three major components: (i) finely ground ion-exchange resin providing the functionality, (ii) polyethylene or polypropylene used as a binder, and (iii) polymeric fibers (polyester or polyamide) providing mechanical strength to the membranes. The membranes are made in the form of large sheets which one produces by laminating slightly melted blend of the ion-exchange resin and the binder. In this process, only control over the composition (how much of each component is in the membrane) is possible. Spatial distribution of ion-exchange resin within the binder is completely random and irregular.

One can measure integral electrochemical characteristics of these membranes after their production and therefore optimize their properties in a trial-and-error method.

In this project we set out to develop techniques that should help to accelerate the trial-and-error method. Our technique allows (i) to analyze the structure of heterogeneous ion-exchange membranes after their preparation and (ii) to characterize these membranes by means of electrochemical measurements. This contribution will focus on micro-computed tomography of swollen membranes. We will present our devised method for scanning of swollen membranes. This technique is based on scanning the membranes in an environment saturated with water vapors to prevent their drying and drying accompanying changes in membrane volume. Simple immersion of membranes in water solutions causes strong attenuation of the signal resulting in poor quality of obtained data. The next part of the presentation will be devoted to a parametrical study investigating the effect of ionic strength of the external solution on swelling/shrinking of the membranes and irreversible changes of the membrane caused by repeatable swelling and shrinkage.

466154 Optimal Design of Continuous Flash Cooling Crystallizers

Abstract:

Crystallization is an important separation process to obtain high value-added chemicals in crystalline form from liquid solution in pharmaceutical, food and fine chemical industries. As in most of the particulate processes, the quality of the solid product is determined by its particle size distribution (PSD), which is the result of the combination of events at the microscopic and macroscopic scale. The microscopic events are governed by complex kinetic interactions between the solute molecules and the crystal lattice, and diffusion mechanisms, whereas the macroscopic events are related to the crystallization operation (solute concentration, temperature profile, mixing).

Traditionally, the crystallization has been operated in batch or semi-batch mode. However, batch operations suffer from disadvantages such as lack of batch to batch reproducibility, long processing times, scale up issues, poor controllability and observability (Porru and Ozkan, 2016). Hence recent research efforts are directed towards the design of continuous crystallization technologies, which may be able to overcome the above mentioned limits of the batch operation. A number of continuous crystallizer types and configuration have been applied, such as the mixed suspension, mixed product removal (MSMPR) crystallizers in single or multiple stage configurations (Alvarez et al., 2011), Plug Flow Crystallizers (PFC), with (Alvarez and Myerson, 2010) or without static mixers, and more recently, PFC with recycle (Cogoni et al, 2015).

This work addresses the optimal design of a continuous flash cooling crystallization system. This technology enhances the crystallization phenomena through the evaporation of the solvent at low temperatures by operating under vacuum and vapour-liquid equilibria. In particular, we are interested in the identification of a suitable configuration of single or multiple MSMPR crystallizers with or without recycle in order to achieve certain process targets, namely production yield and relevant attributes of the PSD. More specifically, we are interested to obtain a process configuration amenable to achieve good compromise between yield and large average dimension size of crystals by optimizing the recycle ratio and pressure of the vessel. To this end, an optimization problem which maximizes the product yield is formulated for the configurations

i) a single MSMPR crystallizer

ii) a single MSMPR crystallizer with recycle

iii) a series of two MSMPRs, without intermediate separation of the solid product and without recycle.

The constraints for the above mentioned optimization problem are derived from the definition of production targets, namely relevant PSD attributes of the final product (in terms of 50^{th}percentile, D50), daily production, and upstream operation (temperature and flow rate of the inlet flow). The optimal design study is performed using a crystallization process model with the kinetics of size independent growth and secondary nucleation due to crystal impeller collision.

The performance of the three configurations are studied through simulations at different pressures and recycle ratios, which were necessary to cast a feasible optimization problem. From this simulation study it has been found that (i) the solute concentration is about at saturation, which depends on the crystallization kinetics used. For the cases C1 and C2 the decreasing of the pressure leads to (ii) an increase of the yield and (iii) a decrease of the D50. (iv) The introduction of a recycle stream is not improving the yield nor the dimension of the crystals. The analysis of the case C3 leads to that (v) the pressure in the first crystallizer (P1) must be higher than the one in the second crystallizer (P2) in order to avoid dissolution, and (vi) the yield only depends on P2 while (vii) the D50 of the final product depends on the combination of P1 and P2.

The optimal design study has shown that in the first configuration, the yield is maximized at the minimum allowed pressure (and thus, minimum allowed temperature). If a quality constrain for the D50 is set, then the optimal yield is obtained at a higher pressure which actively satisfy the above mentioned constrain. In the second configuration, it is not possible to obtain an optimal solution since the recycle ratio has no influence on the yield and the D50. In case of the third configuration, it has been found that at a fixed P2 (and thus at fixed yield), the D50 of the final product has a quadratic functionality with P1. Thus, the optimization problem can be reformulated in order to find P1 which maximize the D50, under the constrain of the minimum allowed P2. This P2 is the optimum of the unconstrained optimization problem casted for the single crystallizer (first configuration). When the single MSMPR and the series of MSMPRs operate at the same yield, the third configuration allows to obtain larger crystals than the first configuration.

The optimal design of three different configurations of MSMPRs for flash cooling crystallization has led to the conclusion that the optimal pressure for the single MSMPR is the result of a compromise between high yield and large D50. Although the addition of a recycle stream does not improve the crystallization performance, the addition of a second crystallization unit does.Â Indeed an optimal setting of the pressure of the first and second crystallizer allows to operate at the maximum yield and obtaining larger crystal compared with the use of a single unit.

An evaluation of the annual expenses associated with the crystallization is under consideration in order to evaluate if the cost of adding a second unit is paid back by the improvements in the dimension of the crystals.

Acknowledgments

This work has been done within the project Improved Process Operation via Rigorous Simulation Models (IMPROVISE) in the Institute for Sustainable Process Technology (ISPT).

References

Alvarez, A.J., Myerson, A.S., 2010. Continuous plug flow crystallization of pharmaceutical compounds, Cryst. Growth Des. 10 (5), 2219â€“2228.

Alvarez, A.J., Singh, A., Myerson, A.S., 2011. Crystallization of cyclosporine in a multistage continuous MSMPR crystallizer, Cryst. Growth Des. 11 (10), 4392â€“4400.

Cogoni G., de Souza B.P., Frawley P.J. 2015 Particle size distribution and yield control in continuous crystallizers with recycle, Chemical Engineering Science 138, 592-599.

Porru, M. and Ozkan, L. Systematic observability and detectability analysis of industrial batch crystallizers, accepted paper at DYCOPS 2016.

466408 Carbon Dioxide Absorption Characteristics in Blended Amine Solutions for Wet Scrubbing Processes

Abstract:

The removal of carbon dioxide using mixed solvents continues to be of interest. A mixed (or blended) solvent can be defined as consisting of chemical-chemical solvents and chemical-physical solvents. In this study, Sulfolane was chosen as the physical solvent because of its stability and outstanding affinity. MEA, DEA, MDEA, DIPA were chosen because they are representative amines and are thermodynamically and kinetically selective for CO2. The solubility of CO2 has been measured from 40 to 130â„ƒ in a various mixing combination of solvents by the static method. In order to overcome non-ideality, activity coefficients and fugacity coefficients were introduced. The activity coefficients also take into account interaction between solute species in the liquid phase. Several reported thermodynamic models have been used to obtain activity coefficient matrices. The fugacity coefficients were also calculated to consider non-ideality of pressure.

467514 Electrokinetics and Electrochemical Characteristics of Ion-Exchange Particles

Abstract:

Ion-exchange particles (resin) find a number of applications in areas such as processing of radioactive waste, water purification or desalination. The last two mentioned examples are associated with electro-membrane separation processes in which finely ground ion-exchange particles embedded in polymeric binder (so called heterogeneous ion-exchange membranes) provide selectivity towards migration of ion caused by external electrical field. This selectivity comes from a so called fixed charge that is bound to polystyrene â€“ divinylbenzene structure of ion-exchange particles and exerts electrostatic attraction and repulsion on mobile counter-ions (ion with opposite charge to that one of the fixed charge) and co-ion (ions with like charge to that one of the fixed charge), respectively.

Electrochemical properties of ion-exchange particles and their behavior in DC electrical field play a very important role in production of heterogeneous ion-exchange membranes with optimized properties for a given process. For instance, ion-exchange membranes for electrodialysis should have limited (ideally no) tendency to split water (they should only exchange ions) while membranes creating bipolar membranes (stack of a cation and anion exchange membrane) should have properties enhancing water splitting reaction. Functional properties of the membranes are mostly given by the properties of ion-exchange particles. Electrochemical characterization of single ion-exchange particles would thus open a new scenario for producing membranes with tailored properties for a given process. However, this characterization presents a difficult task due to technical challenges in the experimental set-up.

In this contribution, we will present our approach to electrochemical characterization of single ion-exchange particles and also to observation of fundamental phenomena (electrohydrodynamic vortices, water splitting) that occur on ion exchange particles provided by a company MemBrain a. s. (Czech Republic). These ion exchange particles are used for production of heterogeneous ion exchange membranes industrially used on a large-scale. We characterize the ion

exchange particles by measuring current-voltage characteristics, and chronoamperometric curves. The electrokinetic behavior is studied by means of fluorescence microscopy when fluorescent markers are used for both imaging of vortices and tracking pH zones that develop on either side of the particles.

468093 Membrane Protein Insertion and Compatibility with Biomimetic Membranes

Abstract:

Cell membrane matrices are mainly composed of lipids and membrane proteins. Membrane proteins, present in the membrane, performing a vast majority of functions including signal reception[1], material transport and biochemical reactions[2,3] that occur at membrane-liquid interfaces. High compatibility between cell membranes and membrane proteins are necessary for living organisms to conduct complicated tasks and thrive in the nature[4]. Inspired by living cell systems, biomimetic membrane-matrices are developed for mimicking cell membrane matrix functions. Biomimetic membrane matrices are composed of membrane proteins (or membrane protein-mimics) and lipids (or lipid-like block copolymers). Similar to nature cell membrane matrices, only compatible biomimetic membranes specific to different membrane proteins will provide desired membrane performance[5]. Highly compatible biomimetic membranes have high membrane protein (or membrane protein-mimic) reconstitution density without compromising membrane protein (or membrane protein-mimic) single molecule properties. However, there are few accurate methods for compatibility evaluation[4].

Here, we propose and test methods to evaluate membrane protein and membrane protein-mimic compatibility with biomimetic membranes. We conducted biophysical activity characterization using stopped-flow spectroscopy, membrane protein density determination using Fluorescence Correlation Spectroscopy (FCS), and evaluation of the effect of matrix hydrophobicity on protein insertion using a new distyrylbenzene chromophore optical characterization technique. Combined the methods utilized provided single protein activity, and for a first time a quantitative measure of the chemical and physical compatibility between proteins and membranes. Water transport protein (aquaporins, AQPs), rhodopsins, and artificial water channels (specifically peptide-appended pillar[n]arene (PAP) channels) were reconstituted into a range of biomimetic membrane matrices to evaluate the proposed platform. Compatibility measurement results showed that both AQPs and PAP[8] channels preserve their single molecule water transport rates in different biomimetic membranes while their reconstitution density changes leading to different membrane permeability. We also show that membrane protein reconstitution density correlates with hydrophobicity mismatch between membrane protein and biomimetic membranes: smaller hydrophobicity mismatch between the membrane matrix and the proteins lead to higher membrane protein (or membrane protein-mimic) reconstitution densities.

References:

1. Broselid, S. *et al.* G protein-coupled Receptor 30 (GPR30) Forms a Plasma Membrane Complex with Membrane-associated Guanylate Kinases (MAGUKs) and Protein Kinase A-anchoring Protein 5 (AKAP5) That Constitutively Inhibits cAMP Production. *Journal of Biological Chemistry* **289**, 22117-22127,(2014).

2.Parker, M. D., Myers, E. J. & Schelling, J. R. Na+-H+ exchanger-1 (NHE1) regulation in kidney proximal tubule. *Cell. Mol. Life Sci.* **72**, 2061-2074,(2015).

3.Krall, J. F. & Korenman, S. G. Soluble Precursor of Membrane-Associated Protein-Kinase in Uterine Smooth-Muscle Cells. *Arch. Biochem. Biophys.* **199**, 479-487,(1980).

4.Shen, Y.-x., Saboe, P. O., Sines, I. T., Erbakan, M. & Kumar, M. Biomimetic membranes: A review. *Journal of Membrane Science* **454**, 359-381,(2014).

5.Nichols, S. P., Koh, A., Storm, W. L., Shin, J. H. & Schoenfisch, M. H. Biocompatible Materials for Continuous Glucose Monitoring Devices. *Chem. Rev.* **113**, 2528-2549,(2013).

6.Erbakan, M., Shen, Y.-x., Butler, P. J., Kumar, M. & Curtis, W. R. Molecular cloning, overexpression and characterization of a novel water channel protein from Rhodobacter sphaeroides. *PLoS ONE* **9**, e86830,(2014).

7.Garner, L. P., J ; Dyar, SM ; Chworos, A ; Sumner, JJ ; Bazan, GC Modification of the Optoelectronic Properties of Membranes via Insertion of Amphiphilic Phenylenevinylene Oligoelectrolytes. *J. Am. Chem. Soc.* **132**, 10042-10052,(2010).

8.Shen, Y. X. *et al.* Highly permeable artificial water channels that can self-assemble into two-dimensional arrays. *Proceedings of the National Academy of Sciences of the United States of America* **112**, 9810-9815,(2015).

469885 Developing a Predictive Form of Mosced for Nonelectrolyte Solids for Extraction and Crystallization Processes

Abstract:

Accurate and efficient models to predict the phase-behavior of nonelectrolyte solids in a wide range of solvents are central to the design of novel separation processes. A promising design tool is the MOSCED limiting activity coefficient model which is parameterized for 133 solvents. However, before predictions may be made for a solute of interest, solute MOSCED parameters are required. For novel compounds of interest, the necessary MOSCED parameters are unavailable and sufficient reference data is likely unavailable to regress the necessary parameters.

In this study we explore the use of electronic structure calculations to generate the reference data necessary to obtain solute MOSCED parameters. Specifically, we use solvation free energies computed using the popular software packages QChem and Gaussian, which employ the SM8 and SMD solvation models, respectively. Results will be shown wherein the method is successfully applied to a wide range of compounds of biological, environmental, and industrial interests. We additionally demonstrate how the limiting activity coefficients provided by MOSCED may be used to parameterize an excess Gibbs free energy model and used to make equilibrium solubility predictions.

470626 Polyacrylonitrile Supported Thin Film Composite Hollow Fiber Membranes for Forward Osmosis

Abstract:

Hollow fiber membranes have long been considered a valuable platform for membrane separations because of their high packing density relative to flat sheet membranes. Recently, the osmotic process community has developed hollow fiber membranes intended for forward osmosis (FO) and pressure retarded osmosis (PRO) applications in order to capitalize on these same advantages. However, this requires rethinking how the hollow fiber membrane should be made. Previous work on FO membranes demonstrated that employing a thin, highly porous, minimally tortuous and hydrophilic supporting structure is essential to mitigating concentration polarization. In this study, we use a systematic approach to fabricating a hollow fiber membrane taking into consideration these particular needs. We use a polyacrylonitrile (PAN) hollow fiber as a hydrophilic support to a thin polyamide film with excellent selective properties. This polyamide layer was formed on the outer surface of hollow fibers via interfacial polymerization. The effect of support layer structure could then be evaluated based on flux performance with some of our membranes exhibiting water flux of 36.6 Lmh using a 1 M sodium chloride draw solution against deionized water in the PRO mode. Thin fiber walls with open and straight pore structures were found to improve water flux substantially and that these structures could be readily tuned by adjusting spinning parameters.

470946 The Role of the Non-Newtonian Behavior of Blood in the Hemodialyzer with Cylindrical Geometry

Abstract:

According to the U.S. Department of Health & Human Services, every day 13 people die waiting for a kidney. Of the more than 122,000 Americans currently on the waiting list for a lifesaving organ transplant, over 101,000 need a kidney, and fewer than 17,000 people receive one each year. The United States Renal Data System states that approximately 468,000 Americans receive dialysis, and approximately 193,000 live with a functioning kidney transplant. While current dialyzers assist patients presenting with pathological kidney conditions, they fail to provide a good quality of life. An implantable artificial kidney would be ideal to allow patients to live a normal, healthy life. As a path to this artificial kidney, the micro-rheological properties of blood in the hemodialyzer are analyzed at the microscopic scale based on continuum mechanic approaches. Once the proper understanding of the microscopic behavior is attained (with the possible impact on the macroscopic scale), the system will be downscaled to a size ideal for potentially designing an artificial kidney. Due to the fact that the blood behaves as a non-Newtonian fluid as one downscales, an analysis needs to be performed to determine how the bloodâ€™s microrheology will affect the mass transfer regime in the smaller-scale hemodialyzer. The analysis also involves comparing bloodâ€™s non-Newtonian regime to that of a Newtonian model, along with the presentation of the hydrodynamics including the mass transfer effects. Furthermore, the analysis considers strategies with illustration of the potential impact of these non-Newtonian properties on dialyzer performance. For example, the roles of these properties in influencing the separation performance of the dialyzer under a variety of operating conditions will be considered in the development of an asymptotic (analytical) solution for the separation (cleaning) efficiency to assess the impact of the bloodâ€™s Non-Newtonian behavior. Parameter bounds for the validity of this asymptotic solution will also be discussed. The future direction of the research will also be illustrated.

Modeling of Hollow Fiber Membrane Module-Performance in Direct Contact Membrane Distillation

Mahdi Mohammadi-Ghaleni, Siamak Nejati
Department of Chemical and Biomolecular Engineering, University of Nebraska-Lincoln

SUMMARY

We constructed a three-dimensional model to study the effect of operating conditions, and physical and geometrical parameters in a hollow fiber membrane module under Direct Contact Membrane Distillation (DCMD) operation.

The normalized permeate flux was calculated for different module arrangements, various hollow fiber physical properties (e.g. length and thickness), and operating conditions.

The results indicated that the permeate flux decreases when we increase the fiber length, thickness, and tortuosity. In contrast, increasing the feed velocity and temperature, and the porosity of fibers result in increased permeate flux. We found that the normalized permeate flux significantly drops once the interspacing parameter of fibers—the ratio of external radius of fibers to the center-to-center distance between neighboring fibers—approached 0.47.

The temperature difference across the membrane decreases with reducing the membrane thickness; however, the effective driving force, pressure gradient over the membrane thickness, for water transport through the membrane increases.

MEMBRANE DISTILLATION (MD)

The requirements for a high performance membrane in membrane distillation are: Non-wetting, low thermal conductivity, and in general low structural parameter.

The main driving force in MD is the vapor pressure gradient between the membrane interfaces.

A multi-physics modeling of the system and module configuration is helpful to guide both membrane and module fabrication.

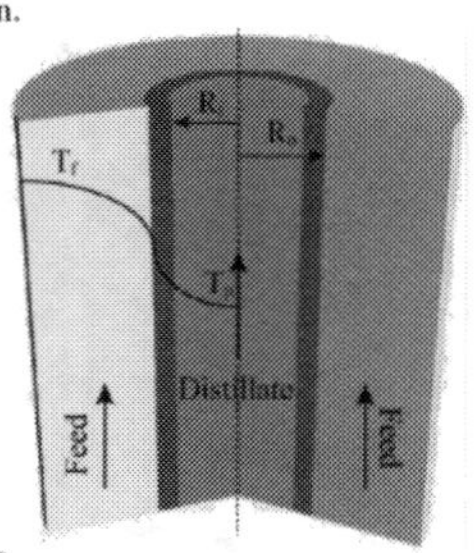

Figure 1. Schematic of DCMD process with a single-layer hollow fiber membrane.

Feed and permeate

Continuity equation:

$$\nabla.(\rho \vec{V_L}) = 0$$

Energy equation:

$$\nabla.[\rho \vec{V_L} h_L] = -\nabla.q + Q_i$$

Equations of motion:

$$\nabla.[\rho \vec{V_L} \vec{V_L}] = \nabla P + \nabla.[\mu_L(\nabla \vec{V_L} + \nabla(\vec{V_L})^T)]$$

Inside the membrane

Heat transfer:

$$-K_m \nabla^2(T) = 0 \quad \text{where, } K_m = (1-\epsilon)K_p + \epsilon K_a$$

Mass transfer:

Mass transfer occurs by Knudsen and molecular diffusions [1]:

$$J = -\frac{D_t}{RT}\Delta p_w$$

where, D_t is total diffusion coefficient: $D_t = \left(\frac{1 - p_w(1-\sqrt{\frac{M_w}{M_a}})}{(\frac{\epsilon}{\tau})P_t D_{w-a}} + \frac{1}{D_{kn}}\right)^{-1}$

$P_t D_{w-a} \quad [pa.m^2 s^{-1}] = (1.895 \times 10^{-5}) T_m^{2.072}$ Molecular diffusion coefficient

$D_{kn} = \frac{4}{3}\frac{\epsilon d_p}{\tau}\sqrt{\frac{RT_m}{2\pi M_w}}$ Knudsen diffusion coefficient

MEMBRANE MODULES

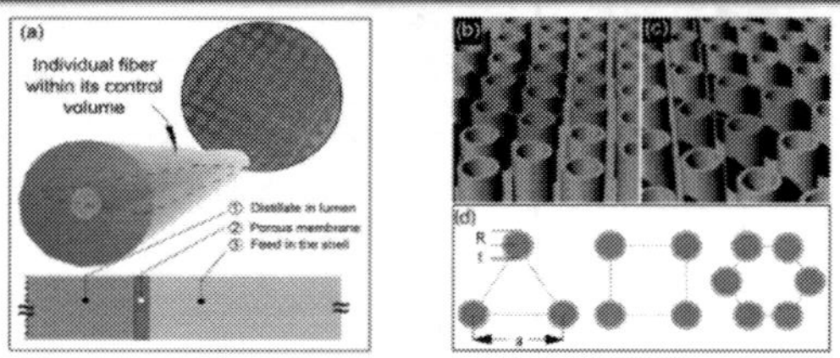

Figure 2. The schematic of the 3-D model used for simulating the hollow fiber membrane distillation process. (a) The fiber stack with an individual fiber in its control volume (the cross section representation is not to scale). (b) and (c) The arrangement of fibers in close-packed and square lattice configurations. (d) Schematic of the unit cell with the symmetry planes for a various lattice geometry; where, a is the center-to-center distance between neighboring fibers, t and R are the membrane thickness and the outer radius of the fibers, respectively.

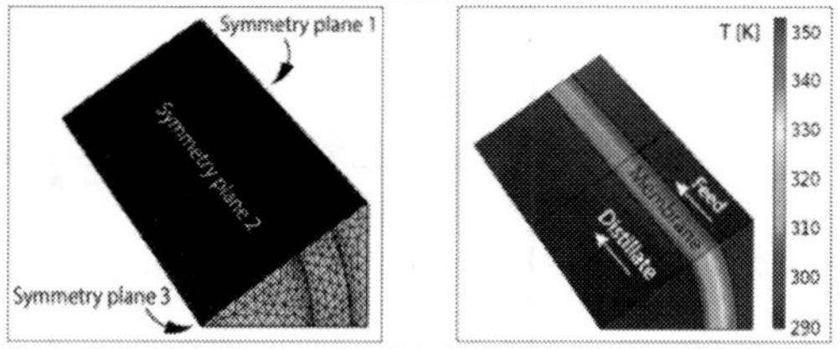

Figure 3. The schematic of the 3-D computational domain (Close-packed configuration). (Left) A sample of mesh, and the symmetry planes used in the simulations. (Right) The steady-state simulation results for temperature distribution inside the computational domain, where the permeate (in the lumen) and feed (in the shell) inlet temperatures are set at 290 K and 353 K, respectively.

MODULE PERFORMANCE

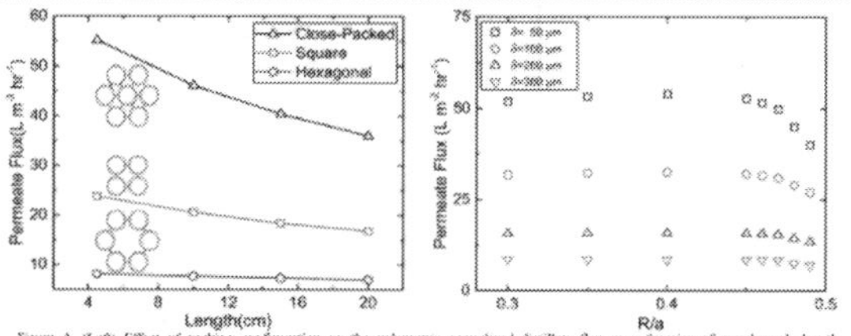

Figure 4. (Left) Effect of packing configuration on the volumetric normalized distillate flux as a function of membrane's length (interspacing parameter=0.45). (Right) Effect of membrane thickness on the normalized distillate flux as a function of fiber-fiber interspacing (Length=4.5 cm). For both cases, the membrane outer and inner radii were kept constant at 400 μm and 300 μm, respectively. The feed and distillate inlet temperatures are 17 °C and 80 °C, respectively. The fiber's porosity and tortuosity are set at 75% and 2.5, respectively.

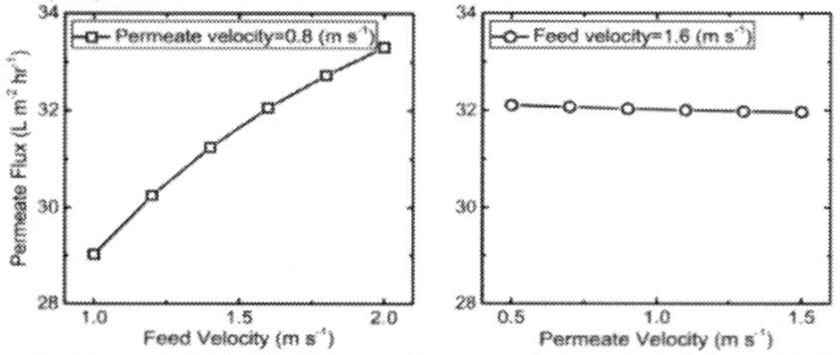

Figure 5. The effect of (Left) feed and (Right) distillate stream velocities on the normalized permeate flux. Here the flow rate in the feed stream (shell-side) is a flux-determining parameter. The membrane outer and inner radii were kept constant at 400 μm and 300 μm. The feed and distillate inlet temperatures are 17 °C and 80 °C, respectively. The fiber's length, porosity, and tortuosity are set at 4.5 cm, 75%, and 2.5, respectively.

PARAMETRIC STUDY

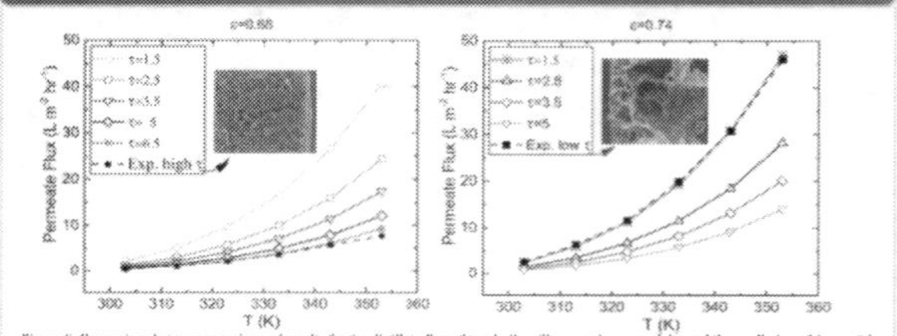

Figure 6. Comparison between experimental results for the distillate flux of two hollow fiber membrane modules and the prediction of the model at comparable porosity (68% and 74 %) and a range of tortuosity factor (τ) for the membranes in a close-packed configuration. All parameters describing the geometry and operating condition are taken from reference [2] and are as follow: feed velocity 1.6 m/s, distillate velocity 0.8 m/s, fiber length, inner and outer diameters are 20 cm, 600 μm and 800 μm, respectively. The distillate temperature and fiber interspacing parameter are 17 °C and 0.45, respectively.

The rate of change of temperature difference across the membrane increases with lowering the thickness.

Lowering the thickness is favorable in terms of permeate flux due to increasing ratio of pressure gradient over the thickness (figure 8-left).

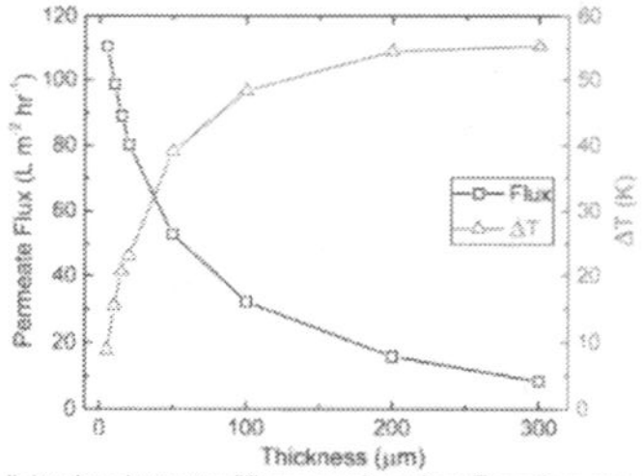

Figure 7. Effect of membrane thickness on the normalized distillate flux and temperature difference across the membrane. The membrane outer and inner radii were kept constant at 400 μm and 300 μm. The feed and distillate inlet temperatures are 17 °C and 80 °C, respectively. The fiber's length, porosity, and tortuosity are set at 4.5 cm, 75%, and 2.5, respectively.

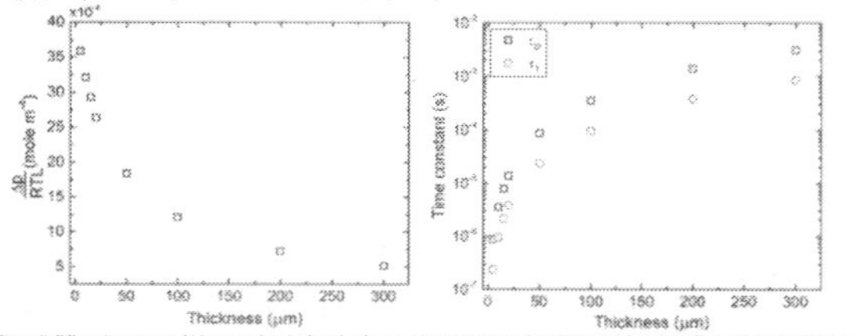

Figure 8. Effect of membrane thickness on the: (Left) ratio of vapor concentration over the thickness of the hollow fiber, and (Right) diffusion and heat conduction time constants, the time required for a unit of mass/heat passes the length of L across the membrane. The hollow fiber outer and inner radii were kept constant at 400 μm and 300 μm. the feed and distillate inlet temperatures are 17 °C and 80 °C, respectively. The fiber's length, porosity, and tortuosity are set at 4.5 cm, 75%, and 2.5, respectively.

CONCLUSION REMARKS

The close-packed module, comparing to the square and hexagonal modules, offers the highest volume normalized distillate flux and packing density.

The permeate flux decreases when fiber length, thickness and tortuosity increases. However, increasing feed temperature and velocity, and fiber porosity result in increased permeate flux.

The temperature difference across the membrane decreases with reducing the membrane thickness; however, the effective driving force, pressure gradient over the membrane thickness, for water transport through the membrane increases.

REFERENCES

[1] Li, Lin, and Kamalesh K. Sirkar. "Influence of microporous membrane properties on the desalination performance in direct contact membrane distillation." *Journal of Membrane Science* 513 (2016): 280-293.

[2] Wang, Kai Yu, Tai-Shung Chung, and Marek Gryta. "Hydrophobic PVDF hollow fiber membranes with narrow pore size distribution and ultra-thin skin for the fresh water production through membrane distillation." *Chemical Engineering Science* 63.9 (2008): 2587-2594.

471669 Theoretical Investigation of Optimized Molecular Separation Across a Nanochannel Membrane

Abstract:

Separation is critically important in various technological, industrial, and medical applications. Although there are many experimental methods that allow to separate molecules, frequently they are expensive and not efficient. Recently, a new method of separation of chemical mixtures based on utilization of channels and nanopores has been proposed and successfully tested in several systems. However, mechanisms of selectivity in the molecular transport during the translocation are still not well understood. We develop a simple theoretical approach to explain the origin of selectivity in molecular fluxes through a nanochannel membrane. Our method utilizes discrete-state stochastic models that take into account all relevant chemical transitions. More specifically, we analyze channels with different binding sites employed for separating mixtures of two types of molecules. The effects of the postion and the strength of the molecular-pore interactions are examined. It is found that the most efficient separation is predicted when the specific binding site is located near the entrance to the nanopore. In addition, the selectivity is higher for large entrance rates into the channel. It is also found that the molecular transport is more selective for repulsive interactions than for attractive interactions. The physical-chemical origin of the observed phenomena is discussed.

471809 Effect of Supersaturation on Morphology in the Crystallization of an Energetic Material

Abstract:

The detonation characteristics of high energetic materials have been strongly affected by not only crystal shape, size distribution, and surface roughness, but also purity and internal structure. Especially, crystal shape and average diameter have determined the packing density and mechanical strengths of the energetic material when formulated with polymeric binder and other ingredients. Among most formulations, polyhedral or spherical particles are needed due to the spatial dissipation of mechanical stresses and increment of flowability exposed at blending, kneading and other processes. In the drowning-out crystallization of 4,10-dinitro-2,4,6,8,12tetraoxa-4,10-diazatetracyclo [5.5.0.0^{5,9}.0^{3,11}]-dodecane(TEX), size and shape of TEX crystal was controlled by initial supersaturation generated at the mixing of the anti-solvent with saturated TEX solution. Shapes of TEX particles were changed by the mixing pattern of the anti-solvent with solvent. Either rapid pulse or continuous injection of anti-solvent produced the irregular or needle-like particles with average diameters ranged 10 to 30 μm. However, polyhedaral TEX particles with smooth surface were obtained at the slow feeding of anti-solvent into TEX solution. According to our experiments, higher supersaturation may not be preferable to obtain polyhedral particles. In the present work, initial supersaturations, saturation temperature and mixing rate were considered. Average diameter of TEX particles was 15 to 45 μm with polyhedral, needle-like and plate depending on crystallization condition. Crystal structure was confirmed by the comparison of XRD patterns based on Karaghiosoff's results.

471982 Membrane-Based, Simple Filtration Device for Rapid Separation of Biological Particles from Complex Matrices

Abstract:

Despite recent developments in polymerase chain reaction (PCR) based assays for identification of infectious diseases, there is still a need for the development of novel viral sample preparation techniques from blood samples for improved detection and identification of rare biological particles. Conventional laboratory diagnostics based on culture or molecular methods need prolonged assay time, complex processing by skilled technicians using costly, large-scale instrumentations in centralized laboratories. This can lead to physicians making treatment decisions based on an incomplete diagnosis contributing to patient morbidity. Ideally a diagnostic platform should be simple, rapid, and reliable that can be processed by non-technical staff even in non-hospital based settings where immediate clinical decision-making can be life-saving. Thus, rapid and reliable detection of bloodstream infections will gain a lot from improved and straightforward isolation of highly purified virus from all blood.

In this discussion, we demonstrate a simple viral or bacterial purification/concentration device that is easy to use, is reliable, and requires no energy for rapid viral separation. The device is comprised of two chambers, a feed chamber and a viral concentration chamber. Each chamber, which are 20 ml in volume, is divided with a poly (ethylene glycol) methacrylate (PEGMA) functionalized membrane to prevent membrane fouling and clogging. High water flux is maintained through driving forces from the packed polymer hydrogels in the concentration chamber and through gravitational pull. Filtration experiments with different sizes of colored-dye particles indicated that the device can exclusively separate small nanoparticles from large particles in mixture with a high filtration rate, up to 1.3 ml/min. In addition, the test results with dengue virus showed high filtration efficiency of the device with 90% of viral particle recovery and a concentration factor of 4.5. Our results suggest that the filtration device may provide a breakthrough toward the development of simple diagnostic platform for rapid detection of bloodstream infections, polymerase chain reaction sample preparation, and viral separation/concentration.

472183 Formation of Spherical Energetic Composites By Crystallization

Abstract:

Compounding of 1,3,5-trinitro-1,3,5-triazacyclohexane (RDX) with ammonium perchlorate (AP) has the advantages of high energy and reduced smoke. However, a short coming of lower burning rate exacerbates the performance of solid rocket motors. Therefore, the formation of composites whose physical distance between components was considerably reduced is very useful because it improves inter-particle mass transfer so that a relatively high decomposition rate of AP could be achieved. In the present work, the crystallization/agglomeration process was introduced for preparation of the spherical Al/RDX/AP energetic composites. Fine RDX and AP particles were obtained by drowning-out and agglomeration was induced by adding a bridging liquid into a solution. Experimental results show that the amount of a bridging liquid, agitation rate, and agitation time have great impacts on the formation of Al/RDX/AP composites. As the amount of bridging liquid increases, the Al/RDX/AP composites become larger. The agitation rate is responsible for the efficiency on a shear force promoting agglomeration between particles leading to a change in the size of Al/RDX/AP composites. Moreover, a sufficient agitation time seems to influence the capillary suction induced by squeezing and followed by compression.

472360 Effect of Ultrasound on Crystallization of Paracetamol

Abstract:

Paracetamol, one of poorly water-soluble active pharmaceutical ingredients (APIs), has some shortcomings such as low dissolution rates and inefficient bioavailabilities when passed through oral of administration. In the present work, micronized paracetamol crystals were achieved by sonication on highly supersaturated solution. To attain highly supersaturated state of paracetamol solution, a nucleation inhibitor was used. To understand the effect of sonication, the measurements of induction time and the statistical approach on the induction time distribution were implemented by using a focused beam reflectance measurement probe (FBRM). Nucleation kinetics was evaluated in terms of supersaturation and nucleation rate was found to greatly increase by sonication. Using a thermodynamic model, the values of the effective interfacial energy for heterogeneous nucleation was estimated. Dissolution profile of produced paracetamol crystals was determined by an UV-visible spectrophotometer. It was found that micronized paracetamol crystals obtained in the present model have higher drug dissolution rates.

472314 Influence of Fixed Charge Group Concentration on Ion Sorption in Ion Exchange Polymers

Abstract:

Controlling ion sorption and transport in polymer membranes is critical for efficient operation of a number of membrane-based technologies such as reverse osmosis, electrodialysis, reverse electrodialysis, and fuel cells. Often, membranes for these processes are made from polymers having ionizable functional groups (fixed charge groups) bound to their backbone (i.e., ion exchange membranes, ionomers, etc.). Sorption of charged species (e.g., ions) in such materials is profoundly influenced by the fixed charge groups, due to Donnan exclusion. Consequently, the concentration of fixed charge groups can be used to tune, to an extent, sorption of ions in charged membranes. However, changes in the concentration of fixed charge groups are typically accompanied by changes in membrane water content (i.e., swelling), which also substantially influences ion sorption in membranes. Decoupling these two effects is critical for developing fundamental understanding of the influence of membrane fixed charge concentration on ion sorption in charged membranes, ultimately leading to strategies for rational design of high performance materials.

In this study, positively and negatively charged membranes having different fixed charge group concentrations but similar water content were synthesized. Ion concentrations in membranes equilibrated with NaCl solutions of varying concentration were measured using a desorption technique. Counter-ion (i.e., ion with opposite charge to that of the fixed charges) sorption increased with increasing fixed charge group concentration, since a larger number of counter-ions are required to electrically balance the fixed charge groups. Co-ion (i.e., ion with same charge as that of the fixed charges) sorption decreased with increasing fixed charge group concentration, presumably due to enhanced Donnan exclusion. However, despite similar changes in membrane fixed charge group concentrations, co-ion sorption in the negatively charged membranes was suppressed to a greater extent than in the positively charged membranes, presumably due to structural differences between the materials. The ion sorption data were interpreted using a model

based on Donnan theory and Manningâ€™s counter-ion condensation theory. The model predicted the experimental ion sorption data in the positively charged membranes with no adjustable parameters. However, agreement between model and experimental data for the negatively charged membranes was poor, presumably due to membrane inhomogeneities introduced during polymerization. Use of the Donnan/Manning model for predicting ion sorption in inhomogeneous materials is discussed.

472485 Surfactant-Aided Extraction of Volatile Organic Acids from Aqueous Solution

Abstract:

The phase equilibrium data for ternary systems containing volatile organic acids (i.e. acetic, propionic, and butyric acids), water and a surfactant (Novel 23-Ethoxylate 2 (N23-E2)) were established at 35°C. The phase equlibrium data were obtained to evaluate the possibility of using the N23-E2 as solvent for the extraction of volatile organic acids from aqueous solutions. The data were fitted to local composition activity coefficient models (i.e. NRTL and UNIQUAC) to identify the more suitable model for establishing the phase diagrams of the ternary systems considered. Results showed that the extractability of volatile organic acids from aqueous solution using N23-E2 increases with the number of carbon in the acid, indicating that this surfactant could be used to selectively extract butyric acid from an aqueous solution containing both acetic and and butyric acids.

472519 Study on the Distillation Sequence with Dividing Wall Column

Abstract:

Study on the Distillation Sequence with Dividing Wall Column

Zengzhi Du[a], Jianhong Wang[a], Chi Zhai[a], Yunlu Zhang[a], Zhexi Zhu[a], Wei Sun[a*],

[a] China Beijing Key Lab of Membrane Science and Technology, College of Chemical Engineering, Beijing University of Chemical Technology, 100029 Beijing, China

Abstract

As an important separation unit, distillation column is widely applied in petrochemical and other process industry. It is reported that there are over 40,000 distillation columns all over the world[1]. Despite its flexibility and widespread use, one important concern is its considerable energy consumption, as distillation can account for more than 50% of plant operating cost. Therefore a large number of researchers have been concentrated on energy-efficient distillations. For the separation of multi-component mixtures, distillation sequence is required in industry. Dividing wall column (DWC), as a thermally coupled distillation column, has been proposed and applied in distillation sequence[2]. It is an atypical distillation column with an internal, vertical partition wall, which effectively accommodates two conventional distillation columns into one shell, for separating a multi-component feed mixture into pure products. For three components separation process, there are two columns, two condensers and two reboilers in conventional distillation sequence, while only one DWC with a condenser and a reboiler can be achieved. Compared with conventional distillation column, about 30% energy and equipment investment cost can be saved by DWC[3]. Until now, it is reported that the number of applications of DWCs have exceeded 100[4].

Distillation sequence has been well studied based on simple column series, by which an overall optimization at both process and equipment levels can be achieved. A distillation sequence with DWC will give more number of feasible sequences, which will increase computation load significantly. DWC may not be the best choice for all three component mixture separations in terms of its physical properties and feed composition. It will be more efficient to eliminate distillation sequences with unnecessary DWC.

In this paper, the influence of mixture and its component compositions on the operation of DWC is systematically investigated based on the simulation implemented by gPROMS on different mixture systems. The unnecessary DWC can be eliminated from distillation sequences. Then, the optimal distillation sequence with DWC can be efficiently identified.

Key Words: Dividing wall column; Distillation sequence; Simulation

Reference:

1. Kiss, A. A., Design, Control and Economics of Distillation. In Advanced Distillation Technologies, John Wiley & Sons, Ltd: 2013, 37-65.

2. Wright, RO., Fractionation apparatus, United States patent, US 2471134, 1949.

3. Triantafyllou, C., Smith, R., The design and optimisation of fully thermally coupled distillation columns: Process design, Chemical Engineering Research & Design, 1992, 70, 118-132.

4. Asprion, N.; Kaibel, G., Dividing wall columns: Fundamentals and recent advances. Chemical Engineering and Processing: Process Intensification, 2010, 49 (2), 139-146.

472784 Toward Radical Improvements in Biosorption Technology for Rare Earth Elements

Abstract:

Toward Radical Improvements in Biosorption Technology for Rare Earth Elements

Bennett Carv, Soumya Srivastava, and James Moberly

Rare earth elements (REE) are essential to many electronic products, "green" energy technologies, and medical imaging and diagnostics. These elements are desirable for their unique properties that include magnetism, phosphorescence, and catalytic ability. While very little quantities are needed to support the technology for which they are used, growing demand for these products is causing concern about economic availability of these critical materials. Though REE are relatively abundant in the Earth's crust, their similar chemical behavior makes separation of these elements an energy intensive process that generates considerable waste. As such, an efficient and economical method is required to be able to recover these elements from end-of-life products and other waste streams toward a "closed" REE cycle. However, currently employed technologies for separations of REE from waste streams is not yet competitive with terrestrial sources. Biosorption is a "green" process which may lend itself to inexpensive selective removal of REEs, but much work is needed to improve this separation processes.

Microfluidic devices are one means that may allow for the rapid separation and later identification of organisms that are capable of selectively accumulating REE, also known as organisms which hyperaccumulate metals (OHMs). Microfluidic devices purposed for OHM separation may also be capable of deployment as a unit process designed for continuous separation. However, a fundamental understanding of the mechanism of metal accumulation in model organisms and their response to conditions that may occur during separation are needed to develop this promising technology. The ability to quickly identify and select relevant OHMs and operating parameters under process stream conditions may enable biosorption to further mature as a separation technology.

The current research focused on *Cupriavidus necator* as a model organism for study. Previous studies with *C. necator* have focused on its capability to biologically precipitate and accumulate different metals including yttrium, palladium, silver, germanium, cobalt, nickel, mercury, zinc, copper, lead, and cadmium. Our data suggest that *C. necator* is able to selectively accumulate light REE compared to heavy REE. Additionally, these studies evaluated changes in REE accumulation and physical properties of *C. necator* with pH and temperature gradients to better understand the behavior of C. *necator* and the accumulated REE during continuous or extended separation processes within microfluidic devices. These results provide valuable information that will assist in the further development of microfluidic devices to be deployed for biosorption processes.

472610 Composite Palladium Membrane Robust below 573 K Prepared with Sputtering Technique

Abstract:

For the last two decades, several ways such as electroless-plating, electroplating, sputtering and chemical vapor deposition have been proposed to prepare a composite type of thin palladium membrane, whereby improving the hydrogen permeability significantly. However, it has been pointed out that the robustness of palladium composite membrane itself especially at low temperatures is not enough for practical use. Thin palladium film is well known to expand and contract by hydrogen absorption and desorption particularly below 573K, that is, the critical temperature, leading to peel-off from its support and crack.

In this study, an approach is explored to avoid such occurrences of defects. One is to prepare a palladium alloyed membrane. The other is to prepare a double-layer palladium membrane, in which between a palladium film and a porous support a metallic interlayer to depress a deformation is introduced. Pd-Cu alloyed and Pd/Pt layered membranes supported on porous alumina tube are successfully completed employing sputtering technique. Both membrane tubes could show a higher hydrogen permselectivity and a robustness even in the range of 473-573 K.

473008 Energy Consumption Minimization in Multi-Pressure Reactive Distillation Networks

Abstract:

Energy consumed by a multi-pressure reactive distillation network involving a ternary azeotropic mixture is investigated in this work. In the synthesis of reactive distillation networks, a promising process intensification technology, different pressure levels are allowed to be used to overcome the azeotropes, which can enhance the separation process. For the multi-pressure reactive distillation of methanol/isobutene/MTBE azeotropic system, a previous work showed that, for the same purity target, when the reactive holdup is reduced, the total flow at the network increases, which impacts the overall energy needs of the network. A methodology for the globally optimal synthesis of a reactive distillation network with minimum energy consumption, under certain conditions, is presented. The proposed methodology employs the Infinite DimEnsionAl State-Space (IDEAS) conceptual framework, which is shown to be applicable to the problem under consideration. The IDEAS framework considers an infinite number of units with all physically possible mixing and splitting streams existing between them. The resulting mathematical formulation is an infinite-dimensional program (ILP). The method is demonstrated on a case study involving the multi pressure reactive distillation of methanol/isobutene/MTBE azeotropic mixture.

475301 Mechanistic and kinetic insight into cocrystallization of APIs in supercritical CO_2

Abstract:

Pharmaceutical cocrystals are crystalline molecular complexes which have been driving much interest from both the academic and industrial environments because their physical and pharmacokinetic properties (e.g. solubility, stability) can be significantly improved as compared to the pure APIs[1]. Cocrystals can be prepared by several (classical) methods such as mechanochemical (e.g. neat and liquid-assisted grinding) or traditional solution crystallization approaches (e.g. solvent evaporation techniques and slurry techniques). However, using these techniques, cocrystals may become electrostatically charged (as may result from the stress applied in classical grinding methods) or suffer undesired interactions with organic solvents (when using classical evaporation or slurries-forming techniques) that may be incorporated into the crystal lattice with the possibility of solvate formation. Supercritical CO_2 provides a novel medium for the cocrystallization between an API and a coformer[2]. This work explores the mechanisms and kinetics of cocrystallization of distinct APIs and coformers in supercritical CO_2. Each API and coformer were placed in stoichiometric amounts inside a custom-built high-pressure vessel (internal volume = 10 cm^3) and compressed with supercritical CO_2 up to the desired pressure and temperature under magnetic stirring. In selected experimental runs, small amounts of distinct organic cosolvents (e.g. ethanol, methanol, acetone) were added to the raw materials inside the high pressure vessel before its pressurization with CO_2. In-situ monitorization of the cocrystallization over time was performed using a Raman probe inserted into the high pressure vessel. The results show that it is possible to control the cocrystallization rate and the form of the API (e.g. cocrystal, polymorph) by tuning process variables such that the API and/or the coformer are more solubilized in the supercritical CO_2.

Density Functional Theory (DFT) calculations have been applied to investigate APIâ€"coformer-solvent interactions

and corresponding binding energies for each system in order to understand how distinct supercritical reaction media (pure CO_2, CO_2 with cosolvents) influence the final solid form outcome of APIs.

1. Fucke *et al. N. J. Chem.* 36 **(2012)** 1969-1977.

1. Padrela *et al. Cryst. Growth Des. 15* **(2015)** 3175-3181.

This work has been supported by Science Foundation Ireland, Grant number: 12/RC/2275.

473821 A Novel Membrane Distillation Response Technology for Nucleation Detection and Crystallization Process Control

Abstract:

A Novel Membrane Distillation Response Technology for Nucleation Detection and Crystallization Process Control

Xiaobin Jiang*, Gaohong He*, Wu Xiao, Xuehua Ruan, Dapeng Lu

State Key Laboratory of Fine Chemicals, Research and Development Center of Membrane Science and Technology, School of Chemical Engineering, Dalian University of Technology, Dalian, P. R. China

*Corresponding author: Email: xbjiang@dlut.edu.cn, hgaohong@dlut.edu.cn

Abstract

Most of researchers considered the mass transfer flux decline and the crystallization on the membrane interface as a fouling behavior. The research was focused on the strategy of avoiding and reducing the crystallization, non-attention had been paid on applying this sensitive nucleation response phenomenon to the crystallization research [1,2]. The inevitable crystallization on the membrane interface under certain operational conditions and solution concentration inspired us that the membrane interface performed as the response medium of the nucleation detection. The responding mechanism is constructed base on the chain effect of ¡®solution concentrated & metastable zone uplimit reached-nucleation on the pores of the porous membrane interface-transmembrane flux sharply decrease¡¯. The inflection point of the transmembrane flux confirms the nucleation timing on the membrane interface. The accurate recording of the transmembrane flux data can be utilized to calculate the solution concentration to determine the metastable zone width (MSZW), which is key dynamic data for the effective control strategy of crystal nucleation and growth [3,4].

Without relying on the optical signal, MDC based nucleation response method would be easy and more meaningful in the case of opaque solutions that laser intensity response (LIR) method is invalid [5]. This advantage will improve the controlling accuracy and flexibility during MSZW measurement. Moreover, with the proposed method, it would be easy to enhance control over the process so that to select the real working conditions to simulate different evaporation rates for the concentrating rate in the isothermal membrane distillation process can be adjusted by the membrane permeability, the partial vapor pressure difference and the membrane area, etc.

While, a strict requirement on the membrane interface structure is obvious to enhance the detection accuracy of this proposed method. With conducted effective work of Trout and Diao, the developed Nanoparticle Imprint Lithography (NpIL) method had successfully prepared the polymer films with nanopores of various, uniform shapes that can hinder nucleation or promoted it [6]. In addition, the had reported the significant impact of polymer interface with nanometer scale pores on inducing crystallization and polymorphism control [7,8]. These foundational work can promote the theoretical research and application of the MDC based nucleation response and crystallization control by providing the qualified microporous membrane as desired nucleation detection and response interface.

Fig 15

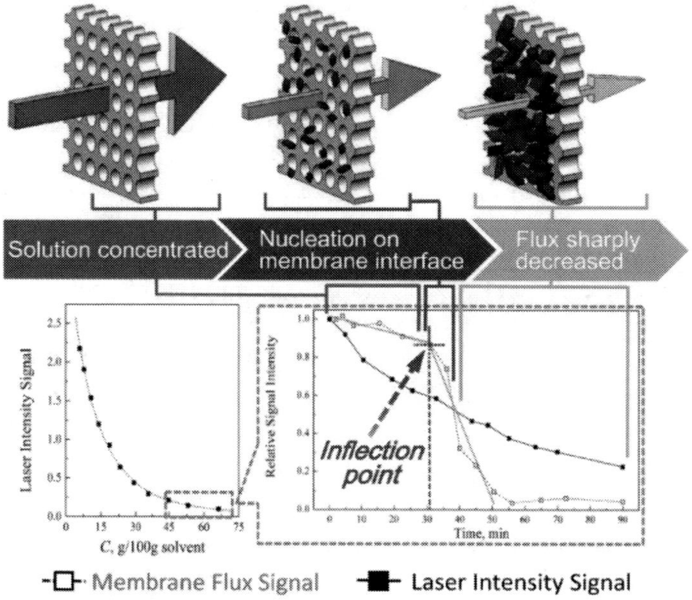

Fig. 1 Schematic diagram of nucleation response and metastable zone width (MSZW) measurement with MDC technology (compared with the laser detection technology)

Acknowledgment

This work is supported by National Natural Science Foundation of China (Grant No. 21306017, 21527812), Program for Changjiang Scholars and Support Project of the China Petroleum and Chemical Corporation (X514001).

References

[1] Chen, G.; Lu, Y.; Krantz, W. B.; Wang, R.; Fane, A. G., Optimization of operating conditions for a continuous membrane distillation crystallization process with zero salty water discharge. *Journal of Membrane Science* **2014,** 450, 1-11.

[2] Meng, S.; Ye, Y.; Mansouri, J.; Chen, V., Fouling and crystallisation behaviour of superhydrophobic nano-composite PVDF membranes in direct contact membrane distillation. *Journal of Membrane Science* **2014,** 463, 102-112.

[3] Ulrich, J.; Jones, M. J., Industrial Crystallization. *Chemical Engineering Research and Design* **2004,** 82, (12), 1567-1570.

[4] Ulrich, J.; Strege, C., Some aspects of the importance of metastable zone width and nucleation in industrial crystallizers. *Journal of Crystal Growth* **2002,** 237, 2130-2135.

[5] Jiang, X.; Ruan, X.; Xiao, W.; Lu, D.; He, G., A novel membrane distillation response technology for nucleation detection, metastable zone width measurement and analysis. *Chemical Engineering Science* **2015,** 134, 671-680.

[6] Diao, Y.; Harada, T.; Myerson, A. S.; Hatton, T. A.; Trout, B. L., The role of nanopore shape in surface-induced crystallization. *Nature Materials* **2011,** 10, (11), 867-871.

[7] Curcio, E.; L¨®pez-Mej¨ªas, V.; Di Profio, G.; Fontananova, E.; Drioli, E.; Trout, B. L.; Myerson, A. S., Regulating Nucleation Kinetics through Molecular Interactions at the Polymer¨CSolute Interface. *Crystal Growth & Design* **2014,** 14, (2), 678-686.

[8] Diao, Y.; Helgeson, M. E.; Siam, Z. A.; Doyle, P. S.; Myerson, A. S.; Hatton, T. A.; Trout, B. L., Nucleation under Soft Confinement: Role of Polymer¨CSolute Interactions. *Crystal Growth & Design* **2012,** 12, (1), 508-517.

南京工業大學 NANJING TECH UNIVERSITY

江苏先进生物与化学制造协同创新中心 THE SYNERGETIC INNOVATION CENTER FOR ADVANCED MATERIALS

Continuous separation of four 5′-ribonucleotides based on ion exchange

AIChE The Global Home of Chemical Engineers

Pengfei Jiao, Yingying Wang, Jinglan Wu, Jingwei Zhou, Wei Zhuang, Pengpeng Yang, Hanjie Ying

Jiangsu National Synergetic Innovation Center for Advanced Materials; College of Biotechnology and Pharmaceutical Engineering, Nanjing Tech University; National Engineering Technique Research Center for Biotechnology, Nanjing, China; State Key Laboratory of Materials-Oriented Chemical Engineering Nanjing, China

Email: yinghanjie@njtech.edu.cn; Jiaopengfei1117@163.com

Summary

5′-ribonucleotides have been widely used in food and pharmaceutical industries. They include mainly uridine 5′-monophosphate (UMP), adenosine 5′-monophosphate (AMP), cytidine 5′-monophosphate (CMP), and guanosine 5′-monophosphate (GMP). In this work, the separation of them from enzymatic hydrolysate of RNA was studied based on ion exchange. **We focused on**

➢ developing a continuous separation process to separate four 5′-ribonucleotides with the strong acid cation exchange resin NH-1 used as exchanger.

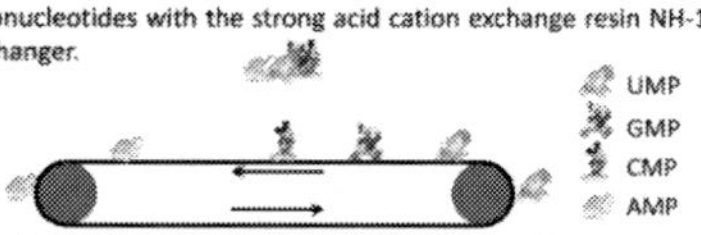

Fig 1. Ideal for separating four 5′-ribonucleotides .

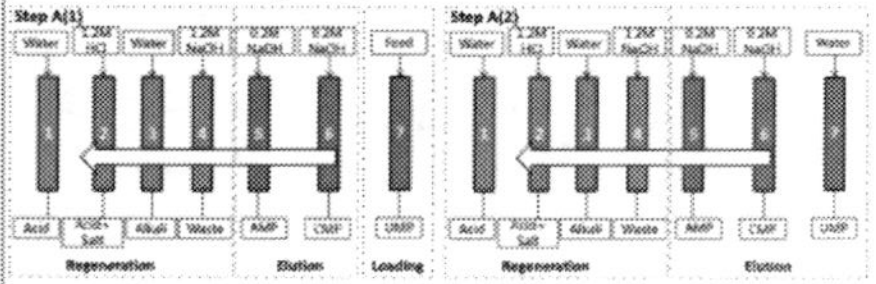

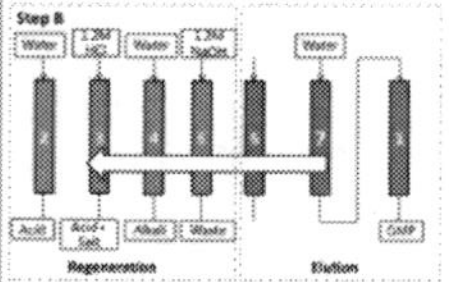

Fig 2. Scheme of continuous separation process of 5′-ribonucleotides.

➢ studying the uptake equilibria and mechanism of 5′-ribonucleotides on the resin.

Outcomes:

➢ The proposed separation process can separate 5′-ribonucleotides with satisfactory results. The productivity is 26% higher than that of fixed bed process. The separation time was diminished by 80%(62.7 h to 12 h).

➢ Modified Helfferich equilibrium model was developed combining ion exchange of cations and adsorption of zwitter ions. The model can predict the uptake equilibria of 5′-ribonucleotides and Na^+ in multicomponent systems. The hydrophobicity of bases of 5′-ribonucleotides determined their relative affinities to the resin.

Introduction

The structural formulas of 5′-ribonucleotides

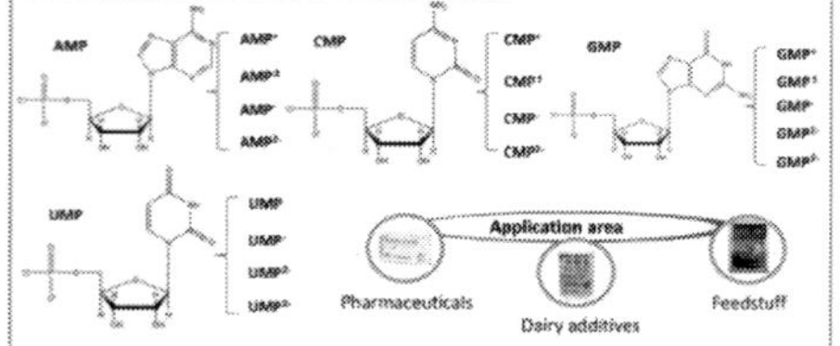

In the industrial manufacturing process, Strong acid cation exchange resins are widely used to separate 5′-ribonucleotides from enzymatic hydrolysate of RNA by 5′-phosphodiesterase.

However, two problems exist in the separation process:

➢ The separation of 5′-ribonucleotides by fixed bed mode is time-consuming and the productivity is very low.

➢ The uptake equilibria and mechanism of 5′-ribonucleotides by ion exchangers were not reported, which is significant for design and scale-up of large-scale ion exchange process.

Therefore, the aim of this work is:

➢ Developing a separation process with a high productivity.

➢ Studying the equilibrium behaviors and mechanism of 5′-ribonucleotides.

Results and discussion

Continuous separation of 5′-ribonucleotides

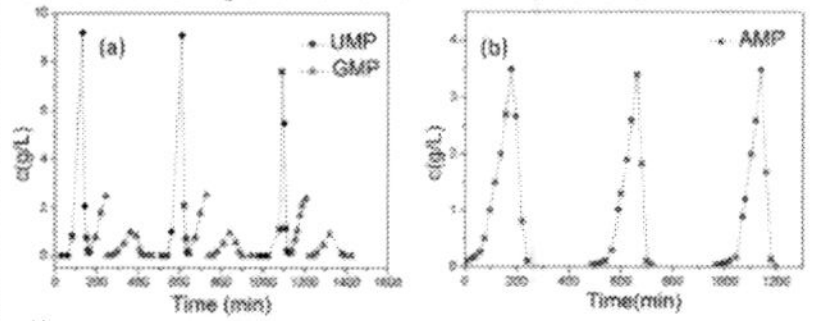

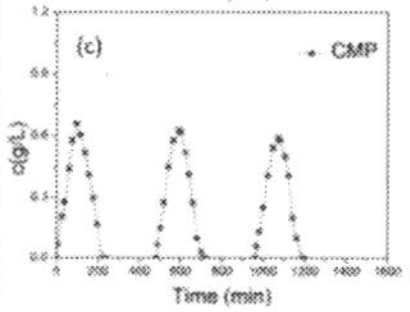

Fig 3. The concentration evolution of four 5′-ribonucleotides with enzymatic hydrolysate of RNA used as feed solution.

Table 1. Comparison of performance between continuous and fixed bed separation processes.

	Continuous separation				Fixed bed separation			
	Concentration (g/L)	Recovery (%)	Purity (%)	Productivity (g/kg/h)	Concentration (g/L)	Recovery (%)	Purity (%)	Productivity (g/kg/h)
UMP	2.42	99.75	85.38	0.50	4.69	100.28	76.72	0.37
GMP	0.66	88.60	98.76		1.78	79.08	97.39	
CMP	0.48	98.32	98.89		1.05	95.14	100	
AMP	0.36	83.97	93.56		0.58	70	99.19	

➢ **The productivity is 26% higher than that of fixed bed process.**

➢ **The separation time was diminished by 80%(62.7 h to 12 h).**

Uptake equilibrium of 5′-ribonucleotides

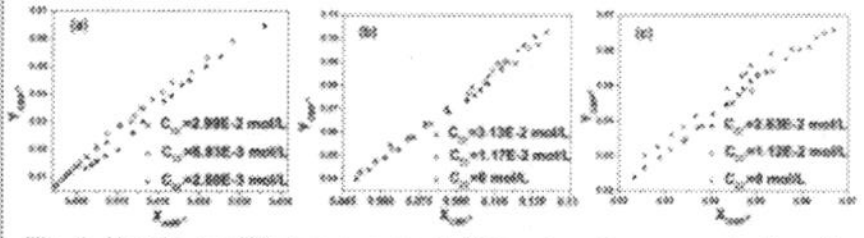

Fig 4. Uptake equilibrium curves at different co-ion concentrations for (a) AMP/H^+ binary system, (b) CMP/H^+ binary system, and (c) GMP/H^+ binary system.

➢ **UMP can not be taken up by the resin.**

➢ **Ion exchange is the predominant uptake form of CMP.**

➢ **AMP and GMP molecules are taken up by NH-1 combining ion exchange of cations and adsorption of zwitter ions.**

Modified Helfferich model

[illegible] (1) [illegible] (2) [illegible] (3) [illegible] (4) [illegible] (5) [illegible] (6) [illegible] (7)

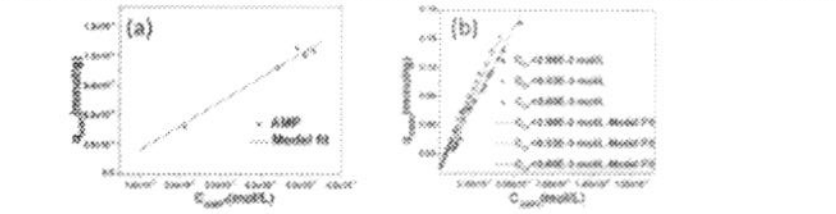

Fig 5. Experimental and fitted uptake equilibrium curves of a) $AMP^{\pm}$, b) AMP combining ion exchange and adsorption.

Table 2. Selectivity coefficients for 5′-ribonucleotides and Na^+ relative to H^+ or distribution coefficients.

	$AMP^{\pm}$	AMP^+	GMP^+	$GMP^{\pm}$	CMP^+	Na^+
Selectivity/Distribution coefficient	2.5	1.95	0.9	2	0.81	1.1

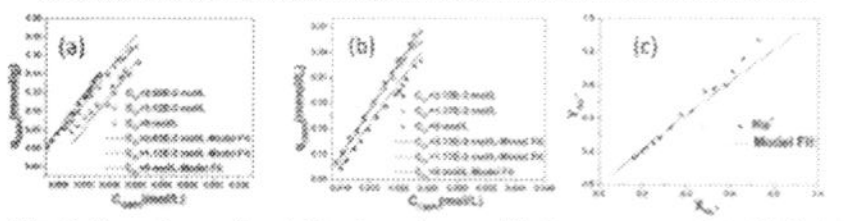

Fig 6. Experimental and fitted uptake equilibrium curves of a) GMP, b) CMP at different concentrations of Cl^-, and c) Na^+.

➢ **The proposed equilibrium model can fit the uptake equilibria of 5′-ribonucleotides and Na^+ in binary systems.**

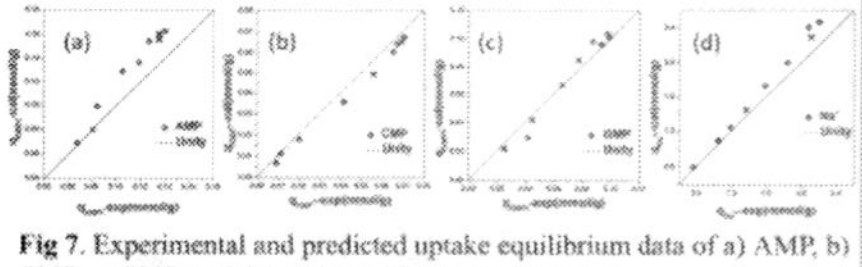

Fig 7. Experimental and predicted uptake equilibrium data of a) AMP, b) CMP, c) GMP, and d) Na^+ in multicomponent system.

➢ **Modified Helfferich model can provide a general prediction to the uptake equilibria of 5′-ribonucleotides and Na^+ in binary systems.**

Uptake mechanism of 5′-ribonucleotide cations

Relative affinity ⇦ Selectivity coefficient ⇦ • Coulombic force ⇦ Valence of cations + hydrated radius; • Van der Waal force ⇦ Polarizability; • Hydrophobic interaction ⇦ log P

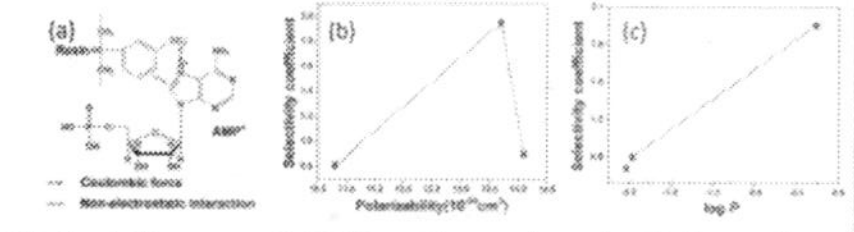

Fig 8. a) The way of AMP^+ uptake on the resin. b) Ion exchange selectivity coefficient versus log *P*, and c) Ion exchange selectivity coefficient versus polarizability.

➢ **The hydrophobicity of bases of 5′-ribonucleotides determined their relative affinities to the resin NH-1.**

Conclusion

➢ The proposed continuous separation process can separate 5′-ribonucleotides with satisfactory results.

➢ Modified Helfferich model was developed considering ion exchange and adsorption equilibrium. The model can provide a general prediction to the uptake equilibria of 5′-ribonucleotides and Na^+ in multicomponent systems.

➢ The hydrophobicity of bases in 5′-ribonucleotides molecules determines the relative affinity of NH-1 to 5′-ribonucleotides.

Methods

➢ **Continuous separation experiments:**

The operating step is shown in Fig 2. All the columns are identical in size (3.17×19 cm). The switch time is 4 h. The loading volume is 96 ml. The feed solution is enzymatic hydrolysate of RNA by 5′-phosphodiesterase. The concentration of feed is 5.72 g/L, 6.21g/L, 10.62 g/L, 7.42 g/L, and 0.137 mol/L for CMP, UMP, GMP, AMP, and Na^+, respectively. The pH of feed solution is 4.86.

➢ **Uptake equilibrium experiments:**

Static method was selected to determine the uptake equilibrium of 5′-ribonucleotides and Na^+.

The concentration and equivalent ionic fraction of 5′-ribonucleotides and Na^+ in the resin and equivalent ionic fraction in the liquid phase can be calculated by the following equations:

$$q_i = \frac{(C_{i,0} - C_i)V}{m} \qquad Y_{CMP^+} = \frac{q_{CMP^+}}{q_{CMP^+} + q_{H^+}} \qquad X_{CMP^+} = \frac{C_{CMP^+}}{C_{CMP^+} + C_{H}}$$

Future work

We are studing the modeling of mass transfer process of 5′-ribonucleotides in a fixed bed, and then we will develop the model of continuous separation process. The optimization of continuous separation process based on modeling is our objective.

Acknowledgement

This work was supported by Program for Changjiang Scholars and Innovative Research Team in University (PCSIRT), National Outstanding Youth Foundation of China (Grant No.:21025625), Major Research plan of the National Natural Science Foundation of China (Grant No.21390204). We would also like to acknowledge the financial support provided by 21306086 and BK20151452.

Mechanism and Kinetics of Polymorphic Transportation: From Amorphous to Hydrate Crystal of Disodium Guanosine 5′-Monophosphate in Ethanol

16AIChE Annual Meeting, San Francisco, CA

Fengxia Zou[1], Qiao Chen[1], Wei Zhuang[1], Jinglan Wu[1] and Hanjie Ying[1,2]
1. College of Biotechnology and Pharmaceutical Engineering, Nanjing Tech University, Nanjing, China
2. State Key Laboratory of Materials-Oriented Chemical Engineering, Nanjing Tech University, Nanjing, China

Motivation/Problem Formulation

Process control for crystallization:

molecules ⟷ (nucleation) nuclei ⟷ (growth) crystals ⟷ (transformation) products

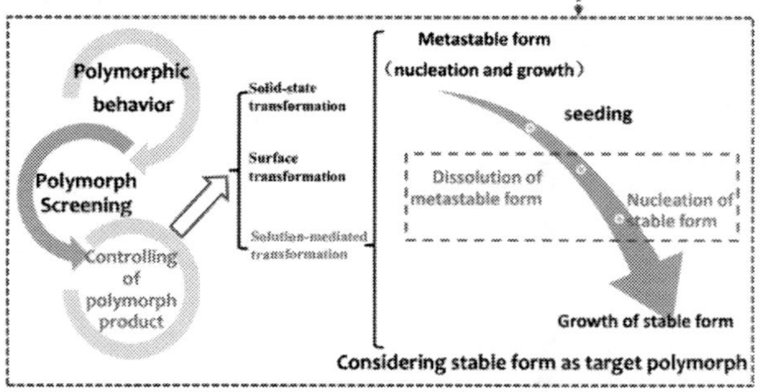

Problem of 5′-GMPNa$_2$ crystallization and products : agglomerate, inhomogenous CSD, poor fluidity and amorphous.

- **How can we acquire a stable polymorphic form?**
  Transformation mechanism (SMT)
- **How can we judge the determining step during the SMT process? And acquire a desired polymorphic form, no matter stable and metastable forms?**
  Transformation kinetics

Experiments and Apparatus

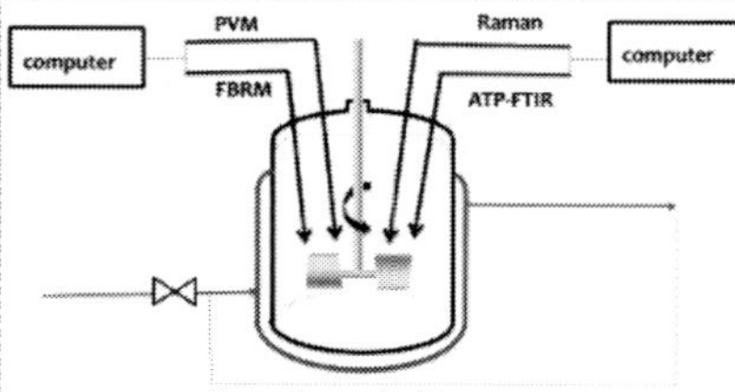

Experiment set up

In-line Monitor
- FBRM
- ATP-FTIR
- PVM
- Raman(lack)

Off-line Monitor
- PXRD
- TG-DSC

Solubility measurement: fundamental data for two polymorphs
Characterization: XRD, DSC -TG,PVM,IR
Seeding experiments: nucleation and growth kinetics of crystal
Dissolution experiments: dissolution kinetics

Results and Discussion

XRD for two sample / TG-DSC

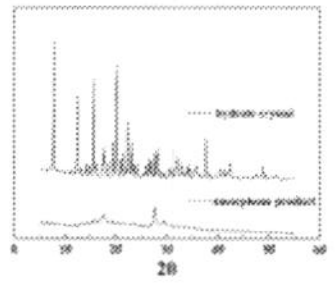

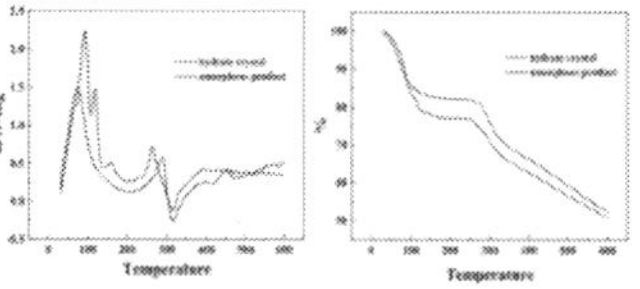

Solubility for two different form

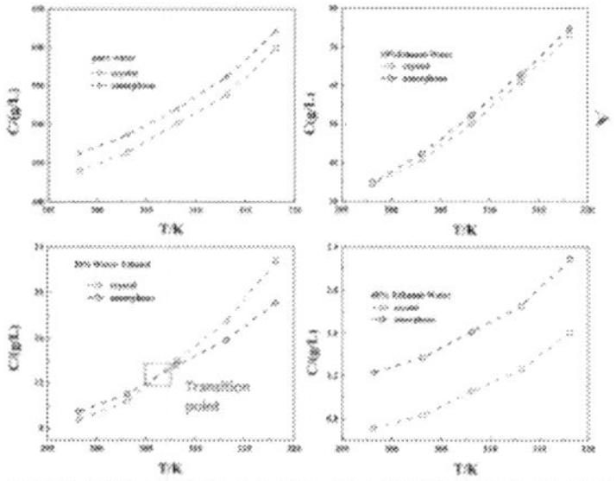

- Amorphous form can turn to hydrate crystal at 303.15K in 20% ethanol-water, which indicates an enantiotropocal transition, however in other binary solution, perform like a monotropical transition.

XRD for offline process samples / PVM image

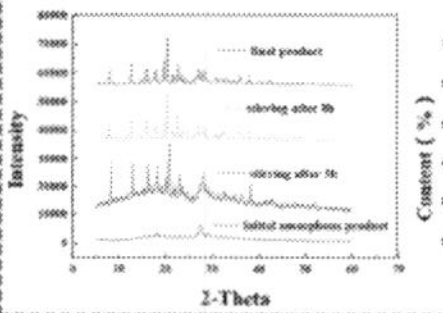

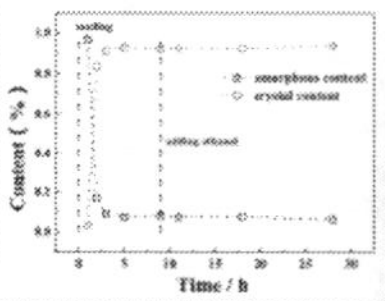

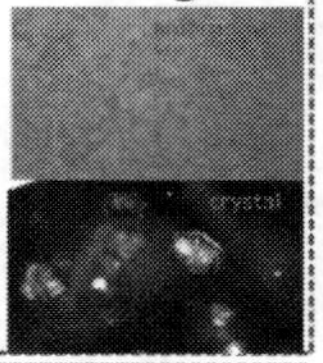

Inline-concentration Data

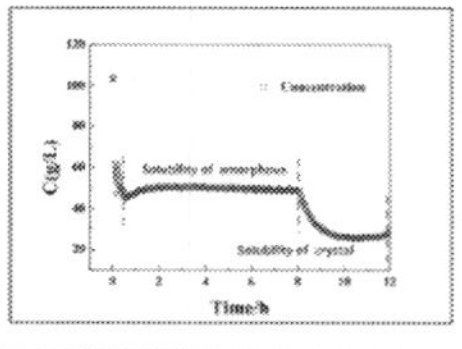

- From XRD data and PVM image, this process goes for a typically transformation. For about 5h, this transformation has been finished.
- From IR data, two straight line appeared, which exists the solubility for amorphous and crystal products, respectively.
- In addition, the first step for a straight line indicate that the dissolution of amorphous and nucleation of hydrate crystal both appeared.
- Which is the determining step for this transformation process?

Kinetics of the transformation process

In this process, the amorphous product produced by burst nucleation, and seeds were added subsequent within 30min, so the nucleation and growth of amorphous were neglect. So the dissolution of amorphous, nucleation and growth of hydrate crystal were studied in this process.

Which is the determining step? The kinetics of the transformation process for dissolution and nucleation show the direct answer and supply the theory guide for future industrial crystallization process.

Dissolution Kinetics:

$$D_{amorphous} = 3.423 \times 10^{-5}(S_{amorphous} - 1)$$

Nucleation Kinetics:

$$B = 1.944 \times 10^{11} m_{amorphous}^{0.3344} \varepsilon^{0.7541}$$

Growth Kinetics:

$$G = 5.360 \times 10^{-10}(S_{crystal} - 1)^{1.103} e^{(1.558/S_{crystal} - 1)}$$

$S_i = C_i^*/C$ C is the solute concentration and C^* is the solubility of polymorph *i*.

m is the mass per volume of amorphous polymorph, ε is stirrer power number.

The determining step in this transformation process is nucleation of hydrate crystal.

Conclusion and Future Direction

I. A typical solution-mediated transformation from amorphous to crystal appeared in 5′-GMPNa$_2$ crystallization process.
II. The determining step is nucleation of hydrate crystal.
III. The results can be used to govern the control for an industrial process of a desired polymorphous structure。

The inline Raman is a key means for the study of transformation process, so later some inline Raman data should be added in this work to verify the XRD data.

References

1. Mangin, D.; Puel, F.; Veesler, S. Polymorphism in Processes of Crystallization in Solution: A Practical Review. *Organic Process Research & Development* 2009, 13, 1241-1253;
2. Trifkovic, M.; Rohani, S.; Sheikhzadeh, M. Kinetics Estimation and Polymorphic Transformation Modeling of Buspirone Hydrochloride. *Journal of Crystallization Process and Technology* 2012, 02 (02), 31-43.
3. Herman, C.; Haut, B.; Douieb, S.; Larcy, A.; Vermylen, V.; Leyssens, T. Use ofin SituRaman, FBRM, and ATR-FTIR Probes for the Understanding of the Solvent-Mediated Polymorphic Transformation of II–I Etiracetam in Methanol. *Organic Process Research & Development* 2012, 16 (1), 49-56.
4. Scholl, J.; Davide Bonalumi; Vicum, L.; Mazzotti, M. In Situ Monitoring and Modeling of the Solvent-Mediated Polymorphic Transformation of L-Glutamic Acid. *Crystal Growth & Design* 2006, 6 (4), 881-891.
5. Cornel, J.; Lindenberg, C.; Mazzotti, M. Experimental Characterization and Population Balance Modeling of the Polymorph Transformation of L-Glutamic Acid. *Crystal Growth & Design* 2009, 9 (1), 243-252.
6. Maher, A.; Croker, D. M.; Seaton, C. C.; Rasmuson, Å. C.; Hodnett, B. K. Solution-Mediated Polymorphic Transformation: Form II to Form III Piracetam in Organic Solvents. *Crystal Growth & Design* 2014, 14 (8), 3967-3974.
7. Kang, J.; Tuan, N. A.; Kim, J. M.; Chang, S. M.; Kim, W. S. Study of phase transformation of guanosine 5'-monophosphate in drowning-out crystallization. *Applied biochemistry and biotechnology* 2010, 160 (2), 561-73.
8. Garcia, E.; Veesler, S. p.; Boistelle, R.; Hof, C. Crystallization and dissolution of pharmaceutical compounds: An experimental approach. *Journal of Crystal Growth* 1999, 1981360—1364.
9. Su, W.; Hao, H.; Barrett, M.; Glennon, B. The Impact of Operating Parameters on the Polymorphic Transformation of D-Mannitol Characterized in Situ with Raman Spectroscopy, FBRM, and PVM. 2010.
10. Févotte, G.; Alexandre; Sheibat Othman Nida, C. A population balance model of the solution-mediated phase transition of citric acid. *AIChE Journal* 2007, 53 (10), 2578-2589.

476638 A Study on Separating Aromatic and Aliphatic Mixtures Using Deep Eutectic Solvents

Abstract:

The separation of aromatic and aliphatic mixtures is an important process in the industry. Due to the formation of azeotropes, traditional distillation method is not suitable for separating these mixtures. Organic solvents such as sulfolane are the most commonly used solvents in the liquid-liquid-extraction (LLE) process. However, these organic solvents are generally toxic and flammable. So it is necessary to develop a new generation of solvents. Recently, a new generation of solvents deep eutectic solvents (DESs) are considered as an environmental benign solvent, which is a combination of a salt as hydrogen bond acceptor (HBA) and a hydrogen bond donor (HBD) through hydrogen bond. DESs have many advantages such as easy preparation, low price, biodegradability, and so on. In this work, we synthesized a series of DESs for the separation of toluene from toluene/heptane mixtures and studied the rules of separating aromatics from aliphatics using DESs, and the obtained DESs with good extraction performance were used to separate aromatics from oil mixtures or naphtha. The results showed that (1) the extraction ability for toluene using DESs was greatly influenced by HBAs and HBDs which formed the DESs. The selectivity of toluene was distinctly enhanced by short side chain, small central atom of cation and large anion of HBA, together with adequate position for alkyl chain and appropriate functional group of HBD. It was also found that an increase of extraction temperature at the studied range could enhance the selectivity for toluene. Fig. 1 shows the effect of HBD on selectivity and distribution ratio of toluene, and effect of selected HBD and different HBA on selectivity and distribution ratio of toluene for the separation of toluene + n-heptane mixtures using DES. (2) The obtained high effective DES namely tetraethylammonium chloride (TEAC)- levulinic acid (LA) was used for separating different systems of aromatics/aliphatics mixtures and the results show that TEAC-LA can efficiently separate benzene from cyclohexane. Besides, the length of the aliphatics chain consisting the oil can remarkably influence the purity of the aromatics which are extracted to the bottom phase. The longer of the aliphatics, the larger of the S values, which means better purity of aromatics. (3) It was also found that when the DES of TEAC-LA is mixed with naphtha at a DES/naphtha mass ratio of 1:1, this kind of DES shows excellent extraction performance. The extraction rate of aromatics is 11.5% and the purity of the aromatics extracted in the bottom phase can reach as high as 95.6%, which is better than the results of sulfolane. The extraction performance of TEAC-LA have not any change after four cycles, indicating that the DES have a good recycling property.

The project is financially supported by the Shanxi Scholarship Council of China (2011-086), the National Basic Research Program of Chin(2011CB201303) and the long-term subsidy mechanism from the Ministry of Finance and the Ministry of Education of PRC

Keywords: Deep eutectic solvents, Extraction, Toluene, n-Heptane, Naphtha, Role, Hydrogen bond donor, Hydrogen bond acceptor

476639 Efficient Absorption of SO_2 in Simulated Flue Gas by Environmentally Benign Functional Deep Eutectic Solvents

Abstract:

The emission of sulfur dioxide (SO_2), which is mainly from the burning of fossil fuels, has drawn much attention all over the world. The emitted SO_2 is harmful to human body and causes serious environmental pollution, so it is essential to remove SO_2 from the flue gas. Up to now, flue gas desulfurization (FGD) is the most efficient way for the capture of SO_2. Liquid absorbents such as aqueous amines have also been applied for the capture of SO_2. However, this method may cause secondary pollution due to the high volatility of amines. As a result, recyclable liquid solvents with low volatility and high capacity are expected as excellent absorbents for the capture of SO_2. Recently, a new generation of solvents deep eutectic solvents (DESs) are considered as an environmental benign solvent, which is a combination of a salt as hydrogen bond acceptor (HBA) and a hydrogen bond donor (HBD) through hydrogen bond. DESs have many advantages such as low vapor pressure, easy preparation, low price, biodegradability, and so on. Based on the properties of DESs, we designed new functional DESs and used them to capture SO_2 in simulated flue gas.

In this work, we designed and synthesized two kinds of biodegradable functional DESs on betaine (Bet) and L-carnitine (L-car) as HBA, and ethylene glycol (EG) or water as HBD with different ratios of HBA to HBD. They were investigated to absorb SO_2 with different partial pressures at various temperatures. The results are showed as follows. (1) At 40 °C with a SO_2 partial pressure of 0.02 atm, SO_2 absorption capacities of the DESs with HBA/HBD mole ratio of 1:3 were 0.332 mol SO_2 / mol HBA for Bet + EG DES and 0.820 mol SO_2 / mol HBA for L-car + EG DES. And SO_2 absorption capacities of the DESs with HBA/HBD mass ratio of 1:1 were 0.155 mol SO_2 / mol HBA for Bet + H_2O DES and 0.606 mol SO_2 / mol HBA for L-car + H_2O DES.

(2) The higher the absorption temperature, the lower SO_2 absorption capacities. (3) The lower the SO_2 partial pressure, the lower SO_2 absorption capacities. At 40 °C and 3700 ppm SO_2 in simulated flue gas, SO_2 absorption capacities of the DESs with HBA/HBD mole ratio of 1:3 were 0.155 mol SO_2 / mol HBA for Bet + EG DES and 0.606 mol SO_2 / mol HBA for L-car + EG DES. (4) The regeneration experiments demonstrated that the absorption capacities of DESs did not change after five absorption and desorption cycles. (5) The absorption mechanism of SO_2 by DESs was studied by FT-IR, ^{1}H NMR and ^{13}C NMR spectra. It was found that there are strong acid–base interactions between SO_2 and $-COO^-$ on HBA for Bet + EG DESs and L-car + EG DESs, and $-COO^-$ on HBA is recovered to $-COOH$ by stronger acid H_2SO_3 for Bet + H_2O DESs and L-car + H_2O DESs. DESs based on quaternary ammonium inner salts as HBA with EG or water are promising absorbents in the application of FGD for SO_2 capture.

The project is financially supported by the National Natural Science Foundation of China (No. 21176020 and 21306007) and the Research Fund for the Doctoral Program of Higher Education of China (No. 20130010120005).

472711 Designer 3D Printed Oleophilic Absorbents and Encapsulants with Tunable Capillary Architectures

Abstract:

Here we show how 3D printing can be leveraged to manufacture absorbents with tailored internal oleophilic capillary networks capable of selectively extracting and sequestering oil dispersed in water. These networks are produced using poly(lactic acid), an inexpensive commodity biopolymer that is degradable and environmentally friendly. 3D printing enables designer absorbent architectures to be crafted that match properties of different petroleum fractions, maximize selectivity against water, and permit either strong encapsulation or easy recovery of the collected oil. We describe fundamental characterization studies aimed at identifying optimal pore network architectures that maximize uptake and rate of absorbance, as well as determining quantitative cost/performance metrics for rational selection of optimal designs.

This platform lays a foundation to manufacture a host of products incorporating sophisticated oil absorption and encapsulation functions. We demonstrate this versatility by highlighting two innovative applications. First, we demonstrate the ability to produce large-sized 3D printed absorbent filters that can be incorporated into an autonomous robotic oil skimmer, enabling continuous routine clean-up of waterways. 3D printing makes it possible to overcome the limitations of current sheet or pad-based materials by achieving on-demand rapid production of rugged, reusable, rigid absorbents with superior uptake, selectivity, and ease of deployment. Second, we employ a new family of flexible filaments to produce 3D printed wearable products containing embedded capillary networks tailored to encapsulate oil-based insect repellents. A core-shell design is embedded in a wristband format to uniquely deliver sustained release of dispersed volatile species at higher doses than possible when applied directly to the skin. This approach may offer a new avenue to help combat the spread of mosquito-borne infectious disease.

459295 Rare Earth Mental Extraction from Geothermal Brine Using Nanofluids

Abstract:

Rare earth metals are critical materials in a wide variety of applications in generating and storing renewable energy and in designing more energy efficient devices. Extracting rare earth metals from geothermal brines is a very challenging problem due to the low concentrations of these elements and engineering challenges with traditional chemical separations methods involving packed sorbent beds or membranes that would impede large volumetric flow rates of geothermal fluids transitioning through the plant. We are demonstrating a simple and highly cost-effective nanofluid-based method for extracting rare earth metals from geothermal brines. Core-shell composite nanoparticles are produced that contain a magnetic iron oxide core surrounded by a shell made of silica or metal-organic framework (MOF) sorbent functionalized with chelating ligands selective for the rare earth elements. By introducing the nanoparticles at low concentration (â‰ˆ0.05 wt%) into the geothermal brine after it passes through the plant heat exchanger, the brine is exposed to a very high concentration of chelating sites on the nanoparticles without need to pass through a large and costly traditional packed bed or membrane system where pressure drop and parasitic pumping power losses are significant issues. Instead, after a short residence time flowing with the brine, the particles are effectively separated out with an electromagnet and standard extraction methods are then applied to strip the rare earth metals from the nanoparticles, which are then recycled back to the geothermal plant. Recovery efficiency for the rare earths at ppm level has now been measured for both silica and MOF sorbents functionalized with a variety of chelating ligands. A detailed preliminary techno-economic performance analysis of extraction systems using both sorbents showed potential to generate a promising internal rate of return (IRR) up to 20%.

461658 New Poly(alkyl imines) for CO_2 Capture from Ambient Air

Abstract:

Research efforts in CO_2 capture technologies have dramatically increased due to interest in reducing the effect of anthropogenic emissions on atmospheric CO_2 concentrations. Much of this research has focused on capture from large point sources such as flue gas. Direct capture of CO_2 from ambient air is proposed as a complementary technology that has the advantage of not needing a flue gas feed stream and thus is more versatile in its deployment. However, at an ultradilute concentration of 400 ppm, the driving force for adsorption is low, necessitating the development of adsorbents that may be different from those used in flue gas capture. Additionally, these adsorbents must be stable under oxidizing conditions due to the presence of oxygen in the feed stream.

In this work, we present work towards the development of a series of new poly(alkyl imines) as well-defined branched molecules for CO_2 capture. By changing the structure and flexibility of the adsorbent aminopolymer, we show that the CO_2 capture capacity and efficiency can be readily tuned. These materials represent a departure from the traditionally-studied poly(ethylenimine) (PEI) for CO_2 capture by using a well-defined molecular species with controlled and varied alkyl linker length. Additionally, these structures only contain primary and tertiary amines, which impart high capture capacity under ultradilute conditions and stability against oxidative degradation.

461888 High-Throughput Screening of Metal-Organic Frameworks for Hydrogen Storage

Abstract:

Metal-organic frameworks (MOFs) are porous materials consisting of metal nodes connected with organic linkers. This class of materials has high surface area and porosity, which make them attractive candidates for gas adsorption applications. Because MOFs can be made from a vast array of nodes, organic linkers, and functional groups, there are a huge number of possible MOF structures. High-throughput computational screening can efficiently characterize large numbers of materials to identify the best candidates for adsorption. [1,2]

In this study, we screened over 130,000 candidate structures [3] for hydrogen storage capacity at cryogenic conditions (77 K), considering loading at 100 bar and desorption at 2 bar using grand canonical Monte Carlo simulations. We also calculated textual properties such as surface area, pore diameter, and void fraction and examined adsorption performance trends with respect to these properties.

Several thousand of the structures exceed the DOE gravimetric adsorption target of 7.5 weight percent; however, none of them meet the volumetric target of 70 g/L. The highest volumetric adsorption we find in this database is 50 g/L.

We will discuss structure-performance trends as well as common traits in the best-performing groups of MOFs. We will also introduce a fast, computationally efficient method of screening MOF structures for gas adsorption performance based on geometry and pore size.

References:

1. Y.J. Colon, R.Q. Snurr, "High-throughput computational screening of metal-organic frameworks," *Chem. Soc. Rev.* 43, 5735-5749 (2014).

2. H. Frost, R.Q. Snurr, "Design requirements for metal-organic frameworks as hydrogen storage materials," *J. Phys. Chem. C* 111 18794-18803 (2007).

3. C.E. Wilmer, M. Leaf, C.Y. Lee, O.K. Farha, B.G. Hauser, J.T. Hupp, R.Q. Snurr, "Large-scale screening of hypothetical metal-organic frameworks," *Nature Chem.*, 4, 83-89 (2012)

464264 Adsorption Separation of N_2 from CH_4 Containing Mixtures Using N_2 Selective Cation Exchanged Clinoptilolite

Abstract:

The application of N_2 selective adsorbent materials within unconventional natural gas and biogas upgrading processes has the potential to achieve a more economically efficient separation of N_2 from hydrocarbon mixtures containing CH_4 than current separation processes. Conventionally, almost all adsorbents are selective for CH_4 over N_2. However, inverse selectivity is desired for this separation. In this study, the potential of cation exchanged clinoptilolite adsorbents were studied for the selective separation of N_2 from CH_4 containing gas mixtures. Single component adsorption equilibrium isotherms and uptake measurements of CH_4 and N_2 confirm the promising N_2 selective properties of the cation exchanged clinoptilolite. Adsorption breakthrough curve experiments and the effects of competitive adsorption of binary CH_4-N_2 mixtures on the studied adsorbents are explored and will be discussed along with the applicability of ion exchanged clinoptilolite in various adsorption technologies for this separation.

465419 Adsorption of Heavy-Metals in Sulfur-Functionalized and Highly Ordered Mesoporous Carbons

Abstract:

In this project, the mesoporous carbons were synthesized from resorcinol-formaldehyde resins templated by Pluronic F127 triblock copolymer. The mesoporous carbons were further treated with a sulfur-bearing compound that introduced sulfur functionalities onto the mesoporous carbons along with simultaneous enhancement of pore textural properties. The mesoporous carbons were characterized with pore textural properties, thermogravimetric analysis (TGA), electron microscopy (SEM and TEM) and x-ray photoelectron microscopy (XPS). In order to understand the heavy metal adsorption on the sulfur-functionalized mesoporous carbons, four metal ions including mercury, lead, nickel and cadmium were adsorbed from their aqueous solutions as a function of the varying pH. Kinetics of adsorption of heavy metals was measured in both individual and in combined form to understand competitive and non-competitive adsorptions. XPS analysis of the metal-loaded mesoporous carbons was performed to understand the binding of metals with sulfur functionalities. The kinetic data of the adsorption were further mathematically modeled to generate the necessary parameters of adsorption.

469103 Removal of Toxic Heavy Metal Cations from Simulated FGD Wastewater Using Functionalized Silica

Abstract:

Flue-gas desulfurization (FGD) process employed by electric power generation utilities produces wastewater containing significant levels of soluble heavy metal ions that are being regulated to parts-per-billion levels due to their toxicity. In order to address this issue, removal of several heavy metal ions was investigated by employing high-surface area amorphous silica grafted with sulfur-containing functional groups, which have been shown to have high affinity towards heavy metal ions. These adsorbents were tested for the removal of single component as well as multiple component heavy metal ions.

The silica support was grafted with 3-mercaptopropyltrimethoxysilane and 2-mercaptobenzothiazole. Thermogravimetric Analysis (TGA) and Potentiometric Mass Titrations (PMT) were performed on functionalized silica to determine surface loadings and the point of zero charge (PZC), respectively. The removal of Hg^{2+} at high concentration (10,000 µg/L) in different matrices was investigated. The effect of FGD matrix ions (Ca^{2+}, Mg^{2+}, Na^{+}, SO_4^{2-}, Cl^{-}, NO_3^{-}) on removal efficiency of Hg^{2+} is also reported. Simultaneous removal of Hg^{2+} along with the several other heavy metal cations has been reported at concentrationsÂ relevant for the FGD wastewater (500 Âµg/L each of Cu^{2+}, Cd^{2+}, Hg^{2+}, Tl^{+}, Pb^{2+}, Ni^{2+}, Zn^{2+}). 99% of Hg^{2+} was captured using these adsorbents for both concentration ranges. The adsorbents were able to selectively capture other heavy metal cations at low concentrations as well. The concentrations of heavy metal ions were determined using the Inductively Coupled Plasma â€“ Mass Spectroscopy (ICP-MS). Preliminary regeneration studies of spent adsorbents were performed using dilute HCl indicating limited regeneration and suggesting their disposal without regeneration.

BTEX REMOVAL FROM SYNTHETIC WASTEWATER USING ORDERED MESOPOROUS CARBON (OMC)

Mas Iwan Konggidinata, University of Louisiana at Lafayette, Lafayette LA, U.S.A
Bing Chao, University of Louisiana at Lafayette, Lafayette LA, U.S.A
William E. Holmes, University of Louisiana at Lafayette, Lafayette LA, U.S.A
Daniel D. Gang, University of Louisiana at Lafayette, Lafayette LA, U.S.A
Rakesh Bajpai, University of Louisiana at Lafayette, Lafayette LA, U.S.A
Ramalingam Subramaniam, University of Louisiana at Lafayette, Lafayette LA, U.S.A

Abstract

Chemical and petrochemical industries produce substantial amounts of wastewater every day. This wastewater contains organic pollutants such as benzene, toluene, ethylbenzene and xylenes (BTEX) that are toxic to human and aquatic life. Benzene, in particular, is carcinogenic to humans while the other compounds can lead to liver and kidney problems. Ordered Mesoporous Carbon (OMC), a novel adsorbent that possesses the characteristics of an ideal adsorbent such as large specific surface area, tunable pore size and stable framework, was investigated to understand its properties and suitability for BTEX removal. Adsorption isotherms, kinetics, the effects of BTEX initial concentrations and temperatures on adsorption were studied. The OMCs were characterized using surface area and pore size analyzer, Transmission Electron Microscopy (TEM), Elemental Analysis, Thermogravimetric Analyzer (TGA) and Fourier Transform Infrared spectroscopy (FTIR). The results suggested that the Langmuir isotherm and pseudo second order models described the experimental data. Furthermore, BTEX adsorption onto OMC was endothermic and physisorption was indicated as the dominant adsorption mechanism. In addition, OMC had 27% higher overall adsorption capacities compared to GAC.

Introduction

Chemical and petrochemical industries produce substantial amounts of wastewater every day. One particular industry that generates a lot of wastewater is oil and gas exploration. In the US, the shale gas boom has been driven primarily by the adoption of hydraulic fracturing technology, or fracking. In the year 2000, approx. 1.6% of total natural gas produced in the United States came from shale gas. This figure had increased rapidly to 8.1% in 2007 and 29% in 2011.[1] In 2007, the estimated total volume of produced wastewater was about 21 billion barrels or an average of 57.4 million barrels per day.[2] Produced wastewater typically contains high total dissolved solids, high salinity, and may contain relatively high levels of metals, organic compounds, and naturally occurring radionuclides.[3] This study focused on the organic compounds commonly found in produced wastewater; primarily benzene, toluene, ethylbenzene, and xylenes, because these compounds pose an immediate health threat to human beings. Benzene, in particular, is carcinogenic. The other compounds can lead to liver and kidney problems.[4]

The adsorbents commonly used for organic removal are activated carbons. The activated carbons can be made from lignite, coal, nut shells, and synthetic polymers.[5] These carbons typically have wide pore-size distributions with a majority of the pores in the microporous range. Therefore, lower adsorption capacity could be due to inaccessibility of some sorption sites.[6] Researchers have explored different materials and modification methods to enhance the adsorption capacities, adsorption kinetics, and regeneration cycles.[7] Different types of carbons that have been studied came from low cost recycled materials such as peat, sawdust, chitosan, waste tire rubbers as well as synthesized materials such as zeolites and carbon nanotubes.[8,9,10,11,12,13] Some of these materials have shown positive results in increasing the adsorption capacity. One of the adsorbents that caught our attention is ordered mesoporous carbon (OMC) because it has the characteristics of ideal adsorbents such as large surface area, high porosity, and high thermal stability.[14] Excellent adsorption performance of OMC for resorcinol removal has also been reported by Guo et al.[15] and Shou et al.[16] In this study, the structure and properties of OMC were investigated to understand its suitability for BTEX removal. Adsorption isotherms, kinetics, the effects of BTEX initial concentrations and temperatures on adsorption were evaluated.

Experimental

Preparation of SBA-15 silica template

SBA-15 silica template was prepared using triblock copolymer surfactant, Pluronic P123 ($EO_{20}PO_{70}EO_{20}$) and tetraethyl orthosilicate (TEOS) as the silica source under acidic condition. 100 mL of concentrated hydrochloride acid (HCl) and 20 g of Pluronic P123 surfactant were added into the 525 mL of distilled water under continuous stirring. After all the surfactant was completely dissolved, 46.5 mL of TEOS was added to the solution and the stirring was kept for another 5 min. The resulting cloudy mixture with white solids was placed in a constant temperature water bath (Premiere Thermostatic water bath HH-4) for 6 h at 50 °C, followed by aging for 24 h at 90 °C. The solid product was filtered, washed with 80 - 90 °C hot distilled water, and dried in an oven at 105 °C overnight. Next, the SBA-15 product was obtained after being calcined in a 550 °C furnace for 8 h.

Fabrication of Ordered Mesoporous Carbon

The Ordered Mesoporous Carbon (OMC) was prepared using hard template method with sucrose as the carbon precursor. 1.5 g of sucrose ($C_{12}H_{22}O_{11}$) and 4 drops of concentrated sulfuric acid (H_2SO_4) were added into 20 mL of water. After dissolved, 2.0 g of SBA-15 was added into the solution and stirred for 1 h. The mixture was then heated to 100 °C for 6 h and 160 °C for another 6 h. Then, additional 0.51 g of sucrose, 4 drops of concentrated sulfuric acid, and 20 mL of water were added into the composite and stirred for 1 h. Following the same heat treatment in the previous step, the composite was then carbonized at 700 °C for 6 h under nitrogen (N_2) flow. Afterwards, the silica template was removed using approx. 50 mL of HF solution at room temperature and stirred for 24 h. The obtained product was filtered and washed with 4 L of DI water and finally OMC-0 was prepared.

Characterizations

The specific surface area, pore volume, average pore size and pore size distribution were obtained from Micromeritics ASAP 2020 surface area and porosimetry analyzer. Elemental composition analysis was obtained using Elementar Vario MICRO Cube CHNS elemental analyzer. The surface functional groups were qualitatively measured with Agilent Cary 630 Fourier Transform Infrared (FT-IR) spectrometer.

TEM images were obtained using a Hitachi 7600 Transmission Electron Microscopy (Hitachi America Ltd., Tarrytown, New York). The acceleration voltage used was 100 kV. The samples were prepared by dispersing a small amount of samples in ethanol and sonicated for 15 min. A few drops of the resulting suspension were placed on a Copper grid for analysis.

The thermogravimetric analysis (TGA) was carried out using a SDT 2960 under oxygen atmosphere from 25 °C to 800 °C temperature range and ramping rate of 5 °C/min.

Batch adsorption study

The adsorption experiments were carried out using 40 mL of different concentration of BTEX solutions with predetermined amounts of OMC. These solutions were placed in an incubator shaker (New Brunswick Scientific) for 48 h at 25 °C and 275 rpm. One sample of the same concentration without the adsorbent (blank) was prepared and treated under the same conditions as the other solutions and served as a reference to establish the initial concentration. The vials were then removed and solutions were filtered using a 0.45 µm syringe filter. 5 mL of the filtered solution were injected into the GC using purge and trap method and the equilibrium concentration of each solution was measured by GC-PID (Varian 3400).

To understand the adsorption equilibrium time, BTEX adsorption kinetics were investigated from 0 to 36 h and the amount of adsorbents used was 10 mg. In addition, the effects of BTEX initial concentrations were also studied. The initial concentrations were varied from 10 to 20 ppm. Furthermore, the effects of temperature on the adsorption capacity of OMC were also inspected. Temperatures were varied from 25 °C to 45 °C.

After addition of adsorbents, the quantity of BTEX adsorbed at equilibrium, q (mg/L) was calculated using the following equation:

$$q = \frac{(C_i - C_f)\,V}{M} \quad (1)$$

Where, q is the adsorption capacity (mg/g). C_i and C_f are the liquid-phase initial and final concentration of BTEX (mg/L), respectively. V is the volume of the solution (L), and M is the mass of adsorbent used (g).

Results and Discussions

SBA-15 and OMC characterizations

The nitrogen adsorption-desorption isotherms in Fig. 1 suggested that both OMC and SBA-15 demonstrated type IV isotherm that is typically observed for mesoporous materials. The SBA-15 capillary condensation step was located in smaller relative pressure range that translated to narrow pore size distributions observed on SBA-15. On the other hand, wider capillary condensation step indicating wide pore size distributions was observed for OMC-0. The structural parameters for both SBA-15 and OMC-0 suggested that the specific surface areas were comparable. As OMC-0 was the inverse replica of SBA-15, the observed pore sizes and pore volumes were dependent on the unit cell size of SBA-15.[17]

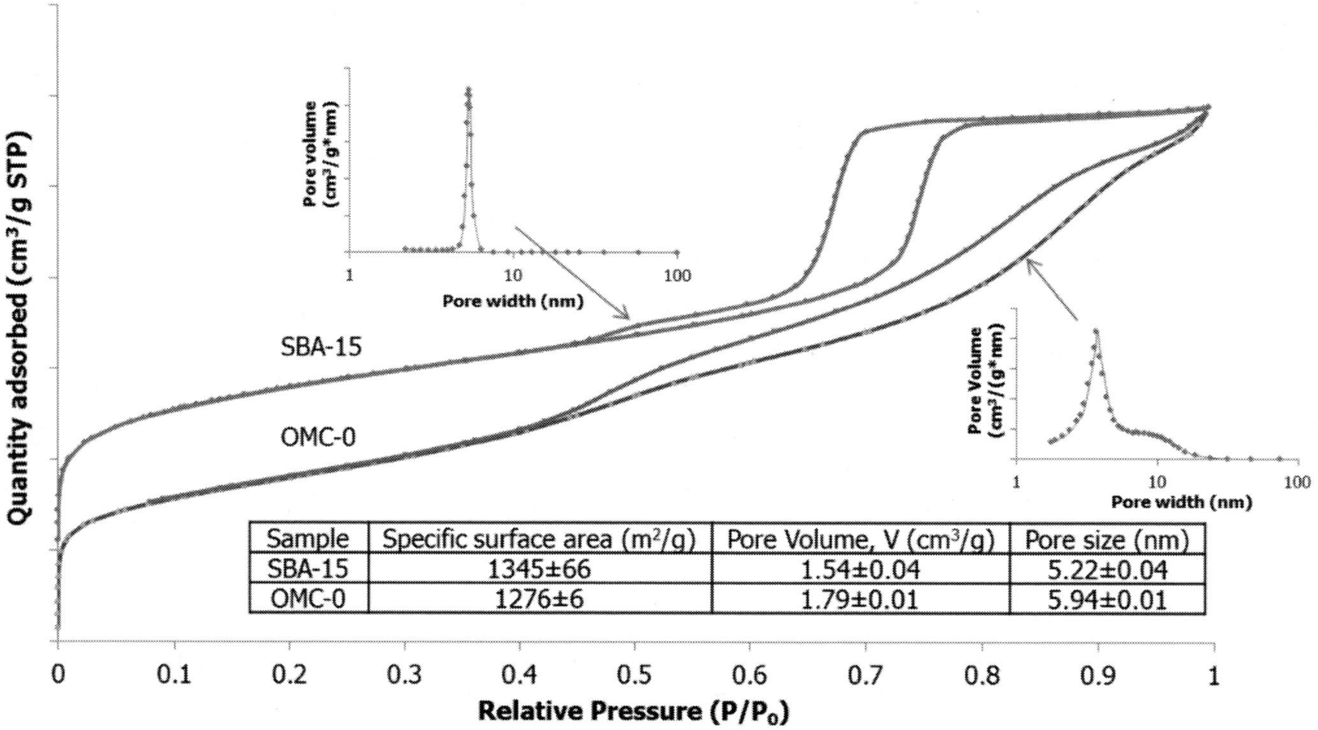

Sample	Specific surface area (m^2/g)	Pore Volume, V (cm^3/g)	Pore size (nm)
SBA-15	1345±66	1.54±0.04	5.22±0.04
OMC-0	1276±6	1.79±0.01	5.94±0.01

Fig.1 Nitrogen adsorption-desorption isotherms, the corresponding pore size distributions and the structural parameters for SBA-15 and OMC-0

The TEM images of SBA-15 below showed the uniformly arranged pores and hexagonally arranged structure that corresponded to the p6mm structure typically observed for SBA-15. On the other hand, the internal structures of OMC-0 were composed of hexagonally arranged carbon rods, which was the inverse replica of SBA-15.

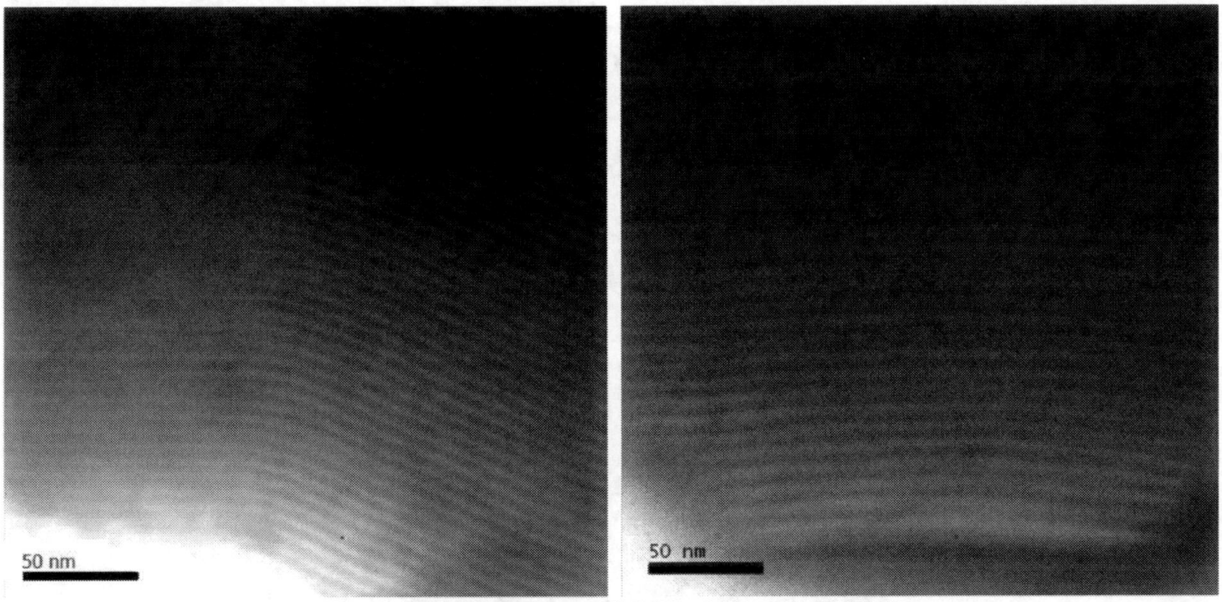

Fig.2 TEM Images of SBA-15 and OMC-0

The FTIR analysis of OMC was shown in Fig. 3. The peak at 1046 cm^{-1} corresponded to C-O bonds from alcohol and carboxylic acids. The peaks observed at 1465 and 1723 cm^{-1} could be associated with $-CH_2-$ bend and C=O bonds from carboxylic acid groups, respectively. A set of peaks in the region of 2000 cm^{-1} to 2300 cm^{-1} could be related to C-O group and the peak observed at 3653 cm^{-1} could be associated with O-H groups.

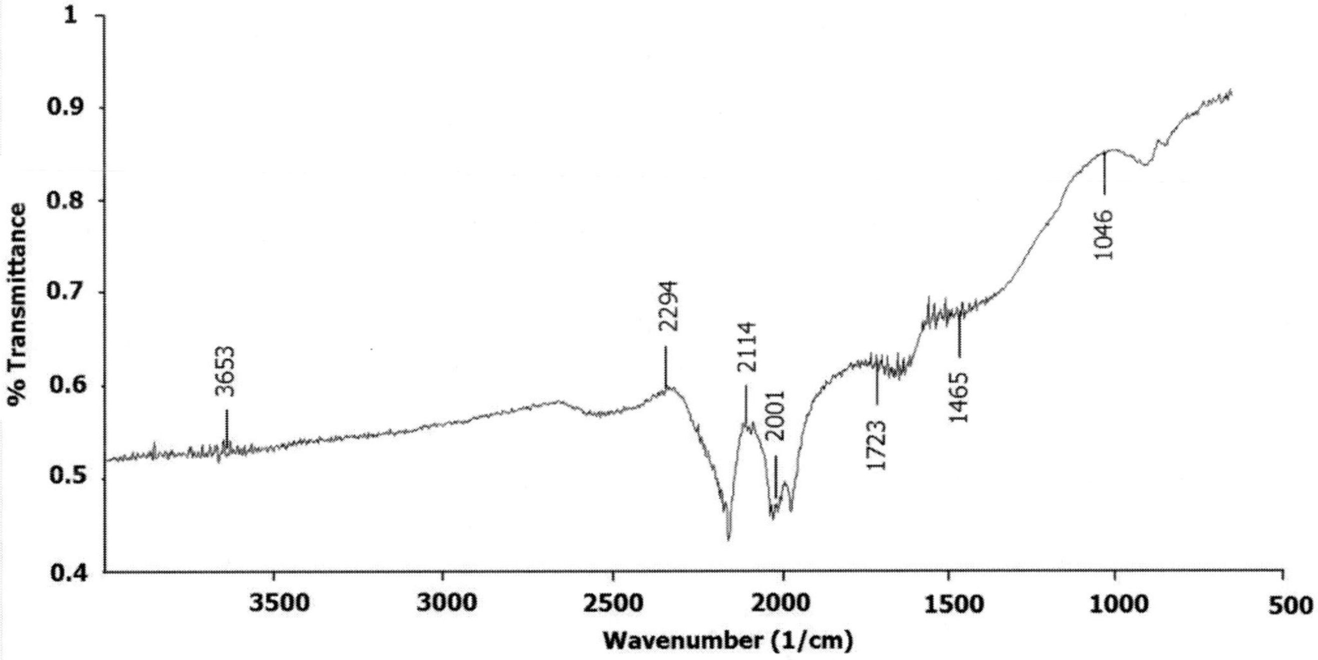

Fig.3 FT-IR Spectra of OMC-0

Thermogravimetric analysis was carried out to investigate the stability of OMC at high temperatures. The result shown in Fig. 4 indicated that the OMC decomposed at 510 – 650 °C temperature range. Similar results were also reported by Ren et al.[6] The elemental analysis indicated that carbon was the most dominant element, followed by oxygen. Sulphur which was found on the OMC could be due to the sulfuric acid used as the catalyst during fabrication process.

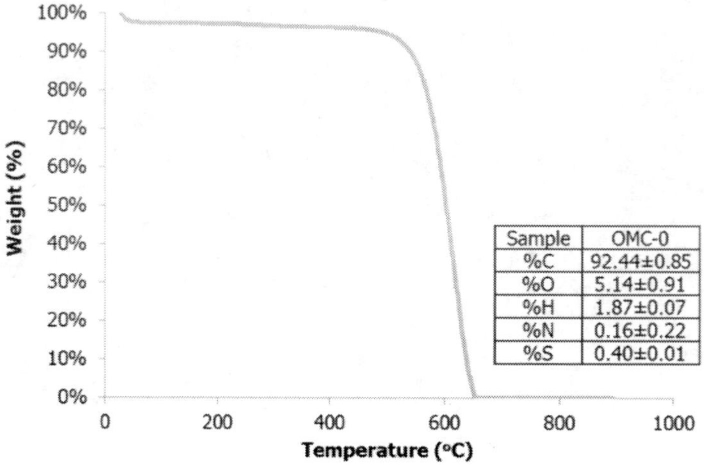

Sample	OMC-0
%C	92.44±0.85
%O	5.14±0.91
%H	1.87±0.07
%N	0.16±0.22
%S	0.40±0.01

Fig.4 Thermogravimetric Analysis (TGA) and Elemental Analysis of OMC-0

Batch Adsorption Study

Adsorption kinetics

BTEX adsorption capacities were investigated as a function of time (0 to 36 h) to determine the adsorption equilibrium time. The investigation revealed that the initial rapid adsorption rate decreased as it approached equilibrium and gradually levelled off. The rapid initial adsorption rate could be due to the high concentration gradient between the adsorbent and adsorbate in the solution and plenty of vacant sites were available at the beginning.[18] It was found that 12 h were the requirement for adsorption to reach equilibrium. To ensure that the equilibrium was achieved, the adsorption experiments were incubated for up to 36 h.

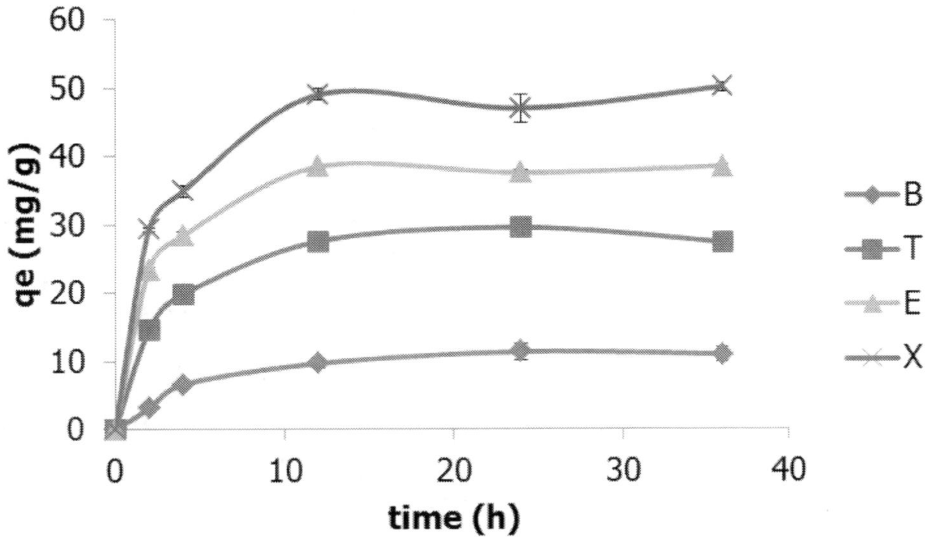

Fig.5 Adsorption kinetics of OMC-0 (T = 25 °C, BTEX conc = 10 mg/L, adsorbent dosage = 0.01 g)

To understand the BTEX adsorption processes on the OMC, kinetic adsorption data were obtained experimentally and was fitted to pseudo-first-order and pseudo-second-order models. From the R^2 values, the pseudo-second-order adsorption mechanism had a better fit to the experimental data compared to pseudo-first-order kinetics indicating the occurrence of chemisorption.[19] Similar kinetic mechanisms were also observed for adsorption of BTEX onto raw and thermally treated lignite, surfactant-modified natural zeolites nanoparticles, and multi-walled carbon nanotubes.[20,21,22]

Table 1. Adsorption kinetics of OMC-0*

	q_e (exp)	Pseudo-first- order			Pseudo-second-order		
		q_e (calc)	k_1	R^2	q_e (calc)	k_2	R^2
B	11.33±1.25	2.476	0.088	0.931	11.960	0.031	0.982
T	29.51±0.16	3.156	0.063	0.611	28.819	0.036	0.994
E	38.43±0.33	4.311	0.180	0.961	39.191	0.031	0.998
X	50.06±0.64	3.975	0.120	0.577	50.763	0.019	0.997

**Units: q_e: mg/g; k_1: 1/h; k_2: g/(mg h)*

Adsorption Isotherm

Based on the experimental data, the removal efficiencies followed the order: Xylenes > Ethylbenzene > Toluene > Benzene. The adsorption isotherm was studied and it was found to follow the Langmuir isotherm model. The Langmuir model relates the concentration of the medium which cover the solid surface at a fixed temperature to the adsorption capacity.

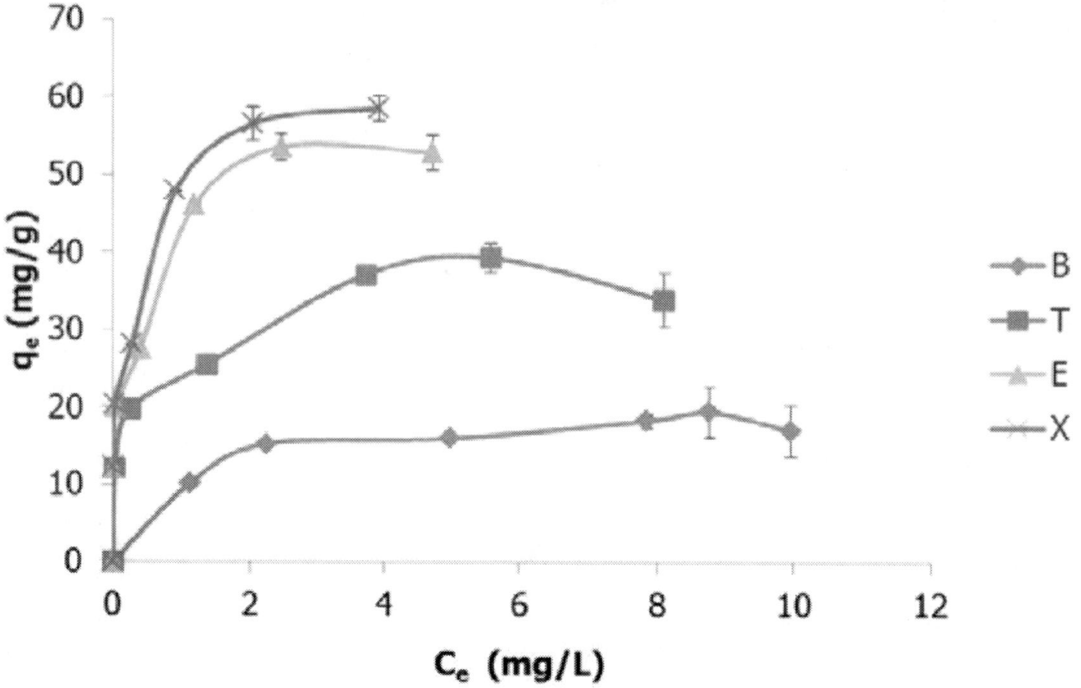

Fig.6 Adsorption isotherm of OMC-0 (T: 25 °C, BTEX conc: 15 mg/L, contact time: 36 h)

Adsorption in Langmuir model is assumed to be monolayer coverage. As the distance from the adsorption surface increased, the intermolecular forces decreased.[18] Similar results were also reported for BTEX removal on organokaolinite, polymeric resin (MN-202), modified clay (Claytone-40).[23,24]

Table 2. Adsorption Isotherms of OMC-0**

	Langmuir			Freundlich		
	q_{max}	b	R^2	k_F	1/n	R^2
B	19.84	0.045	0.981	10.92	0.243	0.832
T	36.36	0.004	0.985	25.39	0.208	0.957
E	54.95	0.003	0.993	39.79	0.217	0.967
X	60.24	0.002	0.994	44.88	0.213	0.974

***Units: q_{max}: mg/g; b: 1/mg; k_F: mg/g*

Effect of BTEX Initial Concentrations, thermodynamic study, and comparison with commercial GAC

Fig. 7 indicated that BTEX adsorption capacities increased with the increase of BTEX initial concentrations in the solution. This could be due to the increase in the driving force between the BTEX compounds and the active sites of the adsorbents at higher concentrations. Similar observation was reported by Nourmoradi et al.[25]

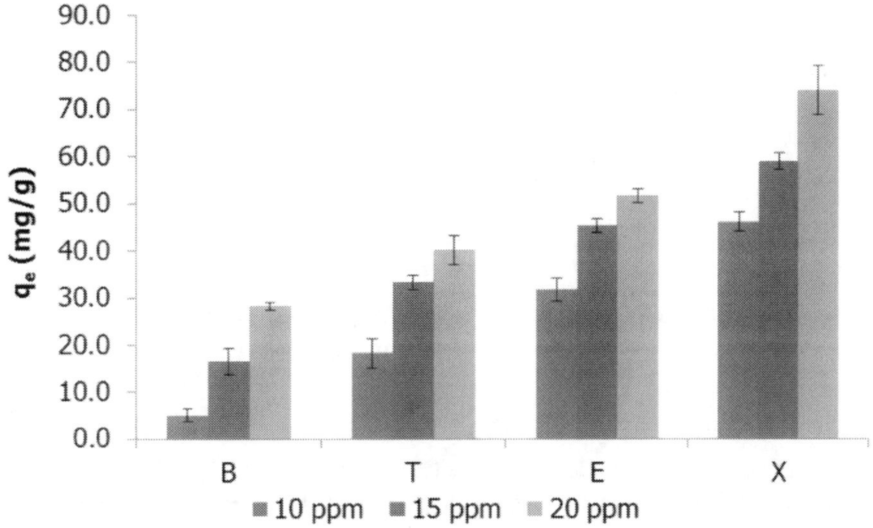

Fig.7 Effects of BTEX Initial Concentrations on OMC-0 (T: 25 °C, adsorbent dosage: 0.01 g, contact time: 36 h)

The thermodynamic study was conducted at three different temperatures (25, 45 and 65°C, respectively). The calculated Gibbs free energy (ΔG^o) values were tabulated in table 3. The values of ΔG^o decreased with increasing temperature. This phenomenon indicated that the adsorption process was spontaneous and more favorable at higher temperatures. The ΔG^o values which were found to be in between -20 and 0 kJ/mol fall in the physisorption range.[26] This indicated that physisorption is the dominant adsorption mechanism for BTEX onto OMC. Next, the positive values of enthalpy (ΔH^o) indicated that the adsorption process was endothermic. This was in line with our experimental results where the adsorption capacity increased with the increase in temperature. Lastly, the positive entropy (ΔS^o) reflects the affinity of OMCs for BTEX and the increasing disorder at the solid liquid interface due to the adsorption process.[27]

Table 3. Effects of temperatures on adsorption capacity and thermodynamic study of OMC-0

	q_e (mg/g)			$-\Delta G^o$ (kJ/mol)			ΔH^o	ΔS^o
	25°C	45°C	65°C	25°C	45°C	65°C	(J/mol)	(J/mol/K)
B	5.55±1.54	9.28±0.13	11.64±0.57	0.01	2.15	3.29	24.7	83.4
T	19.85±3.42	24.88±0.81	28.59±0.34	2.70	4.26	5.35	17.1	66.7
E	34.50±2.62	36.09±0.84	37.07±0.59	5.27	6.31	7.02	7.8	44.0
X	50.09±2.26	50.23±0.87	50.55±0.67	6.31	7.11	7.70	4.0	34.9

For BTEX removal, OMC and commercial GAC (Darco® 12x20 DC) adsorption capacities were compared. As shown in table 3, OMC-0 had 27% higher overall adsorption capacities compared to GAC. This could be due to the higher specific surface area observed on OMC (1276 m^2/g) compared to GAC (890 m^2/g). In addition, since majority of the pores of GAC were in the microporous range, inaccessibility of some sorption sites could lead to lower adsorption capacity.

Table 4. Adsorption capacity comparison for GAC and OMC-0

Sample	B	T	E	X	Total adsorption capacity (mg/g)
GAC	4.62±0.55	17.72±0.07	26.20±0.16	31.05±0.07	79.6
OMC-0	5.10±1.41	18.22±3.15	31.71±2.41	46.05±2.08	101.1

Conclusions

The removal of BTEX from synthetic wastewater onto OMC was studied. Based on the experimental data, the removal efficiencies followed the order: Xylenes > Ethylbenzene > Toluene > Benzene. Adsorption kinetics and adsorption isotherms for the BTEX removal were evaluated. Adsorption equilibrium was established after 12 h and the process was observed to follow the pseudo second order and the Langmuir isotherm models. The calculated values of enthalpy (ΔH^{o}) and Gibbs free energy (ΔG^{o}) suggested that BTEX adsorption was endothermic, spontaneous and physisorption was indicated as the dominant adsorption mechanism. It was also found that OMC-0 had 27% higher overall adsorption capacities compared to GAC.

References

1. Yu, S. (2015). Evaluation of socioeconomic impacts on and risks for shale gas exploration in China. Energy Strategy Reviews, 6, 30-38

2. Clark, C. E., & Veil, J. A. (2009). Produced water volumes and management practices in the United States (No. ANL/EVS/R-09-1). Argonne National Laboratory (ANL).

3. Rahm, B. G., Bates, J. T., Bertoia, L. R., Galford, A. E., Yoxtheimer, D. A., & Riha, S. J. (2013). Wastewater management and Marcellus Shale gas development: Trends, drivers, and planning implications. Journal of environmental management, 120, 105-113.

4. Costa, A. S., Romão, L. P. C., Araújo, B. R., Lucas, S. C. O., Maciel, S. T. A., Wisniewski, A., & Alexandre, M. D. R. (2012). Environmental strategies to remove volatile aromatic fractions (BTEX) from petroleum industry wastewater using biomass. Bioresource technology, 105, 31-39.

5. Chiang, Y. C., Chiang, P. C., & Huang, C. P. (2001). Effects of pore structure and temperature on VOC adsorption on activated carbon. Carbon, 39(4), 523-534.

6. Ren, H., Shou, W., Ren, C., & Gang, D. D. (2016). Preparation and post-treatments of ordered mesoporous carbons (OMC) for resorcinol removal. International Journal of Environmental Science and Technology, 13(6), 1505-1514.

7. Carvalho, M. N., Da Motta, M., Benachour, M., Sales, D. C. S., & Abreu, C. A. M. (2012). Evaluation of BTEX and phenol removal from aqueous solution by multi-solute adsorption onto smectite organoclay. Journal of hazardous materials, 239, 95-101.

8. Uraki, Y., Tamai, Y., Ogawa, M., Gaman, S., & Tokura, S. (2008). Preparation of activated carbon from peat. BioResources, 4(1), 205-213.

9. Garg, V. K., Gupta, R., Yadav, A. B., & Kumar, R. (2003). Dye removal from aqueous solution by adsorption on treated sawdust. Bioresource Technology,89(2), 121-124.

10. Gang, D. D., Deng, B., & Lin, L. (2010). As (III) removal using an iron-impregnated chitosan sorbent. Journal of Hazardous Materials, 182(1), 156-161.

11. Gupta, V. K., Gupta, B., Rastogi, A., Agarwal, S., & Nayak, A. (2011). A comparative investigation on adsorption performances of mesoporous activated carbon prepared from waste rubber tire and activated carbon for a hazardous azo dye—Acid Blue 113. Journal of Hazardous Materials, 186(1), 891-901.

12. Hunger, B., Matysik, S., Heuchel, M., Geidel, E., & Toufar, H. (1997). Adsorption of water on zeolites of different types. Journal of Thermal Analysis and Calorimetry, 49(1), 553-565.

13. Li, Y. H., Wang, S., Wei, J., Zhang, X., Xu, C., Luan, Z., Wu, D. & Wei, B. (2002). Lead adsorption on carbon nanotubes. Chemical Physics Letters, 357(3), 263-266.

14. Ramaswamy, S., Huang, H. J., & Ramarao, B. V. (Eds.). (2013). Separation and purification technologies in biorefineries. John Wiley & Sons.

15. Guo, R., Guo, J., Yu, F., & Gang, D. D. (2013). Synthesis and surface functional group modifications of ordered mesoporous carbons for resorcinol removal. Microporous and Mesoporous Materials, 175, 141-146.

16. Shou, W., Chao, B., Ahmad, Z. U., & Gang, D. D. (2016). Ordered mesoporous carbon preparation by the in situ radical polymerization of acrylamide and its application for resorcinol removal. Journal of Applied Polymer Science, 133(19).

17. Kim, S. S., Karkamkar, A., Pinnavaia, T. J., Kruk, M., & Jaroniec, M. (2001). Synthesis and characterization of ordered, very large pore MSU-H silicas assembled from water-soluble silicates. The Journal of Physical Chemistry B, 105(32), 7663-7670.

18. Arshadi, M., Shakeri, H., & Salvacion, J. W. L. (2016). A green adsorbent for the removal of BTEX from aqueous media. RSC Advances, 6(17), 14290-14305.

19. Mohanty, K., Das, D., & Biswas, M. N. (2005). Adsorption of phenol from aqueous solutions using activated carbons prepared from Tectona grandis sawdust by ZnCl 2 activation. Chemical Engineering Journal, 115(1), 121-131.

20. Aivalioti, M., Pothoulaki, D., Papoulias, P., & Gidarakos, E. (2012). Removal of BTEX, MTBE and TAME from aqueous solutions by adsorption onto raw and thermally treated lignite. Journal of hazardous materials, 207, 136-146.

21. Seifi, L., Torabian, A., Kazemian, H., Bidhendi, G. N., Azimi, A. A., Farhadi, F., & Nazmara, S. (2011). Kinetic study of BTEX removal using granulated surfactant-modified natural zeolites nanoparticles. Water, Air, & Soil Pollution, 219(1-4), 443-457.

22. Yu, F., Ma, J., & Wu, Y. (2011). Adsorption of toluene, ethylbenzene and m-xylene on multi-walled carbon nanotubes with different oxygen contents from aqueous solutions. Journal of hazardous materials, 192(3), 1370-1379.

23. Egbuchunam, T.O, Obi, G., Okieimen, F.E., Tihminlioglu, F. (2016). Removal of BTEX from Aqueous Solution using Organokaolinite. International Journal of Applied Environmental Sciences, 2(11), 505-513.

24. Hackbarth, F. V., Vilar, V. J., De Souza, G. B., de Souza, S. M. G. U., & de Souza, A. A. U. (2014). Benzene, toluene and o-xylene (BTX) removal from aqueous solutions through adsorptive processes. Adsorption, 20(4), 577-590.

25. Nourmoradi, H., Nikaeen, M., & Khiadani, M. (2012). Removal of benzene, toluene, ethylbenzene and xylene (BTEX) from aqueous solutions by montmorillonite modified with nonionic surfactant: Equilibrium, kinetic and thermodynamic study. Chemical Engineering Journal, 191, 341-348.

26. Bhatt, A. S., Sakaria, P. L., Vasudevan, M., Pawar, R. R., Sudheesh, N., Bajaj, H. C., & Mody, H. M. (2012). Adsorption of an anionic dye from aqueous medium by organoclays: equilibrium modeling, kinetic and thermodynamic exploration. RSC Advances, 2(23), 8663-8671.

27. Su, F., Lu, C., Johnston, K. R., & Hu, S. (2010). Kinetics, thermodynamics, and regeneration of BTEX adsorption in aqueous solutions via NaOCl-oxidized carbon nanotubes. Environanotechnology, Elsevier, 71-97.

464380 Adsorption-Based Heating and Cooling – Metal-Organic Frameworks over Zeolites?

Abstract:

Increased tangibility of (anthropogenic) climate change and fossil fuel scarcity demand a transition to sustainable energy sources in the near future. The heating and cooling of residential areas requires roughly a third of the world's primary energy consumption. Thus significantly reducing the energy expenditures for heating and cooling will have a large impact on total energy consumption. Preferably one would like to use sustainable solar or low-grade (< 100 °C) waste energy for this purpose. With adsorption heat pumps and chillers (AHP/ACs) this is possible. These devices us the reversible adsorption of a working fluid onto a porous adsorbent to generate heating or cooling and can be operated with sustainable low-grade energy. Furthermore, with water as preferred working fluid, these devices are environmentally benign.

Market penetration of AHP/ACs has been only minor up to now. This because AHP/ACs are currently bulkier and less energy-efficient than their electrically operated vapor-compression counterparts. The moderate performance of AHP/ACs can be largely attributed to poor to moderate adsorptive properties of current state-of-the-art materials (mainly zeolites). To improve performance and thus commercialization of AHP/ACs, we have embarked on a quest to find more suitable adsorbent materials. Especially a relatively novel set of materials, the Metal-Organic Frameworks (MOFs), holds great promises for AHP/ACs. MOFs consist of inorganic clusters linked together by organic ligands to create porous crystalline materials. More than 20.000 MOFs are known currently, spanning a large variety of topologies and material properties. Not surprisingly thus, MOFs have received attention for a plethora of applications.

In this work we critically explore the promise that Metal-Organic Frameworks (MOFs) hold for adsorption heating and cooling by means of a critical comparison to commercially available materials, with special reference to zeolites, using either water [1, 2] or alcohols [1, 3] as working fluid. Based on both experimental and theoretical efforts we conclude that:

(i) MOFs that are sufficiently stable towards water [1, 2] exist.

(ii) Of these subset, multiple MOFs show the desired adsorption behaviour with either water or alcohols as working fluid [1, 2, 3].

(iii) These working pairs (MOF-working fluid) show enhanced energy efficiency and working capacity compared to benchmark materials and require a lower driving temperature.

(iv) The optimal working pair for specific operating windows depends on the MOF's pore structure (adsorption sites and pore size).

(v) MOFs may hold an added benefit for application, as they can be coated (without binder) directly on heat-exchanger surfaces [4, 5].

In short, the future is bright for the application of MOFs in adsorption heat pumps and chillers.

References

[1] De Lange, M.F.; Verouden, K.J.F.M.; Vlugt, T.J.H.; Gascon, J.; Kapteijn, F.; *Chemical Reviews*, **2015**, 115, 12205-1250

[2] Cadiau, A.; Lee, J. S.; Damasceno Borges, D.; Fabry, P.; Devic, T.; Wharmby, M. T.; Martineau, C.; Foucher, D.; Taulelle, F.; Jun, C.-H.; Hwang, Y. K.; Stock, N.; De Lange, M. F.; Kapteijn, F.; Gascon, J.; Maurin, G.; Chang, J.-S.; Serre, C.; *Advanced Materials*, **2015**, 27, 4775-4780.

[3] De Lange, M.F.; Van Velzen, B.L.; Ottevanger, C.P.; Verouden, K.J.F.M.; Lin, L.-C.; Vlugt, T.J.H.; Gascon, J.; Kapteijn, F.; *Langmuir*, **2015**, 31, 12783-12796

[4] De Lange, M. F.; Ottevanger, C. P.; Wiegman, M.; Vlugt, T. J. H.; Gascon, J.; Kapteijn, F.; *CrystEngComm*, **2015**, 17, 281-285

[5] De Lange, M. F.; Zeng, T.; Vlugt, T. J. H.; Gascon, J.; Kapteijn, F.; *CrystEngComm*, **2015**, 17, 5911-5920

450721 Influence of Different Adsorbates on the Efficiency of Energy Storage Using Adsorption

Abstract:

Influence of different adsorbates on the efficiency of energy storage using adsorption

Tobias Kohler, Karsten Müller, Wolfgang Arlt

University of Erlangen-Nuremberg, Erlangen/Germany;

The efficiency and energy density of energy storage using adsorption depends mainly on the degree of desorption that was achieved before, which is determined by desorption temperature, desorption time and the working pair (adsorbate/adsorbent). The state of the art working pair for adsorptive energy storage is water on zeolite 13X. This working pair needs high desorption temperatures and long desorption times, e. g. 500K for 6 hours, to reach its maximum efficiency. If energy at higher or lower temperature levels has to be stored, the efficiency of the storage is not sufficient to be competitive in the energy storage market.

In order to extend the operation range to different temperature levels, the only possibility is to change the adsorption pair, since desorption temperature and time are fixed by the application.

A huge number of groups work on the optimization of new adsorbents, but most of them aim at maximizing the nominal energy density or the amount of water adsorbed. In this work the goal is to reach high efficiencies of the process at different desorption temperatures. A further peculiarity is the fact that not only the adsorbent is changed but also the adsorbate.

A first examination of the efficiency of an adsorptive energy storage based on literature data of the adsorption of alcohols on BPL activated carbon (Taqvi et al. 1999) shows that the ideal desorption temperature rises, under the assumption that the desorption lasts for six hours, with rising chain length of the alcohol from 70°C for methanol to 147°C for propanol. The efficiencies calculated for these desorption temperatures are significantly higher than for the working pair water on zeolite 13X at the same desorption temperatures.
In order to further underline the effect of different adsorbates on the efficiency of thermochemical energy storage, the possible use of the homologeous series of alcohols on further adsorbents like SAPO and AlPO (Ng,Mintova 2008) are examined. SAPO and AlPO have become more common in recent years and are characterized by a high water and methanol uptake at fixed relative pressures, which corresponds to a high potential in thermochemical energy storage.

In order to examine the efficiencies and energy densities of different working pairs, adsorption isotherms at different temperatures are measured with a Magnetic Suspension Balance. These isotherms are transformed to a temperature independent form, the so called "characteristic curve of adsorption" according to Dubinin and Polányi. This enables the simulation of the strongly non-isothermal process.

References

Ng, E.-P., Mintova, S.: Nanoporous materials with enhanced hydrophilicity and high water sorption capacity. Microporous and Mesoporous Materials 114(1-3), 1-26 (2008).

Taqvi, S.M., Appel, W.S., Le Van, M.D.: Coadsorption of Organic Compounds and Water Vapor on BPL Activated Carbon. 4. Methanol, Ethanol, Propanol, Butanol, and Modeling. Ind. Eng. Chem. Res. 38, 240-250 (1999)

454090 Adsorption Prediction and Modeling of Thermal Energy Storage Systems: A Parametric Study

Abstract:

Thermal energy storage allows for the storage of energy from intermittent sources to correct for the variable supply and demand. The current work investigates adsorption technology for thermal energy storage through the development of a theoretical model, which describes the material and energy transfers in the system. The theoretical model was used to conduct a parametric study which examines the effect of column dimension, particle diameter, adsorption activation energy, flow rate, column void fraction, and adsorbent heat of adsorption on the thermal energy storage system performance. It was found that an optimal column length to column diameter ratio of 1.4, a column diameter to particle diameter ratio of 14.7, a flow rate of 24 LPM, and a void fraction of 0.4 gave the best thermal energy performance for a column volume of 6.275 10^{-5} m^3. Also, a low activation energy and a high heat of adsorption represent the best adsorption parameters for optimal temperature outputs, breakthrough behavior, and energy densities.

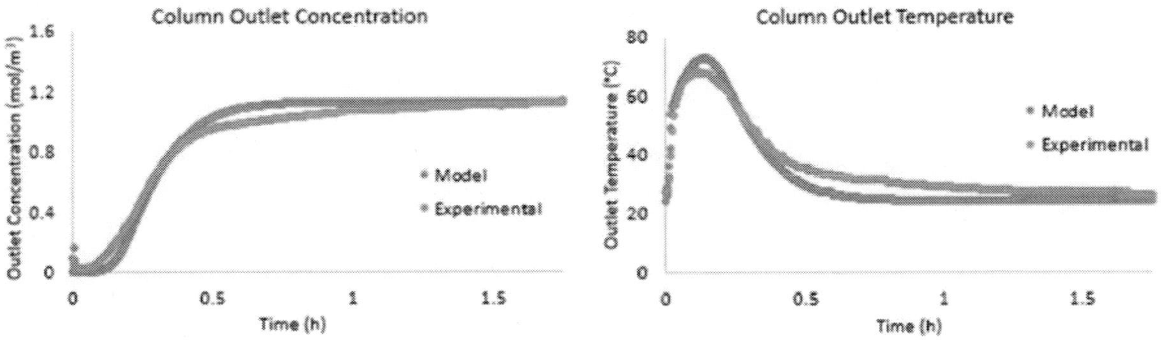

464273 Evaluation of the Performance of Chemically Modified Agrowaste for Solar Refrigeration Applications in Developing Countries

Abstract:

The refrigeration of food and medicine is a necessity for all human settlements. The traditional refrigeration systems use electrical energy and refrigerants that could cause ozone layer depletion and global warming. Thus, solar refrigeration systems have been developed as an alternative to provide cooling by using thermal energies from solar irradiation. The simple solar refrigeration system consists of an evaporator, a condenser, a throttling valve, an adsorption bed and a solar collector. The adsorption bed comprises of solid material that adsorb a vapor (e.g., water, methanol, ethanol), which is desorbed later by solar heating. The desorbed vapor is condensed and collected into the condenser. After, the throttling valve is open causing a sudden pressure reduction that turns the liquid into vapor. The evaporation heat required is taken from the evaporator box causing the cooling effect and then the cycle begins again.

The affinity between the adsorbent (solid material) and the adsorbate (vapor) is an important factor for the refrigeration efficiency. Several researches have investigated the performance of different working pairs (adsorbent-adsorbate) in solar refrigeration systems. The most common working pairs include water-silica gel, water-zeolite, and methanol-activated carbon, among others.

The present paper explores the potential application of agrowastes as adsorbents for solar refrigeration applications since these materials are issue of concern in developing countries such as Mexico. Additionally, Mexico has a high solar irradiation (6 kWh/m^2-day) that is a natural resource that could be used for refrigeration in communities with lack of electricity. In order to evaluate the affinity between agrowastes-vapor working pairs, maximum adsorption capacity are determined using the volumetric method.

461715 Bio-Alcohol Recovery from Aqueous Solutions with a Resistant and Formulated MOF Material

Abstract:

Alcohols, such as biobutanol, form a promising alternative to fossil-based chemicals and fuels, since they can be produced from renewable sources. The economical production of biobutanol, however, still faces several challenges, including its separation from the aqueous mixture in which it is produced by fermentation. Among the different separation techniques adsorption was identified as the most energy-efficient candidate. In a previous study, we studied a wide range of porous materials, like zeolites, active carbons and metal-organic frameworks (MOFs), for their capability to isolate biobutanol from its fermentation medium. We found that the MOF ZIF-8, a member of the Zeolite Imidazolate Frameworks (ZIF) family, shows promising properties, such as a large butanol adsorption capacity, a high selectivity towards butanol over side-products, the exclusion of water from its pores and, easy regeneration and desorption [1].

The ZIF-8 MOF material offers perspectives for the economical and low energy intensive production of biobutanol. A key aspect towards the implementation of ZIF-8 into an actual separation process is its formulation into a resistant structure while maintaining its promising features. As a consequence, in this work, various resistant pellets were produced, by combining polymer binders with ZIF-8 powder under different conditions. The stability of the pellets under acid-base, hydrothermal and long-term storage (3 years) was tested as well as their adsorptive properties under dynamic and static conditions in liquid and vapor phase. Furthermore the pellets were characterized with argon and mercury porosimetry. The results of these different experiments will be shown in this presentation and compared to the binder-free ZIF-8 pellets and powder material. It will be demonstrated that the formulation of the ZIF-8 material resulted in resistant pellets without altering its promising adsorptive properties.

[1] Cousin Saint Remi, J., Remy, T., Van Hunskerken, V., Van de Perre, S., Duerinck, T., Maes, M., De Vos, D.,

Gobechiya, E., Kirschhock, C.E.A., Baron, G.V. and Denayer, J.F.M. Biobutanol Separation with the Metal-Organic Framework ZIF-8. *ChemSusChem* **4**, 1074-1077 (2011).

466024 Understanding the Influence of Carbon Dioxide on the Adsorption of Butanol from the Off-Gas from the ABE Fermentation

Abstract:

The production of bio-alcohols has gained immense interest as an alternative to petroleum-based fuels, as a result of increasing fossil fuel prices and environmental issues. Among bio-alcohols, bio-butanol has shown superior properties such as higher energy and lower volatility and water miscibility. [1] However, the economical production of bio-butanol remains a major challenge as the fermentation production of butanol, ABE (acetone/butanol/ethanol) fermentation, produces a mixture of the three components diluted in water. In the recent years, adsorption on molecular sieves has been posted as the cheapest and most energy-efficient technique for the recovery of butanol from fermentation broth. It is for this reason that many studies have been developed on this regard using activated carbons, zeolites, and Metal-Organic Frameworks (MOFs). [2, 3, 4] Most studies focus on the separation of the main mixture from water. However, other gases such as hydrogen or carbon dioxide are produced as well during the process. Carbon dioxide is responsible for the increment of the amount of butanol present at the head space of the fermentation chamber, however not many studies deal with this phase. Working with the vapor phase has some advantages such as preventing clogging of the adsorbents or the influence of pH. This paper studies the effect of the presence of carbon dioxide on the adsorption of butanol at a molecular level. With this aim in mind, competitive adsorption isotherms and multicomponent vapor phase (water/ABE) breakthrough curves were determined on two zeolitic materials (CHA and LTA) and on a hydrophobic metal-organic framework ZIF-8 in absence and presence of CO_2. In a computational section, Monte Carlo simulations and Molecular Dynamics were employed to study the adsorption and diffusion of binary mixtures containing carbon dioxide and butanol. Prior to the analysis of the mixtures, a model for butanol that reproduces experimental properties of the molecule such as its polarity and vapor-liquid coexistence has been developed. Pure component isotherms and heats of adsorption have been computed and compared to experimental data to check the accuracy of the interacting parameters. Finally, to get a better understanding of the molecular mechanism

that governs the adsorption of the targeted mixture (CO_2/butanol) in the selected materials we also analyze the distribution of the molecules inside the structures. A combination of these features together with the computation of self-diffusion coefficients allow us to get a global picture of the process and to identify the role of carbon dioxide in the butanol purification process.

[1] Garcia, A., Sanz, S., and Beltran, J., J. Chem. Technol. Biotechnol. 84 (2009) 1873-1882.

[2] Cao, Y., Wang, K., Wang, X., Gu, Z., Gibbons, W., and Vu, H., App. Surf. Sci. 349 (2015) 1-7.

[3] Cousin Saint Remi, J., Baron, G., and Denayer, J. Adsorption. 18 (2012) 367-373.

[4] Cousin Saint Remi, J., Remy, T., Van Hunskerken, V., van der Perre S., Duerinck, T., Maes, M., De Vos, D., Gobechiya, E., Kirshhock, C. E. A., Baron, G. V., and Denayer, J. F. M., ChemSusChem. 4 (2011) 1074-1077.

472477 Diffusion and Adsorptive Desulfurization Studies on Bifunctional Zeolites

Abstract:

Diffusion and Adsorptive Desulfurization Studies on Bifunctional Zeolites

Kevin X. Lee and Julia A. Valla

Department of Chemical & Biomolecular Engineering, University of Connecticut, 191 Auditorium Road, Unit 3222, Storrs, CT 06269-3222, USA,

Phone: +1-860-486 4602, e-mail: ioulia.valla@.uconn.edu

Zeolite Y has shown to be an effective sorbent material for the adsorptive removal of sulfur from transportation fuels.[1] In their natural state, Y zeolites are microporous and non-selective. To meet the mandated sulfur standards for gasoline and diesel, several functionalities need to be added to the zeolites to improve the sulfur uptake. Firstly, the introduction of mesoporosity enables the adsorption of refractory sulfur compounds, which could not be achieved by conventional hydrodesulfurization without excessive energy. The bigger pores improve mass transfer and minimize diffusion limitations. Furthermore, adsorption sites for sulfur can be increased through the addition of metals by ion-exchanging the zeolite with active metals such as Ce or Ce. As a result, the adsorption process becomes more selective to thiophenic molecules via the interaction of a π-complexation or a direct S-M σ bond.[2,3] The incorporation of metal ions increases the selectivity for sulfur, especially in commercial fuels. The objective of this work is to fundamentally understand the adsorption mechanism of various sulfur compounds on the zeolite active sites. To achieve this goal, IR spectroscopy experiments were performed using model thiophenic compounds, and parent Y and metal-modified Y zeolites.

Bifunctional zeolites were first prepared by introducing mesoporosity via two top-down methods: (1) desilication (DS) and (2) surfactant-assistant technique.[4,5] The mesoporous zeolite was subjected to an ion-exchange procedure to replace the protons with either Cu or Ce metals. The modified zeolites were characterized using various techniques. The pore size distribution and crystal structure were identified using N_2 adsorption/desorption isotherm and x-ray diffraction (XRD), respectively. Surface acidity and metal distributions were determined using pyridine and CO adsorptions by diffuse reflectance infrared spectroscopy (DRIFT-FTIR), respectively. DRIFTS-FTIR experiments were also used to study diffusion limitations and adsorption mechanism of thiophenic compounds on various modified zeolites. The metal loadings on the zeolite were quantified using ICP. Isosteric heat of adsorption calculations were conducted via a batch method to study the strength of interaction between each sulfur component and the sorbent material.[6] The adsorption pattern was assumed to follow the Langmuir isotherm.

Breakthrough experiments were conducted in a fixed-bed column to study the dynamic adsorption of model fuel (thiophene (TP), benzothiophene (BT) and dibenzothiophene (DBT) in octane). The desulfurized effluent was collected periodically until saturation had been reached and analyzed using a gas-chromatograph-sulfur chemiluminescence detector (GC-SCD). Results indicated that the best materials for the removal of TP and BT were the metal-exchanged zeolites (CeY and CuY). Larger pores, however, were required for the adsorption of sulfur compounds with a larger kinetic diameter, such as DBT. Metal-exchanged mesoporous Y (ie. CuSAY) has shown to be the most effective sorbent with an increase in DBT uptake by about 75 mg/L compared to the parent Y, as demonstrated in Figure 1(a). These results suggest that the adsorption of DBT is perturbed by diffusion limitations, which can be overcome by bifunctional zeolites. Figure 1(b) shows that in a mixture of model fuel containing all three thiophenic compounds, the CeSAY material proved to be the best sorbent, suggesting a preference to adsorb refractory sulfur compounds with a larger kinetic diameter.

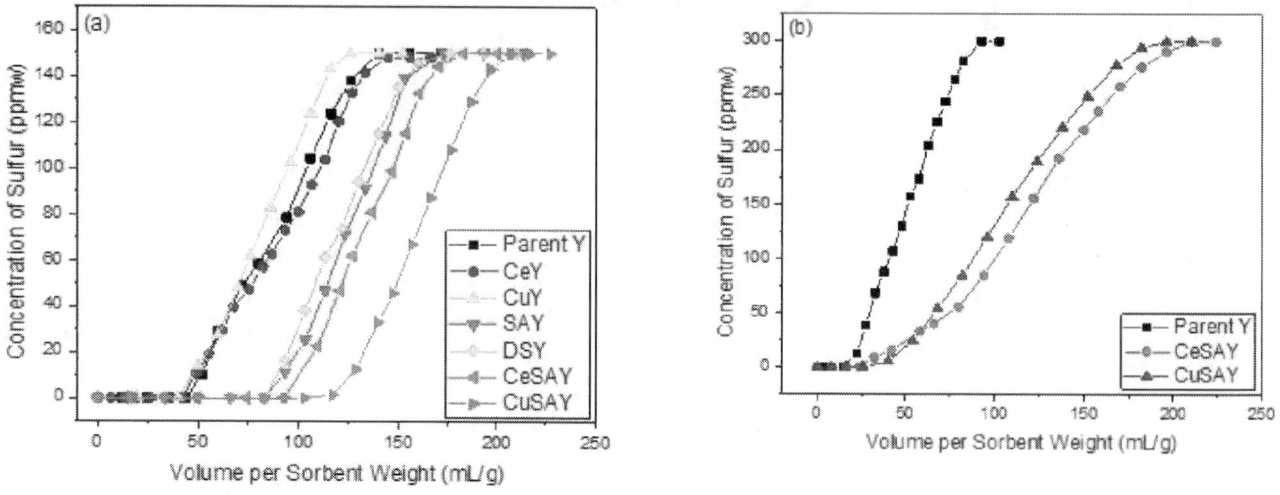

Figure 1: Breakthrough Curves of (a) DBT and (b) model fuel on various Y-derived zeolites.

In all cases, the tendency for zeolites to adsorb sulfur compounds increases in the order of TP < BT < DBT. The isosteric heat of adsorption calculations revealed that DH_{ads} was the highest for BT adsorbed on CuY. This is likely due to the strong π-stacking of the BT ring atop the Cu metal. The CuSAY, despite assisting in overcoming diffusion challenges, showed the lowest DH_{ads}. This shows that the isosteric heat of adsorption is only a descriptor of the interaction strength.

The adsorption mechanism of sulfur on Y zeolites can be revealed by FTIR studies. About 25 mg of zeolite was weighed into the DRIFT cell and activated in-situ with either H_2 or N_2 for 1 hour. Upon activation, benzothiophene vapor was introduced into the cell via a heated line until saturation was achieved. Temperature programmed desorption (TPD) studies were carried out to investigate the sites at which BT is adsorbed. Figure 2 shows the BT adsorption spectra on CuY and parent Y zeolites. The OH region on the left shows that a higher temperature is required to completely desorb the BT ring from CuY compared to parent Y. This is also apparent in the 2000-1000 cm^{-1} region where most of the C-C and C-H vibration modes have disappeared from the parent Y spectra by 360 °C, but the same peaks are still visible on CuY. As suggested by the breakthrough curves and heats of adsorption, Cu-exchanged zeolites can form π-complexations with sulfur, shifting the C=C band vibration of 1640 cm^{-1} to lower frequencies. As a result, more energy is needed to remove the strongly held BT molecule from CuY.

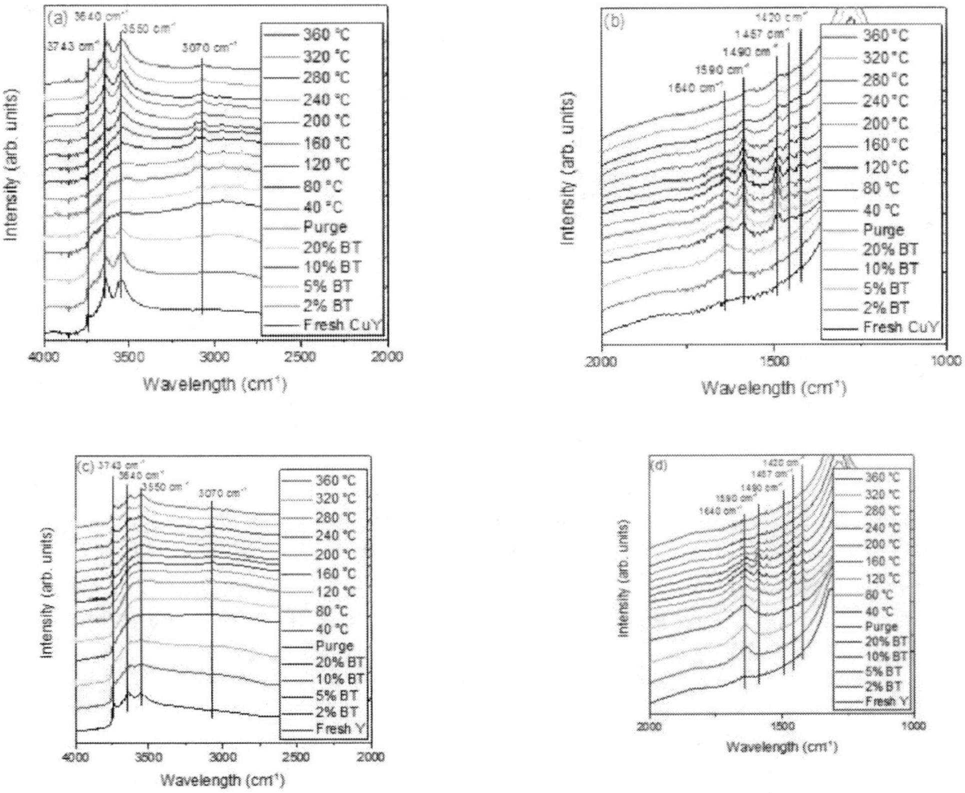

Figure 2: O-H vibration region of BT adsorption on (a) CuY and (c) Y and C-C vibration region of BT adsorption on (b) CuY and (d) Y.

Breakthrough studies revealed that CuSAY has the highest capacity for refractory sulfur compounds such as DBT due to strong π-complexation interaction between the sorbate and the sorbent. In a model fuel containing various thiophenic compounds, CeSAY proved to be the best sorbent in selectively removing DBT via the direct σ-bond interaction. Both isosteric heat of adsorption calculations and FTIR studies suggested that higher energy is required to break the bonds between CuY and BT. Collectively, our findings suggests that determining the optimum balance between the pore structure and metals in a bifunctional zeolite is essential for maximizing capacity and selectivity for sulfur compounds.

References

1. Song, C.; Ma, X. New design approaches to ultra-clean diesel fuels by deep desulfurization and deep dearomatization. *Appl. Catal. B Environ* **2003**, 41, 207 – 238.

2. Hernandez-Maldonado, A. J.; Yang, F. H.; Qi, G.S.; Yang, R. T. Desulfurization of transportation fuels by π-complexation sorbents: Cu(I)-, Ni(II)-, and Zn(II)-zeolites. *Appl. Catal. B Environ.* **2005**, 56, 111 – 126.

3. Velu, S.; Ma, X.; Song, C. Selective adsorption for removing sulfur jet fuel over zeolite-based adsorbents. *Ind. Eng. Chem. Res.* **2003**, 42, 5293 – 5304.

4. Verboekend, D., Perez-Ramirez, J. Design of hierarchical zeolite catalysts by desilication, *Catal. Sci. Technol.* **2011,** 1, 879-890.

5. Garcia-Martinez, J., Johnson, M., Valla, J., Li, K., Ying, J. Y. Mesostructured zeolite Y-high hydrothermal stability and superior FCC catalytic performance. *Catal. Sci. Technol.* **2012,** 2, 987-994.

6. Ma, L., Yang, R. T. Selective adsorption of sulfur compounds: isotherms, heat, and relationship between adsorption from vapor and liquid solution. *Ind. Eng. Chem. Res.* **2007,** 46, 2760-2768.

458613 Study of COS Removal from Gases on Doped or Undoped Aluminas

Abstract:

Gaseous sulfur-containing molecules are considered as harmful contaminants from several points of view in most industrial refining or energy production processes. Their toxicity, strong ability to cause catalyst poisoning and corrosion are the main negative issues induced by these molecules.

The need for light hydrocarbon feedstocks (C1-C4) ultrapurification is presently becoming more and more acute in order to increase the lifetime of catalysts in refineries or petrochemical plants and to achieve the increasingly stringent sulfur content specifications.. Besides the removal of H_2S, which is the main sulfur contaminant, this supposes that carbonyl sulfide (COS), the second most abundant sulfur contaminant in light hydrocarbons, has to be also removed. COS is mainly formed during the thermal or catalytic conversion process of natural gas, oil or biomass, either by reaction of H_2S with CO_2, or CH_4 with SO_2. In refineries, COS is most likely to be found in LPG (liquefied petroleum gases) streams such as C3 cuts (propane-propylene) or C3-C4 cuts (such as propylene/butenes streams issued from FCC); Moreover, in certain cases, COS removal has to be totally removed form process streams as its presence can be very detrimental for downstream processes. For instance, COS is a poison for propylene polymerization catalysts.

Existing desulfurization processes (guard bed sorbents, solvents-based Acid-Gas Removal processes) are often dedicated to the removal of H_2S which generally represents the major sulfur impurity. But yet, COS being less reactive than H_2S, these processes are not or little effective for the removal of COS. For example, ZnO (H_2S capture mass) does not react with COS (Aboulayt et al. 1996), COS removal kinetics using chemical solvents such as amines are not favorable (about 100 times lower than CO_2) (Vaidya et Kenig 2009), and COS removal using physical solvents is effective at very low temperatures (around -40°C for washing with methanol-based solvents) which would make it very costly process. Thus, the preferred industrial COS removal processes are based on the catalytic hydrolysis ($COS + H_2O + CO_2 \leftrightarrow H_2S + CO_2$) and/or the adsorption of COS (P. K. T. Liu 1988).

In both cases, the adsorbent or catalyst generally consists of a basic solid (Bulushev et Ross 2011; A. Aboulayt et al. 1996). The most used materials are simple or mixed metal oxides eventually doped by alkali or alkaline earth metals, by transition metals or even by rare-earth elements. According to the literature, adsorption of COS on metal oxides occurs through interactions with the surface hydroxyl groups (OH), and is more favored at low temperatures ($T < 100°C$) and without water. Hydrolysis of COS is thermodynamically favored but its kinetic is quite slow (Rhodes et al. 2000). Thus it is rather favored at high temperatures ($T > 100°C$) and requires the presence of water (H. M. Huang et al. 2006). Industrially, the choice between adsorption or hydrolysis based method thus depends on the operating conditions imposed in practice. The interaction of COS with basic oxides, such as alumina, has been studied in the literature for widely varying operating conditions. Nevertheless, a number of points are not described in the literature and published data are scarce, especially regarding interaction mechanisms of COS with the oxide surfaces and adsorption competitions with species also involved in the hydrolysis reaction (either CO_2 or H_2S).

The present work has been focused on the study of the adsorption of COS on doped or undoped oxides (Al_2O_3, TiO_2 and ZrO_2). For all of them, we showed that COS is partially hydrolyzed when adsorbed, but only CO_2 is adsorbed on the same sites than COS and may thus compete with it for its adsorption. The formation of CO_2 is less pronounced for more basic oxides that suggests a role of Lewis acid sites. In parallel, the regeneration of the adsorbent and the stability of their capacities have also been investigated. It appears than the most basic oxides show stronger interactions with COS. COS is not fully desorbed during regeneration, explaining a decrease of adsorption capacities over cycling. Moreover, in the case of less basic oxides, some elemental sulfur is observed, whose formation has been assigned to the presence of Lewis acid sites. However, this phenomenon does not seem to be responsible for the decrease of COS adsorption capacity.

In brief, this study gives us a better comprehension of the overall system showing the specific role of Lewis acid sites and also the benefic effect of increasing the basicity of the adsorbent.

Bibliography

A. Aboulayt; F. Mauge; P. E. Hoggan; J. C. Lavalley (1996) Combined FTIR, reactivity and quantum chemistry investigation of COS hydrolysis at metal oxide surfaces used to compare hydroxyl group basicity. In : Catalysis Letters, vol. 39, n° 3-4, p. 213–218.

Aboulayt, A.; Mauge, F.; Hoggan, P. E.; Lavalley, J. C. (1996) Combined FTIR, reactivity and quantum chemistry investigation of COS hydrolysis at metal oxide surfaces used to compare hydroxyl group basicity. In : Catalysis Letters, vol. 39, n° 3-4, p. 213–218.

Bulushev, D. A.; Ross, J. R. H. (2011) Catalysis for conversion of biomass to fuels via pyrolysis and gasification: A review. In : Recent Developments in Operando Spectroscopy, vol. 171, n° 1, p. 1–13.

H. M. Huang; N. Young; B. P. Williams; S. H. Taylor; G. J. Hutchings (2006) High temperature COS hydrolysis catalysed by gamma-Al2O3. In : Catalysis Letters, vol. 110, n° 3-4, p. 243–246.

P. K. T. Liu (1988) Adsorption of carbonyl sulfide from liquid hydrocarbons with activated alumina and other adsorbents. In : AIChE Symposium Series, vol. 84, n° 264, p. 109–118.

Rhodes, C.; Riddel, S. A.; West, J.; Willams, B. P.; Hutchings, G. J. (2000) The low-temperature hydrolysis of carbonyl sulfide and carbon disulfide: a review. In : Recent Developments in Operando Spectroscopy, vol. 59, n° 3-4, p. 443–464.

Vaidya, P. D.; Kenig, E. Y. (2009) Kinetics of carbonyl sulfide reaction with alkanolamines: A review. In : Chemical Engineering Journal, vol. 148, n° 2-3, p. 207–211.

472208 Evaluation of the Cactus Based-Mucilage As an Alternative Natural Dispersant on Toxicity, Effectiveness, and Surface Tension of Oil-in-Water Emulsion

Abstract:

Cactus based-mucilage has the advantage of being biodegradable and can be obtained through sustainable agriculture. This functional natural surfactant was extracted from the *Opuntia ficus indica* cactus plant, and the extraction process yielded two types of mucilage: the non-gelling extract (NE) and the gelling extract (GE). The properties of the cactus plant-based mucilage were studied on the toxicity, surface tension, droplet size and stability of dispersed crude oil in water emulsions (O/W emulsions). The toxicity of the mucilage and conventional dispersants (Corexit 9500A) were evaluated by a standard EPA toxicity test using Daphnia magna colonies exposed to both NE and GE mucilage extracts in concentrations ranging from 0 to 2000 mg/L for 48 hours. It has shown that mucilage can be classified as practically non-toxic to the Daphnia magna colonies with a LC_{50} above 2000 mg/L. The LC_{50} of Corexit 9500A was below 2 mg/L, which is moderately toxic to the Daphnia magna colonies. Cactus mucilage extracts and conventional dispersants has been evaluated by testing the surface tension and measuring the droplet size of O/W emulsions under different concentrations, including three concentrations of oil (3, 6, 30% v/v) and several dispersant to oil ratios (1:1, 1:3, 1:10). Synthetic seawater and surrogate oil from BP were used. The results of the dispersion obtained using mucilage extracts were compared with using Corexit 9500A. The surface tensions of emulsions with cactus mucilage were similar as with Corexit 9500A. The average droplet size in the systems with cactus mucilage (1.5% of NE has generated 5 μm droplets) were smaller when compared with the systems using the Corexit 9500A (1.5% of Corixit 9500A has formed 6.2 μm droplets). Smaller droplet size implies higher emulsion stability, and may improve the biodegradation of oil. An improved laboratory dispersant testing protocol developed by EPA, baffled flask test, was performed to measure the dispersion effectiveness at DOR 1:20, 1:50, and 1:100. The effectiveness of DOR at 1:20 and 1:50 were around 35%. Cactus mucilage is readily available through sustainable agriculture, it is inexpensive to

extract, and has exhibited a long shelf live. Therefore, cactus mucilage can be an alternative technology to mitigate the damage that oil may cause to the aquatic ecosystem and minimize undesired effects associated with the use of synthetic dispersants in oil spills.

462488 Optimization of UF Backwash with Real-Time Control of Coagulant Dosing

Abstract:

Over the past two decades, reverse osmosis (RO) has emerged as the leading technology for seawater and brackish water desalination. However, membrane fouling remains a major challenge for robust RO operation. Hence, effective RO feed pre-treatment is vital to long-term operation of RO desalination plants. Amongst existing RO pre-treatment technologies, ultrafiltration (UF) has proven to be effective in producing consistently high quality filtrate for RO feed. One of the major advantages of UF membrane pre-filtration is the ability to reverse flow through an UF module (a process known as backwash) and thereby removing the foulant layer. After UF backwash, a portion of the foulant may remain on the membrane (i.e., not removed by backwash) as irreversible foulant layer. Irreversible fouling is typically removed through chemical washing which can significantly increase RO feed pretreatment. UF filtration performance can be increased via coagulant dosing of the raw feed water. Studies have also shown that UF backwash effectiveness improves as coagulant dose increases up to a certain threshold, beyond which UF backwash effectiveness no longer improves via additional coagulant dosing. Optimization of coagulant dosing is therefore critical in order to reach both optimal UF and backwash effectiveness while reducing coagulant consumption. However, real-time coagulant dose optimization is a major challenge when feed water quality is temporally variable both in the short term as well as seasonally. Under such conditions coagulant dose optimization via static jar tests is inappropriate. Therefore, in the present work, a novel method is introduced to allow real-time adjustment of coagulant dose so as to ensure effective UF backwash even under conditions of changing water feed quality. In the present method, the change in initial UF filtration resistance (post backwash) is used as a metric of UF backwash effectiveness. The above information is utilized by a coagulant dose controller designed to utilize the above metric to determine whether the coagulant is being underdosed or overdosed, and correspondingly adjust the coagulant dose to the UF feed. The controller was implemented in an integrated UF-RO seawater desalination pilot plant (capable of 18,000 GPD production of fresh water). The controller performance was evaluated, in a field study in California, with respect to its performance, including during a storm event in which raw seawater feed quality rapidly worsened, and its ability to reduce coagulant consumption. Results of the study demonstrated that the present approach enabled system operation that was able to adapt to worsening feed water quality conditions while reducing coagulant dose by up to ~29% relative to operation without real-time coagulant dose optimization.

451292 Performance Analysis of Texas Eagle Ford Shale Oil Hydro Fracturing Produced Water Treatment Process

Abstract:

Performance Analysis of Texas Eagle Ford Shale Oil Hydro Fracturing Produced Water Treatment Process

Chris Skrivanos, Nico Dwarica, Cameron McKay, John Allen Floyd and Mahbub Uddin

Trinity University, San Antonio, Texas

Extended Abstract

With the rise of horizontal drilling, hydraulic fracturing has become increasingly common and it is estimated that 90 percent of new wells drilled in the United States are hydraulically fractured. Hydraulic fracturing involves the injection of a very large quantity of high pressure liquid to fracture the rock formation and pry open the fracture. This fracturing allows natural gas/oil to flow freely from the formation into the well for collection. The components used in the hydraulic fracturing process consist primarily of water, sand, salts, gel, diluted acid, surfactant corrosion inhibitor, scale inhibitor, and other additives to aid in the hydro fracturing process. During the hydro fracturing operation when the pressure used to inject the fracturing fluid into the well is released, some of the fluid returns to the surface; this fluid is called "flowback" water. Once gas production begins at the well, all wastewater emerging from the well is called "produced water". Both types of wastewater – flowback and production phase water – contain potentially harmful constituents. These constituents can be broadly grouped into several principle categories: salts, organic hydrocarbons, metals, chemical additives and some naturally occurring radioactive material. Because of these constituents, shale oil/gas wastewater should not be released into the environment without adequate treatment. The purpose of this paper is to report the performance analysis of a water treatment process of Texas Eagle Ford Shale oil hydro fracturing produced. The treatment process consists of four primary unit operations: Adsorption column, Reverse osmosis, Ion-Exchange and Pump. The produced water is first pumped through the adsorption column, then a second high-pressure pump which then forces the water through the reverse osmosis membrane. The permeate of the membrane then flows through two ion exchange columns and received in a collection bin.

Adsorption is the adhesion of atoms, ions, or molecules from a fluid or dissolved solid to an insoluble surface and is used to remove suspended organics like oil and grease as well as dissolved organics like benzene and toluene. Our adsorption column consists of a cartridge and housing unit rated for operating pressures of 90 psi and temperatures no higher than 125 °F. The cartridge was 10" tall and accepted 10" x 4.5" cartridges. Reverse osmosis filters work on the principle of membrane divergence. An extremely high pressure (800 psi in our process) is applied at the housing entrance that forces the fluid through the fine membrane. Due to the pressure spike, particles that are small enough pass through the fine holes in the filter transfer, while the larger particles are rejected and thus removed from the membrane. In our application, the reverse osmosis membrane rejected the ions and other larger particles like metals, minerals, and organics, allowing only water and smaller particles to remain. The membrane we used is a FilmTec Sea Water Membrane designed to handle increased salinity (greater than 10,000 ppm). The reverse osmosis unit operation used had a flow ratio of permeate to retetante of 73 to 27, which was determined by a series of mass balances across the unit operation. We used a mixed ion exchange in order to remove any remaining positive or negative ions. Our process consists of two ion exchange columns in parallel, both of which were 20" vertical columns housed in separate units. The process uses two pumps. The first pump transferred the water from the tank to the adsorption column at a low pressure. The second pump primed the system to the operating pressure, 800 psi, to transfer the produced water through the reverse osmosis membrane and the ion exchange.

Each water sample analyzed was obtained from one of 4 major positions. Position 1: the untreated water, Position 2: after the activated carbon operation, Position 3: after the reverse osmosis operation and Position 4: after the ion exchange operation. Produced water processed by the system once is defined as first iteration treated water. Produced water processed by the system twice is defined as second iteration treated water. The first iteration sample collection began after flowing untreated produced water through the system for 60 seconds. The second iteration was run identically to the first iteration except produced water that had previously been treated by the system was used instead of untreated produced water. Once collected, samples were labeled according to the iteration, position and the trial number of extraction.

In order to assess the effectiveness of each unit operation and iteration in the wastewater treatment system, a series of experiments were designed to monitor the flow ratio, total dissolved solids present in solution, pH, and change in organic content. Inductively Coupled Plasma Optical Emission Spectrometry (ICP-OES) was implemented as a means to measure the metallic and salt content present in the water at various purification steps. The initial metallic content was determined to be 184.45 ppm, and the initial salt concentration was determined to be 2262.26 ppm for the produced water. A simple pH test using a probe was conducted on samples to monitor the change in acidity of the water as it passed through each unit operation, and the initial pH of the produced water was determined to be 4.25. Gas chromatography measurements were implemented in order to monitor the changing levels of total organic compounds present in the water.

Analysis of Results

First iteration ICP-OES results showed that the initial salt content of the water was 2262.26 ppm and the heavy metal content was 184.45 ppm. The carbon adsorption column removed 99.01 percent of the initial salt content and 98.5 percent of the initial heavy metallic content. The reverse osmosis membrane removed an additional 0.14 percent and 0.16 percent of the initial salt and heavy metal content respectively. After the ion exchange column approximately 100 percent of the total salt content was removed and 99.84 percent of the total heavy metals were removed, leaving a heavy metal concentration of 0.29 ppm and the salt concentration of essentially 0 ppm. Analysis of the second iteration treatment indicated the carbon adsorption column added 0.37 ppm of heavy metal and 2.09 ppm of salt to the water. The reverse osmosis membrane removed 0.29 ppm of heavy metals and added 0.85 ppm of salt content. The ion exchange column removed 0.04 ppm of heavy metals and 1.55 ppm of salts, bringing the final concentration of the second iteration treated water to 0.32 ppm of heavy metals and 1.39 ppm of salts. The first iteration pH results showed that the produced water had an initial pH of 4.25. The carbon adsorption column increased the pH to 8.26, the reverse osmosis membrane lowered the pH to 7.24, and the ion exchange column lowered the pH to the final value of 5.31. Second iteration treated water pH results showed that the carbon adsorption column increased the pH from 5.31 to 8.27. The reverse osmosis membrane decreased the pH to 7.24. The ion exchange column brought the pH to a final value of 6.49. The produced water contained a plethora of unknown organic constituents and it would be unpractical to identify each individual constituent to generate a calibration curve. So instead of measuring the changing concentrations of each constituent, the GC measurements were used to observe the change in the overall organic content in the water, relative to position 1 of the first iteration which is representative of the initial organic content of the produced water. The first iteration GC results showed that the carbon adsorption column removed 48.4 percent of the initial organic content, the reverse osmosis membrane removed an additional 28.7 percent of the initial organic content, and the ion exchange column removed 6.4 percent of the total organic content; the system overall removed 83.5 percent of the initial organic content.

Conclusions and Recommendations

The produced water treated in our process in a single iteration met the EPA standards for acceptable levels of metallic and salt content for reclaimed water, however the pH is at the lower end of the standard. The first iteration treatment reduced the metallic content by 99.84 percent, reduced the salt content by essentially 100 percent, raised the pH from 4.25 to 5.31 and reduced organic content by 83.5 percent. With a second iteration, the treated water pH is within the acceptable range for reclaimed water. The second iteration was shown to be unnecessary since the metallic and salt concentration changed negligibly, and the organic content increased. The pH was raised to acceptable levels, but the addition of a controlled injection pH adjustment unit operation would allow the pH to be raised without the need for a second iteration, making the produced water treatment process more efficient. This change to the process would generate treated water that meets the EPA drinking water standard. Our analysis indicates that the wastewater treatment processes has applications in commercial and residential areas where hydraulic fracturing is common, to treat wastewater that meets EPA standards.

461844 Macroalgae and Chitosan-Macroalgae Biocomposite As Potential Adsorbents of Water-Soluble Hydrocarbons: Effect of pH, Organic Matter and Ionic Strength

Abstract:

Macroalgae and Chitosan-macroalgae biocomposite as potential adsorbents of water-soluble hydrocarbons: Effect of pH, organic matter and ionic strength.

Abstract

Nowadays the recovery of low-molecular aromatic hydrocarbons (HC) released into natural bodies of water continues to be a challenging task. These contaminants bring sever consequences to the environment and to the human health. The oil spill cleanup strategies are primarily design to deal with the heavy fractions accumulated in the water surface. Unfortunately, very limited information is available regarding the treatment of dissolved fractions. Biosorption on macroalgae (Ma) biomass seem to be a potential alternative to overcome disadvantages associated to high-priced costs and complexity of activated carbon production. Three seaweeds of a representative sample of brown, green and red macroalgae (*Macrocystis pyrifera*, *Ulva expansa* and *Acanthophora spicifera*, respectively) were evaluated to remove benzene and toluene from water, which are the most soluble hydrocarbons. Toward this objective, the influences of temperature and ionic strength were also determined. Raw biomasses were characterized by different physical and chemical techniques to assess their potential as biosorbents and the mechanisms involved. The organic and inorganic dissolved fractions were also considered. The characterization techniques mainly identified carbohydrates, lignin and proteins. *M. pyrifera* biomass registered the highest removal capacities: about 112 and 28 $mg \cdot g^{-1}$ for benzene and toluene, respectively, probably attributed to alginate and lignin content. It was found that the hydrocarbons biosorption affinity was not affect up to an ionic strength of 0.6 M, due to the biosorbents pore occlusion by water clusters. For

benzene and toluene nonpolar solutes, the biosorption mechanism is an addition of simultaneous interactions between the sorbents and the chemical constituents of the cell wall, presumably by London (dispersion) forces onto â€"OH and â€"NH active sites from polysaccharides and proteins and hydrophobic interactions from lignin and lipid fractions.

Despite these advantageous properties, the physical characteristics (small particle size, low strength and density) of such biomaterials are not viable for a continuous process operation and make biomass difficult to apply. Therefore the development of innovative low-cost techniques of immobilized biosorbents with special attention to increase their effectiveness in the biosorption process is mandatory. The biopolymer-biopolymer interactions between positive and negative macromolecules can enhance the compatibility and stability of biosorbents, and these new structures are called polyelectrolytes complexes (PEC). The linear amino polycationic compound of chitosan (Ch) possess remarkable quantity of amine and hydroxyl groups that interacts with negatively charged substances with excellent chelation behavior and high affinity toward pollutants like metals and dyes. On the other hand pectin (Pe) is a negatively charged polysaccharide composed mostly by esterified D-galacturonic acid residues linked together by Î±-(1-4) chain. Pectine molecule is also used in the food industry for its well-known thickening, stabilization and gelling properties.

Â The main objective of this research was to explore the performance of Ma and PEC complexes of different proportions of Ma/Ch/Pe on biosorption capacity of the two main soluble hydrocarbons (benzene and toluene) in water. The biocomposite synthesis was optimized by application of the factorial design and response surface methodology. Moreover, a through chemical and physical characterization analysis, by textural properties, potentiometric titrations, elemental content, chemical stability, KBr-FT-IR and TGA analysis, were performed to explain the adsorption mechanisms.

458236 Concentrating 17 α-Ethinyl Estradiol (EE2) Using PES Hollow-Fibre Membranes: Experimental Exploration and Molecular Modelling

Abstract:

The 17Î±-ethinyl estradiol (EE2) is a synthetic estrogen derived from 17Î²-estradiol (E2), the major endogenous estrogen in humans. EE2 is widely used as an active ingredient for birth control and other hormonal treatments. EE2 is excreted into the sewage and the environment mainly via urine. Wastewater treatment plants are usually ineffective in removing EE2 from water which results in the release of EE2 into the environment. The concentration of EE2 in surface water bodies are typically at sub-ppb levels. Even at such low concentration, several studies have reported negative environmental effects of EE2 including feminization of aquatic animals. At present, it is not known if EE2, at such low concentration, has any adverse effect on human or other mammals over short and long term exposure. Developing novel strategies for the efficient removal and treatment of EE2 from water has become an imperative in water reclamation and reuse. Here, we report a novel approach whereas a widely used polyethersulfone (PES) membrane was used to concentrate EE2 through adsorption and solvent elution. The concentrated EE2 can then be further treated with high efficiency using approaches such as biodegradation. Molecular mechanisms underlying the interaction between EE2 and PES membrane was elucidated using molecular dynamic simulation.

454263 Natural Organic Matter Fouling Behaviors on Superwetting Nanofiltration Membranes

Abstract:

Natural organic matter fouling behaviors on superwetting nanofiltration membranes

Linglong Shan, Hongwei Fan, Hongxia Guo, Shulan Ji and Guojun Zhang

Beijing Key Laboratory for Green Catalysis and Separation, College of Environmental and Energy Engineering, Beijing University of Technology, Beijing, China. Email: shanlinglong@126.com

In natural water, natural organic matter (NOM) components have been identified as precursors of disinfection by-products. Nanofiltration has been widely recognized as a promising technology to remove NOM from natural water. Currently, the most challenging aspect is the often inevitable occurrence of membrane fouling, which is caused by NOM accumulation on the membrane surface, as in NOM removal in drinking water treatment and production. NOM fouling is considered to be governed by the interface relationship between NOM and the membrane surface in aqueous solution. Much work has been carried out, but the effect of hydrophilicity/-phobicity of NOM and the membrane surface on fouling behavior is still a controversial issue. [1-7] The controversy originates from the unilateral consideration of membrane wettability or foulant hydrophilicity/-phobicity, when in fact both need to be considered. In our study, both superhydrophilic and superhydrophobic membranes with stable structure have been successfully obtained using two facile hybrid approaches (see Fig. 1). The fouling behavior of NOM on superwetting nanofiltration membranes has been extensively investigated using humic acid (HA: model foulant) and real natural water. The extended Derjaguin-Landau-Verwey-Overbeek (XDLVO) approach and nanoindentor scratch tests suggested that the superhydrophilic membrane had the strongest repulsion force to HA due to the highest positive ¦¤G^{TOT} value and the lowest critical load. Excitation emission matrix analyses of natural water also indicated that the superhydrophilic membrane showed resistance to fouling by hydrophobic substances and therefore high removal thereof. Conversely, the superhydrophobic membrane showed resistance to fouling by hydrophilic substances and therefore high removal thereof (see Figs. 2 and 3). Long-term operation suggested that the superhydrophilic membrane had high stability due to its anti-NOM fouling capacity, which is crucial for practical application. Based on the different anti-fouling properties of the studied superwetting membranes, a combination of superhydrophilic and superhydrophobic membranes was examined to further improve the removal of both hydrophobic and hydrophilic pollutants. The R_{UV254} and R_{DOC} could be increased to 83.6% and 73.3%, respectively, by a combination of superhydrophilic and superhydrophobic membranes. In view of the hydrophilic and hydrophobic NOM fouling behaviors on superwetting membrane surfaces, the superwetting strategy can be regarded as a very important approach for obtaining a membrane with both high NOM rejection and good anti-NOM fouling properties.

Keywords: Superhydrophilic, Superhydrophilic, Nanofiltration, NOM, Fouling

References

[1] L. Shan, H. Fan, H. Guo, S. Ji, G. Zhang, Natural organic matter fouling behaviors on superwetting nanofiltration membranes, Water res., 93 (2016) 121-132.

[2] L. Shan, H. Guo, Z. Qin, N. Wang, S. Ji, G. Zhang, Z. Zhang, Covalent crosslinked polyelectrolyte complex membrane with high negative charges towards anti-natural organic matter fouling nanofiltration, RSC Adv., 5 (2015) 11515-11523.

[3] M. Ng, S. Liu, C.W.K. Chow, M. Drikas, R. Amal, M. Lim, Understanding effects of water characteristics on natural organic matter treatability by PACI and a novel PACI-chitosan soagulants, J. Hazard. Mater., 263 (2013) 718-725.

[4] J.A. Nilson, F.A. DiGiano, Influence of NOM composition on nanofiltration, J. Am. Water Works Assoc., 88(5) (1996) 53-66.

[5] H. Yamamura, K. Okimoto, K. Kinura, W. Yoshimasa, Hydrophilic fraction of natural organic matter causing irreversible fouling of microfiltration and ultrafiltration membranes, Water Res., 54 (2014) 123-136.

[6] L. Zhang, G. Chowdhury, C. Feng, T. Matsuura, R. Narbaitz, Effect of surface-modifying macromolecules and membrane morphology on fouling of polyethersulfone ultrafiltration membranes, J. Appl. Polym. Sci., 88 (2003), 3132-3138.

[7] A. Tiraferri, Y. Kang, E.P. Giannelis, M. Elimelech, Superhydrophilic thin-film composite forward osmosis membranes for organic fouling control: fouling behavior and antifouling mechanisms, Environ. Sci. Technol., 46 (2012) 11135-11144.

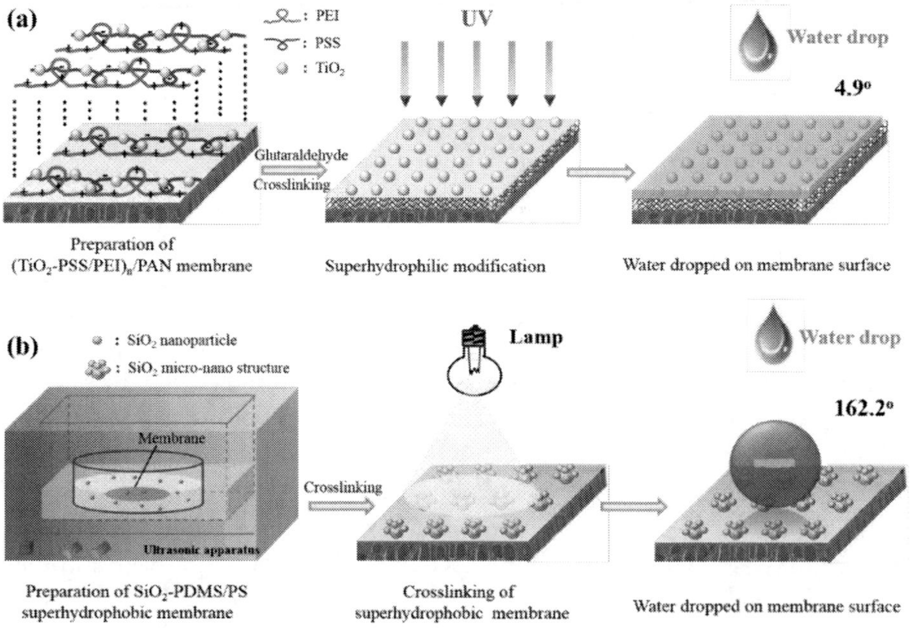

Figure 1. (a) Schematic illustration of the preparation of the superhydrophilic membrane and its surface wettability; (b) schematic illustration of the preparation of the superhydrophobic membrane and its surface wettability.

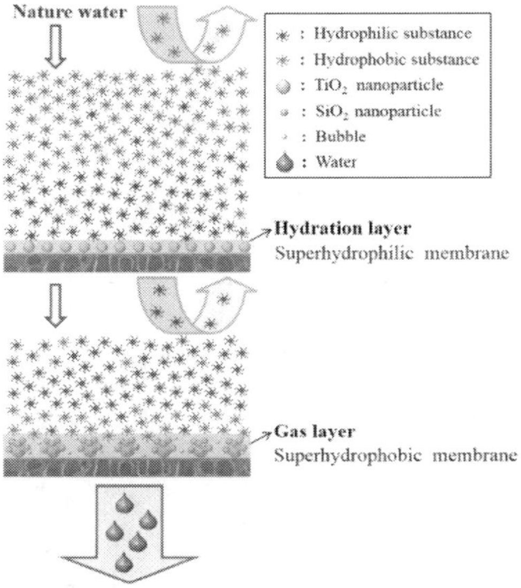

Figure 2. Schematic illustration of different anti-fouling properties of the studied superwetting membranes. Superhydrophilic membrane showed resistance to fouling by hydrophobic substances, while the superhydrophobic membrane showed resistance to fouling by hydrophilic substances.

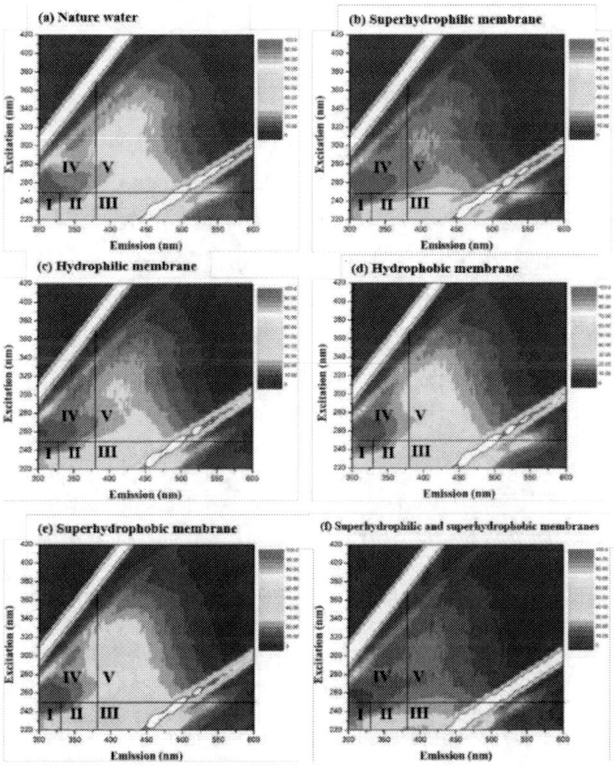

Figure 3. EEMs for (a) natural water; (b) natural water nanofiltrated by superhydrophilic membrane; (c) natural water nanofiltrated by hydrophilic membrane; (d) natural water nanofiltrated by hydrophobic membrane; (e) natural water nanofiltrated by superhydrophobic membrane; (f) natural water nanofiltrated by superhydrophilic membrane-superhydrophobic membrane.

Experimental and Simulation Analysis for Optimization of Solar Desalination System Using Humidification-Dehumidification Method: Open and Close Air Cycle

Nickyar Ghadirinejad [1], Farshad Farshchi Tabrizi [2], Mohammmad Mehdi Zarei [3], Taher Chegini [4], Roberto Moreno-Atanasio [5]

[1] Chemical Engineering Department, University of Sistan and Baluchestan, Zahedn, Iran
[2] Chemical Engineering Department, Shiraz University, Shiraz, Iran
[3] Chemical Engineering Department, Khazar University, Baku, Azerbaijan
[4] Civil Engineering Department, Khazar University, Baku, Azerbaijan
[5] Chemical Engineering Department, Newcastle University, Callaghan, Australia

Abstract:

Over the last decades, desalination of brackish water and seawater has been studied extensively for producing fresh water. Desalination technologies are divided into two major categories: Thermal and Membrane. Humidification-Dehumidification (HDH) which is the subject of the present research, is one of the thermal technologies. It is inspired by the rain cycle and is based on evaporation and consecutive condensation.

In a series of experiments, the processes of production of fresh water has been investigated under various operational parameters such as temperature and flowrate of cooling water, humidifier intake from water boiler, air flowrate and height of packed bed column. Subsequently, an appropriate mathematical model of the system has been developed and validated using the collected data. Furthermore, based on the developed model a series of simulations has been carried out in ASPEN Plus in order to obtain optimum physical parameters of various parts of the desalination system.

The experimental results showed that under constant heat flux and optimum air flowrate, three parameters increase the production of distilled water the most: increasing flowrate of cooling water, decreasing flowrate of water inlet and increasing temperature. Moreover, although increasing flowrate of cooling water leads to an augmentation in the distilled water flowrate, but this effect is mitigated at high rates.

Furthermore, based on the obtained overall heat transfer coefficients of the humidifier and condenser, the most reliable Height Equivalent to a Theoretical Plate (HETP) model of random packing is found to be Mangers & Ponter. The estimation error of the model in predicting the height of packed bed was less than 4%. Based on the results of this model, closed air system increases the efficiency of the production of distilled water by 15% in comparison to open air. This increase is the result of prevention of wasting of air humidity in the condenser outlet.

Finally, optimum parameters of operational units are obtained by applying a Differential Evolution (DE) strategy. The optimum values of flowrate and temperature of cooling water are 0.0298 m^3/hr, 13 °C, respectively. The best value of flowrate of closed air is found to be 0.008075 m^3/hr and of water intake to be 0.021497 m^3/hr. These values produced fresh water with flowrate of 0.3777 m^3/hr.

Keywords: Solar Desalination, Dehumidification, Humidification, Heat Transfer

1- Introduction:

The rapid growth of industries and population, rivers and lakes pollution problems caused by industrial wastewater and flowing municipal wastewater into freshwater resources have caused a shortage of fresh water [1]. However, it should be noted that the only sources of never-ending water reservoir are oceans; water salinity reduction to overcome the shortage of fresh water, is a more applicable solution [2]. Desalination includes many different methods and required energy for this procedure also encompasses a wide range. The heat produced by fossil fuels and/or the excess heat-released of nuclear reactions, renewable energies such as ocean waves or solar energy, all can supply heat power for water desalinisation plants. Desalination technologies are divided into two categories: Thermal and membrane methods. Water is heated in thermal processes and after evaporation, fresh water is produced via condensation. In membrane methods, thin permeable membranes are used to separate feed into both fresh water flow and more saline flow [3].

In recent years, Humidification-Dehumidification (HDH) system, as a new method has been investigated; through this system, air flows as the intemediate fluid. In fact, this method is a combination of membrane processes and thermal distillation. It operates based on the mass diffusion principles and uses cooling air due to water evaporation and humidification. Fresh water is resulted via dehumidification (condensation of humid air). The obvious benefit of HDH is to be one of the most cost-effective desalination mechanism regarding to terms of low temperature and pressure [4]. The investigated method includes solar collector/s, humidification tower/s (evaporation unit), dehumidification tower (condensation unit) which even to achieve higher efficiency, it could include heat recovery unit as well. In order to increase the efficiency of heat transfer, at first, water is passed through the condenser coil, by heat absorption from hot air, water is pre-heated. Afterward, the pre-heated water is led into a solar heater and an additional heater (auxiliary heater) subsequetly to reach a higher temperature. In the next step, hot water is showered into the humidification tower.

2- Closed-air, open-water (CAOW) water-heated systems:

Alluded humidification tower including random packings/ Bubble Columns/ Wetted-wall Columns [5], showered with hot water, gets totally moist using counter-current air flow from the botton; subsequently, after heat absorption, hot air is guided through the condenser to produce condnesate-water (fresh water). In closed air cycle, after passing the condenser, the air is recycled into the system. HDH method has been studied by Al-Hallaj et al. [6,7], Ben-Bacha et al. [8], Garg et al. [9] ،Nafey et al. [10] ،Nawayseh et al. [11], [12] Muller-Holst.

In order to achieve optimum operational conditions and achieve the best yield, many research projects have been studied [13,14,15,16,17] and various mathematical models have been proposed considering measurement of mass and heat transfer coefficients [13], impacts of input parameters into the HDH system, mass and energy balance equations, Consideration of waste heat during HDH process and finding a solution due to heat recovery [18]. To simulate a system including both a flat-plate water-heater and a double-pass solar air heater, assuming some simplified assumptions, Yildirim and Solmus [19] proposed an optimized mathematical model by using fourth order Runge-Kutta solution method. Cost-estimation of produced fresh water via an elaborated, multiple solar desalination system [20] and via a novel solar HDH system [21] has been investigated.

3-HDH Apparatus:

In the study, the humidifier consisted of emedded random packing (Pall ring type with 25 mm nominal diameter). Water was showered from the top and air wsa flown counter-currently. The coil-condensor was a single 8-meter-long cupper pipe. For circulation of air in side the HDH system, two fans were used, without any direct contact between motors and humid air.

As it shown in figure 1, intake salt water in low temperature (as feed water into the HDH system) entered into the condenser's shell, and after relative pre-heating subsequently entered into the solar heat collector. During experimental research, to achieve higher temperatures an electrically additional heater was used. Heated saline water was showered inside the humidifier. Water flow rate was regulated by passing through some valves and rotameters to reach desired amounts. Air-flow rates adjusted through entire HDH system using fans' regulators. By directly counter-current contact between hot saline water and air in random-packed bed, the air temperature got augmented. Consequently, it tended to embrace greater capacity to absorb more moisture. Passing through the dehumidifier, fresh water was by-passed and collected; outlet air was recycled in closed air system and was blowed from the bottom into the humidifier.

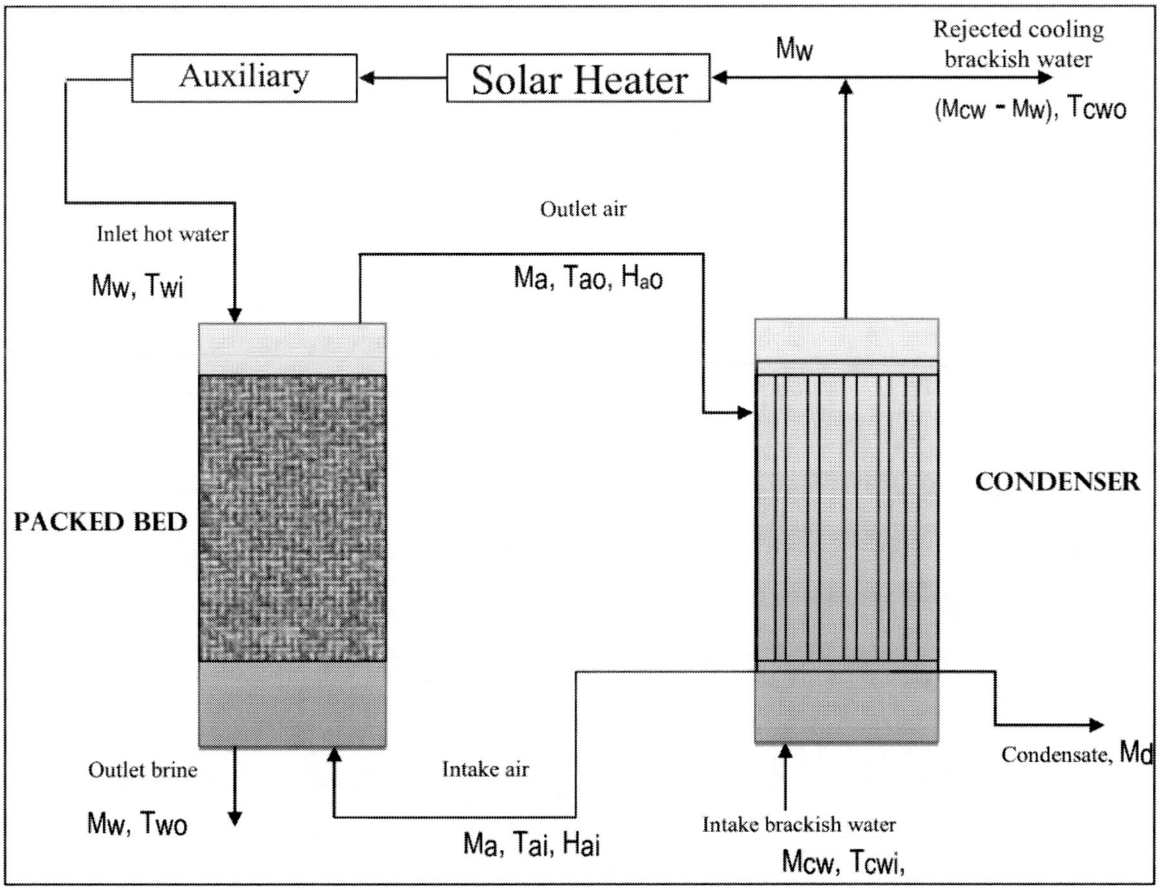

Fig. 1: Layout of HDH system

For experimental data collection, the system was ran continuously for 15 to 20 minutes in the same mode, reaching considered steady-state temperatures and humidity amount in each test; then tests were performed. Each of the experiments were repeated at least three times to ensure their correction.

4- Modeling of humidification-dehumidification system:

In order to determination of diameter and Height Equivalent to a Theoretical Plate (HETP), based on the obtained overall heat transfer coefficients of for the random-packed-bed humidifier and condenser [22], Considering given data about Rasching ring and Pall ring packings [5] following models have been investigated: Shulman et al. [23], Shulman & de Grouff [24], Van Krevelen & Hoftijzer [25,26], Mangers & Ponter [27], Shi & Mersmann [28], Billet and Schultes [29], Onda et al. [30].

The estimation HTEP model of **Mangers & Ponter**:

$$k_L = \frac{(3.22\times10^{8})Sc_L^{0.5}D_L}{a_e(1-cos\theta)}\left(\frac{\rho_L u_L}{\mu_L}\right)^{\alpha}\left(\frac{\rho_L^2\, g\, h_{pe}}{\mu_L^2}\right)^{0.27} \qquad (1)$$

As α in the above equation is calculated by:

$$\alpha = 0.49\left[(1-cos\theta)^{0.6}\left(\frac{\rho_L\sigma_L^3}{\mu_L^4 g}\right)^{0.2}\right]^{0.108} \qquad (2)$$

$$k_G = 0.2\,\frac{D_G}{d_c}\left[\frac{\rho_L u_L}{u_L a_p}\right]^{0.8} Sc_G^{\frac{1}{3}} \qquad (3)$$

Additionaly, for comparison the results obtained from the mathematical modeling and practical experiments, both physical and chemical operational behaviors of HDH units -at steady state conditions- have been investigated by modeling in ASPEN Plus. Initial feed's conditions to HDH system, constituted 3884.8 ppm salinity, Ca^{2+}, Mg^{2+}, K^{+}, Cl^{-}, HCO_3^-, SO_4^{2-}, NO_3^- Na^{+} aqueous solutions were imported to ASPEN Plus model. Enick's Redlich-Kwong-UNIFAC equations [31] and Pall ring packings were selected for the mentioned processes.

In preliminary calculation, the capacity of existing packing in the dumidifier was assumed 10 HETP. Regrading experimental results, output data from run of each stage in ASPEN Plus were compared.

4-1- Modeling of HDH system in open and close air cycle:

Effects of intake water flow rate and its temperature entering the humidifier, air and cooling water flow rates, and effect of packed-bed height on the yield of desalinated water amounts have been studied by developing mass and energy transfer balances for overally and even on every sinlge unit in the HDH system.

$$Mcw.Cp.Tcwi - (Mcw - Mw).Cp.Tcwo + Q_{Heater} = Mw.Cp.Two + Md.Cp.Ta \quad (4)$$

Saturated-air, as a main simplifying assumption was considered to solve formulated equations. On the contrary, unlike precedent models, the term of air solubility in outlet saline water, as well as unsteady air and water flow rates recycling the entire HDH system was considered [32].

To improve derived equations and to reach more realistic modeling in the range of $0 \leq S_p \leq 16000$ ppm and $10 \leq T \leq 180$ °C, correlations of seawater density (ρ_{sw}), specific heat (C_p), seawater thermal conductivity (K_{sw}) [32] were applied:

$$\rho_{sw} = 10^3(A_1.F_1 + A_2.F_2 + A_3.F_3 + A_4.F_4) \quad (5)$$

Which A_n and F_n are specific variables proposed by Isdale and Morris [33].

Evaporation latent heat (λ) (which is T in K degree)

$$\lambda = 2501.89 - 2.407 \times T + 1.1922 \times 10^{-3} \times T^2 - 1.586 \times 10^{-5} \times T^3 \quad (6)$$

Specific heat coefficient of air (T in K degree)

$$C_{pa} = 1.03409 - 2.84887 \times 10^{-4}.T + 7.816818 \times 10^{-7}.T^2 - 4.970786 \times 10^{-10}.T^3 + 1.077024 \times 10^{-13}.T^4 \quad (7)$$

To calculate air density with respect to Al-Sahali & Ettouney's proposal [13] for HDH desalination plant, air assumed as an ideal gas.

$$\rho_a = \frac{P.M_{wei}}{R.T} \quad (8)$$

Thermal conductivity coefficient of air:

$$K_a = -2.276501 \times 10^{-3} + 1.2598485 \times 10^{-4}.T - 1.4815235 \times 10^{-7}.T^2 + 1.73550646 \times 10^{-10}.T^3 - 1.066657 \times 10^{-13}.T^4 + 2.47663035 \times 10^{-17}.T^5 \quad (9)$$

In order to calculate the amount of dissolved air in the water (dependent on its pressure and temerature) in the humidifier [34] Henry's constants for air solubility in water (K_{sol}) were considered too.

Newton's method was used to solve mentioned non-linear equations, using some most probable intial guesses via experimental data. For optimization of outlet fresh water dependent on Q, Mw, Ma, Tcwi and Mcw values, Differential Evolution (DE) strategy was applied to find optimum values. DE strategy [35], one of genetic algorithm methods, via successive tries to generate correct interpolations was coded in MATLAB software.

In the same method, calculations of open-air cycle in the HDH system were performed by rewriting similar equations as above.

5- Results and discussion:

Keeping constant given thermal energy to the inlet water (including solar heater and additional heater), several sets of experimental data were collected in various conditions. By changing the initial independent parameters, trends of intermediate variables and the final variable (amount of produced fresh water) were evaluated.

5-1 Effect of inlet water flowrate and its temperature on deslination production rate:

At constant thermal energy, inlet water flow rate and its temperature were realized as two dependent vaiables. In this series of tests, parameters of air flow, cooling water flow and packed-bed height were kept constant; and consequently, effects of two mentioned parameters of inlet water entering to the humidifier on the yield of HDH system were investigated (Fig.2). As it is shown, at constant conditions, by increasing the temperature of inlet water, production rate of fresh water increases. It is noticible that during experiments, relative humidty of circulating air was checked continuously and it was near to saturation state (97.33%).

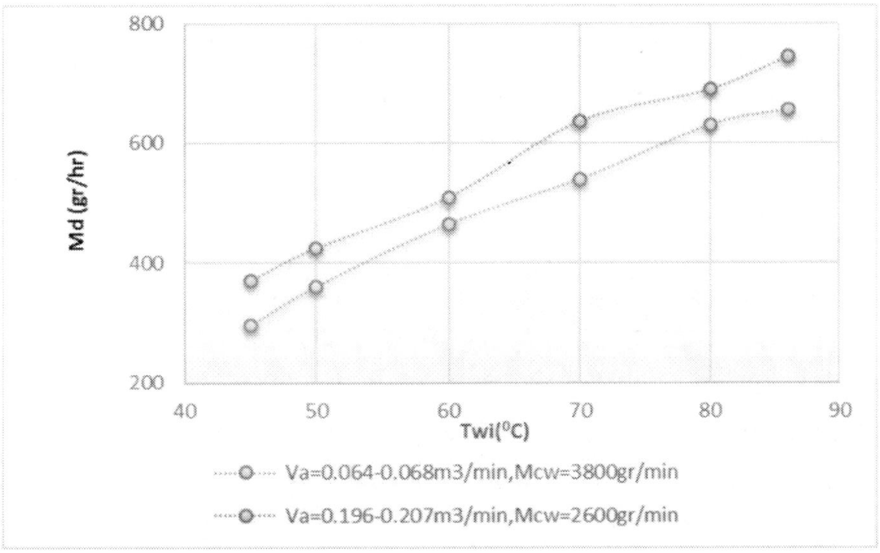

Fig. 2: Effect of temperature of inlet water on produced fresh water

5-2 Survey on air flow rate, considering effects of cooling water flowrate on yield of the HDH system:

As it presented in Figure 3 for three flow rates of circulated-air, the temperature of the exhausted air from the tower is a dependent variable of cooling water flow rate. At constant air flow rate, increasing in water temperature leads to augmentation of air temperature and consequently fresh water production.

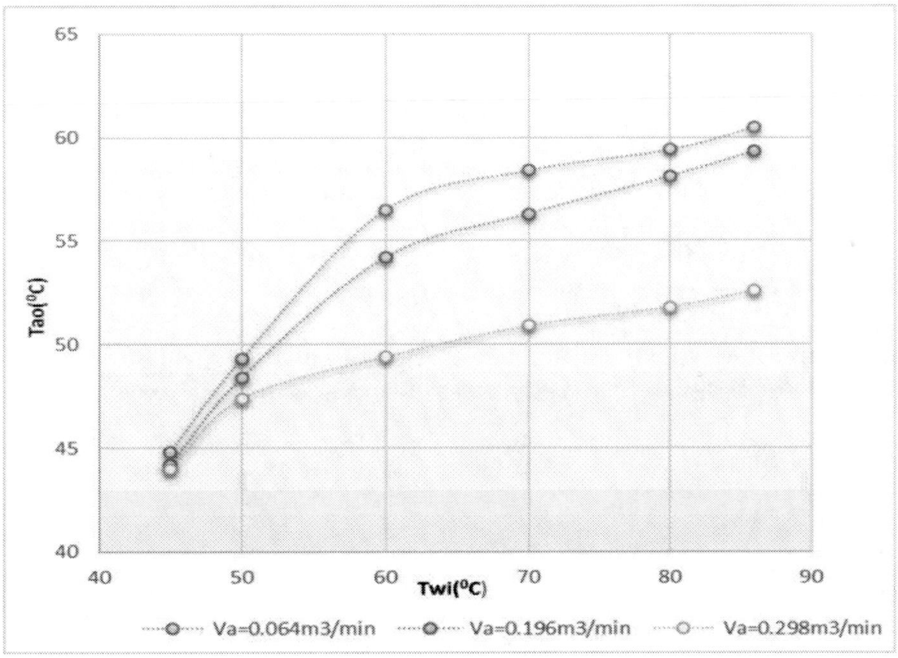

Fig. 3: effects of changes in Twi on Tao

In the present HDH design, to avoid wasting moisture, transfered by exhaust air flow from the condenser, and to spend less energy, close air cycle was contrived. It is noteworthy that the validity of this assumption has been proved by comparing output-results of close and open air models. Experimental tests were conducted in three different ranges (0.064-0.068 $\frac{m^3}{min}$, 0.196-0.207 $\frac{m^3}{min}$ and 0.298-0.310 $\frac{m^3}{min}$). Regarding Fig. 3, 4 and 5, range of 0.196-0.207 $\frac{m^3}{min}$ was determined as the optimal one for the air flow. Obviously, increase of air velocity causes decline of outlet air temperature in the HDH system. Besides, it leads to reduction in air settelment time in the humidifier and finally shrinkage of distilled water.

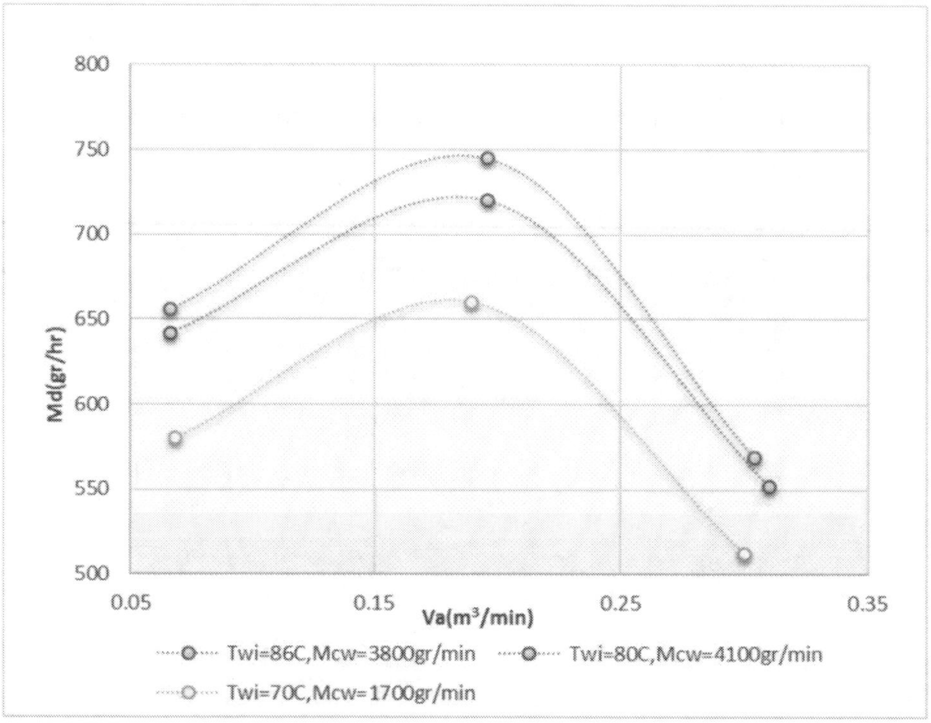

Fig. 4: Trends of M_d in three different air velocities at 70, 80 and 86 °C.

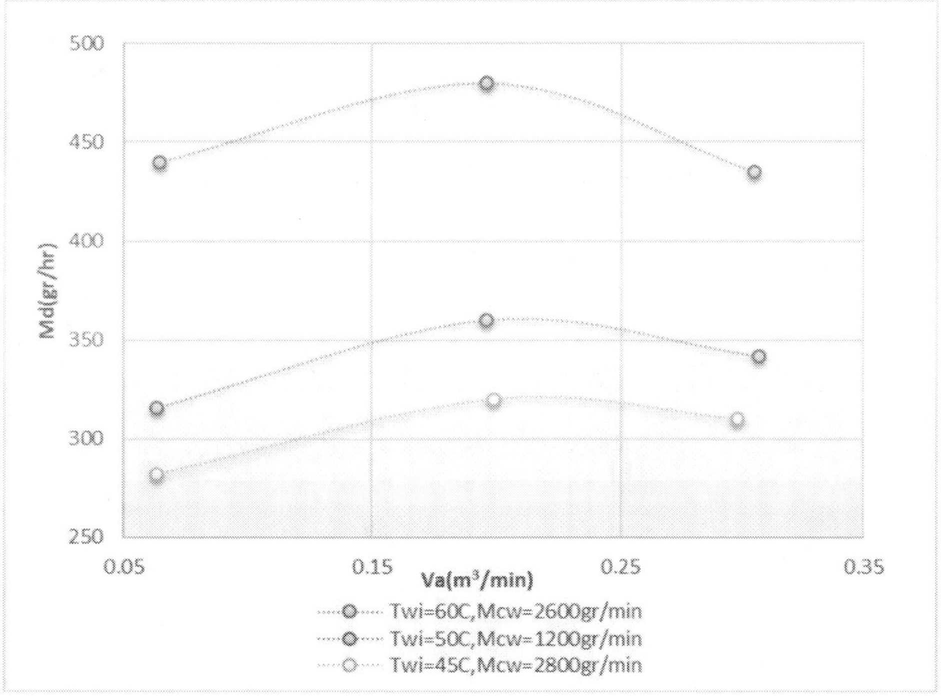

Fig. 5: Trends of M_d in three different air velocities at 45, 50 and 60 °C.

5-1 Distinguishing between results of open and clos air cycle in the investigated HDH system:

This comparibility is shown in table 1. It shows 15% higher productivity of fresh water in close air cycle than production rate in open air cycle. The most probalbe justification for this conlcusion could be waste-heat recovery and lack of moisture loss in the exhaust air from the condenser and recycling to the processes.

Table 1: comparability of 4 samples for open and closed air cycle

Closed cycle				**Open cycle**			
Twi (°C)	Tao (°C)	Tcwo (°C)	Md (m3/hr)	Twi (°C)	Tao (°C)	Tcwo (°C)	Md (m3/hr)
71.6	60.4	32.3	1.70E-04	70.4	56.2	32.1	1.01E-04
91.3	66.5	32.6	1.95E-04	88.4	61.4	31.1	1.50E-04
40.4	41.6	24.9	1.28E-05	37.6	35.1	25.6	1.23E-05
80.2	44.6	33.8	1.71E-04	79.4	39.0	33.2	1.27E-04

5-2 HETP investigation and HDH optimization:

To achieve optimal HETP for humidification tower, several experiments have been executed (Fig. 6). At constant practice conditions, using k_L and k_G equations, the most reliable HETP model has been proposed; Mangers & Ponter's model was determined as the most precise one (Table 2).

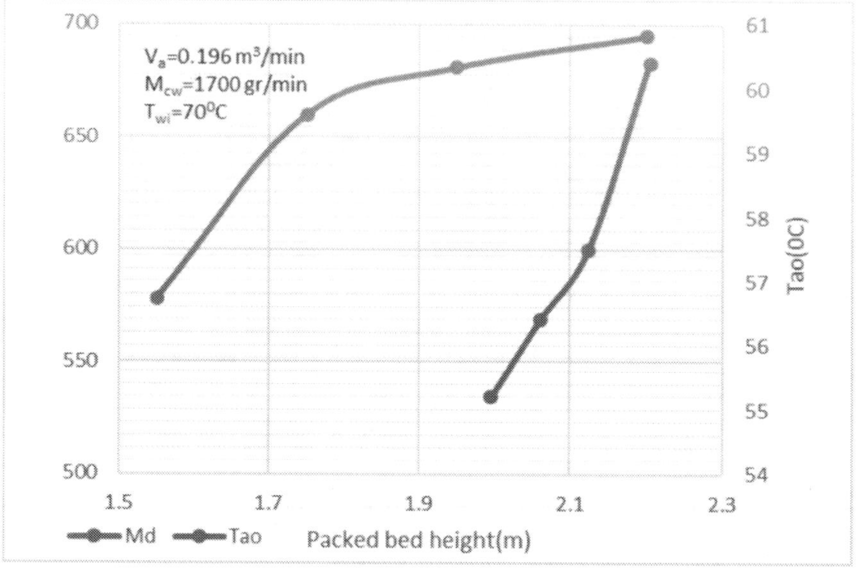

Fig. 6: effects of Packed bed height on M_d ($\frac{gr}{hr}$)

Table 2: estimating HETP models

Model	HETP (m)
Shulman et al.	0.042601
Shulman & de Grouff	0.047809
Van Krevelen & Hoftijzer	0.394546
Mangers & Ponter	0.511026
Shi & Mersmann	0.437404
Billet and Schultes	0.407311
Onda et al.	0.660333

Furthermore, due to admit validity of collected data modeling and simulation in ASPEN Plus have been done. At first, 10-equivalent-step was presumed. After every run of functions implemented in ASPEN Plus and comparison between outcome data and experimental values, it was observed that a tower including two or three steps can help us to meet our expected moisture balance (one single run is shown in table 3).

Table 3: ASPEN Plus simulation for 5-equivalent-setps of HETP

Exp. Cond.	Stages	Twi (°C)	Two (°C)	Tao (°C)	Tcwo (°C)	Md (mol/s)	Mw (mol/s)	Md/Mw
Mw=0.00305 (m3/hr)	5	82.31	42.23	66.48	32.9	0.04489	0.6095	0.07365
Ma=0.00108	4	82.17	43.77	65.9	32.9	0.04334	0.6095	0.07112
(m3/hr) Tcwi=28(°C)	3	81.95	46.18	64.92	32.9	0.04089	0.6095	0.06709
Tai=32.9(°C)	2	81.56	50.26	62.97	32.9	0.03647	0.6095	0.05985
	REAL	80.4	36.2	58.1	30.2			0.06011

According to the Eckert's proposed coefficients [36] and the use of Pall ring packing with 25 mm nominal diameter, the effective height of humidifier was suggested about 170 cm.

Using Differential Evolution (DE) strategy and via differing the number of dependent variables (such as air flow rate, T_{cwi}, M_w and finally M_{cw}) individually and also simultaneously, oprimized model was correlated successfully. Practical values which were resulted optimum M_d=0.3777 $\frac{m^3}{hr}$ were M_{cw}=0.0298 $\frac{m^3}{hr}$ at T_{cwi}=13 °C, wiht M_a=0.008075 $\frac{m^3}{hr}$ and M_w= 0.021497 $\frac{m^3}{hr}$.

6- Conclusion:

One of the most applicable methods for desalination brackish water is humidification-dehumidification system. This thermal method has been studied on a pilot system. After experimental data collecetion due to data validity proof and also finding an optimum fresh water rate, two mathematical models, one on open air cycle and another one on close air cycle have been derived. Furthermore, HETP for the humidifier was studied by comparing experimental data resulted from various tower's height and data from simulation of HDH system in ASPEN Plus; Estimation error of the HETP modelfor Pall Ring packed bed was less than 4%. Yield in closed air system was shown 15% better than open air system.

Convincingly, using of DE strategy, optimal parameters to obtain the choisest yield (0.3777 m^3/hr) have been provided: 0.0298 m^3/hr, 13 °C respectively for flow rate and temperature of cooling water, 0.008075 m^3/hr air flow rate (in closed air system) and 0.021497 m^3/hr for water inake.

Nomenclatures:

D (m)	diameter	C_{pa} $(\frac{KJ}{Kg.K})$	Specific heat coefficient of air
G(kg/s)	Gas mass flow rate	λ $(\frac{KJ}{Kg})$	Evaporation latent heat
L (kg/s)	Liquid mass flow rate	K_a $(\frac{W}{m.K})$	Thermal conductivity coefficient of air
M (gr/min)	Mass flow rate	μ $(\frac{kg}{m.s})$	*viscosity*
Ma (kg/s)	Intake air mass flow rate at the bottom of humidification tower	ρ $(\frac{Kg}{m^3})$	*density*
Mcw (kg/s)	Mass flow rate of saline water Inlet to cooling tower	ρ_a $(\frac{Kg}{m^3})$	Air density
Md (kg/s)	Fresh water (condensed water) mass flow rate	σ	*Thickness*
Mw (kg/s)	Mass flow rate of rejected water from condenser	***Subscripts***	
Q (w)	Heat received /transferred	*a*	*air*
R.H. (%)	Relative Humidity percentages	*ai*	*Inlet air*
T (K)	Temperature	*ao*	*Outlet air*
Tai *(K)*	Intake air temperature at the top of Humidifier	*cw*	*Cooling water*
Tao *(K)*	Outlet air temperature at the top of humidifier	*cwi*	*Inlet Cooling Water*
Tcwi *(K)*	Cooling water inlet temperature	*cwo*	*Outlet Cooling water*
Tcwo *(K)*	Water temperature of rejected water from condenser	*d*	*Distilled water*
Twi *(K)*	Water temperature of inlet water to humidifier	*G*	*Gas*
Two *(K)*	Water temperature of outlet brine water at the bottom of humidifier	*L*	*Liquid*
V_a $(\frac{m3}{min})$	Volumetric flow rate of air	*w*	*water*
R $(\frac{Pa.m^3}{Kmol.K})$	Universal Gas Constant	*wi*	*Inlet water to Humidifier*
M_{wei}	*Molecular weight*	*wo*	*Outlet water from Humidifier*

References:

[1] Malik, M., Tiweri, G.N., Kumer, A., Sodha, M.S., Solar distillation, New York, pergamon, pp. 3-4,1985

[2] Kalogirou, S.A., Solar thermal Collectors and applications, Progress in Energy and Combustion science, Vol. 30, pp. 231-295, 2004.

[3] Watson, I.C., Morin, O.J., Henthorne, L., Desalting handbook for planners, Aqua Resources International Inc., 2003.

[4] Qiblawy, H.M., Banat, F., solar thermal desalination technologies, Desalination, Vol. 220, pp. 633-644, 2008.

[5] Treybal R.E. Mass transfer operations, 3rd ed., NY: McGraw-Hill; 1980.

[6] Al-Hallaj S, Farid MM, Tamimi AR. Solar desalination with humidification–dehumidification cycle: performance of the unit. Desalination 1998;120:273–80.

[7] Farid MM, Al-Hallaj AW. Solar desalination with humidification–dehumidification cycle. Desalination 1996;106:427–9.

[8] Ben-Bacha H, Damak T, Bouzguenda M. Experimental validation of the distillation module of a desalination station using the SMCEC principle. Renewable Energy 2003;28:2335–54.

[9] Garg HP, Adhikari RS, Kumar R. Experimental design and computer simulation of multi-effect humidification (MEH)–dehumidification solar distillation. Desalination 2002;153:81–6.

[10] Nafey AS, Fath HES, El-Helaby SO, Soliman AM. Solar desalination using humidification–dehumidification processes. Part II. An experimental investigation. Energy Conversion Management 2004;45(7–8):1263–77.

[11] Nawayseh NK, Farid MM, Al-Hallaj S, Tamimi AR. Solar desalination based on humidification process. Part I. Evaluating the heat and mass transfer coefficients. Energy Conversion Management 1999;40:1423–39.

[12] Muller-Holst H. Solar thermal desalination using the multiple effect humidification (MEH) method. In: Book chapter, solar desalination for the 21st century; 2007. pp. 215–25.

[13] Al-Sahali M, Ettouney HM. Humidification dehumidification desalination process: Design and performance evaluation, Chemical Engineering Journal, 143, 257–26, 2008.

[14] Farid M.M., Parekh S., Selman J.R., Al-Hallaj S., Solar desalination with humidification–dehumidification cycle: mathematical modeling of the unit. Desalination; 151:153–64, 2002.

[15] Yamali C, Solmus I. Theoretical investigation of a humidificationdehumidification desalination system configured by a double-pass flat plate solar air heater. Desalination 205, 163–177, 2007.

[16] Zamen M, Amidpour M, Soufari SM. Cost optimization of a solar humidification–dehumidification desalination unit using mathematical programming. Desalination 239, 92–99, 2009.

[17] Dai Y.J, Wang RZ, Zhang HF. Parametric analysis to improve the performance of a solar desalination unit with humidification and dehumidification. Desalination; 142:107–18, 2002.

[18] W.F. He, L.N. Xu, D. Han. Parametric analysis of an air-heated humidification-dehumidification (HDH) desalination system with waste heat recovery. Desalination 398 (2016) 30–38.

[19] Cihan Yıldırım,Ismail Solmus. A parametric study on a humidification–dehumidification (HDH) desalination unit powered by solar air and water heaters. Energy Conversion and Management 86 (2014) 568–575.
[20] S.W. Sharshir, Guilong Peng, Nuo Yang, Mohamed A. Eltawil, Mohamed Kamal Ahmed Ali, A.E. Kabeel. A hybrid desalination system using humidification-dehumidification and solar stills integrated with evacuated solar water heater. Energy Conversion and Management 124 (2016) 287–296.
[21] Nabil A.S. Elminshawy, Farooq R. Siddiqui, Mohammad F. Addas. Experimental and analytical study on productivity augmentation of a novel solar humidification–dehumidification (HDH) system. Desalination 365 (2015) 36–45.
[22] Wang GQ, Yuan XG, Yu KT, Review of mass transfer correlation for packed columns. Ind. Eng. Chem. Res. 2005, 44, 8715-8729.
[23]Shulman, HL, Ullrich CF, Proulx AZ, Zimmerman JO, Performance of Packed Columns. 2. Wetted and Effective Interfacial Areas, Gas- and Liquid-Phase Mass Transfer Rates.AIChE J. 1955, 1, 253.
[24]Shulman, H. L.; de Gouff, J. J. Mass Transfer Coefficients and Interfacial Areas for 1-Inch Raschig Rings. Ind. Eng. Chem.1952, 44, 1915.
[25] Van Krevelen, D. W.; Hoftijzer, P. J. Studies of Gas Absorption. 1. Liquid Film Resistance to Gas Absorption in Scrubbers. Recl. Trav. Chim. Pays-Bas. 1947, 66, 49.
[26] Van Krevelen, D. W.; Hoftijzer, P. J. Kinetics of Simultaneous Absorption and Chemical Reaction. Chem. Eng. Prog. 1948, 44, 529.
[27]Mangers, R. J.; Ponter, A. B. Effect of Viscosity on Liquid Film Resistance to Mass Transfer in a Packed Column. Ind. Eng. Chem. Process Des. Dev. 1980, 19, 530.
[28]Shi, M. G.; Mersmann, A. Effective Interfacial Area in Packed Columns. Ger. Chem. Eng. 1985, 8, 87.
[29] Billet, R.; Schultes, M. Advantage in Correlating Packing Column Performance. Inst. Chem. Eng. Symp. Ser. 1992, 128, B129.
[30] Onda, K.; Takeuchi, H.; Okumoto, Y. Mass Transfer Coefficients between Gas and Liquid Phases in Packed Columns. J Chem. Eng. Jpn. 1968, 1, 56.
[31] Eckert, J. S.: Chem. Eng., 82, 70. Apr. 14, 1975.
[32] Mostafa H. Sharqawy, John H. Lienhard, Syed M. Zubair. Thermophysical properties of seawater: a review of existing correlations and data. Journal of Desalination and Water Treatment. 16 (2010) 354–380.
[33] J.D. Isdale and R. Morris, Physical properties of sea water solutions: density, Desalination, 10(4) (1972) 329–339.
[34] Francis L. Smith and Allan H. Harvey (September 2007). "Avoid Common Pitfalls When Using Henry's Law". CEP (Chemical Engineering Progress). ISSN 0360-7275.
[35] Feoktistov, V. (2006). Differential Evolution: In Search of Solutions. Springer. ISBN 978-0-387-36895-5.
[36] H. Bahmanyar, Mass Transfer. Tehran University's Jahad Daneshgahi Publication, 2003.

460667 Multiblock Ionomers for Membrane Applications

Abstract:

The water-energy nexus represents the confluence of two pressing engineering challenges facing todayâ€™s generation. The demand for clean and potable water continues to increase with the expansion of the global population and the rapid industrialization of developing nations. Correspondingly, new technologies to harvest, store, and discharge the energy required for the production of this valuable resource are critical. Membranes prepared from block polymers are useful as designer materials for both of these applications. For example, block polymers microphase separate to form task-specific phases: one that imparts mechanical strength and the other facilitates ion, electron, or water transport. Ionomers are increasing in popularity based on recent work that highlights their stable and unique morphologies, tailorable ionic conductivity, and high water flux with good salt selectivity. This presentation will highlight work from our lab on membranes fabricated from imidazolium-based triblock copolymers prepared using polymerizable ionic liquids or post-polymerization functionalization strategies for energy applications. Specifically, the influence of (i) charge placement and density in the central and outer blocks; (ii) thermomechanical properties; (iii) ionic liquid incorporation; and (iv) chemical composition of the cation and anion on the properties of imidazole-containing polymers will be discussed. Additionally, recent work has utilized these same materials for salt-selective water purification membranes. The structure-property-diffusion relationships that have been developed contribute toward a better understanding of the interplay between polymer morphology, water and salt sorption and transport characteristics, and thermomechanical properties.

469314 Realizing the Potential of Micro-Phase Separated Block Copolymer Electrolytes: Ion Domain Connectivity Plays a Prominent Role in Ion Conduction

Abstract:

Single ion conducting polymer electrolyte membranes (PEMs) are found at the heart of many electrochemical devices (e.g., fuel cells, flow batteries, electrolyzers, electrodialysis etc.).(*1*) The key requirements for such materials include high ionic conductivity, electron insulation, mechanical resilience, selectivity (i.e., high transference #), and chemical, thermal, and mechanical stability.(*2, 3*) Ionic conductivity is particularly important because it strongly influences the ohmic overpotential in the said electrochemical devices affecting overall device efficiency.(*1, 4*)

Block copolymer electrolytes represent an attractive set of PEM materials because their micro-phase architecture yields greater ion conduction over their random copolymer counterparts.(*5-7*) Furthermore, facile and controllable polymerization reactions allow researchers to access a variety of morphological architectures, while engineering of the block copolymerâ€™s self-assembly influences domain alignment.(*8-12*) However, there is lack of systematic studies correlating molecular level structural design to bulk material properties like ion transport.

In this work, a model lamellae-forming diblock copolymer electrolyte system was manipulated to examine the extent of ion domain connectivity on ion conduction.(*13, 14*) Careful control of the model systemâ€™s volume fraction gave diblock copolymer structures with slightly rich electrolyte domains or slightly rich hydrophobic domains. The slightly rich electrolyte domains had greater contiguous area fraction while simultaneously demonstrating substantially lower terminal defects. Having a contiguous area fraction from 0.95 to 1.0 resulted in a 2x improvement in ionic conductivity over a non-micro-phase separated block copolymer electrolyte or a micro-phase separated block copolymer electrolyte with very poor connectivity (i.e., a system with larger volume fraction of hydrophobic blocks). Incremental adjustment in the extent of connectivity revealed an exponential growth curve for ion conductivity as a function of contiguous area fraction. Furthermore, the benefits that a micro-phase separated block copolymer electrolyte affords in terms of ion conductivity were not realized when the block copolymer electrolyte had poor connectivity. The results of this work have far reaching implication into the rationale design of PEM materials based upon block copolymer designs. This talk will emphasize the importance of maximizing ion domain connectivity while taking great strides to minimize terminal defects to boost ion transport in PEM materials. This talk will close with future directions on how molecular level engineering of block copolymer electrolytes offer the potential to shed light on how other structural features (e.g., tortuosity(*15*), grain boundaries(*5*), and counterion condensation) alter ion transport in PEM materials(*16, 17*).

References

1. H. Strathmann, A. Grabowski, G. Eigenberger, Ion-Exchange Membranes in the Chemical Process Industry. *Industrial & Engineering Chemistry Research* **52**, 10364-10379 (2013).

2. H. Strathmann, *Ion-Exchange Membrane Separation Processes, Volume 9*. Membrane Science and Technology (Elsevier Science, Amsterdam, The Netherlands, 2004), vol. 9.

3. T. Sata, *Ion Exchange Membranes: Preparation, Characterization, Modification and Application*. (Royal Society of Chemistry, Cambridge, UK, 2004).

4. A. Z. Weber, J. Newman, Modeling Transport in Polymer-Electrolyte Fuel Cells. *Chemical Reviews* **104**, 4679-4726 (2004).

5. Y. A. Elabd, M. A. Hickner, Block Copolymers for Fuel Cells. *Macromolecules* **44**, 1-11 (2011).

6. N. Li, M. D. Guiver, Ion Transport by Nanochannels in Ion-Containing Aromatic Copolymers. *Macromolecules* **47**, 2175-2198 (2014).

7. Y. Schneider *et al.*, Ionic Conduction in Nanostructured Membranes Based on Polymerized Protic Ionic Liquids. *Macromolecules* **46**, 1543-1548 (2013).

8. H. Hu, M. Gopinadhan, C. O. Osuji, Directed self-assembly of block copolymers: a tutorial review of strategies for enabling nanotechnology with soft matter. *Soft Matter* **10**, 3867-3889 (2014).

9. M. Luo, T. H. Epps, III, Directed Block Copolymer Thin Film Self-Assembly: Emerging Trends in Nanopattern Fabrication. *Macromolecules* **46**, 7567-7579 (2013).

10. S. Ji, L. Wan, C.-C. Liu, P. F. Nealey, Directed self-assembly of block copolymers on chemical patterns: A platform for nanofabrication. *Progress in Polymer Science*, **54-55**, 76-127 (2016).

11. S.-J. Jeong, J. Y. Kim, B. H. Kim, H.-S. Moon, S. O. Kim, Directed self-assembly of block copolymers for next generation nanolithography. *Materials Today* **16**, 468-476 (2013).

12. M. P. Stoykovich, P. F. Nealey, Block copolymers and conventional lithography. *Materials Today* **9**, 20-29 (2006).

13. I. P. Campbell, G. J. Lau, J. L. Feaver, M. P. Stoykovich, Network Connectivity and Long-Range Continuity of Lamellar Morphologies in Block Copolymer Thin Films. *Macromolecules* **45**, 1587-1594 (2012).

14. C. G. Arges, Y. Kambe, H. S. Suh, L. E. Ocola, P. F. Nealey, Perpendicularly Aligned, Anion Conducting Nanochannels in Block Copolymer Electrolyte Films. *Chemistry of Materials* **28**, 1377-1389 (2016).

15. X. Feng *et al.*, Scalable Fabrication of Polymer Membranes with Vertically Aligned 1 nm Pores by Magnetic Field Directed Self-Assembly. *ACS Nano* **8**, 11977-11986 (2014).

16. K. M. Beers, D. T. Hallinan, X. Wang, J. A. Pople, N. P. Balsara, Counterion Condensation in Nafion. *Macromolecules* **44**, 8866-8870 (2011).

17. K. M. Beers, N. P. Balsara, Design of Cluster-free Polymer Electrolyte Membranes and Implications on Proton Conductivity. *ACS Macro Letters* **1**, 1155-1160 (2012).

466765 Structure/Property Relationships in Polymer Membranes for Water Purification and Energy Applications

Abstract:

Providing sustainable supplies of purified water and energy is a critical global challenge for the future, and polymer membranes will play a key role in addressing these clear and pressing global needs for water and energy. Polymer membrane-based processes dominate the desalination market, and polymer membranes are crucial components in several rapidly developing power generation and storage applications that rely on membranes to control rates of water and/or ion transport. Much remains unknown about the influence of polymer structure on intrinsic water and ion transport properties, and these relationships must be developed to design next generation polymer membrane materials. For desalination applications, polymers with simultaneously high water permeability and low salt permeability are desirable in order to prepare selective membranes that can efficiently desalinate water, and a tradeoff relationship between water/salt selectivity and water permeability suggests that attempts to prepare such materials should rely on approaches that do more than simply vary polymer free volume. One strategy is to functionalize hydrocarbon polymers with fixed charge groups that can ionize upon exposure to water, and the presence of charged groups in the polymer influences transport properties. Additionally, in many emerging energy applications, charged polymers are exposed to ions that are very different from sodium and chloride. Specific ion effects have been observed in charged polymers, and these effects must be understood to prepare charged polymers that will enable emerging energy technologies. This presentation discusses research aimed at further understanding fundamental structure/property relationships that govern water and ion transport in charged polymer films considered for desalination and electric potential field-driven applications that can help address global needs for clean water and energy.

460616 Fundamental Water and Salt Transport Properties in Zwitterionic Polymers and Their Use for Membrane Surface Modification to Enhance Antifouling Properties

Abstract:

Fouling by suspended solids and dissolved organic matters presents a great challenge for polymeric membranes for desalination and wastewater treatment. With superior hydrophilicity, zwitterionic polymers have been explored for membrane surface modification to enhance antifouling properties. However, there is a lack of understanding of relationship between polymer structure and water and salt transport properties. In this study, two series of zwitterionic polymers were prepared from sulfobetaine methacrylate (SBMA) and 2 methacryloyloxyethyl phosphorylcholine (MPC). These polymers were thoroughly characterized in terms of sol-gel fraction, density, glass transition temperature, contact angle, and water and salt transport properties. Interestingly, the zwitterionic polymers exhibit water sorption and permeability similar to non charged poly(ethylene glycol) (PEG) based materials. These zwitterionic polymers exhibit lower NaCl diffusivity and permeability and thus higher water/NaCl selectivity than the non charged PEG based materials. Moreover, we have modified ultrafiltration (UF) membranes using a solution containing dopamine and sulfobetaine methacrylate (SBMA). The dopamine and SBMA react and form a thin film on the surface, while dopamine provides adhesion to the membrane surface and SBMA provides superhydrophilic properties. The coating enhances the surface hydrophilicity, which is confirmed by contact angle measurement. The coated membranes demonstrate better stability and higher water flux than those of dopamine-coated and uncoated ones, when tested with water containing bovine serum albumin (BSA). For instance, the membrane coated with dopamine and SBMA exhibits water permeance 80% higher than the uncoated one, after 3 hour filtration using a 3g/L BSA solution in a crossflow system.

460548 Water and Salt Transport in Polymer Membranes Prepared By Solvent-Free Melt Processing

Abstract:

In industry today, polymer membranes are formed via solution processing (i.e., phase inversion techniques) that require large amounts of hazardous organic solvents. While effective from a processing standpoint, these membrane formation methods are highly material-specific and present challenges in controlling pore architecture (i.e., pore size, shape and distribution), which enormously affects the transport of small molecules and ions through the resultant membrane. Moreover, by avoiding the use of toxic and hazardous solvents for membrane formation, many benefits could be realized in terms of cost and environmental impact, and advancements in processing could also provide further insight on the next generation of membrane materials. Herein, work is presented on a new method of membrane formation using charged polymers on porous thin supports via a state-of-the-art multi-nanolayered technology, mainly aimed at producing membranes for desalination and water purification.

Robust, ion-exchange membranes based on sulfonated polymers were prepared by solvent-free melt processing, providing the first example of melt-processed membranes for applications in desalination. To investigate the polymer physics underlying the effects of processing history on these systems, transport of small molecules and ions has been investigated.

Interestingly, melt processed membranes (sulfonated polysulfone, BPS) prepared by different routes (i.e., plasticizer composition, processing temperature, and post-processing treatment) show different water and salt transport properties, compared to solution processed membranes. To understand the origin of this difference, ^{1}H NMR analysis and FT-IR analysis were conducted, but no chemical change was detected. Rather, differences in mass transport behavior result from processing methods having a significant effect on the polymer structures. These results imply that, in addition to varying the chemical structure of the polymer, varying processing methods also provides extra freedom to modify

polymer properties.

472275 Influence of Fixed Charge Group Concentration on Ion Sorption in Ion Exchange Membranes

Abstract:

Controlling ion sorption and transport in polymer membranes is critical for efficient operation of a number of membrane-based technologies such as reverse osmosis, electrodialysis, reverse electrodialysis, and fuel cells. Often, membranes for these processes are made from polymers having ionizable functional groups (fixed charge groups) bound to their backbone (i.e., ion exchange membranes, ionomers, etc.). Sorption of charged species (e.g., ions) in such materials is profoundly influenced by the fixed charge groups, due to Donnan exclusion. Consequently, the concentration of fixed charge groups can be used to tune, to an extent, sorption of ions in charged membranes. However, changes in the concentration of fixed charge groups are typically accompanied by changes in membrane water content (i.e., swelling), which also substantially influences ion sorption in membranes. Decoupling these two effects is critical for developing fundamental understanding of the influence of membrane fixed charge concentration on ion sorption in charged membranes, ultimately leading to strategies for rational design of high performance materials.

In this study, positively and negatively charged membranes having different fixed charge group concentrations but similar water content were synthesized. Ion concentrations in membranes equilibrated with NaCl solutions of varying concentration were measured using a desorption technique. Counter-ion (i.e., ion with opposite charge to that of the fixed charges) sorption increased with increasing fixed charge group concentration, since a larger number of counter-ions are required to electrically balance the fixed charge groups. Co-ion (i.e., ion with same charge as that of the fixed charges) sorption decreased with increasing fixed charge group concentration, presumably due to enhanced Donnan exclusion. However, despite similar changes in membrane fixed charge group concentrations, co-ion sorption in the negatively charged membranes was suppressed to a greater extent than in the positively charged membranes, presumably due to structural differences between the materials. The ion sorption data were interpreted using a model based on Donnan theory and Manningâ€™s counter-ion condensation theory. The model predicted the experimental ion sorption data in the positively charged membranes with no adjustable parameters. However, agreement between model and experimental data for the negatively charged membranes was poor, presumably due to membrane inhomogeneities introduced during polymerization. Use of the Donnan/Manning model for predicting ion sorption in inhomogeneous materials is discussed.

457973 Mixed Ion Sorption and Transport in a Cation Exchange Polymer Based on Sulfonated Polystyrene

Abstract:

Sorption and transport of sodium chloride, sodium fluoride, sodium nitrate, calcium chloride and their mixtures in a commercial cation exchange membrane based on cross-linked polystyrene with fixed sulfonate ions was investigated. For mixed salt sorption and permeation experiments, binary salt mixtures 50:50 wt were considered.

Pure and mixed ion transport parameters were correlated to the ion size and valence, as well to the membrane water content. The effects of the second component on salt transport parameters were investigated. When considering mixtures of salts having the same counter-ion (Na^{+}), mixed salt sorption and permeability coefficients do not change significantly relative to the values measured in the case of single salt, if they are reported as a function of salt activity coefficient in the contiguous, external solution. When considering mixtures of salts having different counter-ions (Na^{+}/Ca^{2+}), mixed salt transport parameters depart significantly from the respective values for single salt. The experimental findings were rationalized based on the Donnan theory [1].

[1] F.G. Donnan, Chem. Rev. 1924 1 73-90

471524 Using Inkjet Printing Devices to Fabricate Charge Mosaics from Chemically Tailored Copolymer Membranes

Abstract:

Charge mosaic membranes, which have distinct anionic and cationic domains that traverse the membrane thickness, allow dissolved salts to permeate rapidly. As such, they can be utilized in a variety of applications that require the selective permeation of ionic solutes (*e.g.*, piezodialysis). In this talk, we will introduce our efforts to prepare charge mosaics from copolymer membrane precursors using ink-jet printing devices. The copolymer precursors are chemically tailored so that well-defined counter-charged domains can be introduced on the surface of the membranes. Specifically, a polyacrylonitrile-based copolymer was synthesized *via* a free radical polymerization mechanism and its chemical composition and molar mass characterized using nuclear magnetic resonance (^{1}H NMR) and gel permeation chromatography (GPC), respectively. Copolymer substrates, which possessed nanopores with a diameter around 3 nm in size, were fabricated using a non-solvent induced phase separation (NIPS) method. The azide moieties incorporated into the copolymer precursor used to fabricate these membranes represent an attractive platform for the straightforward chemical functionalization of the membrane in the solid state. In particular, copper-catalyzed azide-alkyne click reactions enable rapid one-to-one covalent attachment of positively charged amine moieties or negatively charged carboxylic acid moieties to pore walls of the membrane. The properties of the charge-functionalized membranes were examined using hydraulic permeability, molecular weight cut-off, and streaming potential measurements. Subsequently, an ink-jet printer was used to generate well-defined domains of positive and negative charges within the pores of the copolymer substrate in a systematic manner. Membrane performance was optimized by coordinating the printing conditions with the membrane morphology. Results from Fourier transform infrared (FTIR) spectroscopy analysis demonstrate that up to 95% of the reactive sites can be converted into charged domains using this technique. The printed domains were distinguished and visualized using fluorescent microscopy. A series of piezodialysis experiments were conducted over a range of ionic strength using membranes functionalized with a single charge (i.e., cationic or anionic domains) as well as patterned mosaics of charge (i.e., cationic and anionic domains). Significantly, these experiments demonstrate that charge mosaic membranes are highly permeable to ionic solutes while single charged membranes reject up to 80% of the co-ions present in solution. The versatile and precise control of substrate chemistry at the nanoscale by printers reveal the potential application of resulted mixed mosaic membranes in targeted ion separation.

468020 Growth and Nucleation Kinetics in Continuous Antisolvent Crystallization Systems

Abstract:

Most crystallizations in the pharmaceutical industry occur as batch processes. However, a shift towards continuous manufacturing is occurring to decrease costs and increase process control. For heat-sensitive compounds or APIs which exhibit weak solubility-temperature dependence, antisolvent crystallization is commonly the purification method of choice. Previous research has not established a clear relationship between nucleation or growth kinetics and antisolvent composition. Current research elucidates the effect of temperature and supersaturation on empirical models of crystal nucleation and growth, either neglecting the effect of antisolvent or by assuming a â€˜model of best fitâ€™ for solvent composition-dependent kinetics, based primarily on batch crystallization studies. In this work,Â we demonstrate the need to consider antisolvent-dependent growth and nucleation kinetics in continuous crystallization processes. Solvent-dependent growth and nucleation kinetics derived from single-stage, mixed-suspension, mixed-product removal (MSMPR) experiments were applied to model continuous, multi-stage MSMPR systems. The effect of solvent composition on final product specifications, such as yield and mean particle size, was also explored. We successfully showed the implementation of our growth and nucleation kinetic models through the rational design of a multi-stage, continuous, combined cooling/antisolvent crystallization process for an industrially-relevant API.
We expect our findings to provide a framework for the design of additional continuous antisolvent crystallization processes in the future.

468648 Evaluation of *In Vivo* Drug Precipitation or Crystallization Potential through a Novel *In Vitro* Approach for Method Development and Initial Risk Assessment

Abstract:

Evaluation of *In Vivo* Drug Precipitation or Crystallization Potential Through a Novel *In Vitro* Approach for Method Development and Initial Risk Assessment

Chenchi Wang[1], Ching Su[2], and Thanga Mariappan[3], et al.

[1]Pharmaceutical Manufacturing Science and Technology, Global Manufacturing and Supply, Bristol-Myers Squibb, New Brunswick, New Jersey; [2] Discovery Pharmaceutics, Bristol-Myers Squibb, Princeton, New Jersey; [3] Biocon-Bristol-Myers Squibb R&D Centre, Syngene International Limited, India.

As poorly water-soluble drug candidates have become prevalent in the discovery and development phases for pharmaceuticals, incidents of in vivo bio-accumulation in animals of drug molecules or related have been reported in some pre-clinical toxicological studies via precipitation or crystallization. These observations could complicate and may prevent further development of these compounds. In the early development of drug substances, it would be beneficial to enable a quick, *in vitro* assessment to differentiate or screen between multiple drug candidates for their comparative risks of potential precipitation *in vivo*, using a method with minimum efforts required if possible.

In this work, we explored a novel in vitro precipitation risk assessment concept and designed experimental methods based on measurable characteristics, such as solubility, in vivo exposure, and the levels of supersaturation of related molecules, etc., with assumptions made in order to simplify the working hypothesis. Multiple test compounds have been examined using this approach empirically to monitor and rank order their relative precipitation potentials in deliberately created, precipitation-prone conditions. The resultant indications *in vitro* are consistent with the observations obtained *in vivo* in the toxicology studies.

452379 Development of New Screening Methodology for the Assessment of the Agglomeration Potential of APIs

Abstract:

Agglomeration of active pharmaceutical ingredients (API's) in agitated filter driers (AFD) is a common occurrence on scale-up and is typically observed while the wet-cake is being stirred above a certain solvent composition. This can lead to snow-balling, excessive sticking, or simply cake hardening during the drying generating various forms of agglomerates. Agglomeration can impact downstream drug product manufacturing as well as bioavailability and in extreme cases can damage the agitator and/or lead to difficulty in removing the batch from the processing equipment, leading to significant material losses.

Therefore it is important to understand the risk of agglomeration potential prior to scale-up as early in development as possible. Mixer torque rheometry (MTR) has been shown in the literature to be a useful tool for studying the agglomeration potential for certain compounds. However, this methodology requires large quantities of API and we have recently demonstrated that in some cases under-predicts the agglomeration potential and does not account for cake hardening.

We have developed a simple screening protocol for assessing the agglomeration potential using resonant acoustic mixing; that minimizes the quantity of API used and can therefore be used early on in development. This methodology improves upon the MTR one as it allows for multiple conditions to be screened in parallel, saving time and in addition to a qualitative (visual) assessment, provides quantitative data after the material has been dried, thus assessing cake hardening as well. This methodology can also be used to assess the friability of the generated agglomerates. The development and reproducibility of this new methodology will be discussed as well as the impact of particle size, particle size distribution, solvent selection, and solvent loading on the agglomeration potential for several Takeda APIs. Finally, a comparison of this new method and the MTR method as well as a pilot plant validation of the revised drying

conditions developed based on this new methodology will also be presented.

453852 Parameter Estimation Methods for Salting out Crystallization

Abstract:

Predictive modeling is increasingly utilized to gain better understanding of processes and reduce the number of experiments required for process development, optimization and understanding. In the pharmaceutical arena, modeling is typically utilized in late stage projects to provide additional data for QbD and for design space verification and confirmation. All models require the knowledge of appropriate kinetic parameters which are typically acquired from experiments. PAT tools are well suited to generate a wealth of data that can be used to estimate kinetic parameters to be utilized in predictive models. In this study we present two approaches to extract crystal growth kinetic parameters from PAT tools, namely FTIR/ATR and FBRM, combined with off-line microscopy. The methods are applied to a salting out crystallization that affords elongated, columnar crystals. Two parameter estimation approaches are utilized: a GSK developed algorithm that returns growth parameters in two growth directions and a parameter estimation routine developed by Scale-up systems Ltd. in collaboration with GSK. The two approaches provided similar growth kinetic parameters that were utilized to perform simulations to predict process behavior under different process conditions.

467244 Electric-Field-Assisted Protein Crystallization in Continuous Flow

Abstract:

The competitive environment in global market for biopharmaceuticals calls for shorter development times, improved manufacturing flexibility, reduction of batch-to-batch variability and reduction of the footprint of production facilities. A key method to potentially achieve these improvements is to switch from traditional batch-wise manufacturing to continuous manufacturing[1], which offers several potential advantages such as a reduction in the number of required processing steps, simplified scale-up, and the ability to implement more effective process control strategies.[2] The separation of a protein from a mother liquor with impurities is a typical biopharmaceutical process. Compared to traditional packed-bed chromatography, crystallization is an important alternative, yielding high purities in a single processing step that can operate over a range of scales and throughputs. Unfortunately, the design and control of crystallization processes involving large biomolecules such as proteins remains notoriously difficult.

The transition of batch-wise to continuous manufacturing offers opportunities to develop new concepts for continuous protein crystallization.[3] Usually, the kinetics of protein crystallization are optimized by manipulating relevant global variables such as concentration, temperature, and solvent composition. However, studies on the use of local variables to manipulate protein crystallization, such as electric fields, have also been reported[4]. All studies of the mechanism of electric-field-assisted crystallization have been conducted batch-wise and mainly with the objective of producing high-quality material for crystallography or biomedical analysis. The use of low-voltage AC electric fields for protein nucleation in a continuous flow setting for manufacturing has, to the best of our knowledge, not been addressed. Consequently, insights into the mechanism and guidelines for design of continuous electric-field-assisted protein crystallization are lacking.

The objective of this work is to investigate different electrode structures that could be implemented in a continuous flow device to control protein nucleation both spatially and dynamically. The long-term objective is to develop guidelines for the optimal design of a cell with microelectrodes to enable electric-field-assisted protein crystallization in continuous flow. To investigate the interaction of different types of electrodes and field properties on protein crystallization, microelectrode chambers (MEC) have been fabricated and tested. ITO coated glasses are used as electrodes and treated with standard lithographic techniques to obtain specific patterns (Figure 1). Two different types of electrodes have been tested, 1) parallel plate electrodes (Figure 1a-h) in which a potential difference is created in vertical direction, and 2) coplanar electrodes[5] in which a potential difference in horizontal direction is created (Figure 1i-j). Lysozyme with a fluorescent label is used as the model protein in this study. In a typical experiment, supersaturated lysozyme solution is mixed with precipitant and injected into the MECs and observed under a microscope. Results demonstrate that under the applied electric field, the induction time for nucleation is reduced with an average induction time of around 20 hours for some electrode patterns. Furthermore, most of the protein crystals are found on the ITO electrode, which is probably due to the effect of local electric field minima. In addition, the number and quality of crystals are improved when using electric fields, especially compared with a reference case using non-conductive glass chambers.

In conclusion, electric fields can induce faster protein crystallization with higher crystal quantity and better quality. These findings can be used as guidelines for fabricating MECs for continuous protein crystallization by aligning the electrodes in flow channels. The advantage of the tested coplanar electrodes over parallel electrodes is that the height of the flow channels can be varied independently to optimize hydrodynamics whereas the height of channels with parallel electrodes also influences the electric field properties.

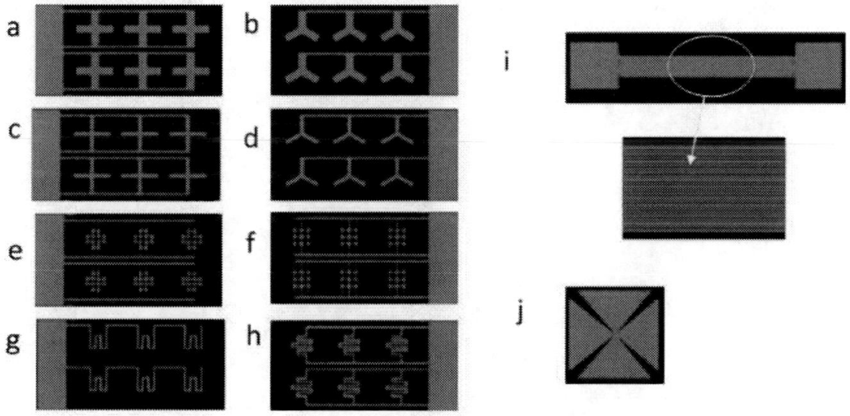

Figure 1. Various electrode structures fabricated with lithographic methods to construct MECs: (a-h) parallel electrode; (i, j) coplanar electrodes.

References

1. Mascia, S., Heider, P. L., Zhang, H. T., Lakerveld, R., Benyahia, B., Barton, P. I., Braatz, R. D., Cooney, C. L., Evans, J. M. B., Jamison, T. F., Jensen, K. F., Myerson, A. S. and Trout, B. L. Angewandte Chemie-International Edition, 2013, 52 (47), 12359-12363.

2. Lakerveld, R., Benyahia, B., Heider, P. L., Zhang, H., Wolfe, A., Testa, C. J., Ogden, S., Hersey, D. R., Mascia, S., Evans, J. M. B., Braatz, R. D. and Barton, P, I., Org. Process Res. Dev., 2015, 19, 1088-1100.

3. Neugebauer, P. and Khinast, J. G., Crystal Growth & Design, 2015, 15, 1089-1095.

4. Taleb, M., Didierjean, C., Jelsch, C., Mangeot, J. P., Capelle, B. and Aubry, A., Journal of Crystal Growth, 1999, 200, 575-582.

5. Hou, D. and Chang, H.-C., Applied Physics Letters, 2008, 92, 223902.

455558 Purification of Nitrophenols Using Complex-Assisted Crystallization

Abstract:

Solution-phase chemical complexation can be used to enhance the purity of crystallized products. Using the separation of nitrophenol isomers as a model, we experimentally demonstrated that complex-assisted crystallization can efficiently remove structural isomers as impurities without affecting product crystal form or decreasing yield. Experiments showed that lattice-impurity incorporation of 3-nitrophenol into 4-nitrophenol can be prevented in an aqueous system using guanidine acetic acid as a complexing agent. Purity enhancements of 70% (mass) with respect to non-complexed controls were observed with no yield reduction. These results serve to further the known applicability of complex-assisted crystallization as a technique for enhancing tradtional crystallization processes.

451215 Rational Pharmaceutical Cocrystallization

Abstract:

Cocrystal formulation plays a crucial role in Pharmaceutical industry by allowing modification of dissolution rate, degree of supersaturation and bioavailability of non-ionizable poor water-soluble drugs; as well as improved chemical and physical stabilities of intermediates and APIs. Since experimental cocrystal screening may be time and cost consuming, various computational approaches were recently proposed for a rational coformer selection to guide screening experiments.[2,3] An overview of the virtual coformer screening approaches will be presented outlying advantages and limitations of each of the methods. A limited number of examples of *in silico* screening applications will be discussed.

References

1. Thakuria, R.; Delori, A.; Jones, W.; Lipert, M.P.; Roy, L.; Rodriguez-Hornedo, N. *Int. J. Pharm.* **2013**, 453, 101–125.

2. Desiraj G.R. *Angew Chem. Int. Ed. Engl.* **1995**, *34*, 2311–2327; Mohammad, M.A.; Alhalaweh, A.; Velaga, S.P. *Int J Pharm* **2011**, *407*, 63–71; Musumeci, D., Hunter, C.A.; Prohens, R.; Scuderi, S.; McCabe, J.F. *Chem. Sci.* **2011**, *2*, 883–890; Wood, P. A.; Feeder, N.; Furlow, M.; Galek, P. T.; Groom, C. R.; Pidcock, E., *CrystEngComm* **2014**, *16*, 5839-5848; Abramov, Y.A.; Loschen, C.; Klamt, A. *J. Pharm. Sci.,* **2012**, *101*, 3687-3697.

3. Abramov, Y.A., Ed. *Computational Pharmaceutical Solid State Chemistry*, John Wiley & Sons, **2016**.

471927 Examining the Partitioning Coefficients of Butanol and Other Fermentation Metabolites Produced By *Clostridium Pasteurianum* into a Lipid Vesicle Solvent

Abstract:

Continuous extractive fermentation is critical to increasing biofuel production. Extracting the fermentation products reduces product inhibition, increases the metabolic activity and drives solvent production, and can increase solvent selectivity. Liquid-liquid extraction using organic solvents has shown to be effective for extractive fermentation, but the solvents employed are often toxic to the microbes and can limit productivity. Self-assemblies such as vesicles and micelles provide an alternative to organic solvents for the extractive fermentation of biofuels. We are currently developing a process where lipid vesicles are used as the solvent phase to recover butanol from fermentation broths and remove toxic impurities present in substrate feedstocks. This approach is based on the high thermodynamic partitioning coefficients observed for butanol from water to lipid membranes, which is one of the main causes of product inhibition as butanol partitions into and fluidizes microbial membranes. This presentation will report on the partitioning behavior of butanol, as well as other metabolites produced by *Clostridium pasteurianum* (acetate, ethanol, 1,3-propanediol), to lipid vesicles as a function of lipid composition, pH, and temperature. We have shown that butanol partitioning increases with the amount of unsaturated lipid present within the vesicles due to an increase in the size of the â€˜binding pocketsâ€™ for amphiphilic butanol to partition into at the vesicle/water interface.

458249 Intensified Liquid-Liquid Extraction with Chemical Reaction By Emulsification

Abstract:

Biobased effluents are commonly defined by undesirable characteristics, such as a low concentration of the target constituent in aqueous solution or separation barriers like azeotropes. The isolation of the constituents from such effluents challenges standard unit operations to their limits. To accomplish this task liquid-liquid extraction has been widely used, but for an increase of efficiency and operability further improvement is necessary. Improvement can be obtained through intensification of the extraction step by combination with chemical conversion of the target constituent. Thus, the separability of the constituent as well as economics can be significantly increased.

Effluents from biorefinery processes like pulping serve as representative separation task. Such effluents often contain a significant amount of carboxylic acids like acetic acid or formic acid. Through spiking an appropriate solvent with an aliphatic alcohol, the extraction rate of the carboxylic acids can be increased by chemical conversion via esterification which yields a favorable change of the physio-chemical properties of the constituents. Admixture of homogeneous catalysts to the solvent provides an additional benefit by increasing the reaction rate of the esterification step. *4*-dodecylbenzenesulfonic acid (4DBSA) is a suitable catalyst. This highly potent catalyst, which additionally acts as a strong anionic surfactant, offers access to two different process routes.

The first route focuses on the prevention of emulsification by adjusting the hydrophilic-lipophilic balance value (HLB value) through adding a second surfactant to the system. Admixture of nonionic surfactants (e.g. SPAN®85) significantly enhances primary phase separation.

The second process route makes use of emulsification in order to increase the interaction between reactants and catalyst, thus increasing the reaction rate although splitting of the emulsion must be applied prior to product isolation. This technology is a promising alternative to state of the art isolation of carboxylic acids from aqueous effluents by liquid-liquid extraction.

447293 Hybrid Shortcut-Modell for Estimating Partition Coefficients in Aqueous Two-Phase Extraction of Therapeutic Proteins

Abstract:

The development of industrial scale bioprocesses has come into high scientific and industrial interest over the last decades. Especially in the red and white biotechnology, process costs are often dominated by 50 - 80 % by the downstream processing. First investigated in the 1896, aqueous two-phase systems (ATPS) were shown to have an enormous potential for the extraction of these biomolecules. In general ATPS show a high biocompatibility and are cheap in comparison to chromatographic methods. The most common ATPS consist of a hydrophilic polymer and a salt serving as phase-forming components. The distribution of the target protein within these systems can either be influenced by the choice of the phase forming components, and additionally by a choice of a displacement salt. The displacement salt influences both the solubility and the distribution of the protein in the two phases. Thus, knowledge of the influence of the different salts potentially useful to enhance the distribution coefficient is necessary.

One promising and fast measure accounting for complex interactions of biological macromolecules is the second osmotic virial coefficient (B_{22}) characterizing the interactions between the biological macromolecules in solution and the cross virial coefficient (B_{23}) characterizing interactions of the biological macromolecules with the salts / impurities. It combines the different influence factors such as pH, Temperature, solvent, salt and salt concentration within one measure.

B_{22} as well as B_{23} data was measured using composition gradient multi angle light scattering (GC-MALS) for different proteins and salts in aqueous solution. The results were then used to characterize the distribution behavior of the target protein in the ATPS dependent on the displacement salt. The results point to the fact, that based on B_{22} and B_{23} the suitability of a salt as displacer for proteins in aqueous two-phase systems can be investigated. One major advantage

using this approach is the fact, that the partition coefficient can be estimated using a hybrid ShortCut Modell based on a minimal set of experimental data. Furthermore the influence of the displacer (salt) on the solubility of the protein can be judged qualitatively serving as an ideal first measure in the development of aqueous two-phase extraction systems. The results will help to improve the development and optimization of ATPS used for efficient separation of processes.

461385 Novel Purification Process for Bio-Derived Citric Acid Using Solvent Extraction and CO_2 Antisolvent Purification

Abstract:

Citric acid (CA) is a chemical commodity directly used in food and confectionery, as well as in the pharmaceutical and chemical industry. CA is used as raw material for calcium supplements to prevent osteoporosis, for the manufacture of biobased plasticizers, and to produce citrate salts used as food preservatives, buffer and chelating agents. Generally, CA is produced from a fungal broth fermentation with *Aspergillus niger* using sucrose as the main substrate. As the microorganism is able to withstand acidic condition, CA is obtained in the acidic form at the end of the fermentation process. Traditional downstream processing involves removal of biomass and suspended solids by filtration, followed by CA precipitation with calcium hydroxide. Impure calcium citrate is acidified with sulfuric acid to release CA in aqueous solution, and precipitating gypsum.

In a typical purification process, CA solution is subjected to ion exchange to remove metal and remaining sucrose, adsorption with activated carbon, followed by evaporation and crystallization. Sucrose traces removal is of major concern as it heavily affects CA crystallization, and prevents from obtaining high CA purity (i. e. higher market value). This process is energy and mass intensive, and generates considerable amounts of liquid and solid waste.

Taking the aforementioned into account, this work describes a novel CA purification process using direct solvent extraction from the clarified fermentation broth and compressed gas (CO_2) as an antisolvent treatment to remove impurities. After the CA is extracted with a medium polarity solvent mixture, precipitation of impurities was carried out by compressing the liquid under CO2 pressure between 10 and 45 Bar, and different contact times (from 2 to 30 h). Under these conditions the impurities, mostly containing sucrose, precipitate out from the solution, allowing high purity CA recovery after solvent evaporation.

461913 Conceptual Process Design and Techno-Economic Analysis of a Hybrid Zeolite Membrane/Distillation Separation System for Butane Isomer Separation

Abstract:

The application of a hybrid separation process integrating zeolite membranes with distillation column was studied for butane isomer (*n*-butane/*i*-butane) separation. Currently, the industrial separation of butane isomers is accomplished using distillation, and due to the close boiling-point of the components, it is considered as one of the most energy-intensive distillation separations. Thus, the implementation of a hybrid process offers a great opportunity for large reductions in energy use and environmental emissions.

In this work, a conceptual process model of a series hybrid configuration, where zeolite membrane is placed before the distillation column to accomplish the initial portion of separation requirements, was developed and analyzed for an industrial case study of butane isomer separation. The permeation through the zeolite membrane was described using the real adsorption solution theory and the Maxwell-Stefan theory for diffusion. The membrane model was simulated in gPROMS v4.0.0 and validated by comparison with results obtained from the experiments. The distillation column was modeled using the McCabe-Thiele method in Aspen Plus v8.0, and was validated with the performance of an industrial *n*-butane/*i*-butane fractionator. The *n*-butane recovery and purity, obtained from the fractionator, were selected as the target for the hybrid system performance, and a techno-economic analysis was performed to this end.

The hybrid process performance was evaluated in terms of the membrane area required and the reduction in energy usage of the distillation column, which represents the additional capital cost and the savings in operational costs, respectively. The membrane area required was found to be one-tenth of that required for stand-alone membrane process, and the reboiler duty was found to be almost half of that required for stand-alone distillation column. This suggests that the hybrid membrane/distillation process has a great potential for industrial implementation. The Net Present Value (NPV) savings projections of the hybrid system were also calculated to generate a set of performance targets in term of the membrane permeance and the membrane cost.

457969 Process Integration for Simulated Moving Bed Reactor for Production of Glycol Ether Acetate

Abstract:

Simulated moving bed reactor (SMBR) has significant advantages over the traditional sequential reactor and separator operation in terms of improved process performance such as separation resolution, increased productivity, and reduced solvent consumption. Academic interest in SMBR operation have manifested through past investigations; however, these studies have focused on the SMBR operation in isolation from the overall process flow, and to the best of our knowledge, it has never been analyzed as a component of a process flowsheet.

This work identifies the optimal SMBR operating parameters for its integration within an overall process. The dynamic mathematical model developed in our previous work is employed in a process flowsheet in this work. The SMBR model was developed for an industrial case study for the continuous production of a solvent, propylene glycol methyl ether acetate (DOWANOL™ PMA[1]) through an acid-catalyzed esterification reaction of 1-methoxy-2-propanol and acetic acid. This overall process consists of the SMBR unit and two downstream units. Specifically, the extract and raffinate streams from the SMBR are further processed to improve product purity, and recycle streams leaving the downstream units return to the SMBR unit to reduce solvent consumption and waste generation.

The proposed process flowsheet is optimized in a deterministic, nonlinear dynamic optimization scheme. The overall operation is formulated as multiple objective optimization problem. The objective functions are to minimize the operating cost of the overall process indirectly through constraints on process variables and to maximize the PMA productivity. The results from this work are compared to those of an isolated SMBR unit, allowing for a detailed comparison between the SMBR operation in the overall process to that of an isolated unit. The work presented offers a more realistic and holistic overall process understanding that is representative of industrial SMBR operations.

1. Trademark of The Dow Chemical Company ('Dow') or an affiliated company of Dow.

464862 Hybrid Membrane-Cryogenic Distillation Processes for Argon Production from Air

Abstract:

Nowadays simultaneous production of oxygen, nitrogen and argon from air on a commercial scale is performed mainly by cryogenic distillation[1]. Although this technology is relatively mature and can deliver high purity products at high capacities, it is highly energy and capital intensive. In particular, separation of oxygen and argon represents a significant challenge due to their similar volatilities.

In recent years, demand for high purity argon has grown, in tandem with increased demand for oxygen[2]. Current distillation technologies can produce extremely high purity argon but many distillation stages are required and energy demand of the separation is high[3]. Consequently, there is a need for development of cost-efficient processes to produce argon from air.

In this work, a novel hybrid process is proposed wherein a membrane unit is coupled with distillation unit to create a novel, energy-efficient process for argon production. A range of membranes at various levels of technological maturity is considered.

Appropriate models are developed to evaluate the performance of these novel processes. Air separation units are modelled in Aspen Plus. Permeation of gas molecules through membranes is described by solution diffusion mechanism. Countercurrent and crossflow membrane models are implemented in Aspen Plus as â€˜user-defined unitsâ€™. The models are validated against published experimental data. An optimisation based solution technique is applied to determine optimum process conditions of the hybrid flowsheet.

A systematic methodology is under development to design, evaluate and optimise alternative hybrid membrane-cryogenic distillation processes for air separation. The performance of hybrid processes will be evaluated in terms of the power demand of the main air compressor and the membrane area required in order to identify the most promising design options offering capital and energy savings compared to conventional cryogenic air separation processes.

Â

[1] Castle, W. 2002. Air separation and liquefaction: recent developments and prospects for the beginning of the new millennium. International Journal of Refrigeration, 25, 158-172.

[2] Transparency Market Research. 2014. Noble Gases Market (Helium, Xenon, Neon, Krypton, and Argon) - Global Industry Analysis, Size, Share, Growth, Trends and Forecast, 2013 - 2019 [Online].

[3] Agrawal, R., Auvil, S., Choe, J. and Woodward, D. 1990. Membrane/cryogenic hybrid scheme for argon production from air. Gas Separation & Purification, 4, 75-80.

471152 Controllability of Reactive Distillation Columns with Single and Double Reactive Sections for Separation of Two-Stage Consecutive Reactions

Abstract:

Process intensification appears as one of the most promising tools in chemical industry, because increasing economic and safety considerations force the industry to focus on smaller scale and more energy-efficient technologies. One of the most glaring examples of this tool is reactive distillation columns which combine both the reaction and separation units into a single vessel. An overview on chemical reaction types shows that single-stage reversible reactions including quaternary systems (two reactants and two products) and ternary systems (two reactants and one product) lead the studies on reactive distillation columns. On the other hand, the third type of reactions widely found in the industry is the two-stage consecutive reversible reactions. However, there are only a few studies reported in the literature for the separation of this type of reactions, and reactive distillation columns with a single reactive section (RDC-SRS) have been used in all of these studies. Recently, a novel reactive distillation column with double reactive sections (RDC-DRS) has been proposed for the separations of the two-stage consecutive reversible reactions [1]. Results of this study indicate better steady-state performance than the RDC-SRC, but the dynamic performance and controllability of the RDC-DRS has not been studied. The aim of this study is designing effective and robust control structures for both RDC-SRS and RDC-DRS. The same hypothetical ideal two-stage consecutive reversible reactions,Â A+B â†" C+D and C+B â†" E+D, and two-stage consecutive trans-esterification of dimethyl carbonate with ethanol are chosen as illustrative examples to evaluate dynamic performance of the proposed control structures.

[1]Â Yu, C., Yao, X., Huang, K., Zhang, L., Wang, S., Chen,H., 2014, A reactive distillation column with double reactive sections for the separations of two-stage consecutive reveresible reactions, Chemical Engineering and Processing: Process Intensification, 79, 56-68.

472130 Process Alternatives for the Direct Esterification of Fusel Oil By Reactive Distillation

Abstract:

Process alternatives for the direct esterification of fusel oil by reactive distillation

César Augusto Sánchez-Correa, Alvaro Orjuela, Iván D. Gil, Gerardo Rodríguez

Grupo de Procesos Químicos y Bioquímicos,

Departamento de Ingeniería Química y Ambiental

Universidad Nacional de Colombia Sede Bogotá

Fusel oil (FO) is a mixture of linear and branched alcohols, ranging from ethylic to isoamylic, obtained as byproduct of bioethanol industry. Typically, around 1-11 L of FO are generated per each 1000 L of anhydrous ethanol produced [1], [2]. In countries where the bioethanol plays an important economic role (e.g USA, Brazil, China, etc), the valorization of FO is an important issue regarding the sustainability of the whole process, and also an opportunity for creating new value-added products for different market niches.

An alternative process to upgrade FO into more-valuable byproducts is the conversion into esters, which are used mainly as flavoring, fragrances and solvents. There are two main processing alternatives for the esterification of fusel alcohols [3], [4]: 1) the indirect process that consists in the separation of the individual FO alcohols followed by their esterification with a carboxylic acid, and 2) the direct process that consists in the simultaneous reaction of the whole FO alcohol mixture with a carboxylic acid. In the first case, only single esterifications (one-acid to one-alcohol at a time) occur and they can be independently studied. However, when directly reacting the fusel mixture, the chemistry of the process is more complex as it is necessary to consider simultaneous esterification reactions.

Because the isoamyl alcohol is the major component in FO, most literature reports dealing with FO upgrading have focused on the production of iso-amyl esters [5]–[7]. In the case of FO direct esterification, the process is less developed partly because the numerous interactions that must be considered in terms of thermodynamic and kinetic modeling. For example, when modeling simultaneous esterification of acetic acid with FO (consisting of a mixture of three alcohols), a higly non-ideal thermodynamic model (e.g. activity-based model) is required for the evaluation of the chemical and phase equilibria, taking into account the 8 components in the mixture (three alcohols, three esters, acetic acid, and water). In such system, several binary and ternary azeotropes and also liquid-liquid behavior are observed.

In this work several process alternatives for the direct esterification of the FO using reactive distillation were synthetized and designed with the aid of the geometrical tools based in the reactive residue curve maps. For this purpose the previous works were updated in three directions: 1) the existent thermodynamic model [4] was improved by the addition of new experimental vapor liquid equilibrium data for the binary mixtures of acetates and the quality assessment for the previous reported data, 2) new experimental kinetic data on the ion exchange resin Amberlyst 70 were used to fit a kinetic model that is useful at the distillation operation conditions and 3) sequences of two and three columns thermally coupled (or coupled in the decanter) were considered during the analysis.

At the end of this work we conclude that is possible convert completely the fusel oil in anhydrous mixtures of acetates that can be used as solvent base in the paint synthesis or for the straightforward production of pure acetates. On the other hand, we contribute with the refinement of the available geometrical techniques (shortcuts) for sequencing of columns in reactive distillation process with multiple esterification reactions. Finally, results also confirmed that the simple sequential procedure, where the synthesis and design are based in the geometrical analysis of the residue curve maps, is a powerful tool that providing good approximations to the rigorous calculations in a simulator.

References

[1] A. G. Patil, S. M. Koolwal, and H. D. Butala, "Fusel oil: Composition, removal and potential utilization.," *Int.Sugar J.*, vol. 104, no. 1238, pp. 51–58, 2002.

[2] M. C. Ferreira, A. J. A. Meirelles, and E. A. C. Batista, "Study of the Fusel Oil Distillation Process," *Ind. Eng. Chem. Res.*, vol. 52, no. 6, pp. 2336–2351, Feb. 2013.

[3] P. Patidar and S. M. Mahajani, "Esterification of fusel oil using reactive distillation – Part I: Reaction kinetics," *Chem. Eng. J.*, vol. 207–208, pp. 377–387, Oct. 2012.

[4] P. Patidar and S. M. Mahajani, "Esterification of Fusel Oil Using Reactive Distillation. Part II: Process Alternatives," *Ind. Eng. Chem. Res.*, vol. 52, no. 47, pp. 16637–16647, Nov. 2013.

[5] B. Saha, H. Teo, and A. Alqahtani, "iso-Amyl acetate synthesis by catalytic distillation," *Int. J. Chem. React. Eng.*, vol. 3, no. A11, pp. 1–14, 2005.

[6] W. Osorio-Viana, H. N. Ibarra-Taquez, I. Dobrosz-Gómez, and M. Á. Gómez-García, "Hybrid membrane and conventional processes comparison for isoamyl acetate production," *Chem. Eng. Process. Process Intensif.*, vol. 76, pp. 70–82, Feb. 2014.

[7] F. Leyva, A. Orjuela, A. Kolah, C. Lira, D. Miller, and G. Rodríguez, "Isoamyl propionate production by reactive distillation," *Sep. Purif. Technol.*, vol. 146, pp. 199–212, May 2015.

481125 Characterizing Convective Heat Transfer Coefficients in Membrane Distillation Cassettes

Abstract:

Accurate characterization of conductive heat transfer in membrane distillation (MD) is essential for determining membrane permeability, calculating thermal efficiency, and describing the effects of process parameters on permeate flux. This work demonstrates the effect of convective heat transfer on common MD performance metrics, reviews existing methods to model and measure convective heat transfer in MD cassettes, and proposes a novel experimental method to validate and or correct estimated convective heat transfer coefficients. This simple method compares trends in experimental permeability to those of theoretical permeability over a series of driving forces and or flow conditions. We first validate this method using experimental data collected from two cassettes using a single membrane. After correcting the heat transfer coefficients for each cassette, the permeability of a single membrane is consistent across two different cassettes. We then validate this method for a single cassette using two different membranes, finding that the experimentally determined cassette correction factor is consistent within the cassette. Consistent and accurate reporting of cassette heat transfer coefficients will promote direct comparison of membrane and module performance across different MD systems.

470263 iCVD Modified Electrospun Membranes for Membrane Distillation

Abstract:

The rising demand for clean water has intensified the necessity for large scale seawater desalination processes. Membrane Distillation (MD) is one method for desalination, which relies on a temperature gradient to allow for water vapor to diffuse through a membrane. While MD is useful for treating highly concentrated feed solutions, the challenge comes with finding an appropriate membrane. Membranes fabricated through electrospinning exhibit many important characteristics for MD including high porosity and low tortuosity. These fibers must also exhibit hydrophobicity to prevent wetting. Many of the most spinnable and strongest polymer fibers are derived from hydrophilic fibers, making their use in MD challenging. Using the most spinnable polymer and then modifying the membrane for hydrophobicity would enable all polymers to be considered for this new type of membrane.

This work investigates the use of initiated chemical vapor deposition (iCVD) as a fiber modification technique for electrospun membranes. PAN and PVDF nanofiber membranes were fabricated using electrospinning to produce flat sheets. The fibers were then coated with a hydrophobic polymer, divinylbenzene (DVB), through iCVD. Since iCVD is a vapor phase, conformal coating technique, depositing a hydrophobic polymer onto the fiber mat can render any polymer fiber hydrophobic and viable for MD. The deposition rate for the iCVD coating was also investigated with the goal of accelerating the coating process. The membrane coupons were tested in a bench top direct contact membrane distillation unit with highly concentrated sodium chloride as the feed solution. It was discovered that the nanofiber membranes exhibited higher flux than commercial PVDF membranes, which can be attributed to the improved properties of the membranes through electrospinning and the conformal hydrophobic coating.

463059 Influence of Microporous Membrane Properties on VMD and DCMD: Experimental and Modeling Studies

Abstract:

A variety of microporous hydrophobic flat sheet membranes of polyvinylidene fluoride (PVDF) and expanded-polytetrafluoroethylene (e-PTFE) were studied to evaluate the influence of membrane properties on their performance in desalination by vacuum membrane distillation (VMD) and direct contact membrane distillation (DCMD) processes. The membrane thickness was varied between 23 µm to 125 µm; the pore size was varied from 0.05 µm to 0.45 µm. The porosity was generally high in the range of 0.7- 0.8. VMD experiments were performed over a hot brine temperature range of 65 °C to 85 °C for various feed flow rates and various vacuum levels in a rectangular chlorinated polyvinyl chloride (CPVC) cell. Membrane properties, such as the maximum pore size, tortuosity as well as the liquid entry pressure (LEP) were experimentally determined. Water vapor fluxes were predicted and compared using two models: the Knudsen diffusion and the dusty-gas model (DGM). The deviation between two models is within 1.3 %. Knudsen diffusion is the dominant regime in VMD transport since Knudsen number, K_n, for all membrane were larger than 1 at all temperatures. The boundary layer heat transfer resistance in the membrane cell and the membrane surface temperature were determined from experimental data via Wilson plot. Good agreements of membrane mass transfer coefficients and water vapor fluxes were found between the DGM simulations and the experimental results (deviation within 5%). Same membranes were studied in a circular stainless steel cell and the CPVC cell for DCMD behavior using the Wilson plot method. Experimental and predicted values of the membrane mass transfer coefficients and water vapor fluxes have been compared for brine temperature range of 65°C - 85°C and distillate temperature at 20°C. Pore size distribution (PSD) did not matter if the membrane pore size variation falls entirely in the Knudsen diffusion or the transition regime. However, for membranes having nominal pore size   0.1 µm, K_n 1, PSD played an important role in DCMD. The effect of membrane thickness on water vapor flux and thermal efficiency were also simulated and compared with the experimental results.

471923 Free Surface Electrospun Polyvinylidene Fluoride Membranes for Direct Contact Membrane Distillation

Abstract:

Water scarcity is a problem that affects 2.7 billion people around the world. By the year 2025, water scarcity is expected to affect nearly 2/3 of the worldâ€™s population, particularly in regions that do not have the economic stability to provide solutions to this problem. The use of current technologies such as reverse osmosis, multistage flash distillation, electrodialysis reversal and membrane distillation are commonly used methods of water purification. In this study, the use of direct contact membrane distillation (DCMD) has been chosen due to its low operating costs in comparison to processes such as reverse osmosis which operates at high pressures. The current process for manufacturing DCMD membranes is considered to be expensive, inefficient, and does not allow for the optimization of desirable membrane properties. Through this research an alternative method of fabricating membranes has been studied through the means of free surface electrospinning. This process allows for controllability of desirable properties such as fiber diameter (shown to be related to pore size), and membrane thickness. Here, DCMD membranes are produced using polymeric solution of 22 wt% Polyvinylidene Fluoride (PVdF) in Dimethylacetamide (DMAc). In the free surface electrospinning setup, a rotating wire spindle was partially submerged in a solution bath and a voltage of 40 kV was applied. The electric field causes charge to accumulate on the free surface, leading to the formation of electrohydrodynamic jets. Electrospinning was performed in an environment with relative humidity ranging from 50-80 RH%. In the presence of high humidity, water absorbs into the electrohydrodynamic jets resulting in the phase separation of the PVdF. The gelled fibers were collected on a grounded rotating drum and were allowed to solidify under room conditions. The performance of the produced membranes was evaluated using an in-house built DCMD apparatus and the membranes were characterized by DCDM performance and scanning electron micrograph (SEM) images.

473623 Optimizing Polysulfone Mixed Matrix Membranes for High Permeate Flux in Membrane Distillation

Abstract:

The development of low-cost membranes specific to Membrane Distillation (MD) remains a major challenge to the progression of the technology towards commercialization. This is because the desirable properties and morphology for an ideal MD membrane are fundamentally different from most commercially available membranes. MD membrane should be hydrophobic and highly porous. The hydrophobicity is so that liquid water does not pass through, and high porosity allows for more space for the water vapor molecules to pass through. Polysulfone is a well-known hydrophobic and thermally resistant polymer, which makes it a good polymer candidate to produce laboratory fabricated membranes.

Additives to the membrane matrix are known to affect the pore-formation mechanism. Morphological studies on GO mixed matrix membranes show that the GO particles tend to aggregate near the surface of the membrane, contributing to a stronger negative surface charge due to the oxygen-containing functional groups. Adding graphene oxide (GO) can increase the hydrophilicity and surface charge of the membrane. This hydrophilicity can be engineered by using thiolated graphene, with sulfur-containing functional groups such as thiol. The sulfur would theoretically contribute less hydrophilicity than the oxygen, due to sulfur having a lower electronegativity oxygen on the Pauling scale.

In this research study, Polysulfone/Graphene Oxide and Polysulfone/Thiolated Graphene mixed matrix membranes were produced using the phase inversion method and tested for permeate flux and salt rejection on MD. Wet phase inversion is a well-known method that can be used to produce membranes in the laboratory without the need for overly sophisticated technology. Morphological studies of fabricated membranes were done using SEM. The effects of membrane thickness, coagulation temperature, and graphene-type additive weight percentage on the permeate flux and salt rejection were quantified.

463411 Bio-Mimetic Hollow Fiber Membrane Contactor Process for Low-Cost High-Purity Oxygen Production

Abstract:

Oxygen production from air has wide applications in oxygen-intensive industries, such as energy production (e.g., IGCC and oxy-combustion systems), fuels, chemicals, and other industries. Cryogenic distillation, pressure swing adsorption, polymeric and ionic transport membrane technologies have been used for air separation. Among these technologies, cryogenic distillation is the most mature technology for large scale and high purity (>99%) O_2 production. However, cryogenic distillation-based air separation is costly and energy-intensive to operate, accounting for up to 15% of the total gasification plant capital cost, and consuming over 35% of in-plant power use.

Supported by the US Department of Energy (No. DE-FE0024080), we are developing a novel technology to produce greater than 95% oxygen from air via a single-stage gas-liquid membrane contactor with oxygen carrier solutions as solvent. In this process, air is sent to a membrane absorber and passes through small-diameter membrane tubes, roughly 15 mil inside diameter, while a lean O_2 carrier solution flows counter currently on the shell side of the membrane. Unlike the other production alternatives, the air stream needs to be compressed to only a few psi and does not require heating for the ion transport through membranes or cooling for cryogenic separations. The O_2 permeates through the membrane pores and is absorbed in the O_2 carrier solution. The O_2-rich carrier solution can be regenerated in a second membrane module (desorber) operated in a reverse process. In that case, the O_2-rich carrier solution is fed to the shell side of the hollow fibers and vacuum is used to draw O_2 into the pore side of hollow fibers. In this bio-mimetic process, a membrane module with hollow fibers and O_2 carrier solution mimics "blood vessels" and "blood".

In our study, low-cost poly(ethyleneimine)-cobalt (PEI-Co) complex solutions with high O_2 absorption capacity (as high as 750 mL O_2 (STP)/L solution), low viscosity (< 3 times of water), and fast O_2 absorption rate have been developed. The O_2 equilibrium absorption capacity increases with the PEI/Co ratio and reaches a plateau at ratio of 10. Two different approaches were employed in the desorption process: 1) vacuum, and 2) using argon as sweep gas. By using the second approach, O_2 concentration as high as 99% was observed in desorbed gas stream on an argon free basis. The production of concentrated O_2 has been validated experimentally in the bio-mimetic process. The estimated cost including capital, operating, maintenance, and energy use for the bio-mimetic O_2 separation technology is \$19.96/ton O_2, which is only about 56% of the benchmark cryogenic distillation (\$35.80/ton O_2). This technology has potential to produce oxygen with purity as high as 99.9% for applications in IGCC, oxy-combustion, and other advanced power generation technologies.

459680 Tri-Bore PVDF Hollow Fibers with a Super-Hydrophobic Coating for Membrane Distillation

Abstract:

Membranes with good mechanical strength, high vapor flux and outstanding anti-wetting properties are essential for membrane distillation (MD) applications. In this work, porous polyvinylidene fluoride (PVDF) tri-bore hollow fiber membranes with super-hydrophobicity are developed to achieve these desired properties. The tri-bore hollow fiber offers better mechanical strength than the conventional single-bore fiber. To improve its anti-wetting properties, Teflon® AF 2400 is coated on the membrane surface. The effects of coating on membrane morphology, performance and anti-wetting properties have been thoroughly investigated. With an optimal coating condition (0.025 wt% of Teflon® AF 2400, 30 s), a super-hydrophobic surface with a contact angle of 151° is achieved. The resultant membrane shows an increase of 109% in liquid entry pressure (LEP) with a slight sacrifice of 21% in flux. Long term direct contact MD tests have confirmed that the Teflon® AF 2400 coated membrane has enhanced stability with an average flux of 21 kg m^{-2} h^{-1} and rejection of 99.99% at 60 C for desalination application.

464065 Design and Economic Analysis of Membrane Crystallization Processes

Abstract:

Design and economic analysis of processes for the crystallization of adipic acid are studied. Two membrane-based process alternatives, reverse osmosis (RO) and porous hydrophobic membranes, are considered, as well as a conventional multi-effect vacuum evaporation crystallization process. For each of these alternatives, a small-scale batch process (2×10^5 kg/yr) and a large-scale continuous process (2×10^7 kg/yr) are designed and a detailed economic analysis is performed. For the batch processes, the seeding policy and optimization of the evaporation rate trajectory are considered. The batch reverse osmosis process is similar to one described by Kuhn et al. (2009). Unlike the situation when using RO membranes, when porous hydrophobic membranes are used for the separation it is necessary to supply the enthalpy of vaporization of the water. However this energy is supplied at a relatively low temperature; it could be provided by waste heat if waste heat is available (Drioli et al., 2015). Therefore, in the economic analysis two scenarios are considered: one in which low-grade waste heat is available at no cost, and one in which no such heat is available.

The results show that when low-grade waste heat is not available, the RO membrane process has the lowest cost because, although pumping costs are significant, these costs are more than offset by the energy that is saved because the enthalpy of vaporization of the water does not need to be provided. When low-grade waste heat is available at no cost, a conventional single-effect vacuum evaporation process has the lowest cost because the low-grade waste heat can also be applied to power the evaporation in the conventional process and the conventional process does not incur the capital cost for the membranes.

References

Drioli, Enrico, Aamer Ali, Francesca Macedonio, Membrane Distillation: Recent Development and Perspectives. *Desalination* **2015**, 356, 56-84

Kuhn, Jelan, Richard Lakerveld, Herman J. M. Kramer, Johan Grievink, Peter J. Jansens, Characterization and Dynamic Optimization of Membrane-Assisted Crystallization of Adipic Acid. *IECR* **2009**, 48, 5360-5369

470546 Optimizing Design Parameters for Thin Film Composite Hollow Fiber Membranes and Modules for Forward Osmosis

Abstract:

Hollow fiber membranes have been considered a valuable platform for osmotic membrane processes because of their high packing density relative to flat sheet membranes. This study evaluates an approach to make thin film composite (TFC) hollow fiber membranes for forward osmosis (FO) by simply employing commercial hollow fiber ultrafiltration (UF) membranes as support material. A thin polyamide film with excellent selective properties was synthesized on the inner surface of hollow fibers via interfacial polymerization. Besides demonstrating the feasibility to be used as good TFC FO membrane supports, the commercial hollow fiber UF membranes also provide a systematic platform with consistent properties to study structure-performance relationship of FO hollow fiber membranes. A series of commercial hollow fiber membranes were used to evaluate how molecular weight cutoff (MWCO), fiber dimensions (inner and outer diameter) and module dimensions (length and width) impacted overall performance of the TFC membranes and module. With the availability of reproducible membranes and modules, we have developed a computational fluid dynamics (CFD) model with COMSOL Multiphysics in order to optimize hollow fiber module design for FO application. The comparison between the model and experimental results provided insight into both how more accurate models can be developed and how these models can be used to design better modules without costly experimental testing.

458145 Monitoring Biofouling Dynamics on Foward Osmosis (FO) Membranes Using a Clsm-Compatible Microfluidic Biofilm Flow Cell

Abstract:

In membrane biofouling studies, the current method for examining the growth of biofilms on water-purifying membranes in membrane testing flow cells has been a â€œmembrane autopsyâ€ followed by microscopic observations. However, during the preparation of membrane samples, uncontrollable and unpredictable disturbances to the biofilms on the membrane surfaces may occur, resulting in artifacts that may lead to misinterpretation of experimental results. Furthermore, the autopsy approach can only provide a snapshot of and the dynamically changing biofouling conditions over time. This limitation is mainly due to the lack of tools that allow us to monitor dynamics of biofouling without the need of dissembling the membrane testing systems. In this study, we developed a novel multichannel microfluidic membrane flow cell that enables non-destructive, real-time monitoring of biofouling dynamics using confocal laser scanning microscopy (CLSM). As a proof of concept, we examined the development of green florescent protein (GFP)-tagged bacterial biofilms on forward osmosis (FO) membranes. The temporal profiles of quantitative biofouling parameters including surface coverage, biovolume and biofilm thickness were obtained without disrupting the continuous operation of the membrane testing system. The microfluidic membrane flow cell developed in this study can be readily applied to evaluate anti-biofouling activities of FO membranes and allows direct comparison of biofouling dynamics between FO membranes with different surface modifications.

462938 Forward Osmosis Using Draw Solutions Manifesting Liquid-Liquid Phase Separation

Abstract:

Over the past decade, Forward Osmosis (FO) technology has emerged as a promising alternative to conventional Reverse Osmosis (RO) to produce fresh water through desalination of seawater. In FO, water is extracted from a saline water source (such as seawater) by a draw solution having a higher osmotic pressure than the saline water; the diluted draw solution is then split into a stream of concentrated draw solution and a stream of essentially pure water in process steps collectively referred to as regeneration. Draw agents whose regeneration is aided via liquid-liquid phase separation have gained much attention in recent years [1-4]. Specifically, aqueous solutions of these draw agents manifest either a lower critical solution temperature (LCST) phase transition or upper critical solution temperature or both. With draw solutions manifesting LCST, the FO step is typically performed at a temperature (T) below (or near) the LCST; for medium to high concentration of the draw agents, the osmotic pressure of the draw solution decreases with increasing T. Modest heating of the diluted draw solution facilitates regeneration by causing liquid-liquid phase separation yielding a water-rich phase and a second phase rich in the draw agent.

In this presentation, we describe the results from an experimental study where mixtures of two different glycol ethers (GE), tripropylene glycol methyl ether (TPM; MW=206 Da) and tripropylene glycol n-butyl ether (TPnB; MW=248 Da), were used as draw agents. These are larger in size than the GEs used in previous studies [2].

Cloud point temperatures of aqueous solutions of these GE mixtures were determined. By adjusting the relative amounts of TPM and TPnB, one can tune the cloud point temperature and the LCST. We also determined the activity of water (in the 288-323K range) at various draw solution compositions, using an Aqualab TDL water activity meter. At high GE wt% and temperature modestly below the cloud point T, the osmotic pressure of the draw solution is much higher than that of seawater, enabling their use as draw agents in FO.

The kinematic viscosities of these draw solution mixtures at different temperatures were found using a micro-Ubbelohde viscometer. Through NMR measurements, the self-diffusion coefficients of the GEs in the draw solutions at different temperatures were determined.

FO experiments were conducted using commercial CTA embedded support FO membrane (Fluid TechnologyÂ Solutions, Inc., USA) to assess water flux, reverse flux of the draw agent, and the effect of orientation of the membrane (support layer facing the draw solution â€“ the so-called FO mode; and, its inverse â€“ the so-called Pressure Retarded Osmosis (PRO) mode). These experiments were performed with fresh water on one side and draw solution on the other side of the membrane. As expected [5], the water flux was higher in the PRO mode than in the FO mode. While the initial water flux was high (~7 lit/m^{2}h), it dropped rapidly to ~ 2 lit/m^{2}h indicating appreciable membrane fouling. Even after thorough washing, the original high flux state could not be recovered.

In the PRO mode, increasing the circulation rate of the draw solution led to measurably large increase in the water flux, clearly revealing the effect of concentration polarization. FO experiments were done at three different temperatures; water flux increased when the operating temperature was increased from 295K to 305K, but decreased when the operating temperature was increased to 315K. This can be rationalized qualitatively as a consequence of the temperature dependence of the osmotic pressure, kinematic viscosity and diffusivity.

Appreciable reverse flux was recorded in the FO experiments (4.91 gr/m^{2}h in FO mode and 2.90 gr/m^{2}h in PRO mode), indicating that molecular weight in the range of 200-250 Da is not sufficiently large to suppress reverse flux. FO experiments performed with aqueous solution of polypropylene glycol (PPG 425; average MW=425 Da) showed lower reverse fluxes (2.53 gr/m^{2}h in FO mode and 1.74 gr/m^{2}h in PRO mode), suggesting that even higher molecular weights are required to suppress reverse flux of draw agent.

In summary, as a result of reverse flux and membrane fouling, using aqueous solutions of glycol ethers directly as draw solutions does not appear promising. Their indirect use, as in the process described by [6], which uses an ionic salt as a draw agent and employs an aqueous polymer solution to recover water from the diluted draw solution, appears to be more realistic.

Keywords: *forward osmosis, lower critical solution temperature, glycol ethers, liquid-liquid phase separation*

References:

1. Mok, Y., et al., *Circulatory osmotic desalination driven by a mild temperature gradient based on lower critical solution temperature (LCST) phase transition materials.* Physical Chemistry Chemical Physics, 2013. **15**(44): p. 19510-19517.

2. Nakayama, D., et al., *Lower critical solution temperature (LCST) phase separation of glycol ethers for forward osmotic control.* Physical Chemistry Chemical Physics, 2014. **16**(11): p. 5319-5325.

3. Noh, M., et al., *Novel lower critical solution temperature phase transition materials effectively control osmosis by mild temperature changes.* Chemical Communications, 2012. **48**(32): p. 3845-3847.

4. Zhao, D., et al., *Thermoresponsive copolymer-based draw solution for seawater desalination in a combined process of forward osmosis and membrane distillation.* Desalination, 2014. **348**: p. 26-32.

5. Cath, T.Y., et al., *Standard methodology for evaluating membrane performance in osmotically driven membrane processes.* Desalination, 2013. **312**: p. 31-38.

6. Rajagopalan, N. and V.A. Patel, *Solvent removal and recovery from inorganic and organic solutions*, 2014, Patent US8852436.

465747 Zwitterions Coated Hollow Fiber Membranes with Enhanced Antifouling Properties for Osmotic Power Generation from Municipal Wastewater

Abstract:

The rapid increases in energy consumption and CO_2 emissions have accelerated the exploration of renewable and green energy. Among them, osmotic energy released from the mixing of water streams with different salinities via pressure retarded osmosis (PRO) has attracted global attention owing to its abundance and renewability. Two factors primarily affect the productivity of a PRO process; namely, the PRO membrane and the feed pair. An ideal PRO membrane requires a high water permeability, high salt rejection, low concentration polarization, high mechanical strength, and minimal fouling tendency. In case of feed pairs, in addition to the conventional pair of river water and seawater, the feed pair of concentrated reverse osmosis (RO) brine and wastewater from municipal recycle plants has received great attention because it has a higher salinity gradient and can potentially produce a higher power density. In addition, this new feed pair may add extra values such as lowering the overall energy consumption for RO, diluting the seawater RO brine for ecologically-friendly disposal, and saving some seawater pretreatment costs.

However, fouling on PRO membranes leads to severe declines in water flux and power density because their porous substrates are facing the wastewater feed. Thus, inorganics, organics and microorganisms in the wastewater are prone to depositing on the substrate surface and even in its pores. In order to reduce the fouling propensity, coating the substrate surface of PRO membranes with zwitterionic materials proves to be an effective way. In our work, 2-methacryloyloxyethylphosphorylcholine (MPC), is modified and grafted onto the polydopamine (PDA) coated poly(ether sulfone) (PES) hollow fiber substrate. Both the synthesis and surface coating of MPC are simple and straightforward. Compared with the pristine PES and PES-PDA substrates, the MPC modified substrate (PES-PDA-MPC) exhibits high resistance to protein adsorption as well as bacteria adhesion. By using a state-of-the-art thin-film composite poly(ether sulfone) (TFC-PES) hollow fiber membrane as the control for power generation, the power density of the TFC-PES-PDA-MPC membrane can achieve as high as 7.7 W/m^2 while the unmodified one has only 6.0 W/m^2 after 3 hâ€™s PRO tests. In conclusion, the osmotic power generation of PRO membranes can be significantly sustained by modifying the membrane surface with zwitterions.

453481 Second Interfacial Polymerization of Thin-Film Composite Polyamide Membrane with Polyethyleneimine for Forward Osmosis Applications

Abstract:

In this work, a fast and simple surface modification-second interfacial polymerization (SIP) of thin-film composite (TFC) membranes using polyethyleneimine (PEI) of various molecular weights, was applied to improve the separation performance for forward osmosis (FO) applications. Compared to the control TFC membranes, PEI -modified membranes exhibit higher water flux and lower fouling propensity. The effects of PEI molecular weight and PEI concentration on the membrane morphology and separation performance were investigated systematically with various characterizations. Post-treatment to surface modified membranes was carried out using four aqueous solutions of different pH values. Meanwhile, the FO performance, intrinsic separation properties and antifouling properties of these post-treated membranes were also studied. In comparison with the surface-modified TFC membranes with PEI, the post-treated TFC membranes show higher water flux and improved anti-fouling properties, especially those post-treated with acidic or alkaline aqueous solutions.

455190 Integrated Seawater Reverse Osmosis and Pressure Retarded Osmosis Processes for Sustainable Water and Energy Generation

Abstract:

Pressure retarded osmosis (PRO) is a promising technology to reduce the specific energy consumption of a seawater reverse osmosis (SWRO) plant. In this study, it is projected that 25.6-40.7 million kWh/day of energy can be recovered globally, if the brines from SWRO are used as the draw solution and diluted to the seawater level in a PRO system. Detailed integrated SWRO-PRO processes are developed in this study with the option to form a closed-loop SWRO-PRO process that can substantially reduce the pretreatment cost of desalination. The governing mathematical models that describe both the transport phenomena on a module level and the energy flow on a system level are developed to evaluate the performances of the SWRO-PRO processes. The model aims to investigate the performance of the hollow fibers as dilution occurs and provides guidelines on hollow fiber module design and process operation. Determining the dilution factor and the corresponding operating pressure of PRO is the key to optimize the integrated process. The specific energy consumptions of three SWRO-involved processes; namely, (1) SWRO without a pressure exchanger, (2) SWRO with a pressure exchanger, and (3) SWRO with pressure exchangers and PRO are compared. The results show that the specific energy consumptions for the above three processes are 5.51, 1.79 and 1.08 kWh/(m^3 of desalinated water) for a 25% recovery SWRO plant; and 4.13, 2.27 and 1.14 kWh/(m^3 of desalinated water) for a 50% recovery SWRO plant, using either freshwater or wastewater as the feed solution in PRO.

472843 Magnetic Nanoparticles on Membrane Surface As Micromixers and Heaters

Abstract:

It is well known that the two most important factors limiting the use of membrane technology are concentration polarization and membrane fouling.[1] Efforts have been made to solve these problems by means of optimizing module design or inducing mixing during operation, pretreatment of the feed and surface modification of the membrane. Feed pretreatment can only postpone the decline of membrane performance while surface modification is not to suppress concentration polarization[2]. Therefore, there is still a need to find an efficient way to improve fouling resistance of membranes and to reduce concentration polarization.

Here, we report a novel approach to solve these problems by attaching magnetically responsive nanoparticles to the membrane surface. Here polymers (polyHEMA and poly NIPAAm), were grafted to membrane surfaces by surface initiated atom transfer radical polymerization. After that, the bromine groups at chain ends were converted to primary amine groups by Gabriel synthesis and then used for attaching carboxyl covered superparamagnetic Fe_3O_4 nanoparticles. The chain/nanoparticle density can be controlled by varying the ratio of active to inactive ATRP initiator in the initiator immobilization step. The modification and nanoparticle immobilization were monitored by XPS and SEM. XPS confirmed each step of modification. SEM images showed that the nanoparticles were successfully attached to the membrane surface and the density can be tuned in a broad range. With nanoparticles attached to the end of polyHEMA, the polymer chains acted as micromixers under oscillating magnetic field and mixing above the membrane surface was observed. This mixing resulted in a significantly improved membrane performance (flux and salt rejection) which can be ascribed to reduced concentration polarization. On the other hand, with polyNIPAAm on the surface, nanoparticles were used as heater in the presence of magnetic field to exert temperature change below and above LCST of this thermo-responsive polymer. Subsequently conformational changes of polyNIPAAm leaded to a dense/loose layer structure transformation on the membrane surface and changed membrane performance. This result demonstrated the possibility of using external magnetic field to control the membrane performance which is promising for industrial applications to realize instantaneous product control without interrupting the production line.

References

1. R.W. Baker, *Membrane Technology and Applications, 2nd ed.*; John Wiley: Chichester, (2004).

2. D. Rana and T. Matsuura, *Chem. Rev.*, **110**, 2448 (2010).

463224 Surface Characterization of Membrane Surfaces Nanostructured with Tethered Hydrophlic Polymers

Abstract:

Membrane surface structuring with end-tethered hydrophilic polymers is an effective approach for reducing the fouling propensity and improving cleaning efficiency of RO and UF membranes. It has been suggested that, in addition to tailoring of surface chemistry, swelling of the polymer surface layer plays a crucial role in surface screening and minimization of foulant attachment. Accordingly, in the present study the impact of solvent conditions (i.e., salinity and pH) on swelling of hydrophilic tethered polymers was investigated focusing on poly(acrylic acid) (PAAc) chains end-grafted onto polysulfone (PSf) membrane. PAAc chains were graft polymerized onto PSf surfaces via atmospheric pressure plasma-induced graft polymerization (APPIGP). Surface topography of the grafted PAAc layers was evaluated in various solvents (i.e., air, deionized water, and saline water). The size of the tethered polymer chains was assessed by determining the rupture length distributions of the tethered polymer chains via single molecule force measurements performed in aqueous solutions of different salinity. Fouling potential of the polymer layers was then evaluated by adhesion force measurements performed with a functionalized colloidal AFM probe. The results of the present study provide a direct measure of the potential effectiveness of tethered chains for membrane fouling reduction. In this regard, direct surface force measurement can guide the optimization of surface structuring with polymer chains for synthesizing high performance low fouling membranes.

469179 Facile Membrane Surface Activation Eliminates Fouling on Demand and Mitigates Concentration Polarization

Abstract:

Fouling of high pressure membranes is a recalcitratnt problem, particularly in reverse osmosis (RO) and nanofiltration (NF) systems. Colloidal, organic and biological fouling are the most commonly encountered challenges in industrial desalination and wastewater treatment processes and there is no convenient way to clean the membrane without interrupting operation or resorting to expensive fouling mitigation infrastructure. Therefore, significant energy savings can be achieved by real time fouling elimination and concentration polarization mitigation. Although several important technological improvements were achieved in membrane design and fouling control in past few years, for example- TiO_2 embedded composite antifouling membranes, photothermal nanoheater membranes or carbon nanotube containing composite electrolytic membranes, which are promising in reducing different types of fouling. However, all the reported methodologies suffer from requiring extensive new infrastructure, lack of scalability and most importantly, efficiency without loss in productivity.

Here, we have demonstrated a simple scalable approach of coating RO/ NF membrane with bioinspired adhesive polymer polydopamine and copper oxide nanoparticles, which decomposed H_2O_2 (0.006% - 0.024%) to oxygen bubbles and thereby, swept foulants away from the surface and at the same time reduced concentration polarization by enhancing mass transfer co-efficient. The polydopamine layer embedded the nanoparticles and acted as the reactive catalytic surface. Additionally, polydopamine layer helped to protect the underlying polyamide layer from hydroxyl radicals by acting as free radical scavenger. Such composite membrane eliminated flux decline with colloidal foulants in repeated runs and even increased the flux level beyond the baseline value with organic humic acid foulants. Enhanced catalytic activity of humic acids in presence of H_2O_2 and reduction of concentration polarization were found to be responsible for such flux enhancements. Embedded copper oxide nanoparticles also retarded *E. Coli* attachment to the membrane surface in stirred cell experiments. On the other hand, such composite membrane was found to eliminate concentration polarization upon H_2O_2 addition by bubble induced mixing. Therefore, such scalable facile approach holds promise to mitigate fouling in industrial membrane systems, reduce concentration polarization and thereby, saving considerable energy.

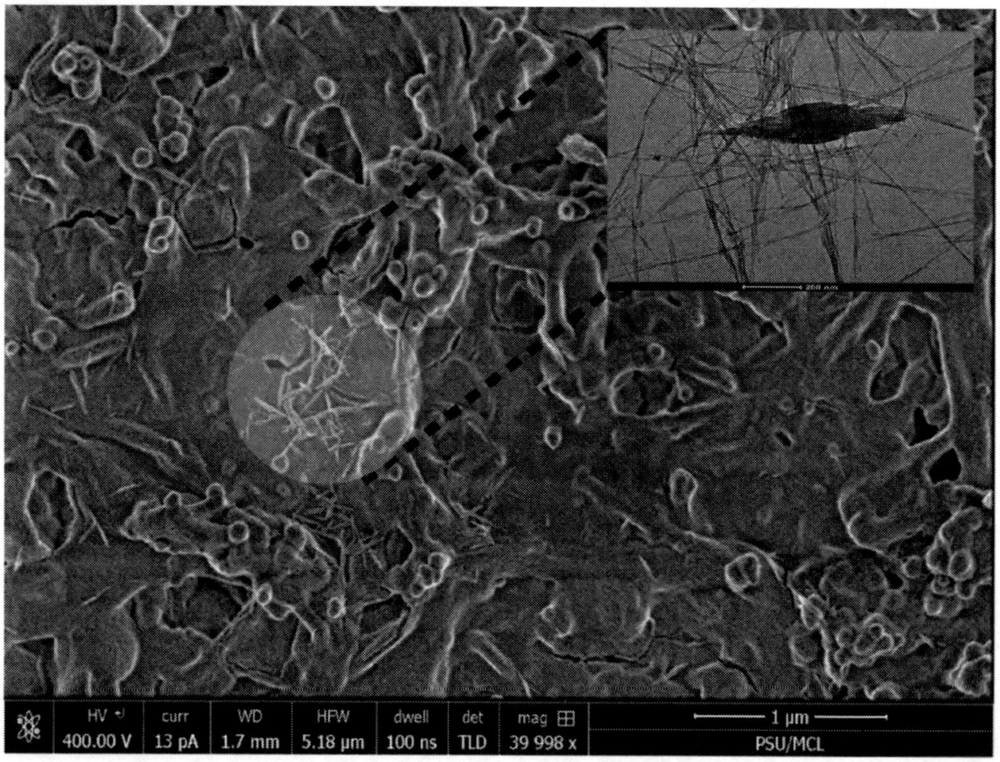

Figure. SEM micrograph of rod shaped CuO nanoparticles on polydopamine coated RO membrane. In the inset, TEM micrograph of petal shaped CuO nanorods.

457389 Development of Anti-Fouling, Anti-Microbial Membranes for Wastewater Treatment

Abstract:

Development of Anti-Fouling, Anti-Microbial Membranes for Wastewater Treatment

Steven Weinman[a], Maria Bass[b], Viatcheslav Freger[b], Moshe Herzberg[c], Scott Husson[a]

[a]Department of Chemical & Biomolecular Engineering, Clemson University, Clemson, SC 29634
[b]Wolfson Department of Chemical Engineering, Technion - Israel Institute of Technology, Haifa, 32000 Israel
[c]Zuckerberg Institute for Water Research, Ben Gurion University of the Negev, Sede-Boqer Campus, Midreshet Ben Gurion, 84990 Israel

Treatment of impaired waters for beneficial use exposes membranes to feed waters that often rapidly foul the membranes with particles (salt, bacteria, biomolecules, etc.). Fouling leads to a loss of membrane productivity over time and is one of the largest costs associated with membrane processes in water treatment. New coatings that limit and reverse fouling would have significant economic and societal impacts. Developing these advanced coatings is the focus of our work.

Prior studies to control biofouling largely have focused either on preventing microorganisms from attaching or killing the microorganisms once they do attach. Based on considerable literature and experience, we know that each of these control strategies alone can be effective but not sufficient for controlling biofouling. Our hypothesis is that membranes coated with dual-mode polymer nanolayers can limit and reverse membrane biofouling by switching reversibly between passive (antifouling) and active (antimicrobial) states of biofouling control. This strategy differs fundamentally from most other surface modification strategies that rely solely on passive control (i.e., using coatings designed only to weaken foulant adhesion and, thus, limit their accumulation) or active control (e.g., adding biocides).

In this presentation, we will introduce the methodologies used to coat surfaces with carboxybetaine zwitterion dual-mode polymer nanolayers designed to combat biofouling. We will explain the chemical switching mechanism and use FTIR measurements to support the reversible switching of the polymer nanolayer between its antifouling and antimicrobial states. Subsequently, the feasibility of the approach will be illustrated by bacterial deposition and membrane biofouling experiments using appropriate model systems. Initial bacterial deposition results show a similarly low bacteria initial attachment rate as compared to other common zwitterion coatings.

- Abstract Text

470814 Unsteady-State Facilitated Diffusive Membrane Systems

Abstract:

Separation processes are vital to the energy and chemicals industries as a means of transforming reaction mixtures into purified products, and increasingly as a tool to mitigate environmental contamination. Membrane separations are often proposed over other separation techniques, such as distillation, because they do not require energy-intensive phase changes to accomplish their goal. In a typical membrane separation, the chemical potential gradient of the solute itself is the force that drives the target solute through a membrane. In the case of dilute solutions, which characterize many environmental contaminants, the chemical potential gradient is low resulting in low fluxes and poor separations.

Membranes that exhibit facilitated transport mechanisms whereby chemically selective reactive "carriers" that promote solute permeation are incorporated within the membrane are often proposed as a means of addressing the low fluxes. Despite its potential, facilitated transport has proven difficult to realize in practical membrane systems. Supported liquid membranes, which are one method for realizing facilitated transport, allow for highly selective processes to be developed; however, because these membranes comprise a thin organic layer suspended between two aqueous solutions, they are prohibitively unstable.In our design, the carriers, which exhibit pH-dependent solute binding, are fixed to the membrane matrix and are used to enhance flux by forcing the membrane into an unsteady process. This system resembles a barrier membrane wherein reactive sites capture target solutes as they attempt to cross the membrane; however by using an external stimulus, such as pH, we manipulate the behavior of the reactive sites over time. The pH dependent reactive sites are periodically forced into a loading state, where a target solute is captured by the membrane's active sites, followed by an unloading state, where the solute is released from the membrane's structure, by varying the pH of the membrane's surroundings. As a result the concentration gradient, and hence the solute's flux across the membrane, are constantly varying. A theoretical model that describes the performance of the system, e.g., the net flow of material across the membrane, is developed.

This model describes the flux-enhancement ability of this membrane system as a function of membrane properties, such as thickness, density of reactive sites, and solute's diffusivity, and as a function of operating parameters such as the solute feed concentration, the loading time, and the unloading time.

An experimental membrane system was assembled to test the validity of theoretical predictions. The experimental system consists of chelation resins as the reactive carrier suspended in a poly(vinyl alcohol) hydrogel. Iminodiacetic acid groups within the resins bind the target solute, divalent metallic ions, reversibly as a function of solution pH. Experiments conducted under a variety of operating conditions demonstrate strong agreement between the model predictions and experimental results. The theoretical model, corroborated by the bench-scale experiments, allows for predictions for the performance of unsteady facilitated diffusion membrane systems as a function of membrane and system properties.

472239 Effect of Monomer and Cross-Linker Concentration on pH Responsive Functionalized Membranes

Abstract:

Polymers and polymer dispersions with different functional groups finds its application as templates for ion exchange, layer-by-layer deposition, and incorporation of different metal nanoparticles. In particular, cross-linked hydrophilic polymers are studied in different functionalization processes especially in preparation of responsive membranes. Membrane functionalization such as surface modification with functional groups; provides us an opportunity to incorporate catalytic nanoparticles in the membrane pores for in situ surface reaction. Functionalized membranes incorporated with Fe/Pd nanoparticles provide a novel platform for organic pollutant degradation reactions and for selective removal of contaminants.

However, to incorporate nanoparticles in membrane pores, monomer and cross-linker concentration plays a significant role during polymerization process. A change in monomer and cross-linker concentration could have effect the whole functionalization of membranes. So far there is no study regarding the effect of changing of monomer and cross-linker concentration on functionalized membranes.

For this study we have changed the monomer concentration from 10 to 20 wt% of polymer solution and cross-linker concentration from 0.5 to 1 mol% of monomer concentration during preparation of different batches of functionalized pH responsive membranes, keeping all other parameters constant. This give us a matrix of monomer and cross-linker concentration to do functionalization of membranes. The synthesized nanoparticles incorporated membranes were characterized by electronic microscopy, X-ray spectrometry and image analysis before-, after- nano particle loading.

Finally, we report the optimum concentration of monomer, cross-linker concentration along with Fe/Pd nano particle loading for making pH responsive membrane. These membranes have both responsive behavior and catalytic properties

to degrade toxic organics (such as, TCE, PCBs) from water.

We like to thank NIH-NIEHS-SRC for funding this research project.

471129 Self-Cleaning, Photoresponsive Thin Film Composite Water Filtration Membranes

Abstract:

Materials with reversibly switchable properties hold great interest in the development of membranes with new functions such as fouling resistance and self-cleaning, adjustable pore size, and tunable flux. Photoresponsive membranes can show these changes without the need for changes to the feed composition. They can also show orthogonal response when combined with pH- or ionic strength responsive functionalities. This research focuses on light-responsive membranes *that can self-clean to remove foulants upon exposure to light.* We show the synthesis of new comb-shaped polymers with photo-regulated zwitterion formation and their use as responsive membrane selective layers. Zwitterions are charged molecules that comprise equal numbers of covalently bound cations and anions. The unique co-existence of opposite charges bridged through alkylene spacers provides zwitterions extremely high polarity. This feature leads to the exceptional anti-fouling properties for zwitterionic materials. This has exciting applications in the membrane field. We synthesized novel comb-shaped graft copolymers at different side-chain lengths with polyacrylonitrile (PAN) backbones and photoreactive side-chains using atom transfer radical polymerization (ATRP). The side-chains undergo a light-induced transition between a hydrophobic neutral state and a zwitterionic state, allowing photo-regulated control over membrane features. We used these graft copolymers to modify a commercially available UF membrane by coating a thin layer of the copolymer solution followed by precipitation in a non-solvent. Prior to any photo-treatment, coated membranes consist mainly of hydrophobic groups that dominate the membrane surface, and they increase upon visible light irradiation. We postulate that the hydrophobic groups promote the adsorption of hydrophobic solutes on channel walls, reducing flow rate. Upon reversing the photochemical response by irradiation with UV light, the hydrophobic groups are converted to zwitterionic groups that release the adsorbed molecules and permit water passage once again. This â€œself-cleaningâ€ behavior has been shown by measuring pre- and post-UV water permeability after fouling with bovine serum albumin. Short chain copolymer shows essentially complete self-cleaning performance, as indicated by the fact that even after three cycles of protein fouling, the water permeability can be increased back to its initial value using UV exposure. In addition, the short chain membrane surface proves antifouling at UV-irradiated state, which is shown by the absence of permeability decline after 2 hours of BSA filtration. The following visible light exposure does not change the permeability, however the visible-irradiated surface shows a significant permeability decline after 2h BSA filtration. Changes in fingerprint IR peaks pertinent to neutral and zwitterionic forms upon light treatment have further confirmed the structural difference between the two forms at a molecular level. This finding is also in agreement with the hydrophilicity and surface roughness results; the UV-irradiated surface has a smoother surface with enhanced hydrophilic wettability. Additional work focuses on determining optimal phototreatment regimes for self-cleaning behavior with minimal energy use, and developing a better understanding of how polymer self-organization controls this responsive behavior.

461684 The Degradation Mechanism of MOFs during Acid Gas Exposure: An Experimental and Computational Study

Abstract:

Developing sorbents for the adsorption of acid gases has garnered much research attention in the field of porous materials. Impregnated activated carbons have been widely explored for acid gas separation applications; however, metal-organic frameworks (MOFs) have emerged as a promising class of materials for acid gas separations. MOFs are characterized by metal clusters and organic linkers, large surface areas, and tunable chemical properties that make them excellent candidates for these gas adsorption applications.[1] Several MOFs using titanium for the metal source have shown promising adsorption properties for CO_2, water, and H_2S.[2] However, MOFs often degrade and are unable to retain their performance in the presence of SO_2, NO_2 or other acid gases that are commonly present in industrial applications of interest.[3-5] In this study, MIL-125 and MIL-125-NH_2 were investigated with SO_2 exposure in dry, humid, and aqueous environments using in situ IR experiments and timed, controlled exposures. MIL-125 was found to be unstable in both humid and aqueous environments while the amine functionalized MIL-125-NH_2 was stable under all conditions for extended time periods, showing no change in textural properties or visual degradation as observed through SEM. Both materials were found to be stable under pure water and pure dry SO_2, thereby suggesting the formation of sulfurous acid and the associated bisulfite ion is likely a key step in the degradation mechanism. In situ IR experiments confirmed the presence of sulfite species supporting the hypothesis that the presence of the bisulfite ion likely leads to the degradation of the MIL-125 structure. Computational investigation of several potential reaction mechanisms in the MIL-125 framework indicated the reaction involving the bisulfite ion is favored over reaction with pure water or pure SO_2. DFT simulations support the observation that MIL-125-NH_2 is stable in humid and aqueous conditions as all reactions are less favorable with the functionalized framework compared to the unfunctionalized framework. This study advances the fundamental understanding of MOF degradation mechanisms

during acid gas exposure.

[1] K. Sumida, D.L. Rogow, J.A. Mason, T.M. McDonald, E.D. Bloch, Z.R. Herm, T.H. Bae, J.R. Long, Chem. Rev., 112 (2012) 724-781.

[2] S. Vaesen, V. Guillerm, Q. Yang, A.D. Wiersum, B. Marszalek, B. Gil, A. Vimont, M. Daturi, T. Devic, P.L. Llewellyn, C. Serre, G. Maurin, G. De Weireld, Chem. Commun., 49 (2013) 10082-10084.

[3] S.G. Han, Y.G. Huang, T. Watanabe, S. Nair, K.S. Walton, D.S. Sholl, J.C. Meredith, Micro. Meso. Mater., 173 (2013) 86-91.

[4] C. Petit, B. Levasseur, B. Mendoza, T.J. Bandosz, Micro. Meso. Mater., 154 (2012) 107-112.

[5] W.P. Mounfield III, U. Tumuluri, Y. Jiao, M. Li, S. Dai, Z. Wu, K.S. Walton, Micro. Meso. Mater., 227 (2016) 65-75.

470067 Post-Synthetic Functionalization of Mg-MOF-74 with Tetraethylenepentamine: Structural Characterization and Enhanced CO_2 Adsorption

Abstract:

Metal-organic frameworks (MOFs) are an emerging class of micro- and mesoporous materials with great potential in gas adsorption and separation. MOFs are composed of organic linkers coordinated to metal centers, possess open-framework structures that rank among the materials with the highest surface areas known. Among the MOFs used for carbon dioxide adsorption, Mg-MOF-74 is an interesting option to study due to the ease of synthesis, very high equilibrium and dynamic gas uptake, the economic feasibility of its reagents. The materialâ€™s cylindrical pores and unsaturated metal centers can adsorb between 21-26% CO_2, which places it among adsorbers with the highest capacity for CO_2. In an effort to improve the adsorption behavior of MOFs under different conditions, we grafted the Mg-MOF-74 with tetraethylene pentamine (TEPA), in order. Multi-unit amines (similar to polyethyleneimines) have been shown to provide high CO_2 uptake capacity as well as regenerability over multiple cycles. We expect that through functionalization by longer-chain amines such as TEPA the MOF adsorption behavior can be enhanced. Gas adsorption testing on our MOF/TEPA materials suggests that CO_2 total uptake capacity is dependent on the amine loading, with the adsorption increase dependent on the functionalization degree. CO_2 uptake is reduced over the parent MOF when amines are loaded in excess, but improved when limited amine fraction is added. The crystal structure from XRD is identical pre- and post-functionalization, as well as pre-post-capture. Neutron diffraction was performed on the functionalized structures to understand the conformation of amines within the pores and draw structure-activity relations. Finally, Dynamic Vapor Sorption (DVS vacuum) is employed to perform competitive co-adsorption of CO with H_2O on Mg-MOF-74. The study shows how the competition of probe molecules for specific adsorption sites

affects the efficiency and selectivity of the process. Water vapor is a major molecular competitor and affects CO separation performance for porous materials in almost all industrial processes. This is one of the few multi-component H_2O/CO_2 studies existent on post-synthetic functionalized Mg-MOFs, and is expected to provide guidelines for future materials design.

461642 Facet-Specific Stability of ZIF-8 in the Presence of Acid Gases

Abstract:

Microporous materials such as metal-organic frameworks (MOFs) and zeolitic imidazolate frameworks (ZIFs) are typically described by their bulk properties, such as crystal structure, internal surface area, and metal-ligand connectivity. However, little consideration is given to the external surfaces of these materials, which are necessarily undercoordinated compared to the bulk structure and become increasingly more significant as the crystal sizes approach the nanoscale. It is currently unclear how these surfaces are terminated and how the specific external termination affects the stability of a given crystal facet and bulk structure as a whole. It is expected that the external surface will play a large role in the crystal stability under exposure of these materials to acid gases such as SO_2.

In this contribution, we examine the stability of the (110) and (100) crystallographic facets of ZIF-8 under mild acidic conditions. We find that though the (110) facet is more thermodynamically stable than (100), it is more susceptible to degradation by acid exposure. We hypothesize that the mechanism of particle degradation follows a shrinking-core model, with surface imidazolates being replaced by hydroxyls, as suggested by XPS and FTIR studies. Computational investigations support this hypothesis and suggest that the activation barrier for insertion of water into the Zn-MeIm bond is higher on the (100) facet than (110) due to steric constraints. Additionally, the barrier for water insertion is lowered to a greater extent in the presence of SO_2 on the (110) facet.

466977 Investigation of the Dynamic Adsorption of CO Using Metal-Organic Frameworks

Abstract:

Carbon Monoxide (CO) is an important industrial gas. Its uses range from being a reducing agent in the smelting metals to being a vital component of synthesis gas (syngas), used during the synthesis of methanol and synthetic oils. Production of CO is accomplished through various processes typically involving partial combustion of hydrocarbon. These processes result in impurities in the gas stream, including nitrogen (N_2), carbon dioxide (CO_2) and water (H_2O). Due to similar size, molecular weight and boiling points, it is particularly challenging to separate CO and N_2. Presently, the most common large scale method for this separation is through cryogenic distillation. This approach is not only costly, but also ineffective in purifying CO with high concentrations of N_2.

This study investigates the use of a class of adsorbents, namely metal-organic frameworks (MOFs), for purification and separation of CO from gas streams containing high levels of N_2. Particularly, several MOFs were synthesized, characterized and then tested for CO adsorption under dynamic conditions. The effect of Cu^+ impregnation on the performance of the materials was investigated as well. The mechanisms of adsorption were studied so as to identify the features controlling the uptake and selectivity and lead to the design of better adsorbents.

468718 Designing MOF-Fiber Composites for Targeted Adsorptive Filtration Applications

Abstract:

Metal-organic frameworks (MOFs) are a class of porous materials that show great promise over a variety of applications including catalysis, separations, sensing, and storage. Even with the large number of MOFs synthesized to date along with an incredible degree of tunability, one challenge facing these materials is their incorporation into an engineering ready form. One promising technique for arranging MOFs is to imbed them in electrospun fibers. Techniques for optimizing the adsorptive properties of MOF impregnated fibers is to be discussed, including fiber crosslinking for applications in secondary growth and selective polymer removal from the fibers to improve the kinetic accessibility of MOFs in fibers.

459380 Solvent Template-Directed Synthesis of Flexible and Thermosensitive Metal-Organic Framework (CPL-1) for Highly Selective C_3H_6/C_3H_8 Separation

Abstract:

The separation of olefins from paraffin is important for the production of industrial chemicals. A flexible and thermosensitive microporous Metal-organic framework (CPL-1) was synthesized by a solvothermal method using various sized solvents as the template. The isotherms of propylene and propane on CPL-1 materials were measured by a volumetric method. Experiments of multiple adsorption-desorption cycles were conducted to estimate reversibility of propylene on CPL-1. Results showed the C_3H_6 isotherms of all samples showed a sudden rise at the threshold pressure and attained a saturated level rapidly, indicating that the structure of CPL-1 materials could expand in response to propylene molecule at 273 K. Whereas, CPL-1 materials were not responding to propane molecule. Therefore, CPL-1 demonstrated a highly selectivity towards C_3H_6 against C_3H_8. The solvents could be divided into two groups based on the kinetic diameters: the kinetic diameters of the solvents inÂ one group are smaller than that of C_3H_6, those in the other are greater than that of C_3H_6. And C_3H_6 gate opening pressure of CPL-1 materials synthesized by different solvents reduced with the decreasing of the kinetic diametersÂ ofÂ two groups of solvents, separately. Among them, CPL-1 synthesized in the solvent of CH_2Cl_2, which molecular size is slightly larger than propylene, performanced the lowest C_3H_6 gate opening pressure about P/P_0=0.37. Itâ€™s worth noting that the structure of thermosensitive CPL-1 could expand in response to propylene molecular at 273 K, but shrink at 298 K. Consecutive cycles of adsorption-desorption experiment showed that the desorption efficiency could reach 90% for five cycles at 298 K, P/P_0=0.9, indicating that CPL-1 had excellent regeneration property of propylene at room temperature.

467201 Effect of Shaping MOFs on Adsorption Performance By Using Adsorption Microcalorimetry

Abstract:

Many laboratory scale studies in the field of gas adsorption deal with as-made samples in the form of powders. However, to be of any applied use in an adsorption based process such as PSA (pressure swing adsorption), the powders need to be shaped, often in the form of spherical beads or cylindrical pellets. Nevertheless, gas separation studies using MOF pellets are still very scarce and the majority of studies concentrate on the separation performance without asking questions about the effect of the binder used or pelletization process on the fundamental adsorption properties.

The effect of pelletization on the adsorption properties of the active materials should ideally be limited. Indeed, the physical properties of the material can be slightly affected as one expects a certain decrease in uptake/working capacity, proportional to the amount of binder used. However, the chemical properties which influence selectivity should remain unchanged in the presence of the binder. Other parameters such as adsorption uptake kinetics should equally remain unchanged.

In this work, we have concentrated on the fundamental gas adsorption properties of several 'topical' MOFs, which not only are hydrothermally stable, cheap and easy to scale-up, but have also recently demonstrated a particular interest for gas separations. The aim here is to compare the initial powder with the shaped materials.

Thus, a series of four Metal-Organic Frameworks (UiO-66(Zr), UiO-66(Zr)_NH_2, MIL-100(Fe) and MIL-127(Fe)) were synthesized as powders and spheres and these were characterized by gas adsorption at 30°C to 20 bars. The adsorption experiments were coupled with microcalorimetry to gain the energetic insight required to highlight any changes in chemistry. Results show that whilst most variations are as expected, interesting effects including partial pore restriction and active site 'protection' were also evidenced for some shaped materials suggesting that the binder may play a more complex role. These effects will be explored further in this communication.

DOI: 10.1002/chem.201602599

Metal–Organic Frameworks

Incorporation of N-Methyl-D-glucamine Functionalized Oligomer into MIL-101(Cr) for Highly Efficient Removal of Boric Acid from Water

Jiawei Wang,[a] Zongbi Bao,*[a] Huabin Xing,[a] Baogen Su,[a] Zhiguo Zhang,[a] Qiwei Yang,[a] Yiwen Yang,[a] Qilong Ren,[a] and Banglin Chen*[b]

Abstract: Polymeric resins are practically important adsorbents in a wide variety of applications, but they generally suffer from low surface areas and limited functionalized adsorption sites owing to their closely compacted and tangled polymeric chains. A metal–organic framework (MOF)–polymer composite with enhanced adsorption capacity against the compacted polymeric resins was reported. The strategy to incorporate functionalized oligomer within the cavities of the MOF was demonstrated by the preparation of MIL-101(Cr) incorporated with N-methyl-D-glucamine-based organosiloxane polymer. The resulting MOF composite shows high efficiency for the removal of boric acid from water because of exceptionally high loading of functional groups responsible for the boron adsorption. This material offers promising perspectives for boron removal applications in seawater desalination.

Introduction

Water security is receiving an increasing attention and will remain a worldwide concern over next decades.[1] It is well-known that seawater desalination using reverse osmosis technology has reinstated boric acid as a major contaminant of concern in water supply.[2] The presence of appreciable quantities of boron species in drinking water deteriorates the quality of the water.[3] The World Health Organization (WHO) has set the tolerance limit in drinking water to 2.4 mg L^{-1} (Drinking-Water Quality Committee regulations published in 2011). Boric acid is a very weak acid with a pK_a of 9.15, and therefore undissociated boric acid [$B(OH)_3$] exists predominantly in aqueous solution at pH < 7 while the metaborate anion $B(OH)_4^-$ becomes the major species in solution at pH > 10. Between these two pH values, highly water-soluble polyborate ions are formed, such as $B_3O_3(OH)_4^-$, $B_4O_5(OH)_4^-$, and $B_5O_6(OH)_4^-$. Many approaches have been explored for the removal of boron species from water including precipitation–coagulation,[4] reverse osmosis,[5] electrodialysis,[6] solvent extraction,[7] membrane filtration,[8] and adsorption.[9] Among these methods, adsorption is a more efficient and economical technique.[10]

Polymer resins have been widely applied as adsorbent materials for water purification; however, most polymer adsorbents suffer from relatively low surface areas and limited functionalized adsorption sites owing to their closely compacted and tangled polymeric chains. For example, Amberlite IRA-743 has possibly been one of few commercial resins for the boron species-selective removal;[11] however, it exhibits unsatisfactory performance for the uptake of boron species (generally less than 1 mmol(B) g^{-1}),[10a] making the resin less economically attractive. Moreover, most polymer-supported resins are based on the copolymers of styrene and divinylbenzene, which show some mass-transfer inefficiency owing to their strong hydrophobic properties. Other types of polymer adsorbents, especially the silica-based ones synthesized by the sol–gel reaction, tend to degrade into nanoscale oligomers in the presence of acid/base, resulting in seriously reduced adsorption capacity. Therefore, there is a need to develop adsorbent materials with high uptakes of boron species.

Metal–organic frameworks (MOFs) or porous coordination polymers (PCPs), emerging as a new class of porous polymer materials, show great potential in adsorptive removal of contaminants from water.[12] MOFs are structurally diverse and provide well-defined nanostructures ranging from 4 to 20 Å, enabling the fabrication of hybrid composites with some special properties by incorporating some functional species into their pores.[13] Herein, we report a MOF composite by incorporating N-methyl-D-glucamine (NMDG) functionalized oligomer into a mesoporous MOF material for the exceptionally high adsorption of boron species and thus the removal of boric acid from water. It is well-established that compounds containing hy-

[a] J. Wang, Dr. Z. Bao, Prof. H. Xing, Dr. B. Su, Dr. Z. Zhang, Dr. Q. Yang, Prof. Y. Yang, Prof. Q. Ren
Key Laboratory of Biomass Chemical Engineering, Ministry of Education
College of Chemical and Biological Engineering
Zhejiang University, Hangzhou 310027 (P. R. China)
E-mail: baozb@zju.edu.cn

[b] Prof. B. Chen
Department of Chemistry, University of Texas at San Antonio
One UTSA Circle, San Antonio (USA)
E-mail: banglin.chen@utsa.edu

The ORCID identification number(s) for the author(s) of this article can be found under http://dx.doi.org/10.1002/chem.201602599.

droxy groups (that is, N-methyl-D-glucamine) are efficient in the selective interaction with boron species by complexation. NMDG is a small molecule and highly soluble in water, it is thus necessary to develop insoluble NMDG derivatives to meet the need. It has been known that NMDG reacts with (3-glycidyloxypropyl)trimethoxysilane (GPTMS) to form an organosiloxane, which will undergo hydrolysis and subsequent polymerization in the acidic solution to produce NMDG-functionalized oligomer. In this work, we employed the nanocavities of the highly porous MIL-101(Cr)[14] as the reaction vessels to host NMDG functionalized oligomer. The resulting MIL-101(Cr) composite decorated with N-methyl-D-glucamine-based organosiloxane polymer, shows exceptionally high efficiency for the removal of boric acid from water.

Results and Discussion

The preparation of the MOF composite (denoted as poly(Si-NMDG)@MIL-101(Cr)) is shown in Scheme 1. MIL-101(Cr) was synthesized and purified following a well-developed procedure.[14] Subsequently, the solution of N-methyl-D-glucamine and (3-glycidyloxypropyl)trimethoxysilane (GPTMS) in N,N′-dimethylformamide (DMF) spontaneously diffuses into MIL-101(Cr) under stirring at room temperature. After sufficient equilibrium of both reactants, NMDG derivatives (denoted as Si-NMDG) were formed within the cavities of MIL-101(Cr) by heating at 75 °C for 8 h. The MOF composite loaded with Si-NMDG was filtered off and continued to polycondensate in the acidic aqueous methanol solution. The synthesized poly(Si-NMDG) within the cavities was interlocked in the porous MIL-101(Cr).

The successful introduction of poly(Si-NMDG) inside the cavities could be evident from FTIR spectra with the appearance of Si–O–Si stretching vibration characteristic band for the polymer at around 1057 cm^{-1} (Figure 1).[10b] All peaks of MIL-101(Cr) can be correspondingly found in the spectrum of poly(Si-NMDG)@MIL-101(Cr). The characteristic band at 1020 cm^{-1}, assigned to the Cr–O vibration,[15] is observed for MIL-101(Cr), poly(Si-NMDG)@MIL-101(Cr), and boron-adsorbed poly(Si-NMDG)@MIL-101(Cr) except for poly(Si-NMDG), indicating reservation of the MOF structure. The additional peak in the spectrum of poly(Si-NMDG)@MIL-101(Cr) at 1057 cm^{-1} could be assigned to the Si–O–Si stretching vibration when compared with the IR pattern for poly(Si-NMDG) only. The adsorption band of poly(Si-NMDG)@MIL-101(Cr) at 1394 cm^{-1} is slightly broader than the same peak of MIL-101(Cr) because of

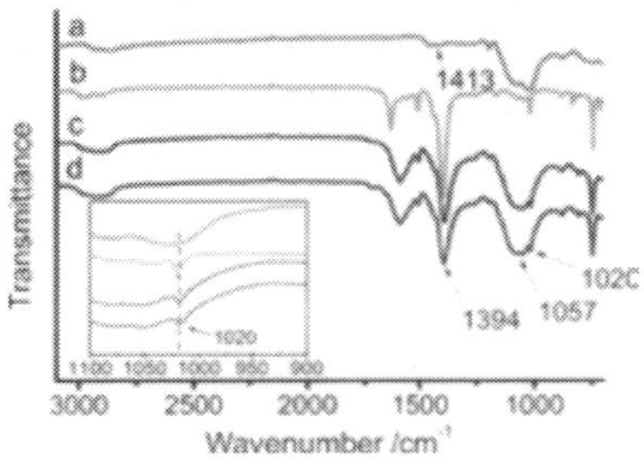

Figure 1. FTIR spectra of a) poly(Si-NMDG), b) MIL-101(Cr), c) poly(Si-NMDG)@MIL-101(Cr) with an NMDG loading of 2.23 mmol g^{-1}, and d) poly(Si-NMDG)@MIL-101(Cr) with adsorbed boron. Inset: partial enlarged view from 1100–900 cm^{-1}.

the combination of the weak C–OH bending vibration band at 1413 cm^{-1} that is observed in poly(Si-NMDG). There is no additional peak emerging in these spectra, indicating that the Si-NMDG has not reacted with terephthalic acid or Cr^{3+} cations in the preparation of the MOF composite. As shown in the PXRD patterns (Figure 2), poly(Si-NMDG)@MIL-101(Cr) with a low loading of poly(Si-NMDG) remained a crystalline structure, but gradually turned amorphous with the increase of poly(Si-NMDG) loading. This might arise from the pore expansion/contraction during the introduction of polymers into the cavities of the MIL-101(Cr) as the Si-NMDG loading up to 2.23 mmol g^{-1} occupied the most of the pore space of MIL-101(Cr) and thus Si-NMDG polymer made the walls between one cavity and another cavity severely deformed. Although

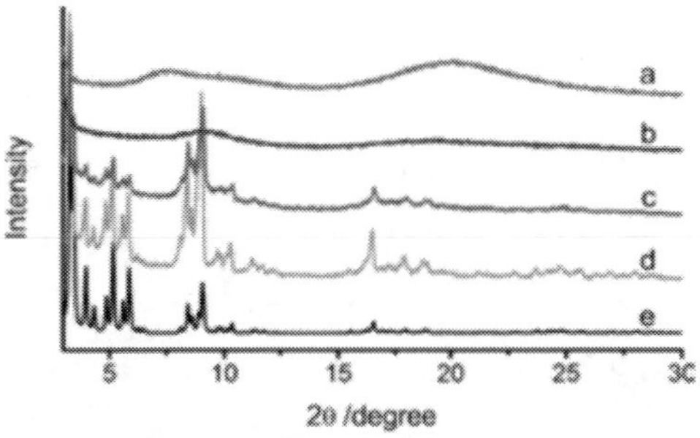

Figure 2. The experimental PXRD patterns for a) poly(Si-NMDG), b) poly(Si-NMDG)@MIL-101(Cr) with NMDG loading of 2.23 mmol g^{-1}, c) poly(Si-NMDG)@MIL-101(Cr) with NMDG loading of 0.60 mmol g^{-1}, d) MIL-101(Cr), and e) the simulated PXRD pattern of MIL-101(Cr).

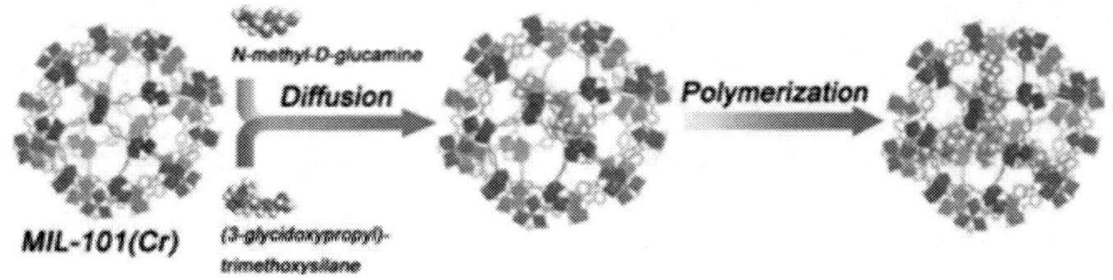

Scheme 1. Illustration of the synthesis of poly(Si-NMDG)@MIL-101(Cr).

very similar changes in PXRD patterns and peak intensities have been observed on the polymer encapsulated MOF materials,[16] it cannot be thoroughly ruled out that the MIL-101(Cr) may arise minor defect during the synthetic process of poly(Si-NMDG)@MIL-101(Cr). ^{29}Si NMR spectra (Figure 3) show the distinct resonances for Si-NMDG ($T^m = RSi(OSi)_m(OH)_{3-m}$, $m = 1$–3) centers. As shown in Figure 3, only the T^3 ($RSi(OSi)_3$) peak was observed on the absolute poly(Si-NMDG), indicating that the hydrolyzed $RSi(OH)_3$ were completely condensed in the form of Si-O–Si bonds. A large T^2 ($RSi(OSi)_2(OH)$) peak was also detected in the sample of poly(Si-NMDG)@MIL-101(Cr), implying that the monomer Si-NMDG mainly generated oligomers during the polymerization process. SEM images (Figure 4) showed that the crystal size, morphology, and dispersity of the host materials did not change after introduction of poly(Si-NMDG), suggesting the survival of MIL-101(Cr) during the synthetic process of the MOF composite.

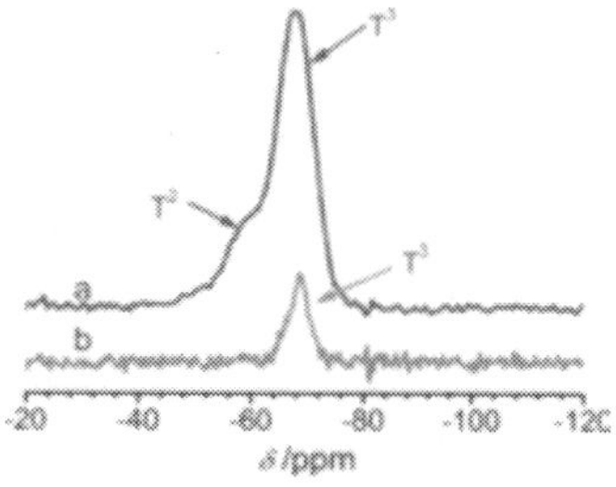

Figure 3. ^{29}Si NMR spectra of a) poly(Si-NMDG)@MIL-101(Cr) and b) poly(Si-NMDG).

According to the ^1H NMR results (Figure 5) of poly(Si-NMDG)@MIL-101(Cr) dissolved in NaOD/D_2O (^1H NMR, 400 MHz: $\delta = 0.24$ ppm (m, 2H; $SiCH_2$), 1.47 ppm (m, 2H; $SiCH_2CH^*_2$), 7.69 ppm (s, 4H; C_6H_4)), the loading of Si-NMDG in poly(Si-NMDG)@MIL-101(Cr) is 2.23 mmol g^{-1} (calculated with the ratio of the peak areas of $SiCH_2$/$SiCH_2CH^*_2$ and C_6H_4 of the ligand). The embedded peaks in the ^1H NMR spectrum were also a clear indication for polymerization of Si-NMDG in the channels of MIL-101(Cr). To the best of our knowledge, it is a highest loading of NMDG ever reported for porous materials. It is expected that boric acid can form 1:1 tetradentate complex with *N*-methyl-D-glucamine group,[17] so 1.0 g of poly(Si-NMDG)@MIL-101(Cr) can theoretically take up 2.23 mmol boric acid. The Si-NMDG loading was further confirmed by XPS characterization. The content of N element in the poly(Si-NMDG)@MIL-101(Cr), which is all from Si-NMDG, was used to estimate the loading of NMDG. The loading of NMDG in poly(Si-NMDG)@MIL-101(Cr) was calculated as 2.49 mmol g^{-1}, which is rather close to the loading calculated by ^1H NMR (2.23 mmol g^{-1}). Additionally, according to the XPS spectra of Cr regions shown in Figure 6, the Cr $2p_{3/2}$ with a binding energy of 577 eV was ascribable to Cr–O.[18] Moreover, the binding energy of Cr 2p regions of poly(Si-NMDG)@MIL-101(Cr) is the same as that in pure MIL-101(Cr), confirming the invariability of the valence and surrounding of Cr^{III}.

Figure 4. SEM images of the activated MIL-101(Cr) (a, b), and activated poly(Si-NMDG)@MIL-101(Cr) (c, d), and poly(Si-NMDG)@MIL-101(Cr) after adsorption (e, f).

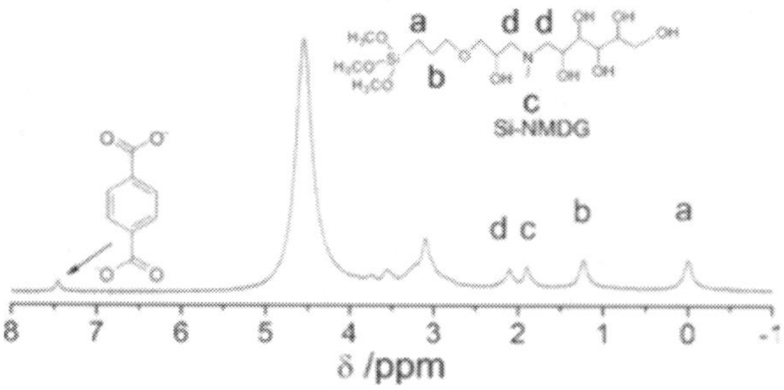

Figure 5. ^1H NMR spectrum of poly(Si-NMDG)@MIL-101(Cr) dissolved in NaOD/D_2O.

The high loading of NMDG in the poly(Si-NMDG)@MIL-101(Cr) prompted us to investigate the adsorption performance for the removal of boron species. We directly employed the packed bed breakthrough experiments to determine the adsorption capacity of poly(Si-NMDG)@MIL-101(Cr)

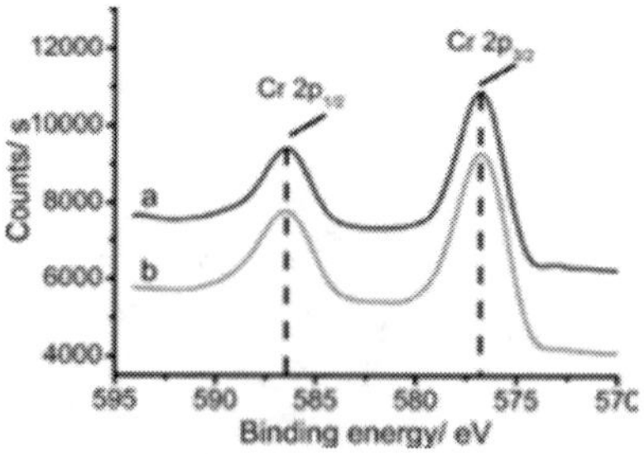

Figure 6. XPS spectra of the 2p regions of the Cr in the (a) poly(Si-NMDG)@MIL-101(Cr), and (b) MIL-101(Cr).

for boron species. A variety of boric acid aqueous solutions with different concentrations (0.1, 0.5, 1.0, 2.0, 4.0 mg mL^{-1}) were used as feed solution. As shown in Figure 7, the break-

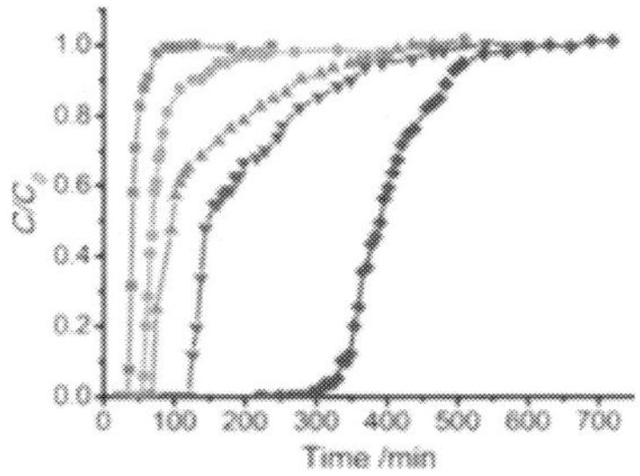

Figure 7. Breakthrough curves for 0.1 (◆), 0.5 (▼), 1.0 (▲), 2.0 (●), and 4.0 mg mL^{-1} (■) boron aqueous solution on poly(Si-NMDG)@MIL-101(Cr).

throughs are rather steep for the different boric acid solutions, indicating a very fast adsorption for boron species in the column. The dynamic saturated adsorption capacities, determined by integrating above the breakthrough curves, are 0.46, 0.79, 1.06, 1.52, 1.90 mmol g^{-1} for 0.1, 0.5, 1.0, 2.0, 4.0 mg mL^{-1} boric acid aqueous solution, respectively. For comparison, the breakthrough curves obtained on MIL-101(Cr) packed column were also investigated as a benchmark using a boric acid solution with concentration of 0.1 mg mL^{-1}, as shown in Figure 8. It

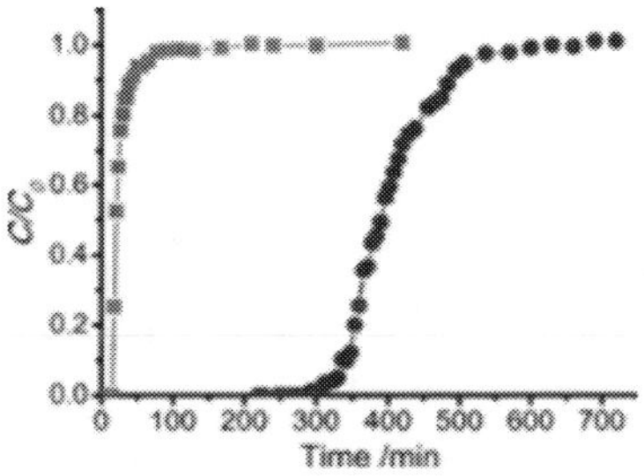

Figure 8. Breakthrough curves for 0.1 mg mL^{-1} boric acid aqueous solution for original MIL-101(Cr) (■) and poly(Si-NMDG)@MIL-101(Cr) (●).

can be seen that the breakthrough point was practically equal to the column hold-up time, indicating that MIL-101(Cr) show no affinity to boron species. Poly(Si-NMDG)@MIL-101(Cr) could prolong the breakthrough time up to 300 min. MIL-101(Cr) shows negligible saturation adsorption capacity of 0.04 mmol g^{-1}, corresponding to only 8% of poly(Si-NMDG)@MIL-101(Cr) at the same feed concentration.

The adsorption isotherm (Figure 9) was plotted on the basis of dynamic saturated adsorption capacities obtained with different concentrations of boric acid solution, and fitted with the Langmuir model (Eq. (2)). The theoretical maximum adsorption capacity was found to be 2.29 mmol(B) g^{-1} (Table 1), which is in good agreement with the NMDG loading in the composite.

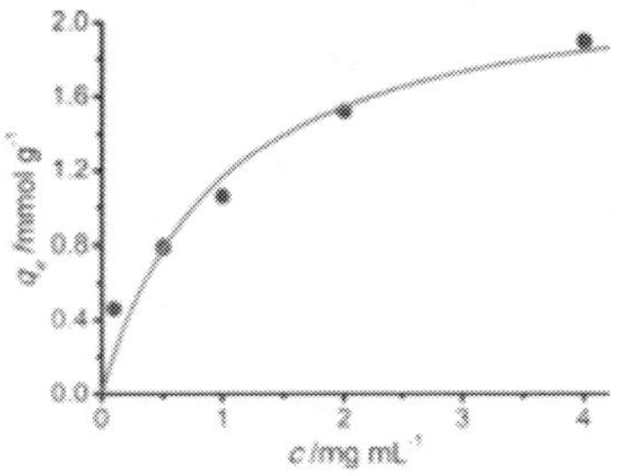

Figure 9. Boron adsorption isotherm of poly(Si-NMDG)@MIL-101(Cr) on basis of adsorption data obtained from breakthrough experiments. The curve represents a fit to the Langmuir equation (see Table 1).

Table 1. Langmuir isotherm parameters for adsorption of boric acid into poly(Si-NMDG)@MIL-101(Cr) at 303 K.

Langmuir	K_L	Q_m	R^2
$q_e = \frac{Q_m K_L c_e}{1+K_L c_e}$	1.0300	2.2946	0.9665

Table 2. Comparison of Poly(Si-NMDG)@MIL-101(Cr) with the best performing boron-selective adsorbents in terms of maximum adsorption capacity.

Adsorbents	Maximum adsorption capacity [mmol B g^{-1}]	pH value
Poly(Si-NMDG)@MIL-101(Cr)	2.29	9
Saccharide grafted UVM-7[19]	1.85	8
Polyol grafted SBA-15[20]	0.63	Not adjusted
Polyol grafted MCM-41[20,21]	0.80	>6
Amberlite IRA 743[7,9,10a]	0.62–1.16	7–8
Diaion CRB series[22]	0.99–1.32	5–8

This exceptionally high adsorption capacity outperforms all other boron-selective silica adsorbents and commercial resins (Table 2).

The pH value always influences boron uptake behavior considerably. The effect of solution pH on the adsorption of boron species by batch equilibration experiments is shown in Figure 10. It demonstrates that the adsorption capacity reaches a climax around pH 9. Theoretically, the boron-selective adsorption process is driven by the formation of tetradentate

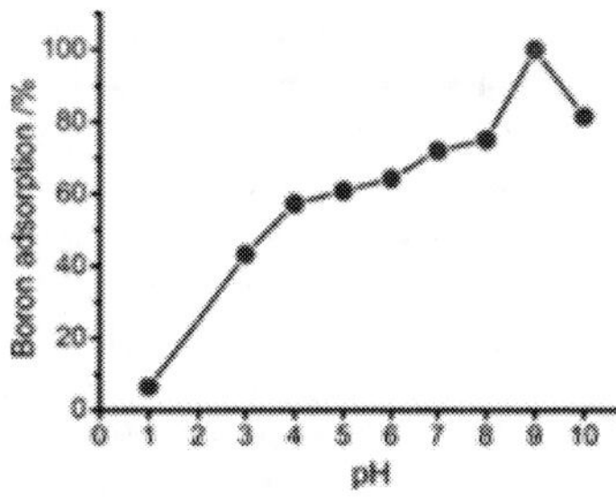

Figure 10. Effect of pH of boric acid solution on the boron adsorption.

complex with NMDG, and increasing pH is favorable for the formation of tetradentate complex. However, at much higher pH the adsorbent surface is negatively charged due to formation of OH^- group when NMDG chelated with boron, and competitive adsorption can be observed between $B(OH)_4^-$ and OH^- groups for the adsorption sites. It reveals the reason why boron uptake decreased as pH was up to 10. An plausible explanation is that repulsion between negatively charged surface and negatively charged ions (OH^-, $B(OH)_4^-$) occurs in this circumstance. It was remarkable that the adsorption capacity at the pH 1 was only 6.6% of the capacity at the pH 9, demonstrating that the adsorbent and boric acid could be regenerated by pH swing effect.

To probe the interference of foreign ions, boron adsorption was also carried out with the presence of anions and cations in the batch experiments. As shown in Figure 11, the breakthrough curve for 0.1 $mg\,mL^{-1}$ boric acid aqueous solution simultaneously with various equimolar anions (NO_3^-, CO_3^{2-}, SO_4^{2-}, PO_4^{3-}, Cl^-, Br^-) for poly(Si-NMDG)@MIL-101(Cr) was similar to the case without anions. Both curves breakthrough almost at the same time although they are slightly different in the shapes that are probably related to the packing properties. Importantly, the adsorption capacity of poly(Si-NMDG)@MIL-101(Cr) in the boron solution with the addition of various anions, calculated to be 0.49 $mmol\,g^{-1}$, was almost the same as the original adsorption capacity of 0.46 $mmol\,g^{-1}$. Thus the anions had a negligible effect on the adsorption of boric acid and that the functionalized MOF composites show high selectivity towards boric acid. Furthermore, the effect of cations (Li^+, Na^+, Mg^{2+}) was also investigated. As illustrated in Figure 12, the uptake of boron was negligibly fluctuated with the increase of the cation concentration, indicating the high selectivity towards boron. It is notable that there was a slight increase in the adsorption capacity when cation was added compared with that without cation. It can be interpreted that there was an increase in $B(OH)_4^-$ concentration with the addition of cations, and the $B(OH)_4^-$ was bound to the protonated tertiary amine groups of the adsorbent by ion pairing.[9] To demonstrate the reason, the speciation of boric acid in cations solution was investigated.[23] As seen in Figure 13a, an evident enhancement of fraction of $B(OH)_4^-$ from 0.37 to 0.58 was observed at pH 9 when the c_{NaCl} was 0.5 $mol\,L^{-1}$. Furthermore, the enhancement of $B(OH)_4^-$ was negligible with an increase of c_{NaCl} from 0.5 to 2 $mol\,L^{-1}$. Similar results were observed in boric acid solution that included $MgCl_2$ (Figure 13b).

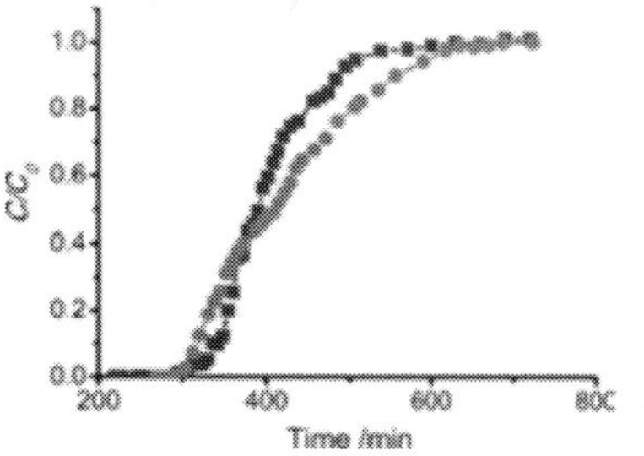

Figure 11. Breakthrough curves for 0.1 $mg\,mL^{-1}$ boric acid aqueous solution with (●) or without (■) the equimolar foreign anions (NO_3^-, CO_3^{2-}, SO_4^{2-}, PO_4^{3-}, Cl^-, Br^-) on poly(Si-NMDG)@MIL-101(Cr).

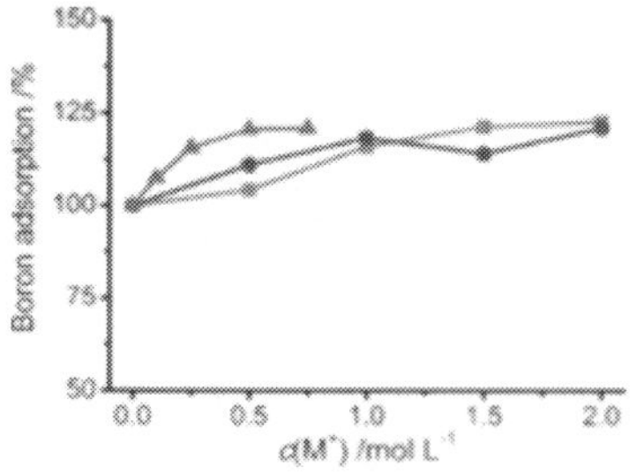

Figure 12. Boron adsorption on poly(Si-NMDG)@MIL-101(Cr) with the presence of different concentrations of foreign ions: Li^+ (■), Na^+ (●), Mg^{2+} (▲). The concentrations of foreign ions range from 0.5–2.0 M.

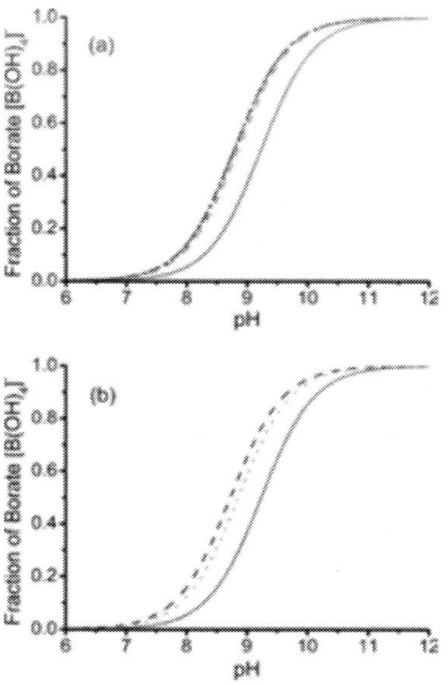

Figure 13. Speciation of boric acid in aqueous solution at different concentrations of NaCl at 25 °C (a), $c_{NaCl}=0$ (——), $c_{NaCl}=0.5$ M (– – –), $c_{NaCl}=1.0$ M (•••••), $c_{NaCl}=1.25$ M (–•–•), $c_{NaCl}=2.0$ M (-----), and (b) $c_{NaCl}=0$ (——), $c_{NaCl}=0.5$ M (•••••), $c_{MgCl_2}/c_{NaCl}=0.05/0.5$ (– – –) at 25 °C. The pK_a data was taken from Hershey et al.[23] and used to determine the speciation of boric acid in foreign ions solution.

Regeneration potential of the adsorbent was also evaluated by measuring its boron adsorption capacity after soaking the boron-loaded material in a 0.1 M HCl solution for 6 h, and neutralized with 0.01 M NaOH. It is noteworthy that nearly 90% boric acid was recovered when regenerating. For convenient comparison, the boron uptake on the fresh MOF composite was set as benchmark, and the same composites were used for consecutive adsorption–regeneration cycles. The adsorption performance for 4 cycles compared with the initial was shown in Figure 14, from which it can be found that the com-

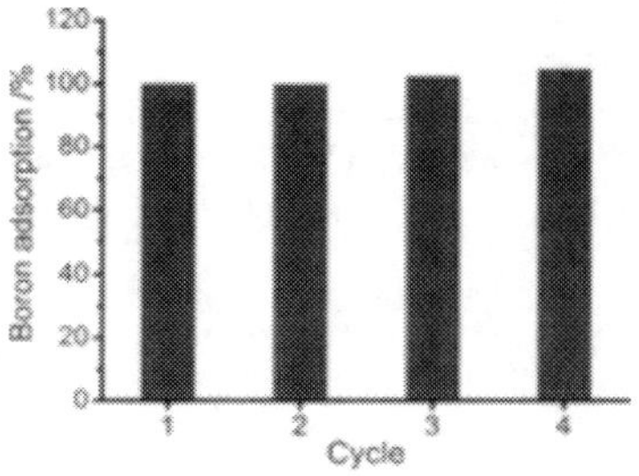

Figure 14. Adsorption cycles of regenerated poly(Si-NMDG)@MIL-101(Cr).

posite remains the constant adsorption capacity. The slight increase in the last two cycles may be attributed to the fluctuation in the boron concentration measurement. The remained adsorption performance after regeneration also indicates the structural stability of the synthesized composite in the acid–base treatments.

Conclusion

In summary, we report a highly selective boron-specific adsorbent condensed in MOF with high hydrophilicity, high stability in aqueous solution, and facile accessibility. The exceptionally high adsorption capacity arises from the ultrahigh surface area of support and consequent loading of functional groups. The results indicated that this MOF supported adsorbent demonstrates great potential for the boric acid removal from aqueous solution. The strategy to incorporate functional groups within the channel of MOF is quite novel compared with the conventional surface-modified or bulk-polymerized boron-selective adsorbents. It not only takes full advantage of the support but also enhances the loading and accessibility of effective functional groups.

Experimental Section

Materials

Chromium(III) nitrate nonahydrate ($Cr(NO_3)_3 \cdot 9H_2O$, 99.95%), *N*-methyl-D-glucamine, (3-glycidyloxypropyl)trimethoxysilane, and boric acid were purchased from Sigma–Aldrich Co. (America), and terephthalic acid (H_2bdc) was obtained from TCI Co. (Japan). All other chemicals were commercially available and used without further purification.

Synthesis of MIL-101(Cr): MIL-101(Cr) was synthesized and purified following the previously reported method.[14] $Cr(NO_3)_3 \cdot 9H_2O$ (4.00 g, 10 mmol), terephthalic acid (1.64 g, 10 mmol), and HF (125 μL, 37 wt%) were dissolved in H_2O (70 mL) and transferred to a Teflon-lined stainless steel autoclave. The autoclave was heated at 220 °C in an oven for 8 h. After completion of reaction, a highly crystalline green solid was produced. To obtain the pure MIL-101(Cr), the as-synthesized product was purified according to the following steps. First the as-synthesized product was treated by hot water at 70 °C for 5 h and then hot ethanol at 60 °C for 3 h, respectively. After the treatment, the green precipitate was collected and further purified by dispersing in 30 mM NH_4F aqueous solution at 60 °C for 10 h. Finally, it was washed five times with water.

Synthesis of poly(Si-NMDG)@MIL-101(Cr): The monomer of organosiloxane within MIL-101(Cr) was synthesized according to Scheme 2. To a 50 mL three-necked flask, activated MIL-101(Cr) (0.50 g) and NMDG (6.5 g) were added, and then DMF (35 mL) and GPTMS (8 mL) were successively added under nitrogen atmosphere. The resulted mixture was stirred for 3 h at room temperature prior to reaction at 75 °C for 8 h with continuous stirring. After the completion of reaction, the product was washed with methanol and dried in the air. It was then transferred to Teflon-lined stainless steel autoclaves. Subsequently, methanol-water (1:1, *v/v*) mixture (5 mL) was added and the pH of solution was adjusted to 3 by addition of hydrochloric acid.[24] Afterwards, the crosslinking polymerization of Si-NMDG within the channels of MIL-101(Cr) was carried out at 80 °C for 12 h, yielding the dark green jelly-like product. With the aid of ultrasound and centrifugation processes, the as-synthesized poly(Si-NMDG)@MIL-101 (Cr) was further purified with deionized water until the eluent was colorless. The purified product was dried at 80 °C overnight, and activated at 150 °C for 5 h. Analytical data of poly(Si-NMDG)@MIL-101 (Cr): ^{1}H NMR (400 MHz, NaOD/D_2O): $\delta = 0.24$ ppm (m, 2H; $SiCH_2$), 1.47 ppm (m, 2H; $SiCH_2CH^*{}_2$), 7.69 ppm (s, 4H; C_6H_4); ^{29}SiNMR (400 MHz): $\delta = -68$ ppm (T^3, $RSi(OSi)_3$), -61 ppm (T^2, $RSi(OSi)_2(OH)$); IR: $v = 3406$ cm^{-1} (OH), 2923 cm^{-1} (CH), 1057 cm^{-1} (Si-*O*-Si), 881 cm^{-1} (Si-OH), 704 cm^{-1} (Si-C).

Characterization

Attenuated total reflection Fourier transform infrared spectra (ATR-FTIR) were performed in a Nicolet iS50 FTIR spectrometer with diamond as an internal reflection element at an incident angle of 45°, using a resolution of 4 cm^{-1} and 32 scans and between 500 and 4000 cm^{-1}. The ^{1}H NMR was acquired on a Bruker Ascend 400 MHz (Switzerland). The ^{29}Si CP/MAS NMR spectrum was obtained on a 400 MHz WB solid-state NMR Spectrometer and a Bruker AVANCE III 400 MHz (Switzerland). Powder X-ray diffraction (PXRD) patterns of the samples were recorded with a XPert diffractometer (Panalytical Corp., Netherlands) using Cu Kα ($\lambda = 0.1543$ nm) radiation at 40 kV from 3° to 30° (2θ angle range) with a step size of 0.02°.

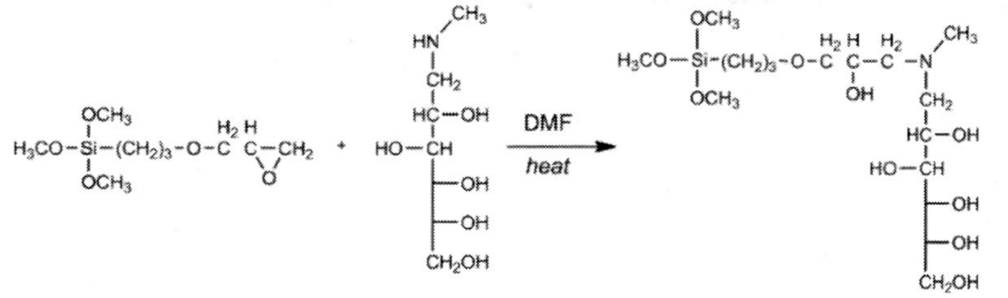

Scheme 2. Synthesis pathway of organosiloxane Si-NMDG.

Scanning electron microscopy (SEM) was carried out on a SIRION-100 instrument (FEI Corp., Netherlands). It was performed on the samples previously dried and sputter-coated with a thin Au layer. XPS results were obtained on a combined VG ESCALAB MARK II instrument at a base pressure of 7.1×10^{-10} mbar, which was operated at Mg Kα: 1253.6 eV, CAE: 50 eV and Step: 0.2 eV, 0.5 eV.

Adsorption experiments

The adsorption capacity of boron on poly(Si-NMDG)@MIL-101(Cr) was evaluated by breakthrough experiments. Different concentrations of boric acid aqueous solution (0.1, 0.5, 1.0, 2.0, 4.0 mg mL^{-1}, pH 9) were separately delivered into a stainless steel column (50 mm *x* 4.6 mm ID) manually packing with the adsorbent. The concentration in the outlet effluent was measured by azomethine-H method[25] on a UV-2550 UV/Vis spectrophotometer (Shimadzu Corp., Japan) with detection wavelength set at 415 nm. The dynamic saturated adsorption capacity of boron at a given equilibrium concentration was calculated according to the breakthrough curve based on Equation (1) in which c_0 and c_e are the initial and exit concentration of boron (mg mL^{-1}), respectively. V is the volume of boron solution delivered (mL), and m is the weight of adsorbents used (g).

$$q_e = \int_0^{V_T} \frac{(c_0 - c_e)\,dV}{m} \quad (1)$$

The adsorption isotherm was investigated on the basis of the dynamic saturated adsorption capacities, and correlated by Langmuir model [Eq. (2)] in which c_e is equilibrium concentration of boric acid (mg mL^{-1}), q_e is amount adsorbed at equilibrium (mmol g^{-1}), Q_m is Langmuir constant (maximum adsorption capacity) (mmol g^{-1}), and K_L is Langmuir constant (mL mg^{-1}).

$$\frac{c_e}{q_e} = \frac{c_e}{Q_m} + \frac{1}{Q_m K_L} \quad (2)$$

The effect of both pH and foreign ions on the adsorption uptake of boric acid was performed in the static adsorption experiments. For pH dependent boron adsorption studies, pH values of 1.0 mg mL^{-1} boric acid ranging from 1 to 10, were adjusted by NaOH solution. The effect of cations (Li^+, Na^+, and Mg^{2+}) was also investigated using 1.0 mg mL^{-1} boric acid solution. The concentrations of the cations were 0.5, 1.0, 1.5, and 2.0 mol L^{-1} except Mg^{2+} (0.1, 0.25, 0.5, and 0.75 mol L^{-1}) due to its low pK_a. Then the boric acid with the cation was adjusted pH to 9 for following adsorption. The effect of the presence of other anions on the boron adsorption was examined by breakthrough experiments. The equimolar concentration of anions (NO_3^-, CO_3^{2-}, SO_4^{2-}, PO_4^{3-}, Cl^-, Br^-) with boric acid (1.62 mmol L^{-1}, equal to 0.1 mg mL^{-1}) were prepared in the deionized water. The mixed boric acid solution with the anions was then adjusted pH to 9 for breakthrough adsorption. It was delivered into a stainless steel column (50 mm×4.6 mm ID) manually packing with the studied MOF composite. The concentration in the outlet effluent was measured by azomethine-H method on a UV-2550 UV/Vis spectrophotometer with the detection wavelength set at 415 nm.

The boron-loaded samples were treated with 0.1 mol L^{-1} HCl solution for 6 h at 30 °C for regeneration and then neutralized with 0.01 mol L^{-1} NaOH solution. The boric acid content of the filtrate collected by centrifugation was measured, and the regenerative adsorbent was dried and activated for the next cycles.

Acknowledgements

This work was supported by the Natural Science Foundation of Zhejiang Province (Grant No. LY13B060002), and the National Natural Science Foundation of China (Grant No. 21376205 and U1407134), and partially supported by the Welch Foundation (AX-1730).

Keywords: boron · composite materials · confined polymerization · metal–organic frameworks · *N*-methyl-D-glucamine

[1] a) K. Bakker, *Science* **2012**, *337*, 914–915; b) C. J. Vörösmarty, P. B. McIntyre, M. O. Gessner, D. Dudgeon, A. Prusevich, P. Green, S. Glidden, S. E. Bunn, C. A. Sullivan, C. R. Liermann, P. M. Davies, *Nature* **2010**, *467*, 555–561; c) M. A. Shannon, P. W. Bohn, M. Elimelech, J. G. Georgiadis, B. J. Marinas, A. M. Mayes, *Nature* **2008**, *452*, 301–310.

[2] K. L. Tu, L. D. Nghiem, A. R. Chivas, *Sep. Purif. Technol.* **2010**, *75*, 87–101.

[3] M. H. Isa, E. H. Ezechi, Z. Ahmed, S. F. Magram, S. R. M. Kutty, *Water Res.* **2014**, *51*, 113–123.

[4] A. E. Yilmaz, R. Boncukcuoğlu, M. M. Kocakerim, *J. Hazard. Mater.* **2007**, *149*, 475–481.

[5] R. Bernstein, S. Belfer, V. Freger, *Environ. Sci. Technol.* **2011**, *45*, 3613–3620.

[6] Z. Yazicigil, Y. Oztekin, *Desalination* **2006**, *190*, 71–78.

[7] Y. Xu, J.-Q. Jiang, *Ind. Eng. Chem. Res.* **2008**, *47*, 16–24.

[8] a) M. Palencia, M. Vera, B. L. Rivas, *J. Membr. Sci.* **2014**, *466*, 192–199; b) Q. Shi, J.-Q. Meng, R.-S. Xu, X.-L. Du, Y.-F. Zhang, *J. Membr. Sci.* **2013**, *444*, 50–59.

[9] H. Mishra, C. Yu, D. P. Chen, W. A. Goddard, 3rd, N. F. Dalleska, M. R. Hoffmann, M. S. Diallo, *Environ. Sci. Technol.* **2012**, *46*, 8998–9004.

[10] a) Y.-T. Wei, Y.-M. Zheng, J. P. Chen, *Water Res.* **2011**, *45*, 2297–2305; b) M. S. Moorthy, D.-J. Seo, H.-J. Song, S. S. Park, C.-S. Ha, *J. Mater. Chem. A* **2013**, *1*, 12485.

[11] N. B. Darwish, V. Kochkodan, N. Hilal, *Desalination* **2015**, *370*, 1–6.

[12] a) Q. Zhang, J. Yu, J. Cai, R. Song, Y. Cui, Y. Yang, B. Chen, G. Qian, *Chem. Commun.* **2014**, *50*, 14455–14458; b) S. H. Jhung, J. H. Lee, J. W. Yoon, C. Serre, G. Férey, J. S. Chang, *Adv. Mater.* **2007**, *19*, 121–124; c) A. J. Howarth, M. J. Katz, T. C. Wang, A. E. Platero-Prats, K. W. Chapman, J. T. Hupp, O. K. Farha, *J. Am. Chem. Soc.* **2015**, *137*, 7488–7494; d) J. H. Wang, M. Li, D. Li, *Chem. Eur. J.* **2014**, *20*, 12004–12008; e) M. Maes, S. Schouteden, L. Alaerts, D. Depla, D. E. De Vos, *Phys. Chem. Chem. Phys.* **2011**, *13*, 5587–5589; f) S.-H. Huo, X.-P. Yan, *J. Mater. Chem.* **2012**, *22*, 7449; g) N. A. Khan, Z. Hasan, S. H. Jhung, *J. Hazard. Mater.* **2013**, *244–245*, 444–456; h) A. Samokhvalov, *Chem. Eur. J.* **2015**, *21*, 16726–16742.

[13] a) C.-C. Wang, J.-R. Li, X.-L. Lv, Y.-Q. Zhang, G. Guo, *Energy Environ. Sci.* **2014**, *7*, 2831; b) Q. L. Zhu, Q. Xu, *Chem. Soc. Rev.* **2014**, *43*, 5468–5512; c) F. K. Shieh, S. C. Wang, C. I. Yen, C. C. Wu, S. Dutta, L. Y. Chou, J. V. Morabito, P. Hu, M. H. Hsu, K. C. Wu, C. K. Tsung, *J. Am. Chem. Soc.* **2015**, *137*, 4276–4279; d) G. Lu, S. Li, Z. Guo, O. K. Farha, B. G. Hauser, X. Qi, Y. Wang, X. Wang, S. Han, X. Liu, J. S. DuChene, H. Zhang, Q. Zhang, X. Chen, J. Ma, S. C. Loo, W. D. Wei, Y. Yang, J. T. Hupp, F. Huo, *Nat. Chem.* **2012**, *4*, 310–316; e) Y. K. Park, S. B. Choi, H. J. Nam, D. Y. Jung, H. C. Ahn, K. Choi, H. Furukawa, J. Kim, *Chem. Commun.* **2010**, *46*, 3086–3088; f) C. Wang, K. E. deKrafft, W. Lin, *J. Am. Chem. Soc.* **2012**, *134*, 7211–7214; g) J. Song, Z. Luo, D. K. Britt, H. Furukawa, O. M. Yaghi, K. I. Hardcastle, C. L. Hill, *J. Am. Chem. Soc.* **2011**, *133*, 16839–16846; h) V. Lykourinou, Y. Chen, X.-S. Wang, L. Meng, T. Hoang, L.-J. Ming, R. L. Musselman, S. Ma, *J. Am. Chem. Soc.* **2011**, *133*, 10382–10385; i) J.-R. Li, J. Sculley, H.-C. Zhou, *Chem. Rev.* **2012**, *112*, 869–932; j) S. Li, F. Huo, *Nanoscale* **2015**, *7*, 7482–7501; k) C. Zlotea, R. Campesi, F. Cuevas, E. Leroy, P. Dibandjo, C. Volkringer, T. Loiseau, G. Férey, M. Latroche, *J. Am. Chem. Soc.* **2010**, *132*, 2991–2997.

[14] a) G. Ferey, C. Mellot-Draznieks, C. Serre, F. Millange, J. Dutour, S. Surble, I. Margiolaki, *Science* **2005**, *309*, 2040–2042; b) D.-Y. Hong, Y. K. Hwang, C. Serre, G. Férey, J.-S. Chang, *Adv. Funct. Mater.* **2009**, *19*, 1537–1552.

[15] S. Sartipi, M. J. Valero Romero, E. Rozhko, Z. Que, H. A. Stil, J. de With, F. Kapteijn, J. Gascon, *ChemCatChem* **2015**, *7*, 3243–3247.
[16] a) T. Uemura, N. Yanai, S. Watanabe, H. Tanaka, R. Numaguchi, M. T. Miyahara, Y. Ohta, M. Nagaoka, S. Kitagawa, *Nat. Commun.* **2010**, *1*, 83; b) L. Bai, P. Wang, P. Bose, P. Li, R. Zou, Y. Zhao, *ACS Appl. Mater. Interfaces* **2015**, *7*, 5056–5060; c) T. Kitao, S. Bracco, A. Comotti, P. Sozzani, M. Naito, S. Seki, T. Uemura, S. Kitagaw, *J. Am. Chem. Soc.* **2015**, *137*, 5231–5238.
[17] K. Yoshimura, Y. Miyazaki, F. Ota, S. Matsuoka, H. Sakashita, *J. Chem. Soc. Faraday Trans.* **1998**, *94*, 683–689.
[18] M. Saikia, L. Saikia, *RSC Adv.* **2016**, *6*, 14937–14947.
[19] a) G. Rodríguez-López, M. Dolores Marcos, R. Martínez-Máñez, F. Sancenón, J. Soto, L. A. Villaescusa, D. Beltrán, P. Amorós, *Chem. Commun.* **2004**, 2198–2199; b) C. Sanfeliu, R. Martinez-Manez, F. Sancenon, J. Soto, V. Puchol, P. Amoros, M. D. Marcos, *J. Mater. Chem.* **2012**, *22*, 25362–25372.
[20] L. Wang, T. Qi, Y. Zhang, *Colloids Surf. A* **2006**, *275*, 73–78.
[21] Ö. Kaftan, M. Açıkel, A. E. Eroğlu, T. Shahwan, L. Artok, C. Ni, *Anal. Chim. Acta* **2005**, *547*, 31–41.
[22] a) L. Xu, Y. Liu, H. Hu, Z. Wu, Q. Chen, *Desalination* **2012**, *294*, 1–7; b) T. Qi, A. Sonoda, Y. Makita, H. Kanoh, K. Ooi, T. Hirotsu, *Ind. Eng. Chem. Res.* **2002**, *41*, 133–138; c) A. Harada, T. Takagi, S. Kataoka, T. Yamamoto, A. Endo, *Adsorption* **2011** *17*, 171–178; d) S. Nishihama, Y. Sumiyoshi, T. Ookubo, K. Yoshizuka, *Desalination* **2013**, *310*, 81–86.
[23] J. P. Hershey, M. Fernandez, P. J. Milne, F. J. Millero, *Geochim. Cosmochim. Acta* **1986**, *50*, 143–148.
[24] H. W. Oviatt, K. J. Shea, J. H. Small, *Chem. Mater.* **1993**, *5*, 943–950.
[25] H. Liu, B. Qing, X. Ye, Q. Li, K. Lee, Z. Wu, *Chem. Eng. J.* **2009**, *151*, 235–240.

Received: May 31, 2016
Published online on ■■ ■■, 0000

FULL PAPER

Stops the tangling: A highly porous metal–organic framework MIL-101(Cr) was loaded with *N*-methyl-D-glucamine-based organosiloxane polymer. The resulting composite exhibits exceptionally high efficiency for the removal of boric acid from water.

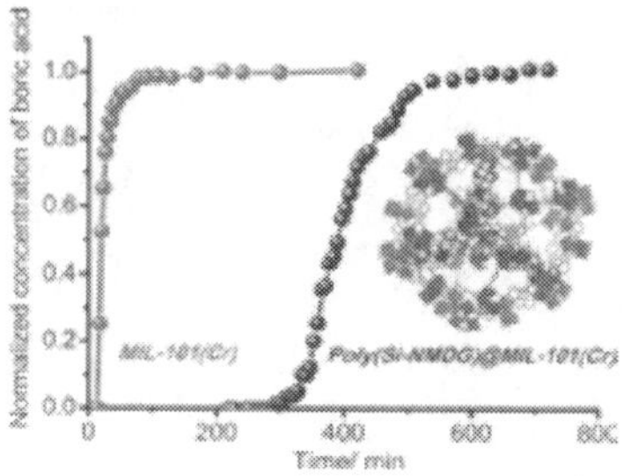

Metal–Organic Frameworks

J. Wang, Z. Bao, H. Xing, B. Su, Z. Zhang, Q. Yang, Y. Yang, Q. Ren, B. Chen**

■■ – ■■

Incorporation of *N*-Methyl-D-glucamine Functionalized Oligomer into MIL-101(Cr) for Highly Efficient Removal of Boric Acid from Water

468367 Halophyte Biochar for Water Desalination Brine Concentrate Management

Abstract:

Disposal of concentrate from brackish water desalination is limited by concerns of salt leaching into and contaminating surface and ground waters. These concerns, and measures to prevent leaching, add substantially to concentrate disposal costs with little or no opportunity to recover those costs. One option being explored is the use of concentrate as irrigation water for producing halophyte biomass for forage and other applications. Continuing development of this option requires more knowledge about what happens to the salt in the concentrate once land applied as irrigation water.

The purpose of this project is to determine how much of the salt taken up by the halophyte crops can be protected from leaching by pyrolysis of the biomass, i.e. to determine the potential of sequestering salt in mineral form. Results from this research will be useful for those managing high salinity biomass streams, such as halophyte crop wastes and animal manures, and those seeking local concentrate disposal options with a potential for producing economic and/or environmental returns. The ability to sequester salt in biochar would decrease the negative impacts of land-applying high-salinity materials, including those used in desalination concentrate management systems.

455179 ^{13}c Pathway Analysis for the Role of Formate in Electricity Generation By *Shewanella Oneidensis* MR-1 Using Lactate in Microbial Fuel Cells

Abstract:

Microbial fuel cell (MFC) is a promising technology for wastewater treatment. Basically, the microorganisms sticking on the anode of the MFC utilize various organic compounds from the wastewater as the electron donors to generate electricity, while in the meantime removing the organic compounds in the wastewater. The type of electron donors fed into MFCs affects the electrical performance, and mechanistic understanding of such effects is important to optimize the MFC performance. In this study, we used a model organism in MFCs, *Shewanella oneidensis* MR-1, and ^{13}C pathway analysis to investigate the role of formate in electricity generation and the related microbial metabolism. Our results indicated a synergistic effect of formate and lactate on electricity generation, and extra formate addition on the original lactate resulted in more electrical output than using formate or lactate as a sole electron donor. Based on the ^{13}C tracer analysis, we discovered decoupled cell growth and electricity generation in *S. oneidensis* MR-1 during co-utilization of lactate and formate (i.e., while the lactate was mainly metabolized to support the cell growth, the formate was oxidized to release electrons for higher electricity generation). To our best knowledge, this is the first time that ^{13}C tracer analysis was applied to study microbial metabolism in MFCs and it was demonstrated to be a valuable tool to understand the metabolic pathways affected by electron donors in the selected electrochemically-active microorganisms.

461312 *Advanced Redox-Based Electrochemical Separations for Wastewater Treatment*

Abstract:

Water scarcity is one of the principal challenges for the 21st century, both from a geographical and an economic standpoint. Developing novel, more energetically efficient technologies for wastewater treatment, retreatment and reuse is crucial for long-term water sustainability. One of the principal challenge for water remediation, both in the U.S. and worldwide, is to target contaminants of emerging concern, such as complex organic molecules including pharmaceuticals, chemical precursors, endocrine inhibitors and pesticides.[1] These originate from chemical processes and organic synthesis, the food industry, electronic manufacturing among others. Many of these pollutants disperse in remote locations, such as agricultural locations, in which large-scale purification systems cannot be implemented. Electrochemical methods are particularly attractive for addressing these problems due to their modularity, fast kinetics and integration with renewable energy sources.[2]

We have developed an redox-mediated separation method based on organometallic-functionalized electrodes (e.g. ferrocene or cobaltocene) for the sorption and release of specific ions from the liquid-phase under oxidizing and reducing conditions.[3] Many of the organic contaminants are anionic in nature (e.g. carboxylates, sulfonates or phosphonates), thus selectivity based on chemical interactions with their functional groups is essential in chemical design. Rather than rely solely on charge or size, our immobilized redox-centers targets these compounds through the activation of a specific chemical binding. Moreover, in the charging and discharging steps, electrochemical charge can be stored by ion-selective adsorption onto the porous electrodes, thus dramatically decreasing energetic costs.

This modular system can then be easily implemented at the point-of-exit in chemical manufacturing and wastewater treatment. Here, we present a complete asymmetric redox-system in both batch and flow for (1) water remediation, through the specific separation of anionic and cationic contaminants in water, (2) tandem energy recovery due to pseudocapacitive charge storage and (3) increasing sustainability in chemical processes, through the recovery of expensive synthesis products from complex organic reactions, thus reducing organic waste and decreasing water usage during separation steps.

References

[1] a) Gilliom R.; Hamilton P. *U.S. Geological Survey*. 2014. (USGS fact sheet: http://pubs.usgs.gov/fs/2006/3028/

). b) Doughton, C.G. and Thomas, A. T. (1999). Pharmaceuticals and personal care products in environment: agents of suble change? *EHP* **107** (6), 907-938.

[2] Suss, M. E.; Porada, S.; Sun, X.; Biesheuvel, P. M.; Yoon, J.; Presser, V. *Energy & Environmental Science* **2015**. Zhao, R.; Satpradit, O.; Rijnaarts, H. H.; Biesheuvel, P. M.; van der Wal, A. *Water Res* **2013**, *47*, 1941.

[3] **Su, X.**; Kulik, H; Jamison, T.F.; Hatton, T. A. 2016. Anion-selective redox electrodes: electrochemically-mediated separation with organometallic interfaces. *Advanced Functional Materials*. Advance Article Online. DOI: 10.1002/adfm.201600079

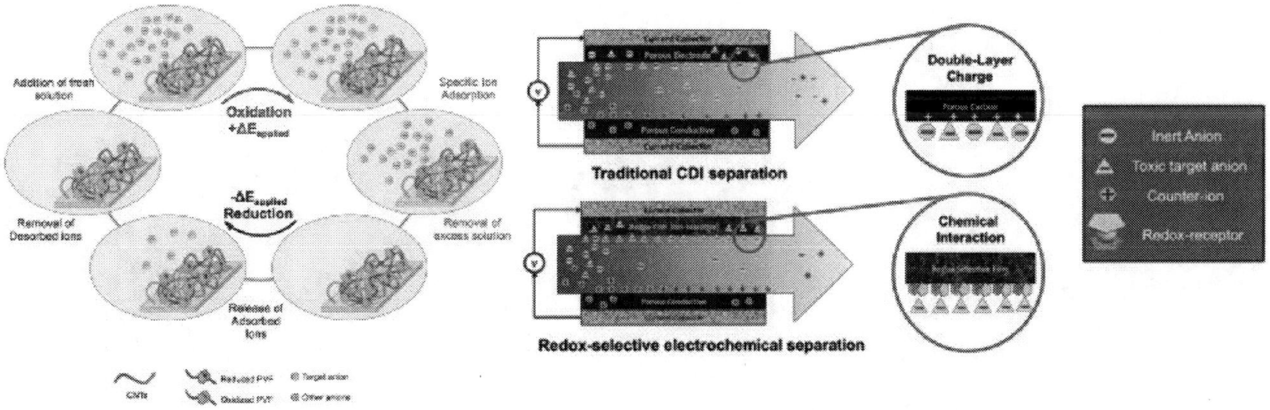

458179 Influence of Elevated Salinity and the Presence of Cytostatic Drugs on Microbial Community in FO Anaerobic Membrane Bioreactors

Abstract:

Wastewater treatment using foward osmotic anaerobic membrane bioreactors (FO-AnMBR) has a number of advantages over traditional wastewater treatment approaches. Previous studies have investigated the overall performance of lab-scale FO-AnMBR in treating synthetic wastewater. During the operation of FO-AnMBR, it is expected that salinity gradually increases, which has also been shown experimentally. In addtion, emerging contaminants such as cytostatic drugs in wastewater would also be accumulated in the FO-AnMBR. Little has been known on whether and how the increased salinity and the presence of cytostatic drugs affect the microbial community in FO-AnMBR. The objective of this study was to investigate the influence of increased salinity and the presence of cytostatic drugs on the structure and function of the microbial community in FO-AnMBR through a metagenomics approach. A lab-scale FO-AnMBR was operated at 25 ^{o}C and fed with synthetic wastewater. Total DNA was extracted from sludge samples taken at pre-determined time points and sequenced using a HiSeq 2500 platform with 250-bp paired-end sequences. Based on the analyses of the sequences, structure and potential functions of the microbial communities were obtained. We found that the increased salinity in FO-AnMBR did not reduce the microbial diversity; however, the relative abunance of methane-producing archaea decreased significantly at high salinity condition, resulting in a reduced methane yield. The general metabolism did not differ significantly under low and high salinity conditions. Intriguingly, genes responsible for cell attachment and biofilm related signalling were enriched at high salinity condition. Transition of the gene clusters for assimilatory and dissimilatory nitrate reduction was also found. The presence of cytostatic drugs only slightly affected the microbial community composition. Biofilm-forming, multidrug resistant and siderophore-producing bacteria were found more dominant after the treatment with drugs.

471970 Batch and Packed-Bed Adsorption Study for the Removal of Dye from Textile Wastewater Using in-House Prepared Silica-Based Nanosorbcats: Kinetics and Computational Modeling

Abstract:

The worldwide shortage of fresh water and the huge competing demands from variety of users stimulates an urgent need for finding alternative cost-effective and environmentally friendly wastewater treatment processes. For instance, synthetic dyes wastewater discharged from the textile and dyestuff industries poses an environmental alert since these organic pollutants are toxic to many aquatic and non-aquatic living organisms. In addition, some of these pollutants are non-biodegradable, and thus, they will exist for a long time in the environment, which may cause a real challenge to the conventional wastewater treatment processes. Accordingly, innovative and environmentally sound techniques are needed. The application of nanoparticle technology as adsorbents and catalysts (nanosorbcats), whether as a standalone or as an enabling technology, in cleaning up wastewater has recently received great attention. This is because of the unique chemical and physical properties of nanoparticles in comparison with their counterparts, which make them superior to the conventional adsorbent/catalysts.

Accordingly, the present study investigated the employment of newly in-house prepared silica-based nanosorbcats functionalized with active species for cleaning up textile wastewater. A facile co-precipitation synthesis route was used to prepare those nanosorbcats, which were characterized by different characterization techniques like XRD, BET, TPD, TPR, HRTEM and zeta potential. The prepared nanosorbcats were employed for the adsorptive removal of cationic and anionic dyes from synthetic and real textile wastewater. Batch adsorption studies were performed for kinetics and equilibrium modeling on the removal of different dyes from wastewater. The operating variables studied were contact time, initial dye concentration, and solution pH. Results showed an optimum pH value, where the maximum removal of dye was achieved. The experimental adsorption data have been modeled using Sips and BET adsorption isotherm models. In addition, a computational modeling on the selected model compounds was carried out to understand the adsorption mechanism on the prepared nanosorbcats. Breakthrough capacities were also investigated in a packed-bed column study. The most successful silica-based nanosorbcats in cleaning up the real textile wastewater in the batch experiments were prepared in extrudate form and used in up-flow packed-bed adsorption experiments. The effects of the influent concentration, flow rate, and bed height were investigated. Breakthrough time and adsorption capacity of the packed-bed increased with increasing bed height, whereas decreased with the increase in influent initial concentration and flow rate. The reliable results of this study will be elaborated and discussed further during the presentation.

463610 Optimization of Operational Parameters for Application of Zirconium/Polyvinyl Alcohol Modified Flat-Sheet Polyvinylidene Fluoride Membrane As Better Adsorptive Material in Treatment of Organic Arsenic Containing Wastewater

Abstract:

Organic arsenic in waters has been a global concern due to its toxicity. In this study, a novel polyvinyl alcohol (PVA)/zirconium modified polyvinylidene fluoride (PVDF) membrane was applied to remove organic arsenic from water. The impregnation of zirconium ions within the modified membrane was attributed to the coordination reactions among the zirconium ions, ether and hydroxyl groups. The synthesized membrane worked better at the acidic conditions and achieved the optimal uptake for both monomethylarsonic (MMA) and dimethylarsinic (DMA) at pH 2.0. The adsorption isotherm study demonstrated that the adsorption of both organic arsenic species was controlled by the mono-layer adsorption process, and the maximum adsorption capacities of 73.04 and 29.78 mg/g for MMA and DMA can be achieved at pH 2.0. The presence of humic acid had a negligible impact on the uptake of organic arsenic, whereas varying impacts on the arsenic adsorption were observed due to the presence of coexisting anions such as fluoride, phosphate, carbonate and silicate. The filtration study was also carried out to evaluate the performance of the modified membrane on the removal of organic arsenic. Based on the XPS analyses, the exchange between chloride ions on the membrane surface and organic arsenic species should be responsible for the removal of both MMA and DMA.

459282 Analysis of Magnetic Bead Separation in Continuous-Flow Magnetophoretic Microsystems for Biomedical Applications

Abstract:

In recent years, there has been a growing interest in the removal of blood-borne toxins through the use of magnetic nanocomposites. These toxins circulate in the bloodstream and can have a detrimental impact on health. Blood detoxification using magnetic beads comprised of an extracorporeal process where the patient's blood is mixed with the magnetic materials with the ultimate goal of selectively removing the toxin while maintaining normal functionality of blood constituents. Once the adsorption of the toxins by the beads surface is completed, the magnetic separation stage takes place and the toxins are removed along with the material, leading to a toxin-free blood solution that returns to the circulatory system of the patient. Selective sequestration of these toxins from the patient's blood using functionalized magnetic particles can improve the effectiveness of treatment while reducing side effects compared to conventional methods [1]. The aim of this work is to contribute to the design of continuous-flow magnetic separators and discuss their application for the recovery of magnetic beads from flowing blood streams.

Numerous microfluidic magnetic separator designs have been proposed in the last few years to recover magnetic beads from fluids; however, the application of continuous-flow systems offers numerous advantages such as higher separation efficacy and greater throughput as compared to batch separators. When using continuous microdevices, the magnetic beads and the adsorbed toxin are continuously injected through one inlet, deflected from the original blood stream and collected into a flowing buffer stream by a magnetic gradient applied perpendicular to the flow direction. In order to employ this separator, two fundamental requirements must be met, i.e. the complete recovery of the magnetic beads from the blood solution and the minimization of intermixing between the blood and buffer streams inside the device.

With regard to the evaluation of using continuous-flow systems for this application, we introduce a combination of magnetic and fluidic computational modelling that describes the bead trajectory inside a continuous microchannel under the influence of an external magnetic field (due to a permanent magnet) and the potential mixing between fluid streams. The model reported here represents a rational design guide for experimentalists in this field of research since it can be used to predict the separator efficacy taking into account key operating variables and impact parameters, such as bead and fluid properties, the characteristics of the magnets employed, the dimension of the microchannel and the selected flow rates.

Additionally, the experimental performance of a microfluidic magnetic separator was evaluated. Aqueous solutions of fluorescein and magnetic particles functionalized with fluorophores have been used in order to quantify both the mixing of both fluids inside the device and the particle separation efficacy according to the analysis of the images taken using an epi-fluorescent microscope. The theoretical and experimental results are compared and discussed.

Finally, it should be noted that although the fundamentals of continuous-flow magnetic separators have been previously discussed in the literature [2,3], a key distinguishing feature of this research is that we study, for the first time, the interaction between two fluids flowing simultaneously in the device while taking into account the effects of particle-fluid interactions in the flow field and potential diffusion between both streams.

ACKNOWLEDGEMENTS

Financial support from the Spanish Ministry of Economy and Competitiveness under the project CTQ2015-66078 is gratefully acknowledged. Jenifer Gómez-Pastora also thanks the FPI postgraduate research grant (BES-2013-064415).

REFERENCES:

1. "Magnetic separation-based blood purification: A promising new approach for the removal of disease-causing compounds?", I.K. Herrmann, A.A Schlegel, R. Graf, W.J. Stark, and B. Beck-Schimmer, *J. Nanobiotechnology* **2015**, 13(1), 49–52.
2. "Magnetic biotransport: analysis and applications", E.P. Furlani, *Materials* **2010**, 3, 2412–2446.
3. "Scalability analysis of magnetic bead separation in a microchannel with an array of soft magnetic elements in a uniform magnetic field", S.A. Khashan, E.P. Furlani, *Sep. Purif. Technol.* **2014**, 125, 311–318.

Visualization of particle movements in the periphery of a deflector wheel classifier

Authors: Christian Spötter[1], Kurt Legenhausen[1], Alfred P. Weber[1]
[1]Technical University of Clausthal, Clausthal – Zellerfeld, Germany

Abstract

Although the deflector wheel classifier is the dominating separation device in the industrial separation of fine-grained particle fractions, the classifying mechanisms at high particle loadings are still not described and understood appropriately. Theoretical models for the calculation of the separation efficiency for deflection wheel classifiers are mostly based on single particle fate and fail to recover the behavior at high particle loading where particle-particle and particle-wall collisions are encountered. To gain a better understanding of the particle motion under real operating conditions a deflection wheel classifier was modified in a way which allows an optical access onto the classifying wheel. Using a high-speed camera analyses it was shown that particle streaks are formed which are influenced by the peripheral wheel speed and the feed mass loading. With increasing peripheral speed the streak form changes from an arc-like to a tangential structure. The occurrence of the streaks is periodical with a frequency which depends on peripheral speed and feed mass loading. While shape, size and density of the streaks change with the peripheral speed, the velocity of the streaks, when they detach and leave the wheel periphery, is quiet independent of the peripheral speed.

1. Introduction

Air classification is an important step for the production of many industrial powders. According to *Rumpf* [1, 10] the basic task of this process is to classify a mixture of particles of different sizes in a fine and a coarse particle fraction. Since product characteristics and quality are closely related to the particle size, products with a narrow particle size distribution are desired e.g. for color pigments and abrasive powders [2, 10].

In contrast to classifiers with unrestricted flow, such as the spiral air classifier with a free vortex the deflection wheel classifier produces a forced vortex due to the rotating paddles which is stated to be less depending on the mass loading. The underlying conception is that the particles experience a centrifugal force F_c in the rotational flow field which is counteracted by the drag force F_w of the air flowing inwards between the paddles (cf. Fig. 1) [1, 10]. On the outer edge of the deflector wheel coarse particles are deflected, while fine particles are following the air flow through the deflector wheel and are separated later from the air flow by an aero cyclone [1-2].

In the last century *Rumpf*, *Molerus*, *Sender*, *Schubert* and *Husemann* developed separation models which are describing the particle behavior and the separation efficiency of a deflector wheel classifier on basis of geometrical and equilibrium considerations. These models, which consider the behavior of a single particle under the influence of the centrifugal and drag force, are applicable for the calculation of the cut point using several empirical coefficients, which can only be obtained from experiments. However, they suffer from several drawbacks. First, the dependency of the empirical parameters on the operating conditions cannot be attributed to a physical sound picture of the separation process. Second, the particle-paddle- and particle-particle-collisions, which become significant at higher loadings, are not taken into account.

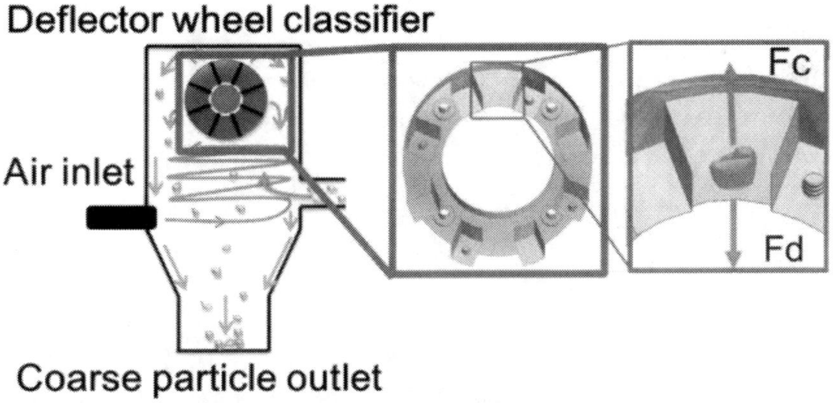

Fig. 1: Schematic diagram of a deflection wheel classifier

As little attention has been payed to the flow field through the paddles of the deflection wheel, *Legenhausen* [3] analyzed the flow behavior between the separation paddles using a Laser Doppler Anemometer (LDA) based on a water model. He found that the flow between two separation paddles detaches forming a flow vortex. Later *Toneva, et al.* [4] determined the gas flow behavior in a deflector wheel classifier using Particle Image Velocimetry (PIV) and Computational Fluid Dynamic (CFD) simulations. In agreement with *Legenhausen* they found a detachment of the gas flow from the paddles accompanied by a gas vortex between the separation paddles. Further, *Toneva et al.* demonstrated that the vortex formation leads to a constriction of the particle flow and an increase in the inward radial gas velocity. Although they neglected the influence of particle-particle-collisions on the separation behavior, their numerical calculations indicated that the particles frequently collide with the separation paddles, where coarse particles are leaving the separation zone faster than fine particles. The calculations of *Tenova et al.* have also shown, that the coarse particles are rebounding in both the inside and the outside direction, which leads to a deterioration of the separation sharpness. Deduced from the calculated trajectories a large proportion of the particle stream is not transported to the inside of the separation wheel. On the contrary their calculations have shown that a particle accumulation in the flow stream near the outside edge of the separation wheel is possible. According to the CFD simulations of *Toneva et al.* this accumulation will lead to a deterioration of the separation sharpness due to long residence times and particle-particle-interactions [5-7].

In order to translate the simulation results of *Toneva et al.* to studies at high loadings, *Stender et al.* [5] reconstructed a deflection wheel classifier type Hosokawa Alpine ATP 50. By replacing the two-sided suspension of the wheel axis by a one-side double suspension they introduced a direct optical access to the front side of the separator wheel. With this optical access and by using a high speed camera, initial studies of the gas flow and particle movement in a real separation process of a deflection wheel classifier were carried out. During their experiments, *Stender et al.* found that at a revolution rate of the deflection wheel of 6000 rpm a vortex is formed near the heading separation paddles. This vortex moves into a separation area between the two separation paddles at an increasing revolution rate and finally reaches the area near the pursuing paddle at a revolution rate of 15000 rpm. With increasing revolution rate, the vortex increasingly constricts the particle flow stream. Through this study *Stender et al.* could confirm the model from *Legenhausen* for the first time under real process conditions. They pointed out that particle-particle- and particle-wall-collisions are affecting the separation process [5-7].

2. Material and method

2.1 Basic construction and separation method of a deflector wheel classifier

To determine the separation characteristics the modified deflector wheel classifier type Turboplex ATP 50 from Hosokawa Alpine with an aero cyclone and a bag filter system was used, as illustrated in Fig. 2.

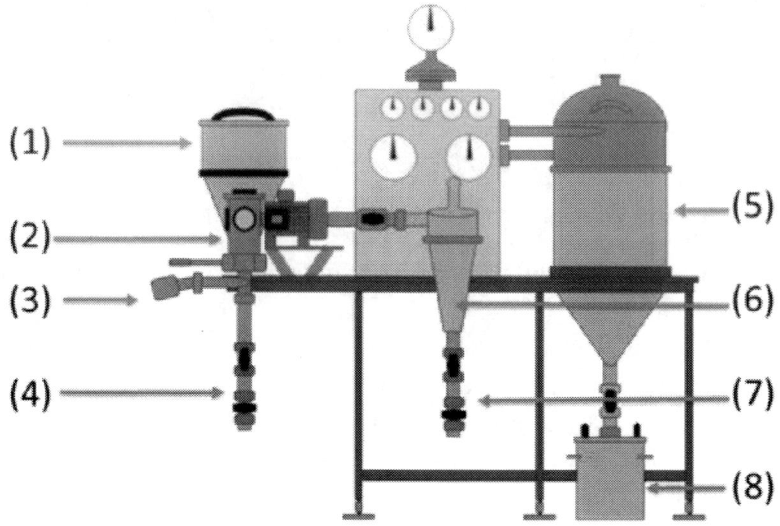

Fig. 2: Scheme of the Hosokawa Alpine ATP 50: (1) feed screw, (2) deflector wheel classifier ATP, (3) air pipe, (4) particle lock system 1: coarse material outlet, (5) bag filter system, (6) aero cyclone (7) particle lock system 2: fine material outlet 1, (8) fine material outlet 2 [7].

As feed material limestone ($CaCO_3$) is introduced into the deflector wheel classifier (2) below the rotor via a screw conveyor which is attached to the side of the deflector wheel classifier (1). The cumulative particle size distribution of the feed material has a $x_{10,3}$ of 5.86 μm, a $x_{50,3}$ of 59.86 μm and a $x_{90,3}$ of 179,54 μm. From the bottom of the housing the classifying air is supplied into the deflector wheel classifier tangential via an air pipe (3) which is equipped with a filter. The aspirator produces a constant air flow of 75 m^3/h. The classifying wheel of the Turboplex ATP 50 has a wheel diameter of 50 mm and is powered by a horizontally positioned shaft. The peripheral speed of the separating wheel was varied over a frequency converter from 7.86 m/s (3000 rpm) to 39.27 m/s (15000 rpm). The rejected coarse material leaves the separator downwards through the central particle lock system outlet (4). The fine material leaves the classifier wheel together with the classifying air flows via a hollow shaft and a subsequent fixed pipe [5-7].

The fine particle fraction is separated from the classifying air by an aero cyclone (6). The fine particles are leaving the cyclone through a particle lock system into fines outlet 1 (7). The ultrafine fraction leaves the aero cyclone via the vortex finder and is finally separated in a bag filter system (5). The bag filters are periodically cleaned with compressed air and the particles are collected in a fine outlet 2 (8) [5-7].

2.2 Reconstruction of the deflector wheel classifier for an optical accessibility

To determine the separation characteristics of a deflector wheel classifier *Stender et al.* [9] reconstructed the ATP 50 so that an optical access onto the deflector wheel was achieved, however, maintaining the flow and classification conditions and the geometry of the classifying wheel and its surrounding components. In order to realize this, the motor of the wheel and the outlet pipe of the fine particle fraction had to be moved to the same side of the deflector wheel. The outlet pipe of the fine particles has been realized using a hollow shaft supported by two high-speed spindle bearings (A), driven by a flat belt (B) which is sealed with an oil-soaked gasket. The drive of the hollow shaft takes place via a 1:3 gearing which is powered by a frequency-controlled three-phase asynchronous motor (D) (see Fig. 3) [5].

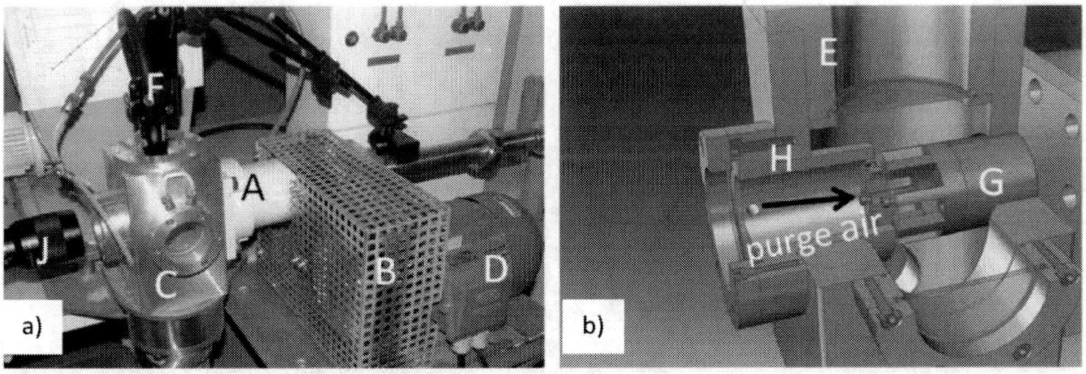

Fig. 3: Remodeled Turboplex ATP 50 deflector wheel: a) One-sided bearing for optical access on the opposite side, b) Installations for an optical access and purge air for keeping the view on the deflector wheel free. (A = high-speed spindle bearings, B = flat belts, C = classifier chamber, D = three-phase asynchronous motor, E = top viewing window, F = light source, G = classifier wheel, H = side viewing window, I = transparent front wall, J = high-speed camera) [5-7]

For the optical access to the classifier wheel specially designed viewing windows were installed on the top (E) and on the side (H). The windows were placed in the side wall and the ceiling of the housing. Special glass discs of the type Mirogard were used for the windows exhibiting anti-reflective properties on both sides leading to a high transparency. A high-speed camera Keyence VW-600M was placed in front of the window (cf. Fig. 3 (J)) to visualize the motions of the particles during the separation process. In order to prevent light reflections, the camera was aligned in a 90° angle to the illumination realized with a tripod lamp across the ceiling window [5-7].

3. Results and discussion

3.1 Visualization of gas and particle flow in the separation zone

The following results present the gas flow and the particle motion in the area between two separation paddles (separation zone). While for the visualization of the gas flow very fine limestone particles were used, the motion of larger particles was visualized with coarser limestone particles ($x_{50,3}$ of 59.86 µm). As tracers fine limestone particles with an $x_{50,3}$ of 2.15 µm were used which are following the gas flow without lag and without classification. To catch the gas flow behavior the discernable contrast points, i.e. due to some inhomogeneous distribution of the fine particles, were followed on subsequent images. Using the software Motion analyzer VW 9000 the motion of these fine particle accumulations was represented by lines of arrows. In Fig. 4 a) it is shown that at a peripheral speed of 7.86 m/s the limestone tracer particles are passing the whole separation area in an arc-like flow. The gas flow starts to be deflected slightly in the left/middle section of the classifying zone, while near the pursuing paddle the gas flows in straight lines. Nevertheless, the gas flow of the very fine limestone particles passes the entire classification area at a peripheral speed of 7.86 m/s. When entering the separation zone the coarse limestone particles (Fig. 4 e) follow the gas flow into the separation zone but later they continue their motion towards the pursuing paddle deviating from the gas flow due to their higher inertia. Only at this low revolution rate the spatial distribution of the particles is relatively homogeneous over the classification area as a consequence of the superposition of largely spreading incoming particles and deflected larger particles (cf. Fig. 4 e) [5-7].

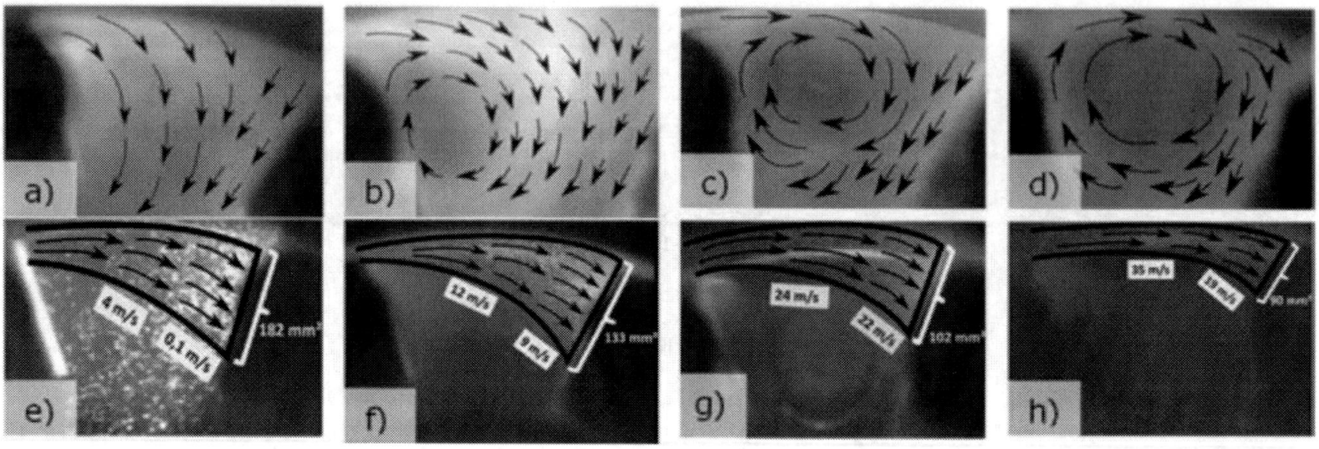

Fig. 4: Visualization of the gas flow (picture a) to d)) and the coarse particle motion (picture e) to h)) between the separating paddles for an increasing peripheral speed from 7.86 to 31.42 m/s. View onto the separation area between two deflection paddles (The separation wheel moves anti-clockwise) [7].

From subsequent images the particle velocity relative to the pursuing paddle was measured using the Motion analyzer VW 9000 software. Interestingly, the particle velocity, which was initially close to the peripheral speed of the paddle, decreased during the approach of the paddle. This deceleration effect could be caused either by bouncing particles colliding with the approaching particles or by peripheral air flow created by the paddle motion, or a combination of both. In any case the deceleration of the approaching particles was observed at all revolution rates, as indicated by the relative velocities given in Fig. 6. At an increased peripheral speed of 15.71 m/s the tracer particle flow (Fig. 4 b) begins to form a vortex on the left side of the separation area. This vortex leads to a constriction of the particle flow thereby focusing it onto a smaller area of the impaction paddle (Fig. 4 f). The vortex begins to move away from the heading separation paddle into the middle of the separation area, if the peripheral speed is increased further to 23.56 m/s (Fig. 4 c). With increasing revolution rate (Fig. 4 d) the vortex moves closer and closer to the pursuing paddle [5-7].

3.2 Visualization of particle streaks in the periphery of the deflector wheel

The high-speed camera investigations of the present study have confirmed and characterized for the first time the formation of particle streaks in the periphery of the deflection wheel of a classifier under real process conditions. For a given mass loading and a given air flow rate, the formation of the particle streaks is mainly influenced by the peripheral speed of the deflection wheel. Experiments have shown that at different peripheral speed rates different streak types with different streak sizes (not discussed here) and residence times are created. The evolution of the streaks is governed by the periodical detachment of small streaks from the deflection wheel which accumulate to larger streaks at a certain distance from the wheel before they are discharged out of the viewing area.

At a peripheral speed of 7.86 m/s the deflected particles form a primary streak at the outer edge of the deflecting paddle which grows to an arc-like structure, as shown in figure 5. With time the growing streak lags increasingly behind the paddle which continues to emit deflected particles. While the paddle pursues its circular motion the streak detaches and follows a tangential trajectory, thereby collecting roving particles.

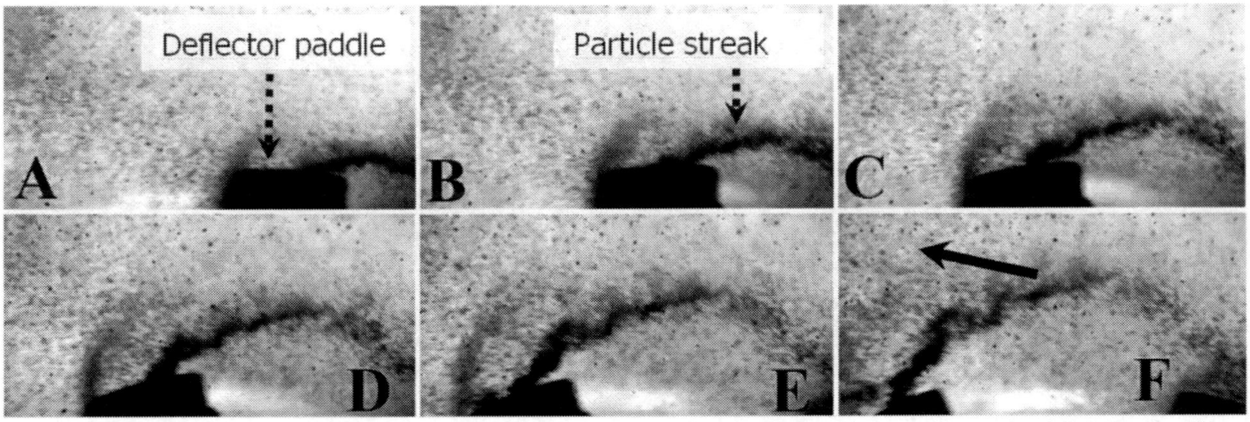

Fig. 5: Formation of a particle streak with an arched form at a peripheral speed of 7.86 m/s. Time between A) and F) 1.8 ms.

At peripheral speeds equal or higher than 15.71 m/s the structure of the streak changes. Instead of the arc-like form the streak approach a more linear structure which is oriented tangentially to the deflection wheel. At these higher peripheral speeds the deflected particles accumulate to form lager streaks, however, which continues to be discharged tangentially (cf. figure 6).

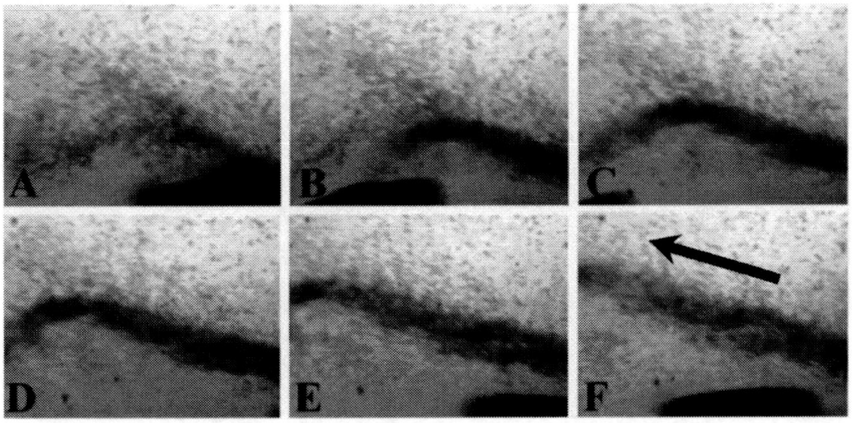

Fig. 6: Formation of a particle streak with a tangential form at a peripheral speed of 15.71 m/s. Time between A) and F) 2.3 ms.

To gain a better understanding of the dynamics of the streak formation the frequency for detachment of the primary streaks and the discharge frequency of the large streaks have been investigated. Generally the number of deflected particles increases with an increasing peripheral speed leading to an increased frequency of detached streaks (cf. figure 7). However, in spite of the increasing number of primary streaks the discharge frequency of the large streaks stays rather constant (cf. figure 7). We hypothesize that the small streaks accumulate to increasingly larger and denser streaks with increasing peripheral speed before being discharged.

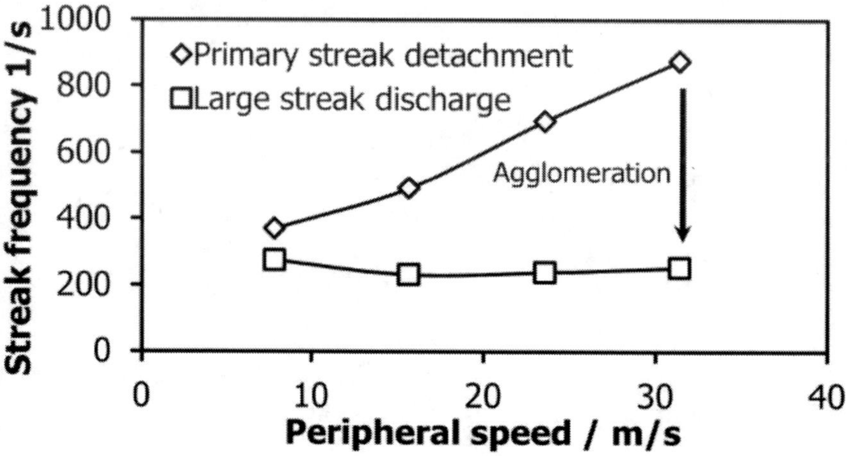

Fig. 7: Frequency of created streaks vs. peripheral speed of the deflector wheel

In figure 8 the discharge velocity of the large streaks is shown as a function of the peripheral speed. As discuss above, the streak mass and density increases with increasing peripheral speed. However, as indicated in figure 8 the peripheral speed has only a minor influence on the discharge velocity of the streaks. This may be a consequence of the acting forces. While with increasing density of the streak the permeability for the classifying air is reduced, leading to a higher drag force towards the center of the wheel, the centrifugal force increases also with increasing density. Nevertheless, the increasing peripheral speed should also influence the centrifugal force. It seems, however, that the large streaks experience only a fraction of the peripheral speed of the deflection wheel as already indicated in figure 5 and 6 where the streak motion lags behind the wheel motion.

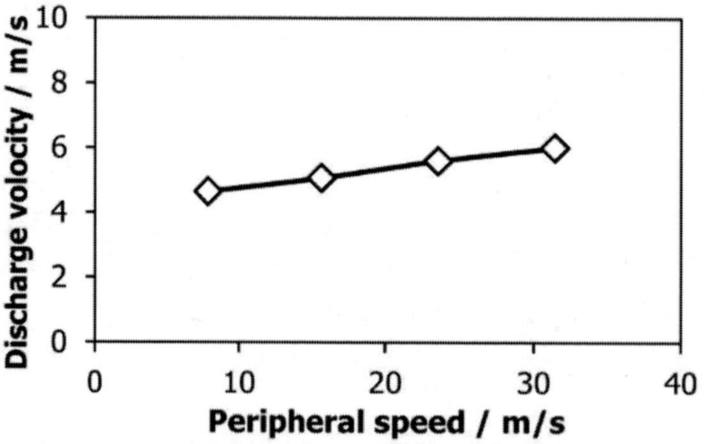

Fig. 8: Discharge velocity of the large streaks vs. peripheral speed of the deflector wheel

4. Conclusions

In this study the particle motion in a deflector wheel classifier was investigated under reals process conditions using a high speed camera. Special attention was payed to the formation of particle streaks around the deflector wheel. Surprisingly the occurrence of streaks is periodical with a frequency which depends on peripheral speed and feed mass loading. Depending on the peripheral speed streaks of different shape, density and size are formed. However, the discharge frequency and velocity of the streaks are nearly independent of the peripheral speed.

From a performance point of view these streaks may have a significant influence on the separation efficiency and the throughput. In particular the streaks may act as filter for counter flowing feed particles. This hypothesis would agree with the observations that with increasing mass loading and increasing peripheral speed the amount of displaced fine particles in the coarse fraction increases. However, this issue will be addressed in future work.

5. Acknowledgement

The authors gratefully acknowledge the German Research Foundation (DFG) for their financial support of this project under the grant no. WE 2331 / 16-2 in context of the priority program SPP 1679 "Dynamic simulation of cross-linked solid processes".

6. References

1. Rumpf H. and Raasch J. (1962), Desagglomeration in Strömungen, Karlsruhe.
2. Leschonski K. (1988), Windsichter, verfahrenstechnische Maschinen zur Herstellung definierter pulverförmiger Produkte, Jahrbuch 1988 der Braunschweigischen Wissenschaftlichen Gesellschaft, Braunschweig.
3. Legenhausen K. (1991), Untersuchung der Strömungsverhältnisse in einem Abweiseradsichter, Dissertation, TU Clausthal.
4. Toneva P., Epple P., Wirth K.E. and Peukert W. (2011), Grinding in an air classifier mill, Powder Technology, Erlangen - Hamburg.
5. Stender M., Legenhausen K. und Weber A.P. (2015), Visualisierung der Partikelbewegung in einem Abweiseradsichter, Chemie Ingenieur Technik, Clausthal.
6. Spötter C., Stender M., Legenhausen K. und Weber A.P. (2015), Dynamische Trenncharakteristik von Abweiseradsichtern, Produktgestaltung in der Partikeltechnologie Band 7, Faunhofer Verlag.
7. Spötter C., Legenhausen K. and Weber A.P., Separation characteristics of a deflection wheel classifier in stationary conditions and at high loadings: New insights by flow visualization, KONA Powder and Particle, Submitted in Mai 2016.

466021 An Efficient Numerical Approach for Transient Simulation of Multiphase Flow Behavior in Centrifuges

Abstract:

Suspensions are intermediate or finished products in a variety of industrial processes. In order to separate the disperse phase of the liquid phase an attached solid-liquid separation step is required. Centrifuges are widely used for this purpose. Despite the wide range of industrial usage only very simplified analytical approaches like the sigma theory are available for dimensioning centrifuges, allowing only the prediction of a separation grain. Therefore the theoretically calculated separation efficiency differs from the actual separation efficiency in operation. The main reason for the inaccuracy of the simplified dimensioning models is the underlying assumption that the flow conditions within centrifuges are well-known and can be described analytically. As the actual flow conditions within centrifuges are time depending and highly complex. In addition to this, the rising sediment influences the flow conditions. The result is a time-depending separation efficiency. Therefore costly experiments in Terms of raw materials and time are currently required to optimize centrifuges.

A more efficient method compared to experiments is the application of numerical simulations. Standard approaches for multiphase flows are Euler-Lagrange and Euler-Euler simulations. To describe the impact of the sediment on the separation efficiency long-term simulations are necessary. With the current simulation approaches these are not feasible to accomplish. In order to realize long-term simulations a new time-efficient simulation procedure for multiphase flows has to be developed.

An efficient method to simulate the multiphase flow in solid bowl centrifuges taking into consideration the flow behavior of the sediment is being developed at the Institute for Mechanical Process Engineering and Mechanics at the Karlsruhe Institute for Technology. The multiphase is modelled as a mixed phase. To consider the flow behavior, the rheology of suspension and sediment are differently modelled and locally and transient calculated within the centrifuge. The flow behavior of the sediment influences the distribution of the sediment and the homogeneity in the sediment height. The focus of the talk will be on the developed simulation approach and results comparing different process and material parameter.

465932 Fractionation of Suspensions of Elongated Particles in a Channel: Influence of Particle Concentration and Reynolds Number

Abstract:

Hydrodynamic fractionation describes a size, and/or shape based particle separation process in channel flow by a side-channel or slotted channel wall. Wall-bounded fluid is removed via the side channels due to reduced pressure in the side channels. Particles entrained in the side channel flow are removed from the main flow. Separation performance depends on a multitude of parameters including upstream flow conditions, slot geometry, and particle size and shape. The influence of the parameter are investigated for analytical and industrial applications ranging from the separation of different types of human and bacterial cells (Sugaya, et.al., Biomicrofludics, 5, 2011, 024103:1-13) to the separation of fibers (Dong, et.al., Tappi Journal, 4, 2004, 3-7).

Reported studies however are limited to dilute particle flow without particle-particle interaction. At increased particle concentration, (i) particles collide, and (ii) might form clusters. An extreme example is the formation of flocs for fibrous, i.e. long aspect ratio particles (Kerekes, Nordic Pulp and Paper Research Journal, 21, 2006, 598-612). Single particle motion is highly restricted and particles move as a bulk.

In this talk we will present results from a recent investigation of the separation of elongated particles at higher concentration. The main channel is constructed from Plexiglas to allow image analysis of the flow conditions. The separation channel is constructed by additive manufacturing to allow sophisticated geometry. Flocculation and thus the formation of a strong network was achieved by using flexible elongated particles. The particle concentration and the Reynolds number and thus the strength of the particle network had been varied in the experimental investigation.

Specifically we will present the (i) upstream condition of the particle network and (ii) and the motion of

single particles at the separation slot, and (iii) link it to the measured grade efficiency, i.e. the fraction of particles per length remaining in the main channel. The results show the dependence of the separation performance on the network strength. With increasing Reynolds number, the number of separated particles per particle length increased and thus the grade efficiency decreased.

Acknowledgement
JDRP, JK, and SR acknowledge funding through the FLIPPR project (www.flippr.at).

References

S. Sugaya, M. Yamada, and M. Seki, "Observation of non-spherical particle behaviors for continuous shape-based separation using hydrodynamic filtration", Biomicrofluidics, 5, 2011, 024103:1-13.

S. Dong, M. Salcudean, and I. Gartshore, "The effect of slot shape on the performance of a pressure screen", Screening, Tappi Journal, 4, 2004, 3-7.

R. Kerekes, "Rheology of fibre suspensions in papermaking: An overview of recent research", Nordic Pulp and Paper Research Journal, 21, 2006, 598-612.

SEDIMENTATION OF NON-SPHERICAL PARTICLES, USING SMOOTH PARTICLE METHOD

Adnan Hamid

Pakistan Institute of Engineering and Applied sciences (PIEAS), Islamabad, Pakistan.

Samia Mehdi

Pakistan Institute of Engineering and Applied sciences (PIEAS), Islamabad, Pakistan.

Atta Ullah

Pakistan Institute of Engineering and Applied sciences (PIEAS), Islamabad, Pakistan.

Ryoichi Yamamoto

Kyoto University, Kyoto 615-8510, Japan.

Abstract

The purpose of this study is to understand the behavior of non-spherical particles in sedimentation and to show that how the settling of non-spherical particles can be qualitatively different from that of spherical particles. Direct numerical simulations are performed in order to investigate the static and dynamic settling properties of non-spherical rod like particles, embedded in Newtonian fluid in Stokes regime, using smooth profile method. Results reveal that average sedimentation velocity of rod like particles is orientation dependent. At lower volume fractions, mean settling velocity is greater than the velocity of a single rod like particle obtained from simulation, as single particle prefers horizontal orientation. Analysis of orientation angle shows that at lower volume fraction particles prefer horizontal orientation but with increase in volume fraction they attempt to align themselves vertically.
Keywords: Sedimentation, direct numerical simulations, smoothed profile method, non-spherical particles

Introduction

In general sedimentation is a natural process of removing suspended particles, heavier than the fluid, in which they are entrained by gravitational settling. Although sedimentation is widely used technique and significant work [1-10] has been done on sedimentation of spherical particles in recent years but less information is available about sedimentation of non-spherical particles. Previous studies [1, 2, 4-9] on sedimentation have been done mostly in Stokes regime, for low volume fractions (ϕ) and mono-dispersed spherical particles. Sedimentation phenomenon is strongly affected by the host fluid properties, particles parameters and by system size, which makes it difficult to tackle computationally, especially in case of non-spherical particles. Sedimentation phenomena is used in many chemical industries such as paints, water purification, powder processing, paper and pulp etc.

Stokes [3] performed first theoretical work and calculated the terminal velocity of a single sphere in an incompressible fluid in Stokes regime, called Stokes velocity. Later, effects of back flow of fluid is explored by Richardson and Zaki [9] and suggested an empirical relation for the hindered settling velocity. Caflisch and Luke [11, 12] estimated that velocity fluctuations diverges with domain size. Guazzelli [4] suggested that the walls of the simulation box acts as sink for the nearby particles, reducing their fluctuations.

Nicolai [8] examined the settling properties of particles, with changing volume fraction, simulation box size and polydispersity, by performing a series of experiments. Segre [10] performed a number of experiments and examined the effects of ϕ on velocity fluctuations at and found $\phi^{1/3}$ scaling of velocity fluctuations. Hinch [13] purposed the scale of velocity fluctuations as $(\phi l/a)^{1/2}$ where ϕ, l and a denote particle volume fraction, size of globule of fluid and particle radius, respectively. Climent et al. [2] and Koch [14] used force coupling method and Lattice Boltzmann method, respectively, to investigate the sedimentation at finite Re. Recently, Hamid et al. [5] suggested a modified form of hindered settling velocity at non-zero Re, using direct numerical simulations (DNS).

In real life, mostly we encounter with non-spherical particles e.g., rods, fibers, plates and many others. The most common example of non-spherical particles is pulp fibers in paper making process. Sedimentation is a challenging phenomenon because of long ranged hydrodynamic interactions (HIs) and effect of these hydrodynamics is even more intense in case of non-spherical particles. Settling of non-spherical particles is orientation dependent and this orientation is largely dependent on the flow pattern in the near vicinity. Change in orientation significantly affects the behavior of non-spherical particles as terminal velocity of an fiber is roughly twice in its vertical orientation as compared to horizontal orientation.

Jeffery [15] purposed a model to find orientation angle for an ellipsoidal fiber, when flow is incompressible, Newtonian and homogeneous, which was later validated by experiments [16]. Bretherton [17] showed that Jeffery's equation can also be used for any axisymmetric particle provided an effective aspect ratio rather than geometric aspect ratio is used. Koch and Shaqfeh [7] investigated the instability of fiber suspensions theoretically and showed rise of particle density in the vicinity of a test particle, caused by HIs. Kumar & Ramarao [18] performed a series of experiments for non-spherical particles and observed the formation of large flocs at low volume fraction compared to spherical particles, settling at higher velocities. Herzhaft et al. [19] and later Guazzelli [20] systematically investigated the behavior of fibers in low to moderate volume fraction regime. They observed that average sedimentation velocity increases in the low volume fraction than an isolated fiber, whereas it decreases in the moderate volume fractions, because of the particle particle hindrance. They also estimated self-diffusion which was strongly anisotropic. Kussela et al. [21, 22] performed simulations by considering the effect of inertia at Re of order one and found that for very low volume fractions, particles tend to align horizontally, because of the fluid inertia, whereas at moderate and high volume fractions, they align vertically because many particle interactions. This behavior is validated in the experiments of of Salmela et al. [23].

Butler and Shaqfeh [24] discovered an increase in the settling velocity with increasing volume fractions fibers through numerical simulations. Their prediction of orientation was in agreement with the experiments of Herzhaft & Guazzelli [20]. Moreover, Saintillan et al. [25] found the formation of clusters even at low volume fractions. Shin et al. [26] purposed that two factors which are "shear induced drift" and "orientation drift", control the lateral migration of fiber.

Although extensive work [1, 2, 4-10, 13-15, 18-23, 25-30] on sedimentation phenomena has been performed by many of the researchers, as briefly described above but still there are many problems which are open to research e.g., orientation of non-spherical particles in concentrated suspensions, effect of polydispersity, system size and particle shape on sedimentation phenomena. The main objective of present study is to explore the behavior of non-spherical particles in sedimentation phenomena in Stokes regime and to show that how the settling of non-spherical particles can be different from spheres, both qualitatively and quantitatively. Furthermore, we investigated the microstructure of the suspension along with temporal & spatial autocorrelation functions.

Simulation Method

We performed DNS with smoothed profile (SP) method [28] in order to study the transport properties of non-spherical rod like particles. Each rod like particle is composed of spherical beads so that SP method can be used. In SP method, a diffuse interface, which has a finite thickness replaces the boundary between particle and fluid, so that simulations of fluid and particle motions can be performed on fixed square grid. SP method also provides coupling between particle and fluid motion. Because of this continuous interface modification, we can accurately characterize the HIs along with many-body interactions, which significantly improved numerical computations.

In this method, Newton and Euler's equation is solved to track the ith particle as:

$$M_i \dot{V}_i = F_i^c + F_i^F + F_i^{ext}, \quad (1)$$

$$\dot{R}_i = V_i \quad , \quad (2)$$

$$I_i . \dot{\Omega}_i = N_i^H + N_i^{ext}, \quad (3)$$

where V_i, $\dot{\Omega}_i$ and $\dot{R}_i$ are translation velocity, angular velocity and position of ith particles. F_i^H and N_i^H are hydrodynamic force and torque, respectively. External torque and force are represented by N_i^{ext}, F_i^{ext} and direct inter-particle interaction by F_i^c. I_i and M_i are tensors for inertia and mass of the ith particle. Navier-Stokes (NS) equations within Eulerian description are used to solve the fluid motion;

$$(\partial_t + \boldsymbol{v}_f.\nabla)\boldsymbol{v}_f = \rho_f^{-1}\nabla.(-p\boldsymbol{I}+\boldsymbol{\sigma}_f), \quad (4)$$

$$\nabla.\boldsymbol{v}_f = 0, \quad (5)$$

here ρ, v_f, σ_f and p are fluid density, fluid velocity, stress tensor and pressure field, respectively.

Modified NS equations are solved on whole computational domain comprising of host fluid and sediments by treating the settling particles as fluid [28]. To sustain the rigidity of particles, a body force has been introduced in modified NS equation [29]. To differentiate fluid and particle domain, a density field $\varphi_i(\boldsymbol{x},t)$ is introduced whose value ranges from 0 to 1, where fluid domain is represented by $\varphi_i = 0$, particle domain by $\varphi_i = 1$, and interface is $0 < \varphi_i < 1$. In SP method, modified NS equation is used which is given as:

$$(\partial_t + \boldsymbol{v}.\nabla)\boldsymbol{v} = \rho_f^{-1}\nabla.(-p\boldsymbol{I} + \boldsymbol{\sigma}_f) + \varphi \boldsymbol{f}_p, \quad (6)$$

incompressibility condition is introduced as:

$$\nabla.\boldsymbol{v} = 0, \quad (7)$$

and total velocity field of entire domain is given as:

$$\boldsymbol{v} = (1-\varphi)\boldsymbol{v}_f + \varphi \boldsymbol{v}_p, \quad (8)$$

$$\boldsymbol{v}_p = \sum_{i=1}^{N_p} \varphi(x,t)\left[\boldsymbol{V}_i(t) + \boldsymbol{\Omega}_i(t) \times (X - \boldsymbol{R}_i(t))\right], \quad (9)$$

here ρ_f, η, $\boldsymbol{v}$ and p are density, viscosity, total velocity field and pressure field. $\varphi \boldsymbol{f}_p$ is body force, $(1-\varphi)\boldsymbol{v}_f$ and $\varphi \boldsymbol{v}_p$ represents the velocity field of fluid and particle, respectively. Further details of SP method along with detailed formulas for the rigid body can be found in previous publications [6, 28, 29, 31].

In this study we performed 3D simulations in Stokes regime for volume fraction $\phi = 0.01$. Aspect ratio of system is 2. System size is constant where $L_x = L_y = 128a$ and $L_z = 256a$. Each rod like particle is composed of 3 spherical beads. The radius of spherical bead a is 3Δ and equivalent radius a_e of rod like particle is 4.33Δ, where Δ is lattice spacing. The ratio of density of particle to fluid is 5 and viscosity of fluid is 5. The particles are randomly distributed as shown in Fig. 1. In order to remain in Stokesian regime we kept $Re = \frac{2\rho_f V_t a_e}{\eta} = 0.1$, where V_t is terminal velocity of a single rod in vertical orientation, calculated from a theoretical correlation [20] which is given as:

$$V_t = \frac{\Delta \rho g d_p^2}{16\eta}\left[\left(2\ln 2r - 1.614 - 0.355(\ln 2r)^{-1} + (\ln 2r)^{-2}\right)\right] \quad (10)$$

The force of gravity is in z direction. Particles are non-Brownian as Peclet number $\mathrm{Pe}=\infty$. Simulations time is $3500t_t$ where $t_t = a/V_t$ is terminal time calculated from terminal velocity V_t.

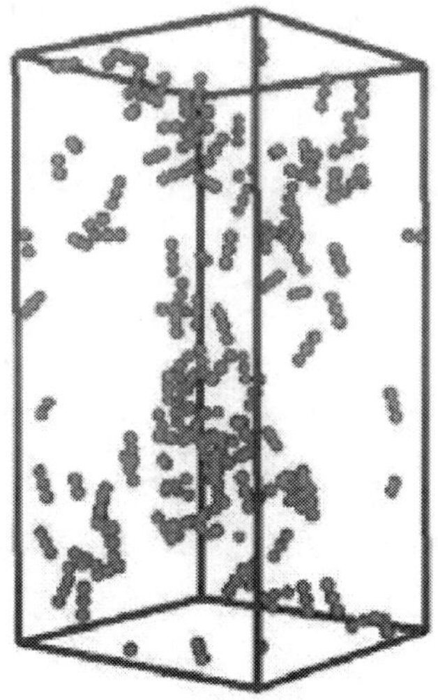

Fig. 1.Random distribution of rod like particles in simulation box

Results and Discussion

Microstructure

Microstructure is a static property of sedimentation phenomena. We calculated radial distribution function (RDF) to study the microstructure of suspension, as transport properties of settling particles are strongly dependent upon microstructure. RDF is used to find the average buildup of pairs of particles in suspension;

$$g(\mathbf{r}) = \frac{2L^3}{N^2}\left\langle \sum_{i<j} \delta\left(\mathbf{r}-\mathbf{r}_{ij}\right)\right\rangle, \qquad (11)$$

where $r_{ij} = \left|\mathbf{R}_i - \mathbf{R}_j\right|$, L is simulation box length, N is for total number of rod like particles, and $\sum_{i<j}$ is ensemble average. RDF is used to characterize the local density of particles around a reference particle as a function of distance in radial direction. It provides information about the particles preference to position themselves around neighboring particle.

Fig. 2 shows the RDF plotted against the distance in radial direction, which is normalized with particle radius. This figure shows two peaks of g(r), where first peak is indication of particle presence around the reference particle at radial distance $r/a = 2$ and second peak also indicates that particle is present but with different orientation at $r/a = 4$. This behavior of rod like particles is different from spherical particles because of change in orientation of particles, which does not matter in case of spheres.

Microstructure of suspension is strongly affected by the HIs. Drafting, kissing and tumbling (DKT) mechanism takes place in sedimentation of rod like particles like spherical particles [32]. During settling, rod like particles keep their distance in z direction, until the leading particle turned over 45 degree, as the leading particle turns more, trailing particle is drawn into wake produced by the leading particle, called drafting. The trailing particle makes a contact with the leading, which is kissing phase, after this contact, the lift force will cause the particles to tumble. In this way, microstructure of suspensions changes because of these interactions.

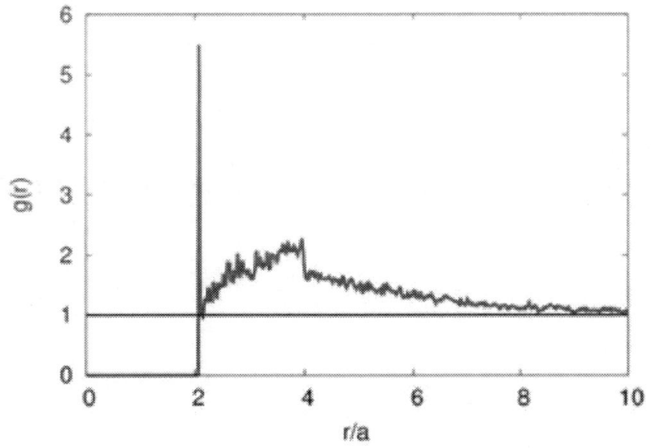

Fig. 2.Radial distribution function $g(r)$ as a function of distance in radial direction r, where $r = |r|$

Results shows that even for dilute suspensions coupling among particles produce inhomogeneity and cluster formation, which cause increase in mean settling velocity over the maximum possible speed of an isolated particle. This behavior of rod like particles is also confirmed by experimental investigation of Herzhaft & Guazzelli [20].

Temporal Autocorrelation Function

We computed temporal autocorrelation functions of mean velocity fluctuations in the horizontal and vertical directions as:

$$C_x(t) = \langle V_{ix}(t) V_{ix}(0) \rangle, \quad (12)$$

$$C_z(t) = \langle \delta V_{iz}(t) \delta V_{iz}(0) \rangle, \quad (13)$$

where $\delta V_{iz} = (V_{iz} - V_{sed})$, $V_{sed} = \langle V_{iz} \rangle$ is the averaged settling velocity, V_{iz} and V_{ix} represent temporal velocities in the vertical and horizontal directions, of ith particle, respectively. These functions provide the information of correlation length, i.e., the time for which velocity fluctuations remained correlated.

Fig. 3 and Fig. 4 show the autocorrelation function plotted against time, normalized with initial value of velocity fluctuations autocorrelation function to visualize the relaxation of velocity fluctuations. Initially correlation function decays rapidly because of collision of particles with surrounding particles and then it decays to zero which indicates the absence of any residual correlation.

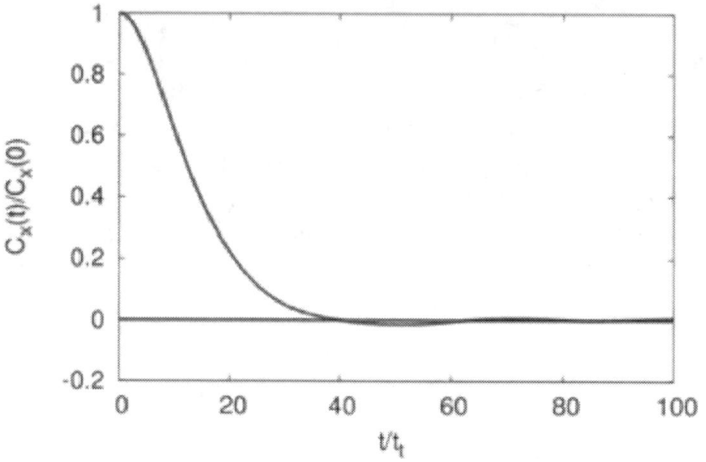

Fig. 3.Temporal autocorrelation function of vertical velocity fluctuations normalized with initial correlation function value as a function of time, where time is normalized by terminal time

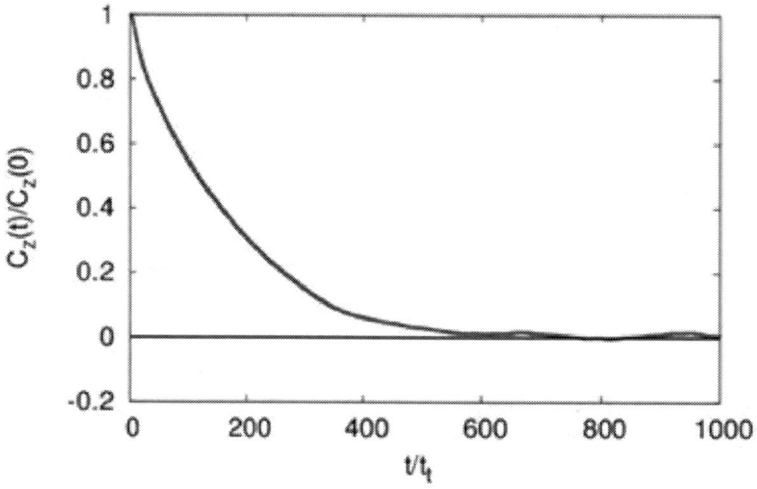

Fig. 4.Temporal autocorrelation function of horizontal velocity fluctuations normalized with initial correlation function value as a function of time, where time is normalized by terminal time

From this temporal autocorrelation function, we calculated relaxation time, which is linked with fluctuating motion of rod like particles. By comparing the relaxation time in x and z direction, we observed that correlations in vertical direction decays very slowly as compared to horizontal because of long ranged effects of memory in gravity direction, induced by the gravity. Values of normalized relaxation time are $\tau_x/t_t = 13.8$ and $\tau_z/t_t = 157$ in x and z directions, respectively. By comparing results with spherical particles we examined that values of relaxation time are comparatively higher in our case, the reason is higher velocity fluctuations in case of non-spherical particles because of continuous deviations in particles orientation.

Spatial Correlation Function

To determine the dependency of results on system size, we find spatial correlation function of fluctuations of vertical velocity component in x and z directions:

$$C_z(x) = \langle \delta V_{iz}(x) \delta V_{iz}(0) \rangle , \quad (14)$$

$$C_z(z) = \langle \delta V_{iz}(z) \delta V_{iz}(0) \rangle . \quad (15)$$

These functions provide information about correlation length, i.e., the space for which velocity fluctuations remained correlated. Different size eddies are form because of HIs, which induces the velocity fluctuations in both directions due to spiral motion. In the x direction, for large system size, correlation function decays in an exponential manner, then it grows into negative direction before becoming zero. Horizontal correlation length is the minima of anti-correlation and vertical correlation length is up to that point where 63.2% of initial correlation function value dies out as defined by Segre et al. [10].

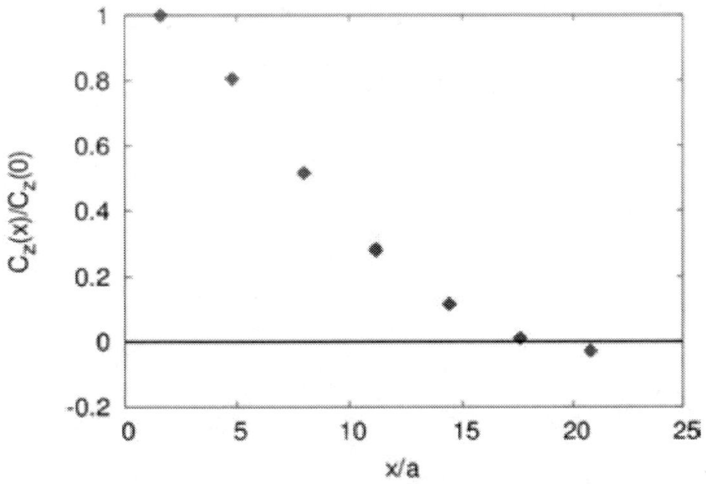

Fig. 5. Spatial correlation function of mean velocity fluctuations of the vertical component of the velocity, which is normalized with initial correlation value as a function of the distance in horizontal direction

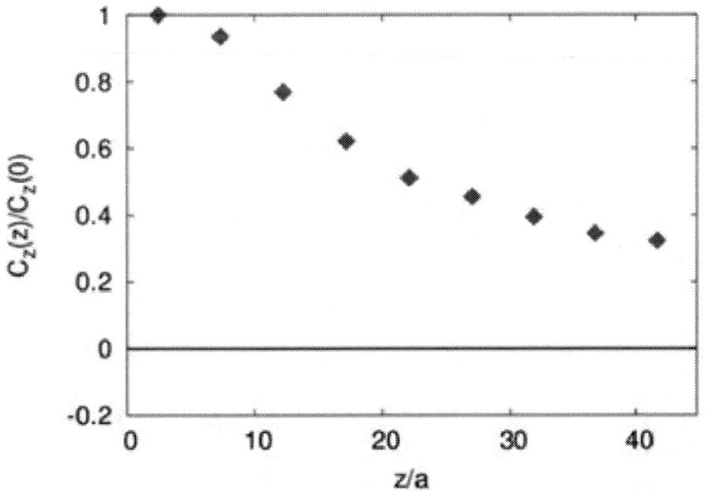

Fig. 6. Spatial correlation function of mean velocity fluctuations of the vertical component of the velocity, which is normalized with initial correlation value as a function of the distance in vertical direction

In this system, we examined that our system size is smaller than the correlation length. In the x direction, we have anti-correlation, but our simulation box is small, therefore a minimum of anti-correlation is not shown in Fig. 5. In the z direction, length of simulation box is not enough to show that point clearly where 63.2% of initial correlation value dies out as shown in Fig. 6. We observed that our system size is small and affected by artefacts generated by finite size effects. These artefacts are stronger at low volume fraction. We are limited to this system size because of the high computational cost.

Conclusion

In this study, we investigated the static and the dynamic transport properties of rod like particles by using DNS with SP method. We observed the cluster formation and larger inhomogeneties even at $\phi = 0.01$, which is low volume fraction, and this result is also validated by previous studies [20, 27]. Vertical relaxation time is higher than the horizontal, because of the long ranged HIs in the gravity direction. Future studies are in process for higher volume fraction to investigate its effects on the properties of settling particles.

References

1. Brenner, M.P., Screening mechanisms in sedimentation. Physics of Fluids (1994-present), 1999. **11**(4): p. 754-772.
2. Climent, E. and M. Maxey, Numerical simulations of random suspensions at finite Reynolds numbers. International journal of multiphase flow, 2003. **29**(4): p. 579-601.
3. Dietrich, W.E., Settling velocity of natural particles. Water resources research, 1982. **18**(6): p. 1615-1626.
4. Guazzelli, É. and J. Hinch, Fluctuations and instability in sedimentation. Annual review of fluid mechanics, 2011. **43**: p. 97-116.
5. Hamid, A., J.J. Molina, and R. Yamamoto, Sedimentation of non-Brownian spheres at high volume fractions. Soft Matter, 2013. **9**(42): p. 10056-10068.
6. Hamid, A., J.J. Molina, and R. Yamamoto, Direct numerical simulations of sedimenting spherical particles at non-zero Reynolds number. RSC Advances, 2014. **4**(96): p. 53681-53693.
7. Koch, D.L. and E.S. Shaqfeh, The instability of a dispersion of sedimenting spheroids. Journal of Fluid Mechanics, 1989. **209**: p. 521-542.
8. Nicolai, H. and E. Guazzelli, Effect of the vessel size on the hydrodynamic diffusion of sedimenting spheres. Physics of Fluids (1994-present), 1995. **7**(1): p. 3-5.
9. Richardson, J. and W. Zaki, The sedimentation of a suspension of uniform spheres under conditions of viscous flow. Chemical Engineering Science, 1954. **3**(2): p. 65-73.
10. Segre, P., E. Herbolzheimer, and P. Chaikin, Long-range correlations in sedimentation. Physical Review Letters, 1997. **79**(13): p. 2574.
11. Luke, J.H., Decay of velocity fluctuations in a stably stratified suspension. Physics of Fluids (1994-present), 2000. **12**(6): p. 1619-1621.
12. Mucha, P.J., et al., A model for velocity fluctuations in sedimentation. Journal of Fluid Mechanics, 2004. **501**: p. 71-104.
13. Gómez, D.C., et al., On stratification control of the velocity fluctuations in sedimentation. Physics of Fluids (1994-present), 2007. **19**(9): p. 098102.
14. Yin, X. and D.L. Koch, Hindered settling velocity and microstructure in suspensions of solid spheres with moderate Reynolds numbers. Physics of Fluids (1994-present), 2007. **19**(9): p. 093302.
15. Jeffery, G.B. The motion of ellipsoidal particles immersed in a viscous fluid. in Proceedings of the Royal Society of London A: Mathematical, Physical and Engineering Sciences. 1922. The Royal Society.
16. Wang, J., J.F. O'Gara, and C.L. Tucker III, An objective model for slow orientation kinetics in concentrated fiber suspensions: Theory and rheological evidence. Journal of Rheology (1978-present), 2008. **52**(5): p. 1179-1200.

17. Bretherton, F.P., The motion of rigid particles in a shear flow at low Reynolds number. Journal of Fluid Mechanics, 1962. **14**(02): p. 284-304.
18. Mackaplow, M.B. and E.S. Shaqfeh, A numerical study of the sedimentation of fibre suspensions. Journal of Fluid Mechanics, 1998. **376**: p. 149-182.
19. Herzhaft, B., et al., Experimental investigation of the sedimentation of a dilute fiber suspension. Physical review letters, 1996. **77**(2): p. 290.
20. Herzhaft, B. and É. Guazzelli, Experimental study of the sedimentation of dilute and semi-dilute suspensions of fibres. Journal of Fluid Mechanics, 1999. **384**: p. 133-158.
21. Kuusela, E., K. Höfler, and S. Schwarzer, Computation of particle settling speed and orientation distribution in suspensions of prolate spheroids. Journal of engineering mathematics, 2001. **41**(2-3): p. 221-235.
22. Kuusela, E., J. Lahtinen, and T. Ala-Nissila, Collective effects in settling of spheroids under steady-state sedimentation. Physical review letters, 2003. **90**(9): p. 094502.
23. Salmela, J., D. Martinez, and M. Kataja, Settling of dilute and semidilute fiber suspensions at finite Re. AIChE journal, 2007. **53**(8): p. 1916-1923.
24. Butler, J.E. and E.S. Shaqfeh, Dynamic simulations of the inhomogeneous sedimentation of rigid fibres. Journal of Fluid Mechanics, 2002. **468**: p. 205-237.
25. Saintillan, D., E. Darve, and E.S. Shaqfeh, Hydrodynamic interactions in the induced-charge electrophoresis of colloidal rod dispersions. Journal of Fluid Mechanics, 2006. **563**: p. 223-259.
26. Shin, M., D.L. Koch, and G. Subramanian, A pseudospectral method to evaluate the fluid velocity produced by an array of translating slender fibers. Physics of Fluids (1994-present), 2006. **18**(6): p. 063301.
27. Fonseca, F. and H. Herrmann, Sedimentation of oblate ellipsoids at low and moderate Reynolds numbers. Physica A: Statistical Mechanics and its Applications, 2004. **342**(3): p. 447-461.
28. Nakayama, Y. and R. Yamamoto, Simulation method to resolve hydrodynamic interactions in colloidal dispersions. Physical Review E, 2005. **71**(3): p. 036707.
29. Yamamoto, R., Y. Nakayama, and K. Kim, A method to resolve hydrodynamic interactions in colloidal dispersions. Computer physics communications, 2005. **169**(1): p. 301-304.
30. Zhang, D., et al. Numerical Evaluation of Single Fiber Motion for Short Fiber Composites Materials Processing. in ASME 2010 International Mechanical Engineering Congress and Exposition. 2010. American Society of Mechanical Engineers.
31. Molina, J.J. and R. Yamamoto, Direct numerical simulations of rigid body dispersions. i. mobility/friction tensors of assemblies of spheres. The Journal of chemical physics, 2013. **139**(23): p. 234105.
32. Rong, X., Numerical study of flow in three-dimensional fiber networks and sedimentation of curved fiber by lattice Boltzmann method. 2007.

Effect of Charge Type and Density of Polyacrylamide on the Electrokinetics Behavior of Flocculated Bentonite Suspensions

Shifa M.R. Shaikh, M. S. Nasser, Ibnelwaleed Hussein, Abdelbaki Benamor*
Gas Processing Center, College of Engineering, Qatar University, P.O. Box 2713 Doha, Qatar

Abstract

The study of electrokinetics behavior of clay suspensions is of fundamental importance in many industrial applications. This paper addresses the electrokinetics behavior of polyacrylamide (PAM)-bentonite suspensions. The impact of surface charge distribution of PAM-bentonite suspensions on the degree of flocculation is examined. The influence of the flocculant properties such as the surface charge type and density, molecular weight and concentrations on the point of zero charge (PZC) of bentonite is investigated. In this study, the effect of different PAM types such as cationic, anionic and amphoteric PAM on the destabilization and flocculation of the systems is studied using zeta potential measurements. Flocculation for each polyacrylamide is conducted using jar tests and the optimum flocculant dosage is defined from turbidity and zeta measurements. This work has provided fundamental understanding of the influence of surface charge types and densities of PAM on the stabilization or destabilization of flocculated bentonite.

Key words: Electrokinetic; Bentonite; Polyacrylamide; Flocculation; Zeta Potential; Surface Charge Density.

Introduction

In the mineral industry, waste residues known as tailings are produced and these must be disposed of in an environmentally and economically acceptable manner. Many of these wastes contain problem minerals such as kaolinite and bentonite, which due to small particle size and charge properties are difficult to settle and consolidate. Bentonite is known chemically as a hydrated aluminum silicate which exists in the form of thin, roughly hexagonal platelets. Bentonite platelets have amphoteric properties. The faces or surfaces carry a permanent negatives charge and depending on the pH, there is a net positives charge on the edges [1]. Bentonite platelets can associate in edge-edge (E-E), edge-face (E-F), and face-face (F-F) configurations. The conventional method used to flocculate colloidal particles (e.g. bentonite) involves high molecular weight polymer–aided flocculation and subsequent sedimentation in thickeners or centrifuges [2-5]. Several types of polymeric flocculants are currently used with clay systems. Of these, polyacrylamides form the most common type of polymeric flocculant and are employed because of their effectiveness as flocculants and ability to produce good settling performance for relatively low cost [1, 6-9]. Flocculation of fine particles may occur by polymer bridging, charge neutralisation, polymer-particle surface complex formation and depletion flocculation, or by a combination of these mechanisms [10]. The bridging mechanism requires that the polymer chains be adsorbed on the particle surfaces, with only a few points of attachment, with the bulk of the chains projecting into the surrounding solution for contact and adherence to other particles [11]. Strong adsorption does not favour the flocculation process, because strong adsorption can cause surface saturation, preventing effective bridging and restabilising fine particles (see Fig. 1).

*Corresponding Author: Tel: + 974-4403-4380; E-mail: m.nasser@qu.edu.qa

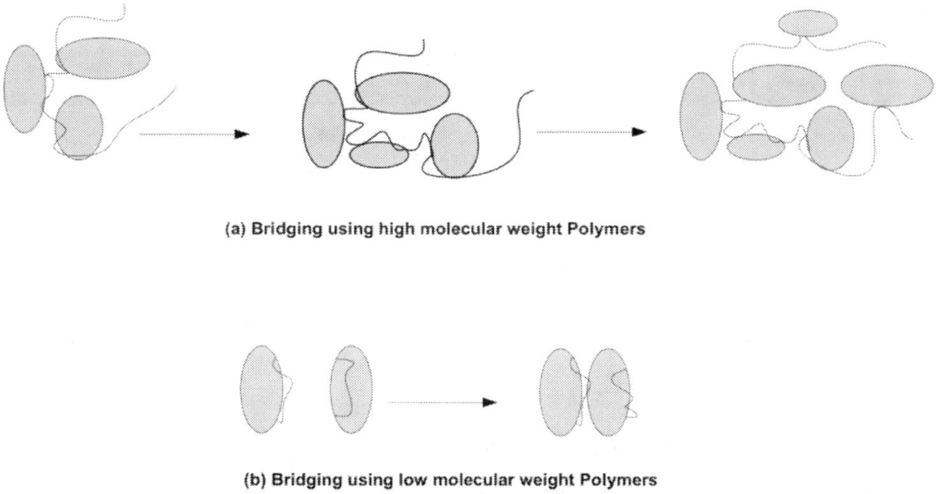

Fig. 1. Differences in bridging mechanism for high and low molecular weight cationic polymers

Extensive research is required to optimize the use of different molecular weight and surface charge polymers as flocculants to improve the settling of fine particle suspensions. Toward this aim, the electrokinetics behavior of polyacrylamide (PAM)-bentonite suspensions is investigated. The impact of surface charge distribution of PAM-bentonite suspensions on the degree of flocculation is examined. The influence of the flocculant properties such as the surface charge type and density, molecular weight and concentrations on the point of zero charge (PZC) of bentonite is investigated.

Experimental and Methods

Material

The bentonite used is a lab grade sample supplied by Sigma-Aldrich Company Ltd., Germany. The chemical analysis of the obtained bentonite is shown in Table 1. In this study, the used flocculants, the polyacrylamides (PAMs), were provided by SNF Floerger, France. FO 4800 SH is high molecular weight cationic with 80 % charge group on chain, FO 4350 SH is high molecular weight cationic with 25% charge group on chain, FO 4800 is Medium molecular weight cationic with 80 % charge group on chain, AN 956 is high molecular weight anionic with 50 % charge group on chain and FAM 4802 is amphoteric high molecular weight with 80 % charge group on chain.

Methods

Turbidity Analysis

Turbidity measurements were performed as a mean to analyze the removal efficiency of bentonite by different types of polyacrylamide. A Hach 2100N turbidity meter was employed to analyze the turbidity of supernatants (measured in Nephelometric TurbidityUnits (NTU)).

Zeta potential

The zeta potential of bentonite suspensions was measured using a Zetasizer 3000 (Malvern Instruments Ltd, UK). The calculation of the zeta potential was assisted with modern software. The equipment measures the electrophoretic mobility of the particles, and converts it to the zeta potential using the von Smoluchowski equation. The von Smoluchowski equation, the most elementary expression for zeta potential, provides a direct relationship between zeta potential and electrophoretic mobility:

$$\xi = \frac{4\pi\mu}{\varepsilon_d}.U \quad (1)$$

where U is electrophoretic mobility, μ is viscosity of the suspending liquid, ε_d is the dielectric constant, π is constant and ξ is the zeta potential. The zeta potential measurements were carried out as a function of polyacrylamide concentration at pH around 8-9.

Table 1. Chemical composition of bentonite

Composition	Wt. %
Na_2O	2.43
K_2O	0.255
Fe_2O_3	3.25
MgO	2.67
Al_2O_3	24.05
SiO_2	58.02
CaO	0.75
FeO	0.31

Results and Discussion

Effect of pH on turbidity and zeta potential of bentonite suspensions

The effect of pH on both turbidity and zeta potential of bentonite without adding any flocculant is shown in Fig. 2. The initial pH value of the bentonite suspensions significantly affects both turbidity and zeta potential. It is apparent from the results shown in Fig. 2 that the zeta potential of the

suspensions increases with increasing pH. This is consistent with background theory on the electrical double layer.
Different modes of particle-particle interaction can occur in different pH and electrolyte environments. Under acidic condition, aluminum exposed at the edges of the plates of the bentonite apparently binds hydrogen ions of water and thus assumes a positive charge. In these circumstances the zeta potential will be decreased (toward the positives sign) and edge and faces will mutually attract giving raise face-edge flocculation and structures referred to as "card-house" flocs. Under alkaline conditions, the edges, become natural or negatively charged, and the particles are deflocculated and hence the turbidity increased substantially.

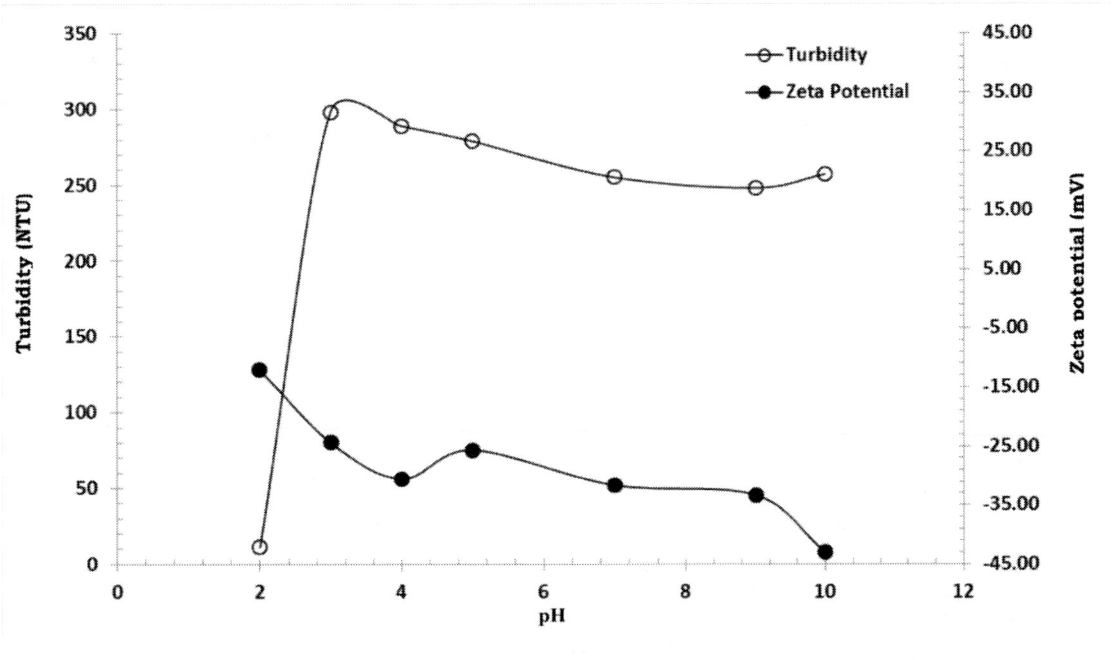

Fig. 2. Turbidity and electrokinetic zeta potentials for pure bentonite suspensions as a function of pH

Effect of PAM Concentration and Type on turbidity and Zeta potential

Table 2 summarizes the turbidity and zeta potential of bentonite as a function of PAM concentration. Prior to flocculant addition, the particles have negative charge and corresponding zeta potential (–43.62 mV). The reduction in the zeta potential to zero by the cationic polyacrylamide arises from charge neutralization through electrostatic attraction between the positively charged polymer and negative surface sites of the particle, and the adsorption of large amount of positively polymer onto the surface of bentonite gives rise to charge reversal. The amount of polymer required to neutralize the particle charge is inversely proportional to charge density and this clearly shown in above Table 1. The amount of PAM required achieving point of zero charge (PZC) when bentonite is flocculated using FO 4800 SH with 80 % charge group on chain is ~ 30 mg L^{-1} and the amount of PAM required achieving point of PZC when bentonite is flocculated using FO 4800 with 25 % charge group on chain is ~ 45 mg L^{-1}. This confirms the charge neutralization mechanism; where the high charge density polymers neutralize the surface charge of bentonite more rapidly than the low charge density polymers, and have require a lesser amount of polymer. On the other hand, the molecular showed great influence on PZC and degree of the bentonite flocculation. For example, when bentonite is flocculated using FO 4350 SH which is medium molecular weight cationic with 80 % charge group on chain, no PZC was achieved in the range of 0-100 mg L^{-1} of tested PAM.

In the case of anionic AN 956 polyacrylamide, the small decrease in the magnitude of the zeta potential with increasing polymer concentration does not arise from charge neutralization but rather from the shift in position of the plane of shear, due to the adsorbed layer, which is caused by the negatively polymer chains.

Similar to cationic PAMs, the amount of PAM required achieving point of PZC when bentonite is flocculated using amphoteric FAM 4802 polyacrylamide is ~30 mg L^{-1}. The reduction in the zeta potential to zero by the amphoteric polyacrylamide arises from charge neutralization through electrostatic attraction between the positive chain group on the amphoteric PAM and negative surface sites of the particle

Table 2. turbidity and zeta potential of bentonite as a function of PAM concentration

Concentration (mg/l)	0	2.5	5	10	15	20	30	35	40	45	50	75	100	
zeta potential (mV)	-43.6	-40.8	-39.3	-39.44	-40.38	-39.96	-15.78	3.79	12.7	15.9	17.7	34.2	40.9	FO 4800 SH
Turbidity (NTU)	373	263	214	134	123	82	1.06	0.527	0.64	1.2	0.98	1.13	1.27	FO 4800 SH
zeta potential (mV)	-43.6	-43.6	-39.7	-39.2	-39.7	-36.0	-38.3	-39.2	-30.2	-36.6	-33.9	-35.7	-15.6	FO 4350 SH
Turbidity (NTU)	373	373	318	293	258	213	181	133	146	67.5	90	77	17.4	FO 4350 SH
zeta potential (mV)	-43.6	-43.6	-36.5	-35.3	-34	-37.7	-38	-21.5	-9.61	0.87	2.65	1.5	1.19	FO 4800
Turbidity (NTU)	373	373	264	222	132	75.7	21	1.04	0.23	0.34	0.57	0.7	1.50	FO 4800
zeta potential (mV)	-43.6	-36.6	-37.18	-24.08	-40.5	-41.86	-48.66	-48	-48.4	-49.5	-53.1	-44.5	-50	AN 956
Turbidity (NTU)	373	329	336	334	336	336	338	342	339	336	333	344	334	AN 956
zeta potential (mV)	-43.6	-34.8	-43.08	-36.84	-27.6	-37.02	-5.08	2.12	2.24	6.85	11.43	28.14	44.82	FAM 4802
Turbidity (NTU)	373	280	240	193	140	82.7	0.714	0.296	0.434	0.607	0.713	1.32	1.71	FAM 4802

Conclusions

The importance of polymer charge-type, density and molecular weight in the flocculation for negatively charged bentonite suspensions was investigated to determine optimal flocculation conditions. The following conclusions have been drawn:

- Polyelectrolyte structure and type has a marked effect on the particle zeta potential and suspension turbidity, whose decrease in the magnitude was much greater in the presence of cationic and amphoteric than anionic polyacrylamides. The decrease in the magnitude of zeta potential was much greater in the presence of cationic and amphoteric than anionic polyacrylamides. Such an increase stems from the charge neutralization through electrostatic attraction between the positively charged polyelectrolytes and the negatively charged surface sites of the particle.

- The result is that the zeta potentials of the bentonite suspensions, flocculated using anionic flocculant, remain negative and this prevents strong adsorption of the polymer.

ACKNOWLEDGMENTS

The authors would like to acknowledge the support provided Qatar University .

References

1. Nasser, M.S. and A.E. James, *Compressive and shear properties of flocculated kaolinite–polyacrylamide suspensions.* Colloids and Surfaces A: Physicochemical and Engineering Aspects, 2008. **317**(1–3): p. 211-221.

2. Nasser, M.S. and A.E. James, *Settling and sediment bed behaviour of kaolinite in aqueous media.* Separation and Purification Technology, 2006. **51**(1): p. 10-17.

3. Al-Shamrani, A.A., A. James, and H. Xiao, *Destabilisation of oil–water emulsions and separation by dissolved air flotation.* Water Research, 2002. **36**(6): p. 1503-1512.

4. Al-Shamrani, A.A., A. James, and H. Xiao, *Separation of oil from water by dissolved air flotation.* Colloids and Surfaces A: Physicochemical and Engineering Aspects, 2002. **209**(1): p. 15-26.

5. Chassagne, C., F. Mietta, and J.C. Winterwerp, *Electrokinetic study of kaolinite suspensions.* Journal of Colloid and Interface Science, 2009. **336**(1): p. 352-359.

6. Nasser, M. and A. James, *The effect of polyacrylamide charge density and molecular weight on the flocculation and sedimentation behaviour of kaolinite suspensions.* Separation and Purification Technology, 2006. **52**(2): p.

241-252.

7. Nasser, M.S. and A.E. James, *The effect of polyacrylamide charge density and molecular weight on the flocculation and sedimentation behaviour of kaolinite suspensions.* Separation and Purification Technology, 2006. **52**(2): p. 241-252.

8. Nasser, M.S. and A.E. James, *Numerical simulation of the continuous thickening of flocculated kaolinite suspensions.* International Journal of Mineral Processing, 2007. **84**(1–4): p. 144-156.

9. Nasser, M.S. and A.E. James, *Degree of flocculation and viscoelastic behaviour of kaolinite-sodium chloride dispersions.* Colloids and Surfaces A: Physicochemical and Engineering Aspects, 2008. **315**(1–3): p. 165-175.

10. Mpofu, P., J. Addai-Mensah, and J. Ralston, *Investigation of the effect of polymer structure type on flocculation, rheology and dewatering behaviour of kaolinite dispersions.* International Journal of Mineral Processing, 2003. **71**(1-4): p. 247-268.

11. Mpofu, P., J. Addai-Mensah, and J. Ralston, *Flocculation and dewatering behaviour of smectite dispersions: Effect of polymer structure type.* Minerals Engineering, 2004. **17**(3): p. 411-423.

463913 Coupling Mechanical Research of Molecule Migration and Stochastic Simulations in Mesoscopic Field

Abstract:

The flow field in micro filtration screening devices can be classified into the mesoscopic flow scale. The research points out the rule of the molecule migration and stochastic simulations in the mesoscopic flow field. In the aspect of molecule migration theory, based on the two fluid theory and dilute fluid theory, we use the perturbation method and Taylor expansion to get the mass transfer equation, momentum equation and constitutive equation. And then we get the stress tensor and concentration results. Besides, the effect of hydrodynamic interaction will be considered. In the stochastic simulations part, Brownian Dynamic Simulation Method and Stochastic Rotation Simulation Method will be used to get the simulation results about stress tensor and concentration to check the agreement with analytical results. Finally, the above methods will be used in the typical flow in micro filtration screening devices to set up the basic migration theory for working principle and process design.

469074 Understanding the Impact of Bile Salts on Crystallization Kinetics of Supersaturated Solutions of a Poorly Water Soluble Compound

Abstract:

Amorphous solid dispersions (ASD) have been one of the most popular formulation strategies to address the issue of inadequate aqueous solubility for small molecules. Although ASD greatly improves oral drug absorption by creating a highly supersaturated solution upon dissolution, addition of a crystallization inhibitor is often required to maintain the supersaturation advantage. In addition to commonly employed polymeric additives, bile salts, which are endogenous surfactants, have attracted the interest of researchers since they show potential as crystallization inhibitors. While the commercial recipe for simulated intestinal fluid contains only one bile salt, sodium taurocholate, there are at least six different bile salts found in human intestinal fluids, and the composition varies from person to person and amongst species. To better understand and maximize oral drug absorption, it is imperative to evaluate the impact of bile salts on key attributes of supersaturated drug solutions.

In this study, a systematic experimental design was carried out to investigate the potential of six biologically relevant bile salts (sodium taurocholate (STC), sodium glycocholate (SGC), sodium taurodeoxycholate (STDC), sodium glycodeoxycholate (SGDC), sodium taurochenodeoxycholate (STCDC), sodium glycochenodeoxycholate (SGCDC)) as crystallization inhibitors for telaprevir, a poorly soluble drug commercialized as an ASD formulation. A side-by-side diffusion cell was used to evaluate solute flux for solutions of telaprevir in the absence and presence of bile salts at various solute concentrations, and to provide an improved estimate of the crystallization driving force in a complex medium. Telaprevir nucleation induction time experiments were performed at the same activity-based supersaturation in the absence and presence of six different bile salts, at monomeric and micellar levels. To decouple crystal growth and nucleation, telaprevir crystal growth experiments with an in-situ common history seeding method were designed and performed at the same activity-based supersaturation. The potential roles of both monomeric and micellar bile salt as crystallization inhibitors were then evaluated.

All bile salts investigated in this study slow down crystallization of highly supersaturated telaprevir solutions. However, the extent and kinetics of crystallization inhibition varies with structural difference amongst the bile salts and their aggregation level. The tri-hydroxyl bile salts (STC and SGC) do not impact solution thermodynamics, and are good nucleation and crystal growth inhibitors at both aggregation level. Di-hydroxyl bile salts solubilize telaprevir at micellar level. In addition, monomeric STDC and SGDC are strong nucleation inhibitors while monomeric STCDC and SGCDC are strong crystal growth inhibitor for telaprevir. Lastly, bile salts are not interchangeable from either a thermodynamic or crystallization inhibition standpoint.

466855 Crystallization Design: A Case Study on an Active Pharmaceutical Ingredient (API)

Abstract:

Crystallization serves as a key separation and purification technique to deliver material of specified quality, form and powder properties for downstream processing in the pharmaceutical industry. This case study will focus on the systematic development of an API crystallization designed to: (1) simplify the current process to deliver the desired form, (2) improve the powder properties through particle morphology manipulation to meet the target material profile (TMP) and (3) control in-process impurities.

Early attempts to generate the desired form relied on a multi-step (three isolation) transformation starting from the free acid, to the addition of the counter-ion and finally hydration by a slurry to slurry conversion. Although this process successfully provided the desired form, the laborious sequence and poor powder properties were prohibitive for long-term use. As a result, further development was necessary to identify a superior alternative.

This case study will review the steps taken to simplify the process through elimination of one isolation, to improve the particle morphology from poor handling thin needles/plates to columnar/rod-like particles and finally to implement control strategies to ensure acceptable purity. The discussion will begin with an analysis of solvent screening results and viable bases and highlight the decision process to achieve suitable form, morphology and yield. The solvent and base system selected was optimized for particle morphology and powder properties through careful mapping of the crystallization space guided by process analytical technology (PAT). Understanding of the super-saturation profile allowed for determination of the optimal seed point, seed type, and anti-solvent addition rates to control growth and nucleation. Finally, strategies implemented to control the formation of in-process impurities will be reviewed.

469143 Design of Continuous Enzymatic Reactive Crystallization for Beta-Lactam Antibiotic Synthesis

Abstract:

Design of Continuous Enzymatic Reactive Crystallization for Beta-lactam Antibiotic Synthesis

Matthew A. McDonald[1], Andreas S. Bommarius[1], and Ronald W. Rousseau[1]

[1]School of Chemical and Biomolecular Engineering,

Georgia Institute of Technology, Atlanta, GA 30332-0100, USA

β-lactam antibiotics, despite a long and storied history, find continued widespread use in our fight against infection. In 2010 the two most common classes of β-lactam antibiotics, those derived from penicillin and cephalosporin, accounted for nearly 60% of world antibiotic consumption, with over 20 billion doses given of penicillin-derived antibiotics alone [1]. Their high rate of use, both historical and contemporary, has contributed to a growing deadly allergy to β-lactams, particularly those derived from penicillin. The rate of allergy is significant enough for the FDA to recently require that all penicillin-derived antibiotics be manufactured, processed, and packaged in facilities separate from those used for any other pharmaceuticals [2]. Traditionally, β-lactam antibiotics have been made through chemical synthesis requiring use of several protection groups and resulting in unsustainable amounts of potentially contaminated waste [3].

An alternative route that has received considerable interest involves using the enzyme penicillin G acylase (PGA) to synthesize semi-synthetic penicillin-derived antibiotics from an acyl donor (e.g. phenylglycine methyl ester for ampicillin) and 6-aminopenicillanic acid (6-APA) [3]. However batch processing, even enzymatic, results in only moderate yields and thus inefficient use of materials. PGA also catalyzes the hydrolysis of the acyl donor (primary hydrolysis) and the hydrolysis of the antibiotic (secondary hydrolysis, ampicillin cleaves into phenylglycine and 6-APA), leading to unnecessary waste.

This work focuses on the design of continuous reactive crystallization of ampicillin catalyzed by PGA, such that the precipitated ampicillin is protected from secondary hydrolysis and is of high purity. The design of such a system requires models for kinetics of reaction and crystallization, solubility of all components, and robust controls for continuous operation. The dependence of the kinetics and solubility on pH promises a means of control in which yield and selectivity can be maximized.

Crystallization kinetics (nucleation and growth) were determined via online chord-length distribution and concentration versus time data in batch crystallizers [4]. Change in pH was used to induce supersaturation, and the effect of crystallization on pH was recorded. Since the goal is steady state operation of a continuous reactive crystallizer, seed crystals should be used in all experiments, the loading of seed crystals being another important design parameter. Crystallization kinetics were combined with the reaction kinetics scheme of Youshko and Svedas [5]. The reaction kinetics were modified to include the effect of enzyme protonation state on the rates of synthesis and primary and secondary hydrolysis. Batch experiments across a range of pH values were used to factor in the equilibrium of the enzyme-substrate complex and enzyme-product complex protonation states. The effect of pH on the rates of ampicillin synthesis, primary hydrolysis, and secondary hydrolysis were determined independent of each other. A combined model was used to determine the yield and selectivity for ampicillin over the hydrolysis product. The largest obstacle to high conversion remains the accumulation and precipitation of the hydrolysis product phenylglycine, an intolerable impurity in the final project.

Model simulations show that continuous operation can increase yield considerably over batch processes. Simulations also show the substantial benefit of crystallization and reaction occurring in the same vessel. Different vessel designs (i.e. stirred tank, tubular reactor) have also been investigated via the model and show significant differences between competing designs. A continuous stirred tank arrangement will be implemented experimentally to validate the model. Preliminary results also suggest controls over important variables including crystal size distribution are possible while maintaining high yield, selectivity, and purity.

[1] Gelbrand, Helen, et al. "The State of the WorldÕs Antibiotics 2015." *Wound Healing Southern Africa* 8.2 (2015): 30-34.

[2] 21 C.F.R ¤ 211.42 2015

[3] Giordano, Roberto C., Marcelo PA Ribeiro, and Raquel LC Giordano. "Kinetics of β-lactam antibiotics synthesis by penicillin G acylase (PGA) from the viewpoint of the industrial enzymatic reactor optimization."*Biotechnology advances* 24.1 (2006): 27-41.

[4] Encarnacion-Gomez, Luis G., Andreas S. Bommarius, and Ronald W. Rousseau. "Crystallization Kinetics of Ampicillin Using Online Monitoring Tools and Robust Parameter Estimation." Industrial & Engineering Chemistry Research 55.7 (2016): 2153-2162.

[5] Youshko, M. I., and V. K. Śvedas. "Kinetics of ampicillin synthesis catalyzed by penicillin acylase from E. coli in homogeneous and heterogeneous systems. Quantitative characterization of nucleophile reactivity and mathematical modeling of the process." *Biochemistry (Moscow)* 65.12 (2000): 1367-1375.

472714 Membrane Assisted Crystallization: Process Model and Antisovlent Crystallization Control for Continuous Pharmaceuticals Manufacture

Abstract:

Membrane assisted crystallization: process model and antisovlent crystallization control for continuous pharmaceuticals manufacture

Xiaobin Jiang*, Gaohong He*, Linghan Tuo

State Key Laboratory of Fine Chemicals, Research and Development Center of Membrane Science and Technology, School of Chemical Engineering, Dalian University of Technology, Dalian, P. R. China

*Corresponding author: Email: xbjiang@dlut.edu.cn, hgaohong@dlut.edu.cn

Abstract

Membrane assisted crystallization (MAC) has furnished a great progress in seawater desalination, brine treatment, wastewater recovery and crystallization control in recent decade, due to its advantages on utilizing low grade energy, promoting nucleation, controlling crystallization, etc [1,2]. Systematically frameworks and methods revolved MAC will be developed that can address the challenges in seawater treatment, brine wastewater recovery, fine crystals manufacture and biomolecular crystallization, etc [3].

In this paper, the process model combined membrane assisted crystallization, which involved the evaluation of the transmembrane flux and nucleation/growth rate will be established. The advantages of MAC on supersaturation degree control, nucleation kinetic and crystal size distribution modification will be revealed and investigated. The impact of the desired membrane property on the nucleation control was discussed in details.

On the basis of authors previous work [4,5], the membrane distillation crystallization is introduced to the antisolvent crystallization process to validate the developed model and the impact of membrane participant mass transfer process on continuous crystallization process control. The equivalent supersaturated degree was introduced to evaluate the controlling accuracy of MAC process and the stability of the control strategy. Due to the micrometer scale porous channel of the membrane, the micro mixing can be significantly improved and the mass tranfer control accuracy can be improved by 1 or 2 order of magnitudes compared to the existing droplet method with the millimeter scale mixing.

As a potential high accuacy control approach of antisolvent crystallizaiton, the MAC assisted antisolvent crystallization can be utilized in continuous and scale-up crystallization. These demands are especially significant in the area that have strict requirements on excellent, steady crystal qualities and robust crystallization control (pharmaceuticals crystallization, fine chemicals preparation, e.g.).

In addtion, these findings can drive the development of the comprehensive understanding on the complex nucleation mechanism under the heterogeneous chemical interface and confined environment. Further studies in this domain will be important for furthering the landscape of MAC application and the potential research results may provide conclusive evidence on the nucleation mechanism in heterogeneous phases[7-8], which is the one of the core concerns in nucleation theory.

Acknowledgment

This work is supported by National Natural Science Foundation of China (Grant No. 21306017, 21527812), Program for Changjiang Scholars and Support Project of the China Petroleum and Chemical Corporation (X514001).

References

[1] Edwie, F., Chung, T.-S.: Development of simultaneous membrane distillation¨Ccrystallization (SMDC) technology for treatment of saturated brine. Chem. Eng. Sci. 98, 160-172 (2013).

[2] Wang, P., Chung, T.-S.: Recent advances in membrane distillation processes: Membrane development, configuration design and application exploring. J. Membrane Sci. 474, 39-56 (2015).

[3] Chabanon, E., Mangin, D., Charcosset, C.: Membranes and crystallization processes: State of the art and prospects. J. Membrane Sci. 509, 57-67 (2016).

[4] Jiang, X., Ruan, X., Xiao, W., Lu, D., He, G.: A novel membrane distillation response technology for nucleation detection, metastable zone width measurement and analysis. Chem. Eng. Sci. 134, 671-680 (2015).

[5] Jiang, X., Lu, D., Xiao, W., Ruan, X., Fang, J., He, G.: Membrane Assisted Cooling Crystallization: Process Model, Nucleation, Metastable Zone, and Crystal Size Distribution. AIChE Journal 3, 829-841 (2016).

[6] Chabanon, E., Mangin, D., Charcosset, C.: Membranes and crystallization processes: State of the art and prospects. J. Membrane Sci. 509, 57-67 (2016).

[7] Diao, Y., Harada, T., Myerson, A. S.; Hatton, T. A.; Trout, B. L.: The role of nanopore shape in surface-induced crystallization. Nat. Mater. 10, 867-871 (2011).

[8] Myerson, A. S.; Trout, B. L., Nucleation from Solution. Science 341, 855-856 (2013).

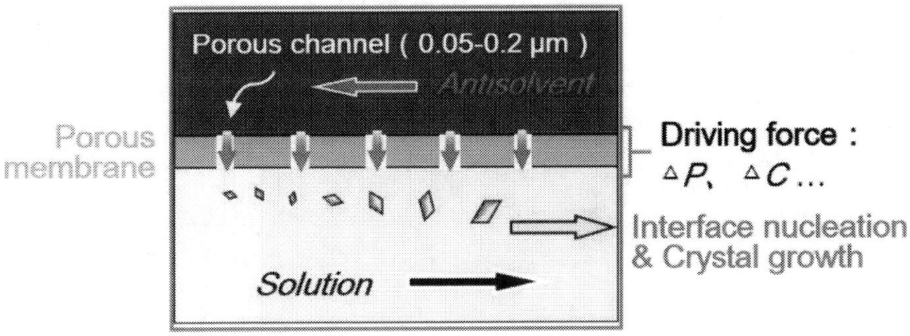

Fig. 1 Schematic diagram of the antisolvent crystallization assisted by MAC technology.

458389 Investigations into the Use of Ultrasounds for Seed Generation for Continuous Cooling Crystallization Process in an Oscillatory Flow Baffled Crystallizer

Abstract:

Investigations into the use of Ultrasound for Seed Generation for Continuous Cooling Crystallization Process in an Oscillatory flow Baffled Crystallizer
One of the challenges associated with tubular crystallizers in general is the high risk of scaling and uncontrolled nucleation due to the large surface area they offer. This poor control over nucleation leads to undesired product attributes which is detrimental for the production process. Oscillatory flow baffled crystallizer (OFBC) is a type of tubular crystallizer which offers the opportunity to manipulate and control the mixing conditions by controlling the amplitude and frequency of the flow oscillations. The control over mixing coupled with high heat transfer area offered by the (OFBC) also allows for better control of the process temperature and hence the super-saturation in the crystallizer, which is promising for controlling nucleation. Furthermore, the very low operating flowrates can be realized as the mixing in the OFBC is independent of the process flowrate. Thus, leading to large residence times in the comparatively compact set-up, when compared to a crystallizer with a simple tubular geometry. However, there is still a lack of understanding in terms of effect of local turbulence in the crystallizer on the nucleation and growth kinetics in a crystallization process. There have been concerns regarding the plugging and scaling of the crystallizers, possibly an influence of the large surface area. Additionally, introduction of seeds in the system, especially at low flowrates, has been challenging and has called for novel feed methods.
Since continuous operation carried out in a single or multiple well-mixed vessels connected in series leads to a broad residence time distribution in the crystallizer and hence results in poor process control, the use of such an infrastructure to generate seeds for feeding into the OFBC seems to be sub-optimal. Therefore we aim at the use of an external field in the form of ultrasound to gain on control of nucleation in a continuous flow set-up, if not to apply directly in the OFBC set-up.
Application of ultrasound (US) has been demonstrated to affect the crystallization process by reducing the induction time, promoting nucleation at relative low supersaturation levels. In this study we investigate the use of US for the continuous generation of small seed crystals in the feed stream of a continuous OFBC crystallizer. Two different reactors will be tested in their efficiency to generate the seed crystals, a tubular flow crystallizer surrounded by piezo electric element which induces ultrasound directly inside the tube and a flow cell in which an ultrasound horn is in direct contact with the flowing supersaturated solution. The number of nuclei entering the OFBC is monitored as well as the control to maintain growth only conditions in the crystallizer. The approach is aimed at the generation of a well-defined feed stream containing small seed crystals at a concentration that can suppress the undesirable (secondary) nucleation in the OFBC, yielding product crystals with a high quality, i.e mono modal with a narrow size distribution.

465865 Microwave-Assisted Evaporative Crystallization for Enhancing Physicochemical Properties of Pharmaceuticals By Size Reduction

Abstract:

Solving bioavailability problems of new active pharmaceutical ingredients is a major challenge for the pharmaceutical industry, since nearly 50% of the new active substances in the pipeline are poorly soluble in water [1-3]. One approach to address this problem is to increase the dissolution rate and achieve a higher bioavailability of drugs by reducing particle size through the selection of proper crystallization conditions [4-7].

Solvent evaporation can be significantly enhanced by microwave irradiation during the crystallization processes, resulting in reduced particle size due to the rapid crystallization induced by the high prevailing supersaturations. In this study the effect of microwave heating on the evaporative crystallization of niflumic acid (NIF) is investigated and compared to conventional heating. NIF is an important active pharmaceutical ingredient, with anti-inflammatory activity accompanied by an analgesic effect [8]. According to the Biopharmaceutics Classification System (BCS), NIF belongs to class II, having a low solubility in water (26 $\mu g\ mL^{-1}$ at 25 °C).

Microwave heating is fundamentally different from conventional (conductive) heating. In particular, microwaves offer rapid and volumetric heating, without the need of heat transfer surfaces or heat transfer fluids. In addition, as microwaves couple directly with the (polar) molecules of the solution, heat transfer stops immediately when the microwave heating is turned off thus minimizing thermal inertia. In principle, microwave irradiation can significantly speed up solvent evaporation in crystallization processes, resulting in rapid crystallization and reduced particle size [9]. The effect of microwaves on the solvent evaporation rate and its effects on crystal size and crystallinity were studied.

A single-mode microwave setup was used for evaporative crystallization of NIF, with and without the polymer excipient, polyvinylpirrolidone (PVP). Clear undersaturated solutions of NIF in ethanol were prepared. When microwave heating is used for solvent evaporation, the 5 mL solution started boling within 4 seconds with maximum microwave power (300 W). The solvent was completely evaporated within 21 seconds at 300 W applied microwave power. Significantly larger times were needed for complete solvent evaporation when conventional heating was used.

Particle sizes decreased to submicron-size range when microwave-assisted evaporative crystallization was applied. Further, in the events of microwave-assisted evaporative crystallization of NIF in the presence of PVP, the drug morphology changed from rectangular to spherical shape and nanoparticles (around 200 nm) were produced. According to the structural analysis, the drug was in crystalline form without the PVP excipient and an amorphous solid dispersion was formed with PVP.

A 2.5-fold increase in the dissolution rate of the produced niflumic acid crystals was observed compared to the dissolution rate of the original drug. When niflumic acid was produced together with the excipient PVP in the microwave system, an amorphous solid dispersion was created with particles in the nano-size range, which showed a 5-fold increase in dissolution rate compared to that of the crystalline niflumic acid samples created by the microwave irradiation in the absence of PVP [10].

Microwave irradiation offers a novel way for drug formulation, and by reducing the particle size the dissolution rate and bioavailability of the active pharmaceutical ingredient can be enhanced.

References

[1] Y. Kawabata, K. Wada, M. Nakatani, S. Yamada, S. Onoue, Formulation design for poorly water-soluble drugs based on biopharmaceutics classification system: Basic approaches and practical applications, Int. J. Pharm 420, (2011), 1–10.

[2] A. A. Noyes, W. R. Whitney, The rate of solution of solid substances in their own solution, J. A. C. S. 19 (1897) 930–934.

[3] M. Lindenberg, S. Kopp, J. B. Dressman, Classification of orally administered drugs on world health organization model list of essential medicines according to biopharmaceutics classification systems, Eur. J. Pharm. Biopharm. 58 (2004) 265–278.

[4] T. Yasuji, H. Takeuchi, Y. Kawashima, Particle design of poorly water-soluble substances using supercritical fluid technologies, Adv. Drug. Dev. Rev. 60 (2008) 388–398.

[5] M. Sarkari, J. Brown, X. Chen, S. Swinnea, R. O. Williams, K. P. Johnston, Enhanced drug dissolution using evaporative precipitation into aqueous solution, Int. J. Pharm. 243 (2002) 17–31.

[6] R. Ambrus, P. Kocbek, J. Kristl, R. Sibanc, R. Rajkó, P. Szabó-Révész, Investigation of preparation parameters to improve the dissolution rate of poorly soluble meloxicam, Int. J. Pharm. 318 (2009) 153–159.

[7] K. Kannan, P. K. Karar, R. Manavalan, Formulation and evaluation of sustained release microspheres of acetazolamide by solvent evaporation technique, J. Pharm. Sci. 1 (2009) 36–39.

[8] P. A. Insel, Analgesic-antipyretics and anti-inflammatory agents; drugs employed in the treatment of rheumatoid arthritis and gout, in: Gilman AG, Rall TW, Nies, AS, Taylor P. (Eds.), Goodman and Gilman's The Pharmacological Basis of Therapeutics, McGraw-Hill, Singapore, 1991. p. 668.

[9] N. Radacsi, J. H. ter Horst, G. D. Stefanidis, Microwave-assisted Evaporative Crystallization of Niflumic Acid for Particle Size Reduction, *Crystal Growth & Design* 2013, *13*, 4186–4189.

[10] N. Radacsi, G. D. Stefanidis, P. Szabó-Révész, R. Ambrus, Analysis of Niflumic Acid Prepared by Rapid Microwave-assisted Evaporation, *Journal of Pharmaceutical and Biomedical Analysis* 2014, *98*, 16-21.

464575 Metal-Organic Framework Membrane Process for High Purity CO_2 Production

Abstract:

The development of the membrane process for CO_2 capture or separation has been of great interest from the energy and environmental perspectives. Current CO_2 separation membrane technologies are dominated by polymeric membranes due to their low cost and easy processibility. However, polymeric membranes suffer from CO_2-induced plasticization and the trade-off between permeability and selectivity. Zeolite membranes have also been extensively studied for CO_2/N_2 and CO_2/CH_4 separation during the past decade. Although some of them have demonstrated high selectivity for CO_2 over CH_4 or N_2, their CO_2 permeance needs to be further improved to be economically viable. Gas separation by metal-organic framework (MOF) membranes is an emerging research field. Their commercial application potential is however still rarely explored due in part to unsatisfied separation characteristics and difficulty in finding suitable applications. Herein, we report "sharp molecular sieving" properties of high quality isoreticular MOF-1 (IRMOF-1) membrane for CO_2 separation from dry, CO_2 enriched CO_2/CH_4 and CO_2/N_2 mixtures. The IRMOF-1 membranes exhibit CO_2/CH_4 and CO_2/N_2 separation factors of 328 and 410 with CO_2 permeance of 2.55×10^{-7} and 2.06×10^{-7} mol m^{-2} s^{-1} Pa^{-1} at feed pressure of 5 atm and 298 K, respectively. High grade CO_2 is efficiently produced from the industrial or lower grade CO_2 feed gas by this MOF membrane separation process.

461364 Tuning of Carbon Molecular Sieve Membranes for Separation of the C_3H_6-C_3H_8 Pair

Abstract:

Carbon molecular sieve (CMS) membranes display exceptional separation properties for gas-pairs such as C_3H_6-C_3H_8, which cannot be addressed using conventional polymers.Â Such CMS membranes are often formed by controlled pyrolysis of asymmetric polymer precursor fibers. This study considers non-standard pyrolysis protocols for an important polyimide, 6FDA:BPDA-DAM that offers attractive CMS properties for the C_3H_6-C_3H_8 pair. Specifically, effects of heating ramp rate, maximum pyrolysis temperature, post-pyrolysis thermal soak time were considered. Fundamental bases for optimization of each of these parameters are offered. The heating ramp profile controls the time spent in each individual stage of the pyrolysis processâ€"especially the period the precursor spends beyond its decomposition temperature. For a given precursor, this critical period sets the micropore and ultramicropore dimensions in the CMS fiber as well as their distributions. Subsequent thermal soak times enable equilibration and rearrangement of the carbon domains, thereby providing tools to tune the CMS fine structure. Somewhat surprisingly, an increase in ramp rates led to a reduction in C_3H_6 permeance, with the highest ramp rate providing a considerable gain in C_3H_6-C_3H_8 selectivity. For a given constant ramp rate, thermal soak time reductions led to discernible improvements in the C_3H_6 permeance with a moderate compromise in C_3H_6-C_3H_8 selectivity. Decreased soak time reduce processing time for CMS membrane fabrication and will be considered in the context of ultimate scale-up of these membranes. Preliminary data on the impact of the above process variables other gas-pair separation will also be considered.

464780 Enhanced Gas Separation Performance of Zeolitic Imidazolate Framework ZIF-8 Membranes Via Post Synthetic Ligand Exchange

Abstract:

Currently propylene/propane separation is performed with a highly energy-intensive cryodistillation process mainly due to their close boiling points. Although a variety of membranes with different materials have been studied for membrane-based propylene/propane separation, there are currently no membranes and membrane processes that are commercially available. This is mainly due to the fact that the majority of new membrane materials proposed failed to meet the separation performance requirements with respect to the propylene permeability and propylene/propane separation factor.[1,2,3]

Zeolitic imidazolate frameworks (ZIFs), a subclass of metal organic framework with zeolite topology, are composed of divalent metal nodes (typically Zn or Co) interconnected with imidazolate-based ligands, forming three-dimensional crystalline structures with pores and cavities in the scale of molecules.[4] ZIFs are considered very promising gas separation membrane materials because of their chemical and thermal stability in combination with their ultra-micropores (less than 0.5 nm). For example, membranes of ZIF-8, made of Zn and 2-methyl-imidazolate with a SOD structure, were reported having excellent propylene/propane separation due to the fact that its effective aperture size lies between propylene and propane.[5] Highly propylene-selective ZIF-8 membranes were synthesized using several methods including *in situ* counter diffusion and microwave-assisted secondary growth methods reported by Kwon et al.[6,7]

For industrial applications of ZIF-8 membranes, their long term stability is critically important. There are, however, only a few such studies have been reported.[8] In our long-term stability studies, we observed that time-dependent propylene/propane permeation and separation behaviors depend on synthesis methods.

Here, we would like to discuss the relationship between the membrane stability and inherent surface defects resulting from different synthesis conditions and to show a post-synthetic treatment as an effective means to stabilize membranes.

References and Notes

1 Koros, William J. and Fleming, G.K. (1993), "Membrane-based gas separation", *Journal of Membrane Science*, 83, pp 1-39

2 Baker RW., (2002) "Future Directions of Membrane Gas Separation Technology" *Industrial & Engineering Chemistry Research.*, 41 (6), pp 1393-1411

3 Tanaka, K., Taguchi, A., Hao, JQ, Kita, H., and Okamoto, K., (1996), "Permeation and separation properties of polyimide membranes to olefins and paraffins", *Journal of Membrane Science*, 121, pp 197-207

4 Park KS, Ni Z, Côté AP, et al., (2006), "Exceptional chemical and thermal stability of zeolitic imidazolate frameworks", *Proceedings of the National Academy of Sciences*, 103 (27), pp 10186-10191

5 Zhang C, Lively RP, Zhang K, Johnson JR, Karvan O, Koros WJ., (2012), "Unexpected Molecular Sieving Properties of Zeolitic Imidazolate Framework-8", *The Journal of Physical Chemistry Letters*., 3 (16), pp 2130-2134.

6 Kwon HT, Jeong H-K., (2013), "In Situ Synthesis of Thin Zeolitic–Imidazolate Framework ZIF-8 Membranes Exhibiting Exceptionally High Propylene/Propane Separation", *Journal of the American Chemical Society*, 135 (29), pp 10763-10768.

7 Kwon HT, Jeong H-K., (2013), "Highly propylene-selective supported zeolite-imidazolate framework (ZIF-8) membranes synthesized by rapid microwave-assisted seeding and secondary growth", *Chemical Communications*, 49, pp 3854-3856

8 Liu, D., Ma, X., Xi, H., Lin, YS., (2014), "Gas transport properties and propylene/propane separation characteristics of ZIF-8 membranes", *Journal of Membrane Science*, 451, pp 85–93

455633 Synthesis and Characterization of Nanoporous Polyethersulfone Membranes As Support for Composite Membrane in CO_2 Separation: From Lab to Pilot Scale

Abstract:

Nanoporous polyethersulfone (PES) membranes were prepared from PES/N-methyl-2-pyrrolidone (NMP)/2-methoxyethanol (2-ME) casting solutions with water as coagulant by both vapor- and non-solvent-induced phase inversion steps successively under various preparation parameters. 2-ME was incorporated into the casting solution as a pore-forming additive because of its high affinity with water, leading to more open membrane morphology. A detailed study of the effects of different parameters, including polymer concentration, 2-ME/NMP ratio, relative humidity, water vapor exposure time, and water coagulation bath temperature, was conducted in lab-scale experiments to determine the operation guidelines for pilot-scale fabrication. 14-inch wide PES membranes were fabricated successfully by using a continuous casting machine. A parametric study in the pilot-scale fabrication was also carried out. The surface morphologies of the PES membranes were characterized by scanning electron microscopy (SEM). The morphological differences between both scales by the same preparation conditions were compared. Based on the guidelines established from the lab-scale experiments, the casting solution composition and coagulation bath temperature were optimized in the pilot-scale continuous casting machine to fabricate the 14-inch wide PES membrane with a desirable morphology suitable for use as the substrate of composite membrane in CO_2 separation. The developed fabrication process also showed the potential for the commercial production of the PES membrane in a larger scale by the vapor- and non-solvent-induced phase inversion technique. In addition, a pilot-scale PES membrane with a pore size of ~ 69 nm and surface porosity of ~ 17% was employed in the continuous Zeolite-Y (ZY) nanoparticle deposition. A thin ZY layer with a controlled thickness and a uniform coverage was prepared. PES membranes deposited with ZY particles were also utilized as the substrate of composite membrane for CO_2 capture

from flue gas in coal-fired power plants. From the gas transport measurement by using a simulated flue gas, the prepared composite membrane with an amine-containing selective cover layer showed a CO_2 permeance of > 800 GPU (1 GPU = 10^{-6} cm^3 (STP)·cm^{-2}·s^{-1}) and a CO_2/N_2 selectivity > 140, indicating the great potential for CO_2 capture from flue gas.

458190 Polymer Gas Separation Membranes with Tailorable Ultrafine Microporosity and Physical Aging Enhanced Transport Propertie

Abstract:

Polymer Gas Separation Membranes with Tailorable Ultrafine Microporosity and Physical Aging Enhanced Transport Properties

Ruilan Guo* (rguo@nd.edu), Jennifer Wiegand, Shuangjiang Luo, Ashish Kushwaha

Department of Chemical and Biomolecular Engineering, University of Notre Dame, Notre Dame, IN 46556-5637

Polymer membranes with well-defined microporosity are highly desired for gas separations, wherein high microporosity enables fast gas transport while the finely tuned pore size distribution regulates selective transport. Central to a successful microporous gas separation membranes are the microcavity size distribution and their long term durability, i.e., resistance to physical aging. We report here a new series of microporous polyheteroarylenes containing shape-persistent iptycene-based building blocks in the backbone structures, several of which show superior gas separation performance outperforming the permeability-selectivity tradeoff upper bounds. Particularly, it is found that the intrinsic microporosity defined by the shape of the iptycene units offers unique opportunity to tailor the microcavity architecture in the membranes, which leads to well-defined yet tailorable bimodal microcavity size distribution. This paper will first focus on the discussions of how to exquisitely tailor the architecture of the microcavities in these iptycene-containing condensation polymers via modifying substituent groups, varying linkage geometry, adjusting backbone rigidity, or varying monomers combination to produce microporous polymer membranes to meet various gas separation needs. The synthesis and membrane properties of these new iptycene-containing polyheteroarylenes will be presented in details. Then the discussions will focus on the study of physical aging behavior of these microporous iptycene-containing membranes. Notably, unlike other reported microporous polymers that are susceptible to physical aging, all the iptycene-containing microporous polymers recently developed in our laboratory show superior resistance to physical aging. In particular, some of the aged membrane samples showed physical aging enhanced separation performance which manifests in marked increases in both gas permeabilities and ideal selectivities for several gas pairs. Microstructure analyses suggest that local chain relaxation in iptycene-polymers facilitates the formation of ultrafine microcavities leading to improved gas transport. The fundamental relationship between the microporous architecture and macroscopic transport properties will be elucidated to provide guidance in new membrane design.

457959 Modeling Gas Sorption in Thermally Rearranged Polymers: A Thermodynamic Approach

Abstract:

Thermally rearranged (TR) polymers, first reported by Park et al. [1], have attracted increasing attention in natural gas sweetening [2]. They offer an unusual combination of high gas permeability and selectivity, which, in some cases surpass the upper bound reported by Robeson in 2008 [3]. TR polymers can be prepared via thermal rearrangement of solvent-soluble polyimides to form insoluble poly-benzoxazoles.

In this work, hydrogen, nitrogen, methane and carbon dioxide sorption isotherms in three TR samples, obtained via thermal conversion of HAB-6FDA polyimide, were theoretically interpreted using the Non-Equilibrium Lattice Fluid (NELF) model [4]. Since PVT data in the rubbery region are not accessible for HAB-6FDA and its thermally rearranged analogs, the polymer characteristic lattice fluid parameters, T^*, p^* and Ï*, were estimated as reported by Galizia et al.Â [5], using a collection of infinite dilution sorption data available at multiple temperatures. The model outcomes show good agreement with the experimental sorption isotherms for both the pristine polyimide and the corresponding thermally rearranged samples. Interestingly, the model confirms by a completely independent perspective that the increase in gas solubility in TR samples is due mainly to entropic factors, i.e. the increase of excess free volume upon thermal rearrangement. Finally, the model results indicate that TR samples have, in the presence of carbon dioxide, greater dimensional stability relative to HAB-6FDA polyimide.

[1] H.B. Park et al., Science 2007 318 254-258

[2] D.F. Sanders et al., J. Membr. Sci. 2012 409-410 232-241

[3] L.M. Robeson, J. Membr. Sci. 2008 320 390-400

[4] F. Doghieri et al., Macromolecules 1996 29 7885-7896

[5] M. Galizia et al., J. Membr. Sci. 2012 405-406 201-211

466795 Tailoring Microcavities in Thermally Rearranged (TR) Polymers for Gas Separation

Abstract:

Rigid and microporous polymers are expected to show high gas permeability with excellent selectivity. Thermally rearranged (TR) polymers are a new class of microporous organic polymers architecting their free volume elements or so-called microcavities during thermal conversion. Microcavities and their distribution in the TR polymers can be manipulated *via* degree of rearrangement, rigidity of the original chain, diverse polymer precursor design. These cavity-tuned TR polymers have been explored to hit over the limits which conventional polymers have faced with including natural gas sweetening or carbon capture.

Herein we report on another cavity-tunable TR polymers for the first time induced by hydrophobic surface coating which enhances build-up of a kinetic layer on the membrane surface. In addition, cavity size and distribution across the coated TR polymer membranes can be easily tailored. Consequently, gas transport of large gas molecules was strongly interrupted due to the tailored cavities, while small gas molecular transport was maintained, resulting in outstanding separation performance as molecular sieves with improved selectivity and good gas permeability for smaller gas molecular pairs such as helium, hydrogen and carbon dioxide or even difficult separation of oxygen and nitrogen.

460488 Ionic Polyimides: New Dimensions in the Design of Polymer Gas Separation Membranes

Abstract:

Ionic liquids (ILs) and polyimides (amond several other classes of materials) have been at the forefront of gas separation membrane materials in recent years. Although ILs and polyimides are seemingly disparate classes of materials, ILs and polyimides can be covalently attached via condensation polymerization reactions to form "ionic polyimides". Based on rational selection of precursors and reagents, there are 10,000's of feasible ionic polyimide structures along with the potential to incorporate elements from PIMs and TR polymers as well. We have already developed several unique ionic polyimide materials based on monomers formed from aminopropylimidazoles with PMDA, 6-FDA and other dianydrides. These diimide monomers are then reacted with dihalides through the imidazole moeities via SN2 reactions to form the ionic polyimides. Our first-generation ionic polyimides show excellent processability and display increased ordering in the presence of ILs. This presentation will discuss our design strategies, polymer characterization and performances as membranes in the separation of CO2, H2, N2 and CH4.

468615 Investigation of Ultrafiltration, Nanofiltration,Ozonation and Biofiltration for the Removal of Microcystin-LR from Water

Abstract:

Investigation of Ultrafiltration, Nanofiltration,Ozonation and Biofiltration for the Removal of Microcystin-LR from Water

On August 2, 2014, the greater Toledo area woke up to a Do Not Drink or Boil Water Advisory. The advisory was due to the presence of a cyanotoxin (algal toxin) produced by cyanobacteria in Lake Erie called microcystin-LR in the drinking water supply that has a WHO provisional guideline of 1 mg/L. Upon entering the City of Toledo Collins WTP crib at the Lake Erie intake, potassium permanganate is added to the water for mussel control. While potassium permanganate is needed to control mussels, it lyses cyanobacteria cells, releasing algal toxins to the water. The water is then pumped nearly three miles to the Low Service Station, where powdered activated carbon (PAC) is added to the water for taste and odor control, and the water is transported approximately six miles to the WTP (High Service Station). PAC is generally effective for removal of algal toxins through adsorption onto its surface. At High Service, alum, lime and soda ash are added to the water for coagulation-flocculation, softening, and removal of metals. The water is then sand filtered, carbonated and chlorinated before being sent to the distribution system. However, traditional physicochemical water treatment processes, such as coagulation-sedimentation-filtration, have been shown to only be partially effective for the removal of whole algal cells and not effective for the removal of algal toxins. Furthermore, chlorination is effective for oxidizing algal toxins at relatively high free chlorine concentration as long as the pH is below 8; however, for corrosion control, Toledo water is kept at a pH above 9. Therefore, the treatment process was not enough of a barrier to prevent microcystin-LR from entering the drinking water supply. By August 4, the water treatment plant increased its PAC dosage by nearly four times, and while the toxin was removed, a significant amount of PAC sludge was produced and the cost of PAC addition was unsustainable. The objective of this project is to study alternative water treatment processes, namely membrane separations, biofiltration and ozonation, for the effective removal of algal toxins.

462400 On the Operational Flexibility of Process Systems for Membrane-Based Water Treatment and Desalination

Abstract:

Membrane-based processes are continuing to be applied for water treatment and desalination of a growing portfolio of underutilized, geographically distributed water sources. With diversified source waters, the conventional paradigm of centralized water treatment via reverse osmosis (RO) membranes using large-scale process systems is increasingly being challenged. The alternative is the use of smaller distributed but more robust RO water treatment/desalination systems. The flexibility of a single RO process system to operate over a wide range of feed water conditions is therefore becoming increasingly important. High operational flexibility that maintains high energy efficiency and compact system footprint, however, remains challenging due to the inherent limitations of conventional membrane system designs. Membrane process systems are composed of integrated process components that are highly interdependent. Optimal operation of each system component is often limited to a narrow flow range, which in turn restricts the operational range of the integrated process system. Pumps (and associated motors) and energy recovery devices, for example, typically have narrow flow ranges at which energy efficiencies are optimal (i.e., best efficiency points). Furthermore, membrane modules are inherently limited with respect to the flow ranges for optimal operation close to the thermodynamic limit, while maintaining conditions that protect physical integrity, maximize membrane area utilization, minimize concentration polarization, and avoid excessive membrane fouling and/or mineral scaling. Such module-level operational limits can significantly restrict water recovery and permeate productivity ranges of the integrated system, which is highly dependent on the specific membrane array design. Existing methods for improving operational flexibility rely on RO concentrate recycling (to RO feed); this provides an additional degree of freedom for steady-state process regulation and also for enabling unsteady-state cyclic operation. At present, however, the impact of operational flexibility on systems energy and membrane footprint has not beem fully explored and remains unclear in the published literature. Accordingly, the present work presents a fundamental analysis framework for quantifying operational flexibility and its realizable improvements. Using the framework, the interdependence between system operational flexibility with energy consumption were explored. The resulting analysis indicated that concentrate recycling, whether applied in a steady-state or cyclic mode of operation, can significantly enhance system operational flexibility. However, increased operational flexibility leads to either higher energy consumption or larger system footprint when compared to a less-flexible system with no concentrate recycling. This finding is consistent with the higher expected axial average osmotic pressure in the membrane array of systems with concentrate recycling. Analysis also shows that an energy recovery device can impose a restriction on operational flexibility; such restriction can be overcome through process decomposition. It is concluded that high operational flexibility is well suited for small- to medium-scale applications or distributed deployments, especially where effective handling of feed water variability to maintain system optimality is critical. In such systems, it is imperative to balance system operational flexibility with the expected increased energy consumption and/or membrane footprint. Less-flexible membrane systems with high efficiencies with respect to energy consumption and membrane area utilization may better serve large-scale, centralized applications with low feed water variability or where feed water equalization tank/basin can be cost effective.

471085 Polyamide Thin Film Nanocomposite (TFN) Membranes for Simultaneous Boron Removal and Desalination of Seawater

Abstract:

Production of potable water has become a worldwide concern. For many countries the available water resources are not adequate due to population growth and industrialization. Desalination of brackish or seawater has emerged as a solution to the limited water supplies. Reverse osmosis (RO) today is the leading technology that is responsible for producing more than half of the worldâ€™s desalination capacity. Polymeric membranes have dominated the RO market since they offer high salt rejection, durability, small footprint and simple operation. An emerging problem is removing boron from seawater in order to meet the World Health Organization (WHO) recommendation of 2.4 ppm for potable water and 0.5 ppm for water used for irrigating crops such as citrus fruits and nuts. Since boron is present in seawater as boric acid whose effective diameter is nearly the same as that of a hydrogen-bonded cluster of water molecules, conventional RO membranes cannot effective achieve the WHO recommendations for boron removal. Recently, a new concept of mixed-matrix composite RO membranes was advanced whereby nanoparticles are embedded into the selective layer of thin film composite polyamide RO membranes fabricated on porous supports via interfacial polymerization. These mixed-matrix RO membranes offer considerably promise for r achieving desalination while concurrently reducing the boron concentration to the WHO recommendations. Incorporating materials such as zeolites and functionalized carbon nanotubes into the selective thin layer of polyamide RO membranes offer the potential for improving the boron rejection as well as the, flux and anti-fouling properties. However, this requires optimizing the conditions for loading, size, type, and functional groups of the nanomaterials as well as the interfacial polymerization and post-treatment protocols. This study reports the first assessment of the performance of thin film nanocomposite (TFNC) desalination membranes for more effectively removing boron from seawater.

463696 Hollow Fiber Lumen Modification Via Environmentally Friendly Poly(zwitterion) Grafting

Abstract:

Membrane fouling is a major obstacle to the practical applications of membrane-based techniques such as ultrafiltration, nanofiltration, reverse osmosis, forward osmosis and pressure-retarded osmosis. To prevent organic fouling, surface modification with hydrophilic polymers has been identified as a potential strategy because many organic foulants are hydrophobic in nature. Efficiency of fouling prevention depends on surface materials and modification techniques. Although grafting poly(ethylene) glycol improves fouling propensity, the layer is susceptible to cleaning solutions, which becomes critical in long-term operations. Zwitterionic polymers are demonstrated as a promising alternative non-fouling material because of their high capacity to generate a stable hydration layer which resists to non-specific protein adsorption. On the other hand, an environmentally friendly grafting technique was developed with a capability of modifying the lumen of hollow fiber membranes, where common techniques such as UV or plasma treatments are unlikely to be feasible and other techniques requiring thermal control are also complicated. This work is an important contribution to reduce fouling and eliminate constrains, which are currently seen as hurdles for the use of hollow fibers in specific membrane-based applications.

469224 Interfacial Transport in Thin Film Nanocomposite Membranes for Water Purification

Abstract:

Current commercial reverse osmosis membranes are limited by the properties of the polymeric semipermeable membranes used. In the past, a variety of nanoparticles have been added in order to improve the membrane flux and selectivity. Some of these thin film nanocomposite membranes have exhibited enhanced membrane properties, allowing for improved reverse osmosis process performance. Previous work in our laboratory has demonstrated that the addition of zwitterion-functionalized carbon nanotubes (CNTs) to the selective layer of thin film composite membranes increases the overall flux of the membranes while maintaining high salt rejection.[1] The CNTs also have also shown improvements in anti-fouling behavior. There is also evidence that the improvement in flux is not necessarily due to transport inside the CNTs, but could be due to transport at the polymer matrix/nanotube interface. Due to the high cost of CNTs, other nanoparticle additives are currently being studied. Cellulose nanocrystals (CNCs) are of interest due to their lower cost and availability as well as their potential for functionalization. In the current work, polyamide/CNC based thin-film nanocomposite membranes have been fabricated on polysulfone supports via vacuum filtration and interfacial polymerization and their water flux and salt rejection have been studied for RO desalination applications. Results have shown that TEMPO-oxidized CNCs may increase the flux of the membranes by up to 40% while maintaining high salt rejection. The flux and salt rejection can be tuned via careful control of the CNC surface functionality and functional group density, and via control of polymer-surface interactions.

[1]Chan, Wai-Fong, Eva Marand, and Stephen M. Martin; *J. Membrane Science* **509** (2016): pp. 125-137.

465759 Fabrication of Polyphenylenesulfone (PPSU) and Sulfonated Polyphenylenesulfone (sPPSU) Membranes for Water Treatment and Reuse

Abstract:

Polymeric membranes fabricated via non-solvent induced phase inversion method have been employed for various water treatment applications including ultrafiltration (UF), nanofiltration (NF), forward osmosis (FO) and reverse osmosis (RO). Polyphenylenesulfone (PPSU) is one of the most widely used materials due to its good chemical resistance, hydrolysis stability and great mechanical strength. Sulfonation of PPSU could further improve the hydrophilicity of the fabricated membranes and therefore enhance their water permeability and anti-fouling performance. In this study, membranes were fabricated with PPSU and sulfonated PPSU (sPPSU) polymers with various degrees of sulfonation. The rheological properties and phase separation behavior of PPSU and sPPSU were investigated. These characteristics of dope solutions greatly influence the morphology and separation performance of the fabricated membranes. sPPSU polymers exhibit enhanced rheological properties compared to PPSU. In addition, the sPPSU polymers show greater non-solvent (i.e., water) tolerance and delayed demixing during phase separation. For the flat-sheet membranes cast by the non-solvent induced phase separation method, sPPSU membranes possess higher water permeability and hydrophilicity than PPSU membranes. This fundamental study may contribute to the design strategies of preparing hydrophilic membranes from sulfonated polymers for water treatment applications.

450977 Removal of Organic Compounds from Aqueous Systems Using Carbon Nanomaterial-Based Aerogels

Abstract:

Continued efforts are underway to develop new and improved techniques for removing contaminants from aqueous environments. Contaminants vary and include organic, metal, and biological compounds. The use of organic compounds (e.g. fertilizers) in agricultural settings continued to cause concern due to the runoff and buildup in downstream environments. Buildup of high concentrations of these compounds can potentially lead to adverse health effects in humans and animals. Adsorbents, such as activated carbon, are well known given their ability to adsorb various contaminants. However, activated carbon is limited in its effectiveness when introduced in environments where pH and temperature swings reduced activated carbon's uptake capacity. Therefore, alternatives to current state-of-the-art adsorbents are desired. In this work, the adsorption capability of a lightweight, carbon-based aerogel is presented. The aerogel is composed of 0 to 2 wt% graphene nanoplatelets and single-walled carbon nanotubes (SWCNT). Two organic compounds, 2,4-dichlorophenoxyacetic acid and 1-pyrenebutyric acid, served as the contaminants. A design of experiment approach was taken to optimize the aerogel for maximum performance. Results show that the aerogel containing 0.2 wt% GnP-SWCNT achieved an uptake capacity of 0.22 mg 2,4-D/mg aerogel and 0.083 mg PBA/mg aerogel. Compared to activated carbon, the aerogels produced in this work demonstrated 39 times higher adsorption of 2,4-D and 5 times higher adsorption of PBA making them a viable candidate as a next generation adsorbent.

471101 VMD Desalination Using Nanocomposite Membranes Prepared By Hydrophilic Nano-Additives

Abstract:

Desalination has attracted many efforts over the past few decades. Lack of fresh water all over the world from south west of the United States to the Middle East and extended to Australia has drawn much attention toward the desalination as an alternative to compensate the water shortage. Among the processes that are used for this purpose, a larger portion of the market is occupied by the membrane-based technologies, i.e. 63.7%, and Reverse Osmosis (RO) has placed on top by managing approximately 60% of the market. In spite of its lower water production cost compared to the other desalination techniques, RO is considered to be a quite energy intensive process, which approximately 40% of the total water cost is spent for electrical energy consumption within the process. However, Membrane Distillation (MD) has demonstrated outstanding characteristics, which is potentially considered an appropriate alternative for RO in global market, although the process has not yet been commercialized, mainly due to the huge thermal energy consumption in MD besides lack of a champion in manufacturing the high performance durable MD membranes at industrial scale. To improve the performance of the MD membranes by enhancing their permeability, hydrophilic nanoparticles have shown a great potential via optimizing the membrane structure. Hydrophilic nanomaterials, such as $CaCO_3$, CuO, and amine modified SiO_2, were able to increase the pure water flux in a Vacuum Membrane Distillation (VMD) process by 102.3%, 153.4%, and 2456%, respectively, at an appropriate concentration, i.e. 2.0 wt.% for $CaCO_3$ and CuO and 7.0 wt.% for silica nanoparticles. Surface hydrophobicity of the MD membranes was not compromised as a result of the addition of the hydrophilic nanoparticles into the polymeric PVDF matrix and nanocomposite membranes possessed adequate Liquid Entry Pressure of water (LEP_w) to be used in seawater VMD desalination. Nanomaterials incorporated membranes demonstrated high selectivity by rejecting more than 99.9% of the sodium chloride when an artificial seawater with a NaCl concentration of 35 g/L was fed to the system.

469778 High-Purity Oxygen Production Using Silver Exchanged Titanosilicates (Ag-ETS-10)

Abstract:

High-purity (>99%) oxygen is required in many medical and industrial applications such as pharmaceutical and aerospace applications. Producing oxygen with purity of higher than 95% from atmospheric air (78% N_2, 21% O_2 and 1% Ar) is challenging because of similar physical properties of oxygen and argon such as the molecule size and the boiling point. Adsorption separation techniques are economically preferable to conventional cryogenic distillation of O_2/Ar for small and medium scale applications but a limited number of commercial adsorbents are selective toward O_2/Ar. Recent multi-component breakthrough experiments in our laboratory have shown that silver exchanged titanosilicates (Ag-ETS-10) have the potential to separate these gases based on their thermodynamic affinities. This opens up the possibility of developing processes for high purity O_2 separations.

In this work, adsorption isotherms of N_2, O_2, and Ar on Ag-ETS-10 extrudates have been measured utilizing a volumetric apparatus and described using a Langmuir isotherm. In order to validate the large-scale performance of the adsorbent, single, binary, and ternary breakthrough profiles were obtained using a laboratory scale dynamic column breakthrough apparatus. These experiments have been described by discretising mass and energy balances using the finite volume technique. The model could predict the experimental profiles to a high level of precision. Various Pressure/Vacuum swing adsorption (P/VSA) cycle configurations including simple Skarstrom cycle and more complicated VSA cycles were simulated using mathematical models to maximize O_2 purity and recovery. Both 95%/5% O_2/Ar and dry air feed were considered in the simulations and a rigorous multi-objective optimization was conducted to maximize O_2 purity and recovery which indicated the ability of Ag-ETS-10 to separate ultra-high purity O_2 with good recovery. Effect of different operating parameters on O_2 purity and recovery and rigorous multi-objective optimization results to maximize oxygen productivity and minimize energy consumption of the P/VSA cycles (while meeting the purity-recovery constraint) were studied. The comparison of the simulation results with single-column experiments will be presented.

464883 Effect of Mass Transfer on Rapid Pressure Swing Adsorption Performance

Abstract:

Rapid pressure swing (rPSA) adsorption processes rely on small diameter particles for fast cycling. High cycling frequencies reduce process size, which is ideal for small portable applications. Conventional wisdom is based on the premise that the rate of transport within an adsorbent particle is the rate limiting step for mass transfer. However, with small diameter adsorbent particles used in rPSA (< 1 mm), the diffusion path within the particle is significantly reduced as well as the flow Reynolds number. Despite the particle size reduction, rPSA processes suffer from a lower overall product recovery as the cycling frequency increases. What causes this product recovery decline remains not well understood. Possible causes include mass and heat transfer resistances and column pressure drop.

At the 2015 AIChE Annual Meeting, a rapid pressure swing adsorption process for air separation was introduced that was capable of cycles below 3 seconds and a BSF around 50 lbs./TPD oxygen. The role of column pressure drop on process performance was investigated using the equipment. It was determined within the operating limits of the process that pressure drop only played a small role in product recovery decline as cycling frequency was increased.

In this study, another possible cause of product recovery decline, mass transfer resistance, is investigated using the same commercial Li-X zeolite (~ 0.5 mm in diameter). Experimental rPSA cycles using different column lengths and diameters will be presented to demonstrate how mass transfer affects process performance in rPSA. The experimental cycles are compared to the results of a PSA simulator to further aid in determining the role of mass transfer in rPSA performance as cycling frequency increases.

455383 Experimental and Modeling Studies on High Purity Propylene Production By PSA Using a Novel Carbon Molecular Sieve

Abstract:

Dow has developed new carbon molecular sieves (CMS) that show a high separation factor for olefin and paraffin. The new CMS was investigated for its propane - propylene separation performance via experimentation and computer simulation. Recent efforts have focused on improving the propylene product purity either by using a two stage PSA or alternate PSA cycle configurations.

Using the new CMS in a 4-step basic PSA cycle, propylene can be enriched to over 80% from a 25/75% propylene and propane mixture with over 95% propylene recovery. Adding co-current blowdown and propylene rinse steps to the 4-step basic PSA cycle increased the propylene product concentration to over 88% from the same mixture. A two stage PSA can readily enrich propylene to >98% when using 4-step PSA cycles and to>99% if the second stage PSA uses a cycle including co-current blowdown and propylene rinse steps.

Aspen models were developed for the base PSA cycles as well as the alternate cycle configurations. Model predictions are compared to PSA experimental results. Simulation results show very good agreement with experimental results.

464701 Advanced Novel Hybrid PSA/TSA Thermal Adsorptive Processes for Air Purification

Abstract:

Pre-purification of air prior to an air separation units (ASU) is a major unit operation responsible for removal of water, CO2, Oxides of Nitrogen and Hydrocarbons. The operation of the pre-purification unit has a major role in stability, operation, overall economics and safety of an ASU. Pre requisite for such adsorption units, C2H2, with low solubility in liquid oxygen has to be fully removed prior to cryogenic distillation system for safety concerns.

First generation (1970's) of the adsorptive air purification utilised conventional thermal swing adsorption (TSA) systems, utilised high proportions or all NaX molecular sieve. The optimal design of a TSA system requires a pre-cooling system, which lowers the water loading duty of the system to minimise the adsorbent utilisation, regeneration power and the vessel size. Such a system typically operate with an on-line time in the range 2-8 hrs. Extensive hardware and regeneration power requirements for a TSA system adversely impact its economics.

More recently, Air Products starting from basic Skarstrom's 2 bed PSA Cycle, has developed and commercialised on over 500 industrial applications worldwide new generations of pioneering PSA, Thermal PSA (TPSA) and Thermally Enhanced PSA (TEPSA) to meet the requirements of the modern Air Separation plants across different product lines and varied applications. Objective throughout has been to reduce capital cost and regeneration energy requirements. Parallel, Air Products has developed new generation of proprietary advanced materials to meet the varied demands and match mass transfer characteristics and requirements of the hybrid cycles. Combined, the new cycles and adsorbents have resulted in significant reduction in both capital and power footprint of these units by around 50% compared with industry standard (TSA) and take advantage of synergies to retrofit older units for improved throughput and power reduction.

A major feature of Hybrid PSA/TSA is use of pioneering advanced proprietary low cost activated alumina for partial removal of water and CO2 and ability to take full advantage of PSA effect removals thus minimise the use of more costly NaX materials. We shall therefore report our progress on modelling H2O and CO2 isotherms and simultaneous breakthrough characteristics.

References:-

1. M.A.Kalbassi et.al, US Patent 9,108,145 B2 (2015)
2. T.C.Golden, et al ,US Patent 8,734,571 B2,(2014)
3. A. Wright, M.A Kalbassi, Pre-purification of Air using an Advanced Temperature-Pressure Swing Adsorption (TPSA) Cycle, AICHE Fall Meeting, Cincinnati USA, 3 November 2005
4. Xiang Jun Liu a, et al, "Water vapor adsorption isotherm expressions based on capillary Condensation", Separation & Purification Technology, China, 116(2013), 95-100
5. Xiang Jun Liu a, et al "comprehensive description of water vapor equilibriums on alumina F-200: Adsorption, desorption, and H2O/CO2 binary adsorption" Separation and Purification Technology , China,133 (2014) 276–281
6. M.A.Kalbassi, T.C.Golden, US Patent 5,846,295, (1998)
7. M.A.Kalbassi, T.C.Golden, US Patent 5,614,000, (1997)
8. T.C Golden, T., et al. U.S. Patent 5,656,064 (1997).
9. T.C.Golden et.al US Patent 5,656,064 (1997)

471271 Two-Stage PSA System for CO_2 Removal during Closed-Loop Human Space Exploration Missions

Abstract:

The development of a novel two-stage PSA system is underway that can be used for enriching and recovering CO_2 from a spacecraft cabin that is suitable for use in a Sabatier reactor. This two-stage pressure swing adsorption (PSA) process utilizes Stage 1 to concentrate metabolic CO_2 from about 2667 ppm (0.2667 vol%) to about 600,000 ppm (60 vol%) and Stage 2 to further enrich the CO_2 product from Stage 1 up to > 97 vol% CO_2, while recovering at least 95% of it, which corresponds to removing 4.0 kg/day of CO_2. Each stage of this PSA system utilizes a combination of equalization, cocurrent depressurization and heavy reflux cycle steps to facilitate heavy component enrichment and recovery. The first generation of this two-stage PSA process utilizes beaded commercial adsorbent, i.e., 13X zeolite, in both stages. Implications from these results potentially reach far beyond the needs of NASA and begin to hint at the extreme concentration of CO_2 from ambient air at relatively high recovery or the extreme concentration of any dilute species from a light gas stream using a simple PSA cycle. This presentation will discuss the most recent modeling and/or experimental results.

463664 Recovery and Purification of Methane in Biogas Using One-Tower Vpsa-Methane with Integrated Pretreatment Adsorbent Layers

Abstract:

From the viewpoints of both energy conservation and alternative energy in the context of high-efficiency power generation, it is very useful to consume highly concentrated methane as a gas-engine fuel, purified from biogas. The authors of the present report designed and fabricated a biogas purifier which is compact and has a low energy power consumption rate. This biogas purifier features one-tower VPSA-methane loaded with pretreatment adsorbent layers (H_2S, Siloxane, and CO_2 from bottom to top). We experimented with bio-gas purification using this one-tower VPSA-methane (inlet gas rate of 400m^{3}N/day, inlet methane concentration of 60vol%), and confirmed the performance thereof in this study as having a methane recovery ratio of 90% and a purified methane concentration of 90vol%. The flow sheet for one-tower VPSA-methane is indicated in Fig.1, while the operating conditions and performance are listed in Table 1.

[Adsorption Stage]

One-tower VPSA-methane consists of an adsorption tower (H_2S adsorbent, Siloxane adsorbent, CO_2 adsorption), a blower/vacuum pump combination rotary mechanism, valves, tubes and pipes, a pressure equalization tank, a product tank, and a raw material tank. The raw material (methane concentration of 60vol%, H_2S concentration of less than 200ppm, Siloxane consisting of 10ppm, and a bio gas rate of 16m^{3}N/h). The raw material (16m^{3}N/h) is supplied at an adsorption pressure of 140kPa-abs and an adsorption period of 60 seconds, such that H_2, Siloxane, and CO_2 are adsorbed. At the outlet, the purified methane (methane concentration of greater than 90vol%, H_2S of less than 1ppm, Siloxane undetected, and a purified 9m methane gas rate of 9m^{3}N/h) is collected.

[Desorption Stage]

The adsorption tower saturated with H_2S, Siloxane, and CO_2 (60 second adsorption period) is connected to the blower and vacuum pump combination rotary mechanism and the internal pressure is reduced to the desorption pressure of 10 kPa-abs. The desorption period is set at 60 seconds, i.e., the same as the adsorption period. During the desorption period, the purified methane can be released from the auxiliary tank containing purified methane product to maintain outlet flow continuity. The above-mentioned one-tower VPSA-methane operation is characterized by a purified methane concentration level of 90vol%, a methane recovery ratio of 90%, and an electric power consumption rate of 0.3kWh/m^3N-methane. With a 1m^3 adsorption tower, purified methane of 173m^3N/h can be recovered. We therefore concluded that the one-tower VPSA-methane in this study demonstrated strong economical performance.

Fig.1 Schematic diagram of VPSA-methane

oTable.1 VPSA-methane performance

Biogas composition

CH_4 70vol%

CO_2 30vol%

H_2O 4vol%

H_2S 100ppm

Siloxane 10ppm

Biogas flow volume

5~1500m^3N/h

CO_2, H_2O, H_2S, CH_3SH,

Siloxane removal

>90%

Methane recovery

>80%

Electric power consumption

1~100kW

Methane recovered

10~100m^3N/h

Methane purity>95vol%

Energy power consumption

0.16~0.2kWh/m^3N-CH_4

Energy/Generation ratio

Approx. 4%

473306 CO_2/N_2 Separation Using Cu-BTC, Uio-66 and MIL-53(Al) Metal Organic Frameworks

Abstract:

CO_2/N_2 Separation Using Cu-BTC, UIO-66 and MIL-53(Al) Metal Organic Frameworks

Satyannarayana Edubilli, Pradip Das, Sasidhar Gumma

Department of Chemical Engineering, Indian Institute of Technology Guwahati

Guwahati - 781039, Assam, India.

Email: s.edubilli@iitg.ernet.in (S.Edubilli), pradip.das@iitg.ernet.in (Pradip.das), s.gumma@iitg.ernet.in (S. Gumma)

In recent years Metal Organic Frameworks (MOFs) have gained much attention for adsorptive separations due to their exceptional capacities, tunable pore Structures and ease of regenarability. MOF materials are being extensively studied in terms of equilibrium measurements, but for the practical application point of view, they have to be tested for their cyclic and thermal stabilities under process conditions.

In this presentation, the results of breakthrough and cyclic adsorption experiments for CO_2 / N_2 separation on different MOFs viz. Cu-BTC, Uio-66 and a flexible metal organic framework MIL-53(Al) will be reported. MOF powders are made into beads using PVA/H_2O as binder. Different ratios of MOF to binder were prepared and the binder content was optimized based on its strength and surface area. The adsorbent column was filled with MOF beads and this column was used for breakthrough as well as the cyclic PSA adsorption experiments. A mixture of 15% CO_2 in N_2

was used as the feed. During the experiments, the column pressure history, feed inlet flow rate, and column outlet flow rate are recorded using data acquisition system. The column outlet stream composition was analyzed using a Gas Chromatograph. The performance of a three step PSA cycle consisting of pressurization, feed and blowdown as well as cycles incorporating N_2 purge and CO_2 rinse steps will be reported. In a typical experiment about 10-12 grams of the adsorbent was used and the feed flow rate was about 200 cc/min. The adsorption step was carried out at about 1.3 bar, while desorption was done at 0.1 bar.

In addition to these results we also report, the effect of framework flexibility on the process performance. In our earlier work on MI-53(Al) we have reported that the pores of parent MIL-53(Al) into narrow pore (**np**) domain at ambient temperature in absence of guest molecules by manipulating its history [1]. Using this procedure, we have studied the performance of MIL-53 (Al) using both **lp** and **np** phased samples. Under the same experimental conditions, the observed CO_2 breakthrough times of **lp** and **np** phase materials are 180 and 510 seconds respectively.

References

[1] P. Mishra, S. Edubilli, H. Uppara, B. Mandal and S. Gumma, *Langmuir*, 29, 12162-12167,2013

457858 Downhole Oil/Water Separation Based on Hydrocyclones

Abstract:

Hydrocyclones were developed as inertial separators at the end of the 19th Century. These separators are widely used to thicken liquid suspensions, clarify liquids, de-gas liquids, and classify solids. The internal flow structure within these confined vortex separators is incredibly complicated. Commercially available hydrocyclones have major diameters ranging from 10mm to more than 300mm. The hydrocyclone is used to separate two immiscible phases with different densities and with a dispersed phase particle size distribution between 5 and 500 microns. It is also used to classify solids suspended in a continuous liquid phase. Hydrocyclones may be operated either as "reverse-flow" separators or as "forward-flow" separators with or without an air core.

Although earlier studies in the 1950s demonstrated the utility (and promise) of hydrocyclones as liquid/liquid separators for concentrated and stable dispersions, the risk of forming an emulsion in high Reynolds number flows has limited the practical use of the hydrocyclone as a liquid/liquid separator to relatively dilute dispersions. Over the past 50 years, a few important practical commercial designs for deoiling produced water on offshore platforms have emerged. During the 1990s, high-capacity deoiling hydrocyclones were field tested as downhole oil/water separators (i.e., DOWS-technology); however, due to the risk associated with this technology, this initiative was abandoned.

This presentation will discuss the potential use of very small diameter, low-capacity hydrocyclones for downhole oil/water separation. The proposed approach aims to mitigate problems associated with single-stage, high-capacity hydrocyclones. Clearly, a viable DOWS technology would have a significant impact on oil and gas production worldwide.

454139 Centrifugal Separation of Flocculated Sludges in Wastewater Treatment

Abstract:

Decanter centrifuges are commonly used in municipal wastewater treatment to dewater sludge cake to a high consistency. Likewise, the liquid centrate has to be free from solids as it is being re-circulated in the plant as make-up water or being discharged periodically to the environment. Flocculants are used to agglomerate the biological solids in the wastewater otherwise it is practically impossible to separate by centrifugal sedimentation due to the small density difference between the biosolids and liquid. Jar tests are used to identify and screen flocculants to determine the most appropriate flocculant for use with the wastewater under centrifugation. It is also known that the flocculated solids (i.e. flocs) would break up in the centrifuge due to intense shear stress during acceleration of the feed to the pool so that the size of flocs observed from jar tests could be much larger than the ones in the centrifuge. This paper presents a model on separation of flocs in a decanter. The Leung number [1] is used to derive the "cut size" in making separation for the decanter. When the cut size is smaller than the floc size, a high solids recovery can be maintained, and vice versa.

Despite it is not possible to determine the floc size in-situ in the machine, the recovery–size cut curve can be determined for a given wastewater application (i.e. with organic-to-inorganic ratio defined and a selected flocculant), from which the performance of the machine can be inferred and projected for other decanter sizes.

The model has been applied in a secondary wastewater treatment plant to several decanters with different configurations in separating flocs from raw mixed sludges (48% Primary Sludge, 52% Waste Activated Sludge). The approach also provides a scale-up of decanter centrifuge performance in separation floc factoring in the scale-up, (a) the operating conditions (pool depth, rotation speed, feed rate) and (b) configuration of the decanter (pool depth, clarifier length, etc.).

Reference:

[1] Wallace Woon-Fong Leung, "Industrial Centrifugation Technology", McGraw-Hill, 1998.

KEYWORDS

Decanter Centrifuge, Flocculated Solids, Wastewater, Separation

462562 Flexible and Low-Energy Approach for RO Desalination

Abstract:

Rigorous optimization in system and process design has enabled reverse osmosis (RO) membrane technology to gain foothold in various water treatment and desalination applications. However, given that water capacity demand and water product quality requirements may change over time, it is critical that the RO membrane system can be easily adapted to the new requirements. Therefore, it is desirable to impart operational flexibility to the RO system to be able to handle temporal variability in feed water quality and satisfy new treatment targets (e.g., recovery) without having to retrofit the system. Recent studies at UCLA have suggested that flexible and low-energy RO (FLERO) operation can be achieved by employing process decomposition to overcome limited operational flexibility imposed by hydraulic components such as energy recovery devices (ERDs). In order to demonstrate the enhanced operational system flexibility (allowing a wide water recovery range of 10-95% with a single system) and the high energy efficiency of the FLERO approach, a small RO/NF system platform was developed capable of autonomous, self-adaptive operation over a wide range of feed water salinity (e.g., 500 mg/L â€" 35,000 mg/L) while maintaining energy-optimal conditions. A key feature of FLERO is the ability to switch (on demand) between various operational modes (e.g. batch, continuous, and semi-batch) depending on feed conditions (e.g., salinity and temperature) and treatment targets using a single system platform. Furthermore, the present unique FLERO system design enables continuous operation (without interruptions of permeate productivity) and without the need for the high pressure feed pump used in single or multi-stage RO systems. System performance was then analyzed via process models that were validated with both laboratory and field data for brackish water and seawater desalination, as well as for nitrate removal from impaired potable groundwater sources. Both experimental and modeling results demonstrated that the innovative FLERO system platform allows for flexible operation over a wide range of product water recovery, while maintaining high energy efficiency (i.e., low energy consumption suitable for use with renewable energy). It is shown that FLERO is particularly attractive for distributed deployment in rural or isolated communities where high water recovery is critical for various water sources of different salinity levels.

449804 Modeling of Moving Bed Granular Filtration Using Eulerian-Eulerian Approach

Abstract:

Modeling of moving bed granular filtration using Eulerian-Eulerian approach

Ravi Kolakaluri

Software Electronics and Mechanical Systems

3M Corporate R&D

Granular filtration is traditionally used for water treatment, but due to the ease of its usage it is also widely used in chemical industries. Granular filtration is mostly used to treat flue gas, combustion products, polymers and petrochemical products. The common process of conducting granular filtration is in the fixed-bed mode, but it can also be conducted in a moving-bed or fluidized-bed mode so that the operation can be carried out continuously.

Granular filtration consists of two solid phases and fluid phase. The first of the solid phases comprises the granules through which the fluid flows, and the other solid phase consists of fine particles that are suspended in the fluid phase. Direct numerical simulation of a full-scale granular bed is prohibitively expensive in terms of the required computational resources. This is due to the range of length and time scales that need to be simulated, hence the Eulerian-Eulerian two-fluid approach is a viable approach to model a full-scale granular bed.

In this study, we develop a computational fluid dynamics (CFD) model where the granules and fluid phase are modeled using an Euler-Euler two-fluid model and the fine particles that are suspended in the fluid phase are

simulated using two User-Defined-Scalars (UDS). One of the UDS represents the number density of particles suspended in the fluid phase and the particle velocity in this UDS is obtained from fluid velocity using the equilibrium Eulerian model suggested by Ferry and Balachandar (2001). The other UDS is for number density of particles deposited on the granules, which are moving with the granules. The validation of the CFD modeling results are done using the experiments.

454000 Transition of Particle Motions in an Evaporating Droplet

Abstract:

In an evaporating water droplet laden with particles, the liquid flow profile varies in time and space. It has been shown by Hu and Larson (2005) that evaporative cooling results in a temperature gradient on the surface of the droplet, thus driving Marangoni currents near the surface. Marangoni currents along with the internal edgeward flows of the droplet result in the formation of circulation eddies, which keep the particles in suspension initially. Near the end of evaporation, the particles will begin to deposit and accumulate at the contact line. This process has been studied both experimentally and computationally by Devlin (2015). However, the transition of particles from suspension to deposition is not clearly understood. This transition affects the separation of different sized particles in the droplet, which has implications in biomedical applications such as diagnostic testing. Here we modelled the evaporating system and used the finite element method to computationally analyze the flows of the water and the motion of the particles. The simulation indicates the disappearance of circulation eddies of water at the end of evaporation, and can be used to predict the time and the nature of which the particles accumulate.

462932 Influence of Particle Size Distribution on the Separation Processes in Decanter Centrifuges

Abstract:

The mechanical fluid separation is widely applied in the processing industry. Almost all processes which are handling finely dispersed particles in the liquid use unit operations, such as the filtration, the thickening or the centrifugation, for the separation of both phases. Centrifuges generate high mass forces. This also allows to separate submicron particles. At this point, especially continuously working decanter centrifuges are widely applied. The screw conveyor system within the centrifuge permits the change in the process parameters in a wide range. The design of centrifuges is limited to simplified assumptions for the machine and the process parameters [1]. The calculation of the separation efficiency takes not into account the sediment build-up, real flow conditions or bulk transport.

We developed a 2D-model for the calculation of the separation efficiency within decanter centrifuges. The current model describes the behaviour of the apparatus by material and machine functions. The material functions consider the behaviour of the particulate system such as, sedimentation velocity distribution, sediment build-up and transport of the sediment. These material parameters are empirical parameters and measured in well-established laboratory equipment. The machine functions represent the time-dependent behaviour of the machine. The 2D-modelling approach can be used for a better prediction of the process behaviour of the machine. The model does not only considering the sedimentation of the particles, but also the whole behaviour of the centrifuge. The results of the numerical simulations are validated with different experiments. The investigations were performed with a lab decanter centrifuge.

The financial support of DFG (German Research Foundation) within the priority program SPP 1679 "Dynamic simulation of interconnected solids processes DYNSIM-FP" is gratefully acknowledged."

[1] M. Gleiss, H. Nirschl: Modeling Separation Processes in Decanter Centrifuges by Considering the Sediment Build-

up, *Chem. Eng. Technol*. 2015, *38*, No. 10, 1873-1882

458462 Bioinspired Antifouling Membranes for Sustainable Water Purification

Abstract:

One of the daunting challenges to the sustainability of modern society is an inadequate supply of clear water. Due to its energy-saving and cost-effective attributes, membrane technology has become an indispensable platform technology for water purification, including seawater and brackish water desalination as well as municipal and industrial wastewater treatment. However, membrane fouling, which arises from the nonspecific interactions between membrane surface and foulants, significantly impedes the efficient application of membrane technology. Preparing antifouling membranes is a fundamental strategy to cope with pervasive fouling problems from diverse varieties of foulants. In recent years, great advancements have been made both in membrane preparation techniques and in elucidating the antifouling mechanisms of membrane processes, spanning from ultrafiltration, nanofiltration, reverse osmosis to forward osmosis. This presentation will first introduce the major foulants and principal mechanisms of membrane fouling, and then highlight the development, current status and future prospects of bioinspired antifouling membranes, including antifouling strategies, preparation techniques and common applications.

471292 Janus Hollow Fiber Membranes with Bio-Inspired Polydopamine Coating for Desalination

Abstract:

Membrane distillation (MD) is a thermally driven water desalination process. Traditionally, hydrophobic membranes are applied to separate the hot feed and cold permeate. The hydrophobic membranes prevent the saline feed permeation and only allow the transport of water vapor. In addition, they also prevent the membrane pore wetting. For the direct contact membrane distillation (DCMD) process, the mass and heat transfer are coupled transport processes, and high mass transfer accompanied with low heat transfer are preferred to maintain the desalination driving force and subsequently permeation flux. However, the reduction of membrane thickness would benefit the mass transfer, but it also reduces the heat resistance simultaneously. Such a phenomenon leads to a dilemma in the DCMD membrane fabrication.

During the DCMD process, the water vapor transport through the membrane pores while the heat is mainly conducted via the membrane matrix due to its higher heat transfer coefficient. As a result, if we can strategically introduce a single-sided hydrophilic layer onto the hydrophobic membrane, we may reduce the mass transfer distance while remain the heat transfer resistance relatively unchanged. Therefore, the Janus membrane with different wettability could be a promising candidate to fulfill the requirement of DCMD process. Conventional approaches to constructing a Janus membrane require complicated fabrication processes or highly specialized equipment, and a facile, scalable method to fabricate a Janus membrane has not been available to date, especially for hollow fiber membranes.

In this work, we applied dopamine (DA) / polyethyleneimine (PEI) co-deposition method to coat the lumen side of a hydrophobic polypropylene hollow fibre membrane (PPHFM, the inner diameter is 1800±150 μm, the wall thickness is 450±50 μm, and the pore size is ~200 nm). The PPHFM was firstly assembled into a membrane module and then the mixture (DA/PEI) were recirculated within the membrane module to allow its gradual deposition on the membrane lumen surface. In order to improve the membrane surface wetting by the deposition, the PPHFM was firstly pre-wetted by ethanol and the residual ethanol in the pores was absorbed by a filter paper.

Due to the capillary effect, single-faced modification of the membrane is challenging through a wet modification process. However, the bio-inspired deposition process enables us a facile process to modify the hydrophobic membranes by an aqueous solution. Initially, only the lumen side was wetted by the deposition solution. Then the solid-liquid-air triple phase contact line moved to the membrane outer surface with the increase of deposition time. As a result, the hydrophilic layer deposition thickness can be regulated by applying different deposition times. After the deposition, the original membrane porous structure was still well preserved.

One of the most remarkable characters for the Janus membrane is the asymmetric wettability. We further investigated the water contact angle of the Janus membrane, and results suggest the successful single-sided deposition of the DA/PEI. The inner side gradually became hydrophilic with the increase of deposition time (contact angle 30 ° for 6 h deposition), while the water contactor angle for the hydrophobic outer side was relatively unchanged. Furthermore, the liquid entry pressure (LEP) test reveals the introduction of a hydrophilic layer only marginally reduced the LEP value, suggesting the Janus membrane could be applied for DCMD process.

We further evaluated the Janus membrane performance for DCMD. With the increase of deposition time, the pure water flux gradually increased for the Janus membrane, and the flux increment was more significant at elevated temperature. The most significant flux improvement of 120 % was observed at 80 °C using 6 h deposited membrane. In terms of the salt rejection test, the Janus membrane exhibited good operational stability. For example, over 99 % of the rejection was obtained over a 72 h testing period with 4 h deposited membrane.

Conclusively, we developed a facile approach to prepare Janus hollow fibre membranes via the DA/PEI deposition approach. The Janus membrane exhibited asymmetric surface components and wettability. During the DCMD testing process, the Janus membrane had high salt rejection and improved permeate flux. This work introduced a novel approach to developing the Janus membrane for various applications.

457305 Living Biofouling-Resistant Membranes with Probiotic Biofilms

Abstract:

Membrane systems are being increasingly used for water treatment, recycling water from wastewater, during food processing, and energy production. They are thus a key technology to ensure water, energy and food sustainability. However, biofouling, the build-up of microbes and their polymeric matrix, clogs these systems and reduces their efficiency. Realizing that a microbial film is inevitable, we engineered a beneficial biofilm that prevents membrane biofouling by limiting its own thickness by sensing the number of its cells that are present via a quorum-sensing circuit. The beneficial biofilm also prevents biofilm formation by deleterious bacteria by secreting nitric oxide, a general biofilm dispersal agent as demonstrated by both short-term dead-end filtration and long-term crossflow filtration tests. In addition, the beneficial biofilm was engineered to produce an epoxide hydrolase so that it efficiently removes the environmental pollutant epichlorohydrin. Thus, we have created a living biofouling-resistant membrane system that simultaneously reduces biofouling while providing a platform for biodegradation of persistent organic pollutants.[1]

References:

1. Wood, T. L.; Guha, R.; Tang, L.; Geitner, M.; Kumar, M.; Wood, T. K.: Living biofouling-resistant membranes as a model for the beneficial use of engineered biofilms. *Proceedings of the National Academy of Sciences* 2016 (doi:10.1073/pnas.1521731113)

472475 Synthetical Recovery of Oil Extraction Wastewater with Combined Membrane Distillation Crystallization Technology

Abstract:

Synthetical recovery of oil extraction wastewater with combined membrane distillation crystallization technology

Xiaobin Jiang*, Dapeng Lu, Gaohong He, Wu Xiao

State Key Laboratory of Fine Chemicals, Research and Development Center of Membrane Science and Technology, School of Chemical Engineering, Dalian University of Technology, Dalian, P. R. China

*Corresponding author: Email: xbjiang@dlut.edu.cn

Abstract

The regeneration of ethylene glycol (EG) aqueous solution used as an anti-freezer and hydration inhibitor from oil extraction wastewater is an important and challenging work due to the combining separation requirement of enrichment of EG and desalination[1]. Membrane distillation (MD) is a recently developed thermally or pressure-driven membrane process combined distillation and crystallization [2,3]. In recent decade, the combined membrane distillation crystallization (MDC) process had significant advantages on the brine treatment, crystals manufacture and crystallization control [4,5,6].

In this work, ethylene glycol (EG) aqueous solution of oil extraction wastewater was concentrated and desalinated by means of combined MDC technology using polypropylene (PP) hollow fiber membrane. The controlling strategy that inspired by the biomineralization control research on the porous membrane platform was developed [6]. Higher feed

EG concentration could lead to a decline in the permeate flux and the best concentration was 60%, where the largest feed treatment equivalent could be obtained with NaCl rejection higher than 98.9 % and water content in the permeate higher than 99%.

The crystal size distributions were quite different. The crystal distribution of MDC method was more uniform than that of evaporation crystallization by introducing the synchronous concentration controlling approach. The average size of NaCl crystals from both methods were about 100 ¦Ìm, while the C.V. value of the crystals from membrane-crystallization was 33.76 lower than that from evaporation crystallization, 46.12. The secondary nucleation inhibited in MDC process led to the desire crystal size distribution. It is obvious that the uniform crystal size distribution is the prominent advantages of MDC, which could be benefit to solid-liquid separation process.

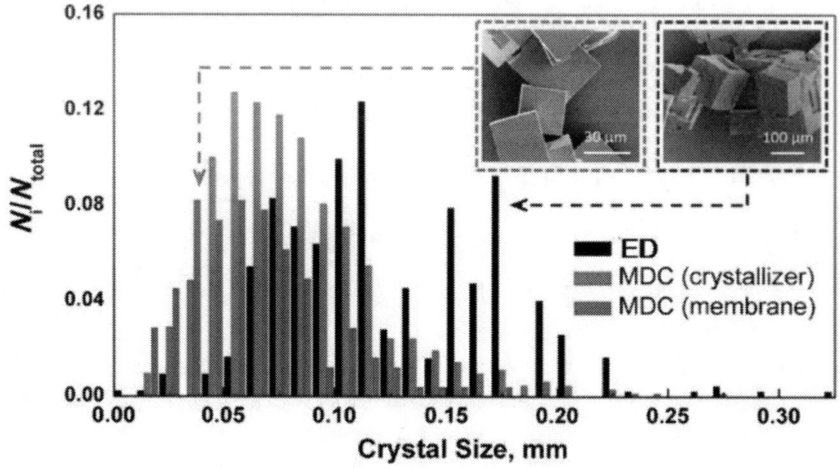

Fig.1 Comparison of crystal size distribution and crystal habit with evaporative crystallization (ED) and membrane distillation crystallizaiton (MDC).

Acknowledgment

This work is supported by National Natural Science Foundation of China (Grant No. 21306017, 21527812), Program for Changjiang Scholars and Support Project of the China Petroleum and Chemical Corporation (X514001).

References

[1] Yong, A. and Obanijesu, E.O.: Influence of natural gas production chemicals on scale production in MEG regeneration systems. Chem. Eng. Sci. 2015. 130, 172¨C182(2015)

[2] Zhang, Y.: Review of thermal efficiency and heat recycling in membrane distillation processes. Desalination 367, 223-239(2015).

[3] Alkhudhiri, A., Darwish N., Hilal N.: Membrane distillation: A comprehensive review. Desalination 287, 2-18 (2012).

[4] Jiang, X., Lu, D., Xiao, W., Ruan, X., Fang, J., He, G.: Membrane Assisted Cooling Crystallization: Process Model, Nucleation, Metastable Zone, and Crystal Size Distribution. AIChE Journal 3, 829-841 (2016).

[5] Chabanon, E., Mangin, D., Charcosset, C.: Membranes and crystallization processes: State of the art and prospects. J. Membrane Sci. 509, 57-67 (2016).

[6] Di Profio, G.; Salehi, S. M.; Caliandro, R.; Guccione, P.; Nico, G.; Curcio, E.; Fontananova, E., Bioinspired Synthesis of CaCO Superstructures through a Novel Hydrogel Composite Membranes Mineralization Platform: A Comprehensive View. Adv. Mater. 28, (4), 610-616 (2016).

471930 Biomimetic Carrier-Mediated Membrane for Transport of Hydrophilic Molecules

Abstract:

The ability of biological membranes to facilitate the selective transport of ions or hydrophilic molecules using carrier proteins inspired the design of carrier-mediated transport membranes for selective separations in aqueous solute systems. Thin film mesoporous silica, which have tunable pore morphology, chemical, thermal stability and ease of organic functionalization, serve as a support for lipid bilayers. Solute carriers are incorporated in lipid bilayers, which acting as barriers. Solute carriers of particular interest are boronic acid derivatives, which can reversibly interact with the 1, 2-diols or 1, 3-diols of molecules (carbohydrates, glycoproteins, nucleosides and dopamine etc.) to form cyclic esters with five or six member rings. Furthermore, the association strength is largely dependent on the structure of solute, which makes this reversible reaction with specific complexation very attractive for selective separation. The combination of high surface area silica materials and selectively permeable lipid bilayers suggests potential applications for efficient, selective dilute aqueous solute separations.

In this work, the transport properties of boronic acid carrier immobilized lipid bilayer which is supported by orthogonally oriented hexagonally close packed (HCP) silica thin films on macroporous support will be examined. The effect of lipid bilayer preparation methods (in the form of pore enveloping, pore filling, tethering) on the barrier properties are first explored, as a good barrier is essential for detecting selective separation with the incorporation of biomimetic carriers in the bilayer. Next, the ability of immobilized boronic acid to facilitate the transport of active molecules through the membrane is demonstrated. The effect of factors like pH gradient, temperature (related to the lipid bilayer diffusivity) and the moleculeâ€™s affinity for boronic acid are explored to improve the selectivity and permeability of this boronic acid-mediated membrane. In addition to boronic acid, other carriers immobilized in lipid bilayers are explored for the construction of selective biomimetic carrier-facilitated membranes.

455892 Characterization and Anti-Fouling Behavior of Bio-Inspired Liquid-Gated Membranes

Abstract:

Recently, a new approach to reducing fouling on surfaces was described. Inspired by the *Nepenthes* pitcher plant, this method relies on immobilizing water-immiscible liquids to a solid surface to create an immobilized liquid layer resistant to adhesion by bacteria, proteins, and salts. These liquid layers have been shown to line membrane pores when pressure is applied, reducing the deposition of salts and proteins on the surface during small-scale filtration. However, no work has been done to determine the effect that liquid-gating has on the filtration characteristics of the membrane, as well as how the presence of a liquid layer effects the development of fouling. In this study, we determine the effects of liquid gating on effective pore size and determine whether immobilized liquid layers reduce the rate of fouling or increase the effectiveness of a backflush step at removing foulants from the membrane surface during filtration of both proteins and bacteria. We also show that the addition of a liquid layer reduces the adhesion of biofilms to the membrane surface by bacterial species known to be strong biofilm formers.

457018 Synthesis of Aquaporin-Based Large-Scale Biomimetic Membrane with Enhanced Water Flux Performance

Abstract:

In recent years, aquaporin-based biomimetic membranes for water treatment have attracted much more attention. Although several achievements have reached and the first biomimetic membrane is commercially available, there are still many challenges for future industrial applications. In the current study, a larger piece of â€œwater channelâ€ embedded membrane for nanofiltration (NF) based on an aquaporin Z (AqpZ) incorporated biomimetic membrane was fabricated by using a simple method. The AqpZ-incorporated vesicles were immobilized onto the sulfonated polyethersulfone (NTR7450) and ultrafiltration cellulose acetate (CA) porous membrane respectively. In situ polymerization without initiator was then conducted to form a relatively dense hydrophobic protection layer and its thickness was adjusted by changing the polymerization duration. The morphology of freeze dried membranes was observed by FESEM and the surface modification was confirmed by FTIR transmission. Membrane performance was characterized by nanofiltration test. The synthesized membrane exhibited high mechanical strength and stability during the water filtration process. Volume water flux and pure water permeability was greatly increased for the AqpZ embedded membrane compared with the results of some reported membranes. This study provides some new possibilities of using large piece of AqpZ based biomimetic membranes for the water-treatment process.

461460 Highly Permeable Artificial Water Channels in Block Copolymer Membranes

Abstract:

Highly permeable artificial water channels in block copolymer membranes

Yue-xiao Shen, Tingwei Ren, Manish Kumar*

Department of Chemical Engineering, Pennsylvania State University, University Park, PA, 16802

*Corresponding author: 155 Fenske Laboratory, Pennsylvania State University, University Park, PA, 16802, (814) 865-7519, manish.kumar@psu.edu

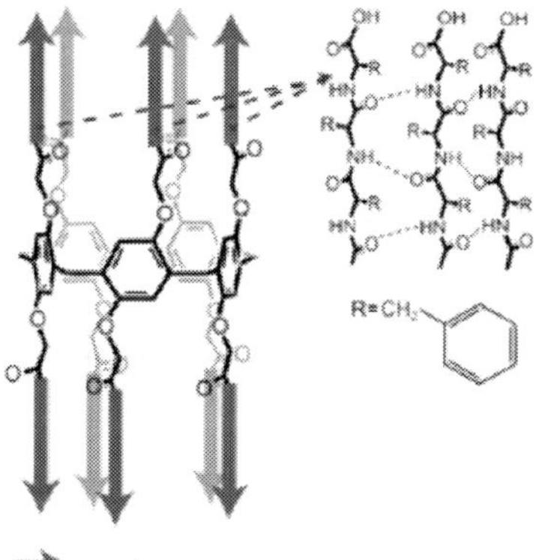

Fig. 1. Structure of the peptide-appended pillar[5]arene (PAP) channel.

We have reported that peptide-appended pillar[5]arene (PAP) artificial water channels (Fig. 1) combine the advantages of biological water channels Aquaporins (AQPs) and carbon nanotubes (CNTs).[1] It can be synthesized using simple chemistry at gram scale in the lab and provide stable structure and good solvent compatibility. The water conductance of PAP is in the range of AQPs. It can also be vertically aligned in lipid membranes, with much higher packing density than current synthetic CNTs-based membranes.

Although the initial testing of the PAP channels was conducted in lipid systems, this is not an ideal format for commercial separation materials. We continue to investigate the functionality of these channels in block copolymer (BCP) membranes. BCPs, with lipid-like structures, are most commonly synthesized for biomimetic applications because of the advantages of BCPs membranes including high mechanical and chemical stability,[2,3] low water and gas permeability[4] and customizable properties (e.g., a larger range of membrane thickness[2] and end groups[5]). We are testing a series of poly(butadiene)-*b*-poly(ethylene oxide) (PB-PEO) diblock copolymers with different block length. These BCPs have different hydrophobicities and thicknesses compared to lipids, and been shown to functionally incorporate water channel proteins aquaporin 0,[6] potassium channel[7] and ¦Á-hemolysin.[8]

The PAP channels can be incorporated into PB-PEO BCP membranes using the film rehydration method. The number of channels per vesicle was measured based on a fluorescence correlation spectroscopy (FCS) technique.[9] We combined the channel number data and permeability data using stopped-flow light scattering technique (Fig. 2A and 2B) and obtained the single channel water permeability in PB_{23}-PEO_{16} vesicles (1.5¡À0.4´10^8 H_2O molecules per second). This value is close to that in lipid system (3.7¡À0.6´10^8 H_2O molecules per second, Fig. 2C).[1] In the future, we will systematically test the PAP channels in different BCPs systems, visually confirm the incorporation in giant unilamellar vesicles and discuss the interaction of different BCPs and the PAP channels. We will also plan to use fluorescence recovery after photobleaching (FRAP) technique to determine the diffusion coefficient of the PAP channels in lipid and BCPs bilayers and expect these results can provide insights into to maximizing the channel¡¯s packing density in BCP membranes.

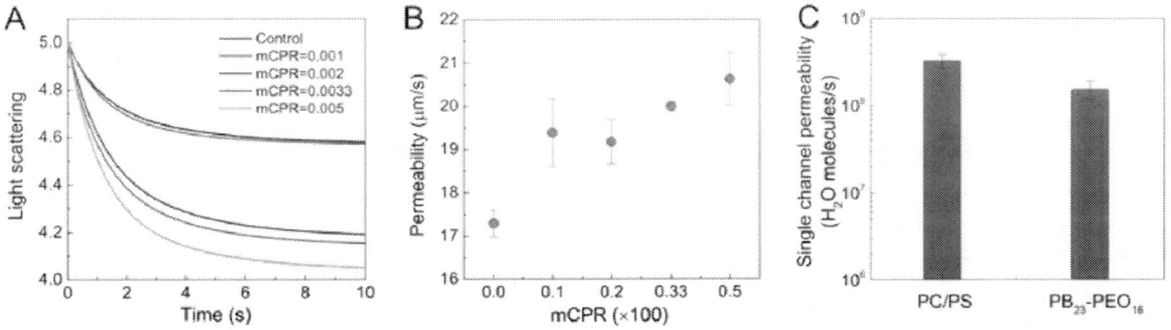

Fig. 2. Functional studies of peptide-appended pillar[5]arene (PAP) artificial water channels in block copolymer vesicles. (A) Representative light scattering traces of PB23 polymersomes with different molar channel-to-polymer ratio (mCPR=0, 0.001, 0.002, 0.0033 and 0.005) after a rapid exposure to a hypotonic solution without 100 mM PEG600 used to form the vesicles on a stopped-flow instrument. (B) The water permeability of PAP channel-containing PB23 polymersomes formed with different mCPRs measured under hypotonic conditions. (C) Single-channel water permeability of the PAP channels in PC/PS liposomes and PB23 polymersomes. Data shown are the average of triplicate measurements with standard deviation.

Reference

1 Shen, Y.-x. *et al.* Highly permeable artificial water channels that can self-assemble into two-dimensional arrays. *Proc. Natl. Acad. Sci. U.S.A.* **112**, 9810-9815, (2015).

2 Discher, D. E. & Eisenberg, A. Polymer Vesicles. *Science* **297**, 967-973, (2002).

3 Discher, B. M. *et al.* Polymersomes: Tough Vesicles Made from Diblock Copolymers. *Science* **284**, 1143-1146, (1999).

4 Kumar, M. *et al.* Highly permeable polymeric membranes based on the incorporation of the functional water channel protein Aquaporin Z. *Proc. Natl. Acad. Sci. U.S.A.* **104**, 20719-20724, (2007).

5 Rakhmatullina, E. & Meier, W. Solid-Supported Block Copolymer Membranes through Interfacial Adsorption of Charged Block Copolymer Vesicles. *Langmuir* **24**, 6254-6261, (2008).

6 Kumar, M. *et al.* High-Density Reconstitution of Functional Water Channels into Vesicular and Planar Block Copolymer Membranes. *J. Am. Chem. Soc.* **134**, 18631-18637, (2012).

7 Kowal, J. Ł. *et al.* Functional surface engineering by nucleotide-modulated potassium channel insertion into polymer membranes attached to solid supports. *Biomaterials* **35**, 7286-7294, (2014).

8 Zhang, X. *et al.* Natural channel protein inserts and functions in a completely artificial, solid-supported bilayer membrane. *Sci. Rep.* **3**, (2013).

9 Erbakan, M. *et al.* Molecular Cloning, Overexpression and Characterization of a Novel Water Channel Protein from *Rhodobacter sphaeroides*. *PLoS ONE* **9**, e86830, (2014).

455879 Construction of Graphene Oxide (GO) Framework on Hollow Fiber Membranes Via Layer-By-Layer Approach for Heavy Metal Removal

Abstract:

A layer-by-layer (LbL) approach was employed to deposit an ultrathin graphene oxide (GO) framework layer on a modified Torlon® hollow fiber support, enabling the composite membrane with superior nanofiltration (NF) performance. The GO framework was molecularly designed following the 2-step method: repeated GO and ethylenediamine deposition and an amine-enrichment modification by hyperbranched polyethylenimine (HPEI). Surface characterizations by SEM, XPS and zeta potential confirmed the formation of the GO framework structure. A 5 to 10 deposition cycle was proven efficient to seal the surface defects, reduce the pore size and narrow down the pore size distribution of the composite membranes. The synergistic combination of the GO framework layer and the Torlon® support can effectively reduce the polymer consumption for the fabrication of traditional integrally skinned NF membranes. The GO/Torlon® composite membrane has rejections higher than 95% towards Pb^{2+}, Ni^{2+}, and Zn^{2+} with a superior water permeability of 4.7 L m$^{âˆ'2}$ h$^{âˆ'1}$ bar$^{âˆ'1}$. The membrane also exhibits excellent long-term performance stability during a 150-h NF test. Thus, the newly developed membrane has great potential for heavy metal removal and may provide useful insights on the fabrication of new generation 2-dimensional (2D) NF membranes.

461795 Analysis of Graphene Membranes and Time-Varying, Stochastic Gas Transport

Abstract:

Single layer graphene is a promising system as a separation membrane. We consider mechanisms and methods of analysis for gas transport through graphene. We also present a detailed analysis of experimental gas permeation data through single layer graphene membranes under batch depletion conditions parametric in starting pressure for He, H_2, Ne, and CO_2 between 100 and 670 kPa. We show mathematically that the observed intersections of the membrane deflection curves parametric in starting pressure are indicative of a time dependent membrane permeance (pressure normalized molecular flow). Analyzing these time dependent permeance data for He, Ne, H_2, and CO_2 shows remarkably that the latter three gases exhibit discretized permeance values that are temporally repeated. Such quantized fluctuations (called â€œgatingâ€ for liquid phase nanopore and ion channel systems) are a hallmark of isolated nanopores, since small, but rapid changes in the transport pathway necessarily influence a single detectable flux. We analyze the fluctuations using a Hidden Markov model to fit to discrete states and estimate the activation barrier for switching at 1.0 eV. This barrier is and the relative fluxes are consistent with a chemical bond rearrangement of an 8â €“10 atom vacancy pore. Furthermore, we use the relations between the states given by the Markov network for few pores to determine that three pores, each exhibiting two state switching, are responsible for the observed fluctuations; and we compare simulated control data sets with and without the Markov network for comparison and to establish confidence in our evaluation of the limited experimental data set.

467795 Graphene and Graphene Oxide Nanoplatelets As Fillers to Reduce Ageing and Improve Permselectivity of PTMSP Membranes

Abstract:

We have determined the effect of physical dispersion of small amounts (1wt%) of nanoplatelets of graphene and graphene oxide to the gas permeation and ageing behavior of poly(1-trimethylsilyl-1-propyne) (PTMSP). PTMSP has extremely large permeability but is subject to ageing, that causes loss of free volume and permeability with time.[1]

The membranes formed with an optimized protocol based on solvent casting with an organic solvent are stable and mechanically robust, although the repeatability of membrane properties is still not optimal, a main factor to be controlled being the solvent evaporation rate.

To analyze the ageing, that is a thickness-dependent mechanism, we fabricated both thin (1-10 microns) supported over a microporous support, and self-standing thick membranes (50-180 microns).

Although the quantitative extent may vary from one sample to another, it has been found that the addition of Graphene Oxide to PTMSP enhances its gas permeability detectably in all cases, despite the small amount added (1 wt%). On the other hand, the effect of the addition of graphene nanoplatelets depends strongly on the preparation protocol and the final sample behavior cannot be controlled.

The ageing was studied with gas permeability tests over time (up to 9 months) and it was concluded that both graphene and graphene oxide nanoplatelets reduce ageing, for both thin and thick samples. The effect of temperature was studied in the range 30-60°C. In the case of He/CO_2 mixture, the samples show a switchable behaviour (He or CO_2 selective) that can be tuned varying the temperature and the filler.

[1]Olivieri, Ligi, De Angelis, Cucca, Pettinau, *Effect of Graphene and Graphene Oxide Nanoplatelets on the Gas Permselectivity and Aging Behavior of Poly(trimethylsilyl propyne) (PTMSP)*, «INDUSTRIAL & ENGINEERING CHEMISTRY RESEARCH», 2015, 54, 11199 - 11211

468193 Chitosan-Graphene Oxide Composite Membranes for Water Filtration

Abstract:

Chitosan (CS) is a natural, biologically-derived polymer in which deacetylated and acetylated units are randomly distributed. CS can be produced by treating crustacean shells such as shrimp crab, and squid shell with alkali sodium hydroxide. The use of CS as a polymer film, membrane, or scaffold has been explored for different applications, including drug delivery, wound healing, tissue engineering, biosensors, and water treatment. Due to its biocompatibility, biodegradability, low toxicity, and antibacterial and hemostatic properties, CS continues to be explored as a key biopolymer and potential low-cost, naturally-occurring alternative to petroleum-based synthetic polymers. Moreover, CS contains amino and hydroxyl functional groups. The functional groups make CS hydrophilic and amenable to adsorption-based dye and heavy metal removal from water. Two critical challenges in CS utilization as a membrane material in water treatment applications are the weak mechanical properties and the solubility of CS in acidic aqueous environments. To overcome these drawbacks, modification methods, including cross-linking strategies and the use of mechanical reinforcement agents (e.g., graphene oxide, hydroxyapatite, cellulose whiskers, clay), can address these two challenges and result in a more robust membrane material. The use of nanoscale additives has shown particular promise. The greater available surface area per unit mass or volume and the unique properties of materials at the nanoscale allow greater mechanical stability and benefit membrane performance, as well as add additional functionality (e.g., reactivity).

Due to its hydrophobic nature, pristine graphene has a tendency to aggregate, which can reduce the available surface area and its usefulness as a nanofiller. To address this challenge, hydrophilic graphene oxide (GO) is produced by chemical modification of graphene, where oxidation causes the addition of hydroxyl, carboxyl, and epoxide functional groups to the basal planes and edges of the graphene sheets. Low cost, high surface area, water solubility, and high negative charge density make GO a notable adsorbent for removal of heavy metal ions and cationic dyes from water.

The oxidative surface modification of GO also enables GO to be a useful, dispersible nanofiller for water filtration membranes, due to the strong interactions between hydrophilic polymer functional groups and GO.

In this talk, we will discuss characterization and performance results for a set of chitosan-graphene oxide composite membranes, where two types of graphene oxide, a nanoscale GO and a granular GO, were evaluated and compared to results for chitosan-only and GO-only membranes. Electron microscopy of membrane surfaces and cross-sections demonstrate significant differences in structural organization and GO layering within the composite membrane materials. Elemental analysis by energy dispersive x-ray spectroscopy identified the presence of aluminum in GO-only membranes and no aluminum in chitosan-GO composites, as is expected from differences in membrane synthesis. X-ray diffraction measurements show that a GO-only membrane has a high level of order with a calculated interlayer spacing of 8.38 Å, which increases slightly to 8.52 Å when wetted with pure water. In comparison, chitosan-GO composite membranes result in broad diffraction peaks and the loss of apparent ordered interlayer spacing upon wetting. X-ray photoelectron spectroscopy results confirmed the presence of aluminum in GO membrane samples, as well as a decrease in amine and an increase in protonated amine and amide contributions in the nitrogen N1s spectrum when GO was added to chitosan. This increase in both amide and protonated amine species suggests oxidation and protonation of the amine groups on the chitosan polymer backbone through interactions with the oxygen-based functional groups on the surface of GO. Membranes were tested for pure water flux and rejection with model water contaminant methylene blue; performance results will be discussed for methylene blue concentrations of 1 â€" 100 mg/L.

469087 Reduced Graphene Oxide (rGO) Membranes for Water Treatment Applications

Abstract:

Graphene oxide (GO) membranes have recently been studied for separations of molecules in nanoscales. With an interlayer spacing of 1.3-1.5 nm, GO nanochannels can reject dyes (rejection> 95%, size: 1.3 nm) depending on the size of dye and can have moderate rejections for divalent ions. It provides a robust active layer for separations as compared to polymeric membranes. However, GO is readily dispersible in water due to the presence of the different oxygen-containing functionalities like carboxylic, hydroxyl, and epoxy on the basal plane and along the edge. GO layer can be peeled off the porous substrate easily under the shear force exerted by water, thus limiting its practical applications. This study focusses on the reduction of graphene oxide membranes and understands its impact on stability in aqueous environment. Catalytic properties of these membranes in advance oxidation are also explored.

GO membranes were synthesized using vacuum filtration for varying loading of GO on porous microfiltration membrane (Nanostone PVDF 200). GO membranes were further chemically reduced using either $NaBH_4$, or ascorbic acid. Reduced graphene oxide membranes (rGO) were more hydrophobic as compared to GO membranes. Characterization of rGO membranes was carried out to understand reduction of oxygen containing functionalities, and to quantify the mechanical strength of the film.

The rGO membrane was observed to be stable in a crossflow membrane setup. The membrane had a rejection of 95% for neutral red dye (1.3 nm diameter) and 35% rejection for $CaSO_4$. GO is known to possess antibacterial properties and thus makes it suitable for applications where biological fouling is severe. We investigated the fouling behavior of these membranes in the presence of Humic acid (HA), and bovine serum albumin (BSA), and compared it with commercial nanofiltration membranes (Nanostone Inc. NF8 and DOW NF270).
GO and rGO also have been reported to have catalytic properties in advanced oxidation processes. This makes these membranes interesting as no transition metal catalyst are involved. We observed that rGO membranes had better efficiency in the decomposition of sodium persulfate (to generate sulfate free radicals). This project is supported by, NSF KY EPSCoR program, and by NIH-NIEHS-SRC, and by Australian Research Council (through Monash University).

469316 Dynamic Microstructure and Compaction of Graphene Oxide Thin Film Membranes in Nanofiltration

Abstract:

Graphene-based materials has attracted a lot of interest for developing ultra-thin and high performance membranes due to their unique two dimensional structure, one atomic thickness and strong mechanical strength. Graphene oxide (GO) membranes made by assembling small GO flakes have shown excellent performance in nanofiltration due to their well-defined interlayer space. However, besides the interlayer distance, the microstructures of GO membranes such as the stacking of GO flakes and the defects within membranes may also affect the membrane performance [1-2]. In our study on GO membranes in nanofiltration, serious membrane compaction was observed, where water permeation flux decreased considerably over time. The compaction of GO membranes could be due to the change in microstructure, where highly order GO laminar microstructure was induced when pressure was applied during the permeation test. However, the disorder of microstructure was reversible and could be restored when GO membranes were dried in air, and compaction reoccurred every time when GO membranes were retested after drying. The change in microstructure was found in GO membranes prepared by most of the synthesis methods (pressure driven or surface tension driven) and can be affected by the conditions where the membranes were stored. Understanding of the dynamic of the microstructures is crucial as it could affect the performance of GO membranes/ thin films in applications where the microstructure is critical. The observation of the change in membrane microstructures also provided important information for the understanding of water transport phenomenon in GO membranes.Â Though the microstructure evolved over time, GO membranes still showed an excellent molecular cut off for nanofiltration, ~250 Dalton after repeated drying-wetting cycles.

[1] Aba, N.F.D., et al., Graphene oxide membranes on ceramic hollow fibers â€" Microstructural stability and nanofiltration performance. Journal of Membrane Science, 2015. **484**: p. 87-94.

[2] Chong, J. Y., et al., UV-enhanced sacrificial layer stabilised graphene oxide hollow fibre membranes for nanofiltration. Scientific Reports, 2015. **5**: 15799.

469607 Tuning the Gas Barrier Properties and Selectivity By Means of Graphene-Based Coatings on Polymeric Film

Abstract:

The peculiar properties of graphene and its derivatives are very attractive for several applications, and significant efforts are devoted to their implementation in various fields. Among the others, the very low gas permeability of the graphene sheet together with the characteristic 2-D structure are very attractive for the fabrication of barrier materials, as well as of gas separation membranes. However, the production of a large and defect-free graphene surface is not practically feasable, and consequently the assembling technique and the method used to couple graphene with e.g. polymeric materials are crucial. Clearly, the resulting material cannot be fully impermeable as a single graphene sheet, and gaseous species are forced to diffuse around graphene sheets. The organization can exploited for gas separation purposes, selectively blocking only some penetrants, as in molecular sieves.

In this work, two different assembly methods are employed to fabricate graphene oxide based coatings on polymer substrates.

First, a low oxygen graphene oxide (G, C/O â‰ˆ8) solution obtained by electrochemical exfoliation was used to produce graphene coatings by vacuum-assisted filtration; the obtained graphene cakes were then transferred, by roll-to-roll or simple press, onto various polymeric substrates, suitable e.g. for packaging, such as polyethylene terephthalate (PET), polyproylene (PP) or polylactic acid (PLA). The coating was firmly attached to the polymer matrix, and the adhesion proved to be very effective resisting also over time, while the distribution of the graphene sheets and their assembly have been observed by Scanning Electron Microscopy measurements. This simple production technique was successfully used to produce low oxygen permeability multilayer films, showing a reduction of the O_2 transfer rate of more than 70% with respect to the bare substrate in case of PET and more than 90% in case of PP and PLA.

A second assembly method involved the use of the layer-by-layer (LBL) technique, to produce alternating ultra-thin multilayer structures to coat the polymeric matrixes. After its proper activation, the substrate is alternatively dipped into polyelectrolyte or graphene oxide water solution, exploiting electrostatic forces to obtain a molecular-thin layer with a high order degree. Highly oxidized graphene oxide (GO, C/O â‰ˆ1) was used as a electronegative-behave species, together with the partially-positive polyethyleneimine (PEI) to built a well ordered, self assembling coating on top of various substrates. The SEM images revealed a quite flat top surface of the coating, in which the GO sheets can be clearly identified, while the analysis of the sample cross section suggested the existence of an alternated structure.

The highly ordered GO structure produced by this method was able to provide very effective barrier systems, leading to a 96% reduction of the O_2 transfer rate on PET, by means of a 25 bilayer translucent coating (whose thickness can be estimated as 200nm). The estimated O_2 permeability of the coating only resulted remarkably low, comparable to the best already-available gas barrier solutions.

The sieving ability of the obtained GO coating was also investigated by means of direct permeability test of CO_2 and smaller gas molecules, such as H_2 and He, allowing the determination of selectivity values. The peculiar assembly produced by the LbL technique, indeed, conferred to the coating a sieving ability based on the probe different molecular sizes, which is suitable for H_2/CO_2 separation, relevant in pre-combustion CO_2 capture applications. Indeed, the evaluation of the coating selectivity returned an astonishing value of 70, proving the validity of such approach. The same method is also applied to another polymer matrix of relevance in membrane field, Matrimid polyimide, characterized by moderate H_2/CO_2 separation performances, obtaining the same results.

Therefore, this technique allows to conveniently tailor the gas transport properties through very thin coating layers characterized by large sieving selectivity values, and moderate resistance towards the diffusion of small probes such as H_2.

470109 Multiscale Graphene Topographies Programmed By Sequential Mechanical Deformation

Abstract:

Multiscale Graphene Topographies Programmed by Sequential Mechanical Deformation

Po-Yen Chen[1,2]*, Robert H. Hurt[2], Ian Y. Wong[2]

[1]*Department of Chemical Engineering, Massachusetts Institute of Technology, Cambridge, MA, 02139*

[2]*School of Engineering, Brown University, Providence, RI 02912*

Email: pychen@mit.edu

Complex surface topographies emerge in ultrathin-layered materials undergoing large mechanical deformations. Yet, the ability to independently engineer feature size and orientational order across multiple length scales remains a challenge. In this study, we demonstrate hierarchical graphene surface architectures generated using various sequences and combinations of extreme mechanical deformation. The method involves multiple cycles of compression driven by thermal actuation of pre-stretched polymer substrates followed by polymer dissolution, graphene film transfer, and recompression. In each generation, the compression can be biaxial (2D) or unidirectional (1D), and the 1D contraction step can be aligned parallel or perpendicular to the previous step, leading to a family of different hierarchical wrinkle/crumple textures (**Figure 1,** the scale bars are 4 micrometers). Analysis of the three-generational genealogy of these graphene films shows that both directionality and sequence play a role in final texture, and that the characteristic length scale of the features increases with subsequent generations. This behavior can be explained through the increase in effective film thickness (which is directly related to feature size) as complex, out-of-plane textures are added in each successive generation. These films show systematic increases in hydrophobicity and electrochemical current density with each successive generation, and the final product of three-generational extreme compression shows superhydrophobicity (static contact angle >160 degree) and high electrochemical activity (20-fold increase in comparison with planar GO film). We believe this texturing concept can be extended to other 2D material films and has potential applications in anti-fouling substrates, stretchable electronics and advanced electrode architectures.

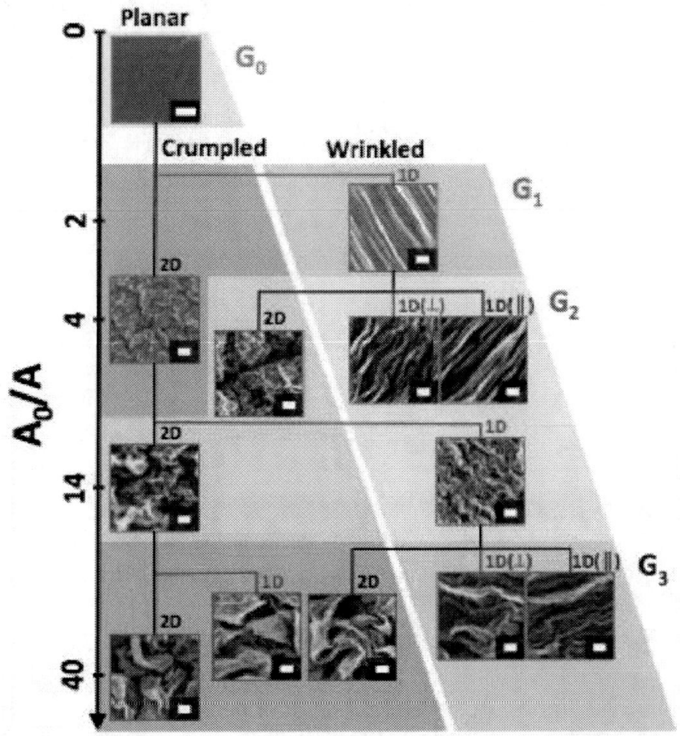

Figure 1. The genealogy of GO multigenerational structures from planar generation 0 (G_0) coatings to multiscale generation three (G_3) structures. A_0 indicates the area of initial planar film; A is the area of multigenerational GO film. All of the scale bars are 4 micrometers.

466704 High-Performance Separation of Cis/Trans Isomers on MIL-125(Ti)

Abstract:

High-performance separation of cis/trans isomers on MIL-125(Ti)

Stijn Van der Perre[1], Ana-Martin Calvo[1], Bart Bueken[2], Dirk E. De Vos[2], Sofia Calero[3], Gino V. Baron[1] and Joeri F.M. Denayer[1]

[1]Department of Chemical Engineering, Vrije Universiteit Brussel, Brussels, Belgium

[2]Centre for Surface Chemistry and Catalysis, Katholieke Universiteit Leuven, Leuven, Belgium

[3]Department of Physical, Chemical, and Natural Systems, Universidad Pablo de Olavide, Sevilla, Spain

In comparison to conventional stationary phases like (functionalized) silica or alumina in High-Performance Liquid Chromatography (HPLC), MOFs can lead to chromatographic separations based on molecular size and shape due to a uniform pore size distribution and ordered pore structure. Hereby, the versatility of organic linkers and metal centers used in MOFs introduce different interactions with the analytes than those obtained with commercial stationary phases in HPLC, which opens possibilities to separate different kind of mixtures.

In this work, the MIL-125 (a Ti-based MOF) framework was used as stationary phase in liquid chromatography to study the separation of different model and stereoisomer mixtures. MIL-125(Ti) has distorted octahedral (12.5 Å) and tetrahedral (6.1 Å) cages connected with triangular windows of around 6 Å in size. Monodisperse MIL-125(Ti) particles were synthesized and slurry packed into a stainless steel column for HPLC measurements.

MIL-125(Ti) allowed separation of various model mixtures in HPLC, but moreover, the structure displays a stereo selectivity for difunctionalized cyclohexane molecules (Fig. 1). This remarkable selectivity could be attributed to the shape selective confinement in the cages of the structure, where the tetrahedral cages act as channels between the octahedral cages, favoring the trans isomer. In a next stage, a Van Deemter curve was measured and intraparticle diffusion coefficients of some components were calculated from peak parking experiments in order to characterize and to determine performance of the crystal packed column, as will be discussed in this work. For a better understanding of the molecular interactions that occur within the chromatographic process, simulation techniques have been used. We performed Monte Carlo simulations to study adsorption properties of the most interesting molecules inside the framework. For this purpose, models of 4-ethylcyclohexanol and 1,3-dimethylcyclohexane isomers have been developed to reproduce characteristic properties of the molecules as well as their interactions with MIL-125(Ti). Adsorption isotherms, selectivity, distribution of the molecules, and Radial Distribution Functions were obtained and discussed.

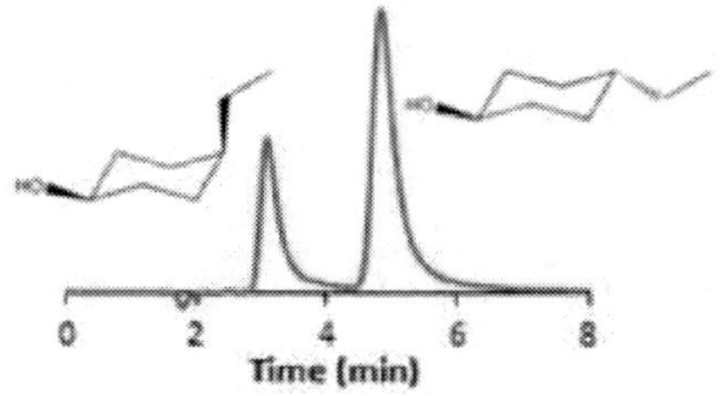

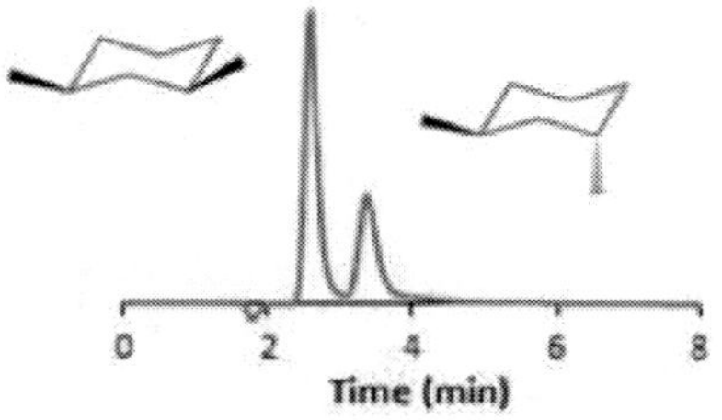

Fig. 1: Chromatographic profiles of cis and trans mixtures of 4-ethylcylohexanol (left) and 1,3-dimethylcyclohexane (right), using a mobile phase of ACN at 303 K and 0.2 ml/min. A preference for the trans isomer was observed.

In conclusion, the separation potential and applicability of MIL-125(Ti) in chromatography was demonstrated via an experimental and theoretical way. This could be a start for the separation of other cis/trans mixtures, giving possibilities for separation of diastereomers, which is of tremendous importance for the production of pharmaceuticals.

Acknowledgements The authors are grateful to IWT Vlaanderen, for financial support for MOFSHAPE "Shaped Metal-Organic Frameworks for Adsorption, Energy and Catalysis Applications".

Separation of Carotenoids by Ultra High Performance Supercritical Fluid Chromatography (UHPSFC)

L. S. Guedes; C. C. Santana; M. M. Beppu; M. C. Breitkreitz; I. C. S. F. Jardim; L. V. Melo

UNICAMP

BRAZIL

2016 AIChE Annual Meeting
November 13-18, 2016

Summary

- Motivation for carotenoids separation.
- Objectives of this study.
- Supercritical fluid and small diameter particles columns : Ultra High Performance Supercritical Fluid Chromatography (UHPSFC).
- Analytical chromatographic separation.
- Experimental design.
- Raw extract chromatography and scaling route.
- Conclusion.

Carotenoids: natural sources and functions

Carotenoids form an important class of natural pigments that are found mainly in plants . Besides the pro-vitamin A activity these compounds have been associated with a wide number of health benefits as antioxidant capacity and help on prevent cancer, diabetes and cardiovascular diseases.

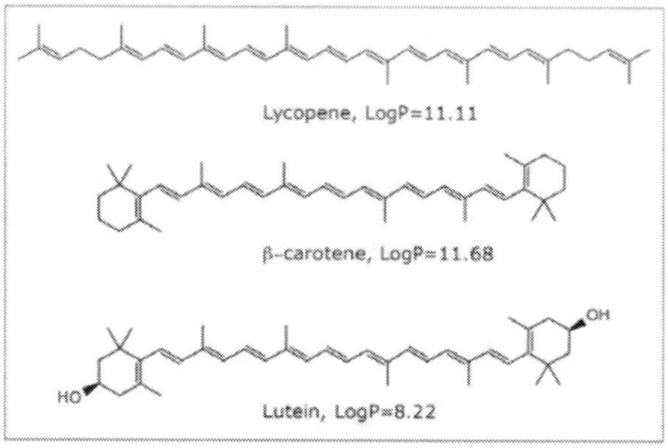

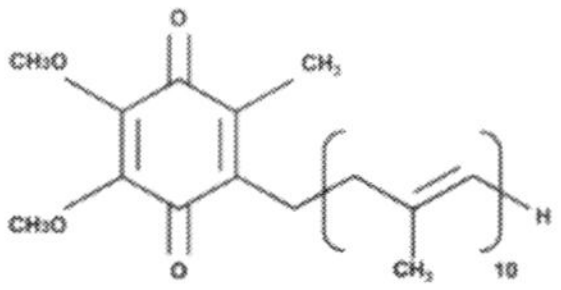

Coenzyme Q10
Log P = 11,00

Strong antioxidant is a fat-soluble, vitamin-like nutrient that is involved in the mitochondrial respiratory chain

Similar solubility parameters P.

Temperature and pressure also play an important role in separation processes.

From palm oil seed to high added value carotenoids: by-product of biodiesel production

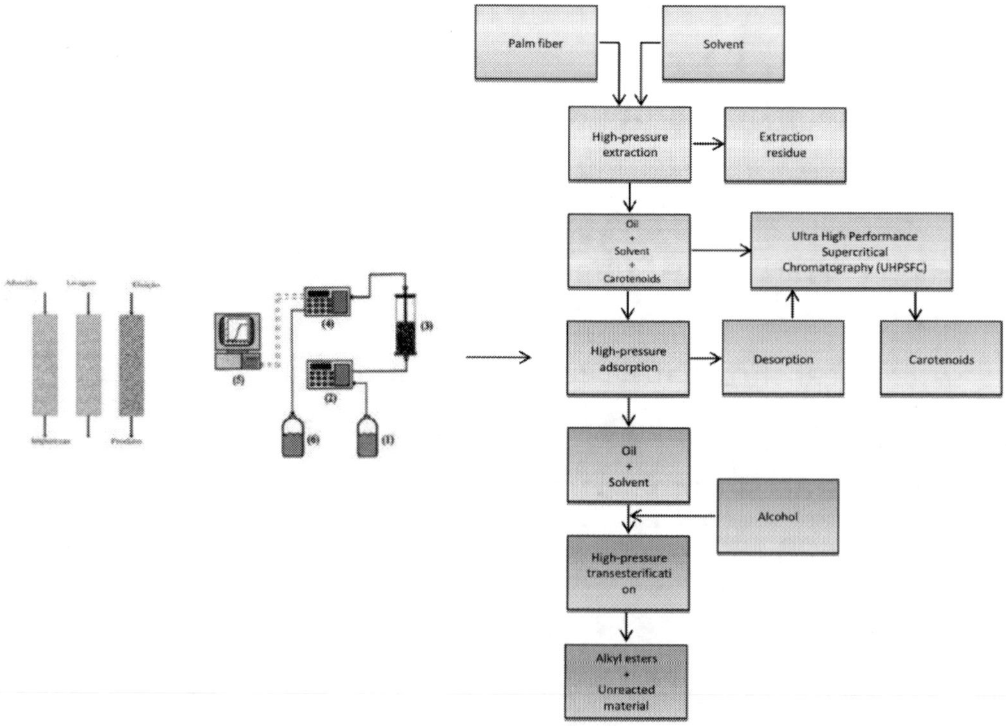

Objectives

- The main objective of this study was to identify the key factors that affect the separation of lycopene, beta-carotene, coenzyme Q10 and lutein by ultra high performance supercritical fluid chromatography.

- Also aimed at identifying the supercritical fluid chromatography operating conditions that could lead to adequate selectivity and retention by using preliminary experiments and response surface methodology.

- Chromatographic aspects of raw extract separation and route to scaling the process.

Supercritical fluid (SF): suitable for separation of thermally unstable compounds

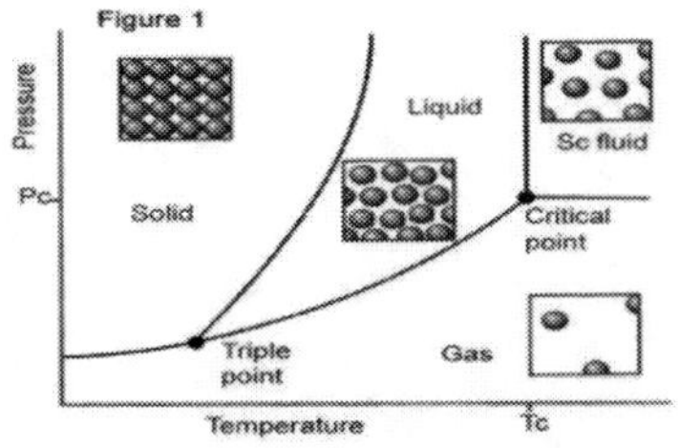

CO_2 critical properties close to room conditions.

Supercritical CO_2 employed as mobile phase in Supercritical Fluid Chromatography (SFC) is environmentally friendly, nontoxic and non-flammable. It is not expensive, has good miscibility with most polar and no polar organic solvents and is easily available. Experiments are very often carried out with a mixture of CO_2 and an organic modifier in order to increase the solubility of polar compounds.

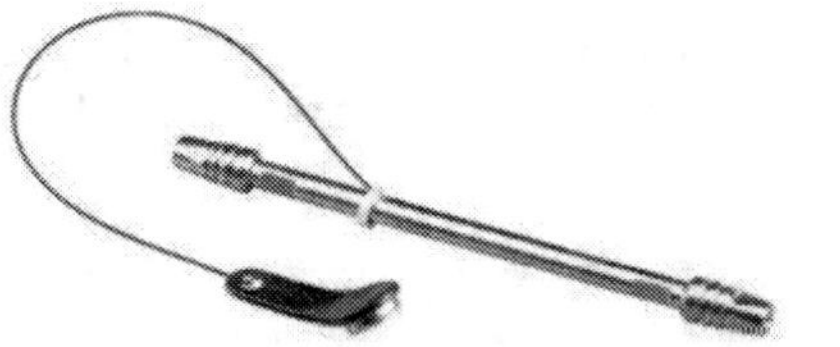

HSS C_{18} SB (3,0 x 100 mm, 1,8 µm) column

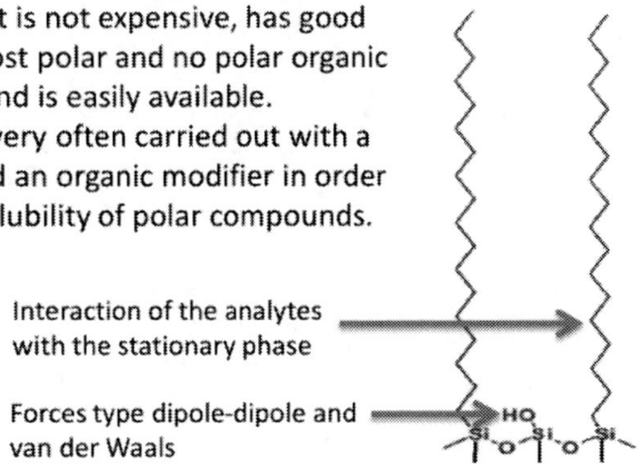

Recently, there has been a revival of interest in SFC due to the introduction of new state-of-the-art instrumentation, in which the pressure control system was improved, proving superior repeatability of the results. This fact, together with the possibility of using sub-2µm particle stationary phases gave rise to the new generation of SFC, named Ultra High Performance Supercritical Fluid Chromatography (UHPSFC).

Advantages of UHPSFC

CO_2 critical properties close to room conditions

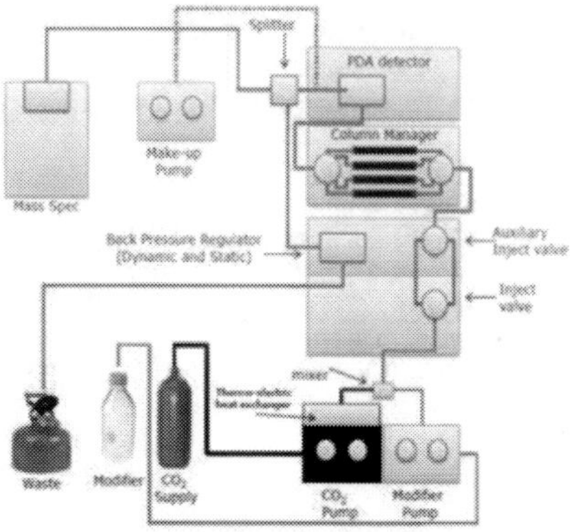

UHPSFC equipment.

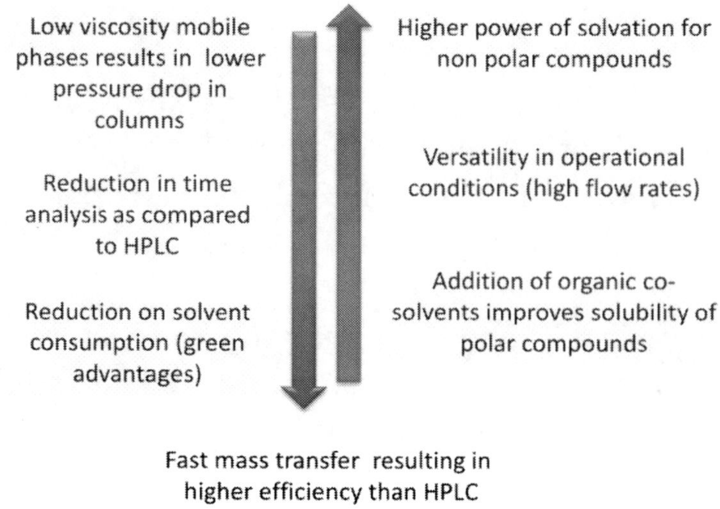

Bernal *et al.* (2013) *J. Chromatogr. A*, 1313, 24
Knöfel (2013) Waters Corporation

Analytical chromatographic separation

$$k = \frac{t_r - t_0}{t_0}$$

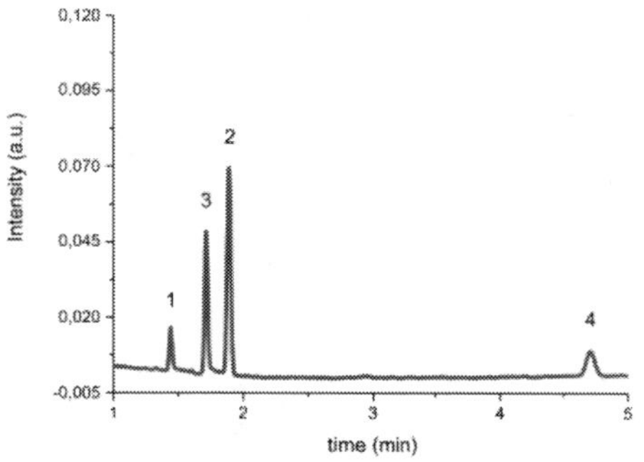

Chromatogram of the mixture (1) lycopene, (2) coenzyme Q10, (3) beta-carotene and (4) lutein. Experimental conditions: 15% of ethanol, 2190 psi and 40 °C.

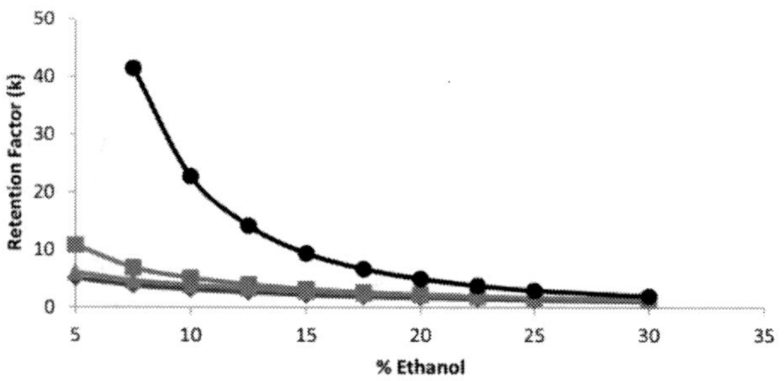

Variation of retention factor for (◆) lycopene, (▲) beta-carotene, (■) coenzyme Q10 and (●) lutein as a function of the ethanol percentage.

Retention factors (k) for the compounds. In the range of 15 to 30% ethanol as co-solvent

Compound	15%	20%	25%	30%
Lycopene	2,17	1,65	1,29	1,09
Coenzyme Q10	3,16	2,21	1,63	1,32
Beta-carotene	2,77	2,21	1,63	1,43
Lutein	9,35	4,94	2,92	1,94

Experimental design: effect of operating parameters

Chromatographic conditions: Waters Acquity UPC[2] system, Acquity UPC[2] HSS C_{18} SB 3.0 x 150 mm 1.8 µm column, sample temperature 5 °C, flow rate 1.5 mLmin^{-1}, injection volume 1 µL. All injections were done in duplicate. Analytes were detected at 275 nm. Statistical analyses were performed by using Design Expert software version 9.0 (Stat-Ease).

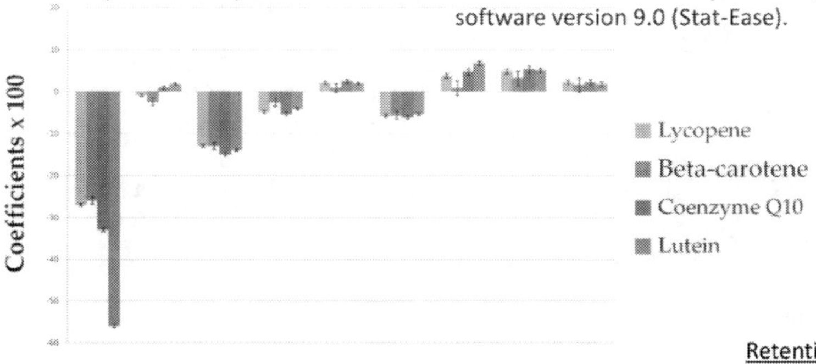

Coefficients of the model fit for the retention factor of lycopene, beta-carotene, coenzyme Q10 and lutein. The error bars represent the 95% confidence interval of the coefficients.

Coded and actual levels of the independent variables used for DOE experiments (Central Composite Design).

Symbol	Variables	Coded levels		
		-1	0	+1
A	Percentage of modifier (%)	15	19,5	24
B	Temperature (°C)	25	37,5	50
C	Pressure (psi)	1500	1850	2200

Retention: The percentage of ethanol (C_E) is the factor that most affects the retention of the compounds, followed by pressure and temperature. Many interaction effects were significant, justifying the multivariate approach used in this study (Figure 1). Some quadratic coefficients were also significant.

Experimental design: effect of operating parameters

It is desirable that all compounds present $k > 2$. The contour maps of the retention factor of lycopene is shown as a function of the temperature and percentage of ethanol. It was observed that lower pressure values favored to find $k > 2$ for all compounds.

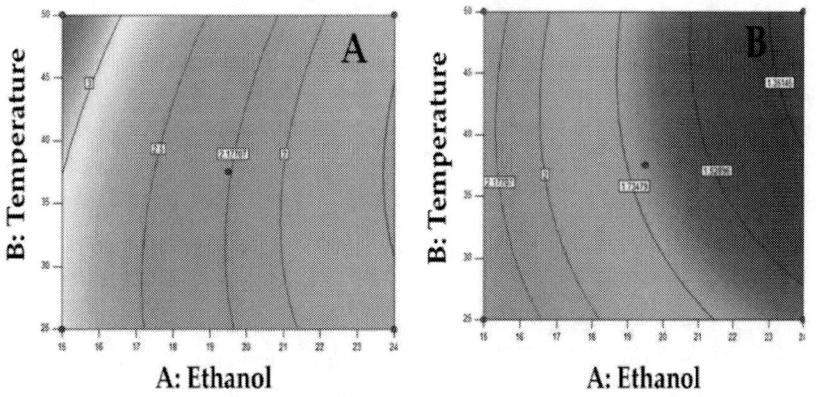

Contour maps of retention factor (k) of lycopene for 1500 (A) and 2200 (B) psi.

Experimental design: effect of operating parameters

Selectivity (α): The separation of the compounds was measured by the separation factor α, which is desirable to be higher than 1.5). The critical separation was of beta-carotene and coenzyme Q10. The contour map showing the separation of these two compounds as a function of the experimental conditions is shown below.

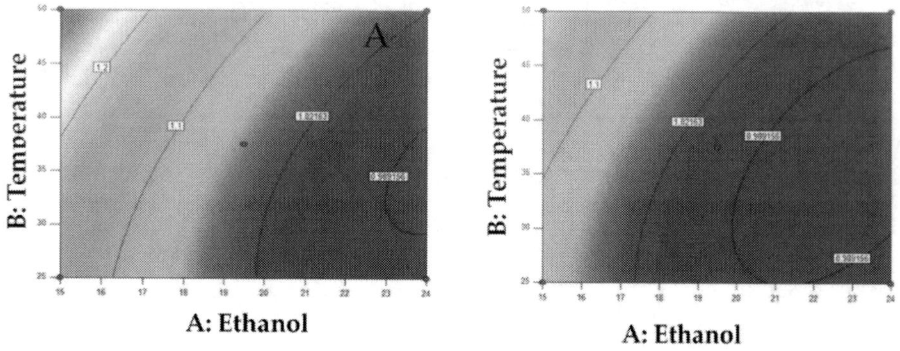

Contour plots for the separation factor of beta-carotene and coenzyme Q10 as a function of temperature and ethanol percentage at 1500 (A) and 2200 (B) psi.

Experimental design: effect of operating parameters

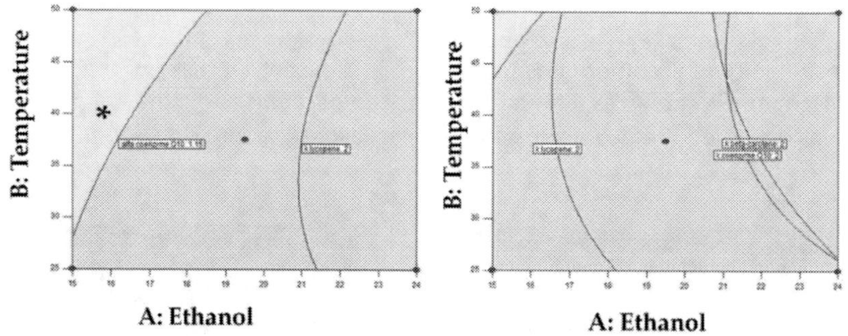

The overlay plot highlighting in yellow the region that fulfils the requirements of retention and selectivity. (A) shows the results at 1500 psi and (B) at 2200 psi. The condition selected for further work is marked with the (*).

Remark - The multivariate approach based on DOE allowed to quantify the influence of the experimental variables on the retention and separation of the bioactive compounds for UHPSFC method development and identify a robust region of work which will be employed for further experiments.

Extraction of raw oil from palm pressed fiber

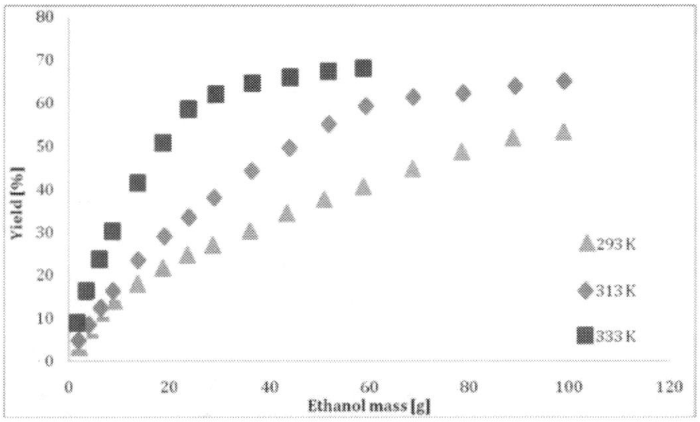

313K, flow rate of 2ml/min and 100 bar

Extraction with compressed propane and ethanol

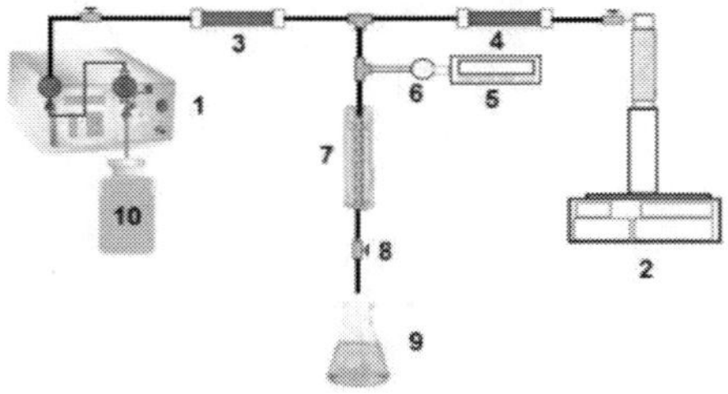

Carotenoids content as beta-carotene (HPLC): 0.63 mg/mL

Oil Extract composition by GC

Fatty acid (as triglycerides)	Concentration (%)
Palmitic	42
Oleic	35.4
Linoleic	9.4
Stearic	4,4
Myristic	0.8
Linolenic	0.3
Lauric	0.1

UHPSFC applied to palm oil extract

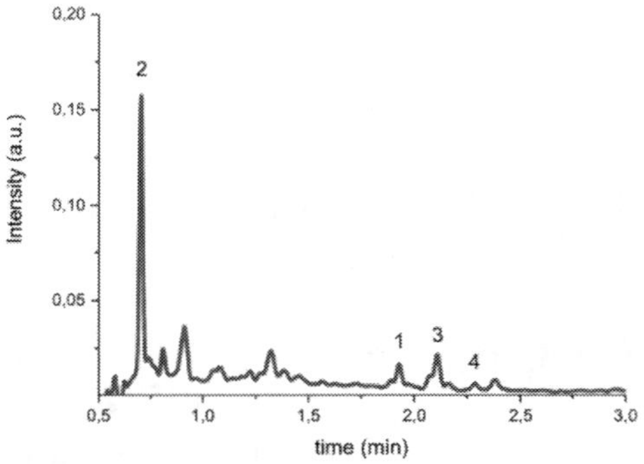

Chromatogram of the palm oil extract (1) lycopene, (2) coenzyme Q10, (3) beta-carotene and (4) lutein. Sample concentration 400 mg/mL.

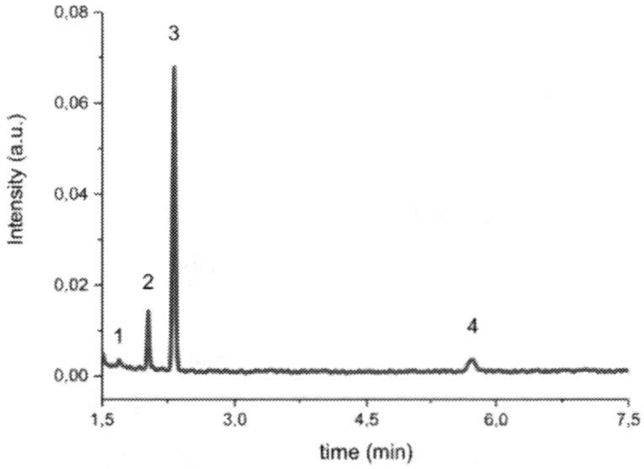

Chromatogram of the mixture (1) lycopene, (2) coenzyme Q10, (3) beta-carotene and (4) lutein. Sample concentration 0.2 mg/mL.

Experimental conditions: 15,5% of ethanol, 1500 psi and 40 °C.

Raw extract chromatography and scaling route

Effect of pressure: variation in the retention factor k with UHPSFC (isopycnic method), (Tarafder and Guiochon, 2011, 2015)

Thermodynamic aspects for scaling: effect of density

$$\left(\frac{\partial \ln \rho k_i}{\partial P}\right)_T = -\lambda_i \left(\frac{\partial \ln \rho}{\partial P}\right)_T$$

$$\ln(k_i) = -(\lambda_i + 1)\ln(\rho) + \ln(f_i(T))$$

Modified Van Deemter equation : Bartmann e Schneider 1973

$$H = B\,\frac{\bar{u} d_p \rho}{\eta} + \frac{1{,}3 D_m}{\bar{u}} + \frac{1{,}5k}{(1+k)^2}\frac{\bar{u} d_p}{D_m} F$$

$u = Q/S$

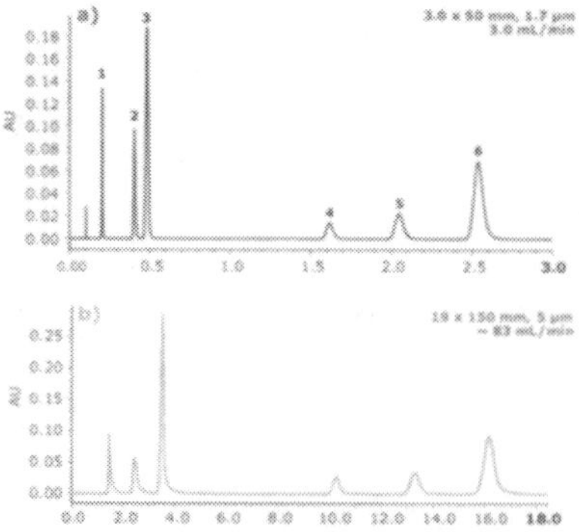

Darcy's law

$$\frac{dP}{dz} = -\frac{\eta}{\rho}\frac{Q}{KS}$$

$$\bar{u} = Q/S$$

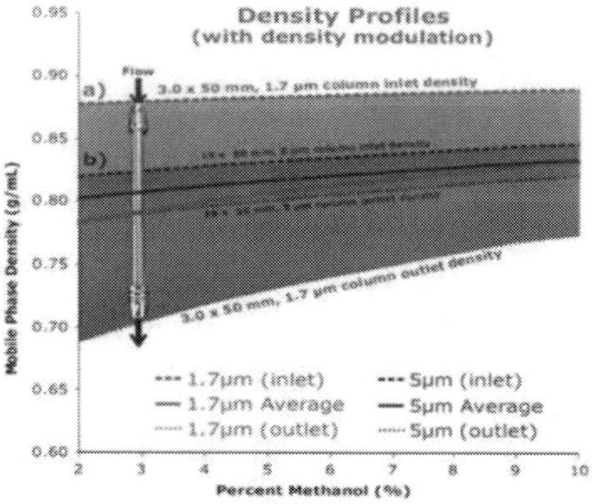

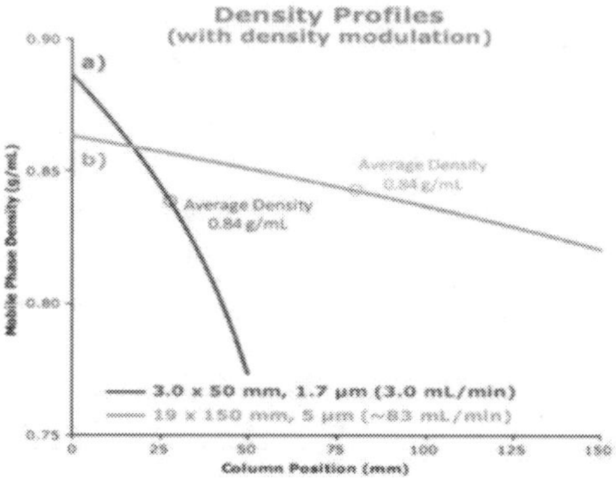

Conclusion

- Ultra High Performance Supercritical Fluid Chromatography (UHPSFC) is na efficient separation technique for carotenoids separation reducing analysis time and solvent consume.

- Co-solvent ethanol plays an important role on separation factors and selectivity.

- A multivariable approach planning allowed to quantify the influence of experimental variables on the retention selectivity of carotenoids separation by UHPSFC method development and to identify the robust region of work.

- Separation of carotenoids from raw extract by UHPLC is a promising technique to get high added value compounds.

Acknowledgments

School of Chemical Engineering
Department of Materials and Bioprocesses

Institute of Chemistry
Department of Analytical Chemistry

UNICAMP

CNPq
Conselho Nacional de Desenvolvimento Científico e Tecnológico

464280 Experimental Realization of Single-Column SMB Chromatography with Recycle Lag

Abstract:

In previous work (AIChE J. 2005;51:1641-1653) we demonstrated that the periodic state of the simulated-moving-bed (SMB) process is reproduced by a single-column chromatographic process with a recycle lag of (N-1)Ã—Ï„ time units, where N is the number of columns of the equivalent SMB unit and Ï„ is the switching interval. The operation of this ideal single-column process is formulated by selecting an arbitrary column of the SMB and following its operation over a complete cycle. To implement the recycle lag in practice, a special type of plug-flow tube was proposed. It includes internal elements to make the flow as close as possible to plug flow, and a piston to compensate for the difference between the inlet and outlet flow rates. This is necessary because the recycle lag is not a multiple of the overall cycle duration. The proposed system is a more compact, less expensive, and simpler-to-operate alternative to the SMB. Depending on the efficiency of the recycle tube, it wa s shown that the single-column process can achieve the same purities as the analogous SMB unit while keeping the specific productivity constant.

Here we present an experimental realization of the single-column process with recycle lag that uses a simplified version of the plug-flow device. The experimental proof of concept uses the separation of nucleosides by reverse phase and protein capture by protein-A affinity chromatography as case studies to demonstrate the effectiveness of the proposed chromatography apparatus.

471539 Speedy Standing Wave Design, Optimization, and Scaling Rules of Simulated Moving Bed Systems with Linear Isotherms

Abstract:

SMBs consume orders of magnitude less solvent, require an order of magnitude less sorbent, take up less space, and require less manpower than batch chromatography operations. These advantages make SMBs economical for large-scale separations. SMBs, however, have not been widely used for large-scale production. One key barrier is the complexity of SMB design and optimization. A four-zone SMB for a binary, linear adsorption system has 23 variables, which include nine material properties and 14 design parameters (two yields, five operating parameters, and seven equipment parameters). Trial and error design of the 14 parameters is not feasible. Furthermore, scaling rules of SMB systems have not been well established. The most commonly used method for scale-up is to design and test the SMB at lab scale and then simply increase the column diameter to meet the production requirements, while keeping the same column length and the fluid zone velocities for the production scale SMB. This method, however, does not guarantee an optimal large-scale process.

In a previous study, we developed for size exclusion SMB (SEC-SMB) systems the analytical expressions of solvent consumption and sorbent productivity as a function of key dimensionless groups. The resulting Speedy Standing Wave Design (SSWD) equations could be used to quickly optimize SEC-SMB design parameters to achieve maximum productivity, minimum solvent consumption, or minimum cost if cost functions are known.

In this study, the SSWD is extended to systems with linear adsorption isotherms, which can result from various separation mechanisms, such as size-exclusion, ion-exclusion, or complexation. Moreover, in deciding whether simulated moving bed (SMB) is technically or economically feasible for a separation, many nonlinear systems are first approximated as linear systems for preliminary studies. Furthermore, general scaling rules based on the key dimensionless groups in the SSWD equations are also developed. The SSWD equations are simplified for diffusion-controlled systems. The separation of acetic acid from glucose in biomass hydrolysate is used as an example to illustrate the effects of various material properties and key dimensionless groups on sorbent productivity and solvent consumption. The SSWD equations are also used in a search routine, along with cost functions, to determine the SMB design with the lowest separation cost for this example.

The SSWD equations also allow estimation of the maximum SMB productivity based on intrinsic material properties and specified yields. For an isocratic system, the desorbent flowrate is always over two times the feed flowrate for the maximum productivity design. The optimal column configuration for either maximum productivity or minimum solvent consumption is controlled by the ratio of the effective diffusivities of the two components and the selectivity. Optimal large-scale SMB process can be found using the SSWD method if the intrinsic parameters are known. For lab-scale testing of a scaled-down process, one can use smaller particles to speed up the testing process, if all other material properties are the same. By keeping the values of key dimensionless groups constant, the lab-scale SMB experiments can be performed four times faster using particles with ½ the diameter, and the lab-scale SMB will have the same dimensionless solvent consumption and productivity as the plant-scale SMB.

465205 Simultaneous Control of Product and Feed Streams for High Performance in Simulated Moving Bed Chromatography

Abstract:

Simulated Moving Bed (SMB) Chromatography has been considered as an important separation process due to its effective separation efficiency. Therefore, it has been widely used in various industries including petrochemicals, fine chemicals, pharmaceuticals and biomaterials. To improve the SMB performance, many operating strategies using stream flow control have been reported. However, they often generate significant flow and pressure fluctuations during a switching period. Since such harsh flow and pressure variation can lead to damaging packed adsorbents and system, some operating conditions canâ€™t be acceptable for real fields even though theoretical separation performance is high.

In the study, a novel operating strategy, SimCon operation, was developed to improve SMB performance. The SimCon operation controlled simultaneously product outlet streams and a feed inlet stream to reduce the flow and pressure fluctuations. It consisted of initial, middle, and last steps in a switching period. The extract and feed ports were closed in an initial step and the raffinate port was closed in a last step. Due to three sub-divided steps in a switching period, two additional operating variables were introduced: middle step time and middle length. In the SimCon operation, the control of feed inlet stream could keep total pressure drop from exceeding the maximum allowable pressure drop due to lessening the internal flow rate fluctuations by the control of product outlet streams. The performance of SimCon operation was compared with the corresponding conventional SMB and other stream control strategies in terms of purity, recovery, productivity and eluent consumption. The SimCon operation showed better performance parameters even under maximum allowable pressure drop condition. The SimCon operation was proved as one of practically applicable operating strategies with high performance.

465219 Transesterification of Propylene Glycol Methyl Ether Using Heterogeneous and Homogeneous Catalyst in Simulated Moving Bed Reactor

Abstract:

Reactive chromatography is an integrated process that combines reaction and separation in a single unit that leads to a higher productivity. This process is especially advantageous when the reaction is equilibrium limited where in-situ separation of product shifts the equilibrium in the direction of conversion increase. A continuous reactive separation process, simulated moving bed reactor (SMBR), utilizes this principle in a train of multiple chromatographic columns that are packed with an adsorbent that also acts as a catalyst.

In the present work, we study SMBR for the catalytic synthesis of propylene glycol methyl ether acetate using two types of catalysts: AMBERLITE™ IRA-904 and sodium alkoxide. Heterogeneous catalyst, AMBERLITE™ IRA-904, has an advantage of not requiring the separation of catalyst from the product. On the other hand, the homogeneous catalyst, sodium alkoxide, shows a higher activity for the transesterification and also deactivation of catalyst is prevented by continuously feeding fresh homogeneous catalyst into the system. The benefits of each catalyst are discussed with comparisons throughout the study.

Case studies of process development for the ester product are discussed using heterogeneous and homogeneous catalysts. For heterogeneous catalysis, AMBERLITE™ IRA-904 is employed that acts both as catalyst and adsorbent. When homogeneous catalyst was used, AMBERLITE™ IRA-904 in an inactive ionic form was used as a packing material that functions as an adsorbent. Equilibria and kinetics of the reaction and adsorption are investigated by carrying out batch reaction experiments and chromatographic pulse tests. Model parameters are obtained from batch experiments by the inverse method, where a transport dispersive model with a linear driving force for the adsorption rate is used for modeling. Using parameters obtained from the experiments, the SMBR is designed and optimized considering multiple objectives—to maximize the production rate of DOWANOL™ PMA glycol ether acetate, to maximize the conversion of the transesterification reaction, and to minimize the consumption of the excess reactant, DOWANOL™ PM glycol ether, which also acts as the desorbent in the chromatographic separation. The two SMBR processes are compared systematically at the optimal operating conditions.

™Trademark of The Dow Chemical Compay

466619 Cellulosic Stabilized Amorphous Solid Dispersions of HMX and CL-20

Abstract:

Amorphous solid dispersions of small organic molecules offer a number of advantages in comparison to their crystalline counterparts for practical application. When considering pharmaceuticals, glassy materials exhibit enhanced bioavailability, making them beneficial both for cost and effectiveness. In the field of energetics, amorphous solid dispersions of explosive compounds are desirable because they display exciting properties such as optical transparency, enhanced reactivity, and glass castability. While possessing useful properties, stability of amorphous solid dispersions is problematic, as the form is necessarily strained, and thus of higher energy than the ordered crystalline state. Stability can be achieved by using polymeric excipients, which inhibit crystallization, and in this application, cellulosics have been found to be particularly effective. In this work, the use of cellulosic polymers to inhibit the crystallization of a mixture of octogen (HMX) and hexanitrohexaazaisowurtzitane (CL-20) is explored. The difference in stability of nitrocellulose, cellulose acetate, and cellulose acetate butyrate stabilized dispersions of HMX and CL-20 is elucidated. Of particular interest is crystallization kinetics and glass transition temperature. These amorphous solid dispersions are characterized with the following techniques: X-Ray diffraction, Raman spectroscopy, thermomechanical analysis, differential scanning calorimetry. Overall, nitrocellulose has been found to be particular effective for inhibiting crystallization, which can be tentatively attributed to its moieties which match that of HMX and CL-20.

V. Stepanov, R. B. Patel, R. Mudryy, H. Qiu, Investigation of Nitramine-Based Amorphous Energetics, Prop. Exp. Pyro. 2016, 41, 142-147. DOI: 10.1002/prep.201500118

R. Knepper, K. Browning, R. R. Wixom, A. S. Tappan, M. A. Rodriguez, M. K Alam, Microstructure Evolution during Crystallization of Vapor-Deposited Hexanitroazobenzene Films, Prop. Exp. Pyro. 2012, 37, 459 â€“467. DOI: 10.1002/prep.201200013

N. Shamin, Y. P. Koh, S. L. Simon, G. B. McKenna, The glass transition of trinitrotoluene (TNT) by flash DSC, Thermochimica Acta, 2015, 36-39. DOI: 10.1016/j.tca.2015.10.003

466105 Spray Drying of Ternary Solid Dispersions of Itraconazole: Amorphous Stability By Design of Experiment

Abstract:

SPRAY DRYING OF TERNARY SOLID DISPERSIONS OF ITRACONAZOLE: AMORPHOUS STABILITY BY DESIGN OF EXPERIMENT

Mark T. Davis[1]*, Catherine B. Kelly[1], Ahmad B. Albadarin[1], Maryam Mohammadpour[1]â€Ž
Mustafa Sajjia[2] & Gavin Walker[1]

[1]Synthesis and Solid State Pharmaceutical Centre (SSPC), Bernal Institute,
University of Limerick, Limerick, Ireland
[2]Pharmaceutical Manufacturing Technology Center (PMTC), University of Limerick, Limerick, Ireland
*Tel.: +353-(0)-61- 213096, E-mail: mark.davis@ul.ie

Abstract

Itraconazole is a BCS Class II drug with very limited aqueous solubility (<1 Âµg/ml) and poor dissolution kinetics. Many studies have been conducted in an effort to produce formulations of this drug with higher bioavailability. In this context, we have examined ternary amorphous solid dispersions (ASD) of itraconazole with excipient polymers Soluplus® and HPMCP. Evidence exists to suggest that ternary formulations can be more stable and have better release profiles than binary systems alone. Soluplus® is a structurally diverse excipient with excellent solubility and melt extrusion characteristics. HPMCP is used as a complimentary excipient, owing to the phthalate groups which help protect against gastric juices.

In this study we have taken a design of experiments (DoE) approach to optimise ternary formulations of itraconazole-Soluplus-HPMCP, producing 17 outcomes. A pre-formulation solubility test was run to ensure complete solubility of each combination. The solutions were then converted into powders by spray drying in a Büchi B-290 mini spray drier, equipped with a pneumatic nozzle, inert loop condenser and high efficiency cyclone. Yields were consistent and could be increased to as high as 83% by varying parameters.

The powders were split into three parts and stored at (i) RT and 0% RH, (ii) 30 °C and 45% RH and (iii) 40 °C and 75% RH for up to one year. PXRD measurements revealed samples to be amorphous at day 0 and importantly still amorphous after a year of storage. Low angle XRD was used to ensure that samples had not formed a meso phase, which is possible with solid state itraconazole. Differential scanning calorimetry (DSC) confirmed that the samples were amorphous and demonstrated how the glass transition temperature varied with content. Particle size distribution measurements was carried out on using laser differaction and the data was analysed by cumulative and histogram plots. A statistical analysis revealed a relationship between PSD and formulation content. This result was rationalised by hydrogen bonding in HPMCP, the only component capable of self-association.

455766 Engineering the Mechanical Properties of Amorphous Solid Dispersion Particles

Abstract:

Spray-dried amorphous solid dispersions (SDDs) is one of the most robust and mature available technologies to improve oral absorption of drug candidates with low bioavailability due to poor aqueous solubility and dissolution rate. As an active intermediate of the drug substance, SDDs are readily integrated into oral solid dosage forms, the most common being tablets. However, due to the dilution of the active ingredient in the SDD and then further dilution in the tablet formulation, pill burden can become a problem for less potent drugs. One possible solution is to engineer improved SDD particle mechanical properties through the spray drying process thereby reducing the amount of tablet fillers and improving the robustness of the tableting process. This concept of SDD particle property optimization is a multifaceted optimization problem between physical state and chemical stability of the drug, downstream processing of the resulting powder, and in vivo performance. Although this talk focuses on one aspect of downstream processing, the mechanical properties specifically related to tableting, there have been several recent reviews on this subject [1-2]. The current pharmaceutical literature is rich with information on engineering SDD particles, but there lacks the connection to how these engineering principles influence the compression-related mechanical properties.

The aim of the current study is to make that connection by engineering the mechanical properties of SDD particles through the spray-drying process. Ketoconazole spray-dried with HPMCAS-M or PVP VA-64 at varying concentrations was used as a model SDD system. Atomization and drying kinetics were adjusted to vary surface area by altering particle size distribution, particle density, and morphology. The mechanical properties were studied using compression profiles, SEM imaging, and yield pressure determination with a modified in-die compression analysis [3]. The results showed that the compression pressures required to reach a 2.0MPa tensile strength compact varied with

SDD composition and were in the range of typical fillers. Yield pressures also varied with drug loading and polymer selection which could be reasonably predicted by the pure material behaviors composing the SDD particles. The deformation and tensile strength mechanisms were further evidenced by the SEM micrographs. The results suggest that a tablet formulation using rationally engineered particles could reduce the amount of fillers needed to decrease pill burden.

[1] Patel, B.B. et al. Revealing facts behind spray dried solid dispersion technology used for solubility enhancement. *Saudi Pharm. J.* **2015**, *23 (4)*, 352-365.

[2] Singh, A. and Van den Mooter, G. Spray drying formulation of amorphous solid dispersions. *Adv. Drug Deliv. Rev.* **2015**.

[3] Katz, J. M., Roopwani, R., Buckner, I. S. A Material-Sparing Method for Assessment of Powder Deformation Characteristics Using Data Collected During a Single Compression-Decompression Cycle. *J. Pharm. Sci.* **2013**, *102 (10)*, 3687-3693.

466729 Investigation of Phase Separation Phenomena in a Supersaturated Ternary Amorphous Solid Dispersion

Abstract:

Investigation of phase separation phenomena in a

supersaturated ternary amorphous solid dispersion

Catherine B. Kelly[1*], Ahmad B. Albadarin[1] & Gavin M. Walker[1]

[1]Synthesis and Solid State Pharmaceutical Cluster (SSPC), Bernal Institute,

University of Limerick, Limerick, Ireland *Tel.: +353-(0)-61-237732, E-mail: Catherine.Kelly@ul.ie Abstract

Many new chemical entities (NCEs) are failing to progress through drug-development pipelines due to a display of insufficient bioavailability, stemming from the extremely low aqueous solubility properties of many of these materials. The overwhelming majority of these NCEs are small-molecule, crystalline solids so lend themselves to formulation in the higher solubility but thermodynamically unstable amorphous form. A common approach to the formulation of amorphous APIs is through the use of an amorphous solid dispersion. In order to stabilise the thermodynamically unstable amorphous form an excipient to which the API exhibits an affinity through molecular interactions is added during formulation.

Recent publications have suggested that even in cases in which interactions between an API and a biocompatible polymer has been detected, in order to produce an acceptable solid dosage form the amorphous API could be present at a loading above that which is thermodynamically stable. At the required higher drug loadings successful long-term stability appears to be achieved through a combination of thermodynamic and kinetic mechanisms.

Additionally, the action of the excipients in an amorphous solid dispersion is twofold. As well as the stabilisation of the amorphous form in the solid state, the excipients should also be capable of maintaining the supersaturated concentration of the API which occurs during dissolution due to the presence of the amorphous form. Due to the dual-action properties required of the polymeric excipient the use of two excipients is proposed, with this study investigating the first mechanism, namely the stabilisation of the amorphous form in the solid state.

Production of the drug-polymer temperature-composition phase diagram

Itraconazole (ITR) was used as a model poorly aqueous soluble drug and Soluplus and hydroxypropyl methylcellulose phthalate (HPMCP) were investigated as compatible excipients for the formulation of a ternary amorphous solid dispersion. The value of the Flory-Huggins interaction parameter and its dependence on temperature was determined for Itraconazole with each polymer as well as a 1:1 blend of both polymers using the dissolution end point method. This method determines the liquid-solid boundary in the API-polymer binary system by measuring the endset of melting during a heating ramp in a differential scanning calorimeter (DSC). The melting point value gained from this method was further investigated using a 16-hour annealing method which indicated that the liquid-solid curve had been underestimated for both polymers. The annealing method involved the use of hyper DSC and a heating rate of 100°C/min to search for the presence of residual crystalline API following a 16 hour hold at or below the temperature gained using the dissolution end point method.

It is known that the dissolution end point method of determining the liquid-solid curve in high viscosity systems such as these can overestimate the drug melting temperature, however the greater accuracy associated with the annealing method should be considered against the much greater period of time required to gain a phase diagram data point.

Investigation of phase separation

Following production of the three phase diagrams a number of amorphous solid dispersions with a range of drug-polymer compositions were produced through ball milling and a subsequent melt-quench. Hyper DSC was used to measure the temperature and width of the glass transition and confirm the absence of crystalline API. Itraconazole is also known to form a liquid-crystal state but the thermal events associated with this state were not apparent in the any of the thermograms of the solid dispersions. The dispersions were stored at 100°C which is within the region of the phase diagram where spontaneous spinodal decomposition should occur (Figure 1). After one week phase separation had occurred in all samples but the development of two distinct demixing mechanisms was apparent. At higher drug loadings demixing occurred by a release of the API by recrystallisation. At lower drug loadings the amorphicity of the formulations was maintained, however two distinct amorphous phases appeared, identified by two separate glass transition events. The change in demixing from recrystallisation to phase separation was observed in all of the three systems examined but shifted in drug-polymer ratio depending on the polymer(s) involved.

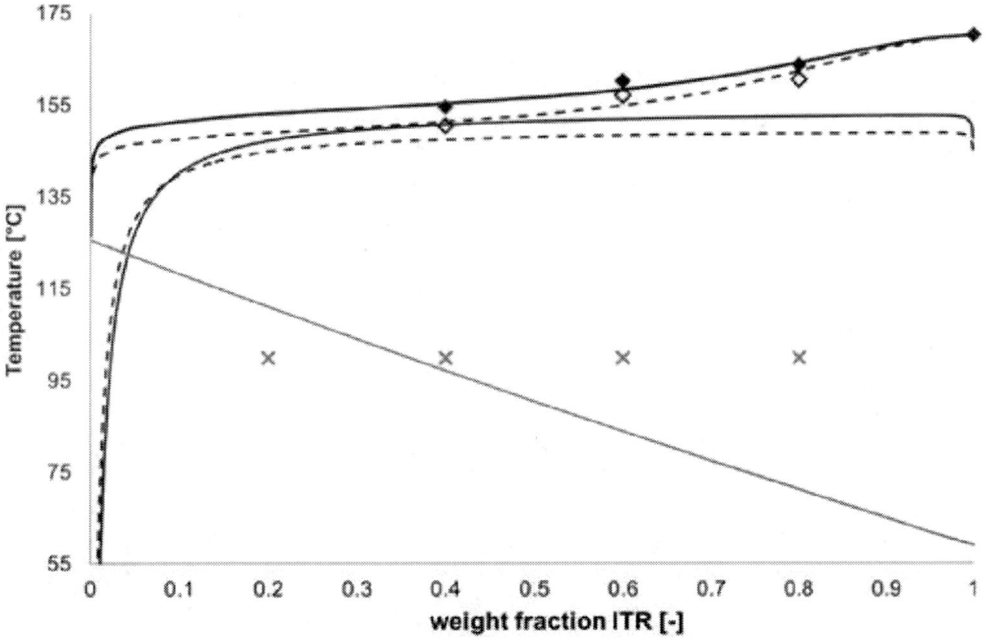

Figure 1: Phase diagram of the ITR-HPMCP-Soluplus system indicating the boundaries identified using the dissolution end point (full line) and annealing (dashed line) methods and the points at which phase separation was investigated (crosses). The upper boundary marks that between liquid and solid API while the lower is the spinodal curve.

473925 Effects of Intermediate Spray-Drying Conditions on Downstream Processing of Pharmaceutical Solid Dosage Formulations

Abstract:

As enabled formulations continue to assume an expanded role in the pharmaceutical industry, a better understanding of how spray-dried particle characteristics influence granulation and tablet compression are necessary to enable reproducible results across spray-drying scales. In this effort, drying kinetics were varied during the spray drying process to produce a spray-dried intermediate (SDI) with varying particle properties. One property, particle density, has been shown to strongly influence the compaction properties of oral solid dosage formulations containing high SDI compositions. It was observed that responses to uniaxial compaction could be successfully employed to project granulation performance at pilot and commercial scale manufacture. Tablet hardness was found to be inversely proportional to SDI particle density, which translates to significant differences in dissolution response. For bilayer tablets compressed with this SDI, extent of granulation and SDI particle density enhance interfacial adhesion. These findings enable flexibility in future spray drying campaigns by establishing how measureable SDI properties impact downstream processing and quality attributes.

456469 A Multidimensional Population Balance Model for Predicting Crystal Size and Aspect Ratio in the Production of Phosphogypsum

Abstract:

A Multidimensional Population Balance Model for Predicting Crystal Size and Aspect Ratio in the Production of Phosphogypsum

Zhilong Zhu

You Peng

Kamal Samrane

Allan S. Myerson

Richard D. Braatz

Crystals of inappropriate crystal size distribution (CSD) and shape can cause problems in downstream processes [1]. Many recent efforts have been directed towards modeling CSD and shape using a multidimensional population balance model (mD-PBM) that can be used to optimize crystal shape [2]–[7]. The model equation is a hyperbolic integro-partial differential equation, which must be solved numerically except for some simple cases. Due to the advection term in the mD-PBM, most discretization methods experience numerical diffusion. Though more advanced schemes such as the high-resolution finite-volume method have demonstrated the ability to solve mD-PBM with much higher numerical accuracy [8], the computational cost of these methods is still not favorable for applications in process optimization and model predictive control. The method of characteristics (MOC) has demonstrated to be computationally efficient and accurate in handling the advection term in the 1D-PBM [9]–[11]. This presentation extends the MOC for solving mD-PBM with both growth and nucleation for the reactive crystallization of phosphogypsum.

During the production of phosphoric acid by the Wet Process, gypsum crystal is produced as a byproduct in the crystallizer. The phosphoric acid is separated from the gypsum by passing the slurry to the filter, which is the bottleneck process due to the undesired gypsum CSD and shape. A two-dimensional PBM with simultaneous growth and secondary nucleation is developed to model the crystal size and its varying aspect ratio in a continuous crystallizer. The solution supersaturation is estimated using the state-of-art Mixed Solvent Electrolyte (MSE) thermodynamic model [12], which models both the solution nonideality and speciation. In the presence of metal impurities, the growth rate in each direction is inhibited differently and results in a change in aspect ratio. The relative change in growth rate due to the added impurity is modeled by growth inhibition model [13]. The presented model is the first to combine the 2D-PBE, an impurity inhibition model, and the MSE model in the study of reactive crystallization under high ionic strength as found in the industrial process.

The proposed MOC method solves the 2D-PBM combined with the inhibition and the MSE model. Compared to existing PBM solvers, the MOC algorithm offers a computationally efficient and highly resolved method, especially in higher dimensions. The simulation results with experimentally measured crystal kinetic parameters show a good agreement with the measured CSD and mean aspect ratio. These simulation results also suggest that the presence of metal ion impurities cannot be ignored as they can modify crystal shape. This information would not be captured in a 1D-PBM for gypsum crystallization.

References:

[1] J. W. Mullin, *Crystallization*. Woburn, MA: Butterworth-Heinemann, 2001.

[2] R. Gunawan, D. L. Ma, M. Fujiwara, and R. D. Braatz, "Identification of kinetic parameters in multidimensional crystallization processes," *Int. J. Mod. Phys. B*, vol. 16, pp. 367–374, Jan. 2002.

[3] Z. K. Nagy and R. D. Braatz, "Advances and new directions in crystallization control," *Annu. Rev. Chem. Biomol. Eng.*, vol. 3, pp. 55–75, Jan. 2012.

[4] F. Puel, G. Févotte, and J. P. Klein, "Simulation and analysis of industrial crystallization processes through multidimensional population balance equations. Part 1: A resolution algorithm based on the method of classes," *Chem. Eng. Sci.*, vol. 58, no. 16, pp. 3715–3727, Aug. 2003.

[5] M. Jiang, X. Zhu, M. C. Molaro, M. L. Rasche, H. Zhang, K. Chadwick, D. M. Raimondo, K.-K. K. Kim, L. Zhou, Z. Zhu, M. H. Wong, D. O'Grady, D. Hebrault, J. Tedesco, and R. D. Braatz, "Modification of crystal shape through deep temperature cycling," *Ind. Eng. Chem. Res.*, vol. 53, no. 13, pp. 5325–5336, Apr. 2014.

[6] C. Borchert, N. Nere, D. Ramkrishna, A. Voigt, and K. Sundmacher, "On the prediction of crystal shape distributions in a steady-state continuous crystallizer," *Chem. Eng. Sci.*, vol. 64, no. 4, pp. 686–696, Feb. 2009.

[7] K. Sato, H. Nagai, K. Hasegawa, K. Tomori, H. J. M. Kramer, and P. J. Jansens, "Two-dimensional population balance model with breakage of high aspect ratio crystals for batch crystallization," *Chem. Eng. Sci.*, vol. 63, no. 12, pp. 3271–3278, Jun. 2008.

[8] R. Gunawan, I. Fusman, and R. D. Braatz, "High resolution algorithms for multidimensional population balance equations," *AIChE J.*, vol. 50, no. 11, pp. 2738–2749, Nov. 2004.

[9] F. Févotte and G. Févotte, "A method of characteristics for solving population balance equations (PBE) describing the adsorption of impurities during crystallization processes," *Chem. Eng. Sci.*, vol. 65, no. 10, pp. 3191–3198, 2010.

[10] E. Aamir, Z. K. Nagy, C. D. Rielly, T. Kleinert, and B. Judat, "Combined quadrature method of moments

and method of characteristics approach for efficient solution of population balance models for dynamic modeling and crystal size distribution control of crystallization processes," *Ind. Eng. Chem. Res.*, vol. 48, no. 18, pp. 8575–8584, Sep. 2009.

[11] S. Qamar and G. Warnecke, "Numerical solution of population balance equations for nucleation, growth and aggregation processes," *Comput. Chem. Eng.*, vol. 31, no. 12, pp. 1576–1589, Dec. 2007.

[12] P. Wang, A. Anderko, and R. D. Young, "A speciation-based model for mixed-solvent electrolyte systems," *Fluid Phase Equilib.*, vol. 203, no. 1–2, pp. 141–176, Dec. 2002.

[13] N. Kubota and J. W. Mullin, "A kinetic model for crystal growth from aqueous solution in the presence of impurity," *J. Cryst. Growth*, vol. 152, no. 3, pp. 203–208, 1995.

459234 Optimal Operation of Batch Enantiomer Crystallization: From Ternary Diagrams to Predictive Control

Abstract:

The compounds making up many useful molecules such as the active ingredients in pharmaceuticals appear in nature as enantiomers [1]. To produce a pure enantiomer, a variety of techniques can be used, including asymmetric synthesis and methods of chiral resolution such as chromatographic methods and crystallization [2]. When a set of enantiomers forms a racemic compound, production of pure crystals of one enantiomer in a batch operation requires that the original liquid mixture purity be above that of the eutectic composition due to thermodynamic restrictions [3]. These thermodynamic restrictions can be visualized through a ternary diagram for the mixture of the two enantiomers and a solvent [3]-[4].

In this work, we investigate the operation and control of a batch crystallization process for a racemic compound forming system with a fines trap and dissolution tank, and seeding of the initial liquid within the system (which is enriched in the desired enantiomer) with crystals of the desired enantiomer. We use the ternary diagram in the development of operating conditions for the system and for the development of constraints for an optimal control design for the system. We develop model equations for the batch crystallization process based on a population balance model for the crystals coupled with mass balances. We then apply a model predictive control strategy to the system to maximize the volume of crystals developed from the seeds while reducing the volume of crystals developed from nucleation, and show how the closed-loop trajectories are related to the ternary diagram thermodynamic data.

[1] Nguyen LA, He H, Pham-Huy C. Chiral drugs: An overview. International Journal of Biomedical Science. 2006;2:85-100.

[2] Lorenz H, Seidel-Morgenstern A. Processes to separate enantiomers. Angewandte Chemie International Edition.

2014;53:1218-1250.

[3] Kaspereit M. Separation of Enantiomers by a Process Combination of Chromatography and Crystallisation. PhD Thesis, Max Planck Institute for Dyanamics of Complex Technical Systems. Aachen, Germany: Shaker, 2006.

[4] Jacques J, Collet A, Wilen SH. Enantiomers, Racemates, and Resolutions. New York: John Wiley & Sons, 1981.

461227 Using Semidefinite Programming to Calculate Bounds on Particle Size Distributions

Abstract:

Many chemical engineering processes involve a population of particles with a distribution of sizes that changes over time. For example, crystallization, colloidal suspension formation, catalyst attrition, polymerization, and aerosol formation all fit this general framework [1]. In each of these processes, the particle size distribution (PSD) can have a large effect on macroscopic properties of engineering interest. For example, for pharmaceutical crystals, the PSD affects the ease with which the crystals can be filtered and compacted into tablets, thereby affecting the cost and processing time of the pharmaceutical product [2, 3]. Moreover, once the drug has been introduced to a patient's system, the PSD affects it's bioavailability. Thus, the PSD is tied to both the pharmaceutical's efficacy and safety [4,5].

Because of the importance of the PSD in these diverse chemical engineering applications, many methods have been developed to model how a PSD changes over time. Usually, this model is a PDE known as a population balance model [1]. In some cases, this PDE can be solved analytically. However, it often must be solved numerically. Solving the PDE has the advantage that the result is a full description of the final PSD in terms of a number density function; the disadvantage is that obtaining this solution numerically can be computationally expensive. For this reason, it is very common to instead model only finitely many *moments* of the PSD, which amounts to solving a system of ODEs [6,7]. Modeling only the moments certainly reduces the computational burden, but this comes at a cost: moments are only a summary description of the PSD, i.e, they do not contain enough information to reconstruct all of its details. This is because there are, in general, many PSDs corresponding to a given finite set of moments [8]. Thus, given only finitely many moments of an unknown distribution, there is no clear answer to industrially relevant questions such as:

- How many particles have size in the range a to b?
- What is the D10 of the distribution?
- What is the qualitative shape of the distribution?

Faced with these questions, one might be tempted to apply one of the various methods available for constructing a number density function with a specified finite set of moments [9,10,11,12]. With the resulting number density function, answering the above questions would be trivial. However, the problem with this strategy should be clear from the foregoing discussion: the calculated number density function describes just one of the many PSDs with the specified moments. Accordingly, it would provide just one of the many possible (valid) answers to each of the above questions, giving us a false sense of certainty in our knowledge of the distribution.

We take a more rigorous approach. Acknowledging that reconstructing a PSD from finitely many moments is an ill-posed inverse problem, we make no attempt to answer the above questions exactly. Instead, we calculate provable *bounds* on the answers. These bounds require no a priori knowledge of the shape of the distribution, no experimental data, and no regularity assumptions on the number density function describing the PSD.

The bounding algorithms we will present are a natural application of results from the mathematical literature regarding moments of positive finite Borel measures (i.e. generalized distributions) [13]. In particular, we will calculate the proposed bounds using Semidefinite Programs (SDPs) [14]. While SDPs have been applied in chemical engineering in the context of optimal control [15,16], to the best of the authors' knowledge, their natural application to particle size distributions has, until now, gone unnoticed.

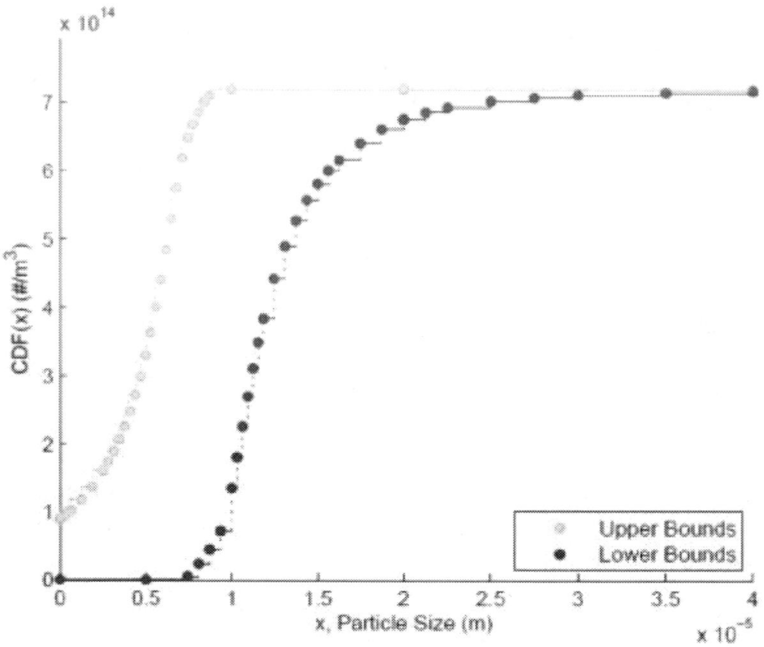

Figure 1: Example of bounds calculated on the PSD cumulative distribution function (CDF)

References

[1] Ramkrishna D. Population Balances. San Diego: Academic Press. 2000.

[2] Myerson A. Handbook of Industrial Crystallization. Woburn: Butterworth-Heinemann. 2002.

[3] Wibowo C, Chang W, Ng K. Design of integrated crystallization systems. AIChE Journal. 2001;47:2474-2492.

[4] Braatz RD. Advanced Control of Crystallization Processes. Annual Reviews in Control. 2002; 26(1):87-99.

[5] Nowee S, Abbas A, Romagnoli J. Model-Based Optimal Strategies for Controlling Particle Size in Antisolvent Crystallization Operations. Crystal Growth Des. 2007;8:2698-2706.

[6] Hulburt H, Katz S. Some problems in particle technology: A statistical mechanical formulation. Chemical Engineering Science. 1963;19:555-574.

[7] Randolph AD, A LM. Theory of Particulate Processes. New York: Academic Press Inc. 1988.

[8] McGraw R, Nemesure S, Schwartz SE. Properties and evolution of aerosols with size distributions having identical moments. Journal of Aerosol Science. 1998;29(7):761-772.

[9] de Souza LGM, Janiga G, John V, Thevenin D. Reconstruction of a distribution from a finite number of moments with an adaptive spline-based algorithm. Chemical Engineering Science. 2010;65(9):2741-2750.

[10] Hutton K, Mitchell N, Frawley PJ. Particle size distribution reconstruction: The moment surface method. Powder Technology. 2012;222:8-14.

[11] Cogoni G, Frawley PJ. Particle Size Distribution Reconstruction Using a Finite Number of Its Moments through Artificial Neural Networks: A Practical Application. Crystal Growth & Design. 2015; Published online.

[12] Woodbury AD. A FORTRAN program to produce minimum relative entropy distributions. Computers and Geosciences. 2004;30(1):131-138.

[13] Lasserre JB. Moments, positive polynomials and their applications, vol. 1. World Scientific. 2009.

[14] Vandenberghe L, Boyd S. Semidefinite programming. SIAM Review. 1996;38(1):49-95.

[15] VanAntwerp JG, Braatz RD, Sahinidis NV. Globally optimal robust process control. Journal of Process Control. 1999;9(5):375-383.

[16] VanAntwerp JG, Braatz RD, Sahinidis NV. Globally optimal robust control of large scale sheet and film processes. In: American Control Conference, 1997. Proceedings of the 1997, vol. 3. IEEE. 1997; pp. 1473-1477.

462030 Characterization of an Integrated Continuous Cooling Crystallization Process with an in-Situ Wet Mill System for Particle Size Reduction

Abstract:

Crystallization is one of the critical unit operations during the production of an active pharmaceutical ingredient or a pharmaceutical intermediate. Control of crystal attributes such as size and size distribution are critical not only to the efficiency of the downstream operations but also to the final drug product. Size reduction can be achieved by dry or wet milling operations. The integration of an insitu wet mill and continuous mixed suspension mixed product removal crystallizer (MSMPR) is studied in this work for cooling crystallization of Paracetamol from an aqueous isopropanol mixture. During this study, the impact of various operational conditions of both continuous *insitu* mill and MSMPR unit on the product properties is evaluated. The paper demonstrates that the MSMPR operational conditions (i.e. residence time, operational temperature) significantly impact the achievable steady state product characteristics. In addition, the influence of insitu mill operating conditions (i.e. rotor speed) on critical product and process qualities attributes such as particle size distribution, mean particle size, and yield. Towards this end, breakage kernels were evaluated and validated through a parameter estimation routine using the experimental data. The specific breakage rate considering magma density and rotor speed showed higher accuracy between the various kernels.

462046 Tailoring the Crystal Size Distribution By Controlling the Crystallization Trajectory in Mass-Count Space

Abstract:

Batch cooling crystallization is widely applied to isolate high-purity chemicals and pharmaceuticals. Control over the size of crystals produced by such operations can be critically important for efficient downstream processes. In this presentation we describe an approach for tailoring the crystal size distribution.

Previously, we proposed an alternative perspective on crystallization dynamics. Instead of understanding crystallization dynamics through the lens of the population balance framework, we viewed crystallization dynamics as movement in a 2D space mapped out by measured properties: the crystal mass and the chord count [1]. This, we found, revealed a simple, empirical model of crystallization dynamics and also facilitated the development of feedback schemes for controlling the mean crystal size [1, 2].

Here, a further use of the mass-count perspective will be described. By carefully controlling the mass-count *trajectory* we can rationally tailor the crystal size *distribution*. This will be demonstrated experimentally for paracetamol crystallization from ethanol.

[1] Griffin, D. J., Grover, M. A., Kawajiri, Y. and Rousseau, R. W. 2015. Mass-Count Plots for Crystal Size Control. *Chemical Engineering Science* 137, 338-351.

[2] Griffin, D. J., Grover, M. A., Kawajiri, Y. and Rousseau, R. W. 2016. Data-Driven Modeling and Dynamic Programming Applied to Batch Cooling Crystallization. *Industrial & Engineering Chemistry Research* 55, 1361-1372.

463021 Attainable Regions of Particle Sizes for Continuous Milling-Crystallization Processes

Abstract:

The properties of particulate products are not only defined by their composition and crystal structure (polymorph, salt form, solvates, etc.), but also by their size and shape. The influence of particle size and shape is particularly felt in downstream processing operations, such as filtration, drying, granulation, and blending of one particulate product with another. For instance, one can surmise that thin needle-like crystals are more prone to break in an agitated dryer than compact crystals of the same material or that the time required to separate mother liquor from crystals by cake filtration must – among other factors – depend on how tightly crystals in the filter cake are packed, which in turn depends on the crystal size and shape. Therefore, crystallization processes should be designed to deliver the high purity and yield, the right crystal form, as well as to deliver a reproducible particle size distribution that avoids undesired effects in downstream processing.

Within the quality by design (QbD) initiative[1], there is an increased drive towards a deeper understanding of manufacturing processes, aiming at identifying a design space for a given combination of product and process in which the product can be reliably produced to specification. Recent works within the crystallization community established a design methodology, which allows to deduce what particle sizes are attainable from a given crystallization process based on knowledge of the thermodynamics and kinetics[2]. This was formulated in regions of mean particle sizes against processing time (residence time for continuous crystallizers or batch time for batch crystallizers), while fulfilling constraints such as yield considerations and maximum/minimum allowed temperatures. The methodology was subsequently extended to take into account additional product characteristics, such as the width of the particle size distribution[3], as well as additional operational constraints[4]. The framework's robustness against uncertainty in the kinetic parameters was assessed as well[5]. However, these efforts were exclusively focused on crystallization processes that are dominated solely by crystal growth and nucleation, i.e., secondary kinetic processes, such as crystal breakage and agglomeration were neglected.

In this work, it will be shown that attainable regions can also be obtained for processes that include extensive breakage processes. The study is motivated by the realization that processes that intricately combine crystallization and milling devices often allow to tune particle size distributions in a more extensive way than crystallization processes alone[6]. For instance, they allow to obtain smaller particles, or allow reaching a desired yield in shorter processing times (through an increase in surface area of the crystals). This work specifically investigates the combination of a continuous mixed suspension mixed product removal crystallizer (MSMPRC) with a high shear wet suspension mill that is connected to the crystallizer in a recycle loop. After determining breakage kinetics for this mill were determined, attainable regions for this process were calculated using a population balance equation model including breakage kinetics. The methodology is applied to the cooling crystallization of paracetamol crystallized from a mixture of isopropanol and water (1:4 w/w).

References:

[1] Food and Drug Administration, "Pharmaceutical Quality for the 21st Century A Risk-Based Approach Progress Report", http://www.fda.gov/aboutfda/centersoffices/officeofmedicalproductsandtobacco/cder/ucm128080.htm, accessed: May 6th, 2016.

[2] Vetter, T., Burcham, C.L., Doherty, M.F., „Regions of attainable particle sizes in continuous and batch crystallization processes", Chem. Eng. Sci., 2014, 106, 167-180.

[3] Yang, Y., Nagy, Z.K., "Advanced control approaches for combined cooling/antisolvent crystallization in continuous mixed suspension mixed product removal cascade crystallizers", Chem. Eng. Sci., 2015, 127, 362-373.

[4] Power, G., Hou, G., Kamaraju, V.K., Morris, G., Zhao, Y., Glennon, B., „Design and optimization of a multistage continuous cooling mixed suspension, mixed product removal crystallizer", Chem. Eng. Sci., 2015, 133, 125-139.

[5] Vetter, T., Burcham, C.L., Doherty, M.F., „Designing robust crystallization processes in the presence of parameter uncertainty using attainable regions", Ind. Eng. Chem. Res., 2015, 54, 10350-10363.

[6] Yang, Y., Song, L., Zhang, Y., Nagy, Z.K., „Application of Wet Milling-Based Automated Direct Nucleation Control in Continuous Cooling Crystallization Processes", Ind. Eng. Chem. Res., 2016, 55, 4987-4996.

471499 Attainable Regions for Critical Quality Attributes in Drug Manufacture and End Performance

Abstract:

The attainable regions for Critical Quality Attributes (CQA) in the manufacture, such as Particle Size Distribution (PSD), and performance, such as fraction absorbed, of Solid Oral Dosage Forms (SODF) can be highly dependent upon the types and operation of the end-to-end production steps. This is particularly the case for low-dose drug products, where other factors, such as Content Uniformity (CU) of SODF can also be of a major concern. The CU of SODF is known to be highly dependent upon the particle size distribution of the Active Pharmaceutical Ingredient (API), with control of API particle size being a common approach to ensuring the content uniformity of drug products. In this work, we examine a continuous cascade cooling crystallization process and explore the impact of varying process configurations, both drug substance and drug product, on the attainable processing regions, subject to various processing and performance constraints. The continuous cooling crystallization of paracetamol and ethanol solutions was employed as an example case. Additional drug substance and drug product manufacturing steps were also considered in a modular fashion, in order to investigate their impact on the attainable regions, including filtration, wet and dry milling. The final performance of the drug product was determined via a segmented model of the Gastro-Intestinal (GI) tract. This model describes the dissolution and permeation of the API in the GI tract, yielding the fraction of the API absorbed in the patient as function of time and hence the bioavailability of the drug *in-vivo*.

A number of optimization objectives were posed, which included increasing throughput of the API and the number of crystallizer stages, whilst maintaining a target CU of the final product, sufficient bioavailability and other processing constraints, in order to evaluate these attainable regions. Once the attainable region of a given CQA versus total production time was determined, the robustness within the attainable region was evaluated and explored via Global Sensitivity Analysis (GSA). This holistic approach to the modelling of API manufacture and end performance, allows the interdependent processing steps of drug substance and drug product manufacture and their impact upon *in-vivo*

drug performance to be considered in a more cohesive way.

456861 Low Temperature Conversion of Biomass into Activated Biochar for Water Treatment in Resource-Constrained Settings

Abstract:

Access to clean water is a major challenge in many parts of the world, especially in remote rural areas such as rain forest communities in Africa. New approaches are needed to ensure adequate water supply and quality, as they are inextricably intertwined with food security and public health. In this presentation, we will focus on the synthesis, characterization, and application of activated biochar for water purification. Unlike conventional gasification reactor systems designed to operate at large scales, not much has been investigated in the past about small-scale batch pyrolysis. Nevertheless, batch systems offer advantages such as increased char yield and net exothermicity. Various types of cellulosic biomass (eg: sawdust, coconut shells) can be used as carbon precursors via pyrolysis. We will describe a process where catalytic activating agents are used in batch setups to lower the pyrolysis temperature and simultaneously activate the carbon produced to BET surface areas in excess of 500 m2/g . A systematic study of process parameters and characterization of the resulting activated carbons is used to optimize the chemistry and minimize contaminants from the produced filter-carbon.

468039 Small-Scale Multiple Effect Distillation (MED) for Brackish Groundwater Desalination

Abstract:

Multiple effect distillation (MED) is the oldest thermal desalination process and has a typical plant capacity of 600 to 300,000 m^3/day. Because MED units do not use membranes, the capital cost, system complexity, and maintenance costs are usually lower than those for reverse osmosis (RO) systems. The purpose of this study was to engineer a small unit that could be operated in the field using renewable energy-sourced heat, namely, heat from the pyrolysis of waste biomass.

A small-scale MED unit was designed and fabricated at NMSU to produce approximately 0.5 m^3/day of very pure water (<10 ppm total dissolved solids (TDS)) from brackish groundwater. The unit is being used to study scaling behavior as a function of water chemistry and identify optimum operating conditions for each water. The fabricated MED unit consists of two falling film evaporators with horizontal smooth copper tubes in a parallel feed arrangement; the flow rates are approximately 60 kg/hr. brackish feed water and 20 kg/hr. of distilled water based on brine concentration factor of 1.5. The feedwater is sourced from the Brackish Groundwater National Desalination Research Facility (BGNDRF) in Alamogordo, NM and has a TDS concentration of 1,000-5,000 ppm. The MED operates under partial vacuum (0.2 bar absolute) and temperatures of 60-70°C. The MED performance was improved by adding a heat exchanger, water heater, and pump to bring the feedwater temperature closer to its boiling point prior to be sprayed into the effects. The aluminum parts in the MED effects were anodized to prevent corrosion. Remote thermocouples were used to monitor the temperature at the surface of copper tubes to study the effects of dry spots. The temperatures and pressures selected in this design are meant to be compatible with local brackish groundwater chemistry in terms of salt solubility to limit scaling on the heat exchanger surfaces.

472093 Bench Scale Investigation of the Biosand Filter Media As the Fluoride, Turbidity, and *E. coli* Removal

Abstract:

Biosand filter (BSF) is a simple and inexpensive facility that can be built as a community-based water treatment technique using locally-available materials. It has been proven that BSF can remove turbidity, *E. coli*, arsenic in water efficiently by modifying the filtration media. Drinking water fluoridation is prevalent in the world. High fluoride intake would cause dental fluorosis and other toxic health effects, especially for children. Therefore, drinking water de-fluoridation is necessary to safeguard public health. There is little information on the use of BSF for fluoride removal. For instance, some previous works reported that aluminum oxide-coated pumice (AOCP) was effective for fluoride removal and had better removal efficiency compared to activated alumina. Other research works found that pumice media BSF can provide similar water quality as conventional sand BSF but 30% more filtered water. Based on these findings, we have investigated the optimization of using AOCP surfaces to serve as media for BSF to clean contaminant water. Preliminary results have shown that AOCP can remove 93% of fluoride in surface water. In our setting, the traditional sand layer has been replaced by AOCP surfaces. We have found that filtration rate, contaminants removal efficacy are prime parameters that enhance BSF performance. It is expected that our findings will be able to predict the behavior of BSFs and their implementation in low-income communities or as point of use (PoU) systems.

463624 Capacitive Deionization of Brackish Waters Using Biomass-Derived Carbons

Abstract:

Irrigated agriculture water represents the bulk of demand for water in most of the countries suffering from acute water scarcity in the Near East and North Africa, as well as Mexico, Pakistan, South Africa, and large parts of China and India. Greenhouses are emerging solution. Current testing in Mozambique and Sierra Leone has demonstrated significantly reduced water use. Yet even a few gallons per day per greenhouse is challenging because many local water sources are brackish. Capacitive deionization (CDI) is a sustainable, energy efficient, and cost effective technology for removal of low or moderate salinity typical of brackish water as compared to technologies that remove the water instead. Indigenous production of the carbon for the electrodes holds the key to such technology being adopted as a source of irrigation water, simultaneously facilitating the use of such using renewable energy sources. Biochars have recently been explored as electrode materials in CD, and can be easily produced from local plant-based agricultural waste materials.

The goal of this study is thus to identify biomass process conditions to enable the best performance as electrode material in CDI. Studies on CDI so far have been targeted towards achieving the best adsorption capacity, which directs towards the use of highly processed and tailored carbons. This is mostly achieved through chemical activation processes using toxic or corrosive and expensive materials which require deionized water for subsequent washing of the carbon, for achieving better CDI performance. Hence, with the effort directed towards making the process viable for use locally in regions with acute water scarcity, a change in the approach to tailor the electrode material for use in CDI is proposed. This approach is to determine the best possible material utilization by ensuring maximum possible contact between the liquid and the biochar surface (wettability) and by generating the least Ohmic loss thus resulting in highest electrical efficiency.

472455 The Use of Electrodeionization for High Water Recovery in Brackish Water Treatment

Abstract:

Currently reverse osmosis dominates desalination because of low cost, reliability, and choices in the market place. However, there are many new desalination opportunities, especially in brackish water applications, where the size and scale make it possible to consider other technology. One of these technologies is electrodeionization. Electrodeionization is a technique in which we use wafers to remove ions to low concentration through an electricially driven process. Because the cost is related to the amount of ions being moved this technology can be cost competitive at low to medium ion concentration and low to medium scale. The purpose of this presentation is to explore places where electrodeionization can be economically feasible and show the fundamental scientific challenges with the technology. This work will specifically focus on the high water recovery possible with electrodeionization with experimental results to show high current efficiencies at concentration concentrations of greater than 10% salt.

466903 Total Phosphorus Removal from Sewage in Microbial Electrochemically Assisted Septic Tanks

Abstract:

Background

Subsurface sewage treatment systems (SSTS), commonly known as septic systems, typically consist of a septic tank and an absorption field (or drainfield or leachfield), to treat sewage from households. The septic tank as pretreatment step decreases total suspended solids (TSS), biochemical/chemical oxygen demands (BOD/COD), and pathogen levels from sewage influent, while the drainfield or soil absorption field captures nutrients (nitrogen and phosphorus) from the tank effluent so that the water leaving SSTS would have minimal environmental impact. By modifying the conventional septic tank through incorporating electrode materials to make use of microbial electrochemical process, this study assessed this novel design in terms of sewage treatment performance. Since phosphorus becomes increasingly threatening to water environment, it was the goal of this novel design to remove it from liquid phase and potentially capture it at electrode surface.

Methods

The experiments were conducted with 1-L scale simulated septic tanks at four different applied voltage levels (0.50 V, 0.63 V, 0.75 V, and 0.88 V, selection based on previous preliminary results) with control of without applied voltage. Tanks were set up and operated at two temperatures (15 and 25 °C), a hydraulic retention time of 8.3 days, two cycles of feeding and withdrawal during each HRT. The treatment conditions of the reactors are listed in REF _Ref450506281 \h Table 1
08D0C9EA79F9BACE118C8200AA004BA90B02000000080000000E0000005F005200650066003400350030003500300036003200380031000000
. ANOVA and pairwise multiple comparisons based on Bonferroni test were conducted for water characteristics with OriginPro Version 8 at significance level of α=0.05.

Table SEQ Table * ARABIC 1. Treatments of simulated novel septic tanks for 171-day operation

	Treatment*		
Reactor #	Temp, oC	Voltage, V	Chlortetracycline (CTC)
Reactor #1	25	0	No
Reactor #2	25	0.50	No

Reactor #3	25	0.63	No
Reactor #4	25	0.75	No
Reactor #5	25	0.88	No
Reactor #6	15	0	No
Reactor #7	15	0	1 mg-CTC/L
Reactor #8	15	0.50	No
Reactor #9	15	0.63	No
Reactor #10	15	0.75	No
Reactor #11	15	0.88	No
Reactor #12	15	0.75	1 mg-CTC/L
Reactor #13	15	0.88	1 mg-CTC/L

*Water characteristics data of reactors #6 and #7 were pooled for data analysis, and the same procedure was applied to reactors #10 and #12, and reactors #11 and #13, for the reason that CTC addition was not detected to affect other water characteristics.

Results and Discussion

For total phosphorus, results suggested that septic tanks partially removed it from influent (REF _Ref450507954 \h Table 2 08D0C9EA79F9BACE118C8200AA004BA90B02000000080000000E0000005F005200650066003400350030003500300037003900350034000000), and the novel design of microbial electrochemical further decreased its concentration ($p=0.0001$) with an influence from temperature ($p=0.0003$), while the interaction of voltage and temperature was not significant ($p=0.2723$). Those performance differences in total phosphorus removal were further elucidated by pairwise multiple comparisons given in REF _Ref450508201 \h Table 3 08D0C9EA79F9BACE118C8200AA004BA90B02000000080000000E0000005F005200650066003400350030003500380032003000310000000 . The results concluded that at 25 °C, 0.63 V or higher voltage at the selected electrode material is practically effective to remove most total phosphorus (>89.7%) from sewage, resulting in an effluent TP concentration less than 0.82 mg/L over 171-day operation; and that at 15 °C, phosphorus removal efficiency was smaller and it required 0.88 V to remove over 90% of total phosphorus from sewage.

Table SEQ Table * ARABIC 2. Total phosphorus concentrations in influent and effluents of different reactors

	Ave	SD	TP removal
	mg/L	mg/L	%
Influent	7.95	1.39	-
Control, 25 °C	6.98	1.72	12.2
0.50 V, 25 °C	1.81	3.67	77.2
0.63 V, 25 °C	0.82	2.92	89.7
0.75 V, 25 °C	0.14	2.30	98.3

0.88 V, 25 °C	0.11	2.31	98.7
Control, 15 °C	7.36	1.60	7.45
0.50 V, 15 °C	6.30	3.84	20.7
0.63 V, 15 °C	4.31	3.90	45.8
0.75 V, 15 °C	2.26	2.48	71.6
0.88 V, 15 °C	0.49	2.36	93.9

Table SEQ Table * ARABIC 3. Pairwise multiple comparisons results for total phosphorus levels in effluents

Comparisons (Bonferroni Test)	Significance at alpha=0.05 or lower
0.50 V, 25 °C vs. Control, 25 °C	*
0.63 V, 25 °C vs. Control, 25 °C	*
0.63 V, 25 °C vs. 0.50 V, 25 °C	NS
0.75 V, 25 °C vs. Control, 25 °C	*
0.75 V, 25 °C vs. 0.50 V, 25 °C	*
0.75 V, 25 °C vs. 0.63 V, 25 °C	NS
0.88 V, 25 °C vs. Control, 25 °C	*
0.88 V, 25 °C vs. 0.50 V, 25 °C	*
0.88 V, 25 °C vs. 0.63 V, 25 °C	NS
0.88 V, 25 °C vs. 0.75 V, 25 °C	NS
Control, 15 °C vs. Control, 25 °C	NS
Control, 15 °C vs. 0.50 V, 25 °C	*
Control, 15 °C vs. 0.63 V, 25 °C	*
Control, 15 °C vs. 0.75 V, 25 °C	*
Control, 15 °C vs. 0.88 V, 25 °C	*

0.50 V, 15 ºC vs. Control, 25 ºC	NS
0.50 V, 15 ºC vs. 0.50 V, 25 ºC	*
0.50 V, 15 ºC vs. 0.63 V, 25 ºC	*
0.50 V, 15 ºC vs. 0.75 V, 25 ºC	*
0.50 V, 15 ºC vs. 0.88 V, 25 ºC	*
0.50 V, 15 ºC vs. Control, 15 ºC	NS
0.63 V, 15 ºC vs. Control, 25 ºC	*
0.63 V, 15 ºC vs. 0.50 V, 25 ºC	*
0.63 V, 15 ºC vs. 0.63 V, 25 ºC	*
0.63 V, 15 ºC vs. 0.75 V, 25 ºC	*
0.63 V, 15 ºC vs. 0.88 V, 25 ºC	*
0.63 V, 15 ºC vs. Control, 15 ºC	*
0.63 V, 15 ºC vs. 0.50 V, 15 ºC	*
0.75 V, 15 ºC vs. Control, 25 ºC	*
0.75 V, 15 ºC vs. 0.50 V, 25 ºC	NS
0.75 V, 15 ºC vs. 0.63 V, 25 ºC	NS
0.75 V, 15 ºC vs. 0.75 V, 25 ºC	*
0.75 V, 15 ºC vs. 0.88 V, 25 ºC	*
0.75 V, 15 ºC vs. Control, 15 ºC	*
0.75 V, 15 ºC vs. 0.50 V, 15 ºC	*
0.75 V, 15 ºC vs. 0.63 V, 15 ºC	*
0.88 V, 15 ºC vs. Control, 25 ºC	*
0.88 V, 15 ºC vs. 0.50 V, 25 ºC	NS

0.88 V, 15 °C vs. 0.63 V, 25 °C	NS
0.88 V, 15 °C vs. 0.75 V, 25 °C	NS
0.88 V, 15 °C vs. 0.88 V, 25 °C	NS
0.88 V, 15 °C vs. Control, 15 °C	*
0.88 V, 15 °C vs. 0.50 V, 15 °C	*
0.88 V, 15 °C vs. 0.63 V, 15 °C	*
0.88 V, 15 °C vs. 0.75 V, 15 °C	*

* indicates significant difference, p-values less than 0.05; NS indicates non-significant difference, p-values equal or larger than 0.05.

Sludge solids were collected from simulated septic tanks (reactors at 25 °C, REF _Ref450512244 \h Figure 1 08D0C9EA79F9BACE118C8200AA004BA90B02000000080000000E0000005F005200650066003400350030003500310032003200340034000000) and quantified for mixed-liquor (ML) total suspended solids (TSS) and total solids (TS), sludge volume index (SVI), and composition. After log transformation, both MLTS and MLTSS displayed a linear increase along with voltage level. At 0.63 V where TP removal achieved close to 90%, the MLTSS and MLTS increased by 2.53 and and 1.17 folds, respectively. However, the corresponding SVI of the sludge was only 52.5% of the control reactor, indicating the formation of denser sludge, which suggested a smaller increase of sludge volume (85.3%) than the weight increase. The composition analysis based on energy dispersive X-ray spectroscopy (EDS) revealed the concentration and amount of phosphorus in sludge samples (REF _Ref450512615 \h Table 4 08D0C9EA79F9BACE118C8200AA004BA90B02000000080000000E0000005F005200650066003400350030003500310032003600310035000000). More phosphorus was captured in sludge sediment when voltage was increased, but the phosphorus balance, assumed to be captured on electrode surface, was higher at voltages levels of 0.50 and 0.63 V than 0.75 and 0.88 V.

Figure SEQ Figure * ARABIC 1. Mixed-liquor total suspended solids (A), mixed-liquor total solids (B), and sludge volume index (C) of the sludge from reactors with different voltage treatments at 25 °C.

Table SEQ Table * ARABIC 4. Phosphorus balance in different reactors.

	Amt of eff P	Amt P removed	Amt P in Sed	P of balance (on electrodes)
	mg	mg	mg	mg
Inf	178	-	-	-
Control, 25 °C	143	35	13	21
0.50 V, 25 °C	37	141	26	114
0.63 V, 25 °C	17	161	37	125
0.75 V, 25 °C	3	175	92	83
0.88 V, 25 °C	2	176	92	83

Conclusions

These results confirmed the phosphorus retention capability of the novel septic tank, suggesting a lower burden on the soil absorption for nutrient removal. The amount of sludge solids was also increased, which may require a more frequent tank cleaning.

462469 Applying Phase-Change Materials (PCM) to Enhance the Performance of Carbon Dioxide Adsorption Using Hollow Fibers

Abstract:

Applying phase-change materials (PCM) to enhance the performance of carbon dioxide adsorption using hollow fibers

Héctor Octavio Rubiera Landa, Matthew J. Realff, Ryan P. Lively, and Yoshiaki Kawajiri.

School of Chemical & Biomolecular Engineering, Georgia Institute of Technology,

311 Ferst Drive N. W., Atlanta, GA, 30332-0100, USA.

The challenges posed by carbon capture and storage (CCS) have motivated an increasing interest in the last years to investigate new materials and process alternatives. Adsorption-based separations have the potential to cope with this task in an economically and technically feasible way. A key factor to the success in the application of these processes relies on the development of suitable adsorbent materials, with high adsorptive capacity and selectivity. It has been demonstrated experimentally that metal-organic frameworks (MOF) display a significant adsorptive capacity for carbon dioxide at cryogenic temperatures in the range of 233 K to 273 K **[1]**.

Hollow fibers such as polyethylenimine (PEI) with embedded sorbents have attracted significant interest for their potential to address current challenges of CCS technologies. Due to their heat transfer properties, an efficient heat management can be achieved by cooling/heating with a fluid that circulates through the fiber bore. The potential application of PEI-impregnated hollow fibers in cyclic gas adsorption processes, such as rapid temperature-swing adsorption (RTSA), has been investigated, see e.g., **[2,3]**.

An attractive improvement of carbon dioxide adsorption may be achieved by operating at cryogenic conditions using hollow fibers impregnated with MOFs. Furthermore, additional heat-management improvements may be brought about by employing a phase-change material (PCM), encapsulated inside the fiber bore, hence eliminating the external cooling/heating fluid requirements.

This work investigates, via a numerical simulation study, the possibility of coupling thermally the inherent heat generated within a hollow fiber due to carbon dioxide adsorption with the latent heat of a PCM, therefore serving as a heat sink/source for cryogenic adsorption-desorption cycles. This concept provides improvements to the adsorption process using hollow fibers by allowing suitable heat integration, henceforth aiming to reduce capital and operation costs. The case study provides further insight into the practical experimental realization of this new thermally-enhanced adsorption process.

References

[1] J. M. Simmons, H. Wu, W. Zhou, and T. Yildirim. Carbon capture in metal-organic frameworks---a comparative study. *Energy & Environmental Science*. 4(6):2177-2185, 2011.

http://dx.doi.org/10.1039/c0ee00700e.

[2] J. Kalyanaraman, Y. Fan, R. P. Lively, W. J. Koros, C. W. Jones, M. J. Realff, and Y. Kawajiri. Modeling and experimental validation of carbon dioxide sorption on hollow fibers loaded with silica-supported poly(ethylenimine). *Chemical Engineering Journal*. 259:737-751, 2015.

http://dx.doi.org/10.1016/j.cej.2014.08.023.

[3] S. Swernath, K. Searcy, F. Rezaei, Y. Labreche, R. P. Lively, M. J. Realff, and Y. Kawajiri. Optimization and Technoeconomic Analysis of Rapid Temperature Swing Adsorption Process for Carbon Capture from Coal-Fired Power Plant. In Fengqi You, editor, *Sustainability of Products, Processes and Supply Chains Theory and Applications, Computer Aided Chemical Engineering*. Vol. 36:253â€"278. Elsevier B. V., 2015.

ISBN 978-0-444-63472-6. http://dx.doi.org/10.1016/B978-0-444-63472-6.00010-0.

465162 Direct Air Capture of CO_2 Using Faujasite Structured Zeolite

Abstract:

In contrast to conventional carbon capture systems that capture CO_2 from large stationary point sources such as power plants, the ability to capture CO_2 from atmospheric air has been studied in this study. This presentation will include an experimental analysis of CO_2 capture and concentration from ambient air using temperature swing adsorption (TSA) and vacuum temperature swing adsorption (VTSA) as a means to capture and effectively reduce the amount of CO_2 in the atmosphere. Commercially available faujasite structured zeolites have been screened using gravimetric analysis and breakthrough experimentation to determine which adsorbent would be the best for a direct carbon capture plant from air.

471137 Carbon Dioxide Capture from Flue Gas By a PSA Process Utilizing a Structured Adsorbent: Full Scale Simulation with Model Validated at the Bench Scale

Abstract:

The primary objective of this work is to carry out a comprehensive modeling study to evaluate the performance of a structured adsorbent for capturing CO_2 from coal fired power plant flue gas at full process scale. This study is being carried out with the Dynamic Adsorption Process Simulator (DAPS) developed at USC and modified to account for pressure drop through a structured adsorbent with a one dimensional pressure drop correlation in place of the packed bed Ergun equation. DAPS will be validated by comparing the modeling results to experimental results obtained from a 1-bed PSA system utilizing a structured adsorbent. The model will then be used to predict the performance of the PSA cycle at the process scale. Challenges were faced in the scale up process, especially due to the high pressure drop that occurred within the structured adsorbent bed during the very low pressure regeneration steps. This presentation will present the latest results obtained by Ritter and his group.

453471 Modeling, Simulation and Optimization of Pressure Swing Adsorption (PSA) Processes for Post-Combustion Carbon Dioxide (CO_2) Capture from Flue Gas

Abstract:

Anthropogenic carbon dioxide (CO_2) emissions are considered as a great threat to the environment because of their contribution to the greenhouse effect and global warming. CO_2 is considered to be responsible for 60% of the global warming caused by greenhouse gases (GHGs) [1]. However, post-combustion CO_2 capture is still required to avoid excess emissions of CO_2 from existing power plants. The process of CO_2 capture and sequestration (CCS), involves CO_2 separation followed by pressurization, transportation, and sequestration. The capture units are expected to concentrate the CO_2 from flue gas with CO_2 purity and CO_2 recovery exceeding 95% and 90%, respectively according to the Department of Energy requirements [2]. Thus it is important to develop energy-efficient industrial technologies for CO_2 capture. A recent study presents a general and critical review of the state of the art emerging CO_2 capture technologies [3]. Pressure or vacuum swing adsorption (PSA/VSA) is a promising technology for carbon capture due to its relatively better separation performance, lower energy consumption and lower cost [4]. In a typical adsorption process, CO_2 is selectively adsorbed onto a solid sorbent while the clean flue gas passes through. The adsorbed CO_2 is released or desorbed by lowering the pressure of the process. Several PSA/VSA cycles specific to CO_2 capture have also been studied in the literature which indicate that cyclic adsorption processes are promising options for CO_2 capture [5]. Adsorption-based CO_2 capture using zeolites and metal-organic frameworks (MOFs) with large internal surfaces are promising technologies to separate CO_2 from power plant flue gases [6, 7].

This work presents a mathematical modeling framework developed from our research group [8,9] for the simulation and optimization of PSA/VSA processes for post-combustion CO_2 capture from dry flue gas (15% CO_2, 85% N_2). The core of the modeling framework represents a detailed adsorption column model relying on a coupled set of mixed partial differential and algebraic equations (PDAEs) for mass, heat and momentum balance at both bulk gas and particle level, equilibrium isotherm equations, transport and thermo-physical properties of the gas mixture and boundary conditions according to the operating steps. The proposed modeling equations have been implemented in the gPROMS™ modeling environment. The modeling framework is first validated against data available from the literature [10], illustrating good agreement in terms of several process performance indicators [11]. Accordingly, the developed modeling framework is employed for a comparative evaluation of three available potential adsorbents for CO_2 capture, namely, zeolite 13X, activated carbon and Mg-MOF-74 [12]. A two-bed configuration (six-step VSA cycle) with light product pressurization has been employed in all simulations in accord with recent studies [13, 14]. The results from systematic comparative simulations demonstrate that zeolite 13X has the best process performance among the three adsorbents, in terms of CO_2 purity and CO_2 recovery. On the other hand, Mg-MOF-74 appears to be a promising adsorbent for CO_2 capture, as it has considerably higher CO_2 productivity compared to the other two adsorbents. Furthermore process optimization studies using zeolite 13X and Mg-MOF-74, have been performed to minimize energy consumption for specified minimum requirements in CO_2 purity and in CO_2 recovery at nearly atmospheric feed pressures. The optimization results indicate that there is a complex relationship between optimal process performance indicators and operating conditions that varies among the different adsorbents and cannot be quantified by simple comparison of CO_2/N_2 adsorption isotherms and selectivity data. Evidently, detailed process modeling, simulation and optimization strategies, provide the most reliable way to evaluate both qualitatively and quantitatively potential adsorbents for CO_2 capture. Finally, the reduction of CO_2 emissions from dry flue gas employing two successive PVSA units, using zeolite 13X and Mg-MOF-74 as adsorbents, is considered and the effects of operating parameters on the separation quality are investigated by simulation and optimization. With the proposed two-stage PVSA process, a CO_2 purity higher than 95% with a relatively high CO_2 recovery unit (greater than 90%) is obtained.

References

[1] Houghton, J. T.; Ding, Y.; Griggs, D. J.; Noguer, M.; Van der Linden, P. J.; Dai, X.; Maskel, K.; Johnson, C. A. Climate Change 2001: The scientific basis; Cambridge, UK 2001.

[2] DOE/NETL Quality guidelines for energy system studies: CO2 impurity design parameters, Technical report; DOE/NETL-341/011212; National Energy Technology Laboratory: 2012.

[3] Abanades, J.C., Arias, B., Lyngfelt, A., Mattisson, T., Wiley, D.E., Li, H., Ho, M.T., Mangano, E., Brandani, S., 2015. Emerging CO2 capture systems. Int. J. Greenh. Gas Control 40, 126–166.

[4] The future of coal: options for a carbon-constrained world. Massachusetts Institute of technology, Cambridge, MA, 2007.

[5] Ebner AD, Ritter JA. State-of-the-art adsorption and membrane separation processes for carbon dioxide production from carbon dioxide emitting industries. Sep Sci Technol. 2009; 44(6):1273–1421.

[6] Hasan, M. M. F.; Baliban, R. C.; Elia, J. A.; Floudas, C. A., Modeling, simulation, and optimization of postcombustion CO2 capture for variable feed concentration and flow rate. 2. Pressure swing adsorption and vacuum swing adsorption processes. Industrial and Engineering Chemistry Research 2012, 51, (48), 15665-15682.

[7] Spjelkavik, A.I., Aarti, Divekar, S., Didriksen, T., Blom, R., 2014. Forming MOFs into spheres by use of molecular gastronomy methods. Chem. Eur. J. 20 (29), 8973–8978.

[8] Nikolic, D.; Giovanoglou, A.; Georgiadis, M. C.; Kikkinides, E. S., Generic modeling framework for gas

separations using multibed pressure swing adsorption processes. Industrial and Engineering Chemistry Research 2008, 47, (9), 3156-3169.

[9] Nikolic, D.; Kikkinides, E. S.; Georgiadis, M. C., Optimization of multibed pressure swing adsorption processes. Industrial and Engineering Chemistry Research 2009, 48, (11), 5388-5398.

[10] Ko, D.; Siriwardane, R.; Biegler, L. T., Optimization of pressure swing adsorption and fractionated vacuum pressure swing adsorption processes for CO2 capture. Industrial and Engineering Chemistry Research 2005, 44, (21), 8084-8094.

[11] Nikolaidis, G. N., Kikkinides, E. S., Georgiadis, M. C., Modelling and Simulation of Pressure Swing Adsorption (PSA) Processes for post-combustion Carbon Dioxide (CO2) capture from flue gas, Computer Aided Chemical Engineering, Elsevier, 37 (2015), 287-292.

[12] Nikolaidis, G. N., Kikkinides, E. S., Georgiadis, M. C., Model-Based Approach for the Evaluation of Materials and Processes for Post-Combustion Carbon Dioxide Capture from Flue Gas by PSA/VSA Processes, Industrial & Engineering Chemistry Research, 55 (3), (2016), 635-646.

[13] Haghpanah, R.; Nilam, R.; Rajendran, A.; Farooq, S.; Karimi, I. A., Cycle synthesis and optimization of a VSA process for postcombustion CO2 capture. AIChE Journal 2013, 59, (12), 4735-4748.

[14] Krishnamurthy, S.; Rao, V. R.; Guntuka, S.; Sharratt, P.; Haghpanah, R.; Rajendran, A.; Amanullah, M.; Karimi, I. A.; Farooq, S., CO2 capture from dry flue gas by vacuum swing adsorption: A pilot plant study. AIChE Journal 2014, 60, (5), 1830-1842.

470758 Optimization of a High Temperature PSA Process for Pre-Combustion CO_2 Capture

Abstract:

In a U.S. Department of Energy (DOE) sponsored project, TDA Research, Inc. is developing a novel sorbent that removes CO_2 well above the dew point of synthesis gas stream generated by commercial gasifiers using a pressure swing adsorption (PSA) process. The relatively strong affinity of the sorbent to CO_2 enables effective operation at temperatures up to 300°C. However, because the sorbent and the CO_2 do not form a true covalent bond, the energy needed to regenerate TDA's sorbent (5.4 kcal per mol of CO_2) is much lower than that observed for either chemical absorbents (e.g., 29.9 kcal/mol CO_2 for sodium carbonate) or amine-based solvents (e.g., 14.2 kcal/mol CO_2 for monoethanolamine). TDA's sorbent can be regenerated isothermally and CO_2 could be recovered at pressures as high as 150 psia. Thus, the energy needed to regenerate the sorbent and compress the CO_2 for sequestration is significantly lower than that for any technology reported to date. In this presentation we will summarize the results of collaboration between TDA Research and the University of Alberta to develop PSA models for TDA's precombustion CO_2 capture sorbent.

High-pressure dynamic column breakthrough experiments have been performed using a bench scale fixed bed reactor system over a wide range of temperatures. The adsorption isotherms and kinetic parameters for CO_2 and H_2 were measured by fitting the experimental breakthrough responses with those simulated using a detailed adsorption model. Based on these parameters, a variety of PSA cycles, that include pressure equalization, blow down and steam purge steps were designed. Using a detailed model coupled with a genetic algorithm toolbox, these PSA cycles were optimized for maximizing CO_2 purity and recovery. Using this approach, we were able to show that many cycle configurations do achieve high CO_2 purity (>95%) and recovery (>90%) that are recommended by the DOE for sequestration. For those cycles that satisfied the high purities and recoveries, the problem of minimizing parasitic energy was studied. Factors such as energy consumption for CO_2 compression, loss of H_2 into the desorption product, steam utilization were all combined into a single objective function. The cycle parameters such as various pressure levels, step time and feed velocities were used as decision variable while the CO_2 purity and recovery were used as a constraint. The results of this optimization study will be discussed at the meeting.

460361 Batch Adsorber Analogs for Rapid Screening of Adsorbents for CO2 Capture

Abstract:

The adsorptive CO_2 capture arena has seen tremendous growth with the ability to synthesize new materials, e.g., MOFs, ZIFs, etc. With a wide range of materials that are now available there is a need for effective tools to rank materials according to their potential to perform at a process scale. Our recent work has indicated that traditional metrics such as selectivity, working capacity, etc. show a poor ability to predict process performance, since they do not take into account the complexities, of adsorption kinetics, isotherm shapes, heat effects and process configurations into consideration [1]. With the access to inexpensive computational power, it is now possible to combine rigorous optimization methods with detailed models to compare material performance in a quantitative manner. However, these approaches are time-consuming and require more sophisticated numerical schemes. Maring and Webley introduced the analog model in which the different steps of a P/VSA process was simulated by considering a batch adsorber without axial gradients [2] . The key advantage of this approach is that the Partial differential equations (PDEs) are converted to an Ordinary differential equation (ODE); and the requirement to repeated simulations to reach cyclic steady state (CSS) is eliminated. They also showed that the prediction of process performance indicators such as purity and recovery when compared to detailed models were reasonable compared to the savings in computational time. In their work, they considerd a simple 4-step cycle which resulted in low value of purity and recovery.

In this presentation, we extend the analysis of Maring and Webley by considering more complex cycles that include a three pressure-level operation; reflux, pressure equalization and light-product pressurization steps. This model is then used to predict the performance of 4 adsorbents, namely Zeolite 13X, Mg-MOF-74, UTSA-16, a type of coconut-shell derived activated carbon for post-combustion CO2 capture. The model assumes instantaneous equilibrium and isothermal operations. The impact of making these assumptions is first verified and then the batch-analog model is first validated by comparing the CO2 purity and recovery with that predicted by detailed models. By varying the operating conditions, the Pareto curve for maximizing purity and recovery is obtained and is comapred to the ones from detailed optimization [1]. The results show that although the Pareto curves from the two modeling approaches are different, the relative ranking of adsorbents is indeed correct. Based on this observation, the effect of operating variables, e.g., intermediate and low pressure on the process performance is evaluated and the specific energy consumption for the different materials is calculated. Finally the by considering the Langmuir adsorption isotherm to represent both CO2 and N2, the effect of the isotherm parametrs, equilibirum constant, saturation capacity and heats of adsorption, on the process performance is mapped in the form of contour plots. These plots provide guidance for the development of new materials that could potentially show superior performance at the process scale.

References

[1] Rajagopalan, A. K.; Avila, A. M. & Rajendran, A.Do adsorbent screening metrics predict process performance? A process optimisation based study for post-combustion capture of CO_2 *Int. J. Greenhouse Gas Control, Elsevier,* **2016**, *46*, 76-85.

[2] Maring, B. J. & Webley, P. A. A new simplified pressure/vacuum swing adsorption model for rapid adsorbent screening for CO_2 capture applications *Int. J. Greenh. Gas Con., Elsevier,* **2013**, *15*, 16-31

468575 Mixed Matrix Membranes for H_2/CO_2 Separation

Abstract:

We verified the feasibility of mixed matrix membranes based on poly-2-6-dimethylpoly(phenylene oxide) (PPO) for H_2/CO_2 separation in the purification of syngas obtained from coal gasification process. PPO shows a very high permeability but a moderate selectivity for the H_2/CO_2 couple at room temperature. We tried to improve the H_2-selective behavior by incorporating different types of fillers. The preparation protocol was optimized to allow the formation of stable, compact and mechanically robust films up to 50% in weight of filler, with a thickness of around 50 micrometers. It is shown that the evaporation rate is crucial to obtain good films. Pure and mixed matrix membranes were tested via DSC and SEM to verify the thermal behavior and morphology; fillers were analyzed to detect the particle size distribution. The gas transport behavior of the mixed matrix membranes with different filler loadings was tested with Helium, CO_2, N_2 and CH_4 at different temperatures and correlated to the membrane structure and preparation protocol. The effect of membrane annealing on the membrane structure and gas transport was studied, indicating a strong effect of membrane history, especially in the case of hydrophilic fillers.

466580 Gas Permeation Behaviors of Highly Permeable Thermally Rearranged (TR) Hollow Fiber Membranes in Flue Gas System

Abstract:

In gas separation processes using membranes, for example CO_2/N_2 or O_2/N_2separation, membrane performance itself as represented by permeability (or permeance) and selectivity is important in deposal capacity and quality of resultant gas (purity). Particularly, process optimization achieved by controlling operation variations such as feed flow rate, stage-cut, pressure ratio, operation temperature, etc. plays a critical role influencing separation factor and overall operation cost in mixed gas separation system. In addition, the understanding of gas transport mechanism in presence of water vapor is essential in a process where water cannot be removed before it is in contact with membranes.

In this study, we fabricated highly permeable thermally rearranged polybenzoxazole (TR-PBO) hollow fiber membranes and their modules to examine gas permeation performance in mixed gas system similar to real flue gas exhausted from post-combustion process. The optimal operation condition for CO_2 capture application was identified from simulation data. Additionally, we proposed a conceptual transport mechanism in TR-PBO hollow fiber membranes under humidified condition. Separation factor showed an increase with higher water vapor content without severe permeation flux drop. Moreover, the TR-PBO membrane modules under permeate vacuum condition showed improved gas separation performance with no permeation flux drop.

460359 CO_2 Facilitated Transport Membranes

Abstract:

Membrane separation is a promising technology for CO_2 separation and capture because of the high energy efficiency, small footprint and operational simplicity. CO_2 facilitated transport membrane is an important group of membranes for CO_2 separation. This work first reviews the state of the art in CO_2facilitated transport membranes, followed by some current results on novel amine-containing membranes.

In this work, the facilitated transport mechanism is described in conjunction with the chemical reactions of $CO_2$and amine carriers. A wide range of facilitated transport membranes is discussed, including amine-based fixed-site carrier membranes, amine mobile carrier-containing membranes, and other facilitated transport membranes with different types of carriers and/or inorganic fillers to further improve the performance of facilitated transport membranes.

High molecular weight polyvinylamine was successfully synthesized in our group, which served as the fixed site carrier and the polymer matrix to incorporate mobile carriers into the membranes. The membranes demonstrated a CO_2 permeance of 757 GPU and a CO_2/N_2 selectivity of 151 under the typical conditions for flue gas CO_2 capture (57 °C, 1 atm, 100% relative humidity, 20% CO_2 in the feed gas balanced with N_2 on a dry basis). As reported that the CO_2 loading capacity of sterically hindered amine can be twice of unhindered amine in aqueous solution, the high molecular weight polyvinylamine was also modified into sterically hindered polyvinylamine in this work and demonstrated up to 24% improvement in CO_2 permeability and 14% increase in CO_2/N_2 selectivity. Sterically hindered polyvinylamine has the potential to become the next-generation fixed site carriers for CO_2 facilitated transport membranes.

464311 High-Performance Intrinsically Microporous Dihydroxyl-Functionalized Triptycene-Based Polyimide for Natural Gas Separation

Abstract:

A novel polyimide of intrinsic microporosity (PIM-PI) was synthesized from a 9,10-diisopropyl-triptycene-based dianhydride (TPDA) and dihydroxyl-functionalized 4,6-diaminoresorcinol (DAR). The unfunctionalized TPDA-*m*-phenylenediamine (*m*PDA) polyimide derivative was made as a reference material to evaluate the effect of the OH groups in TPDA-DAR on its gas transport properties. Pure-gas permeability coefficients of He, H_2, N_2, O_2, CH_4, and CO_2 were measured at 35 °C and 2 atm. The BET surface area based on nitrogen adsorption of dihydroxyl-functionalized TPDA-DAR (308 m^2 g^{-1}) was 45% lower than that of TPDA-*m*PDA (565 m^2 g^{-1}). TPDA-*m*PDA had a pure-gas CO_2 permeability of 349 Barrer and CO_2/CH_4 selectivity of 32. The dihydroxyl-functionalized TPDA-DAR polyimide exhibited enhanced pure-gas CO_2/CH_4 selectivity of 46 with a moderate decrease in CO_2 permeability to 215 Barrer. The CO_2 permeability of TPDA-DAR was 30-fold higher than that of a commercial cellulose triacetate membrane coupled with 39% higher pure-gas CO_2/CH_4 selectivity. The TPDA-based dihydroxyl-containing polyimide showed good plasticization resistance and maintained high mixed-gas selectivity of 38 when tested at a typical CO_2 natural gas wellhead CO_2 partial pressure of 10 atm

464132 Influence of Surface-Treated SiO_2 on the Transport Behavior of O_2 and N_2 through Polydimethylsiloxane Nanocomposite Membrane

Abstract:

Mixed matrix membranes were prepared with polydimethylsiloxane (PDMS) and different weight percent surface-treated fumed silica (SiO_2) to investigate the influence of SiO_2 on transport behaviour of O_2 and N_2 gases in the nanocomposite membranes. Fourier transform infrared spectroscopy (FTIR) showed that OH functional group on the surface of SiO_2 was consumed upon incorporation of the silica into the polymer matrix. Thermogravimetric analysis (TGA) results revealed that SiO_2-PDMS has improved thermal property over neat PDMS, supporting the argument that there is good interaction between the polymer and the fumed silica. Scanning electron microscopy (SEM) images of SiO_2-PDMS membranes showed uniform dispersion of SiO_2 nanoparticles in PDMS matrix. SiO_2 nanoparticles disrupted and altered the PDMS polymer chains packing arrangement resulting in different membrane transport behaviour of both O_2 and N_2 gases in SiO_2-PDMS compared to the neat PDMS membrane. While the O_2 flux through SiO_2-PDMS membranes was observed to increase with time, N_2 flux decreased with time before attaining steady state. The 10wt%SiO_2-PDMS membrane exhibited improved performance compared to neat PDMS membranes with O_2/N_2 selectivity and O_2 permeability increased from 2.43 to 3.46 and 520 Barrer to 640 Barrer, respectively, at 30 psig. This improvement is attributed to the influence of the well dispersed SiO_2 nanoparticles in the PDMS matrix.

469603 Characterization and Gas Separation Performance of Two High Performance Polyimide Blend Membranes

Abstract:

Characterization and gas separation performance of two high performance polyimide blend membranes

Saeed Mazinani[a][*], Roozbeh Ramazani[b], Siavash Darvishmanesh[c], Bart Van der Bruggen[a]

[a] ProcESS - Process Engineering for Sustainable Systems, Department of Chemical Engineering, KU Leuven, Celestijnenlaan 200F, B-3001 Leuven, Belgium

[b] Department of Chemical Engineering, Chemical and Petroleum Engineering School, Shiraz University, Shiraz 71345, Iran

[c] Department of Chemical and Biological Engineering, Princeton University, Princeton, New Jersey 08544, USA

Abstract

The emission of greenhouse gases, generated by human industrial activities, is perhaps the most prominent environmental problem today. In recent years, CO_2 has attracted much attention because it is the most important human induced greenhouse gas and is believed to be responsible for approximately 64% of the enhanced greenhouse effect. Carbon Capture and Sequestration (CCS) is considered as a promising way to reduce the CO_2 concentration in the atmosphere. According to the IEAâ€™s 2Â Â°C scenario report, the CO_2 emissions should be decreased to 14 billion tons per year in 2050, and CCS should be responsible for approximately 20% of this reduction [1].

Among various carbon capture technologies, polymeric membrane-based technology is a potential separation method due to small operation units, low energy consumption, low capital and operating costs, lack of mechanical complexity, environmentally friendly, and compact process design. However, the major drawback of membrane gas separation which limits its application is the low CO_2 concentration and pressure of the flue gas, which requires the use of membrane with high selectivity to meet the specification of the International Energy Agency, a CO_2 recovery and purity of around 80% [2].

Generally, properties of membrane desired in gas separation are high transport performance, high mechanical strength, strong thermal and chemical stability, plasticization resistance, commercial availability, reasonable cost, along with high permselectivity. In general, polymeric materials do not simultaneously meet all of these criteria. For example, highly permeable polymers exhibit moderate to low selectivity values while materials with high resistance to harsh chemical environments or plasticizing gases are either difficult to process or are very expensive [3].

In recent years, gas separation research has focused on aromatic, glassy polymers and particularly on rigid, polymers with high glass transition temperature (Tg). Within several classes of aromatic glassy polymers, polyimides have been attracting much attention due to their outstanding set of physiochemical properties including superior thermal stability, remarkable chemical resistance, excellent thermo-oxidative stability, and good mechanical toughness [4]. Due to these unique characteristics, they are widely used as adhesives, coatings, films, fibers, foams. However, due to their high crystallinity (brittleness) and high cost, only few of them (P84, Matrimid, Ultem) are suitable for membrane manufacturing.

Blending of polymers with different features is considered an applicable method to develop new polymer with desired properties [5]. Synergistic interaction of different polyimides can impart superior properties to the blend over the constituting polymers [6]. In the study of polymer blends, polymer-polymer miscibility is crucial because the new properties depend on the degree of miscibility of the polymers at a molecular level. It is well known that in most miscible polymer blends, specific interactions such as dipole-dipole forces, hydrogen bonding and charge transfer complexes are mainly responsible for the miscibility [7]. It is interesting to note that polymer blends are usually incompatible because of unfavorable interactions between polymer segments. The interest in the use of polymeric blends as gas separation membranes is associated with the possibility of gas transport management through the morphological control of the cross-section of the material, which is mainly dependent on blend components, phases and interface characteristics [8, 9]. In this work, the miscibility behavior and intermolecular interactions of two high performance polyimides in compositions of 100/0, 80/20, 50/50, 20/80, 0/100 have been evaluated. The polymer blend systems have been characterized by different analytical techniques including optical microscopy, Differential Scanning Calorimetry (DSC), Fourier Transform Infrared Spectroscopy (FTIR), X-ray Diffraction (XRD) and rheological measurements. DSC results of two blend systems showed the existence of a single glass transition temperature (Tg) in each composition, suggesting the miscibility of the blends. In order to study the specific interactions between polymer blend systems, the Tgs of the polymer blends were estimated by theoretical equations and compared with experimental data. The empirical Tg values exhibited a positive deviation from the linearity, indicating the presence of specific interactions between polymer chains; this was confirmed by FTIR spectra. The interactions between studied polymer systems and four aprotic solvents were assessed on the basis of the difference between their solubility parameters. DMSO and DMAC had the highest and lowest affinity respectively. Moreover, the single and mixed gas separation properties of the blend films were measured for separation of CO_2 from N_2 by a lab-made gas permeation test setup at pressures 2, 4 and 6 bar for temperatures between 303 and 333 K. Comparison of the results with that of the pure membranes revealed that the blend membrane had a higher permeability to CO_2 and a lower permeability to the N_2 and therefore had a higher selectivity value of CO_2/N_2. Gas permeance of N_2, increased with increase in feed pressure irrespective of blend composition.

References

[1] M.H. El-Naasa, M. Marzouqia, S.A. Marzoukb, N. Abdullati, Evaluation of the removal of CO_2 using membrane contactors: Membrane wettability, Journal of Membrane Science, 350 (2010) 410-416.

[2] Polymeric Gas Separation Membranes, Yuri Yampolskii, Macromolecules 2012, 45, 3298-3311

[3] High performance polymer membranes for CO_2 separation Seungju Kim1 and Young Moo Lee, Current Opinion in Chemical Engineering 2013, 2:238-244.

[4] Gupta. R.K., Gaba, ID., Pande, C.P., Singh, Y.P., Sing, R.P., 1987. Polyimides, Pleunm Press: New York, 547-554.

[5] DeMeuse, M., 2014. High Temperature Polymer Blends. Technology & Engineering.

[6] Handbook of Polymer Blends and Composites, Volume 3, Anand K. Kulshreshtha, Cornelia Vasile, iSmithers Rapra Publishing, 2003.

[7] Walsh, D.J., Rostami, S., 2005. The miscibility of high polymers: The role of specific interactions. Springer Berlin Heidelberg. 70, 119-169.

[8] Coleman, M.M., Graf, J., Painter, P.C., 1991. Specific interactions and miscibility of polymer blends. Technomic publishing Co., Pensilvania, USA.

[9] Paul, D.R., Newman, S., 1978. Polymer Blends, Academic Press: New York, NY, USA.

---

[*]E-mail: saeed.mazinani@kuleuven.be

469549 Tuning Separation Properties of Carbon Molecular Sieve Membranes with Diamine Dopants

Abstract:

Tuning Separation Properties of Carbon Molecular Sieve Membranes with Diamine Dopants

Graham B. Wenz and William J. Koros

Georgia Institute of Technology

Atlanta, GA 30332

Natural gas separations have become increasingly important as worldwide natural gas production has increased in the past decade. Common natural gas impurities, such as CO_2, must be removed in order to protect equipment and meet pipeline specifications for transmission. Carbon molecular sieve (CMS) membranes, formed by high temperature pyrolysis of polymers under controlled conditions, have emerged as a new class of materials that can surpass the polymer "upper bound". Similar to their glassy polymer precursors, CMS membranes display a physical aging phenomenon where productivity loss couples with a mild increase in selectivity when monitored over time. This behavior is thought to be cause by the relaxation of "free volume" trapped within the CMS structure during formation. This work outlines a new post-synthetic modification technique where CMS membranes are exposed to a solution containing PPM levels of amine dopants, with the original goal of preventing the "free volume" relaxation within CMS materials. While long term CO_2 and CH_4 permeation experiments showed aging-resistant membranes did not result, attractive performance resulted from the *amine doped materials*. The results of these experiments suggest *amine doping* as new technique to selectively modify the large pores present within the CMS microporous structure, resulting in increased size and shape based selectivity.

466025 *Sod*-Zmof/Polyimide MIXED-Matrix Membranes for CO_2 Separation

Abstract:

Polymer/MOF mixed matrix membranes offer enhanced gas sepration performance compared to neat polymeric membranes. In designing such membranes, the selection of polymer/MOF pairs is of crucial importane. Polyimides have been of interest as polymeric gas separation membranes for a long while due to their thermal and mechanical properties. On the other hand, eeolite-like metal organic frameworks (ZMOFs), which are a subclass of MOFs, are promising candidates for separation and purification applications. They have extra-framework ions in the pores of molecular dimensions increasing the interactions with guest molecules and improving their separation, storage, or ion-exchange capability. The charge-compensating extra-framework ions inside the cavities of ZMOFs can be ion-exchanged with alkali metals and this is expected to increase their affinity for CO_2 and improve the CO_2 separation performance of polyimide based mixed matrix membranes for gas mixtures like natural gas (CO_2/CH_4) and flue gas (CO_2/N_2).

In this work, as-synthesized and alkali ion-exchanged *sod*-ZMOF nanoparticles were used to develop polyimide/ZMOF mixed matrix membranes. Two different types of polyimides (Matrimid® 5218 and 6FDA-DAM) and one copolyimide (6FDA-BTDA-pBAPS) were chosen for this purpose. After the analyses of structural and thermal characteristics of the prepared membranes; CO_2/CH_4 and CO_2/N_2 separation performance of the membranes were tested in terms of CO_2/CH_4 and CO_2/N_2 separation performances via single and mixed gas permeability measurements.

This work is part of the project supported by the Scientific and Technological Research Council of Turkey

(TUBITAK) through the grant no. 113M776.

458430 Efficient Modification of Outer Selective P84 Nanofiltration (NF) Hollow Fiber Membranes for Cadmium Removal

Abstract:

Efficient Modification of Outer Selective P84 Nanofiltration (NF) Hollow Fiber Membranes for Cadmium Removal

Jie Gao, Shi-Peng Sun, Wen-Ping Zhu, Tai-Shung Chung*

Department of Chemical and Biomolecular Engineering

National University of Singapore, Singapore 117576

*Corresponding author

Tel: +65-65166645; fax: +65-67791936; Email: chencts@nus.edu.sg

Abstract:

Cadmium is a common heavy metal present in industrial effluents from pesticide, smelting, dye

manufacturing and electroplating industries. Like other heavy metals, cadmium is non-biodegradable. Accumulation of cadmium can occur in living systems and results in harmful environmental effects. Side effects, such as nausea, bone lesions, renal disturbances and cancer, may happen if cadmium is accumulated in human beings. Thus, the removal of cadmium from wastewater before discharge has important meanings to the human society. One method to remove those heavy metals is nanofiltration (NF) because it is able to offer good rejections to multi-valent heavy metal ions without much compromising water fluxes as compared to those of reverse osmosis (RO) membranes.

There is a growing interest in thin-film composite (TFC) NF hollow fiber membranes. The TFC layer can be formed on both inner and outer surfaces of hollow fiber membranes. The outer selective TFC hollow fiber membranes may have less mass transfer resistance and lower chances being blocked by foulants compared to the inner ones. Thus, the outer selective TFC fibers can be fabricated into smaller fiber dimensions with a higher package density in modules. Despite these advantages, the molecular design of an ideal TFC layer on the outer surface of hollow fiber membranes remains a challenge, making the mass production difficult. Since most TFC layers are made from interfacial polymerization of m-phenylenediamine (MPD) and trimesoyl chloride (TMC), removing excess solution after the MPD coating is the major obstacle in the process. Vacuum assist interfacial polymerization or wipe drying are required to make a defect free polyamide layer. Another obstacle is that a large amount of alkane solvents has to be used during interfacial polymerization, which brings extra costs and environmental issues. Eliminating alkane solvents is critical to lower manufacture costs and yield a green production process.

In this study, green methods to prepare the outer selective TFC hollow fiber membranes were explored by firstly modifying the membrane substrate with polyethyleneimine (PEI) and then by water soluble small molecules such as glutaraldehyde (GA) and epichlorohydrin (ECH). Using P84 polyimide as the substrate, not only do these modifications decrease substrate's pore size, but also vary surface charge by making the membranes less positively charged. As a result, the resultant membranes have higher rejections against salts such as Na_2SO_4, NaCl and $MgSO_4$. The PEI and then GA modified membrane has the best separation performance with a NaCl rejection over 90 % and a pure water permeability of 1.74 ± 0.01 $Lm^{-2}bar^{-1}h^{-1}$. It also shows an impressive rejection of 94 % for $CdCl_2$ removal during long-term stability tests. The $CdCl_2$ rejection remains higher than 90% when testing it from 5 to 60 °C. This study may provide useful insights for green manufacturing of next-generation NF hollow fiber membranes.

Acknowledgements:

The authors would like to thank Chinese Gansu province for the research project entitled "Membrane development to removal heavy metals" with the grant number of R-279-000-360-597 and GSK-EDB Trust Fund for the project entitled "New membrane development to facilitate solvent recovery and pharmaceutical separation in pharmaceutical syntheses" with the grant number of R-706-000-019-592. The authors also acknowledge Mr. Z.L. Cheng, Dr. X. Li, Ms, Y. Zhang, Dr. Z.Z. Zhou and Dr. Y.K. Ong for all their helps and suggestions on the experimental work. Thanks are also due to Mr. K.P. Ng for his help on the NF set-up.

462463 Self-Generated RO Concentrate Pulse Backwash of Ultrafiltration Module in an Integrated Seawater Desalination UF-RO Pilot Plant

Abstract:

The effect of self-generated RO concentrate pulse backwash was evaluated in a novel integrated UF-RO seawater desalination pilot plant (with RO permeate production of up to 45.2 m^3/day at 35% feed water recovery). Direct usage of RO concentrate to backwash the UF module was further enhanced with pulse backwash (PBW) generated using bladder-type hydraulic accumulators. The accumulators were charged directly from the RO concentrate stream within a period of 30-40 s. Experimental and theoretical model analysis of the hydraulic accumulator operation (i.e., both pressure and flux time profiles), validated via a series of field experiments, demonstrated the ability to discharge the pulse over a short period along with continuous UF backwash (directly from the RO brine stream). This combined backwash strategy enabled peak UF backwash flux up to 4.2 - 4.6 times higher than the normal UF filtration flux. Along with the above backwash approach, the study successfully demonstrated automated control of the frequency and intensity of the PBW system in seawater field study. Given the rapid accumulator charging and discharge periods, UF backwash was feasible with multiple consecutive backwash pulses that was more effective than with a single pulse. Field testing of UF PBW with adaptive triggering of UF backwash and variable PBW instances per filtration/backwash cycle enabled highly effective UF operation over a wide range of water quality conditions and without the need for chemical cleaning. The results also showed that the UF system was capable of providing consistently high quality RO feed and the RO system operation was stable without deterioration in performance even during severe fouling conditions that have occurred during stormy weather.

458899 Membrane Fouling and Electrochemical Regeneration at a Sub-Stoichiometric TiO_2 Reactive Electrochemical Membrane

Abstract:

In this study, a non-invasive and non-destructive electrochemical impedance spectroscopy (EIS) characterization method was used to study membrane fouling at a novel ultrafiltration (UF) reactive electrochemical membrane (REM). The REM was fabricated from a titanium dioxide UF precursor and converted to a substoichiometric titanium dioxide using a high temperature reduction method. The EIS method was used to spatially characterize membrane fouling by interpreting the EIS data using a theoretical transmission line model. Model simulations indicate that different modes of membrane fouling (e.g., pore blocking, pore constriction, and cake layer formation) produced distinct EIS spectra. The model was used to interpret experimental EIS data collected during filtration experiments in the presence of two model membrane foulants, humic acid and polystyrene microspheres. Results indicate that the transmission line model could accurately characterize membrane fouling and was used to aid in the development of a chemical free electrochemical regeneration method. Electrochemical regeneration was able to recover the flux of a fouled REM from 3% to between 76% and 99% of the initial flux over 5 continuous fouling/regeneration cycles for humic acid fouling and full flux recovery for polystyrene microsphere fouling. The operating cost of the electrochemical regeneration method is estimated at 1.3% of that for NaOH cleaning, which indicates the great promise of this new method for separation processes.

462886 Molecular Dynamics Simulation of Water Desalination Via Functionalized Covalent Organic Framework Membranes

Abstract:

The scarcity of fresh water has become a great concern with a rapidly growing global water gap. As an energy-efficient desalination technique, reverse osmosis (RO) accounts for more than half of the existing desalination capacity. In order to screen suitable nanoporous membrane materials, and to investigate the effects of pore size and functionality, we design a series of single layer covalent organic framework (COF) membranes based on an experimentally synthesized COF structure TpPa-1. They consist of 1,3,5-triformylphloroglucinol (Tp) and *p*-phenylenediamine (Pa) with various functional groups, namely TpPa-AM2, TpPa-AMC_2NH_2, TpPa-OC_3OH, TpPa-OC_4H_9, TpPa-AMCOOH, TpPa-OBn and TpPa-AM3. The molecular dynamics (MD) simulation results indicate that water permeances in these COF membranes are affected by both pore size and functionality. Water traverse across the pore in AMC_2NH_2, TpPa-OC_3OH and TpPa-AMCOOH is enhanced, due to hydrophilic functional groups pointing inward the pores. When the polarity of functional groups is similar, the pore size becomes a dominant factor exhibiting a positive correlation with water permeance. Nevertheless, all the COF membranes show exceptionally high water permeance ranging from 1216 to 3375 kg/(m^2·h·bar), more than three orders of magnitude higher than commercial RO membranes. Furthermore, water density distribution within the membrane pores are found to be highly dependent on hydrophilicity of the pores, resulting in higher water density for hydrophilic membranes. The salt rejection is calculated to be 95.8% in TpPa-AMCOOH, and above 98.3% in other TpPa membranes. Among all the TpPa membranes designed, TpPa-OC_3OH has high water permeance (2816 kg/(m^2·h·bar)) with excellent salt rejection (99.4%). This simulation study provides a microscopic insight into the key factors governing water desalination, and suggests that TpPa-OC_3OH might be a potential candidate for high performance water desalination.

461747 Novel Amphiphilic Zwitterionic Copolymer Membranes of ~ 1nm Pore Size for Industrial Wastewater Treatment and Reuse

Abstract:

Membranes with ~1 nanometer effective pore size that allow removal of high oil/organic content (dyes, industrial chemicals, surfactants etc.) have numerous applications in the treatment of wastewater from textile plants, oil refineries, fracking and oil extraction. The small effective pore size of such membranes can potentially provide better effluent quality and prevent the environmental release of a wider range of pollutants. However, there are very few commercial membranes in this pore size range, and those that are available exhibit poor selectivity, charged surfaces which lead to complex separation mechanisms, and fouling due to the accumulation of feed components on the membrane surface. In addition to high flux, sharp selectivity and exceptional fouling resistance to common foulants in wastewaters, membranes for these applications should also have high chemical resistance and be easily fabricated into large-area membranes. Zwitterionic groups have been shown to strongly resist biomacromolecular fouling due to their high degree of hydration, which makes them promising for membrane applications. We have introduced a new class of membranes whose selective layers are made of self-assembling zwitterionic amphiphilic random copolymers. These membranes derive not only fouling resistance but also their permeability and selectivity from this self-assembled nanostructure. In this study, we evaluate the potential and performance of these new membrane materials in wastewater treatment. We synthesized random copolymers of a hydrophobic monomer (trifluoroethyl methacrylate) and two zwitterionic monomers (sulfobetaine methacrylate, phosphobetaine methacrylate) by free radical polymerization. These copolymers self-assemble into bicontinuous networks of ~1 nm nanochannels, driven by the strong electrostatic interactions between the zwitterionic groups, and documented by transmission electron microscopy (TEM). Membranes manufactured by coating these copolymers on commercially available support membranes exhibit size-based selectivity with a ~1 nm size cut-off demonstrated by filtering negatively charged dyes. We have studied how feed composition such as ionic strength and presence of foulants in varying wastewater streams affects membrane performance. Specifically, we have tested the performance of these membranes by filtering single foulants, and

simulated wastewater feeds. These membranes are extremely resistant to fouling by proteins and oil emulsions, fully retaining their permeability even during the filtration of the foulant solution in dead-end mode for up to 7 days. They are also stable in a wide range of chemical solutions: salt solutions (upto 1M NaCl), acidic and basic buffers, as well as chlorine. They exhibit a sharp, size-based cut-off for organic molecules such as dyes found in textile wastewater. Their low salt rejection allows operation at lower pressures. These results indicate the promise of this new family of membrane materials for industrial wastewater treatment, reclamation and reuse.

464906 Analysis of Air Scouring Effect in Flat Sheet Type MBR

Abstract:

MBR (Membrane Bioreactor) is a water treatment process to treat the sewage and the waste water. Membrane module is immersed in an activated sludge tank and filtrated by microfiltration membrane or ultrafiltration membrane. As compared with conventional process, activated sludge method, MBR has advantages of space saving and treated water quality, but inferior in much power consumption caused by the aeration for restraining the membrane fouling. Reduction of power consumption is the biggest challenge in MBR process. There are two methods to reduce aeration quantity, one is developing the low fouling membrane, and the other is increasing washing efficiency for membrane surface without increasing power consumption. To develop the method for increasing washing efficiency for membrane surface, it is important to understand the scouring effect for membrane surface caused by aerated bubbles. Several researches studied washing effect for membrane surface caused by aeration, but there are few researches focusing on quantifying the washing effect for membrane surface caused by aerated rising bubbles.

This study is focused on analyzing the scouring effect of rising bubbles in flat sheet type MBR. The relationship between aerated bubbles and shear stress at the membrane surface was analyzed using PIV analysis. We prepared the tank which modeled the zone between adjacent membranes in flat sheet type MBR and filmed the motion of tracer particles around a rising bubble. Velocity distribution of liquid around a rising bubble was analyzed, and then shear rate and shear stress at the membrane surface was derived. It was confirmed that the aeration condition affect the shear stress at the membrane surface by using PIV analysis. And also, relationship between shear stress at the membrane surface and washing effect for membrane surface caused by aeration was studied in model experiment. From the experiment, it was studied what kind of shear stress, such as average and maximum, has more relation with the washing efficiency. These results showed that PIV analysis was a useful method to quantify the washing effect of rising bubbles restraining membrane fouling in MBR. Also it was confirmed that the increase of shear stress at the membrane surface could improve the detachment of foulants and washing efficiency for membrane surface.

472351 Novel Composite Adsorptive Membranes for Ammonia Removal from Wastewaters

Abstract:

Natural zeolite is a superior and inexpensive adsorbent material for ammonia removal from municipal, industrial, and aquaculture wastewaters. However, zeolite used directly as an adsorbent material leads to its undesirable leaching and dissolution water. Thus, it is important to incorporate zeolite particles in a host matrix in order to improve their stability.

Novel composite fiber membranes were developed via incorporation of zeolite particles into polymeric matrices. These composite membranes demonstrate high porosity, high capacity and uniform particle distribution. It was shown that the capacity for ammonia removal increases with the increase of zeolite loading with fast ion exchange kinetics. The ammonia captured can be recovered and concentrated as fertilizers and fuels. These membranes are easy to regenerate and can be used repeatedly.

471622 Nucleation of Entropic Crystals

Abstract:

Many hard polyhedra crystallize into ordered structures, driven by directional entropic forces. Here, we use Monte Carlo simulations to investigate the nucleation behavior of polyhedra that all crystallize in the same structure - the face-centered cubic close sphere packing. We describe how the nucleation of anisotropic shapes, with varying amounts of faceting and multiple symmetries, differs from the behavior of spheres. We quantify the nucleation behavior by measuring crystallization probabilities and nucleation energy barriers, as well as cluster size distributions and local ordering in the dense fluid. Our results elucidate the impact that faceting and particle symmetry have on the nucleation behavior in entropically stabilized crystal structures.

459248 Two-Step Crystal Nucleation Is Selected Because of Lower Surface Free Energy Barrier

Abstract:

Nucleation of crystals from solution underlies multitudinous natural and engineered processes. The formation of a crystal nucleus in a supersaturated environment creates an energy-rich interface. According to classical theory, the associated surface free energy barrier is overcome by random fluctuations leading to direct assembly of monomers into a crystalline embryo. Discrepancies between classical predictions and experimental observations suggested a two-step mechanism of crystal nucleation, whereby nuclei are hosted by disordered solute-rich precursors. A fundamental controversy of two-step nucleation is that whereas precursor populations often occupy a minor volume, nucleation confined to precursors is faster than direct nucleation. Here we demonstrate that nucleation of insulin crystals, which occurs with a highly regulated rate in mammalian pancreases as a part of insulin biosynthesis, is assisted by protein-rich clusters of radius 160 â€“ 200 nm that occupy ca. 10^{-6} of the solution volume. We show that the surface free energy encountered by newly nucleated crystals is ca. 80-fold lower than the independently determined surface free energy of the crystal-solution interface; the latter value dictates the barrier for direct crystal nucleation in the solution. This comparison indicates that two-step nucleation is faster than the direct pathway and is selected as a phase transformation route due to a lower surface free energy barrier and suggests that nucleation of crystals can be controlled by manipulating the properties of the disordered precursors.

455080 Modeling the Step Velocity of Non-Centrosymmetric Growth Units and Accounting for Stable/Unstable Layers

Abstract:

The velocities of step edges on crystal faces are anisotropic. This must be accurately captured for a mechanistic model of crystal growth to be successful, since both a spiral growth regime and a birth-and-spread two-dimensional nucleation and growth regime depend on step velocities. The key components to the step velocity are the density of kinks along the step edge (the kink density) and the net rate of growth unit incorporation into kinks (the kink rate).

Non-centrosymmetric growth units represent the majority of industrially relevant molecules, but introduce significant complexity into this problem. The pattern of different growth units along the step edge produces multiple kink sites, and so the simple centrosymmetric expressions for kink density [1] and kink rate [2] are no longer applicable. Kuvadia and Doherty [3] introduced a framework for the mechanistic modeling of crystals with non-centrosymmetric growth units: they considered hypothetical transformations to form an analogous Boltzmann distribution for the kink density and determined an expression for the kink rate for a general number of growth units. Importantly, they also introduced the concept of stable/unstable step edge layers, which are unique to non-centrosymmetric growth units.

We present several improvements and generalizations of this framework that enable a more universal application and provide the ability to better capture the effect of stable/unstable layers. First, we use surface energies around each step edge site to form the Boltzmann-distributed kink density, which enables calculations of specific site densities. This also ensures the required collapse to the centrosymmetric case under appropriate limits. Second, we consider that specific kink sites are unable to move at their native kink rate velocity indefinitely; they can become limited following collision events with other kink sites. This effect can be captured by considering characteristic timescales for possible collisions and a characteristic timescale (based on kink annihilation) for edge reorganization. Each kink is allowed to advance at its native velocity

for a fraction of the reorganization timescale and is then appropriately limited. Physically, this corresponds to the exclusion of overhang growth along the step surface. To calculate the step velocity, the individual contributions from each type of kink on the surface are summed (i.e., specific kink densities and specific native kink rates, which may become limited).

We have tested this expression for the case of a 2-layer step with alternating layers of 'A' and 'B' growth units against kinetic Monte Carlo simulations, varying the edge energy of B. The predicted step velocities are in excellent agreement with simulations and can capture the trend that as the B edge energy is increased, double kinks become dominant on the step edge surface. The resulting models enable accurate morphology prediction for a wide class of molecular solutes.

References

1. Frenkel, J. On the surface motion of particles in crystals and the natural roughness of crystalline faces. *J. Phys. (Moscow)* 1945, *9*, 392-398.
2. Voronkov, V. The movement of an elementary step by means of the formation of one-dimensional nuclei. *Sov. Phys. Crystallogr.* 1970, *15*, 8-13.
3. Kuvadia, Z. B.; Doherty, M. F. Spiral Growth Model for Faceted Crystals of Non-Centrosymmetric Organic Molecules Grown from Solution. *Cryst. Growth Des.* 2011, *11*, 2780-2802.

471664 The Use of Ionic Liquids for API Purifications and Its Effect on Nucleation Kinetics

Abstract:

The Use of Ionic Liquids for API purifications and its effect on nucleation kinetics

Samir A. Kulkarni[a,b*], Cameron C. Weber[a,c], and Allan S. Myerson[a]

[a] Department of Chemical Engineering, Massachusetts Institute of Technology, 77 Massachusetts Avenue, Cambridge, Massachusetts 02139, USA

[b] SSCI, A Division of Albany Molecular Research, Inc. (AMRI), 3065 Kent Avenue West Lafayette, IN 47906-1076

[c] Department of Chemistry, Imperial College London, London SW7 2AZ, UK

email: samir.kulkarni@amriglobal.com; cameron.weber@imperial.ac.uk; myerson@mit.edu

Keywords: Crystallization, Nucleation, Ionic Liquids

The source of this interest arises from the favorable properties some Ionic Liquids can exhibit, such as low flammability, low vapor pressures, tunable physicochemical properties through appropriate ion selection, large liquidus ranges and good thermal stability. The tunability of Ionic Liquids has been one of their most widely touted and exploited benefits as solvents, as the ability to manipulate solution interactions can lead to the design of Ionic Liquids that are good solvents for chosen target compounds. This has been explored with regards to the solubility of pharmaceutical substances, and their crystallization. These approaches

typically involve the use of antisolvent crystallization or a combination of liquid-liquid extraction and antisolvent crystallization to recover the target [1]. Surprisingly, studies into the use of Ionic Liquids as solvents for the purification by crystallization of organic solids, particularly pharmaceuticals, remains relatively limited despite the industrial significance of this methodology. In order to address these concerns we have considered the use of cooling crystallization in an Ionic Liquid as a purification methodology to reduce the complexity of the resultant solution. The Ionic Liquid possesses a relatively high viscosity over a wide range of temperature and the high initial viscosity can act as kinetic barrier to both nucleation and subsequent crystallization [2].

To ascertain the generality of our approach and to probe the limits of miscibility and thermal stability, we have examined a wide range of different active pharmaceutical compounds (APIs) as model systems for the cooling crystallization approach [3]. To demonstrate proof of principle and examine the efficacy of this purification approach, the cooling crystallization of acetaminophen with various impurities has been conducted and compared with results obtained for antisolvent processes and with cooling crystallizations conducted in water. The results in all cases indicated that impurity inclusion was comparable or improved compared to antisolvent crystallizations from either water, organic solvent systems or Ionic Liquids with substantially improved yields. Such a straightforward crystallization approach taking advantage of the low vapor pressure and wide ranging miscibility of ionic liquids with various solvents suggests a route towards a general purification procedure for many small molecule APIs in Ionic Liquids. The ionic liquid displayed a diverse effect on nucleation kinetics, as ionic liquid was more effective than ethanol-water system to reduce nucleation rate and to enhance growth time. Specific ionic liquids thus provide an additional tool in nucleation control in solution crystallization.

References:

1. CC Weber, AJ Kunov-Kruse, RD Rogers, AS Myerson, Manipulation of ionic liquid anion–solute–antisolvent interactions for the purification of acetaminophen, Chem. Commun., 2015,51, 4294-4297.

2. Kulkarni SA, Kadam SS, Meekes H, Stankiewicz AI, ter Horst JH (2013) Crystal Nucleation Kinetics from Induction Times and Metastable Zone Widths. Crystal Growth & Design 13 (6):2435-2440.

3. Cameron C Weber, Samir A Kulkarni, Andreas J Kunov-Kruse, Robin D Rogers, Allan S Myerson The Use of Cooling Crystallization in an Ionic Liquid System for the Purification of Pharmaceuticals. Cryst. Growth Des., 2015, 15 (10), pp 4946–4951.

458624 Homogeneous Bubble Nucleation in Liquid Carbon Dioxide from a Hybrid of Molecular Dynamics Simulation and Density Gradient Theory

Abstract:

Bubble nucleation is the first stage in cavitation. This phenomenon is well known in industrial processes and generally unwanted. It may occur, if a liquid is brought into the metastable range of the vapor-liquid equilibrium. It is usually described by classical nucleation theory (CNT), which relies on extreme simplifications, e.g. assuming that the usually nanoscopic bubbles have the same properties as the saturated bulk vapor and that their surface tension is the same as that of the planar interface[1–3]. In many cases CNT fails drastically[4–7], while in other cases, it gives fair predictions. This behavior is not fully understood, but it is argued sometimes that in nucleation different phenomena combine and sometimes errors cancel out[8,9].

Another method used for studies of nucleation in literature[10,11] is density functional theory (DFT) or simplifications thereof like the density gradient theory (DGT) first applied by Cahn and Hilliard to droplet nucleation in demixing incompressible liquids[12]. DFT and DGT have the advantage of correct limiting behavior approaching both binodal and spinodal. They do also inherently consider the size dependence of surface tension. Using full DFT, however, the free energy of the heterogeneous system must be known as a functional of density, which is most often not the case. Also, both theories only give information on the thermodynamic barrier, but not on the kinetics of bubble nucleation. This is usually addressed by using expressions for the kinetic pre-factor obtained in the framework of CNT. This may be dangerous, since CNT is usually developed for describing the whole process so that using only the pre-exponential factor of CNT may lead to poor results. We propose here a different method with a clean separation between kinetics and thermodynamics.

In order to properly introduce kinetics into the DGT based nucleation model, we hybridize it with molecular simulation. As an example, CO_2 is studied. All our computations are based on the molecular model of Merker et al.[13] of CO_2. The thermodynamic and interfacial properties of this model are used to fit the parameters of the PC-SAFT[14] equation of state (EOS), as well as the influence parameter from DGT. This way, a thermodynamic link between molecular simulation and thermodynamic model is achieved, which is further confirmed by comparison to molecular simulation data in the metastable range. In a next step, the homogeneous bubble nucleation of the molecular model is simulated for two temperatures using molecular dynamics simulation following an approach first introduced by Yasuoka and Matsumoto[15] for droplet nucleation and later on applied by Diemand et al.[16] for bubble nucleation. The kinetic pre-factor of the proposed hybrid theory is then fit to the direct simulation data at two temperatures, where we assume an Arrhenius type dependence on temperature. This way, our hybrid method describes bubble nucleation in the whole metastable range from triple point to critical point without relying on any further assumptions. The results compare favorably to our direct simulations of the nucleation rates and to available experimental data for CO_2[17,18], while CNT shows the wrong temperature dependence and is off several orders of magnitude in terms of nucleation rates.

(1) Blander, M.; Katz, J. L. The Thermodynamics of Cluster Formation in Nucleation Theory. *J Stat Phys* **1972**, *4* (1), 55–59.

(2) Vehkamäki, H. *Classical Nucleation Theory in Multicomponent Systems*; Springer: Heidelberg, 2006.

(3) Blander, M.; Katz, J. L. Bubble Nucleation in Liquids. *AIChE J.* **1975**, *21* (5), 833–848.

(4) Baidakov, V. G.; Bobrov, K. S. Spontaneous Cavitation in a Lennard-Jones Liquid at Negative Pressures. *The Journal of Chemical Physics* **2014**, *140* (18), 184506.

(5) Gonzalez, M. A.; Abascal, J. L. F.; Valeriani, C.; Bresme, F. Bubble Nucleation in Simple and Molecular Liquids via the Largest Spherical Cavity Method. *The Journal of Chemical Physics* **2015**, *142* (15), 154903.

(6) Römer, F.; Kraska, T. Homogeneous Nucleation and Growth in Supersaturated Zinc Vapor Investigated by Molecular Dynamics Simulation. *The Journal of Chemical Physics* **2007**, *127* (23), 234509.

(7) Sinha, S.; Bhabhe, A.; Laksmono, H.; Wölk, J.; Strey, R.; Wyslouzil, B. Argon Nucleation in a Cryogenic Supersonic Nozzle. *The Journal of Chemical Physics* **2010**, *132* (6), 64304.

(8) Römer, F.; Fischer, B.; Kraska, T. Investigation of The Nucleation and Growth of Methanol Clusters from Supersaturated Vapor by Molecular Dynamics Simulations. *Soft Materials* **2012**, *10* (1–3), 130–152.

(9) Wedekind, J.; Hyvärinen, A.-P.; Brus, D.; Reguera, D. Unraveling the ``Pressure Effect'' in Nucleation. *Phys. Rev. Lett.* **2008**, *101* (12), 125703.

(10) Laaksonen, A.; Talanquer, V.; Oxtoby, D. W. Nucleation: Measurements, Theory, and Atmospheric Applications. *Annual Review of Physical Chemistry* **1995**, *46* (1), 489–524.

(11) Oxtoby, D. W. Density Functional Methods in the Statistical Mechanics of Materials. *Annual Review of Materials Research* **2002**, *32* (1), 39–52.

(12) Cahn, J. W.; Hilliard, J. E. Free Energy of a Nonuniform System. III. Nucleation in a Two Component Incompressible Fluid. *The Journal of Chemical Physics* **1959**, *31* (3), 688–699.

(13) Merker, T.; Engin, C.; Vrabec, J.; Hasse, H. Molecular Model for Carbon Dioxide Optimized to Vapor-Liquid Equilibria. *The Journal of Chemical Physics* **2010**, *132* (23), 234512.

(14) Gross, J.; Sadowski, G. Perturbed-Chain SAFT: An Equation of State Based on a Perturbation Theory for Chain Molecules. *Ind. Eng. Chem. Res.* **2001**, *40* (4), 1244–1260.

(15) Yasuoka, K.; Matsumoto, M. Molecular Dynamics of Homogeneous Nucleation in the Vapor Phase. I. Lennard-Jones Fluid. *The Journal of Chemical Physics* **1998**, *109* (19), 8451–8462.

(16) Diemand, J.; Angélil, R.; Tanaka, K. K.; Tanaka, H. Direct Simulations of Homogeneous Bubble Nucleation: Agreement with Classical Nucleation Theory and No Local Hot Spots. *Phys. Rev. E* **2014**, *90* (5), 52407.

(17) Straub, J. In *Proceedings of the 3rd International Conference on Chemical Thermodynamics and Physico-Chemical Techniques at High Temperatures, Baden, Austria*; Butterworth: London, 1973; Vol. 2.

(18) Huang, J. S.; Goldburg, W. I.; Moldover, M. R. Observation of Anomalously Large Supercooling in Carbon Dioxide. *Phys. Rev. Lett.* **1975**, *34* (11), 639–642.

466555 Simulation of Homogeneous Nucleation of Co-Crystals from a Binary Hard Sphere Fluid Mixture

Abstract:

Co-crystal formation from fluid-mixtures is quite common in a large number of systems. The simplest systems that shows co-crystal (also called substitutionally ordered solids) formation are binary hard sphere mixtures. In this work we study the nucleation of AB_2 type solid compounds using Monte Carlo molecular simulations in a binary hard sphere mixtures with size ratio of 0.55. The conditions chosen for the study lie in the region where nucleation of AB_2 type solid competes with that of pure A solid with FCC (face-centered-cubic) structure. The fluid phase compositon is kept equal to that of the AB_2 type solid. The nucleation free-energy barriers are computed using the seeding technique of Sanz *et al.* [*J. Am. Chem. Soc.* **135**, 15008 (2013)]. Our computed free energy barriers for homogeneous nucleation show that the AB_2 type solid is always favored even under conditions where the pure A solid is more stable. This is primarily due to similarity in the composition of fluid-phase and the AB_2 type solid which in turn leads to much lower interfacial tension between the crystal nucleus and the fluid phase. This system is an example of how the fluid phase composition affect the structure of the nucleating solid phase during crystallization and has relevance to solvent-induced polymorphism seen in molecular crystals.

456379 Crystallization of Calcium Sulfate Dihydrate during Phosphoric Acid Production in the Presence of Magnesium Ions

Abstract:

Crystallization of Calcium Sulfate Dihydrate during Phosphoric Acid Production in the Presence of Magnesium Ions

You Peng

Zhilong Zhu

Kamal Samrane

Richard D. Braatz

Allan S. Myerson

The production of phosphoric acid from phosphate mineral rock involves the addition of phosphate rock to a concentrated sulfuric acid solution. The induced reactive crystallization process produces a side product of calcium sulfate hydrates which become the filter media in the subsequent acid separation process. For most industrial processes, the dihydrate form of calcium sulfate crystals (gypsum) precipitates and its shape and size distribution are key factors in determining the downstream filtration efficiency [1]. Particularly, the metal ion impurities coming from raw phosphate rock plays an important role as shape modifiers. The presence of impurities in the acid mixture has an impact both thermodynamically and kinetically, although most of the available literature focuses on their role as growth inhibitors and has neglected their potential impact on solution speciation [2].

Past studies on gypsum crystallization in phosphoric acid solutions usually involve the study of crystal growth and nucleation kinetics [3, 4]. However, past works did not use the correct definition of supersaturation when fitting kinetic parameters. The high concentrations in this multicomponent electrolyte system implies that supersaturation which be written in terms of the solubility product ratio, as governing by nonideal thermodynamics, which requires the com- putation of activity coefficients as well as free ion concentrations. For this purpose, the mixed solvent electrolyte (MSE) model is utilized to capture the solution speciation in order to properly quantify supersaturation at any given condition. The MSE model is a first-principles model that determines solid-liquid equilibrium by calculating excess Gibbs energy from additive pairwise interactions [5]. When impurities are present, additional binary interactions need to be included in the databank, which is carried out by regression analysis using solubility measurements.

Continuous reactive crystallization experiments are carried out with and without additives using a mixed-suspension, mixed-product removal (MSMPR) crystallizer. Crystal size dis- tribution and supersaturation are measured once the process reaches steady state. Different conditions are imposed to acquire both the temperature and supersaturation dependency of the crystallization kinetics [6]. A two-dimensional growth model is developed in order to capture the needle-like crystal morphology and the varying crystal aspect ratio with temperature, which is made possible by performing multi-scale image segmentation [7] and edge detection using the Canny method [8]. Growth inhibition models such as developed by Kubota and Mullin [9] are used for numerical quantification of step advancement retardation in the presence of impurities.

Experimental and numerical results are obtained for the base system and in the presence of Mg^{2+} ions. The results indicate that, while retarding crystal growth, Mg^{2+} ions tend to increase the crystal aspect ratio and lead to elongated needle-like crystals. The experimental results are consistent with experimental observation reported in the literature [10]. This study goes beyond past studies by providing a full two-dimensional population balance model for a highly concentrated ionic system that includes crystallization kinetics and thermodynamically correct driving force accounting non-ideality as well as the effects of an impurity.

References

[1] Pierre Becker. *Phosphates and Phosphoric Acid: Raw Materials, Technology, and Eco- nomics of the Wet Process*. Marcel Dekker, New York, 2nd edition, 1989.

[2] Gilles Févotte, Nesrine Gherras, and Jacques Moutte. Batch cooling solution crystalliza- tion of ammonium oxalate in the presence of impurities: study of solubility, supersatura- tion, and steady-state inhibition. *Crystal Growth & Design*, 13(7):2737–2748, 2013.

[3] E.T. White and S. Mukhopadhyay. Crystallization of gypsum from phosphoric acid so- lutions. In Allan S Myerson and Ken Toyokura, editors, *Crystallization as a Separations Process*, chapter 23, pages 292–315. American Chemical Society, Washington, DC, 1990.

[4] M El Moussaouiti, R Boistelle, A Bouhaouss, and JP Klein. Crystallization of calcium sul- phate hemihydrate in concentrated phosphoric acid solutions. *Chem. Eng. J.*, 68(2):123–130, 1997.

[5] P Wang, A Anderko, and RD Young. A speciation-based model for mixed-solvent elec- trolyte systems. *Fluid Phase Equilib.*, 203(1):141–176, 2002.

[6] You Peng, Zhilong Zhu, Richard D Braatz, and Allan S Myerson. Gypsum crystalliza- tion during phosphoric acid production: Modeling and experiments using the mixed- solvent-electrolyte thermodynamic model. *Industrial & Engineering Chemistry Research*, 54(32):7914–7924, 2015.

[7] J Calderon De Anda, XZ Wang, and KJ Roberts. Multi-scale segmentation image anal- ysis for the in-process monitoring of particle shape with batch crystallisers. *Chemical Engineering Science*, 60(4):1053–1065, 2005.

[8] John Canny. A computational approach to edge detection. *Pattern Analysis and Machine Intelligence, IEEE Transactions on*, (6):679–698, 1986.

[9] Noriaki Kubota and JW Mullin. A kinetic model for crystal growth from aqueous solution in the presence of impurity. *Journal of Crystal Growth*, 152(3):203–208, 1995.

[10] M.M. Rashad, M.H.H. Mahmoud, I.A. Ibrahim, and E.A. Abdel-Aal. Crystallization of calcium sulfate dihydrate under simulated condition of phosphoric acid production in the presence of aluminum and magnesium ions. *J. Cryst. Growth*, 267:372–379, 2004.

464898 Modifiers of Calcium Oxalate Monohydrate Crystallization: Tailoring Modifier–Crystal Interactions for Rational Drug Design

Abstract:

Modifiers of calcium oxalate monohydrate crystallization: Tailoring modifier–crystal interactions for rational drug design

Bryan G. Alamani, Jihae Chung, and Jeffrey D. Rimer

Chemical and Biomolecular Engineering, University of Houston, Houston, TX, 77204 USA

Calcium oxalate monohydrate (COM) is a major constituent of kidneys stones. Current therapeutics block solute (Ca^{2+} and $C_2O_4^{2-}$) attachment to crystals through either complexation in solution or binding to specific sites on crystal surfaces.[1] Despite advancements in drug design, the incidence rate of stones is on the rise, thus emphasizing the need for alternative therapies. When designing new modifiers as drugs, it is important to consider the underlying fundamental physical interactions of modifiers (ions or molecules) with COM crystal surfaces. These interactions may be elucidated through a combination of experimental and theoretical studies. In this talk, we will describe systematic studies of bulk crystallization and in situ characterization of surface growth in the presence and absence of modifiers.[2] We have explored the effect of metal ions (both monovalent and divalent), small organics, and native proteins that are commonly observed in human urine and kidney stone matrix. The majority of modifiers act as growth inhibitors with varying efficacy.[3] Interestingly, we have identified modifiers that function as COM growth promoters.[4]

Understanding the interactions between growth modifiers and crystal interfaces requires a combination of experimental techniques capable of probing a wide range of length and time scales.[5] Analysis of macroscopic properties (bulk crystal growth) provides information on modifier changes to crystal size and habit. Kinetic studies were employed to assess the effect of modifiers on the rate of COM growth through ion-selective electrochemistry (ISE). To gain molecular insights, it is important to study the interface that captures surface phenomena leading to changes in crystal morphology. For this task, we use *in situ* atomic force microscopy (AFM) to observe time-resolved dynamics of surface growth in the presence of modifiers.[6] The culmination of these studies provide insight on how we can potentially engineer modifiers with tailored properties to take advantage of specific modifier-crystal interactions to suppress growth, or even lead to crystal dissolution. To this end, we will present our discovery of modifiers that are viable candidates to replace current therapeutics of pathological COM crystallization.

References1. Wesson, J. A.; Ward, M. D., *Elements* **2007,** *3*(6), 415-421.

2. Farmanesh, S.; Chung, J.; Chandra, D.; Sosa, R. D.; Karande, P.; Rimer, J. D., *Journal of Crystal Growth* **2013,** *373*, 13-19.

3. Farmanesh, S.; Alamani, B. G.; Rimer, J. D., *Chemical Communications* **2015,** *51*(73), 13964-13967.

4. Farmanesh, S.; Ramamoorthy, S.; Chung, J.; Asplin, J. R.; Karande, P.; Rimer, J. D., *Journal of the American Chemical Society* **2014,** *136*(1), 367-376.

5. Rimer, J. D.; An, Z.; Zhu, Z.; Lee, M. H.; Goldfarb, D. S.; Wesson, J. A.; Ward, M. D., *Science* **2010,** *330*(6002), 337-341.

6. De Yoreo, J. J.; Chung, S.; Friddle, R. W., *Advanced Functional Materials* **2013,** *23* (20), 2525-2538.

464204 Combined Cooling and Anti-Solvent Crystallization of L-Asparagine Monohydrate

Abstract:

Crystallization is a key unit operation especially in fine chemical and pharmaceutical industries to produce high value solids with desired purity and size distribution [1]. The cooling and anti-solvent crystallization are common operational modes used in industry and the optimization in such processes is traditionally sought with respect to optimal cooling profile, anti-solvent feed-rate profile, and seeding characteristics. Essentially, the control is implemented by manipulating the supersaturation trajectory during the course of crystallization due to its relationship with the fundamental crystallization phenomena such as nucleation, growth, etc. Cooling crystallization is advantageous when the solubility of the compound strongly depends on temperature. Anti-solvent crystallization is particularly advantageous when the solute to be crystallized has a solubility that is a weak function of temperature and/or the solute's thermal stability is low. Sometimes the solubility of a compound is significantly affected by both temperature and addition of an anti-solvent. Therefore, combining cooling and anti-solvent crystallization can be advantageous as it offers two manipulative variables that can be varied to exert greater control on the supersaturation profile and thereby on the outcome of the crystallization process [2]-[3].

The objective of this research is to perform a systematic study on how the cooling and anti-solvent addition should be combined using three open-loop conventional cooling policies and three feeding policies that are inspired by these well-known cooling policies [4]. Specifically, we perform a series of combined cooling/anti-solvent crystallization of L-asparagine monohydrate (LAM), an important amino acid, from its aqueous solution using isopropanol as anti-solvent. First, solubility of LAM in water-isopropanol mixture is determined gravimetrically as a function of temperature (25 to 45 °C) and weight fraction of isopropanol (0 to 80%). Next, combined cooling/anti-solvent crystallization experiments are conducted where a saturated

aqueous solution of LAM at 45 °C is cooled using a given cooling policy from 45 to 25 °C in 2 h while isopropanol is added simultaneously with a given feeding policy such that its weight fraction changes from 0 to 40 wt% by the end of the experiment. All combined cooling and anti-solvent crystallization experiments recorded improvement in crystal size distribution characteristics and significant enhancement in the yield compared to only cooling crystallization.

Experiments with various combinations of cooling and feeding policies generate a wealth of data for understanding the dynamics of the process and estimating kinetic parameters for developing a predictive model. The concentration of the solute is measured at regular sampling intervals by high performance liquid chromatography and the final crystal size distribution is measured by laser diffraction. We develop a dynamic model based on population balance equation coupled with mass balance equation; and nucleation and growth kinetics are simultaneously estimated from experimental data using a nonlinear parameter estimation technique. The fully validated model is then used to determine the optimal temperature and anti-solvent feeding profiles. A comparison is then made among the optimal policy and conventional policies to understand the convenient and optimal path of combining the two modes of operation.

References:

[1] J. W. Mullin, Crystallization. Woburn, MA: Butterworth-Heinemann, 2001.

[2] Z.K. Nagy, M. Fujiwara, and R.D. Braatz, J. Process Control 18, 856–864, 2008.

[3] C. Lindenberg, M. Krattli, J. Cornel, and M. Mazzotti, Cryst. Growth Des. 9, 1124–1136, 2009.

[4] H. Hojjati, M. Sheikhzadeh, and S. Rohani, Ind. Eng. Chem. Res. 46, 1232-1240, 2007.

462129 A Semi-Empirical CLD-to-Psd Model for Crystallization Process Monitoring

Abstract:

Controlling the particle size distribution of the final API is often one of the main objectives of a pharmaceutical crystallization process since API particle properties impact the subsequent formulation process and the ability to produce a drug product with the desired quality attributes. Thus, API crystallization process monitoring and control have become more and more important as part of the increasing QbD efforts in the pharmaceutical industry.

Among the possibilities to characterize particle populations in situ and in real-time, the Focused Beam Reflectance Measurement (FBRM) has become broadly accepted mainly due to its ease of process implementation as well as due to its performance robustness. However, one major limitation of the FBRM technique is that its measurement signal, the Chord Length Distribution (CLD), cannot be easily transferred into the particle size distribution (PSD) despite the application of several different resolution approaches over the last 15 years. Thus, the CLD data is often used qualitatively rather than quantitatively and generally provides limited value for advanced crystallization process modeling purposes.

In this work a semi-empirical method is introduced that translates CLD to PSD data based on a training set of several particle sieve cuts at increasing particle concentrations. Unlike matrix-based methods, this approach is not limited to compact particle morphologies but holds as long as changes to the particles shape distribution do not exceed certain ranges. This optimization-based method consists of two steps. First, the â€œspectral decompositionâ€ of the

experimental CLD is generated which yields an idealized particle size distribution. In the subsequent optimization step, fuzzy-trained sets are used to correlate the idealized PSD with the actual experimentally measured PSD. This work highlights the application of this model to several relevant compounds and discusses the possibility of using in situ FBRM data for crystallization process modeling.

468457 Optimizing a Distillative Crystallization and Evaluating Form Risk

Abstract:

Distillative crystallizations represent a class of particle formation with unique challenges in establishing control. In contrast to antisolvent addition rate-controlled or temperature ramp rate-controlled crystallizations, the evolution of supersaturation with time is both scale and equipment dependent for distillative crystallizations. Thus, controlling the nucleation and growth processes that contribute to particle size and form is more challenging. This presentation provides a case study of a distillative crystallization which exhibited poor and unreliable filtration rates upon scale up, leading to a process bottleneck. The variable nature of this unit operation also resulted in the appearance of several polymorphs. Further process understanding was needed to establish a robust process with respect to filtration rate and polymorph control to a desired form. The crystallization was characterized through monitoring trends in nucleation and growth via FBRM, evaluating particle size via photomicroscopy, and determining relative supersaturation via HPLC. Water content, temperature, and age time were identified as critical parameters in maximizing the particle size of the crystals. Competitive slurry experiments and measurement of relative solubilities were used to develop the thermodynamic phase map and establish optimized conditions to enable generation of the desired form.

466961 Application of Ehrfeld® Valve Assisted Micromixer for Generation of Micron Sized Particles of Poorly Water Soluble Drug Compounds

Abstract:

Production of micron and submicron size particles for hydrophobic compounds finds a great application in the pharmaceutical industry due to their enhanced surface properties which aid in overcoming dissolution and bioavailability issues. The key to obtaining fine particles is inducing homogeneous precipitation in supersaturated systems. However, batch systems prove to be inefficient for homogenous nucleation due to non-uniform mixing conditions and continuous reduction in driving force as the process progresses over time. Favorable conditions for homogenous nucleation can be easily established in micromixers, due to presence of microchannels they are able to achieve superior mixing conditions with small mixing times (τ_{mix}). Over the past few years, micromixers have played an instrumental role in the transition of processes from batch to continuous systems in several sectors, yet very few studies focus on their applications for carrying out liquid antisolvent precipitative reactions of pharmaceutical compounds.

The current work investigates the application of Ehrfeld® valve assisted micromixer (VAMM) technology for generation of micronsized drug compounds which are poorly water soluble. The VAMM has emerged as a promising technology due to its unique construction which splits the fluid stream into multiple microchannels triggering impingement at multiple sites. In addition the unidirectional restrictor valve excludes reverse flow of fluid stream (therefore having an advantage over conventional T-mixers) and the passage of the thin fluid lamellae through the body of the micromixer ensures a well-mixed and highly supersaturated solution. First the mixing efficiency of the VAMM was characterized by developing an empirical correlation for mixing time (τ_{mix}) [1], followed by experimentation using a set of competitive parallel reactions (whose kinetics have been well studied) and establishing conversion of key reactant as a function of varying jet velocity. These values were then used, to estimate Damköhler number (Da) and τ_{mix}

which enables a direct comparison of mixing efficiencies between VAMM and other well established micromixer systems [2].

Model Biopharmaceutical classification system (BCS) II compounds were identified and highly supersaturated solutions of same were subjected to anti-solvent crystallization process via VAMM in the presence of suitable stabilizer system [3]. Similar crystallizations were also carried out in batch process, where the results clearly indicated the superior performance of VAMM in generation of fine particles when compared with the batch . In both the cases particles as small as $d_{90} < 5\mu m$ were achieved, therefore establishing the Ehrfeld® Valve Assisted Micromixer as a feasible platform for precipitation of fine pharmaceutical compounds.

References:

1. Wolf A., etl al., Precipitation in a Micromixer –From Laboratory to Industrial Scale. Chemical Engineering & Technology (2015), 2017-2024

2. Johnson K., and Prudhomme R., Chemical processing and micromixing in confined impinging jets, AIChE J, 2003, 2264-2282

4. Flash nanoprecipitation of organic actives and block copolymers using a confined impinging jets mixer, Australian Journal of Chemistry (2003), 1021-1024.

IPDO communication number: IPDO IPM-00468

455229 Linking Particle Size and Shape Distributions to Filtration Performance Using Measurements and Models

Abstract:

Chemical products sold as solids typically undergo at least one crystallization step in their respective production process[1]. Crystallization processes are primarily used for purification, but also as a means to establish the downstream processing properties of the generated crystals, for instance their filterability, flowability and ultimately – if relevant, such as for pharmaceuticals – their tabletability. These properties are thus of vital importance when moving from a suspension towards a dry product.

It is widely accepted that the particle size and shape distribution (PSSD) originating from a crystallization process is chiefly responsible for a product's downstream processing properties[2]. Considering the processing step immediately following crystallization, filtration, the PSSD affects the efficiency and effectiveness of the filtration step, i.e., how fast and how complete the separation of solids from the liquid is achieved. For example, there is ample experimental evidence that small particles and wide particle size distributions lead to longer filtration times (lower filtration efficiency)[3], while it is generally observed that non-isotropic particle shapes (needles, platelets) lead to higher mother liquor retention (lower filtration effectiveness)[4].

Current state of the art approaches describing the relationship between particle size and shape and filtration performance make strong simplifications with respect to the particles' size and shape, i.e., the crystals are often assumed to be monodispersed, all of the same shape or both. In this work, we investigate whether (and when) more accurate information about the particles, such as the full PSSD, is beneficial to predict filtration performance. To this end, several different particle size and shape distributions of β-L-glutamic acid crystals were prepared by different crystallization, milling and sieving operations. The particles were generally needle-like, but with varying aspect ratios and sizes. On these materials two types of measurements were carried out: first, we report measurements of complete, quantitative PSSDs obtained with a custom-built stereoscopic measurement device[5]. Second, we report measurements of the mother liquor flow rate through cakes of these materials at constant pressure drops up to 5 bar. By correlating the full PSSDs as well as simplified particle characterizations (such as 1D particle size distributions and their moments/quantiles), with filtration performance, we aim at providing insight which level of particle characterization allows for a (reasonably) quantitative link.

References:

[1] Variankaval, N., Cote, A.S., Doherty, M.F., "From form to function: Crystallization of active pharmaceutical ingredients", AIChE J., 2008, 54, 1682-1688.

[2] Wakeman, R., Tarleton, S., "Solid Liquid Separation: Principles of Industrial Filtration", Elsevier, Amsterdam, Netherlands, 2005.

[3] Wakeman, R., "The influence of particle properties on filtration", Sep. Purif. Technol., 2007, 58, 234–241.

[4] Garside, J. (ed.), "Separation Technology: The next ten years", Institution of Chemical Engineers, Rugby, UK, 1994, p. 81.

[5] Schorsch, S., Ochsenbein, D.R., Vetter, T., Morari, M., Mazzotti, M., "High accuracy online measurement of multidimensional particle size distributions during crystallization", Chem. Eng. Sci., 2014, 105, 155-168.

471285 Dynamic Data Reconciliation and Model Validation of a CO_2 Capture Process Using Pilot Plant Data

Abstract:

The post-combustion CO_2 capture systems would be subjected to fluctuating flue gas flow rate and composition as the power plants follow the load. In addition, future CO_2 capture plants can vary the extent of CO_2 capture in response to the electricity price. Therefore optimal servo control and disturbance rejection characteristics are highly desired while designing the control system for CO_2 capture processes. One essential element for designing the control system is a high fidelity dynamic model that is validated with the experimental data . However, there is a lack of validated dynamic models for solvent-based CO_2 capture processes. With this incentive, a rigorous dynamic model of a MEA-based CO_2 capture process is developed and validated using steady-state and dynamic data obtained from the US DOEâ€™s National Carbon Capture Center (NCCC) in Wilsonville, Alabama.

Rigorous thermodynamic and transport models with quantified uncertainties are first developed and validated using large number of experimental data. A novel approach is developed to obtain a rigorous mass transfer model by considering the wetted wall column and packed column data simultaneously while regressing the parameters of the interfacial area model, mass transfer coefficient models, and diffusivity models together. In addition, rigorous kinetic model, holdup model, and pressure drop model are developed. The steady-state model is validated using large number of data collected from NCCC.

One issue with the limited dynamic data that are available in the open literature is that they span a limited range of operating conditions and often only contain a single step change. In this work, the dynamic test runs were conducted by designing a test protocol that helps to reflect the process nonlinearity and ensures persistence of excitation

approximately. The dynamic data include solvent composition and loading data that were manually collected every 5 minutes and analyzed in the NCCC laboratory. The dynamic data collected from the NCCC pilot plant were found to have significant measurement noise and some inaccurate measurements. In addition, there were a number of key variables that were unmeasured. To handle these issues, a dynamic data reconciliation (DDR) framework is developed to ensure closure of material and energy balances prior to model validation. As the rate-based models are not currently supported in Aspen Plus Dynamics, the software platform of our choice, an approximate method that matches the results from the rate-based Aspen Plus model closely is developed and implemented using scripts in Aspen Plus Dynamics. It is observed that the results from the dynamic model match very well with the experimental data that span more than three days. The dynamic model is then leveraged to perform a number of transient studies by simulating expected disturbances and setpoint tracking problems.

463128 Rigorous Modeling of a MEA-Based CO_2 Capture Process with Uncertainty Quantification and Validation Using Large Scale Pilot Plant Data

Abstract:

The US DOE's Carbon Capture Simulation Initiative (CCSI) has a strong focus on the development of state of the art process models to accelerate the development and commercialization of post-combustion carbon capture system technologies. One of CCSI's goals is the development of a methodology for a "gold standard" process model that will serve as a definitive reference for benchmarking the performance of steady-state and dynamic CO_2 capture systems. Although this methodology is intended to be applicable for various solvent types and for a wide range of operating conditions at different scales, an aqueous monoethanolamine (MEA) system is treated as a baseline in this work due to its widespread use as the industrial standard. This work is focused on developing the "gold standard" process model in Aspen Plus® and validating the model with pilot plant data.

The MEA model in this work is developed through a compilation of submodels, including physical property, reaction kinetics, mass transfer, and hydraulic models. The model parameters are regressed with experimental data that span the temperature, pressure, and composition ranges of interest. For each submodel, a Bayesian inference methodology is used to quantify the parametric uncertainty.

This model is validated with high quality steady-state data (23 total trials) collected from the National Carbon Capture Center (NCCC) in Wilsonville, Alabama. While most of the steady-state test runs reported in the literature are focused on narrow operating range, the key manipulated variables such as solvent flowrate and reboiler steam flowrate and disturbance variables such as the flowrate and CO_2 concentration of the inlet flue gas were varied widely in these test runs. Moreover, the number of beds, and thus packing height, and the presence of intercooling in the absorber are variable in the test runs. For each of the 23 cases, uncertainty in the key variables of interest, including percentage of CO_2 capture in the absorber and required reboiler duty in the regenerator, are determined by propagating parametric uncertainty through the process model. This rigorous methodology for obtaining uncertainty ranges on key operating variables is not only novel, but also provides insight into the effects of the operating variables on output uncertainty and helps to identify process submodels that have significant impact on output uncertainty. The insights gained through this developed methodology can play a key role in reducing scale up uncertainty of CO_2 capture processes.

471729 Immobilizing Metal Organic Frameworks on Active Substrates for CO_2 adsorption

Abstract:

Scaling up gas separation processes requires novel approaches for manufacturing of adsorbent materials into scalable contactors such as monolithic structures that offer a cost-effective platform for practical applications in comparison to traditional packing systems. In this study, several metal-organic frameworks (MOFs) are supported on a variety of commercial and synthesized monolithic substrates and their dynamic adsorption behavior are assessed in a rapid pressure swing adsorption (RPSA) system for post-combustion CO_2 capture. To gain control over crystal nucleation and growth, various bottom-up growth techniques are employed and optimized with respect to uniformity, thickness, and adsorption characteristics of the MOFs films. To further improve the equilibrium capacity, post-functionalization of composite adsorbents are carried out using several aminopolymers and aminosilanes. In addition, long-term stability of the structured adsorbents are evaluated by performing cyclic operation. Furthermore, to assess the influence of water on adsorption behavior of coated monoliths, both dry and humid runs are performed under real conditions. Finally, the influence of SO_x/NO_x impurities are determined by exposing the structured adsorbents to a simulated flue gas containing SO_2 and NO.

467239 Incorporation of 1-Butyl-3-Methylimidazolium Hexafluorophosphate into ZIF-8: Elucidation of Interactions and Their Consequences on Gas Separation Performance

Abstract:

Incorporation of 1-butyl-3-methylimidazolium hexafluorophosphate into ZIF-8: Elucidation of Interactions and Their Consequences on Gas Separation Performance

Fatma Pelin Kinik, Seda Keskin,* and Alper Uzun*

Chemical and Biological Engineering Department, Koç University, Istanbul 34450, Turkey

Metal organic frameworks (MOFs) have been popular recently in adsorption-based gas separation applications due to their configurable pores, large surface areas, robust framework structures and high chemical and thermal stabilities.[1] Lately, ionic liquids (ILs) have been used to enhance the gas separation performances of MOFs. ILs are liquid salts at room temperature which consist of organic cations and inorganic or organic anions.[2] Because of their favorable physical and chemical properties such as non-flammability, negligible volatility, high thermal stability and remarkable gas solubility (especially towards CO_2), ILs have been used in gas adsorption and separation processes.[3] In this work, an IL was used to improve gas adsorption and separation properties of a MOF. The influence of IL confinement into MOF structure was examined for CO_2/CH_4, CO_2/N_2 and CH_4/N_2 gas separations. ZIF-8 (2-methylimidazole zinc salt) was selected as the MOF as it is a thermally and chemically stable zeolite-type material with large pore size of 11.6 Å and small pore size of 3.4 Å.[4] Because of having high surface area, significant porosity, water resistance and commercial availability, ZIF-8 has been commonly used in gas separation applications.[5] As IL, 1-butyl-3-methylimidazolium hexafluorophosphate, [BMIM][PF_6], was selected because of its hydrophobic nature, low cost, and notable affinity for CO_2.[6]

Crystalline structures, thermal stabilities and interionic interactions were investigated by combining various characterization techniques. XRD and SEM results showed that the crystalline structure of the MOF is preserved upon the incorporation of IL. Moreover, XRF and BET measurements indicated that impregnation of the IL was successfully achieved at an IL loading of 24 wt.%. TGA and FTIR results proved that there is an interaction between IL and MOF. Adsorption measurements of CO_2, CH_4, and N_2 gases in ZIF-8 and [BMIM][PF_6]/ZIF-8 were performed using the volumetric method at room temperature up to 35 bar and ideal selectivities were calculated for CO_2/CH_4, CO_2/N_2, and CH_4/N_2 separations. It was shown that the CO_2 selectivities of IL/MOF samples at low pressures significantly increased compared to the pristine ZIF-8, while CH_4/N_2 selectivity of IL/MOF samples remained lower than the one of pristine ZIF-8. There is a considerable improvement in the CO_2/N_2 selectivity of IL-incorporated MOF, which is almost three times of pristine ZIF-8 up to 0.3 bar and two times of pristine ZIF-8 up to 1 bar. Also IL-incorporated MOF has approximately two times more CO_2/CH_4 selectivity than pristine ZIF-8 up to 1 bar. As a conclusion, IL-modified ZIF-8 is a promising material for gas separation applications, especially for CO_2 capturing processes such as natural gas sweetening and removing CO_2 from power plant flue gas. Tunable physicochemical properties of ILs and MOFs offer great potential for optimizing these structures for better gas separation performance.

Acknowledgement

This work is supported by the Scientific and Technological Research Council of Turkey (TUBITAK) under 1001-Scientific and Technological Research Projects Funding Program (project number: 114R093) and by Koç University Seed Fund Program. Financial support provided by the KUTEM (Koç University TUPRAS Energy Center) is gratefully acknowledged. S.K. acknowledges TUBA-GEBIP and A.U. acknowledges the support by the Science Academy of Turkey under the BAGEP Award Program. Authors thank Koç University Surface Science and Technology Center (KUYTAM) for providing help with the sample characterization.

*Corresponding authors: skeskin@ku.edu.tr and auzun@ku.edu.tr

References:

[1] Li, J.-R.; Kuppler, R. J.; Zhou, H.-C., Selective Gas Adsorption and Separation in Metal-Organic Frameworks. *Chemical Society Reviews* **2009,** 38, 1477-1504.

[2] Marsh, K.; Boxall, J.; Lichtenthaler, R., Room Temperature Ionic Liquids and Their Mixtures—a Review. *Fluid Phase Equilibria* **2004,** 219, 93-98.

[3] Hasib-ur-Rahman, M.; Siaj, M.; Larachi, F., Ionic Liquids for CO_2 Capture—Development and Progress. *Chemical Engineering and Processing: Process Intensification* **2010,** 49, 313-322.

[4] M. Zhu; Venna, S. R.; Jasinski, J. B.; Carreon, M. A., Room-Temperature Synthesis of ZIF-8: The Coexistence of Zno Nanoneedles. *Chemistry of Materials* **2011,** 23, 3590-3592.

[5] Venna, S. R.; Carreon, M. A., Highly Permeable Zeolite Imidazolate Framework-8 Membranes for CO_2/CH_4 Separation. *Journal of the American Chemical Society* **2009,** 132, 76-78.

[6] Chen, Y.; Hu, Z.; Gupta, K. M.; Jiang, J., Ionic Liquid/Metal–Organic Framework Composite for CO_2 Capture: A Computational Investigation. *The Journal of Physical Chemistry C* **2011,** 115, 21736-21742.

470543 Strontium-Based Bimetallic Metal Organic Frameworks (MOFs) for CO_2 Separation

Abstract:

Metal organic frameworks (MOFs) are promising candidates for separation and purification applications due to their tunable chemistry, high surface area, and porosity [1]. An interesting class of nanoporous bimetallic compounds known as strontium-based metal-organic frameworks have been synthesized on the organic basis of malonate. Cobalt and copper were employed to form MOFs with the general formula of $[SrM(C_3H_2O_4)_2 (H_2O)_n]$, (M=Co, n=7 and M=Cu, n=4). These MOFs are designated with the refcodes of FOHQUO and VAMQOP in the Cambridge Structural Database, respectively.

Haldoupis and co-workers [1] examined a new category of MOFs using atomic simulation for gas sorption including FOHQUO. According to their results, FOHQUO with pore limiting diameter of 3 Å has a binary adsorption selectivity above 190 with water-free molecular lattice [1]. Having removed all the guest water molecules between 70°C to 190°C, they report that FOHQUO's framework integrity remain stable up to 350°C [1]. A molecular simulation study on mixed matrix membranes with FOHQUO as the dispersive phase was conducted by Yilmaz and Keskin [2] for CO_2/N_2 separation. According to their predictions, FOHQUO has a CO_2 permeability of about 500000 Barrer and CO_2/CH_4 selectivity of about 5000.

One of the challenges in MOFs nature is their thermal, chemical and physical stability under heat treatment and in presence of acids and moistures. The crystalline lattice of as-synthesized MOFs are typically filled with guest molecules of solvents [3]. Solvent removal can pose obstacles in crystallinity of MOFs and collapse the framework integrity and convert to amorphous materials [4].

In this study, we have synthesized FOHQUO and VAMQOP to investigate their potential for use in CO_2 separation and mixed matrix membrane formation. We examined their thermal and structural stability, surface area, and gas sorption capacity. Interestingly, X-ray diffraction (XRD) analysis proved that FOHQUO framework was not stable above 80°C. However, TGA analysis reflected the dehydration step at 80°C. On the other hand, VAMQOP has a stable structure up to 140°C. XRD analysis revealed that, after an amorphous period above 170°C, phase transformation occurs.

This work is supported by TUBITAK (Turkish Scientific and Technological Research Council) through Project No. 113M776.

References:

1. Haldoupis, E., Nair, S., Sholl, D. S. (2012), Finding MOFs for Highly Selective CO2/N2 Adsorption Using Materials Screening Based on Efficient Assignment of Atomic Point Charges, J Am Chem Soc, 134, 4313-4323.
2. Yilmaz, G., Keskin, S. (2014). Molecular Modeling of MOF and ZIF-Filled MMMs for CO2/N2 Separations, Journal of Membrane Science, 454, 407-417.
3. Qiu, S., Xue, M., Zhu, G. (2014), Metal-Organic Framework Membranes: From Synthesis to Separation Applicatio,. Chem Soc Rev, 43, 6116-6140.
4. Mondloch, J. E., Karagiaridi, O., Farha, O. K., Hup, J. T. (2013,. Activation of Metal–Organic Framework Materials, CrystEngComm, 15, 9258.

474890 Oxygen Selectivity of RPM3-Zn Gate-Opening Metal-Organic Frameworks

Abstract:

Commercial cryogenic distillation process is applied to air separation, based on different normal boiling points of nitrogen (77.4 K) and oxygen (90.2 K), but the process comprises of 30% of capital costs due to significant energy consumption. This study begins to explore the potential for kinetic separations of air via exploiting the gate-opening (GO) process of various metal organic frameworks (MOFs). Here, we report the gas adsorption capacities of nitrogen and oxygen on the GO-MOF RPM3-Zn (a.k.a. $Zn_2(bpdc)_2(bpee)$; bpdc = 4,4â€™-biphenyldicarboxylate; bpee = 1,2-bipyridylethene) at temperatures ranging from 77 K to 273 K up to 20 bar. Given sufficient equilibration time, the nitrogen and oxygen GO pressures are indistinguishable at 77 K, 87 K, 195 K, and 273 K. Yet, somewhat unexpectedly, RPM3-Zn exhibits higher oxygen adsorption capacity than nitrogen at low temperatures. The heats of adsorption in gate opening RPM3-Zn with oxygen and nitrogen are calculated as 12 kJ/mol and 8 kJ/mol, respectively, further evidence for oxygen selectivity. The selectivity is enhanced significantly when kinetic effects are also included. Efforts to deduce the role of various adsorption sites on oxygen selectivity will be discussed.

461656 PIM-1/Poly(ethylene imine) Composites As Solution-Processable "Molecular Baskets" for CO_2 Capture from Dilute Streams

Abstract:

Research for CO_2 capture from dilute streams such as flue gas or ambient air has received increased attention within the last several decades due to rising atmospheric CO_2 levels. Recently, much research has focused on the use of solid sorbent materials composed of mesoporous oxides with tethered or impregnated polymeric amine materials. However, these materials are difficult to process and implement at the scales required for real CO_2 capture applications due to difficulties associated with heat and mass transport and high pressure drops through the powder materials.

Polymers of intrinsic microporosity (PIMs) are highly porous, solution-processable materials that theoretically could act as supports for amines such as poly(ethylene imine) (PEI) for CO_2 capture. These materials can be shaped into a variety of morphologies such as monoliths, hollow fibers and thin film membranes, which can have beneficial implications for heat and mass transfer properties of the final material.

In this contribution, we demonstrate the use of a PIM-1/PEI composite system for dilute CO_2 capture. We demonstrate that these composites, in powder, monolith, and dense film form, exhibit comparable performance to their inorganic powder counterparts, exhibiting good CO_2 capacity, uptake kinetics, and stability under simulated flue gas and air capture conditions and rapid temperature swing operation. Spin-diffusion solid-state NMR experiments were performed to demonstrate that PEI is well-dispersed within the PIM-1 pores at all loadings, allowing for rapid CO_2 uptake kinetics. Our results suggest that other solution-processable, intrinsically porous polymers may be applicable as supports for amine sorbents.

456419 CO_2 Adsorption from Intra Versus Inter Molecular Interactions on Aminosilane Grafted SBA-15

Abstract:

Most of amine adsorbents studied for CO_2 capture have high amine loadings on the various supports, such that it may be presumed that CO_2 adsorption most often occurs via cooperative interactions of two amines in different molecules or chains. However, a fundamental molecular level understanding of CO_2 adsorption requires insight into the propensity or probability for intramolecular vs intermolecular cooperative adsorption when using amine molecules that contain multiple amine species. Very limited research to date has targeted intramolecular amine–amine interactions during CO_2 capture.

In this study, ultralow loadings (0.45 mmol silane/g) of four different organosilane molecules were grafted onto a mesoporous silica SBA-15 support so that CO_2 could be adsorbed by primarily or only intramolecular amine interactions.[1] The four grafted molecules studied contain four different functional moieties [propylamine, (MONO); propylethylenediamine, (DI); propyldiethylenetriamine, (TRI); propyltriethylenetetramine, (TREN)] including organosilanes that can only bind to form alkylammonium carbamates intermolecularly (MONO) as well as amines that can bind with inter- or intramolecular interactions (DI, TRI, TREN). For a series of these samples, hexamethyldisilazane (HMDS) was used to cap accessible surface silanols and exclude the potential for CO_2 capture by amine interactions with silanols on the surface, which is hypothesized to give carbamic acid species.[2] The CO_2 uptake over various CO_2 partial pressures was measured over these adsorbents, and simultaneously, the isosteric heats of adsorption of CO_2 were measured via calorimetry. Thus, the potential for intramolecular CO_2 adsorption has been explored and the effective distances between two amine molecules for CO_2 adsorption via intramolecular interactions was estimated for each silane.

The insights from this work may be utilized in the design of new amine molecules that might have improved sorption properties, while the collected data provide new understanding of the mechanism and kinetics of CO_2 adsorption at low amine loadings.

References

1) C.-J. Yoo, L.-C. Lee, C. W. Jones, "Probing Intramolecular vs. Intermolecular CO_2 Adsorption On Amine-grafted SBA15." *Langmuir* (2015) 31, 9356-9365

2) A. Danon; P.C. Stair, E. Weitz, "FTIR Study of CO2 Adsorption on Amine-Grafted SBA-15: Elucidation of Adsorbed Species." *J. Phys. Chem. C* (2011) 115, 11540–11549

472599 Structure-Function Relationship of Type I CO_2 Adsorbents Via Coupled Calorimetric and Volumetric Measurements

Abstract:

In this study, a series of linear and branched molecular and polymeric amines including triethylenetetramine (TETA), tris(2-aminoethyl)amine (branched TETA), tetraethylenepentamine (TEPA), pentaethylenehexamine (PEHA), linear polyethylenimine (PEI, M_n=423), and branched polyethylenimine (PEI, M_n=600) on silica material were used to synthesize amine-impregnated silica (Type I sorbents). The number of secondary amines in the amine structure increased in the order of b-TETA, TETA, TEPA, PEHA, and PEI amine materials. The experiments were performed in the equilibrium and isothermal conditions at various CO_2 concentrations for different temperatures and amine weight loadings. A variety of information from VA/DSC combination that includes differential and integral heat of adsorption, CO_2 adsorption capacity, distribution of adsorption site energies, energetic heterogeneity and lateral interaction, and thermodynamic parameters for the aminosilica materials have been investigated as a function of secondary amine. By combining an automated volumetric adsorption (VA) instrument and a commercial differential scanning calorimetry (DSC), we simultaneously evaluated the differential and integral heats of adsorption and adsorption isotherms of CO_2 adsorption on solid aminosilica materials. The volumetric adsorption instrument is used to measure CO_2 adsorption capacity at various CO_2 partial pressures while heat of adsorption at each equilibrium partial pressure is directly measured in DSC concurrently. The results demonstrated that heat of adsorption demonstrated non monotonic behavior with experimental variables, such as temperature and weight loading. At constant weight loading with increases in the ratio of secondary to primary amines, the heat of adsorption decreased. In general, branched amine molecules and PEI demonstrated larger heats of adsorption and lower capacities than linear versions.

452030 Titanium Pillared MCM-36 As a Capture Platform for CO2 Capture Applications

Abstract:

Rising global carbon dioxide levels have been heavily implicated in global climate change, requiring the capture and conversion/storage of this greenhouse gas. In order to achieve this result many groups have looked towards a variety of solid adsorbents, including metal oxide particles, metal organic frameworks, and zeolite structures. One of the most promising of these platforms are metal oxide sorbents, which show very high capacities but become deactivated upon regeneration, resulting in sorbents that cannot be used in a cyclic capture system. This has led many groups to begin investigating novel nanostructured metal oxides, including those containing magnesium, calcium, and titanium, for cyclic carbon capture. In this work we present work our group has recently performed on the use of a novel titanium-oxide pillared MCM-36 nanosheet sorbent for cyclic carbon dioxide capture. The material shows good regenerability, high capture capacity, and evidence of a chemisorption mechanism on this traditional physisorbent. These results suggest that the titanium pillared MCM sorbent can be a strong candidate for further study as a CO_2 sorbent or catalyst, especially once functionalized with secondary organic or inorganic groups.

461218 Consequences of 1-Butyl-3-Methylimidazolium Tetrafluoroborate Incorporation on Gas Separation Performance of ZIF-8

Abstract:

Consequences of 1-butyl-3-methylimidazolium tetrafluoroborate Incorporation on Gas Separation Performance of ZIF-8

Burak Koyutürk, Seda Keskin,* Alper Uzun*

Chemical and Biological Engineering Department, Koç University, Istanbul 34450, Turkey

Metal organic frameworks (MOFs), which consist of metals and organic linkers, have gained significant attention for gas storage and separation applications specifically thanks to their high surface area and tunable pore sizes.[1] ZIF-8 is one of the MOFs which is used in membrane-based and adsorption-based gas separations.[2,3] The windows of ZIF-8 allow the passage of small gas molecules, such as hydrogen and carbon dioxide (having kinetic diameters of 2.89 Å and 3.3 Å, respectively) as well as larger molecules, such as methane and nitrogen (having kinetic diameters of 3.80 Å and 3.60 Å, respectively) due to the flexibility of its structure. [4,5]

In this study, we confined 1-butyl-3-methylimidazolium tetrafluoroborate ([BMIM][BF_4]) into ZIF-8 at a loading of 30 wt.% by impregnation and characterized the resulting materials in terms of its crystalline structure, thermal stability, surface area and bond interactions. Volumetric adsorption measurements were performed for CO_2, CH_4, and N_2 gases and selectivities of binary gas pairs were calculated using the single-component gas adsorption isotherms. Data illustrate that IL incorporation directly improve the selectivities of CO_2/CH_4, CO_2/N_2 selectivity by 50 % and 102% at low pressures as a results of interactions of the ionic liquid with the ZIF-8 framework. Results offer opportunities to

tune the adsorption and separation performance of MOFs by the incorporation of ionic liquids with tunable physicochemical properties.

Acknowledgement

This work is supported by the Scientific and Technological Research Council of Turkey (TUBITAK) under 1001-Scientific and Technological Research Projects Funding Program (project number: 114R093) and by KoÃ§ University Seed Fund Program. Financial support provided by the KUTEM (KoÃ§ University TUPRAS Energy Center) is gratefully acknowledged. S.K. acknowledges TUBA-GEBIP and A.U. acknowledges the support by the Science Academy of Turkey under the BAGEP Award Program. Authors thank KoÃ§ University Surface Science and Technology Center (KUYTAM) for providing help with the sample characterization.

References:

1. Zhou, H.-C., J.R. Long, and O.M. Yaghi, Introduction to metalâ€"organic frameworks. Chemical Reviews, 2012. 112(2): p. 673-674.

2. Bux, H., et al., Ethene/ethane separation by the MOF membrane ZIF-8: molecular correlation of permeation, adsorption, diffusion. Journal of Membrane Science, 2011. 369(1): p. 284-289.

3. McEwen, J., J.-D. Hayman, and A.O. Yazaydin, A comparative study of CO_2, CH_4 and N_2 adsorption in ZIF-8, Zeolite-13X and BPL activated carbon. Chemical Physics, 2013. 412: p. 72-76.

4. Fairen-Jimenez, D., et al., Opening the gate: framework flexibility in ZIF-8 explored by experiments and simulations. Journal of the American Chemical Society, 2011. 133(23): p. 8900-8902.

5. Venna, S.R. and M.A. Carreon, Highly permeable zeolite imidazolate framework-8 membranes for CO2/CH4 separation. Journal of the American Chemical Society, 2009. 132(1): p. 76-78.

451416 Evaluation of Physical Adsorbents for Post-Combustion CO_2 Capture

Abstract:

CO_2 capture using physical adsorbents such as activated carbon, zeolites, and metal-organic frameworks (MOFs) have attracted a lot of attention because of the technical capability of adsorption technology using such separation agents to retrofit the energy demanding liquid amine scrubbing process. Nevertheless, this will be conceivable only if the separation agents (adsorbents) in question fulfil many unambiguous attributes in terms of porosity, CO_2 affinity, kinetics, energetics, stability, throughout the right capture mechanism, in addition to the adsorbents cost.

In this paper we report recent study about the evaluation of physical adsorbents for CO_2 capture from simulated flue gas.

Extensive CO_2/N_2:10/90 mixed gas adsorption at 298 and 328 K, in the presence of water vapor and other impurities, were carried out using both temperature and pressure swing regeneration modes. For comparison purpose, absorption in liquid amines and adsorption on 13X zeolites are considered as the reference technologies for post-combustion CO_2 capture.

465588 The Importance of Nitrogen Co-Adsorption on Effectiveness of Post-Combustion CO2 Capture Materials: A Process Optimization Study

Abstract:

Adsorption-based carbon capture is considered to be a potential alternative to current technologies such as absorption. Research in this area has concentrated on the development of new materials and novel processes[1]. On the materials front, many forms of metal-organic frameworks, zeolites and activated carbons have been synthesized. On the process front, many novel processes have been developed and pilot scale demonstrations have proven the capability of adsorption to achieve targets set by US Department of Energy[2].

Most adsorbents prescribed for CO_2 capture show significantly higher affinity for CO_2 compared to N_2 (the major component in post-combustion flue gas). Hence, during adsorbent development it has been customary to ignore the affinity of N_2 while focussing on enhancing CO_2 capacity. However, from the process perspective, the high concentration of N_2 in the flue gas makes it challenging to design processes to achieve high purity CO_2, without sacrificing CO_2 recovery. Hence, understanding the impact of CO2 and N2 affinities on any sorbent is hence key to developing materials that can perform well at a process level.

In this paper we study the trade-off between improving CO_2 adsorption and reducing N_2 adsorption by considering several hypothetical materials with varying CO_2 and N_2 affinity and understanding how the performance of a vacuum swing adsorption process is impacted by these parameters. These hypothetical materials are subject to detailed process modeling and multi-objective optimization which take into account process configuration, constraints, temperature and mass transfer effects. The multi-objective optimization framework[3], which has been validated using our pilot-plant results[2], provides Pareto plots of key performance indicators such as CO purity vs. recovery, and energy vs.

2

productivity. These Pareto plots provide an objective comparison of the performance of various adsorbents.

The study provides many interesting results. Firstly, it appears that minimum selectivity required for achieving CO_2 purity and recovery in excess of 95% is much lower than what is exhibited by conventional materials such as Zeolite 13X. Secondly, increasing the CO_2 affinity of an adsorbent shows minimal impact on the process performance. However, reducing the N_2 adsorption has a significantly higher impact on process performance. In summary, it appears that the efficacy of an adsorbent is limited by its ability to release N_2 rather than by its ability to capture CO_2. These results are illustrated using a graphical approach that provides contours of CO_2 purity, recovery and specific energy consumption as a function of CO_2 and N_2 affinity. The graphical approach provides a direction for the design of new adsorbents that can satisfy process requirements and result in lower parasitic energy.

REFERENCES:

1. M.E. Boot-Handford, et al, *Energy & Env. Sci.* 2014, **7**, 130-189.
2. S. Krishnamurthy et al., *AIChE J.* 2014, **60**, 1830-1842.
3. R. Haghpanah et al., *AIChE J.* 2013, **59**, 4735-4748.

461533 Development of an Evaluation Metric from Pressure Swing Adsorption Simulations for Rapid Material Screening

Abstract:

Development of an Evaluation Metric from Pressure Swing Adsorption Simulations for Rapid Material Screening

K.T. Leperi,[1] Y.G. Chung,[2] F. You,[1] R.Q. Snurr[1]

[1]Northwestern University, Evanston, IL, USA

[2]Pusan National University, Busan, South Korea

In 2010, 30.6 gigatons of CO_2 were emitted into the atmosphere globally.[1] A significant portion of the emitted CO_2 is due to the burning of fossil fuels for the generation of electricity in coal-fired power plants. One way to reduce the CO_2 emission to the atmosphere is to equip existing power plants with Carbon Capture and Sequestration (CCS) technology.

Of all the technologies available for CCS, there has been an increased interest in using pressure swing adsorption

(PSA) to capture the CO_2 due to its higher performance and lower energy requirements compared to the other available technologies, such as amine scrubbing, or membrane-based separations.[2] However, in the majority of recent publications on CCS, there has been a division between research focused on new materials, where simple metrics based on the isotherms are used, and process-level research, where only a few materials are investigated and incorporated into the design. In previous work, we simulated a two-stage Skarstrom cycle and investigated several zeolites and MOFs to determine the optimal adsorbent material for post-combustion separation application.[3] However, full scale process simulations on all possible adsorbent materials would be computationally expensive. The objective of this work is to develop an evaluation metric from process-level simulations of PSA cycles to allow for the rapid screening of adsorbent materials for CO_2 capture.

In this work, a one stage Fractionated Vacuum Pressure Swing Adsorption cycle is used to select and evalute hundreds of metal-organic frameworks (MOFs) from the Computation-Ready Experimental (CoRE) MOF database.[4] In order to select the top 400 candidates from the 5000+ MOFs in the database, generic adsorbents were created and tested using PSA simulation to determine ideal ranges of the energies of adsorption of CO_2 and N_2, densities and metal composition. Process level simulations and optimization were then performed for each of the top candidate MOFs in order to determine the minimum cost of CO_2 capture. Finally, utilizing the process and economic data, we have developed two evaluation metrics for ranking adsorbents. One metric ranks the adsorbents based on their expected maximum CO_2 product purity. The other metric ranks the adsorbents based on the expected cost of CO_2 capture. The goal of these evaluation metrics is to facilitate rapid screening of adsorbents for CCS application.

References

1. IEA. Prospect of Limiting the Global Increase in Temperature to 2°C Is Getting Bleaker; 2011. http://www.iea.org/newsroomandevents/news/2011/may/name,19839,en.html

2. Zhao M, Minett AI, Harris AT. A review of techno-economic models for the retrofitting of conventional pulverised-coal power plants for post-combustion capture (PCC) of CO2. *Energy Environ. Sci.* 2013;6(1):25-40.

3. Leperi KT, Snurr RQ, You F. Optimization of Two-Stage Pressure/Vacuum Swing Adsorption with Variable Dehydration Level for Postcombustion Carbon Capture. *Ind. Eng. Chem. Res.* 2016;55(12):3338-3350

4. Chung YG, Camp J, Haranczyk M, Sikora BJ, Bury W, Krungleviciute V, Yildirim T, Farha OK, Sholl DS, Snurr RQ. Computation-Ready, Experimental Metal-Organic Frameworks: A Tool To Enable High-Throughput Screening of Nanoporous Crystals. *Chem. Mater.* 2014;26(21):6185-6192.

472776 Improving Single Use Bioreactor Design and Process Development, New Research Towards Intensifying Seed-Train and Scale-up Methods Using 5:1 Turn-Down

Abstract:

Operating bioreactor vessels at low working volumes (low turn down ratio) is often desirable but brings about challenges in regard to mixing, mass transfer, and process control. Research done towards optimizing cell culture has provided methods to improve performance and control when operating under these conditions. Implementing bottom heat exchange, making changes to impeller angle and height, taking advantage of the unique Thermo Fisher drilled hole sparge design, and implementing a new cross flow sparge into the headspace have yielded reliable mass transfer and cell culture results with the additional benefit of lower turn down mixing when draining the reactor.

462181 Characterization and Design of a 3L Single-Use Bioreactor to Support Process Development and Tech Transfer

Abstract:

The design and characterization of a 3L rigid single-use bioreactor (SUB), to replace existing 3L glass bioreactors, was pursued to further enhance the efficiency of development operations in Merckâ€™s Kenilworth biologics facility.

A myriad of bench-scale, rigid, single use bioreactors are commercially available at this time. We required one to replace our current 3L glass bioreactor vessels and meet the diverse requirements of our biologics facility, which supports a continually increasing number of cell culture projects. The new SUB needed to perform like our current bioreactors, so that process development activities would not be impacted. Overall, the SUB was expected to be versatile, robust, and simple to use.

A multi-faceted, iterative approach was used in the development of a new custom 3L SUB, with the following requisites:

- Similar configuration to larger vessels (e.g. drilled tube sparger and rigid baffles)
- Magnetic bottom drive agitation system
- Perfusion ready
- Expanded number of sensor ports and addition lines for flexibility
- Compatible with Automated sampling systems

To characterize the designs, we performed gas transfer ($k_L a$) studies, computational fluid dynamics (CFD) modeling, and cell growth studies with both fed-batch and perfusion processes. The data generated from these studies was used to refine the design and confirm comparability to the glass vessel controls.

Implementation of the customized 3L SUBs has enabled enhanced process development capability by introducing technology to improve operational efficiency.

461864 Foam Control in the Mammalian Cell Culture Process through Extrinsic and Intrinsic Methods

Abstract:

During biopharmaceutical manufacturing, the formation of foam within bioreactors has been an industry-wide issue for decades. As the product throughput has increased over the years, higher viable cell densities have increased oxygen demand. To support this higher uptake, agitation and aeration rates have also increased, leading to more foam generation. This results in a myriad of problems, including cell entrapment within foam, cell damage caused by bubbles bursting at the surface, interruption of gas transfer across the air/liquid interface, and equipment failure due to clogged vent filters. To optimize bioreactor conditions, Biogen has been working on different methods of monitoring, removing, and controlling foam in bioreactors to dramatically reduce antifoam addition and prevent the regeneration of foam. While focusing on extrinsic methods of monitoring and removing foam through automation, the intrinsic characteristics of foam generation and stability are also being studied to find alternative methods of eliminating the foam in the cell culture process. New methods for monitoring and controlling the foam in bioreactors include using a digital camera to image the culture surface within the vessel. Antifoam is then programmed to be fed automatically based on the percentage of foam coverage. Studies of the delivery of antifoam have also shown that spraying antifoam as a mist reduces the required dose to remove the foam layer, while not significantly interrupting the mass transfer. It not only reduces the foam regeneration rate, but also protects cells from shear damage by allowing for a lower gas entrance velocity from the sparger. The foam generated during cell culture manufacturing processes has been characterized as stable Pickering foam, with slower foam decay. The foam stability was correlated to the accumulation of biomass within the culture. This stability reduces the efficacy of the traditional antifoam remediation strategy over time. Therefore, alternative methods of foam removal are now being investigated to improve the cell culture processes. Overall, the systematic ways of controlling foam have notably reduced the likelihood of equipment failure and improved bioreactor success rates.

460729 Morphological and Life Cycle Changes Exhibited at the up-Scaling of a *Saccharomyces Cerevisiae* High-Cell Density Fed-Batch Biopharmaceutical Process

Abstract:

In a previous study (Fu et al, 2014), we observed a clear difference in yeast cell volumes (determined by wet cell weight (WCW)) grown at 10L laboratory scale versus 10,000L manufacturing scale. That study also included metabolite analysis which suggested hypoxia during scale up. We hypothesized that hypoxia weakens the yeast cell wall during the scale up, leading to changes in cell permeability, and/or cell mechanical resistance.Â Cells treated with the zymolyase showed a more sensitive cell wall later in the production process at manufacturing scale, although cell viability from samples obtained at both scales were similar. In the current study we further examine the relationship between relative volumes of yeast cells to cell viability, size, morphology, and life cycle at laboratory scale (10L) and manufacturing scale (10,000L). Vi-cell and flow-cam methodologies demonstrate a larger sub-population of cells with aspect ratio 0.5 and â‰¥6 micron-diameter at 10,000L scale, corresponding to more one- and two-budded mothers in the larger scale process. In addition, results from a cell cycle study using flow cytometry and DAPI fluorescence microscopy demonstrated multi-budded cells having increased chromosomal content with a long residence in the M phase without completing cytokinesis. In summary, these results collectively suggest a possible relationship between the level of cell wall compromise with cell volume and the presence of a major sub-population of larger budding cells reflecting a longer residence in M phase.

456526 Accelerating Manufacturing in a Clinical Drug Substance Plant through Application of Plant Simulation

Abstract:

Given the major investment required to build or expand biopharmaceutical manufacturing facilities, intensification of existing assets is a highly valuable proposition. However, drug substance processes have many complex operational constraints. Increasing the run rate of a facility should only be undertaken when a thorough evaluation of all operational constraints has been performed.

In this case study, plant simulation was applied to an existing clinical drug substance manufacturing facility executing a wide variety of processes to evaluate opportunities to increase facility run rate. The modeling effort identified an opportunity to increase capacity by 40% with minimal capital investment. At current time, with partial implementation of the identified changes, a 20% increase in run rate has been realized.

461029 Model Predictive Control: A Process Development Tool for Mammalian Cell Culture Processes

Abstract:

In recent years, there has been a growing necessity to increase the speed, efficacy and information content associated to the development and scale-up of biopharmaceutical processes, particularly given the regulatory authorities current shift in philosophy towards quality-by-design (QbD). This QbD requirement for increased process understanding intensifies the work involved in the earlier phases of the drug-product lifecycle. The benefit, however, is that the increased process knowledge can speed the technical transfer from development into manufacturing, deliver a more optimised, robust process with higher titres and greater reproducibility and aid in troubleshooting and root-cause analysis of deviations during production.

A model predictive control (MPC) technology platform incorporating Raman spectroscopy was developed as a tool to streamline upstream bioprocess development and applied to a fed-batch CHO culture. In this talk, the core elements of the MPC will be discussed, namely the development of the Raman PAT method for the online quantification of cell density, substrate and metabolite concentrations, the process model and the MPC control algorithm.

The application of the MPC to process development will be demonstrated through a case study of its application to a CHO fed-batch culture. The MPC enabled the transition from traditional bolus to a PAT-enabled continuous feeding strategy in order to prevent nutrient depletion and deliver a stable macro-environment for the cells. Overall, application of the glucose set-point control strategy resulted in a 1.5 fold increase in peak viable cell density (VCD) and the integral of the viable cell density (IVC) which is directly related to increased titre.

466474 Accurate Determination of B_{22} Data Multivariate Analysis to Improve B_{22} As a Predictive Tool

Abstract:

ABSTRACT

Protein-protein molecular interactions are known to be involved in protein solution aggregation behaviour and are a common issue for the manufacturing of therapeutic proteins such as mAbs. Much effort has been employed to gain a better understanding of aggregation, however the mechanisms leading to protein aggregation are still not fully understood. The osmotic second virial coefficient (B_{22}) is a fundamental physiochemical property that describes protein-protein interactions in solution, which can be a useful tool for not only predicting the aggregation propensity of proteins but also protein solubility, crystallisation and phase behaviour.

A rapid and high-throughput way of determining B_{22} is self-interaction chromatography (SIC). However, the method is still critiqued to be less accurate than its predecessors such as static light scattering. This paper is evaluating the impact of the different experimental strategies and theoretical estimations on the parameters involved in the second virial coefficient determined by SIC. This has been done to help focus on the small variations and combinations that have the largest impact on B_{22} and can directly determine a critically different value.

A number of journal papers recently published describe important experimental considerations for more accurate B_{22} estimations. Many of these small differences in the experimental procedure could result in wrongly estimated B_{22}

values and the interpretation that the protein-protein interactions are repulsive when they are actually attractive and an instable condition for the protein. To easily identify what initial parameters that have the greatest impact an extensive multivariate and sensitivity analysis has been performed to report the impact of a single or multiple parameters on B_{22}. The values that were found to be the most accurate representations of B_{22} were compared to aggregation data to identity that B_{22} was a useful and powerful predictor of aggregation propensity.

Apart from proving to be an accurate predictor for stability in pre-formulation and formulation studies, this tool has also shown to be a good indicator for suitable conditions that can be used for purification and manufacturing of mAbs to enable better product stability while not compromising recovery and quality.

In summary, this work is providing a good insight into important considerations when determining protein-protein interactions and virial coefficients to obtain accurate values that can be used as a powerful tool for biopharmaceutical development.

REFERENCES

Hedberg, S. H. M., Heng, J. Y. Y., Williams, D. R. & Liddell, J. M. (2015). Self-Interaction Chromatography of MAbs: Accurate Measurement of Dead Volumes. *Pharmaceutical Research,* 32, 3975-3985.

Quigley, A., Heng, J., Liddell, J. & Williams, D. (2013). The Accurate Measurement of Second Virial Coefficients Using Self Interaction Chromatography: Experimental Considerations. *European Journal of Pharmaceutics and Biopharmaceutics,* 85, 1103-1111.

458703 Multivariate Data Analysis (MVDA) for Cell Culture Process Development

Abstract:

Multivariate Data Analysis (MVDA) for Cell Culture Process Development

John Bowers, Maria Khouzam, Louis Obando, John Higgings, and Balrina Gupta.

Analyzing mammalian cell culture results has traditionally been done after batch completion by creating and reviewing univariate plots of each variable separately. This approach results in a large amount of data that is not easily interpreted. MVDA Multivariate Data Analysis (MVDA) resolves these because it analyzes parameters and their covariance of large data sets to summarize the performance of the process. Real-time MVDA allows for online process monitoring, and provides immediate understanding.

To apply MVDA during cell culture development, several hurdles were overcome. The number and variety of bioreactors generated our first challenge. Experiments were performed in systems ranging from an automated set of twenty four 250 mL mini-bioreactors to a manufacturing skid with two 500 L stainless steel bioreactors. Since batch context was not available in a consistent format across the bioreactor controllers, special software was developed to enter it into PI. Furthermore each reactor had its own configuration, so a universal configuration structure was develop and the actual configuration of each bioreactor entered into it.

In addition, continuous process data were collected into the data archiving system while off-line results were collected into a separate electronic notebook (ELN) system, making it harder to use combined results to holistically analyze new

batches. To combine offline measurements with continuous bioreactor data, SIMCA-Online was configured to extract the daily offline measurements from E-Notebook BioAssayÂ® (PerkinElmer) database. Batch information and the continuous data were taken from a PI data historian. A â€œgrab and holdâ€ of the last entered value was sufficiently representative of daily offline measurements to account for the sampling rate difference between the daily and continuous measurements in the model. Models based on these combined modes significantly better described culture behavior, and were much better predictors of batch quality measures, such as titer or charge variants.

Furthermore, modeling approaches were developed to account for scale changes and intentional parameter adjustments. Â Most parameters used were intrinsic, such cell and metabolite concentrations, pH, DO, temperature etc. To account for scale effects in useful extrinsic parameters, such as gas flows or feed amounts, a volume based scaling was introduced. With these, we were able to use results from automated 250 mL mini-bioreactor system Ambr® 250 (Sartorius Stedim Biotech) to build models for 3-L and larger scale bioreactors.

In order to account for intentional changes in parameters, such as pH or temperature settings, variables were tracked as deviations from setpoints. The deviation from setpoint allowed for diagnostics of equipment performance, without the multivariate statistics being dominated by an intentional setpoint change. We applied this approach to process ranging studies to determine the impact of changing a parameter on overall process performance live, giving immediate feedback on what a parameter change impacted. When needed, the setpoints were then included in the models to quantify the magnitude of the change to performance or quality attributes.

We successfully applied MVDA using SIMCA-13 and SIMCA-Online® (Umetrics) to several development activities, such as assessing the impact of the parameter changes on the process performance, confirming similarity between master and working cell banks, and demonstrating process comparability across scales. Applying the techniques described here, as much information as possible was obtained from development activities, even in the context of continuing process changes. Six to eight initial similar batches can be used to create models for future work. These models, in turn, can be used to improve process understanding to guide future work. The new results can then be included in models to further improve understanding and model performance.

461208 Template Particle & Solvent Evaporation Assisted Phase Inversion: Easy and Robust Control of Filtration Membrane Morphologies for Various Applications

Abstract:

We present a robust platform allowing the controlled nanoparticle assisted formation of porous polymer membranes.[1] Applications for porous membranes range from drinking water purification,[2] gas separation,[3] biotechnological downstream processing,[4] battery separators,[5] reverse osmosis,[6] food separation processes[7] to porous scaffolds for tissue engineering.[8]

Depending on the membraneâ€™s specific application, different morphologies are required. For example, the convenient separation of particles such as tiny parvoviruses from pharmaceutical goods, requires membranes thick and flexible enough to be mechanically stable while exhibiting appropriate pore sizes. Furthermore for maximizing the flux rate, asymmetric membranes are mostly preferred, as they comprise thin selective layers containing small pores.

Our filtration membrane fabrication procedure allows the simple and exact shaping of porous polymer membranes with previously unmaintained asymmetric properties.[1] The membrane formation is enabled by solvent evaporation induced spinodal decomposition of particles containing liquids and the usage of template particles[9] as such. Membranes produced with varying concentrations of ingredients were analyzed on their morphology at different steps during membrane synthesis. This allowed deciphering mechanistic aspects of the membrane formation, as well as the key parameters for controlling the pore size and pore size gradient. By only varying the concentration ratio of non-solvent and ZnO nanoparticles within a specific polymer dispersion, selective pore sizes form sub-20 nm up to 100 nm and various pore size gradients could be reproducibly created. A selection of membranes with different pore sizes, was finally tested on pure water flux rates as well as silica nanoparticle retention rates, and compared to commercially available high-end virus filters.

Very recently, we discovered the benefit of pore directed ZnO nanoparticles for the synthesis of MOF-polymer composite membranes. The synthesis of MOF-polymer composite membranes depends on MOF-precursors which are ideally in close contact with the membrane. For synthesis of MOF-polymer composite membranes the ZnO-nanoparticles, instead of being dissolved, were maintained within the pores and used as zinc source for the solvothermal synthesis of mechanically flexible ZIF-8 composite membranes. ZIF-8-polymer composite membranes were tested on H2, N2 and CO2 single gas permeation values.

In addition to the methodâ€™s simplicity and the large scale availability of all the ingredients, it has great upscale-potential via roll-to-roll coating.[10]

1. Hess, S. C.; Kohll, A. X.; Raso, R. A.; Schumacher, C. M.; Grass, R. N.; Stark, W. J. Template-Particle Stabilized Bicontinuous Emulsion Yielding Controlled Assembly of Hierarchical High-Flux Filtration Membranes. ACS applied materials & interfaces 2015, 7, 611-617.

2. Shannon, M. A.; Bohn, P. W.; Elimelech, M.; Georgiadis, J. G.; Marinas, B. J.; Mayes, A. M. Science and technology for water purification in the coming decades. Nature 2008, 452, 301-310.

3. Qiu, S. L.; Xue, M.; Zhu, G. S. Metal-organic framework membranes: from synthesis to separation application. Chem Soc Rev 2014, 43, 6116-6140.

4. Rathore, A. S.; Shirke, A. Recent Developments in Membrane-Based Separations in Biotechnology Processes: Review. Preparative Biochemistry & Biotechnology 2011, 41, 398-421.

5. Liu, H. Y.; Liu, L. L.; Yang, C. L.; Li, Z. H.; Xiao, Q. Z.; Lei, G. T.; Ding, Y. H. A hard-template process to prepare three-dimensionally macroporous polymer electrolyte for lithium-ion batteries. Electrochimica Acta 2014, 121, 328-336.

6. Kosutic, K.; Kastelan-Kunst, L.; Kunst, B. Porosity of some commercial reverse osmosis and nanofiltration polyamide thin-film composite membranes. J Membrane Sci 2000, 168, 101-108.

7. Makardij, A.; Chen, X. D.; Farid, M. M. Microfiltration and ultrafiltration of milk: Some aspects of fouling and cleaning. Food and Bioproducts Processing 1999, 77, 107-113.

8. Madihally, S. V.; Matthew, H. W. T. Porous chitosan scaffolds for tissue engineering. Biomaterials 1999, 20, 1133-1142.

9. Kellenberger, C. R.; Luechinger, N. A.; Lamprou, A.; Rossier, M.; Grass, R. N.; Stark, W. J. Soluble nanoparticles as removable pore templates for the preparation of polymer ultrafiltration membranes. J Membrane Sci 2012, 387, 76-82.

10. Kellenberger, C. R.; Hess, S. C.; Schumacher, C. M.; Loepfe, M.; Nussbaumer, J. E.; Grass, R. N.; Stark, W. J. Roll-to-Roll Preparation of Mesoporous Membranes by Nanoparticle Template Removal. Ind Eng Chem Res 2014, 53, 9214-9220.

466886 Intrinsically Microporous Polyimides Incorporating TröGer's Base (TB) for Membrane Gas Separation

Abstract:

Tröger's Base (TB), which is a rigid, V-shaped, and bridged bicyclic amine, has been studied during more 130 years especially in organocatalysis and stereochemistry and recently, McKeown introduced the TB units into polymers of intrinsic microporosity (PIMs) for gas separation membranes. The resultant PIM-TBs exhibited an improvement in gas transport behaviors, indicating that incorporation of the TB units into polymer backbones endows polymer rigidity and microporosity.

In this study, rigid TB moieties were incorporated in polyimides (PI-TBs) *in-situ* polycondensation reactions using dimethoxymethane (DMM). Polyimides are commonly used for membrane gas separation due to their advantages such as good thermal and chemical stabilities, and processability. Solution-processable PI-TB membranes in this work exhibited an enhancement of mechanical properties and gas transport behaviors compared to conventional polyimide membranes, indicating that TB units imparted their microporosity and improved free volume elements of polyimide chains.

Six copolyimides incorporating Tröger's Base (CoPI-TBs) were also investigated for the first time. The CoPI-TBs were prepared *in-situ* TB formation with dimethoxymethane (DMM) and two kinds of diamines, imide-containing diamines and three kinds of flourene-based diamines. Molecular architecturing of CoPI-TBs was tailored of the ratio of two diamines and species of flourene moieties and the structure-property relationship including the physical properties and gas separation properties of the corresponding CoPI-TBs was evaluated.

469370 Comparison of Performance Between Laboratory Scale and Industrial Scale Membranes

Abstract:

Abstract

Cellulose acetate (CA) membranes are among the most researched polymers for separation and barrier applications because of their many useful characteristics including relative low cost, good toughness, high biocompatibility, good desalting, high potential flux. There are different methods used to fabricate porous polymeric membranes, and these can lead to irreproducible results, especially during scale up. The performance of the membrane is greatly influenced by not only the material but also the processing methods and conditions, because they affect the morphology of the membrane. Generally, a laboratory scale approach, such as knife over blade or solution casting, is used to cast a thin sheet of the polymeric membrane solution. Then, the final form of the membrane is prepared by sintering, stretching, or phase inversion. In many studies, the primary film casting process has been limited to doctor blade extrusion, however, this method is not always the most suitable for the industrial scale, because the solution exposed to the environment during the processing phase. Alternatively, slot die extrusion is a well-developed pre-metered approach that allows for continues casting of liquid film in a controlled environment, which also minimizes changes in liquid properties, such as rheology. The slot die coating process is a commonly used process for the manufacturing of plastics and polymer films, as well as coatings and optical films for liquid crystal displays in industry. For the scale up fabrication of hollow fiber CA membranes, a hollow fiber spin-line was used. The feasibility of processing cellulose acetate using a laboratory-scale doctorâ€™s blade versus a scale up slot die extrusion vs hollow fiber membrane was examined. The effects of processing methods, conditions, and substrate on the morphology and on the flux of CA membranes were studied.Â CA membranes were tested for their performance under the same operational conditions of pressure, feed volume, and feed concentration.

472794 Interfacial Polymerization for Sugar / Ionic Liquid Separations

Abstract:

Ionic Liquids (IL) are tunable green solvents for a variety of chemical and biochemical processes. However, the economic viability of utilizing these green solvents depends on developing processes for their recycle and reuse. ILs are outstanding solvents for biomass fractionation and sugar conversion processes. However, separating saccharides from ILs is generally very difficult due to the similarity of their molecular weight. Here a selective skin layer containing 3-aminophenylboronic acid has been successfully synthesized via interfacial polymerization on the surface of asymmetric polyethersulfone (PES) membranes. The fabricated composite membranes show superior ionic liquid/sugar selectivity at competitive permeability. Moreover, the selectivity can be tuned by varying the monomer composition, the polymerization time and post treatment conditions. While [EMIM]Cl and [BMIM]Cl are able to selectively permeate through the membranes, over 90% of mono- and oligosaccharides are rejected. The compositional, chemical and morphological properties of these nanofiltration membranes are characterized and correlated with their separation characteristics.

455188 Development of High Performance Carboxylated PIM-1/P84 Blend Membranes for Pervaporation Dehydration of Isopropanol and CO_2/CH_4 Separation

Abstract:

By taking the advantages of high permeability characteristics of carboxylated PIM-1 (cPIM-1) and high selectivity properties of P84, a new type of cPIM-1/P84 blend membranes has been designed in this study for pervaporation dehydration of isopropanol and CO_2/CH_4 separation. P84 is known to be a highly selective polyimide with low permeability for alcohol dehydration and gas separation, while cPIM-1 is a highly permeable PIM-1 modified via hydrolysis reaction. Polarized light microscopy was employed to study their homogeneity. It was found that the P84/cPIM-1 blends were fully miscible when either cPIM-1 or P84 content was at or lower than 10wt%. The molecular-level interactions were further confirmed by FTIR and DSC. The cPIM-1 loading has significant impact on sorption and diffusion of water and IPA in blends, as indicated by vapor sorption tests. The pervaporation dehydration of isopropanol (IPA) shows that the incorporation of only 10wt% cPIM-1 in the P84 matrix enhances the water permeability by 87% without compromising its selectivity. It has an IPA/water selectivity of 6263 and a water permeability of 0.0864 mg m^{-1} h^{-1} Kpa^{-1}. Similarly, the addition of 30wt% cPIM-1 into P84 increases the CO_2 permeability by 430% from the original 1.3 to 6.98 Barrers. The P84/cPIM-1 membrane containing 90% cPIM-1 has the CO_2/CH_4 separation performance surpassing the Robson upper bound. The newly developed compatible cPIM-1/P84 blend membranes exhibit great potential for the applications of alcohol dehydration and gas separation.

458504 Effects of Fabrication Conditions on the Microstructures and Performances of Smart Gating Membranes with *in Situ* assembled Nanogels As Gates

Abstract:

ABSTRACT

Smart gating membranes with *in situ* assembled poly(*N*-isopropylacrylamide) (PNIPAM) nanogels as gates are successfully prepared via vapor induced phase separation (VIPS) with different exposure time periods, temperatures and relative humidities of the water vapor. Effects of the fabrication conditions on the microstructures as well as the thermo-responsive and mechanical performances of the membranes are investigated. Both the membrane microstructure and the movement of blended PNIPAM nanogels in the membrane forming solution can be controlled by adjusting the fabrication conditions. With increasing the exposure time, the membrane microstructure undergoes a transition from typical liquid-induced phase separation (LIPS) structure (unsymmetrical finger-like porous structure) to typical VIPS structure (symmetric cellular-like porous structure). The critical time periods for the microstructure transition of membranes prepared with vapor temperature and relative humidity of 25 °C/90%, 25 °C/70% and15 °C/70% are about 1.5 min, 2 min and 10 min respectively. The performances of membranes are heavily dependent on the microstructures. The membranes with unsymmetrical finger-like porous structures show large thermo-responsive factor ($R_{39/20}$, the ratio of water flux at 39 °C to that at 20 °C), with the maximum $R_{39/20}$ value being 43.2. All the membranes with symmetric cellular-like porous structures exhibit strong mechanical property and large water flux at 39 °C, which is higher than the volume phase transition temperature (VPTT) of PNIPAM nanogels. The results provide valuable guidance for rational design and fabrication of smart gating membranes with desirable performances.

453480 Novel Thermally Cross-Linked Polyimide Membranes for Ethanol Dehydration Via Pervaporation

Abstract:

In this work, two novel carboxyl-containing polyimides, 2,2′-bis(3,4-dicarboxyphenyl) hexafluoropropane dianhydride-4,4′-diaminodiphenylmethane/3,5-diaminobenzoic acid (6FDAMDA/DABA, FMD) and 3,3′,4,4′-benzophenone tetracarboxylic dianhydride-4,4′-diaminodiphenylmethane/3,5-diaminobenzoic acid (BTDA-MDA/DABA, BMD), are synthesized via chemical and thermal imidization methods, respectively, and employed as pervaporation membranes for ethanol dehydration. Chemical structures of the two polyimides are examined by FTIR and TGA to confirm the successful synthesis. A post thermal treatment of the polyimide membranes with the temperature range of 250 to 400 °C is applied, and its effects on the membrane morphology and separation performance are studied and characterized by FTIR, TGA, WXRD, solubility and sorption test. It is believed that the thermal treatment of the carboxyl-containing polyimide membrane at a relative low temperature only leads to the physical annealing, while it may cause the decarboxylation-induced crosslinking at a higher temperature. In addition, the operation temperature in pervaporation is also varied and shown to be an important factor to affect the final membrane performance. Performance benchmarking shows that the developed polyimide membranes both have superior pervaporation performance to most other flat-sheet dense membranes. This work is believed to shed useful insights on polyimide
membranes for pervaporation applications.

452769 Recent Advances in Evapoporometry for Determining the Pore-Size Distribution of Membranes

Abstract:

Evapoporometry (EP) is a novel means for characterizing the pore-size distribution (PSD) of membranes. It is based on the phenomenon described by the Kelvin equation whereby the vapor pressure is reduced owing to the interfacial curvature of a volatile wetting liquid contained in small pores. EP employs a specially designed test cell that permits establishing a partial pressure in the gas phase at the surface of the membrane that corresponds to the vapor pressure of the largest pores that are filled with the volatile wetting liquid. A small hole in the lid of the test cell controls the rate at which evaporation of the volatile wetting liquid occurs from the pores. Owing to the implications of the Kelvin equation, the gas phase adjacent to the membrane is supersaturated with respect to all pores that are smaller than those from which liquid is evaporating. This supersaturation condition causes pores to empty via evaporation progressively from the largest to the smallest. Measuring the instantaneous mass of the test cell containing the membrane sample via a high resolution microbalance permits determining the instantaneous evaporation rate. This can be related to the partial pressure of the volatile wetting liquid at the surface of the membrane via a straightforward analysis of the mass transfer in the test cell. The Kelvin equation then permits determining the instantaneous pore diameter from the partial pressure at the surface of the membrane. The limitations of the Kelvin equation imply that EP can determine the PSD for flat sheet and hollow fiber ultrafiltration membranes.

EP offers several advantages relative to other methods for determining the PSD. Since the mass can be measured with microgram accuracy, the measurement error in EP is much less than for techniques that require measuring a flow rate, pressure, volume, or heat input. Implementing EP does not require any specialized equipment or expertise. EP also has a relatively small laboratory footprint since the required equipment consists of the test cell, high resolution microbalance, environmental chamber, and an anti-vibration table. Since EP characterization is carried out at the ambient temperature and pressure, it does not alter the morphology of the membrane sample. EP also can be used to determine the PSD of fully hydrated viable biofilms that cannot be done via other PSD characterization methods. A particular advantage of EP is that it can be used to measure the effect of internal pore fouling on the PSD that cannot be done using any method that involves liquid displacement via an inert gas or immiscible liquid.

This presentation will provide an overview of the EP characterization method and discuss recent advances in this technology. The latter include an improved EP test cell design, improving the resolution of EP, extending EP to larger pore sizes, accounting for the effect of the residual adsorbed layer on the pore walls, and adapting EP for characterizing the PSD on both the outer surface and lumen side of hollow fiber membranes.

461196 Development of Highly Compatible and High-Performance Mixed Matrix Membranes for CO_2 Separation

Abstract:

Development of highly compatible and high-performance mixed matrix membranes for CO_2 separation

Zhi Wang*, Xiaochang Cao, Zhihua Qiao, Jixiao Wang, Shichang Wang

Chemical Engineering Research Center, School of Chemical Engineering and Technology, Tianjin University, Tianjin 300350, PR China Tianjin Key Laboratory of Membrane Science and Desalination Technology, State Key Laboratory of Chemical Engineering, Collaborative Innovation Center of Chemical Science and Engineering, Tianjin University, Tianjin 300350, PR China wangzhi@tju.edu.cn

*Corresponding author

Global energy and environmental problems create an urgent need for new and environmental friendly strategies that can be used to capture CO_2 energy-efficiently. Membrane-based CO_2 separation technology exhibits potential applications for greenhouse gas control (CO_2 capture from flue gas) and clean energy supply (natural gas purification and hydrogen production). Currently, great efforts have been devoted to membranes that preferentially permeate CO_2 with high permeances and selectivities for CO_2/N_2, CO_2/CH_4 and CO_2/H_2 gas pairs. Mixed matrix membranes (MMMs) offering a number of benefits, become one of the most popular membrane morphologies for efficient CO_2 separation applications for the past few years. MMMs, consisting of organic polymers as the continuous phase and filler particles as the disperse phase, not only could circumvent the trade-off limit of the continuous phase as well as the inherent obstacles of brittleness associated with the disperse phase, but also could combine their advantages such as good processability of polymers and excellent gas separation property of fillers.

One of the critical factors that restrict achieving high separation performance of MMMs is the poor compatibility between the filler and polymer matrix leading to non-selective interface voids formation. To systematically explore polymer-filler interface and its influence, a series of MMMs have been fabricated by coating the mixture comprised of same polymer and different inorganic fillers. The size of interface voids is connected to the property of fillers[1]. Under the direction of this work, a series of nanofillers with good interface compatibility towards polymer chains and high permselectivity for CO_2 separation have been synthesized and incorporated into polymers to fabricate highly compatible and high-performance MMMs by our group (presented in Fig. 1).

Polymeric nanofillers possess apparent favorable affinities toward polymer continuous phase based on the theory of similarity and intermiscibility. Two kinds of polymeric nanofillers have been designed. Polyaniline (PANI) nanorod[2] containing amino carrier, constructs the CO_2-facilitated transport highway in the MMM (presented in Fig. 1(a)). Covalent organic framework (COF)[3] containing amino carrier, obtains CO_2 preferential adsorption channels in the MMM (presented in Fig. 1(b)).

Coupling of polymer chains and nanofillers is an effective way to improve the interface compatibility between polymer continuous phase and coupled nanofillers. Therefore, our group has coupled hydrotalcite (HT)'s channels in a polyethyleneimine-epichlorohydrin copolymer (PEIE)[4], in which contains abundant amino-groups and moderate hydroxyl groups, to establish high-speed CO_2 facilitated transport channels in the MMM (presented in Fig. 1(c)).

The as-developed MMMs display outstanding CO_2 separation performance.

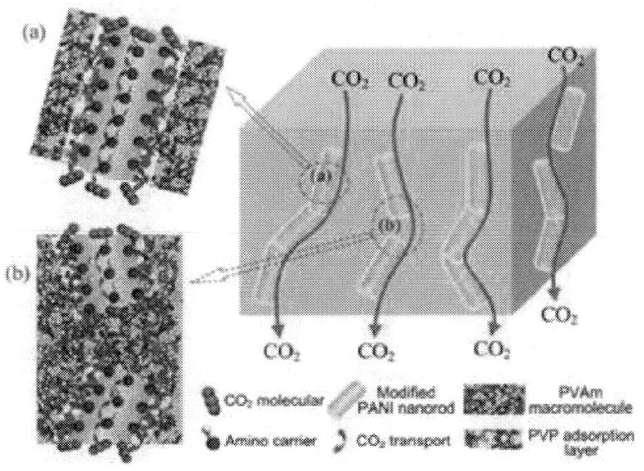

(a)

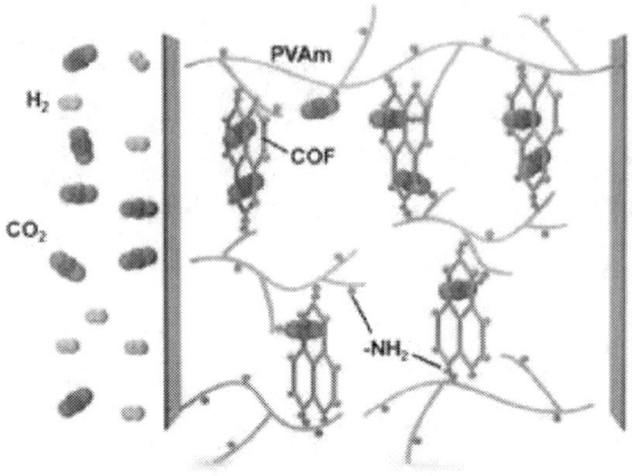

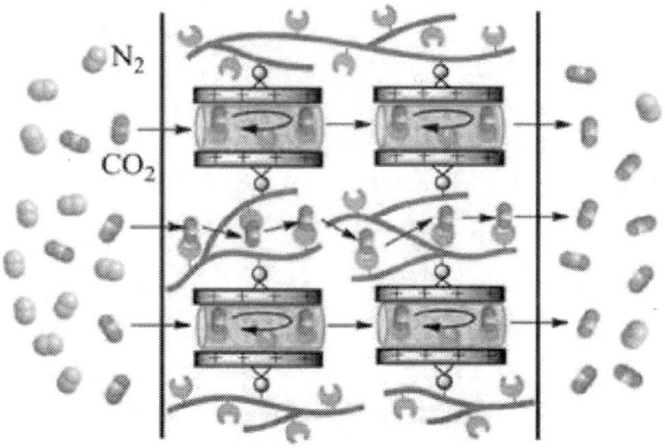

(b)

(c)

Figure 1. Scheme illustration of highly compatible and high-performance MMMs for CO_2 separation: (a) membrane with PANI nanorod; (b) membrane with COF; (c) membrane with HT-PEIE.

References (1) Wang, M.; Wang, Z.; Li, N.; Liao, J. Y.; Zhao, S.; Wang, J. X.; Wang, S. C. *J. Membrane Sci.* **2015**, *495*, 252. (2) Zhao, S.; Wang, Z,; Qiao, Z. H.; Wei, X.; Zhang, C. X.; Wang, J. X.; Wang, S. C. *J. Mater. Chem. A*, **2013**, *1*,246. (3) Cao, X. C.; Qiao, Z. H.; Wang, Z.; Zhao, S.; Li, P. Y.; Wang, J. X.; Wang, S. C. *Int. J. Hydrogen Energy*. **2016**. http://dx.doi.org/10.1016/j.ijhydene.2016.01.137. (4) Liao, J. Y.; Wang, Z.; Gao, C. Y.; Li, S. C.; Qiao, Z. H.; Wang, M.; Zhao, S.; Xie, X. M.; Wang, J. X.; Wang, S. C. *Chem. Sci*. **2014**, *5*, 2843. ADDIN EN.REFLIST

460290 Steric Hindrance Effect Demonstrated in Polyvinylamine Facilitated Transport Membranes for CO_2 Separation

Abstract:

CO_2 capture from flue gas (main component: N_2) is essential for eliminating greenhouse gas emission, and CO_2 separation from synthesis gas is effective for hydrogen purification. Due to the inherent advantages of membrane separation, membrane technology is receiving more and more attention for CO_2 separation. For most conventional polymer membranes, CO_2/N_2 separation follows the solution-diffusion mechanism. The key challenge of solution-diffusion membranes is the trade-off between permeability and selectivity, i.e., the Robesonâ€™s Upper Bound. However, for facilitated transport membranes, the non-reactive gas component, e.g., N_2, diffuses through the membrane via the solution-diffusion mechanism, whereas CO_2 follows the facilitated transport mechanism via its reversible chemical reaction with carriers (often amines). Therefore, CO_2 transport is facilitated, and both high CO_2 permeability and high CO_2/N_2 selectivity can be achieved.

Amine carriers are the key for the CO_2 separation performance of a facilitated transport membrane. Amine carriers exhibiting a high CO_2 loading capacity and reaction rate are always desirable. It has been reported that the CO_2 loading capacity of sterically hindered amine can be twice of unhindered amine in aqueous solution. In this work, high-molecular-weight polyvinylamine (PVAm) was successfully synthesized and modified into sterically hindered polyamines with different degrees of steric hindrance via alkylation reactions. The membranes were synthesized by coating the sterically hindered polyamine solutions with suitable viscosity onto a nanoporous polyethersulfone substrate. Under the typical conditions for CO_2 capture from flue gas (57 °C, 1 atm, 100% relative humidity, 20% CO_2 in the feed gas balanced with N_2 on a dry basis), polyvinylamine membranes showed an average CO_2 permeability of

214 Barrers and CO_2/N_2 selectivity of 48.5. Although the content of amino groups reduced by incorporation of alkyl groups, CO_2 permeability was improved by 24% and CO_2/N_2 selectivity was increased by 14% by modifying polyvinylamine into poly(N-methyl-N-vinylamine). These results have demonstrated the effect of steric hindrance with the incorporation of the methyl group for enhancing CO_2 separation performance in the solid membrane phase. This is one of very few pieces of work demonstrating the steric hindrance in the solid membrane phase. Poly(N-isopropyl-N-vinylamine) and poly(N-tert-butyl-N-vinylamine) membranes showed 15% and 11% improvements on CO_2 permeability, respectively, with no significant changes in CO_2/N_2 selectivity. Comparing the CO_2 separation performances of the above polyamine membranes and their corresponding steric hindrance degrees, we found that the CO_2 permeability increased first significantly with increasing the steric hindrance degree and then less pronouncedly with further increasing the steric hindrance degree. This could be explained by the reduced amino group content in the membranes and the lower reaction rate due to the incorporation of bulky groups onto amino groups. Based on the gas transport results, poly(N-methyl-N-vinylamine) has demonstrated the largest improvement on membrane performance, making it a promising candidate as the next-generation fixed-site carrier for CO_2 facilitated transport membranes.

464312 Thermally-Rearranged Highly Selective Gas Separation Membranes from Hydroxyl-Functionalized Triptycene-Based Polyimide for Natural Gas Upgrading

Abstract:

Recently, a new polyimide of intrinsic microporosity (TDA1-APAF) prepared from 9,10-dimethyl-2,3,6,7-triptycene tetracarboxylic dianhydride (TDA1) with commercially available 2,2-bis(3-amino-4-hydroxyphenyl) hexafluoropropane (APAF) diamine monomer exhibited excellent gas-pair selectivities for O_2/N_2 and CO_2/CH_4 but with moderately high gas permeabilities. To enhance gas permeabilities, conversion of hydroxyl-containing polyimide into polybenzoxazole was used by thermal rearrangement of the aromatic polymer chain with decarboxylation at elevated temperature. TR-TDA1-APAF membrane displayed excellent O_2 permeability of 311 Barrer coupled with O_2/N_2 selectivity of 5.4 and CO_2 permeability of 1328 Barrer coupled with a CO_2/CH_4 selectivity of 27. Gas mixture permeation experiments with a 1:1 CO_2/CH_4 feed mixture demonstrated higher mixed- than pure-gas selectivity (~33) with CO_2 permeability of 1270 Barrer. Interestingly, physical aging over 180 days resulted in an enhanced O_2/N_2 selectivity of 6.3 with O_2 permeability of 185 Barrer and CO_2/CH_4 selectivity of 35 coupled with CO_2 permeability of 699 Barrer. These results suggest that thermally-rearranged membranes from hydroxyl-functionalized triptycene-based polyimides are promising candidate membrane materials for CO_2removal from natural gas and air separation applications.

470062 Clustering Effects in Mixed Matrix Membranes Formed By Nano-Sized ZIF-8 and Pebax

Abstract:

Separation of carbon dioxide from combustion flue gases using selective membranes shows promise to be a low energy carbon capture option. [1] Polymeric membranes exhibit a trade-off between permeability and selectivity which limits their viability as an economically feasible option for use in post-combustion carbon capture. [2] A great deal of attention has been focused on mixed matrix membranes (MMMs) due to the potential improvements that can be made to the selectivity and permeability relative to pure polymer membranes as well as offering a lower cost alternative and improved handling properties relative to pure inorganic membranes. [3]

Whilst mixed matrix membranes show promise there are several challenges slowing their widespread implementation. These challenges largely revolve around minimising performance-hindering defects at the polymer-filler interface. The synthesis route and material selection are key in minimising these defects. [4] Thin membranes are also desirable for greater gas permeances and economical separation processes. Hence small particles are of particular interest but agglomeration of the particles can become an issue.

Nano-sized ZIF-8 crystals were synthesised at Deakin University. ZIF-8 is a zeolitic imidazolate framework containing cavities connected by small windows be which may allow for a molecular sieving effect to occur. [5]

Membranes composed of PEBAX MH1657 and nano-sized ZIF-8 particles were prepared and their gas transport properties tested using the constant-volume method. Significant clusters were observed using SEM at loadings of above 5 wt. %. The presence and prevalence of clusters corresponded with an increase in permeability of all gases tested (helium, nitrogen and carbon dioxide) with no appreciable loss of selectivity. A novel strategy was taken to improve the polymer-filler interface via surface modification of the ZIF-8. ZIF-8 particles were treated with ammonia

gas plasma to substitute methyl groups on the surface of the particles with amine groups in order to improve the adhesion between polymer and particles and investigate the role the clusters play in the gas transport.

[1] T.C. Merkel, H. Lin, X. Wei, R. Baker, J. Memb. Sci. 359 (2010) 126.

[2] L.M. Robeson, J. Memb. Sci. 320 (2008) 390.

[3] C.M. Zimmerman, A. Singh, W.J. Koros, J. Memb. Sci. 137 (1997) 145.

[4] T.-S. Chung, L.Y. Jiang, Y. Li, S. Kulprathipanja, Prog. Polym. Sci. 32 (2007) 483.

[5] D. Fairen-Jimenez, S.A. Moggach, M.T. Wharmby, P.A. Wright, S. Parsons, T. Düren, J. Am. Chem. Soc. 133 (2011) 8900.

451074 Advanced Composite Membranes with Layered Fillers for Gas Separation

Abstract:

Advanced Composite Membranes with Layered Fillers for Gas Separation

Dan Zhao*

Department of Chemical and Biomolecular Engineering,

National University of Singapore, 117585 Singapore

E-mail: chezhao@nus.edu.sg

Gas separation technologies, especially for CO_2 separation, have attracted increased attention recently because of their important applications in clean energy and environmental sustainability. Compared to other CO_2 separation technologies, membrane-based CO_2 separation has been greatly developed in recent years. In order to combine the merits of both polymeric and inorganic membranes, mixed matrix membranes (MMMs) were invented by dispersing porous fillers into continuous polymeric matrices hoping to increase the permeability and selectivity of the resulting membranes while preserving the properties of good mechanical strength and processability of the polymeric matrices. In this talk, I will introduce our work in preparing MMMs containing lamellar MOF or COF fillers with high-aspect-ratio, which exhibit good gas separation performance because of the increased tortuosity of gas permeation paths imposed by lamellar fillers. In addition, I will also discuss the direct fabrication of ultrathin membranes composed of 2D nanosheets. Because of their ultra-small thickness and precisely tunable pore size and functionality, these novel membranes may demonstrate unprecedented gas separation performance.

References

1. Kang, Z. X.; Peng, Y. W.; Qian, Y. H.; Yuan, D. Q.; Addicoat, M. A.; Heine, T.; Hu, Z. G.; Tee, L.; Guo, Z. G.; Zhao, D.* Mixed matrix membranes (MMMs) comprising exfoliated 2D covalent organic frameworks (COFs) for efficient CO_2 separation. *Chem. Mater.* **2016**, *28*, 1277-1285.

2. Kang, X. Z.; Peng, Y. W.; Hu, Z. G.; Qian, Y. H.; Chi, C. L.; Yeo, L. Y.; Tee, L.; Zhan, D.* Mixed matrix membranes comprising two-dimensional metal-organic framework nanosheets for pre-combustion CO_2 capture: a relationship study of filler morphology versus membrane performance. *J. Mater. Chem. A* **2015**, *3*, 20801-20810.

467568 Comparison of the Permeabilities and Selectivities of Polybenzoxazole Membranes Thermally Rearranged from Hydroxyl-Polyimides Formed from Different Dianhydride Precursors

Abstract:

Membrane separation is an excellent method for purification of natural gas. It is much cheaper and easier to implement than the alternatives. Multiple polymers have been studied for use in natural gas separation, and two stand above the rest: polyimide and polybenzoxazole. Originally, polybenzoxazole hadn't been discovered yet, so researchers focused on polyimides. They continuously worked to overcome the limitations of membrane separation, which are the tradeoff between selectivity and permeability, physical aging, and plasticization. It was found that the selection of monomer precursors could influence the separation properties of the resultant polyimide membranes. Other methods of increasing resistance to plasticization are thermal annealing and cross-linking.

Recently, it was discovered that polyimides containing hydroxyl groups in a specific position could be thermally rearranged into polybenzoxazole, which has even better separation properties than polyimide due to the size and distribution of its free volume elements. Methods to influence these properties are similar to those used with polyimides. Polybenzoxazole is also very easy to mold into different forms, such as hollow fibers, because the polyimides are easy to make into these shapes and retain their shape when thermally rearranged.

The intent of this research is to address the egregious lack of permeation testing performed to compare different dianhydride precursors. Many studies have been done to determine the effects of the structure of diamine precursor on the separation properties of polybenzoxazole, but none regarding the effect of the structure of the dianhydride. In this work we compare the permeabilities and selectivities of

polybenzoxazole membranes thermally rearranged from hydroxyl-polyimides formed from three dianhydride precursors.

Thus far, we have synthesized samples of polyimides from each of the following three dianhydrides: 1) 4,4'-oxydiphthalic anhydride (ODPA), 2) 3,3',4,4'-benzophenone tetracarboxylic dianhydride (BTDA), and 3) 3,3',4,4'-biphenyl tetracarboxylic dianhydride (BPDA). Each dianhydride was combined with 2,2'-bis(3-amino-4-hydroxyphenyl)-hexafluoropropane (APAF) to form the polyimide. Multiple membranes have been made from the polyimides made from ODPA and BTDA, and work is on-going to make membranes from polyimides made from BPDA. A few of the formed membranes were successfully thermally rearranged as well.

465996 Hydrogen Permeable Membrane with Improved Separation Property

Abstract:

Hydrogen permeable membrane with improved separation property

Yanying Wei,[1,2] Haihui Wang,[2] Jürgen Caro[1]

[1] Institute of Physical Chemistry and Electrochemistry, Leibniz University of Hannover, Callinstrasse 3A, 30167 Hannover, Germany

[2] School of Chemistry and Chemical Engineering, South China University of Technology, 510640 Guangzhou, China

ceyywei@scut.edu.cn

High-temperature mixed proton and electron conductors (HT-MPECs) have attracted increasingly interest due to their wide potential applications as hydrogen sensors, hydrogen pumps, hydrogen permeable membranes for gas separation and catalytic membrane reactors, solid oxide fuel cells (SOFCs) etc. Materials with both high protonic and electronic conductivity, as well as adequate stability are required to be developed. $BaCe_{0.95}Tb_{0.05}O_{3-\delta}$ (BCT) is popular HT-MPECs, which exhibits the highest hydrogen permeation flux till now, but the instability due to the phase segregation hinders its applications. Considering the doped $BaZrO_3$ exhibits a much higher chemical stability compared with the doped $BaCeO_3$, and the two oxides can form easily a solid solution, the mixed proton and electron conductor material $BaCe_{0.85}Tb_{0.05}Zr_{0.1}O_{3-\delta}$ (BCTZ) is prepared through the glycine-nitrate combustion process, which exhibits enhanced stability for H_2 separation. Moreover, in order to improve the H_2 permeability of such material, metallic Ni is added as another phase for electronic conductor aimed to increase the electronic conductivity. Thus a Ni-BCTZ cermet membrane is synthesized by mixing the BCTZ oxide and the metallic Ni powder followed by sintering, which indicates good homogeneity of the perovskite oxide phase for protonic conductivity and the metallic Ni phase for electronic conductivity. After doping Zr into the BCT oxide, the chemical stability of both the BCTZ ceramic membrane and the Ni-BCTZ cermet membrane have been enhanced remarkably. The XRD and SEM results also prove the outstanding chemical stability, which promotes the Zr-doped $Ba(CeTb)O_{3-\delta}$ ceramic and cermet material as a proper candidate of an MPEC material for different potential applications.

Keywords:

Hydrogen separation/ Proton conductor/ Cermet membrane/ Perovskite/ Stability

References

1. X. Meng, N. Yang, J. Song, X. Tan, Z. Ma, K. Li, Synthesis and characterization of terbium doped barium cerates as a proton conducting SOFC electrolyte, Int. J. Hydrogen Energy 36 (2011) 13067-13072.

2. X. Meng, J. Song, N. Yang, B. Meng, X. Tan, Z. Ma, K. Li, Ni–$BaCe_{0.95}Tb_{0.05}O_{3-\delta}$ cermet membranes for hydrogen permeation, J. Membr. Sci. 401-402 (2012) 300-305.

3. X. Tan, J. Song, X. Meng, B. Meng, Preparation and characterization of $BaCe_{0.95}Tb_{0.05}O_{3-\alpha}$ hollow fibre membranes for hydrogen permeation, J. Eur. Ceram. Soc. 32 (2012) 2351-2357.

467506 Hematin Crystallization Mechanisms Suggest How Antimalarial Drugs Operate

Abstract:

Hematin is a byproduct of heme detoxification in malaria parasites. Current antimalarial compounds are thought to inhibit the growth of hematin crystals, referred to as hemozoin (*in vivo*) or b-hematin (*in vitro*). Identification of antimalarial drug action on this phase transformation could provide a foundation for drug design to overcome parasite resistance to antimalarials. Continued resurgence of antimalarial drug resistance drives research effort to better characterize drug-crystal interactions from a fundamental perspective. To this end, we aim to better understand the molecular mechanism(s) of hematin crystallization and antimalarial drug action within physiologically-relevant environments as the basis for elucidating growth sites on hematin crystal surfaces and design more effective growth inhibitors that selectively bind to these sites.

We employ *in situ* atomic force microscopy (AFM) in parallel with bulk solution and crystallization studies as a platform to study the mechanisms of crystal growth and inhibitor-crystal interactions in biomimetic growth solutions. Time-resolved *in situ* AFM measurements reveal that hematin crystallization occurs by the classical pathway of single molecule incorporation, specifically involving two-dimensional nucleation of layers. We quantified the rate of layer generation and the velocities of anisotropic step advancement as a function of supersaturation in the absence and presence of antimalarials. We identified unique modes of binding for current antimalarials as inhibition pathways: step pinning, kink blocking, and step bunching. In parallel, we studied the prominent functional groups that comprise antimalarial compounds by *in situ* AFM. Identifying fundamental principles governing the molecular recognition of antimalarial drugs to specific sites could lead to the rational design of new compounds with improved efficiency to overcome parasite resistance to current antimalarials.

461662 Atomistic Simulation of the Aqueous-Phase Synthesis of Au Nanoparticles

Abstract:

The catalytic, electronic, and optical properties of metal nanoparticles strongly depend on their three-dimensional atomic structure. In particular, Au nanoparticles have demonstrated a wide range of properties, which can be tuned for different catalytic and biomedical applications. In this work, we have adopted a kinetic Monte Carlo (KMC) simulation approach for modeling the atomistic growth behavior of Au nanoparticles, in order to provide a direct link to the experimentally-observed growth behavior. Such a model can provide information about the impact of individual system events on the final atomic-scale structural features of the Au nanoparticles, and this can help guide the experimental synthesis process. The advantage of our approach is that, compared to traditional molecular dynamics simulations or electronic structure calculations, we can track the atomistic nanoparticle structural evolution on time scales that approach the actual experiments. This has allowed us to perform several different comparisons against experimental benchmarks, and it has helped transition our KMC simulations from a hypothetical toy model into a more experimentally-relevant test-bed. In addition, while previous analytical models have been used for predicting nanoparticle growth, the KMC approach can track atomistic-scale details and account for neighbor-neighbor interactions, structural defects, and structural heterogeneity. These aspects can be particularly influential during nanoparticle growth, leading to a wide variety of interesting nanoparticle geometries.

Here, we report the effects of synthesis temperature and initial precursor concentration on the Au nanoparticle growth behavior, and we find that our model can be trained (via automated feedback) to adequately reproduce the experimental growth curves at the same conditions. The fitting procedure results in reasonable parameter values, including activation energy barriers that are consistent with related experimental measurements. In addition, since our KMC simulations preserve the atomistic details of our growing Au nanoparticle, this allows for detailed structural analyses. Although our nanoparticles are roughly spherical, the maximum/minimum dimensions deviate from the average by approximately 12.5%, which is consistent with the corresponding experiments. Also, our surface texture analysis highlights the changes in the surface structure as a function of time. While the nanoparticles show similar surface structures throughout the growth process, there can be some significant differences during the initial growth at different synthesis conditions. To the best of our knowledge, this is one of the first attempts at modeling the atomistic growth behavior of realistic metal nanoparticles in solution. In the future, we intend to perform additional benchmarking of more atomistic growth features, including an extension of our model to capture anisotropic nanoparticle growth.

470151 A Mechanistic Growth Model for Organic Salt Morphology Prediction

Abstract:

There is growing interest in the pharmaceutical sector to develop strategies for digital drug product design, starting with the molecular structure of a target molecule and creating good initial estimates for all the solid form characteristics, including crystal morphology. For morphology prediction, a detailed understanding of crystal growth is vital. In the past several decades various growth models have been proposed to predict the crystal morphology of organic molecules, which laid the foundation to design schemes to manipulate and tailor the crystal morphology in accordance with the application requirements. Recent developments include [1-2].

However, the prediction of growth rates for organic salts still remains blank, due to their complicated crystal structures and complex atomic interactions. For example, organic salts usually consist of nonstoichiometric ions as the growth units, which makes it difficult to find the periodic bond chains (PBCs) within the crystal lattice; moreover, the short-range intermolecular interactions, and the long-range electrostatic interactions in organic salts may need to be considered; the kink sites in organic salts are complicated due to the existence of positive and negative ions, etc. With these considerations in mind, we propose a mechanistic growth model for organic salt morphology prediction, which identifies periodic bond chains (PBCs) using the concept of building units, and accurately captures the short-range/long-range electrostatic interactions and partial charges. We illustrate the method on ammonium acetate grown from the vapor phase. The electrostatic interaction energies in the kink sites were calculated using a space partitioning method that is computationally efficient. This algorithm for studying the solid-state interactions and building units is based on a mechanistic spiral growth model, which may be useful for predicting the morphology of a wide variety of organic salts.

References:

1. Li J, Tilbury CJ, Kim SH, Doherty MF, A design aid for crystal growth engineering. Progress in Materials Science 82, 1-38 (2016).

2. Li J, Tilbury CJ, Joswiak MN, Peters B, Doherty MF, Rate expressions for kink attachment and detachment during crystal growth. Crystal Growth & Design DOI:10.1021/acs.cgd.6b00292 (2016).

453851 Growing Macroscopic Hydrates Using a Bioinspired Approach and Investigating Dehydration Induced Polymorphism

Abstract:

Growing of macroscopic single crystals of organic materials is not necessarily a straight forward procedure. Carbamazepine dihydrate is one of these cases. Traditional approaches of crystallization of this, particular, compound using mixtures of alcohols with water lead to crystals with high aspect ratios (needle shaped or high aspect ratio plates)[1]. These crystals are not suitable for wettability or XPS measurements on individual facets[2]. In this work, a bioinspired crystallization method is employed for the growth of macroscopic carbamazepine dihydrate crystals.

In a work on the interactions of surfactants with carbamazepine in aqueous solutions, investigators have reported the formation of prismatic carbamazepine dihydrate crystals in the micron scale[3]. From the surfactants employed, it seems that the carbonyl groups of sodium taurocholate influence the crystal habit. Unfortunately, sodium taurocholate proved not to be adequate for the growth of crystals of larger sizes, that can be used as seeds for the growth of macroscopic crystals. Also, other studies suggest that the growth of large protein crystals is favored by biomimetic solutions, imitating the design of the digestive vacuole of malarian parasites such as *Plasmodium falciparum*, comprising of an aqueous and an organic phase[4,5].

A two phase system with cyclohexanone in the organic phase and water in the other phase was developed. An adequate amount of carbamazepine was added in the solution under stirring and the mixture was heated using a water bath. When all the material was dissolved, stirring stopped and a temperature gradient of $1^{o}C*d^{-1}$ was applied. After a few days, depending on the amount of carbamazepine added, crystals started to appear inside precursor droplets formed on the interface between the two phases. When the temperature of the water bath reached 4°C the crystals were harvested. They were quite large with the (0 2 0) facet to be the dominant one, having sizes up to 2x1 cm. The harvested crystals were examined using XRD and found to be carbamazepine dihydrate crystals. Similar experiments were performed with heavy alcohols and butanone, but the crystals obtained were not that promising, even though they were in all cases dihydrate crystals, in terms of crystal habit. The shift from the needle like shape to a more prismatic one can be thus attributed to the carbonyl chemistry.

Molecular dynamics simulations suggest that in such systems droplets of the aqueous phase are formed in the organic phase[6]. This droplets were tracked using Dynamic Light Scattering and found to be about 500 nm in diameter. Nonetheless, these droplets do not seem to act as nucleation sites. Only the main water cyclohexanone interface operates as a nucleation site. This preferential nucleation can be explained on the ground of heterogeneous nucleation. The shape factor for nucleation on an interface should result to a much higher decrease on the Gibbs free energy of nucleation rather than the shape factor for nucleation on a droplet[7,8]. Furthermore, molecular dynamics simulations propose the formation of pockets of enhanced water concentration on the interface. These pockets is speculated to promote even further nucleation.

Drying experiments performed on these macroscopic crystals. From SEM images obtained, it was shown that dehydration is driven by well-ordered cracks formed on the surface of the dominant facet. Two types of cracks appear on the (0 2 0) facet: 1) macroscopic cracks vertical to the (0 2 0) facet with crack spacing a few microns big and 2) cracks with smaller spacing inclined, both clockwise and counterclockwise, with respect to the primary cracks of type 1. These two types of cracks correspond to the following main cleavage planes: (h 0 0), (-1 ±1 1) and (0 ±1 1). Furthermore, macroscopic crystals were sliced with razor blade and the examination of their core revealed an additional cleavage plane, (0 k 0). These findings are important, as previous studies reported only the (h 0 0) cleavage plane. It should, also, be mentioned that the crack appearance is independent of the application of vacuum upon drying.

The anhydrous form nucleates, during dehydration, on the surface of the dihydrate crystals in the form of small islands. The dehydration mechanism for macroscopic crystals results solely to the triclinic polymorph of anhydrous carbamazepine. The triclinic polymorph was depicted both with XRD and SEM. The SEM images are quite interesting showing the needle shaped triclinic crystals growing vertically on the surface of the dihydrate. Comparing the results of the dehydration induced polymorphism for macroscopic dihydrate crystals and micron-sized crystals it can be deduced that the crystal size and habit influence the final polymorph. It seems that for cases where the need for heat dissipation is greater (i.e. in macroscopic crystals), the triclinic polymorph nucleating during drying is higher. This is because the needle-shaped triclinic crystals have a large surface area to volume ratio providing better heat dissipation. This concomitant nucleation scheme has been reported computationally via Monte Carlo simulations[9,10]. However, the suggested drying mechanism can be better understood, mechanistically, on the same ground Machon described phase transformations on nanoparticles using the one dimensional Ginzburg-Landau equation[11, 12].

This study expands the current knowledge on nucleation of organic materials using bioinspired approaches. It is quite important to notice the nucleation of dihydrate and the absence of any other polymorphs. This shows that for such systems, the chemical potential associated with the aqueous phase and with the small amount of water dissolved in cyclohexanone drive the nucleation. It is, also, of particular importance that the drying mechanism is revealed; suggesting that the heat dissipation mechanism drives the dehydration induced polymorphism. The image analysis revealed also the nucleation mechanism of the anhydrous polymorphs resembling to the anisotropic nucleation model proposed in the past[13,14].

Disclosure:

Imperial College London and AbbVie jointly participated in study design, research, data collection, analysis and interpretation of data, writing, reviewing, and approving the publication. Eftychios Hadjittofis is a graduate student at Imperial College London; Jerry Y. Y. Heng is a professor at Imperial College London. They all have no additional conflicts of interest to report. Geoff G. Z. Zhang is an employee of AbbVie and may own AbbVie stock.

References:

1. Khoo JY, Shah UV, Schaepertoens M, Williams DR, Heng JYY. Process-induced phase transformation of carbamazepine dihydrate to its polymorphic anhydrates. *Powder Technol.* 2013;236:114-121.

2. Ho R, Hinder SJ, Watts JF, Dilworth SE, Williams DR, Heng JYY. Determination of surface heterogeneity of d-mannitol by sessile drop contact angle and finite concentration inverse gas chromatography. *Int J Pharm.* 2010;387(1–2):79-86.

3. Rodriguez-Hornedo N, Murphy D. Surfactant-facilitated crystallization of dihydrate carbamazepine during dissolution of anhydrous polymorph. *Journal of Pharmaceutical Sciences.* 2004;93(2):449-460.

4. Olafson KN, Rimer JD, Vekilov PG. Growth of large hematin crystals in biomimetic solutions. *Crystal Growth and Design.* 2014;14(5):2123-2127.

5. Vekilov PG, Rimer JD, Olafson KN, Ketchum MA. Lipid or aqueous medium for hematin crystallization? *Crystal Engineering Communications.* 2015;17:7790-7780.

6. Wick CD, Chang TM. Computational observation of pockets of enhanced water concentration at the 1 Octanol/water interface. *The Journal of Physical Chemistry B.* 2014;118(7785-7791).

7. Fletcher NH. Size effect in heterogeneous nucleation. *Journal of Chemical Physics.* 1958;29:572-576.

8. Johnson WC, White CL, Marth PE, et al. Influence of crystallography on aspects of solid-solid nucleation theory. *Metallurgical transactions A.* 1975;6(4):911-919.

9. Sanders DP, Larralde H, Leyraz F. Competitive nucleation and the ostwald rule in a generalized potts model with multiple metastable phases. *Physical Review B.* 2007;75:132101.

10. Hedges LO, Whitelam S. Limit of validity of ostwald's rule of stages in a statistical mechanical model of crystallization. *Journal of Chemical Physics.* 2011;135:164902.

11. Machon D, Piot L, Hapiuk D, et al. Thermodynamics of nanoparticles: Experimental protocol based on a comprehensive ginzburg-landau interpretation. *Nano Letters.* 2014;14(1):269-276.

12. Machon D, Melinion P. Size-dependent pressure-induced amorphization: A thermodynamic panorama. *Physical Chemistry Chemical Physics.* 2014;17(2):903

13. Bianchet MA, Rigotti G, Blesa MA. Surface nucleation and anisotropic growth model for solid state reactions. *Solid State Ionics.* 1990;42(1):21-27.

14. Bianchet MA, Rigotti G, Blesa MA. Surface nucleation and anisotropic growth model for the dehydration of 11-aminoundecanoic acid dihydrate. *Solid State Ionics.* 1995;78(1–2):49-53.

468938 A Microfluidic Droplet Platform for Investigating Nucleation Kinetics of Hydrocarbons for Application to Wax Crystallization

Abstract:

Crystallization of wax present in crude oil is a major issue for its storage, processing and transportation. As a result, there is a significant interest in identifying inhibitors that can mitigate the nucleation of wax crystals in crude oil. To address this need, droplet-based microfluidics is used to quantify the nucleation kinetics of n-alkanes - a simple model system for wax. In nucleation studies, it is useful to have control of the droplet volume as it dictates the number of impurities (like dust) present per droplet that can act as nucleation site. Therefore, a valve-based flow-focusing device is used to produce droplets of controllable volume. A downstream storage chamber is included to incubate the produced droplets so that nucleation studies can be conducted. The modular design of the microfluidic system provides flexibility, allowing adjustment to specific experimental requirements. Hexadecane-in-water emulsions are produced using Tween 20 and sodium dodecyl sulfate as surfactants. The effect of droplet size and surfactant type on nucleation rate of n-hexadecane is investigated for the first time in a droplet-based microfluidic platform. The results show that the degree of nucleation in hexadecane could be controlled by selecting an appropriate surfactant and controlling the droplet size. The experimental data is used to test current models of nucleation kinetics. This work will enable establishment of a microfluidic platform for future studies of nucleation in paraffins and their mixtures with inhibitors.

463932 Sonocrystallization of L-Asparagine Monohydrate: Modeling and Optimization.

Abstract:

Producing crystals with target attributes consistently is a great challenge in any crystallization process due to stochastic nature of the nucleation event and the limited number of manipulated variables. The use of power ultrasound (defined as frequencies from 20 to 100 kHz) provides an interesting approach to induce nucleation at low supersaturation in a controlled way [1]. Several studies have demonstrated the benefits of applying ultrasound during a cooling or anti-solvent crystallization process such as faster and uniform primary nucleation, higher yield, reduced agglomeration, narrow crystal size distribution, control of polymorphism, and generation of internal seed [2]-[3].

The development of an effective mathematical model describing the crystallization dynamics is a crucial issue towards finding the optimal process performance and to control the crystal product properties. Here we develop and validate a mathematical model based on population balance framework for the ultrasound-assisted unseeded batch cooling crystallization of L-asparagine monohydrate, an important amino acid, from its aqueous solution. The population balance model considers the relevant fundamental events of the crystallization process, such as nucleation, growth, and breakage phenomena. An additional kinetic expression is introduced for induced nucleation due to ultrasound irradiation [4]. Application of ultrasound induces energy into the system and this in turn leads to temperature rise. The modeling considers this rise in temperature appropriately. The population balance model coupled with mass balances is solved by high resolution finite volume discretization with a flux limiter.

A series of ultrasound-assisted batch cooling crystallization experiments have been conducted systematically and the solute concentration data measured by high performance liquid chromatography and final mean crystal size measured by laser diffraction are used to simultaneously extract all kinetic parameters using the population balance model and a nonlinear optimization technique. The model has been fully validated with new experimental data and the good agreement between model predictions and experimental observations clearly demonstrates the predictive ability of the model. Experimental data and model simulations show that the ultrasound significantly reduces the metastable zone width for L-asparagine monohydrate, enhances the nucleation rate, and induces early nucleation. The developed model is then used to solve a set of dynamic optimization problems related to particle engineering. The model is also used to determine the region of attainable size for such a sonocrystallization process.

References:

[1] G. Ruecroft, D. Hipkiss, T. Ly, N. Maxted, and P. W. Cains, Org. Proc. Res. Dev. 9, 923-932, 2005.

[2] O. Narducci and A. G. Jones, Cryst. Growth Des. 12, 1727–1735, 2012.

[3] M. Kurotani and I. Hirasawa, Chem. Eng. Res. Des. 88, 1272–1278, 2010.

[4] A. Kordyllaa, T. Krawczyka, F. Tumakakab and G. Schembecker, Chem. Eng. Sci. 64, 1635–1642, 2009.

461649 Improving the Search for Polymorphs

Abstract:

Improving the search for polymorphs.

I. Sugden, C.Gatsiou, M. Vasiliadis, C.S. Adjiman, and C.C. Pantelides

Centre for Process Systems Engineering, Department of Chemical Engineering, Imperial College London, South Kensington Campus, SW7 2AZ, London, UK

The CrystalPredictor I[1,2] and II[3] codes used to explore the space of crystal structures have been used successfully in several crystal structure prediction (CSP) investigations in recent years[4], including in the series of blind tests organised by the Cambridge Crystallographic Data Centre[5,6] and in the prediction of the crystal structures of pharmaceutically-relevant molecules[7,8]. CSP methods can have an important role to play in the polymorph screening of pharmaceutically relevant molecules, by identifying potential low energy crystal structures.

We present a summary of CrystalPredictor, focussing on improvements to the lattice energy evaluation that the most recent blind test[9], as well as our own investigations into a variety of flexible polymorphic molecules, has prompted. These improvements aim to achieve greater accuracy in the initial ranking of potential crystal structures, while managing computational cost so that a thorough exploration of the search space is possible. Firstly we discuss non-uniform LAMs; an innovation in CrystalPredictor II that allows the most efficient use of computational effort to cover a flexible moleculeâ€™s conformational space. We use blind test molecule 26 as an example.

Secondly, we discuss the smoothing of the intramolecular potential, which improves accuracy in CrystalPredictor II by collating data from *ab initio* calculations. The impact of this approach is investigated based on a CSP study for

flufenamic acid[10].

(1) Karamertzanis, P. G.; Pantelides, C. C. *J Comput Chem* **2005**, *26*, 304.

(2) Karamertzanis, P. G.; Pantelides, C. C. *Mol Phys* **2007**, *105*, 273.

(3) Habgood, M.; Sugden, I. J.; Kazantsev, A. V.; Adjiman, C. S.; Pantelides, C. C. *J Chem Theory Comput* **2015**, *11*, 1957.

(4) Pantelides, C. C.; Adjiman, C. S.; Kazantsev, A. V. *Top Curr Chem* **2014**, *345*, 25.

(5) Day, G. M.; Cooper, T. G.; Cruz-Cabeza, A. J.; Hejczyk, K. E.; Ammon, H. L.; Boerrigter, S. X. M.; Tan, J. S.; Della Valle, R. G.; Venuti, E.; Jose, J.; Gadre, S. R.; Desiraju, G. R.; Thakur, T. S.; van Eijck, B. P.; Facelli, J. C.; Bazterra, V. E.; Ferraro, M. B.; Hofmann, D. W. M.; Neumann, M. A.; Leusen, F. J. J.; Kendrick, J.; Price, S. L.; Misquitta, A. J.; Karamertzanis, P. G.; Welch, G. W. A.; Scheraga, H. A.; Arnautova, Y. A.; Schmidt, M. U.; van de Streek, J.; Wolf, A. K.; Schweizer, B. *Acta Crystallogr B* **2009**, *65*, 107.

(6) Bardwell, D. A.; Adjiman, C. S.; Arnautova, Y. A.; Bartashevich, E.; Boerrigter, S. X. M.; Braun, D. E.; Cruz-Cabeza, A. J.; Day, G. M.; Della Valle, R. G.; Desiraju, G. R.; van Eijck, B. P.; Facelli, J. C.; Ferraro, M. B.; Grillo, D.; Habgood, M.; Hofmann, D. W. M.; Hofmann, F.; Jose, K. V. J.; Karamertzanis, P. G.; Kazantsev, A. V.; Kendrick, J.; Kuleshova, L. N.; Leusen, F. J. J.; Maleev, A. V.; Misquitta, A. J.; Mohamed, S.; Needs, R. J.; Neumann, M. A.; Nikylov, D.; Orendt, A. M.; Pal, R.; Pantelides, C. C.; Pickard, C. J.; Price, L. S.; Price, S. L.; Scheraga, H. A.; van de Streek, J.; Thakur, T. S.; Tiwari, S.; Venuti, E.; Zhitkov, I. K. *Acta Crystallogr B* **2011**, *67*, 535.

(7) Vasileiadis, M.; Pantelides, C. C.; Adjiman, C. S. *Chem Eng Sci* **2015**, *121*, 60.

(8) Vasileiadis, M.; Kazantsev, A. V.; Karamertzanis, P. G.; Adjiman, C. S.; Pantelides, C. C. *Acta Crystallographica Section B-Structural Science* **2012**, *68*, 677.

(9) Reilly, A. M. e. a. *Acta Crystallographica Section B: Structural Science, Crystal Engineering and Materials* **2016**.

(10) Lopez-Mejias, V.; Kampf, J. W.; Matzger, A. J. *J Am Chem Soc* **2012**, *134*, 9872.

463153 Polymorph Selection By Continuous Crystallization

Abstract:

Polymorphic solids are ubiquitous in both nature and industry and generally exhibit unique mechanical, electrical, and physicochemical properties, as well as unique solubility curves in solution. Polymorph selective processing is necessary in these systems due to the typical material and pharmacokinetic property variation among polymorphs of the same chemical composition. The study of polymorphism in batch systems has a long history that has led to deep understanding of the concepts of solvent-mediated phase transformations and Ostwaldâ€™s rule of stages. These fundamental ideas are applied in many papers and processes to steer batch processes towards a desired polymorph. Here, we discuss the implications of continuous processing on polymorphism, and outline a methodology that enables polymorph selection and simplifies the necessary process control strategies in applicable systems.

It is well known in batch crystallizer design that the polymorph distribution is governed by the induction time of the most stable form, and that in the limit of very long batch times the most stable polymorph is obtained. Continuous devices do not operate under a similar constraint. After an initial start-up phase, a continuous device can operate at a dynamically stable operating point indefinitely, regardless of the thermodynamic stability of the effluent crystals. Interestingly, essentially pure, thermodynamically metastable steady-states exist in systems in which the thermodynamically stable solid cannot nucleate and grow on the time scale of the crystallizer residence time. These conditions depend on design choices (such as solvent choice, temperature, residence time, feed supersaturation, etc.), and are therefore accessible in many systems. This concept has been formalized with the use of a bi-polymorph population balance model, the method of moments, and a linear stability analysis. The analysis gives simple functions of parameters (dimensionless groups) for which one can continuously produce thermodynamically metastable products based only on the relative polymorph dynamics.

This work was motivated by the remarkable experiments recently reported by Lai, Trout, and Myerson et al [1] on continuous crystallization in the L-glutamic acid system. We demonstrate agreement with their L-glutamic acid results as well as another set of data describing the continuous crystallization of p-aminobenzoic acid [2]. For many polymorphic compounds, engineering a process to produce a desired polymorph is as simple as finding a reasonable operating point for the continuous mixed-suspension mixed-product removal crystallization process (temperature, residence time, initial supersaturation, etc.).

References

[1] Lai, T. T. C., Ferguson, S., Palmer, L., Trout, B. L., & Myerson, A. S. (2014). Continuous Crystallization and Polymorph Dynamics in the l-Glutamic Acid System. *Organic Process Research & Development*, *18*(11), 1382-1390.

[2] Lai, T. T. C., Cornevin, J., Ferguson, S., Li, N., Trout, B. L., & Myerson, A. S. (2015). Control of Polymorphism in Continuous Crystallization via Mixed Suspension Mixed Product Removal Systems Cascade Design. *Crystal Growth & Design*, *15*(7), 3374-3382.

457919 Coarse-Grained Drug Molecule Simulation and Polymorph Discovery

Abstract:

We describe a coarse-graining method used to study crystallization and polymor- phism in a pharmaceutically relevant drug molecule, all via computer simulation. Our coarse-grained model is capable of capturing the stability of the experimentally- known orthorhombic crystal (space group Pna21), the crystal lattice parameters, and the preferred growth direction (the [100] direction) from the melt. By applying meta-dynamics to the coarse-grained model, a new crystalline form of the drug was discovered (monoclinic, space group P21), which contains two molecules per unit cell. Atomistic simulations and quantum calculations then showed the polymorph to be meta-stable at ambient temperature and pressure, as well as thermodynamically more stable than the conventional orthorhombic crystal at high pressure. The strategy described in this study could be an efficient route to polymorph prediction in other types of small molecules.

469230 Nanocrystal Dissolution Kinetics and Solubility Increase from Molecular Dynamics â€" the Case of Alpha and Beta Glycine

Abstract:

The physics of crystal active pharmaceutical ingredient (API) dissolution is an essential step in the ingestion of a solid oral API tablet. Poor dissolution, and hence bioavailability, can result in insufficient amount of active drug being transported to the target, resulting in suboptimal treatment for the patient. It is estimated that as high as 60% of newly discovered APIs (active pharmaceutical ingredients) possess poor aqueous solubility (<0.1g/ml). Nanocrystal API synthesis offers an intriguing method to combat this issue. Due to the increased surface area to volume ratio, nanocrystals demonstrate an increased solubility, dictated by the interfacial free energy of the nanocrystal, when compared to micron-sized particles. This solubility increase is encapsulated in the Ostwald-Freundlich equation. It is essential to understand that the parameters, most importantly the interfacial free energy , in the Ostwald-Freundlich equation are polymorph specific quantities for a given solvent. Therefore, the polymorph solubility rankings in a given solvent at bulk length scales need not apply for nanoparticles. The incorporation of the polymorph specific size dependent solubility term in classical dissolution models, and the competition between decreasing surface area (smaller particle) and solubility increase, remains largely unaddressed.

Molecular dynamics (MD) provides a first principles simulation procedure capable of simulating the dissolution process for nanoparticles on the size of a critical cluster (~2nm). In this work, a combination of first principles molecular dynamics and classical dissolution models are used to determine (1) the polymorph specific dissolution kinetics and (2) the multiplicative increase in the polymorph specific nanocrystal solubility relative to the bulk solubility for alpha and beta glycine. Alpha and beta glycine seed clusters, with varying radii, are embedded in a MD simulation domain consisting of water. This simulation set up is used to generate crystal size and polymorph specific dissolution data. The effects of the various polymorph specific model parameters (mass transfer coefficient, diffusion layer thickness, and interfacial free energy) are used to analyze the dissolution rates of the polymorphs, as well as the

increase in the solubility, as a function of the particle sizes simulated.

457914 Exploring Self-Assembly Mechanism from Solution: Tolfenamic Acid in Toluene and in DMF

Abstract:

The detailed molecular pathway of nucleation of crystalline materials largely remains unknown. The formation of solution-state assembly is suggested to be the first step of nucleation pathway, and the structure of pre-nucleation aggregates may shed lights on crystal nucleation mechanism. Herein, this study presents a systematic investigation on structure, conformation, size and shape of solution aggregates during the self-assembly process of TFA molecules from both nonpolar (toluene) and polar (n,n-dimethylformamide, DMF) solutions.

Crystallization experiments were conducted by using crash cooling method at various supersaturations and different temperatures. It was shown that both Form â… and Form â…¡of TFA with carboxylic acid homo-dimer synthons can be crystallized from toluene dependent on applied supersaturations while DMF solvate containing carboxylic acid formamide synthon is exclusively crystallized from DMF independent of supersaturations.

The solution phase self-assembly processes are investigated by FTIR spectroscopy, one- and two-dimensional ^{1}H-NMR spectroscopy, and *ab initio* computations. NMR chemical shift changes of TFA in DMF indicate two events, which are likely associated to conformation change and further dimerization through weak aromatic rings interactions. By contrast, dimerization is mediated by H-bonding in toluene. FTIR spectroscopic measurements and *ab initio* calculations of the stability of small clusters in solution support the conclusions. Based upon concentration and temperature dependences of proton chemical shifts of TFA, the reaction constant, cooperativity, and thermodynamic parameters (enthalpy and entropy) of self-association process were determined. The detailed structures of pre-nucleation aggregates are unveiled by chemical shift changes and further conformed by 2D-NOESY NMR. The results show that the solution pre-nucleation dimers bear similar structures as those found in X-ray crystal structure. The size and shape of solution aggregates were revealed by diffusion ordered spectroscopy (DOSY) NMR measurements. It was

found that TFA aggregates take on the shapes of prolate ellipsoid.

466753 Crystallisation of Pyroglutamic Acid: Jumping Crystals and Polymorphic Transformation

Abstract:

L-pyroglutamic acid, an active pharmaceutical material, has attracted intensive research interest both in academic and industrial communities due to its importance in pharmaceutical and food industry[1-2]. Our previous studies investigated the enantiotropic polymorphs of L-Pyroglutamic acid α', α and β and their phase transformations sequence[3-4]:

$$\alpha' \xleftrightarrow{-130°C} \alpha \xleftrightarrow{68°C} \beta \xleftrightarrow{162°c} Liquid$$

The structural changes at the transitions were studied by Raman spectroscopy and variable temperature powder x-ray diffraction, and compared with single crystal structure determination. Differences in Raman spectra were attributed to differences in intermolecular N-H···O interactions in the three polymorphs. L-pyroglutamic acid racemized spontaneously when heated above its melting temperature, 162 °C. This had a major effect on the subsequent composition-dependent crystallization kinetics. Thus, crystallization of L-pyroglutamic acid from undercooled melt was particularly slow and required long-range counter-diffusion of L-pyroglutamic acid molecules in the viscous, hydrogen-bonded melt.

In this study, a novel approach to monitor the α↔β phase transformation was achieved using in house Wide Angle X-ray Scattering (WAXS) with a linkam stage. By optical microscopy, crystals frequently jumped around the microscope slide on passing through the α/β transition; this is attributed to discontinuous changes in the unit cell dimensions leading to a spring effect in the crystals. This thermosalient feature of the α→β transformation was also confirmed by high speed camera and sequential weight losses during repeated thermogravimetric cycling.

References

[1] Tan S.W., et al., *J Org Chem.*, **2015**, 80 (5), 2661-2675.

[2] Heaton, A.L., et al., US Patent US9066953, **2015.**

[3] Wu, H., et al., *Cryst. Growth Des.*, **2010,** 10, 3141-3148.

[4] Wu, H and West, A.R., *Cryst. Growth Des.*, **2011**, 11, 3366-3374.

460732 A Novel Preparative Method and Solid-State Characterization of the Theophylline-Benzoic Acid Co-Crystal

Abstract:

Co-crystals are molecular complexes that contain two or more components together and can be prepared through many methods, such as slow evaporation, cooling crystallization, slurry conversion crystallization, mechanochemical synthesis and so on. Furthermore, mechanochemical synthesis includes neat grinding and solvent-assisted grinding. The theophylline–benzoic acid co-crystal have be reported to be prepared via slow evaporation and neat grinding. But each method has its disadvantage in the process of preparing co-crystals. It takes a long time to get the product by slow evaporation and it is too difficult to make a large amount of co-crystals through neat grinding. In our study, we succeed in synthesizing the theophylline–benzoic acid co-crystal via slurry conversion crystallization in acetonitrile to solve these problems. The solid phase of the theophylline–benzoic acid co-crystal is characterized by powder X-ray diffraction (PXRD), Fourier transform infrared (FTIR), Differential scanning calorimetry (DSC), Raman spectroscopy and Scanning electron microscopy(SEM). The DSC curve of the co-crystal presents melt point of co-crystals. The hydrogen bond in the structure can be confirmed via FTIR and Raman spectrum. The result of SEM could show the difference of morphology between the co-crystal and the initial substance.

In the course of preparing the theophylline–benzoic acid co-crystal through slurry conversion crystallization, the initial conditions are changed and the purity of products which got with different conditions draws our attention. We find that sometimes the final product is a mixture of our target co-crystal and one raw material. A explanation for this observation is that because the benzoic acid solubility in acetonitrile is much higher than that of theophylline (Form , the stable form), the co-crystal dissolves in that incongruently. The region in the phase diagram where the co-crystal is the only solid stable phase is evidently shifted toward the benzoic acid side and is also narrow. To solve the problem, the pure solvent could be replaced by the binary mixed solvent. We measure the solubility of theophylline and benzoic acid in water/ methanol mixture at 298K and prepare the much purer co-crystal in methanol/water mixture by slurry conversion crystallization. In methanol/water mixture the theophylline–benzoic acid co-crystal can dissolve congruently.

To investigate the process of mechanochemical synthesis to prepare co-crystals, a solvent-assisted grinding approach is used to synthesize the title compound. The products are compared with that of neat grinding. The result indicates that the reaction rate is faster and the product of higher crystallinity can be got via solvent-assisted grinding. In conclusion, solvent is conductive to promoting the mechanochemical synthesis of our target co-crystal. Through changing the grinding time, the amount of solvent, the kind of solvent, the effect of every factors in the mechanochemical method can be got with comparing the powder X-ray diffraction data of the co-crystal product in different condition and it has some reference value for preparation of the co-crystal.

459222 Structure and Basic Adsorption Properties of Tcpb, a High Porosity Hydrogen-Bonded Organic Frameworks (HOFs)

Abstract:

A highly porous, molecular crystal has been recently introduced in literature by Zentzer et.al. (**Chem. Comm.** *51*, 11642-11645, 2015). 1,3,5-Tris(4-carboxyphenyl)benzeneÂ (tcpb) crystallizes into a complex 8-fold polycatenated assembly of (6,3) hexagonal nets formed through hydrogen bonds and Ï€-stacking. tcpb is part of a new class of materials in which hydrogen bonds stabilize an organic framework (HOF) in an ordered fashion to produce a porous structure of the molecule itself. The self-assembly of the structure occurs by simple temperature crystallization which is rather different than the synthesis of other adsorbents. The solid can be dissolved easily in a suitable solvent, and re-crystallized without any loss in adsorptionÂ characteristics. Controlling the self-assembly can conceptually result in development of adsorbents with many different morphologies, particle size distributions, on surfaces, in cavities, in micro-scale which may enable new adsorption technologies.

Compared to other adsorbents, the number of hydrogen-bonded organic frameworks (HOFs) examined in literature is very few and almost no adsorption equilibrium data for these structures exist, especially for gases. A possible explanation for this lack of interest may have been the very low external porosity of molecular crystals in general. With BET surface area around 1,000 square meter per gram, tcpb provides a system where large porosity (hence adsorption capacity) is provided in a hydrogen-bonded network in 3D. In tcpb, the porosity is external to the molecule. Not to be confused with internal porosity of molecules (many in crystal form) which have been quite extensively studied for pharmaceutical applications. The porosity of tcpb (and HOFs) is formed by the assembly of the molecules rather than the shape of each individual molecule.

In this presentation, we will describe the structure, describe the several methods to recrystallize the solid, and present basic properties including BET surface area and adsorption of H_2 and CO_2.

462229 Chiral MOFs: Building Blocks, Features and Design of New Frameworks

Abstract:

Recent advances in the design and synthesis of metal-organic frameworks have demonstrated the potential of this material class. De novo synthesis of new framework types for specific applications requires a sound understanding of not only the synthesis chemistry but also of desirable framework features. In this study, the existing literature on chiral metal-organic frameworks is analyzed by using the CoRE MOF database. This database contains most of the reported MOF structures from the literature [1]. Chiral frameworks are identified either by topology (e.g. srs, unc, lcy...) or the presence of chiral linkers. Chiral topologies are identified using structural analysis of their 3D nets. Interestingly, only 19 out of the 362 chiral nets in the Reticular Chemistry Structure Resource (RCSR) database are found within reported MOF structures. MOF structures with chiral linkers (or introduced by post synthetic modification) are identified by data mining of article contents.

Analysis of the reported frameworks gives insight into the organic and inorganic building blocks needed to form chiral frameworks. There are 3 main factors mentioned in the literature that contribute to chirality: asymmetric metal nodes, chiral linkers, or helix formation. Interestingly, analysis of the chiral topologies shows that in many cases chirality is the result of symmetry distortion at the metal node by chelation or mixed O- and N-terminated organic linkers. This essentially means that no chiral building blocks are required to create some degree of chirality in the framework. Several asymmetric metal clusters are identified for Zn, Cu, In, Co, Ni, and Cd that are of interest in the design of novel frameworks (no need for chiral linkers). The formation of helixes or warped pore architectures is more likely if a three-connecting organic building block is present rather than 2- or 4-connecting linkers. On the basis of these insights, several new hypothetical structures are proposed and evaluated for chiral separations. The analysis of structural components is complemented with molecular simulations on the adsorption of R- and S-isomers (organic acid, alcohol, amine, amino acids) to establish relationships and compare the impact of key

features (linker, node, helix) on the separation potential.

[1] Chung et al. *Chem. Mater.*, **2014**, *26* (21), pp 6185–6192

462098 Impregnation of SBA-15 with Task Specific Ionic Liquids for Targeted Gas Capture

Abstract:

Six task-specific ionic liquids (TSILs), comprised of the taurinate anion paired with five tetraalkylammonium cations (where alkyl = methyl, ethyl, propyl, butyl, and hexyl) and a tetrabutylphosphonium cation, were impregnated in SBA-15 to capture CO2. Additionally, a set of sorbents built from SBA-15 and a task specific ionic liquid that contains a sulfonic acid group were synthesized and investigated as a novel means of capturing ammonia from air. The results show that the sorption process is complex with the capacity of some samples being sensitive to preadsorbed water and some ionic liquids exhibiting crystallinity in the pores of SBA-15. The composite materials designed to capture ammonia show an increase in ammonia capacity below 2 kPa, while maintaining nearly the same capacity for ammonia as the parent silica at 101 kPa. More broadly, the results illustrate that it is possible to functionalize the pores of an adsorbent with an ionic liquid and that the functional group of the ionic liquid can be used to capture specific gases in a fixed-bed.

470255 Tailoring the Electric Resistivity of a Molecular Sieve Adsorbent for Electrothermal Regeneration

Abstract:

Electrothermal regeneration is an adsorbent regeneration technique whereby electric current is passed through a material with sufficiently low resistivity (typically, 0.1 â€“ 10 Î©.m) and heat is generated by the Joule effect. This allows for rapid adsorbent regeneration that is decoupled from the purge gas flow. Molecular sieves are effective adsorbents because they have tailored and structured pore size distributions. Unlike carbonaceous materials, however, molecular sieves have high electrical resistivity that prevents their electrothermal regeneration. In this study, carbon nano tube (CNT) was added to zeolite Y (high resistivity, >E+7 Î©.m) to decrease its resistivity. CNT was added by CH_4 decomposition on cobalt impregnated zeolite Y. The effect of CNT growth condition on the properties of zeolite Y was investigated to find the optimum condition to improve its electrical conductivity without compromising its adsorption properties. Specifically, the effect of calcination temperature and duration, cobalt cations reduction, and methane decomposition were studied. Modified samples were characterized using bulk elemental analysis, scanning electron microscopy (SEM), X-ray diffraction, and resistivity measurements. The results indicated that reduction temperature has the most significant effect on the size and amount of CNT deposited on the surface of zeolite Y during each step. The reduction temperature could significantly alter properties of prepared sample, as its CNT content decreased from 37 to 9 %C and its conductivity decreased from 6.25 to 0.71 S/m, once the temperature decreased from 500 to 400 ℃. SEM images revealed smaller CNT for samples prepared at lower reduction temperature. Although reducing the temperature resulted in less CNT on the zeolite, it led to less change in zeolite adsorption properties while providing suitable conductivity for electrothermal regeneration. The developed low-temperature procedure could be applied to tailor the electric resistivity of other adsorbents.

465875 Advancement in High Capacity Adsorbent Materials for Lithium Recovery from Secondary Resources

Abstract:

Lithium (Li) has been one of the most sought-after precious metals by various industries. It has been popularized by its unique electrochemical property and exemplary energy density, which makes it an attractive key component in energy storage devices and next generation electric vehicles. Thus with its surging demand, securing Li^+ supply is an utmost priority.

One of the alternative ways offered to ensure a steady flow of Li supply is through its recovery from all viable resources such as brine pools, seawater, coal ash, and industrial wastes. However, existing technologies are not designed to recover Li from alternative resources, which typically contain low Li^+ levels in the presence of more abundant interfering cations. Among the known processes, adsorption is considered as one of the simplest and most promising methods. The issue on highly selective Li^+ capture can be addressed by using adsorbents with Li ion (Li^+) sieving capability and/or ion recognition property.

While the potential of inorganic lithium ion sieves (LIS) and organic crown ethers (CEs) as Li^+ adsorbents have been known for several decades, their application in actual systems are yet to be fully materialized. The hurdle of employing these materials primarily stems from their physical forms (i.e. either as powder or viscous fluids), which are prone to physical loss if directly used as adsorbents. Herein, the research status and progress on the development of advanced materials with high capacity and selectivity for Li^+ capture are presented. Particularly, various synthesis routes and solutions to render the LIS and CEs stable, recyclable, and applicable for long term use (i.e. continuous Li recovery application in actual systems) are discussed. Immobilization and multi-step functionalization routes using different supports (i.e. polymers, inorganic nanomaterials, etc.) processed into different configurations (i.e. nanofibers, membranes, forms, etc.) as composite Li^+ adsorbents are highlighted. Their potential application in actual systems are also demonstrated.

This research was supported by Basic Science Research Program through the National Research Foundation of Korea (NRF) funded by the Ministry of Education (2009-0093816) and Ministry of Science, ICT and future Planning (2015R1A2A1A15055407).

466930 Investigation of the Formation of Porous Boron Nitride and Its Application for Oil/Water Separation

Abstract:

Very recently, porous boron nitride (BN) has appeared as a new class of adsorbents, one that is complementary to carbonaceous materials owing to its higher thermal and chemical stability as well as the polarity of its bonds. This enables to extend the range of applications of adsorbents, especially those related to extreme conditions. Important aspects of adsorbents include their porous structure and the mechanisms by which they interact with a given adsorbate.

This presentation will report the results of an in-depth investigation of the effects of the synthesis conditions on the formation of porous BN. Towards this goal, porous BN was synthesised via a bottom-up approach and then analyzed using a range of analytical techniques (i.e. N2 sorption, XRD, FT-IR, XPS, thermogravimetric analysis). It was found that the intermediate preparation had a large effect on the physical features of the resulting BN, and particularly its porosity with surface areas ranging from 400 to 1000 m^2/g depending on the conditions. This dramatic change was assigned to the varying crystallinities of the intermediates which in turn impacted their thermal decomposition pattern and hence the formation of BN. Owing to it hydrophobic nature, BN can be used for the separation of oil and water. In order to provide a fundamental understanding of the nature of the interactions between hydrocarbons and the surface of BN, 2D BN nanosheets (BNNSs) were produced and molecular interactions were investigated using Scanning Tunnelling Microscopy (STM).

462925 Nanoporous Boron Nitride: Synthesis, Stability and Gas Separation Characteristics

Abstract:

Boron nitride is a type of ceramic-based material and a structural analogue of graphene. In a single boron nitride layer, the boron and nitrogen atoms are held together in a hexagonal fashion. In this work, nanoporous boron nitride was synthesized in a dynamic templating strategy. The synthesized boron nitride was characterized with pore textural properties, thermogravimetric analysis (TGA), X-ray diffraction (XRD), X-ray photoelectron spectroscopy (XPS) and electron microscopy. The boron nitride sample has BET surface area of over 1300 m^2/g and total pore volume around 1 cm^3/g. Thermogravimetric analysis followed by further characterization studies revealed that it is highly stable in nitrogen and in air upto a very high temperature. In order to examine its gas separation characteristics, different gases, including nitrogen (N_2), carbon dioxide (CO_2), methane (CH_4), ethane (C_2H_6), ethylene (C_2H_4), propane (C_3H_8) and propylene (C_3H_6) were adsorbed in boron nitride in ambient temperature and pressure upto 1 bar. Based on the Ideal adsorbed Solution Theory (IAST), selectivity of separation of different gas pairs, including CO_2/N_2, CO_2/CH_4, CH_4/N_2, C_2H_6/CH_4, C_2H_6/C_2H_4 and C_3H_8/C_3H_6 were calculated. Furthermore, the breakthrough times for the simulated pressure swing adsorption were also calculated from the isotherms and selectivity values. The results suggest that nanoporous boron nitride can be employed for gas separation purposes.

459781 Adsorption, Structural and Dynamic Properties of Ethanol-Water Mixtures in Graphene and Hexagonal Boron Nitride Slit Pores

Abstract:

The industries discharge a variety of pollutants, such as heavy metals, organic toxins, and oils, in water resources. Exposure of these contaminants in water causes adverse health effects on various forms of life [1-3]. Novel materials are needed for the effective removal of pollutants from industrial wastewater. Graphene and hexagonal boron nitride (hBN) sheets are promising materials for removal of organic pollutants [4]. In this work, the suitability of the sheets for the separation of ethanol-water mixture is investigated by studying the adsorption and structural behavior of ethanol-water mixtures in slit pore with variable width (7 to 13 Å) using molecular dynamics. The simulation results indicate that the selectivity of ethanol varies with pore width and with the nature of the surface. The selectivity of ethanol is highest for 9 Å pore and lowest for 7 Å pore, irrespective of the nature of the substrate. However, selectivity of ethanol is relatively higher for hBN compare to the graphene surface, for all slit-pore widths. At lower pore width, molecular sieving plays an important role for separation of ethanol; whereas at higher pore width separation is governed by the adsorption behavior. The diffusion coefficients of water and ethanol molecules substantially decrease with decrease in pore width for both graphene and hBN surfaces. The resident time of water and ethanol molecules decreases with increase in the slit-width. Furthermore, water and ethanol molecules confined in hBN pores show higher residence time and lower diffusion coefficient values compare to graphene slit pores.

References

[1] M. A. Shannon, P. W. Bohn, M. Elimelech, J. G. Georgiadis, B. J. Mariñas, A. M. Mayes, "Science and technology for water purification in the coming decades" Nature, 2008, 452: 301-310.

[2] M. A. Montgomery and M. Elimelech, "Water and sanitation in developing countries: including health in the equation" Environmental Science and Technology, 2007, 41: 17-24.

[3] F. Zhang, WB. Zhang, Z. Shi, D. Wang, J. Jin, L. Jiang, "Nanowire-haired inorganic membranes with superhydrophilicity and underwater ultralow adhesive superoleophobicity for high-efficiency oil/water separation" Advanced Materials, 2013.

[4] W. Lei, D. Portehault, D. Liu, S. Qin, Y. Chen, "Porous boron nitride nanosheets for effective water cleaning" nature communications, 2013.

466478 Quantitative Monitoring of Microphase Separation Behaviors in Cationic Liposomes Using Multiple Fluorescent Probes

Abstract:

Microphase separation behaviors of cationic liposomes have been investigated using a pH-sensitive fluorescent probe with 4-heptadecyl-7-hydroxycoumarin (HHC), 1,6-diphenyl-1,3,5-hexatriene, and 6-lauroyl-2-dimethylaminonaphthalene, and to estimate localized electrostatic potentials. Shifts of the apparent pKa values of HHC were observed in cationic liposomes in proportion to the amount of cationic lipids. Two pKa values were obtained with 1,2-dioleoyl-*sn*-glycero-3-phosphocholine (DOPC)/3ß-[*N*(*N'*,*N'*-dimethylaminoethane)-carbamoyl] cholesterol hydrochloride (DC-Ch) liposomes, while only one pKa value was generated with either DOPC/1,2-dioleoyl-3-trimethylammonium-propane (DOTAP) or DOPC/dimethyldioctadecylammonium-bromide (DODAB) liposomes. The physicochemical membrane property analyses, focusing on membrane fluidity and membrane polarity, revealed heterogeneity among DOPC/DC-Ch liposomes. By analyzing the pH titration curves using sigmoidal fitting, the localized electrostatic potentials were estimated. For DOPC/DOTAP=(7/3), the membrane was in the liquid-disordered phase and the density of cationic molecules was 0.41 cation/nm^2. For DOPC/DC-Ch=(7/3), the membrane was heterogeneous and the densities of cationic molecules in liquid-disordered and liquid-ordered phases were 0.25 and 1.24 cation/nm^2, respectively. We thereby conclude that the DC-Ch molecules can form nanodomains when these molecules are concentrated to 59%.

458670 A Multi-Fluid Model of Phase-Inversion Membrane Formation

Abstract:

Industrial processes for producing polymer-based materials operate away from equilibrium, making the final microstructure – and thus the properties of the material – dependent on process history. Describing such processes by traditional simulation methods has proven difficult because of the confluence of complex thermodynamics and non-trivial fluid mechanics. We have recently developed a "multi-fluid" simulation framework for examining the dynamics of multiphase polymer solutions that combines a field-theoretic approach to thermodynamics with continuum fluid mechanics. In this framework, we use a ternary polymer solution model to investigate the phase inversion process for making polymer membranes. After characterizing the model and methods, we investigate a number of important phenomena that occur across several length and time scales including (i) mass-transfer between the polymer film and the non-solvent bath, (ii) phase separation between polymer lean and polymer rich regions and (iii) long-time coarsening behavior. We will further discuss the implications of these results on the formation mechanisms of various microstructures found in polymer membranes.

470502 Charged Nanochannels By Random Copolymer Micelle Assembly

Abstract:

Membranes that can separate solutes by factors other than size (e.g. charge, hydrophobicity, chemical functionality) would enable the use of this energy-efficient, green technology in new applications. However, achieving membrane selectivity based on parameters other than size has been a challenge, especially when the molecules to be separated are small. Biological pores such as ion channels achieve exceptional selectivity by confining flow into very small pores lined with functional groups. Therefore, incorporating nanostructures that confine permeation into membrane selective layers can significantly enhance selectivity. In this study, we created membrane selective layers with nanopores lined with carboxyl groups using the deposition of micelles formed in methanol onto porous membranes. To prepare these membranes, we prepared an amphiphilic random copolymer that combines highly hydrophobic, fluorinated repeat units of 2,2,2-trifluroethyl methacrylate (TFEMA) with repeat units of methacrylic acid (MAA), PTFEMA-*r*-PMAA, by free radical polymerization, because it a simple, robust and scalable method. The copolymer contained 55 wt% TFEMA, and was insoluble in water. We found that this copolymer forms micelles in methanol with a multimodal size distribution, as shown by dynamic light scattering (DLS). When these micelles are coated onto the surface of a porous support membrane whose pores are smaller than the micelles, a selective layer of micelles packed together is formed in between which permeation occur. If the membrane pore size is too large, micelles go inside the membrane, clogging inner pores and resulting in a membrane with very low flux. Also, if the solvent evaporation time is too long, the micelle merge and create a dense selective layer with a very low permeability. By proper selection of conditions (i.e. copolymer concentration, solvent evaporation time, base membrane MWCO, etc.), we achieved membranes with pure water permeabilities up to ~6 $L/m^2.h.bar$ and high selectivity (separation coefficient up to 24) between small molecules with similar size (~1nm) but opposite charges. For example, these membranes can retain a negatively charged dye by > 85% whereas a positively charged dye with a similar size was retained by < 20%. Membrane rejection can be modeled by Donnan exclusion theory. This is the first demonstration of a new approach to forming membranes from easily synthesized copolymers by micelle assembly. We believe this approach can be adapted to address various pore sizes

and pore surface chemistries, leading to novel membranes with promising selectivity.

469664 Development of Polybenzimidazole Nanofiltration Membranes with Biomimetic Surfaces

Abstract:

Aquaporins are membrane protein water channels present in cells, and they restrict the passage of contaminants without preventing the passage of water. Small molecules such as urea and boric acid, the removal of which is inefficient by conventional membranes are also rejected. Biomimetic membranes, an innovation in water filtration technology attempts to replicate this natural process of highly selective and efficient transport of different molecules across a semi-permeable membrane occurring at the cell level. Therefore, aquaporins have received worldwide attention because of their potential to form biomimetic membranes with high flux and selectivity for water reuse and desalination. However, challenges involved in the incorporation of aquaporin proteins in membranes limit their applicability. One of them is to attach aquaporins to the membranes without chemically altering or damaging the aquaporins during the binding to the membrane. The second challenge is to design and prepare an assembly that allows biomimetic membranes with aquaporins to sustain hydraulic water pressure gradients without losing their integrity and performance. The overarching objective of this project is to form a biomimetic membrane made of unaltered aquaporins dispersed in a polymeric membrane selective layer and capable of operation under high hydraulic pressure. Polybenzimidazole (PBI) membranes were surface modified with treated aquaporins in order to achieve higher water flux and selectivity. Membranes modified with aquaporins displayed lower flux declines and higher flux recovery values after backwash as compared to unmodified PBI membranes. Also, modified membranes showed improved rejection values for both protein and salt solutions of different concentrations. However, there was leakage of ions between the channels. Therefore, in order to improve the rejection of protons, ions and other impurities, aquaporin channels were aligned with the direction of water flow. Functional groups were installed on AqpZ for covalent attachment to the polymer matrix so that the proteins could be immobilized to the membranes and aligned in the direction of the flow.

465129 Bijel Derived, Fouling Resistant and Catalytic Ultra-/Micro-Filtration Membranes for Advanced Water Treatment Applications

Abstract:

Bicontinuous interfacially jammed emulsions (bijels) are a new class of soft materials [1] with promising applications in various areas ranging from healthcare, food, energy and reaction engineering due to their unique structural, mechanical and transport properties. We recently presented a new method to prepare bijels via solvent transfer-induced phase separation (STRIPS), which enables continuous fabrication of hierarchically structured composite fibers and membranes with a manifold of organic/inorganic building blocks [2]. These asymmetric membranes are composed of a porous polymer scaffold with inorganic nanoparticles on their surface and introduce a variety of important features: (i) Hollow fibers can be used as ultra- and micro- filtration membranes for the separation of model contaminants in water and organic solvents due to their controllable pore size and chemical stability. (ii) The membranes are resistant to biofouling due to the simplicity of chemical surface modifications. (iii) Stimuli-responsive membranes that can open/close their pores are fabricated, facilitated by the fact that broad ranges of polymeric building blocks can be integrated. And (iv) photocatalytic titanium dioxide membranes for organic contaminant oxidation are realized; other functionalities can be readily imparted due to the wide range of material choices for the nanoparticles on the membrane surface. Our work demonstrates a novel and versatile fabrication route for separation membranes with potential applications in water treatment, food processing or as separation membranes in the chemical industry.

[1] Herzig, E. M., et al. (2007). Nature materials, 6(12), 966-971.

[2] Haase, M. F., Stebe, K. J., & Lee, D. (2015). Advanced Materials, 27(44), 7065-7071.

461733 Boosting the Performance of Random Zwitterionic Copolymers Using Ionic Liquids during Membrane Formation

Abstract:

Polymer self-assembly is a promising tool for scalable manufacture of membranes while maintaining high permeability and controlled pore size. Tuning copolymer composition (monomers, additives) and processing methods can change copolymer behavior which can dramatically affect the self-assembly and hence the membrane performance (permeability, selectivity, fouling resistance). Studies based on block copolymers demonstrate the influence of additives (solvent, homopolymer etc.) on the copolymer self-assembly and membrane performance. But studies on additives in casting solutions of random copolymers and how they affect membrane performance have not been reported to our knowledge. Recently, we have introduced a new class of membranes with ~ 1 nm effective pore size whose selective layers are made of self-assembling zwitterionic amphiphilic random copolymers. These membranes derive not only their excellent fouling resistance but also their permeability and selectivity from this self-assembled nanostructure. These membranes have numerous applications in the biochemical and pharmaceutical industries, as well as wastewater treatment processes. These membranes are prepared simply by coating random copolymers of hydrophobic and zwitterionic monomers onto a porous support membrane. In this study, we have used ionic liquids and other additives in the coating solutions to boost and alter the performance of these membranes. Membranes prepared by coating copolymer solutions with sufficient amounts of selected additives on commercial ultrafiltration membrane supports exhibit permeances as high as 50 L/m^2.hr.bar, up to 10 times higher than membranes formed without additives. These membranes also exhibit a narrow pore size distribution, retaining the same size-based selectivity with a ~1 nm size cut-off demonstrated by filtering negatively charged dyes. Performance of these membranes depends on the amount of additive as well as the membrane manufacturing method (non-solvent, drying time etc.). This is attributed to phase separation and solvent diffusion kinetics that are in play during membrane formation by SNIPS (self-assembly and non-solvent induced phase separation) mechanism. These structure-property relationships serve as a guideline to develop new materials with high fluxes and sharp selectivity for various applications such as textile wastewater treatment, pharmaceutical purification and bioseparation applications.

470812 Optimizing Process Yields of a Legacy 25,000L Mammalian Cell Culture Process Using Integrated Upstream and Downstream Process Yield Models

Abstract:

Process monitoring is defined as a continuous effort to evaluate manufacturing parameters across multiple lots to ensure commercial manufacturing processes are operating in-control within specifications. As it becomes increasingly important to reduce the cost of goods for cell culture processes, scientists and engineers require digital data-acquisition and aggregation systems to develop a robust process monitoring approach for the entire process. These automated data acquisition tools assist in rapidly contextualizing data over the course of a process lifecycle and can identify correlations across an entire process (cell culture and purification). In particular, advent of digital plant combined with development of informatics tools to generate, acquire, analyze and visualize large data sets for process monitoring has led to increased process understanding, reduced cycle time, and implementation of process improvements based on analysis performed with the tool.

We will present a case study where predictive models of production bioreactor and chromatography steps were used in tandem to identify sources of variability in overall process yield for a 25,000L fed-batch CHO process. From this analysis, bioreactor culture duration was identified to negatively impact glycosylation levels of the recombinant fusion protein produced from the culture. By implementing calibrated, data-driven modifications to the upstream manufacturing schedule, we were able to realize tighter control of upstream protein quality, optimal downstream process performance, and improved overall process yield.

470647 Intelligent Control Strategies Towards the Intensification of Monoclonal Antibody Production Via Continuous Operation

Abstract:

Current industrial trends encourage the development of sustainable, environmental friendly processes with minimal energy and material consumption. Meanwhile, the increasing market demand as well as the tight regulations in product quality, necessitate efficient operation procedures that guarantee products of high purity. In this direction, process intensification via continuous operation paves the way for the development of novel, eco-friendly processes, characterized by higher productivity compared to batch.
This shift could benefit the market of high value biologics, such as monoclonal antibodies (mAbs), as it would lead to shorter production times, decreased costs, as well as significantly less energy consumption compared to batch [1, 2]. In particular, mAb production comprises two main steps: the culturing of the cells (upstream) and the purification of the targeted product (downstream). Both processes are highly complex and their performance depends on various parameters. In particular, the efficiency of the upstream depends highly on cell growth and the longevity of the culture, while the product quality can be jeopardized in case the culture is not terminated timely. Similarly, downstream processing, whose main step is the chromatographic separation, relies highly on the setup configuration, as well as on the composition of the upstream mixture. Therefore, in order to be able achieve a sustainable and efficient continuous operation, it is of vital importance to design intelligent computational tools that form a solid basis for the: (i) execution of cost-free comparisons of various operating strategies, (ii) design of optimal operation profiles and (iii) development of advanced, intelligent control systems that can maintain the process under optimal operation, rejecting disturbances.

This work focuses on the development of advanced control strategies for: (i) a cell culture system in a bioreactor and (ii) a semi-continuous purification process. More specifically, we consider fed-batch culturing of GS-NS0 cells producing a chimeric IgG4 antibody (based on the feeding of five key nutrients) and the semi-continuous Multicolumn Countercurrent Solvent Gradient Purification (MCSGP) [3-5] process for the purification of the upstream mixture. For the controller development we follow the generic framework & software platform PAROC that considers: (i) development of a high-fidelity process model, (ii) approximation of the complex, process model, (iii) design of the multi-parametric controller, (iv) 'closed-loop', in-silico validation of the controller against the process model [6]. PAROC allows 'in-silico' testing of the advanced controllers against the high-fidelity process model and evaluate their performance before operating them online. The results from this study indicate that the in-silico upstream/downstream integration can be achieved and the designed controllers succeed in maintaining both systems under optimal operation. In particular, the upstream system is characterized by a significantly increased mAb titer under the operation of the controller. Moreover, the controller indicates the time point that the culture needs to be terminated in order to minimize the risk of product degradation. Similarly, the downstream control scheme manages to maintain a stable operation throughout the process cycle, tracking efficiently the predefined setpoints. In addition, the controller is immunized against the uncertainty that results from the composition of the upstream mixture by considering the latter as measured disturbance. Finally, the suggested controller input indicates decreased use of utilities, while maintaining process efficiency.

Acknowledgements

The authors would like to thank Mr R. Oberdieck and Miss A. Quiroga-Campano for their contribution in the assessment of the controller validation. The authors would also like to thank Mr. Fabian Steinebach & Prof. M. Morbidelli from ETH Zurich, as well as Dr. Thomas Mueller-Spaeth from ChomaCon AG for their valuable input on the understanding of the MCSGP process. Financial support from the European Commission (OPTICO/G.A. No.280813) & Texas A&M University are also gratefully acknowledged.

References

1. Konstantinov, K.B. and C.L. Cooney, *White Paper on Continuous Bioprocessing May 20-21, 2014 Continuous Manufacturing Symposium.* Journal of Pharmaceutical Sciences, 2015. **104**(3): p. 813-820.

2. Xenopoulos, A., *A new, integrated, continuous purification process template for monoclonal antibodies: Process modeling and cost of goods studies.* J Biotechnol, 2015. **213**(0): p. 42-53.

3. Aumann, L. and M. Morbidelli, *A continuous multicolumn countercurrent solvent gradient purification (MCSGP) process.* Biotechnology and Bioengineering, 2007. **98**(5): p. 1043-1055.

4. Krättli, M., F. Steinebach, and M. Morbidelli, *Online control of the twin-column countercurrent solvent gradient process for biochromatography.* Journal of Chromatography A, 2013. **1293**: p. 51-59.

5. Papathanasiou, M.M., et al., *Advanced control strategies for the multicolumn countercurrent solvent gradient purification process.*AIChE Journal, 2016.

6. Pistikopoulos, E.N., et al., *PAROC—An integrated framework and software platform for the optimisation and advanced model-based control of process systems.* Chemical Engineering Science, 2015. **136**: p. 115-138.

470938 Development of a Scalable Process for Production of a Broadly Protective Influenza Antigen

Abstract:

While most influenza infections cause only mild illness, the World Health Organization (WHO) estimates that, on average, 3 to 5 million people worldwide develop respiratory or heart complications related to seasonal influenza infections each year. The WHO also estimates 250,000-500,000 influenza associated deaths yearly, with a large percentage occurring in developing world children with inadequate access to vaccines. Such mortality would be much worse for a virulent pandemic influenza strain. The current standard of care for influenza immunization is a trivalent whole-inactivated virus vaccine consisting of two influenza A strains (H1N1 and H3N2) and one influenza B strain. Due to antigenic drift, the influenza virus is constantly changing and experts must predict the dominant strains for the upcoming flu season roughly 7-9 months in advance of distribution. As a result, the actual circulating viruses may differ from those included in the vaccine, causing an increase in the number of influenza cases and the number of influenza-related deaths as was the case during the 2014-15 flu season. Once the three strains are selected, the majority of the vaccine is produced in eggs and treated for inactivation. This egg-based method takes roughly 6 months. As a result of the lengthy production times, vaccines cannot be updated to more accurately reflect the circulating strains in the case of a poor prediction.

A broadly protective influenza vaccine that would elicit strong protective responses for most or all possible circulating strains would dramatically reduce influenza-related hospitalizations and deaths by conferring life-long immunity. Our proposed vaccine would address many deficiencies in the current vaccine strategy by targeting a conserved region of the influenza hemagglutinin protein to protect against both seasonal and pandemic influenza strains. We will use a stabilized virus-like particle scaffold to present the antigen in a multivalent manner which more closely mimics the presentation of the antigen on the natural virus. By providing broadly protective immunity worldwide against seasonal and pandemic influenza, we would decrease the cost of medical care for influenza related illnesses, and dramatically reduce the number of influenza related deaths.

We have developed a stabilized hemagglutinin stem antigen which has the potential to provide broad protection against seasonal and pandemic strains of influenza. Here we discuss the development of a scalable refolding and formulation process for the production of our stabilized hemagglutinin stem antigen. By evaluating the kinetic and molecular forces required for the proper folding and trimerization of the hemagglutinin stem antigen, we have transitioned a number of lengthy, industrially unattractive refolding steps to a more reliable, scalable process. We are also working to further stabilize our hemagglutinin stem antigen for improved conformational stability with the ultimate goal of producing a cold-chain independent vaccine which would allow for improved distribution worldwide. We plan to evaluate both lyophilization and spray drying techniques for producing solid drug substance compatible with long term room temperature storage.

467571 Application of Process Monitoring Data Management Systems to Make Multiple Incremental Improvements to Cell Cultures Processes at the Manufacturing Scale

Abstract:

Process monitoring data management systems have been important tools to capture the large number of parameters and conditions that influence both the product quality and productivity of cell culture processes. Recent improvements in process analytical technologies and refinements in process control strategies have improved the biopharmaceutical industryâ€™s ability to gather large, multidimensional data sets that can be used for statistical process monitoring. In this work, we discuss the implementation of a data management platform and its application to build understanding around mammalian cell culture processes at the manufacturing scale. The database comprises of the end to end collection of process information from vial thaw to drug substance release for several commercial biologics produced using mammalian cell culture.

We explore the use of this database to make several incremental improvements to the cell culture process through a series of case studies including responses to health authority requests, technology transfers between multiple facilities, and evaluations of upstream parameters to understand their impact on final drug substance output. These case studies highlight the gains in process knowledge that can be achieved using these data management tools and illustrate some of the challenges in developing these capabilities. While data management tools facilitated the rapid retrieval of available data, there were several obstacles associated with data structure, data acquisition, and the subsequent data analysis. The current work demonstrates the successful application of data management tools to improve the performance and robustness of existing cell culture processes at the manufacturing scale.

471374 Scalable Production of Glioblastoma Cancer Stem Cells in 3 Dimension Thermoreversible Hydrogels for Drug Discovery

Abstract:

There is growing interest in developing drugs that specifically target glioblastoma cancer stem cells. Current cell culture methods, however, cannot cost-effectively produce the large numbers of glioblastoma cancer stem cells required for drug discovery and development. In this paper we report a new method that encapsulates patient-derived primary glioblastoma cancer stem cells and grows them in 3 dimension thermoreversible hydrogels. Our method allows long-term culture (~ 50 days, 10 passages tested, accumulative ~>10^{10}-fold expansion) with both high growth rate (~ 20-fold expansion / 7 days) and high volumetric yield (~ 2.0 x10^7 cells/ml) without the loss of stemness. The scalable method can be used to produce sufficient, affordable glioblastoma cancer stem cells for drug discovery.

470399 Levers to Modulate Product Quality Attributes of Product-a in CHO Cell Culture

Abstract:

Levers to modulate two product quality attributes were investigated as part of process development of Product-A produced in Chinese Hamster Ovary (CHO) cell culture. The addition of a chelating agent to the cell culture fluid was identified as a lever to minimize product fragment formation which indicates that metallo-proteases could potentially be contributing to fragment formation. A study to understand the nature and formation of product size variants identified that aggregate formation was predominantly intracellular and occurred prior to secretion and the fragments were formed in the cell culture fluid after secretion. In addition, media additives and process conditions were evaluated for potential levers to modulate product quality attribute-2. Process parameter-1 was determined to be a robust lever for controlling product quality attribute-2 across various production scales.

468324 Data-Driven Optimization Using an Evolutionary Design of Dynamic Experiments for Biopharmaceutical Processes

Abstract:

Biopharmaceutical processes are substantially complex and are not understood well enough to enable the development of an accurate knowledge-driven model for process optimization purposes. In such a case, a data-driven modelling approach provides a very attractive alternative, which, if designed properly, can quickly optimize the process with a small set of experiments. The recently introduced Design of Dynamic Experiments (DoDE) [1], a generalization of the traditional Design of Experiments (DoE) [2, 3], has demonstrated that it is an effective data-driven modeling and optimization approach [1, 4, 5].

Here we apply the DoDE method in an evolutionary manner to optimize biopharmaceutical processes while satisfying budgetary and developmental time constraints on the number of experiments in two representative biopharmaceutical processes; Penicillin fermentation [6] and Hybridoma cell culture [7]. We maximize the amount of product produced by determining the optimal feeding policy and batch duration.

To minimize the number of experiments we initially consider only a few dynamic sub-factors and the simplest Response Surface Model (RSM). After the initial results are analysed and an approximate RSM is estimated, additional experiments are added sequentially to further improve the data-driven model and the process performance. We demonstrate that with a reasonable amount of evolutionary experiments one can come very close to the model-based optimum. The model-based optimization [8], however, requires a very accurate knowledge-driven model that is rarely at hand.

References

1. Georgakis, C., *Design of Dynamic Experiments: A Data-Driven Methodology for the Optimization of Time-Varying Processes*. Industrial & Engineering Chemistry Research, 2013. **52**(35): p. 12369-12382.

2. Box, G.E.P. and N.R. Draper, *Response Surfaces, Mixtures, and Ridge Analysis*. 2007, Hoboken, NJ: Wiley.

3. Montgomery, D.C., *Design and Analysis of Experiments*. 8 ed. 2013, New York: Wiley. 729.

4. Fiordalis, A. and C. Georgakis, *Data-driven, Using Design of Dynamic Experiments, versus Model-driven Optimization of Batch Crystallization Processes*. Journal Of Process Control, 2013. **23**(2): p. 179-188.

5. Makrydaki, F., C. Georgakis, and K. Saranteas. *Dynamic Optimization of a Batch Pharmaceutical Reaction Using the Design of Dynamic Experiments (DoDE): The Case of an Asymmetric Catalytic Hydrogenation Reaction*. in *Proceedings of the 9th International Symposium on Dynamics and Control of Process Systems*. 2010. Oude Valk College, Belgium.

6. Bajpai, R.K. and M. Reuss, *A Mechanistic Model for Penicillin Production*. Journal Of Chemical Technology And Biotechnology, 1980. **30**(6): p. 332-344.

7. De Tremblay, M., et al., *Optimization of Fed-Batch Culture of Hybridoma Cells Using Dynamic Programming Single and Multifeed Cases*. Bioprocess Engineering, 1992. **7**(5): p. 229-234.

8. Biegler, L.T., *An Overview of Simultaneous Strategies for Dynamic Optimization*. Chemical Engineering And Processing, 2007. **46**(11): p. 1043-1053.

AUTHOR INDEX

AUTHOR INDEX

AUTHOR INDEX

AUTHOR INDEX

AUTHOR INDEX

AUTHOR INDEX

AUTHOR INDEX

AUTHOR INDEX

AUTHOR INDEX

AUTHOR INDEX

AUTHOR INDEX

AUTHOR INDEX

AUTHOR INDEX

AUTHOR INDEX

AUTHOR INDEX

AUTHOR INDEX